Methods in Enzymology

Volume 235
BACTERIAL PATHOGENESIS
Part A
Identification and Regulation of
Virulence Factors

METHODS IN ENZYMOLOGY

EDITORS-IN-CHIEF

John N. Abelson   Melvin I. Simon

DIVISION OF BIOLOGY
CALIFORNIA INSTITUTE OF TECHNOLOGY
PASADENA, CALIFORNIA

FOUNDING EDITORS

Sidney P. Colowick and Nathan O. Kaplan

Methods in Enzymology

Volume 235

Bacterial Pathogenesis

Part A
Identification and Regulation of Virulence Factors

EDITED BY

Virginia L. Clark
Patrik M. Bavoil

SCHOOL OF MEDICINE AND DENTISTRY
UNIVERSITY OF ROCHESTER
ROCHESTER, NEW YORK

ACADEMIC PRESS
A Division of Harcourt Brace & Company
San Diego   New York   Boston   London   Sydney   Tokyo   Toronto

Academic Press, Inc.
525 B Street, Suite 1900, San Diego, California 92101-4495

United Kingdom Edition published by
Academic Press Limited
24–28 Oval Road, London NW1 7DX

International Standard Serial Number: 0076-6879

International Standard Book Number: 0-12-182136-6

PRINTED IN THE UNITED STATES OF AMERICA
94 95 96 97 87 99   MM   9 8 7 6 5 4 3 2 1

Table of Contents

CONTRIBUTORS TO VOLUME 235 ix

PREFACE xv

VOLUMES IN SERIES xvii

1. Recommendations for Working with Pathogenic   STEPHEN A. MORSE AND
   Bacteria   JOSEPH E. MCDADE   1

Section I. Evaluations of Virulence in Animal Models

2. Determination of Median Lethal and Infectious   SUSAN WELKOS AND
   Doses in Animal Model Systems   ALISON O'BRIEN   29

3. Experimental Keratoconjunctivitis (Sereny) As-   DENNIS J. KOPECKO   39
   say

4. Mouse Respiratory Infection Models for Pertussis   ROBERTA D. SHAHIN AND
   JAMES L. COWELL   47

5. Chinchilla Model of Experimental Otitis Media for   BRUCE A. GREEN,
   Study of Nontypable *Haemophilus influenzae*   WILLIAM J. DOYLE, AND
   Vaccine Efficacy   JAMES L. COWELL   59

6. Animal Models for Ocular Infections   ROGER G. RANK AND
   JUDITH A. WHITTUM-HUDSON   69

7. Animal Models for Urogenital Infections   ROGER G. RANK   83

8. Animal Models for Meningitis   MARTIN G. TÄUBER
   AND ANDRÉ ZWAHLEN   93

9. Animal Models for Periodontal Disease   THERESA E. MADDEN AND
   JACK G. CATON   106

10. Animal Chamber Models for Study of Host–Para-   CAROLINE ATTARDO GENCO
   site Interactions   AND ROBERT J. ARKO   120

11. Animal Models for Immunoglobulin A Secretion   DAVID F. KEREN AND
   MARK A. SUCKOW   140

Section II. Species and Strain Identification

12. Serogroup and Serotype Classification of Bacterial   CARL E. FRASCH   159
   Pathogens

13. Analysis of Genetic Variation by Polymerase Chain Reaction-Based Nucleotide Sequencing — KIMBERLYN NELSON AND ROBERT K. SELANDER — 174

14. Determinations of Restriction Fragment Length Polymorphism in Bacteria Using Ribosomal RNA Genes — RIVKA RUDNER, BARBARA STUDAMIRE, AND ERICH D. JARVIS — 184

15. DNA Fingerprinting of *Mycobacterium tuberculosis* — DICK VAN SOOLINGEN, PETRA F. W. DE HAAS, PETER W. M. HERMANS, AND JAN D. A. VAN EMBDEN — 196

16. Phylogenetic Identification of Uncultured Pathogens Using Ribosomal RNA Sequences — THOMAS M. SCHMIDT AND DAVID A. RELMAN — 205

Section III. Bacterial Cell Fractionation

17. Isolation of Outer Membranes — HIROSHI NIKAIDO — 225

18. Isolation and Purification of Periplasmic Binding Proteins — GIOVANNA FERRO-LUZZI AMES — 234

19. Isolation and Characterization of Lipopolysaccharides, Lipooligosaccharides, and Lipid A — MICHAEL A. APICELLA, J. MCLEOD GRIFFISS, AND HERMAN SCHNEIDER — 242

20. Isolation of Peptidoglycan and Soluble Peptidoglycan Fragments — RAOUL S. ROSENTHAL AND ROMAN DZIARSKI — 253

21. Purification of Streptococcal M Protein — VINCENT A. FISCHETTI — 286

22. Isolation and Assay of *Pseudomonas aeruginosa* Alginate — THOMAS B. MAY AND A. M. CHAKRABARTY — 295

23. Purification of *Escherichia coli* K Antigens — WILLIE F. VANN AND STEPHEN J. FREESE — 304

Section IV. Microbial Acquisition of Iron

24. Deferration of Laboratory Media and Assays for Ferric and Ferrous Ions — CHARLES D. COX — 315

25. Detection, Isolation, and Characterization of Siderophores — SHELLEY M. PAYNE — 329

26. Effects of Iron Deprivation on Outer Membrane Protein Expression — J. B. NEILANDS — 344

27. Identification and Isolation of Mutants Defective in Iron Acquisition — J. B. NEILANDS — 352

28. Identification of Receptor-Mediated Transferrin-Iron Uptake Mechanism in *Neisseria gonorrhoeae*   CYNTHIA NAU CORNELISSEN AND P. FREDERICK SPARLING   356

29. Isolation of Genes Involved in Iron Acquisition by Cloning and Complementation of *Escherichia coli* Mutants   SUSAN E. H. WEST   363

Section V. Genetics and Regulation

30. Bacterial Transformation by Electroporation   JEFF F. MILLER   375

31. Analysis and Construction of Stable Phenotypes in Gram-Negative Bacteria with Tn*5*-and Tn*10*-Derived Minitransposons   VÍCTOR DE LORENZO AND KENNETH N. TIMMIS   386

32. Use of Transposons to Dissect Pathogenic Strategies of Gram-Positive Bacteria   ROBERT A. BURNE AND ROBERT G. QUIVEY, JR.   405

33. Identification of Bacterial Cell-Surface Virulence Determinants with Tn*phoA*   MELISSA R. KAUFMAN AND RONALD K. TAYLOR   426

34. Temperature-Sensitive Mutants of Bacterial Pathogens: Isolation and Use to Determine Host Clearance and *in Vivo* Replication Rates   ANNE MORRIS HOOKE   448

35. Use of Conditionally Counterselectable Suicide Vectors for Allelic Exchange   SCOTT STIBITZ   458

36. Gene Replacement in *Pseudomonas aeruginosa*   DEBBIE S. TODER   466

37. Systems of Experimental Genetics for *Campylobacter* Species   PATRICIA GUERRY, RUIJIN YAO, RICHARD A. ALM, DONALD H. BURR, AND TREVOR J. TRUST   474

38. *In Vivo* Expression Technology for Selection of Bacterial Genes Specifically Induced in Host Tissues   JAMES M. SLAUCH, MICHAEL J. MAHAN, AND JOHN J. MEKALANOS   481

39. Regulation of Alginate Gene Expression in *Pseudomonas aeruginosa*   NICOLETTE A. ZIELINSKI, SIDDHARTHA ROYCHOUDHURY, AND A. M. CHAKRABARTY   493

40. Regulation of Expression of *Pseudomonas* Exotoxin A by Iron   DARA W. FRANK AND DOUGLAS G. STOREY   502

41. Regulation of Cholera Toxin by Temperature, pH, and Osmolarity   CLAUDETTE L. GARDEL AND JOHN J. MEKALANOS   517

42. Posttranslational Processing of Type IV Prepilin and Homologs by PilD of *Pseudomonas aeruginosa*   MARK S. STROM, DAVID N. NUNN, AND STEPHEN LORY   527

Section VI. Enzyme and Toxin Assays

43. Bacterial Immunoglobulin A₁ Proteases   MARTHA H. MULKS AND
                                          RUSSELL J. SHOBERG   543

44. Elastase Assays   LYNN RUST,
                     CALVIN R. MESSING,
                     AND BARBARA H. IGLEWSKI   554

45. Zymographic Techniques for Detection and Char-   MARILYN S. LANTZ AND
    acterization of Microbial Proteases   PAWEL CIBOROWSKI   563

46. Assays for Bacterial Type I Collagenases   JOHN D. GRUBB   594

47. Purification and Assays of Bacterial Gelatinases   JOHN D. GRUBB   602

48. Assays for Hyaluronidase Activity   WAYNE L. HYNES AND
                                        JOSEPH J. FERRETTI   606

49. ADP-Ribosylating Toxins   LUCIANO PASSADOR AND
                              WALLACE IGLEWSKI   617

50. Photoaffinity Labeling of Active Site Residues in   STEPHEN F. CARROLL AND
    ADP-Ribosylating Toxins   R. JOHN COLLIER   631

51. Activation of Cholera Toxin by ADP-Ribosylation   JOEL MOSS,
    Factors   S.-C. TSAI,
             AND MARTHA VAUGHAN   640

52. Toxins That Inhibit Host Protein Synthesis   TOM G. OBRIG   647

53. Assays of Hemolytic Toxins   GAIL E. ROWE AND
                                 RODNEY A. WELCH   657

54. Identification and Assay of RTX Family of Cytoly-   ANTHONY L. LOBO AND
    sins   RODNEY A. WELCH   667

55. Assay of Cytopathogenic Toxins in Cultured Cells   MONICA THELESTAM AND
                                                        INGER FLORIN   679

56. Use of Lipid Bilayer Membranes to Detect Pore   BRUCE L. KAGAN AND
    Formation by Toxins   YURI SOKOLOV   691

57. Uptake and Processing of Toxins by Mammalian   CATHARINE B. SAELINGER
    Cells   AND RANDAL E. MORRIS   705

AUTHOR INDEX 719

SUBJECT INDEX 754

Contributors to Volume 235

Article numbers are in parentheses following the names of contributors.
Affiliations listed are current.

RICHARD A. ALM (37), *Centre for Molecular Biology and Biotechnology, University of Queensland, St. Lucia, Queensland 40723, Australia*

MICHAEL A. APICELLA (19), *Department of Microbiology, University of Iowa College of Medicine, Iowa City, Iowa 52242*

ROBERT J. ARKO (10), *Division of Sexually Transmitted Diseases, National Center for Infectious Diseases, Centers for Disease Control and Prevention, Atlanta, Georgia 30333*

ROBERT A. BURNE (32), *Department of Dental Research, University of Rochester Medical Center, Rochester, New York 14642*

DONALD H. BURR (37), *Division of Virulence Assessment, Center for Food Safety and Applied Nutrition, Food and Drug Administration, Washington, D.C. 20204*

STEPHEN F. CARROLL (50), *Department of Biological Chemistry, XOMA Corporation, Berkeley, California 94710*

JACK G. CATON (9), *Department of Periodontology, Eastman Dental Center, Rochester, New York 14620*

A. M. CHAKRABARTY (22, 39), *Department of Microbiology and Immunology, University of Illinois College of Medicine, Chicago, Illinois 60612*

PAWEL CIBOROWSKI (45), *Department of Periodontics, University of Pittsburgh School of Dental Medicine, Pittsburgh, Pennsylvania 15261*

R. JOHN COLLIER (50), *Department of Microbiology and Molecular Genetics, and the Shipley Institute of Medicine, Harvard Medical School, Boston, Massachusetts 02115*

JAMES L. COWELL (4, 5), *Department of Bacterial Vaccine Research, Lederle-Praxis Biologicals, Inc., West Henrietta, New York 14586*

CHARLES D. COX (24), *Department of Microbiology, College of Medicine, University of Iowa, Iowa City, Iowa 52242*

PETRA E. W. DE HAAS (15), *Unit Molecular Microbiology, National Institute of Public Health and Environmental Protection, Bilthoven, 3000 LL, The Netherlands*

VÍCTOR DE LORENZO (31), *Centro de Investigaciones Biológicas, Consejo Superior de Investigaciones Científicas, Madrid 28006, Spain*

WILLIAM J. DOYLE (5), *Department of Otolaryngology, University of Pittsburgh School of Medicine, Children's Hospital of Pittsburgh, Pittsburgh, Pennsylvania 15213*

ROMAN DZIARSKI (20), *Northwest Center for Medical Education, Indiana University School of Medicine, Gary, Indiana 46408*

JOSEPH J. FERRETTI (48), *Department of Microbiology and Immunology, University of Oklahoma Health Sciences Center, Oklahoma City, Oklahoma 73190*

GIOVANNA FERRO-LUZZI AMES (18), *Department of Molecular and Cell Biology, Division of Biochemistry and Molecular Biology, University of California at Berkeley, Berkeley, California 94720*

VINCENT A. FISCHETTI (21), *Laboratory of Bacterial Pathogenesis and Immunology, The Rockefeller University, New York, New York 10021*

INGER FLORIN (55), *Department of Bacteriology, Karolinska Institute, S-104 01 Stockholm, Sweden*

DARA W. FRANK (40), *Department of Microbiology, Medical College of Wisconsin, Milwaukee, Wisconsin 53226*

CARL E. FRASCH (12), *Division of Bacterial Products, Center for Biologics Evaluation and Research, Food and Drug Administration, Bethesda, Maryland 20892*

STEPHEN J. FREESE (23), *Division of Bacterial Products, Center for Biologics Evaluation and Research, Food and Drug Administration, Bethesda, Maryland 20892*

CLAUDETTE L. GARDEL (41), *Department of Microbiology and Molecular Genetics, Harvard Medical School, Boston, Massachusetts 02115*

CAROLINE ATTARDO GENCO (10), *Department of Microbiology and Immunology, Morehouse School of Medicine, Atlanta, Georgia 30310*

BRUCE A. GREEN (5), *Department of Bacteriology Research, Lederle-Praxis Biologicals, Inc., West Henrietta, New York 14586*

JOHN D. GRUBB (46, 47), *Organon-Technica, Westchester, Pennsylvania 19380*

PATRICIA GUERRY (37), *Enterics Program, Naval Medical Research Institute, Rockville, Maryland 20852*

PETER W. M. HERMANS (15), *Pediatric Laboratory, Erasmus Universiteit/Sophia Kinderziekenhuis, Rotterdam, 3015 GD, The Netherlands*

WAYNE L. HYNES (48), *Department of Microbiology and Immunology, University of Oklahoma Health Sciences Center, Oklahoma City, Oklahoma 73190*

BARBARA H. IGLEWSKI (44), *Department of Microbiology and Immunology, University of Rochester, School of Medicine and Dentistry, Rochester, New York 14642*

WALLACE IGLEWSKI (49), *Department of Microbiology and Immunology, University of Rochester, School of Medicine and Dentistry, Rochester, New York 14642*

ERICH D. JARVIS (14), *Laboratory of Animal Behavior, The Rockefeller University, New York, New York 10021*

BRUCE L. KAGAN (56), *Department of Psychiatry and Brain Research Institute, University of California at Los Angeles School of Medicine, West Los Angeles Veterans Affairs Medical Center, Los Angeles, California 90024*

MELISSA R. KAUFMAN (33), *Department of Biological Sciences, Hopkins Marine Station, Stanford University, Pacific Grove, California 93950*

DAVID F. KEREN (11), *Department of Pathology, Warde Medical Laboratory, Ann Arbor, Michigan 48108*

DENNIS J. KOPECKO (3), *Department of Bacterial Diseases, Walter Reed Army Institute of Research, Washington, D.C. 20307*

MARILYN S. LANTZ (45), *Department of Periodontics, University of Pittsburgh School of Dental Medicine, Pittsburgh, Pennsylvania 15261*

ANTHONY L. LOBO (54), *Department of Medical Microbiology and Immunology, University of Wisconsin–Madison, Madison, Wisconsin 53706*

STEPHEN LORY (42), *Department of Microbiology, University of Washington, Seattle, Washington 98195*

THERESA E. MADDEN (9), *Department of Periodontology, Eastman Dental Center, Rochester, New York 14620*

MICHAEL J. MAHAN (38), *Department of Biological Sciences, University of California at Santa Barbara, Santa Barbara, California 93106*

THOMAS B. MAY (22), *Hospital Products Division, Abbott Laboratories, Abbott Park, Illinois 60064*

JOSEPH E. MCDADE (1), *National Center for Infectious Diseases, Centers for Disease Control and Prevention, Atlanta, Georgia 30333*

J. MCLEOD GRIFFISS (19), *Department of Laboratory Medicine, University of California at San Francisco, San Francisco, California 94121*

JOHN J. MEKALANOS (38, 41), *Department of Microbiology and Molecular Genetics, Harvard Medical School, Boston, Massachusetts 02115*

CALVIN R. MESSING (44), *Diagnostics Division, Eastman Kodak Company, Rochester, New York 14650*

JEFF F. MILLER (30), *Department of Microbiology and Immunology, University of California at Los Angeles School of Medicine, Los Angeles, California 90024*

RANDAL E. MORRIS (57), *Department of Anatomy and Cell Biology, University of Cincinnati, Cincinnati, Ohio 45267*

ANNE MORRIS HOOKE (34), *Department of Microbiology, Miami University, Oxford, Ohio 45056*

STEPHEN A. MORSE (1), *National Center for Infectious Diseases, Centers for Disease Control and Prevention, Atlanta, Georgia 30333*

JOEL MOSS (51), *Laboratory of Cellular Metabolism, National Heart, Lung, and Blood Institute, National Institutes of Health, Bethesda, Maryland 20892*

MARTHA H. MULKS (43), *Department of Microbiology, Michigan State University, East Lansing, Michigan 48824*

CYNTHIA NAU CORNELISSEN (28), *Department of Medicine, Division of Infectious Diseases, The University of North Carolina at Chapel Hill, School of Medicine, Chapel Hill, North Carolina 27599*

J. B. NEILANDS (26, 27), *Division of Biochemistry and Molecular Biology, University of California at Berkeley, Berkeley, California 94720*

KIMBERLYN NELSON (13), *Institute of Molecular Evolutionary Genetics, Pennsylvania State University, University Park, Pennsylvania 16802*

HIROSHI NIKAIDO (17), *Department of Molecular and Cell Biology, University of California at Berkeley, Berkeley, California 94720*

DAVID N. NUNN (42), *Department of Microbiology, University of Illinois, Champaign–Urbana, Illinois 61801*

ALISON O'BRIEN (2), *Department of Microbiology, Uniformed Services University of the Health Sciences, Bethesda, Maryland 20814*

TOM G. OBRIG (52), *Department of Microbiology and Immunology, University of Rochester, School of Medicine and Dentistry, Rochester, New York 14642*

LUCIANO PASSADOR (49), *Department of Microbiology and Immunology, University of Rochester, School of Medicine and Dentistry, Rochester, New York 14642*

SHELLEY M. PAYNE (25), *Department of Microbiology, The University of Texas at Austin, Austin, Texas 78712*

ROBERT G. QUIVEY, JR. (32), *Department of Dental Research, University of Rochester Medical Center, Rochester, New York 14642*

ROGER G. RANK (6, 7), *Department of Microbiology and Immunology, University of Arkansas for Medical Sciences, Little Rock, Arkansas 72205*

DAVID A. RELMAN (16), *Departments of Medicine and of Microbiology and Immunology, Stanford University, Stanford, California 94305*

RAOUL S. ROSENTHAL (20), *Department of Microbiology and Immunology, Indiana University School of Medicine, Indianapolis, Indiana 46202*

GAIL E. ROWE (53), *Department of Medical Microbiology and Immunology, University of Wisconsin–Madison, Madison, Wisconsin 53706*

SIDDHARTHA ROYCHOUDHURY (39), *Cell Culture Research and Development, Lilly Research Laboratories, Eli Lilly & Company, Indianapolis, Indiana 46285*

RIVKA RUDNER (14), *Department of Biological Sciences, Hunter College of the City University of New York, New York, New York 10021*

LYNN RUST (44), *Department of Microbiology and Immunology, University of Rochester, School of Medicine and Dentistry, Rochester, New York 14642*

CATHARINE B. SAELINGER (57), *Department of Molecular Genetics, University of Cincinnati, Cincinnati, Ohio 45267*

THOMAS M. SCHMIDT (16), *Department of Microbiology, Michigan State University, East Lansing, Michigan 48824*

HERMAN SCHNEIDER (19), *Department of Bacterial Diseases, Walter Reed Army Institute for Research, Washington, D.C. 20307*

ROBERT K. SELANDER (13), *Institute of Molecular Evolutionary Genetics, Pennsylvania State University, University Park, Pennsylvania 16802*

ROBERTA D. SHAHIN (4), *Division of Bacterial Products, Center for Biologics Evaluation and Research, Food and Drug Administration, Bethesda, Maryland 20892*

RUSSELL J. SHOBERG (43), *Department of Periodontics, University of Texas Health Science Center at San Antonio, San Antonio, Texas 78284*

JAMES M. SLAUCH (38), *Department of Microbiology, University of Illinois, Urbana, Illinois 61801*

YURI SOKOLOV (56), *Department of General Physiology of the Nervous System, A. A. Bogomolets Institute of Physiology, Kiev 252024, Ukraine*

P. FREDERICK SPARLING (28), *Department of Medicine, The University of North Carolina at Chapel Hill, School of Medicine, Chapel Hill, North Carolina 27599*

SCOTT STIBITZ (35), *Division of Bacterial Products, Center for Biologics Evaluation and Research, Food and Drug Administration, Bethesda, Maryland 20892*

DOUGLAS G. STOREY (40), *Department of Biological Sciences, University of Calgary, Calgary, Alberta, Canada T2N 1N4*

MARK S. STROM (42), *Utilization Research Division, Northwest Fisheries Science Center, NMFS, NOAA, Seattle, Washington 98112*

BARBARA STUDAMIRE (14), *Department of Virus Inactivation, New York Blood Center, New York, New York 10021*

MARK A. SUCKOW (11), *Laboratory Animal Program, Purdue University, West Lafayette, Indiana 47907*

MARTIN G. TÄUBER (8), *Infectious Disease Laboratories, San Francisco General Hospital and Department of Medicine, University of California at San Francisco, San Francisco, California 94143*

RONALD K. TAYLOR (33), *Department of Microbiology, Dartmouth Medical School, Hanover, New Hampshire 03755*

MONICA THELESTAM (55), *Department of Bacteriology, Karolinska Institute, S-104 01 Stockholm, Sweden*

KENNETH N. TIMMIS (31), *Department of Microbiology, National Research Center for Biotechnology, Braunschweig, W-38124, Germany*

DEBBIE S. TODER (36), *Department of Pediatrics, University of Rochester, School of Medicine and Dentistry, Rochester, New York 14642*

TREVOR J. TRUST (37), *Department of Biochemistry and Microbiology, University of Victoria, Victoria, British Columbia, Canada V8W 3P6*

S.-C. TSAI (51), *Laboratory of Cellular Metabolism, National Heart, Lung, and Blood Institute, National Institutes of Health, Bethesda, Maryland 20892*

JAN D. A. VAN EMBDEN (15), *Unit Molecular Microbiology, National Institute of Public Health and Environmental Protection, Bilthoven, 3720 BA, The Netherlands*

WILLIE F. VANN (23), *Division of Bacterial Products, Center for Biologics Evaluation and Research, Food and Drug Administration, Bethesda, Maryland 20892*

DICK VAN SOOLINGEN (15), *Laboratory of Bacteriology and Microbial Agents, National Institute of Public Health and Environmental Protection, Bilthoven, 3720 BA, The Netherlands*

MARTHA VAUGHAN (51), *Laboratory of Cellular Metabolism, National Heart, Lung, and Blood Institute, National Institutes of Health, Bethesda, Maryland 20892*

RODNEY A. WELCH (53, 54), *Department of Medical Microbiology and Immunology,*

University of Wisconsin–Madison, Madison, Wisconsin 53706

SUSAN WELKOS (2), *Bacteriology Division, United States Army Medical Research Institute of Infectious Diseases, Frederick, Maryland 21072*

SUSAN E. H. WEST (29), *Department of Pathobiological Sciences, School of Veterinary Medicine, University of Wisconsin–Madison, Madison, Wisconsin 53706*

JUDITH A. WHITTUM-HUDSON (6), *Department of Ophthalmology, The Johns Hopkins University School of Medicine, Baltimore, Maryland 21287*

RUIJIN YAO (37), *Enterics Program, Naval Medical Research Institute, Rockville, Maryland 20852*

NICOLETTE A. ZIELINSKI (39), *Department of Pathobiological Sciences, School of Veterinary Medicine, University of Wisconsin—Madison, Madison, Wisconsin 53706*

ANDRÉ ZWAHLEN (8), *Department of Medicine, Hôpital de Zone St. Loup/Orbe, Pompaple, Switzerland*

Preface

This volume contains contributions that present an overview of methods involved in the identification and regulation of bacterial virulence factors. Recent advances in recombinant DNA technology have enabled a study at the molecular level of genes from pathogenic bacteria. We have attempted to provide a series of articles that will enable researchers to determine if cloned genes or isolated gene products are involved in virulence.

After an introductory chapter on safety considerations in working with pathogenic bacteria, the volume is divided into six parts. First, a variety of animal model systems to determine bacterial virulence is presented, including how virulence, defined as LD_{50} or ID_{50}, is quantitated. This is followed by a section covering basic epidemiological techniques for the identification of bacterial strains and species. The third section presents protocols for the purification of subcellular bacterial products frequently associated with virulence. Section four presents methods involved in the determination of the means by which bacteria acquire iron, an important nutritional requirement normally restricted in the host. This is followed by a section on genetics and regulation of bacterial virulence factors, which includes methods to generate mutants, to construct isogenic strains by allelic exchange, to identify bacterial genes only expressed during infection of the host, and to study the regulation of selected bacterial virulence factors. The last section presents a variety of methods for the assay of destructive bacterial enzymes and bacterial toxins.

We would like to express our appreciation to the contributors to this volume for their generosity in sharing their research expertise.

VIRGINIA L. CLARK
PATRIK M. BAVOIL

METHODS IN ENZYMOLOGY

VOLUME I. Preparation and Assay of Enzymes
Edited by SIDNEY P. COLOWICK AND NATHAN O. KAPLAN

VOLUME II. Preparation and Assay of Enzymes
Edited by SIDNEY P. COLOWICK AND NATHAN O. KAPLAN

VOLUME III. Preparation and Assay of Substrates
Edited by SIDNEY P. COLOWICK AND NATHAN O. KAPLAN

VOLUME IV. Special Techniques for the Enzymologist
Edited by SIDNEY P. COLOWICK AND NATHAN O. KAPLAN

VOLUME V. Preparation and Assay of Enzymes
Edited by SIDNEY P. COLOWICK AND NATHAN O. KAPLAN

VOLUME VI. Preparation and Assay of Enzymes (*Continued*)
        Preparation and Assay of Substrates
        Special Techniques
Edited by SIDNEY P. COLOWICK AND NATHAN O. KAPLAN

VOLUME VII. Cumulative Subject Index
Edited by SIDNEY P. COLOWICK AND NATHAN O. KAPLAN

VOLUME VIII. Complex Carbohydrates
Edited by ELIZABETH F. NEUFELD AND VICTOR GINSBURG

VOLUME IX. Carbohydrate Metabolism
Edited by WILLIS A. WOOD

VOLUME X. Oxidation and Phosphorylation
Edited by RONALD W. ESTABROOK AND MAYNARD E. PULLMAN

VOLUME XI. Enzyme Structure
Edited by C. H. W. HIRS

VOLUME XII. Nucleic Acids (Parts A and B)
Edited by LAWRENCE GROSSMAN AND KIVIE MOLDAVE

VOLUME XIII. Citric Acid Cycle
Edited by J. M. LOWENSTEIN

VOLUME XIV. Lipids
Edited by J. M. LOWENSTEIN

VOLUME XV. Steroids and Terpenoids
Edited by RAYMOND B. CLAYTON

VOLUME XVI. Fast Reactions
Edited by KENNETH KUSTIN

VOLUME XVII. Metabolism of Amino Acids and Amines (Parts A and B)
Edited by HERBERT TABOR AND CELIA WHITE TABOR

VOLUME XVIII. Vitamins and Coenzymes (Parts A, B, and C)
Edited by DONALD B. MCCORMICK AND LEMUEL D. WRIGHT

VOLUME XIX. Proteolytic Enzymes
Edited by GERTRUDE E. PERLMANN AND LASZLO LORAND

VOLUME XX. Nucleic Acids and Protein Synthesis (Part C)
Edited by KIVIE MOLDAVE AND LAWRENCE GROSSMAN

VOLUME XXI. Nucleic Acids (Part D)
Edited by LAWRENCE GROSSMAN AND KIVIE MOLDAVE

VOLUME XXII. Enzyme Purification and Related Techniques
Edited by WILLIAM B. JAKOBY

VOLUME XXIII. Photosynthesis (Part A)
Edited by ANTHONY SAN PIETRO

VOLUME XXIV. Photosynthesis and Nitrogen Fixation (Part B)
Edited by ANTHONY SAN PIETRO

VOLUME XXV. Enzyme Structure (Part B)
Edited by C. H. W. HIRS AND SERGE N. TIMASHEFF

VOLUME XXVI. Enzyme Structure (Part C)
Edited by C. H. W. HIRS AND SERGE N. TIMASHEFF

VOLUME XXVII. Enzyme Structure (Part D)
Edited by C. H. W. HIRS AND SERGE N. TIMASHEFF

VOLUME XXVIII. Complex Carbohydrates (Part B)
Edited by VICTOR GINSBURG

VOLUME XXIX. Nucleic Acids and Protein Synthesis (Part E)
Edited by LAWRENCE GROSSMAN AND KIVIE MOLDAVE

VOLUME XXX. Nucleic Acids and Protein Synthesis (Part F)
Edited by KIVIE MOLDAVE AND LAWRENCE GROSSMAN

VOLUME XXXI. Biomembranes (Part A)
Edited by SIDNEY FLEISCHER AND LESTER PACKER

VOLUME XXXII. Biomembranes (Part B)
Edited by SIDNEY FLEISCHER AND LESTER PACKER

VOLUME XXXIII. Cumulative Subject Index Volumes I–XXX
Edited by MARTHA G. DENNIS AND EDWARD A. DENNIS

VOLUME XXXIV. Affinity Techniques (Enzyme Purification: Part B)
Edited by WILLIAM B. JAKOBY AND MEIR WILCHEK

VOLUME XXXV. Lipids (Part B)
Edited by JOHN M. LOWENSTEIN

VOLUME XXXVI. Hormone Action (Part A: Steroid Hormones)
Edited by BERT W. O'MALLEY AND JOEL G. HARDMAN

VOLUME XXXVII. Hormone Action (Part B: Peptide Hormones)
Edited by BERT W. O'MALLEY AND JOEL G. HARDMAN

VOLUME XXXVIII. Hormone Action (Part C: Cyclic Nucleotides)
Edited by JOEL G. HARDMAN AND BERT W. O'MALLEY

VOLUME XXXIX. Hormone Action (Part D: Isolated Cells, Tissues, and Organ Systems)
Edited by JOEL G. HARDMAN AND BERT W. O'MALLEY

VOLUME XL. Hormone Action (Part E: Nuclear Structure and Function)
Edited by BERT W. O'MALLEY AND JOEL G. HARDMAN

VOLUME XLI. Carbohydrate Metabolism (Part B)
Edited by W. A. WOOD

VOLUME XLII. Carbohydrate Metabolism (Part C)
Edited by W. A. WOOD

VOLUME XLIII. Antibiotics
Edited by JOHN H. HASH

VOLUME XLIV. Immobilized Enzymes
Edited by KLAUS MOSBACH

VOLUME XLV. Proteolytic Enzymes (Part B)
Edited by LASZLO LORAND

VOLUME XLVI. Affinity Labeling
Edited by WILLIAM B. JAKOBY AND MEIR WILCHEK

VOLUME XLVII. Enzyme Structure (Part E)
Edited by C. H. W. HIRS AND SERGE N. TIMASHEFF

VOLUME XLVIII. Enzyme Structure (Part F)
Edited by C. H. W. HIRS AND SERGE N. TIMASHEFF

VOLUME XLIX. Enzyme Structure (Part G)
Edited by C. H. W. HIRS AND SERGE N. TIMASHEFF

VOLUME L. Complex Carbohydrates (Part C)
Edited by VICTOR GINSBURG

VOLUME LI. Purine and Pyrimidine Nucleotide Metabolism
Edited by PATRICIA A. HOFFEE AND MARY ELLEN JONES

VOLUME LII. Biomembranes (Part C: Biological Oxidations)
Edited by SIDNEY FLEISCHER AND LESTER PACKER

VOLUME LIII. Biomembranes (Part D: Biological Oxidations)
Edited by SIDNEY FLEISCHER AND LESTER PACKER

VOLUME LIV. Biomembranes (Part E: Biological Oxidations)
Edited by SIDNEY FLEISCHER AND LESTER PACKER

VOLUME LV. Biomembranes (Part F: Bioenergetics)
Edited by SIDNEY FLEISCHER AND LESTER PACKER

VOLUME LVI. Biomembranes (Part G: Bioenergetics)
Edited by SIDNEY FLEISCHER AND LESTER PACKER

VOLUME LVII. Bioluminescence and Chemiluminescence
Edited by MARLENE A. DELUCA

VOLUME LVIII. Cell Culture
Edited by WILLIAM B. JAKOBY AND IRA PASTAN

VOLUME LIX. Nucleic Acids and Protein Synthesis (Part G)
Edited by KIVIE MOLDAVE AND LAWRENCE GROSSMAN

VOLUME LX. Nucleic Acids and Protein Synthesis (Part H)
Edited by KIVIE MOLDAVE AND LAWRENCE GROSSMAN

VOLUME 61. Enzyme Structure (Part H)
Edited by C. H. W. HIRS AND SERGE N. TIMASHEFF

VOLUME 62. Vitamins and Coenzymes (Part D)
Edited by DONALD B. MCCORMICK AND LEMUEL D. WRIGHT

VOLUME 63. Enzyme Kinetics and Mechanism (Part A: Initial Rate and Inhibitor Methods)
Edited by DANIEL L. PURICH

VOLUME 64. Enzyme Kinetics and Mechanism (Part B: Isotopic Probes and Complex Enzyme Systems)
Edited by DANIEL L. PURICH

VOLUME 65. Nucleic Acids (Part I)
Edited by LAWRENCE GROSSMAN AND KIVIE MOLDAVE

VOLUME 66. Vitamins and Coenzymes (Part E)
Edited by DONALD B. MCCORMICK AND LEMUEL D. WRIGHT

VOLUME 67. Vitamins and Coenzymes (Part F)
Edited by DONALD B. MCCORMICK AND LEMUEL D. WRIGHT

VOLUME 68. Recombinant DNA
Edited by RAY WU

VOLUME 69. Photosynthesis and Nitrogen Fixation (Part C)
Edited by ANTHONY SAN PIETRO

VOLUME 70. Immunochemical Techniques (Part A)
Edited by HELEN VAN VUNAKIS AND JOHN J. LANGONE

VOLUME 71. Lipids (Part C)
Edited by JOHN M. LOWENSTEIN

VOLUME 72. Lipids (Part D)
Edited by JOHN M. LOWENSTEIN

VOLUME 73. Immunochemical Techniques (Part B)
Edited by JOHN J. LANGONE AND HELEN VAN VUNAKIS

VOLUME 74. Immunochemical Techniques (Part C)
Edited by JOHN J. LANGONE AND HELEN VAN VUNAKIS

VOLUME 75. Cumulative Subject Index Volumes XXXI, XXXII, XXXIV–LX
Edited by EDWARD A. DENNIS AND MARTHA G. DENNIS

VOLUME 76. Hemoglobins
Edited by ERALDO ANTONINI, LUIGI ROSSI-BERNARDI, AND EMILIA CHIAN-CONE

VOLUME 77. Detoxication and Drug Metabolism
Edited by WILLIAM B. JAKOBY

VOLUME 78. Interferons (Part A)
Edited by SIDNEY PESTKA

VOLUME 79. Interferons (Part B)
Edited by SIDNEY PESTKA

VOLUME 80. Proteolytic Enzymes (Part C)
Edited by LASZLO LORAND

VOLUME 81. Biomembranes (Part H: Visual Pigments and Purple Membranes, I)
Edited by LESTER PACKER

VOLUME 82. Structural and Contractile Proteins (Part A: Extracellular Matrix)
Edited by LEON W. CUNNINGHAM AND DIXIE W. FREDERIKSEN

VOLUME 83. Complex Carbohydrates (Part D)
Edited by VICTOR GINSBURG

VOLUME 84. Immunochemical Techniques (Part D: Selected Immunoassays)
Edited by JOHN J. LANGONE AND HELEN VAN VUNAKIS

VOLUME 85. Structural and Contractile Proteins (Part B: The Contractile Apparatus and the Cytoskeleton)
Edited by DIXIE W. FREDERIKSEN AND LEON W. CUNNINGHAM

VOLUME 86. Prostaglandins and Arachidonate Metabolites
Edited by WILLIAM E. M. LANDS AND WILLIAM L. SMITH

VOLUME 87. Enzyme Kinetics and Mechanism (Part C: Intermediates, Stereochemistry, and Rate Studies)
Edited by DANIEL L. PURICH

VOLUME 88. Biomembranes (Part I: Visual Pigments and Purple Membranes, II)
Edited by LESTER PACKER

VOLUME 89. Carbohydrate Metabolism (Part D)
Edited by WILLIS A. WOOD

VOLUME 90. Carbohydrate Metabolism (Part E)
Edited by WILLIS A. WOOD

VOLUME 91. Enzyme Structure (Part I)
Edited by C. H. W. HIRS AND SERGE N. TIMASHEFF

VOLUME 92. Immunochemical Techniques (Part E: Monoclonal Antibodies and
General Immunoassay Methods)
Edited by JOHN J. LANGONE AND HELEN VAN VUNAKIS

VOLUME 93. Immunochemical Techniques (Part F: Conventional Antibodies,
Fc Receptors, and Cytotoxicity)
Edited by JOHN J. LANGONE AND HELEN VAN VUNAKIS

VOLUME 94. Polyamines
Edited by HERBERT TABOR AND CELIA WHITE TABOR

VOLUME 95. Cumulative Subject Index Volumes 61–74, 76–80
Edited by EDWARD A. DENNIS AND MARTHA G. DENNIS

VOLUME 96. Biomembranes [Part J: Membrane Biogenesis: Assembly and
Targeting (General Methods; Eukaryotes)]
Edited by SIDNEY FLEISCHER AND BECCA FLEISCHER

VOLUME 97. Biomembranes [Part K: Membrane Biogenesis: Assembly and
Targeting (Prokaryotes, Mitochondria, and Chloroplasts)]
Edited by SIDNEY FLEISCHER AND BECCA FLEISCHER

VOLUME 98. Biomembranes (Part L: Membrane Biogenesis: Processing and
Recycling)
Edited by SIDNEY FLEISCHER AND BECCA FLEISCHER

VOLUME 99. Hormone Action (Part F: Protein Kinases)
Edited by JACKIE D. CORBIN AND JOEL G. HARDMAN

VOLUME 100. Recombinant DNA (Part B)
Edited by RAY WU, LAWRENCE GROSSMAN, AND KIVIE MOLDAVE

VOLUME 101. Recombinant DNA (Part C)
Edited by RAY WU, LAWRENCE GROSSMAN, AND KIVIE MOLDAVE

VOLUME 102. Hormone Action (Part G: Calmodulin and Calcium-Binding Proteins)
Edited by ANTHONY R. MEANS AND BERT W. O'MALLEY

VOLUME 103. Hormone Action (Part H: Neuroendocrine Peptides)
Edited by P. MICHAEL CONN

VOLUME 104. Enzyme Purification and Related Techniques (Part C)
Edited by WILLIAM B. JAKOBY

VOLUME 105. Oxygen Radicals in Biological Systems
Edited by LESTER PACKER

VOLUME 106. Posttranslational Modifications (Part A)
Edited by FINN WOLD AND KIVIE MOLDAVE

VOLUME 107. Posttranslational Modifications (Part B)
Edited by FINN WOLD AND KIVIE MOLDAVE

VOLUME 108. Immunochemical Techniques (Part G: Separation and Characterization of Lymphoid Cells)
Edited by GIOVANNI DI SABATO, JOHN J. LANGONE, AND HELEN VAN VUNAKIS

VOLUME 109. Hormone Action (Part I: Peptide Hormones)
Edited by LUTZ BIRNBAUMER AND BERT W. O'MALLEY

VOLUME 110. Steroids and Isoprenoids (Part A)
Edited by JOHN H. LAW AND HANS C. RILLING

VOLUME 111. Steroids and Isoprenoids (Part B)
Edited by JOHN H. LAW AND HANS C. RILLING

VOLUME 112. Drug and Enzyme Targeting (Part A)
Edited by KENNETH J. WIDDER AND RALPH GREEN

VOLUME 113. Glutamate, Glutamine, Glutathione, and Related Compounds
Edited by ALTON MEISTER

VOLUME 114. Diffraction Methods for Biological Macromolecules (Part A)
Edited by HAROLD W. WYCKOFF, C. H. W. HIRS, AND SERGE N. TIMASHEFF

VOLUME 115. Diffraction Methods for Biological Macromolecules (Part B)
Edited by HAROLD W. WYCKOFF, C. H. W. HIRS, AND SERGE N. TIMASHEFF

VOLUME 116. Immunochemical Techniques (Part H: Effectors and Mediators of Lymphoid Cell Functions)
Edited by GIOVANNI DI SABATO, JOHN J. LANGONE, AND HELEN VAN VUNAKIS

VOLUME 117. Enzyme Structure (Part J)
Edited by C. H. W. HIRS AND SERGE N. TIMASHEFF

VOLUME 118. Plant Molecular Biology
Edited by ARTHUR WEISSBACH AND HERBERT WEISSBACH

VOLUME 119. Interferons (Part C)
Edited by SIDNEY PESTKA

VOLUME 120. Cumulative Subject Index Volumes 81–94, 96–101

VOLUME 121. Immunochemical Techniques (Part I: Hybridoma Technology and Monoclonal Antibodies)
Edited by JOHN J. LANGONE AND HELEN VAN VUNAKIS

VOLUME 122. Vitamins and Coenzymes (Part G)
Edited by FRANK CHYTIL AND DONALD B. McCORMICK

VOLUME 123. Vitamins and Coenzymes (Part H)
Edited by FRANK CHYTIL AND DONALD B. McCORMICK

VOLUME 124. Hormone Action (Part J: Neuroendocrine Peptides)
Edited by P. MICHAEL CONN

VOLUME 125. Biomembranes (Part M: Transport in Bacteria, Mitochondria, and Chloroplasts: General Approaches and Transport Systems)
Edited by SIDNEY FLEISCHER AND BECCA FLEISCHER

VOLUME 126. Biomembranes (Part N: Transport in Bacteria, Mitochondria, and Chloroplasts: Protonmotive Force)
Edited by SIDNEY FLEISCHER AND BECCA FLEISCHER

VOLUME 127. Biomembranes (Part O: Protons and Water: Structure and Translocation)
Edited by LESTER PACKER

VOLUME 128. Plasma Lipoproteins (Part A: Preparation, Structure, and Molecular Biology)
Edited by JERE P. SEGREST AND JOHN J. ALBERS

VOLUME 129. Plasma Lipoproteins (Part B: Characterization, Cell Biology, and Metabolism)
Edited by JOHN J. ALBERS AND JERE P. SEGREST

VOLUME 130. Enzyme Structure (Part K)
Edited by C. H. W. HIRS AND SERGE N. TIMASHEFF

VOLUME 131. Enzyme Structure (Part L)
Edited by C. H. W. HIRS AND SERGE N. TIMASHEFF

VOLUME 132. Immunochemical Techniques (Part J: Phagocytosis and Cell-Mediated Cytotoxicity)
Edited by GIOVANNI DI SABATO AND JOHANNES EVERSE

VOLUME 133. Bioluminescence and Chemiluminescence (Part B)
Edited by MARLENE DELUCA AND WILLIAM D. MCELROY

VOLUME 134. Structural and Contractile Proteins (Part C: The Contractile Apparatus and the Cytoskeleton)
Edited by RICHARD B. VALLEE

VOLUME 135. Immobilized Enzymes and Cells (Part B)
Edited by KLAUS MOSBACH

VOLUME 136. Immobilized Enzymes and Cells (Part C)
Edited by KLAUS MOSBACH

VOLUME 137. Immobilized Enzymes and Cells (Part D)
Edited by KLAUS MOSBACH

VOLUME 138. Complex Carbohydrates (Part E)
Edited by VICTOR GINSBURG

VOLUME 139. Cellular Regulators (Part A: Calcium- and Calmodulin-Binding Proteins)
Edited by ANTHONY R. MEANS AND P. MICHAEL CONN

VOLUME 140. Cumulative Subject Index Volumes 102–119, 121–134

VOLUME 141. Cellular Regulators (Part B: Calcium and Lipids)
Edited by P. MICHAEL CONN AND ANTHONY R. MEANS

VOLUME 142. Metabolism of Aromatic Amino Acids and Amines
Edited by SEYMOUR KAUFMAN

VOLUME 143. Sulfur and Sulfur Amino Acids
Edited by WILLIAM B. JAKOBY AND OWEN GRIFFITH

VOLUME 144. Structural and Contractile Proteins (Part D: Extracellular Matrix)
Edited by LEON W. CUNNINGHAM

VOLUME 145. Structural and Contractile Proteins (Part E: Extracellular Matrix)
Edited by LEON W. CUNNINGHAM

VOLUME 146. Peptide Growth Factors (Part A)
Edited by DAVID BARNES AND DAVID A. SIRBASKU

VOLUME 147. Peptide Growth Factors (Part B)
Edited by DAVID BARNES AND DAVID A. SIRBASKU

VOLUME 148. Plant Cell Membranes
Edited by LESTER PACKER AND ROLAND DOUCE

VOLUME 149. Drug and Enzyme Targeting (Part B)
Edited by RALPH GREEN AND KENNETH J. WIDDER

VOLUME 150. Immunochemical Techniques (Part K: *In Vitro* Models of B and T Cell Functions and Lymphoid Cell Receptors)
Edited by GIOVANNI DI SABATO

VOLUME 151. Molecular Genetics of Mammalian Cells
Edited by MICHAEL M. GOTTESMAN

VOLUME 152. Guide to Molecular Cloning Techniques
Edited by SHELBY L. BERGER AND ALAN R. KIMMEL

VOLUME 153. Recombinant DNA (Part D)
Edited by RAY WU AND LAWRENCE GROSSMAN

VOLUME 154. Recombinant DNA (Part E)
Edited by RAY WU AND LAWRENCE GROSSMAN

VOLUME 155. Recombinant DNA (Part F)
Edited by RAY WU

VOLUME 156. Biomembranes (Part P: ATP-Driven Pumps and Related Transport: The Na,K-Pump)
Edited by SIDNEY FLEISCHER AND BECCA FLEISCHER

VOLUME 157. Biomembranes (Part Q: ATP-Driven Pumps and Related Transport: Calcium, Proton, and Potassium Pumps)
Edited by SIDNEY FLEISCHER AND BECCA FLEISCHER

VOLUME 158. Metalloproteins (Part A)
Edited by JAMES F. RIORDAN AND BERT L. VALLEE

VOLUME 159. Initiation and Termination of Cyclic Nucleotide Action
Edited by JACKIE D. CORBIN AND ROGER A. JOHNSON

VOLUME 160. Biomass (Part A: Cellulose and Hemicellulose)
Edited by WILLIS A. WOOD AND SCOTT T. KELLOGG

VOLUME 161. Biomass (Part B: Lignin, Pectin, and Chitin)
Edited by WILLIS A. WOOD AND SCOTT T. KELLOGG

VOLUME 162. Immunochemical Techniques (Part L: Chemotaxis and Inflammation)
Edited by GIOVANNI DI SABATO

VOLUME 163. Immunochemical Techniques (Part M: Chemotaxis and Inflammation)
Edited by GIOVANNI DI SABATO

VOLUME 164. Ribosomes
Edited by HARRY F. NOLLER, JR., AND KIVIE MOLDAVE

VOLUME 165. Microbial Toxins: Tools for Enzymology
Edited by SIDNEY HARSHMAN

VOLUME 166. Branched-Chain Amino Acids
Edited by ROBERT HARRIS AND JOHN R. SOKATCH

VOLUME 167. Cyanobacteria
Edited by LESTER PACKER AND ALEXANDER N. GLAZER

VOLUME 168. Hormone Action (Part K: Neuroendocrine Peptides)
Edited by P. MICHAEL CONN

VOLUME 169. Platelets: Receptors, Adhesion, Secretion (Part A)
Edited by JACEK HAWIGER

VOLUME 170. Nucleosomes
Edited by PAUL M. WASSARMAN AND ROGER D. KORNBERG

VOLUME 171. Biomembranes (Part R: Transport Theory: Cells and Model Membranes)
Edited by SIDNEY FLEISCHER AND BECCA FLEISCHER

VOLUME 172. Biomembranes (Part S: Transport: Membrane Isolation and Characterization)
Edited by SIDNEY FLEISCHER AND BECCA FLEISCHER

VOLUME 173. Biomembranes [Part T: Cellular and Subcellular Transport: Eukaryotic (Nonepithelial) Cells]
Edited by SIDNEY FLEISCHER AND BECCA FLEISCHER

VOLUME 174. Biomembranes [Part U: Cellular and Subcellular Transport: Eukaryotic (Nonepithelial) Cells]
Edited by SIDNEY FLEISCHER AND BECCA FLEISCHER

VOLUME 175. Cumulative Subject Index Volumes 135–139, 141–167

VOLUME 176. Nuclear Magnetic Resonance (Part A: Spectral Techniques and Dynamics)
Edited by NORMAN J. OPPENHEIMER AND THOMAS L. JAMES

VOLUME 177. Nuclear Magnetic Resonance (Part B: Structure and Mechanism)
Edited by NORMAN J. OPPENHEIMER AND THOMAS L. JAMES

VOLUME 178. Antibodies, Antigens, and Molecular Mimicry
Edited by JOHN J. LANGONE

VOLUME 179. Complex Carbohydrates (Part F)
Edited by VICTOR GINSBURG

VOLUME 180. RNA Processing (Part A: General Methods)
Edited by JAMES E. DAHLBERG AND JOHN N. ABELSON

VOLUME 181. RNA Processing (Part B: Specific Methods)
Edited by JAMES E. DAHLBERG AND JOHN N. ABELSON

VOLUME 182. Guide to Protein Purification
Edited by MURRAY P. DEUTSCHER

VOLUME 183. Molecular Evolution: Computer Analysis of Protein and Nucleic Acid Sequences
Edited by RUSSELL F. DOOLITTLE

VOLUME 184. Avidin–Biotin Technology
Edited by MEIR WILCHEK AND EDWARD A. BAYER

VOLUME 185. Gene Expression Technology
Edited by DAVID V. GOEDDEL

VOLUME 186. Oxygen Radicals in Biological Systems (Part B: Oxygen Radicals and Antioxidants)
Edited by LESTER PACKER AND ALEXANDER N. GLAZER

VOLUME 187. Arachidonate Related Lipid Mediators
Edited by ROBERT C. MURPHY AND FRANK A. FITZPATRICK

VOLUME 188. Hydrocarbons and Methylotrophy
Edited by MARY E. LIDSTROM

VOLUME 189. Retinoids (Part A: Molecular and Metabolic Aspects)
Edited by LESTER PACKER

VOLUME 190. Retinoids (Part B: Cell Differentiation and Clinical Applications)
Edited by LESTER PACKER

VOLUME 191. Biomembranes (Part V: Cellular and Subcellular Transport: Epithelial Cells)
Edited by SIDNEY FLEISCHER AND BECCA FLEISCHER

VOLUME 192. Biomembranes (Part W: Cellular and Subcellular Transport: Epithelial Cells)
Edited by SIDNEY FLEISCHER AND BECCA FLEISCHER

VOLUME 193. Mass Spectrometry
Edited by JAMES A. McCLOSKEY

VOLUME 194. Guide to Yeast Genetics and Molecular Biology
Edited by CHRISTINE GUTHRIE AND GERALD R. FINK

VOLUME 195. Adenylyl Cyclase, G Proteins, and Guanylyl Cyclase
Edited by ROGER A. JOHNSON AND JACKIE D. CORBIN

VOLUME 196. Molecular Motors and the Cytoskeleton
Edited by RICHARD B. VALLEE

VOLUME 197. Phospholipases
Edited by EDWARD A. DENNIS

VOLUME 198. Peptide Growth Factors (Part C)
Edited by DAVID BARNES, J. P. MATHER, AND GORDON H. SATO

VOLUME 199. Cumulative Subject Index Volumes 168–174, 176–194 (in preparation)

VOLUME 200. Protein Phosphorylation (Part A: Protein Kinases: Assays, Purification, Antibodies, Functional Analysis, Cloning, and Expression)
Edited by TONY HUNTER AND BARTHOLOMEW M. SEFTON

VOLUME 201. Protein Phosphorylation (Part B: Analysis of Protein Phosphorylation, Protein Kinase Inhibitors, and Protein Phosphatases)
Edited by TONY HUNTER AND BARTHOLOMEW M. SEFTON

VOLUME 202. Molecular Design and Modeling: Concepts and Applications (Part A: Proteins, Peptides, and Enzymes)
Edited by JOHN J. LANGONE

VOLUME 203. Molecular Design and Modeling: Concepts and Applications (Part B: Antibodies and Antigens, Nucleic Acids, Polysaccharides, and Drugs)
Edited by JOHN J. LANGONE

VOLUME 204. Bacterial Genetic Systems
Edited by JEFFREY H. MILLER

VOLUME 205. Metallobiochemistry (Part B: Metallothionein and Related Molecules)
Edited by JAMES F. RIORDAN AND BERT L. VALLEE

VOLUME 206. Cytochrome P450
Edited by MICHAEL R. WATERMAN AND ERIC F. JOHNSON

VOLUME 207. Ion Channels
Edited by BERNARDO RUDY AND LINDA E. IVERSON

VOLUME 208. Protein–DNA Interactions
Edited by ROBERT T. SAUER

VOLUME 209. Phospholipid Biosynthesis
Edited by EDWARD A. DENNIS AND DENNIS E. VANCE

VOLUME 210. Numerical Computer Methods
Edited by LUDWIG BRAND AND MICHAEL L. JOHNSON

VOLUME 211. DNA Structures (Part A: Synthesis and Physical Analysis of DNA)
Edited by DAVID M. J. LILLEY AND JAMES E. DAHLBERG

VOLUME 212. DNA Structures (Part B: Chemical and Electrophoretic Analysis of DNA)
Edited by DAVID M. J. LILLEY AND JAMES E. DAHLBERG

VOLUME 213. Carotenoids (Part A: Chemistry, Separation, Quantitation, and Antioxidation)
Edited by LESTER PACKER

VOLUME 214. Carotenoids (Part B: Metabolism, Genetics, and Biosynthesis)
Edited by LESTER PACKER

VOLUME 215. Platelets: Receptors, Adhesion, Secretion (Part B)
Edited by JACEK J. HAWIGER

VOLUME 216. Recombinant DNA (Part G)
Edited by RAY WU

VOLUME 217. Recombinant DNA (Part H)
Edited by RAY WU

VOLUME 218. Recombinant DNA (Part I)
Edited by RAY WU

VOLUME 219. Reconstitution of Intracellular Transport
Edited by JAMES E. ROTHMAN

VOLUME 220. Membrane Fusion Techniques (Part A)
Edited by NEJAT DÜZGÜNEŞ

VOLUME 221. Membrane Fusion Techniques (Part B)
Edited by NEJAT DÜZGÜNEŞ

VOLUME 222. Proteolytic Enzymes in Coagulation, Fibrinolysis, and Complement Activation (Part A: Mammalian Blood Coagulation Factors and Inhibitors)
Edited by LASZLO LORAND AND KENNETH G. MANN

VOLUME 223. Proteolytic Enzymes in Coagulation, Fibrinolysis, and Complement Activation (Part B: Complement Activation, Fibrinolysis, and Nonmammalian Blood Coagulation Factors)
Edited by LASZLO LORAND AND KENNETH G. MANN

VOLUME 224. Molecular Evolution: Producing the Biochemical Data (in preparation)
Edited by ELIZABETH ANNE ZIMMER, THOMAS J. WHITE, REBECCA L. CANN, AND ALLAN C. WILSON

VOLUME 225. Guide to Techniques in Mouse Development
Edited by PAUL M. WASSARMAN AND MELVIN L. DEPAMPHILIS

VOLUME 226. Metallobiochemistry (Part C: Spectroscopic and Physical Methods for Probing Metal Ion Environments in Metalloenzymes and Metalloproteins)
Edited by JAMES F. RIORDAN AND BERT L. VALLEE

VOLUME 227. Metallobiochemistry (Part D: Physical and Spectroscopic Methods for Probing Metal Ion Environments in Metalloproteins)
Edited by JAMES F. RIORDAN AND BERT L. VALLEE

VOLUME 228. Aqueous Two-Phase Systems
Edited by HARRY WALTER AND GÖTE JOHANSSON

VOLUME 229. Cumulative Subject Index Volumes 195–198, 200–227 (in preparation)

VOLUME 230. Guide to Techniques in Glycobiology
Edited by WILLIAM J. LENNARZ AND GERALD W. HART

VOLUME 231. Hemoglobins (Part B: Biochemical and Analytical Methods)
Edited by JOHANNES EVERSE, KIM D. VANDEGRIFF, AND ROBERT M. WINSLOW

VOLUME 232. Hemoglobins (Part C: Biophysical Methods) (in preparation)
Edited by JOHANNES EVERSE, KIM D. VANDEGRIFF, AND ROBERT M. WINSLOW

VOLUME 233. Oxygen Radicals in Biological Systems (Part C)
Edited by LESTER PACKER

VOLUME 234. Oxygen Radicals in Biological Systems (Part D) (in preparation)
Edited by LESTER PACKER

VOLUME 235. Bacterial Pathogenesis (Part A: Identification and Regulation of Virulence Factors)
Edited by VIRGINIA L. CLARK AND PATRIK M. BAVOIL

VOLUME 236. Bacterial Pathogenesis (Part B: Integration of Pathogenic Bacteria with Host Cells) (in preparation)
Edited by VIRGINIA L. CLARK AND PATRIK M. BAVOIL

VOLUME 237. Heterotrimeric G Proteins (in preparation)
Edited by RAVI IYENGAR

VOLUME 238. Heterotrimeric G Protein Effectors (in preparation)
Edited by RAVI IYENGAR

VOLUME 239. Nuclear Magnetic Resonance (Part C) (in preparation)
Edited by THOMAS L. JAMES AND NORMAN J. OPPENHEIMER

VOLUME 240. Numerical Computer Methods (Part B) (in preparation)
Edited by MICHAEL L. JOHNSON AND LUDWIG BRAND

VOLUME 241. Retroviral Proteases (in preparation)
Edited by LAWRENCE C. KUO AND JULES A. SHAFER

[1] Recommendations for Working with Pathogenic Bacteria

By STEPHEN A. MORSE and JOSEPH E. MCDADE

Introduction

Microbiology laboratories are special, often unique, work environments that may pose infectious disease risks to persons in or near them. Personnel have contracted infections in the laboratory throughout the history of microbiology. The nature and basis of these laboratory-associated infections have been reviewed in a series of publications,[1-4] which have served to highlight specific laboratory hazards and procedures that increase the risk of infection. As a general rule, all bacteria should be treated as if they are pathogenic. Nevertheless, it is recognized that specific microorganisms are more infectious and more pathogenic than others and that certain laboratory activities pose a greater hazard than other activities. To facilitate safe laboratory practices, a guidebook was prepared by the Centers for Disease Control and Prevention (CDC) and the National Institutes of Health (NIH), with the help of an expert committee, which provides recommendations for working with pathogenic microorganisms and infected vertebrate animals.[5]

Laboratory Biosafety Levels

There are four laboratory biosafety levels (designated Biosafety Level 1 through 4) that are combinations of laboratory practices and techniques, safety equipment, and laboratory facilities appropriate for the operations performed and the hazard posed by the infectious agents and for the laboratory function or activity. Etiologic agents have been classified on the basis of hazards as Class 1 through Class 4 agents.[6] Virtually all bacterial pathogens of humans are Class 2 and Class 3 agents. These classes are comparable to the cohort biosafety level. Thus, studies involving pathogenic bacteria require either the Biosafety Level 2 or Biosafety Level 3 practices, equipment, and facilities described in Table I.

---

[1] R. M. Pike, *Health Lab. Sci.* **13**, 105 (1976).
[2] R. M. Pike, S. E. Sulkin, and M. L. Schulze, *Am. J. Public Health* **55**, 190 (1965).
[3] S. E. Sulkin and R. M. Pike, *N. Engl. J. Med.* **241**, 205 (1949).
[4] S. E. Sulkin and R. M. Pike, *Am. J. Public Health* **41**, 769 (1951).
[5] U.S. Dept. of Health and Human Services, Biosafety in microbiological and biomedical laboratories, HHS Pub. No. (CDC) 88-8395 (1988).
[6] *Federal Register* **51**, 16958 (1986).

Animal Biosafety Levels

Many studies on pathogenic microorganisms require the use of vertebrate animals. If vertebrate animals are used, institutional management must provide facilities and staff and establish practices that reasonably assure appropriate levels of environmental quality, safety, and care. As a general principle, the biosafety level (practices, equipment, and facilities) recommended for working with infectious agents *in vivo* and *in vitro* are comparable. These recommendations are described in Table II and presuppose that the laboratory animal facilities, operational practices, and quality of animal care meet applicable standards and regulations and that appropriate species and numbers have been selected for animal experiments.[7] Animal Biosafety Levels 2 and 3 are applicable for work with agents assigned to the corresponding Biosafety Levels 2 and 3. Facility standards and practices for invertebrate vectors and hosts are not specifically addressed in the standards for commonly used vertebrate animals. However, a useful reference in the design and operation of facilities using arthropods was prepared by the subcommittee on Arbovirus Laboratory Safety of the American Committee on Arthropod-Borne Viruses.[8]

Recommendations for Working with Selected
Pathogenic Microorganisms

Selection of an appropriate biosafety level for work with a particular pathogenic bacterium or animal study depends on a number of factors. Some of the most important are the infectivity, virulence, pathogenicity, biological stability, route of spread, and communicability of the agent; the nature or function of the laboratory; the procedures and manipulations involving the agent; the quantity and concentration of the agent; and the availability of effective vaccines or therapeutic measures. The biosafety levels are also applicable to laboratory operations other than those involving the use of bacterial pathogens of humans. For example, microbiological studies of animal host-specific pathogens post substantially lower risks of laboratory infection. Nevertheless, the laboratory practices, containment equipment, and facility recommendations for working with human pathogens are of value in developing operational standards for laboratories working with animal host-specific pathogens.

---

[7] U.S. Dept. of Health and Human Services, Guide for the care and use of laboratory animals, HHS Pub. No. 86-23 (1985).
[8] Subcommittee on arbovirus laboratory safety for arboviruses and certain other viruses of vertebrates, *Am. J. Trop. Med. Hyg.* **29,** 1359 (1980).

Recommended biosafety levels for working with selected pathogenic bacteria and infected animals are presented in Table III. In addition, recommendations for the use of vaccines and toxoids have been included when such products are available, either as licensed or investigational new drug (IND) products, and are specifically targeted to at-risk laboratory personnel and others who must work in or enter laboratories. It is important to remember that the recommended biosafety levels presuppose that the laboratory personnel are immunocompetent; those with altered immunocompetence may be at increased risk when exposed to infectious agents. Immunodeficiency may be hereditary, congenital, or induced by a number of neoplastic or viral diseases, by therapy, or by radiation. The risk of becoming infected or the consequences of infection may also be influenced by host factors such as age, sex, race, pregnancy, surgery (e.g., splenectomy, gastrectomy), predisposing disease [e.g., diabetes, lupus erythematosus, human immunodeficiency virus (HIV) infection], and altered physiologic function. Each laboratory director should consider these and other variables when deciding on biosafety levels for specific activities.

If possible, studies with pathogenic bacteria should be conducted with antimicrobial-susceptible strains. However, it is recognized that this is not always possible. Therefore, the antimicrobial susceptibility pattern of the strains should be determined by standard methodologies[9] in order to guide appropriate prophylaxis or therapy in the event of a laboratory accident.

The basic biosafety level assigned to a pathogenic bacterium is based on the activities typically associated with the growth and manipulation of small quantities and concentrations of the particular agent. If activities involve large volumes or highly concentrated preparations or manipulations which are likely to produce aerosols or droplets, or which are otherwise intrinsically hazardous, additional personnel precautions and increased levels of containment should be used. "Production quantities" refers to large volumes or concentrations of infectious bacteria as occurs in large-scale fermentations or batch cultures, antigen [e.g., capsule, lipopolysaccharide (LPS), toxin] and vaccine production, and a variety of other research and commercial activities which require significant masses of infectious agents. However, it is difficult to define "production quantities" in finite volumes or concentrations for any given pathogenic bacterium. This must be done by the laboratory director after making a risk assessment of the activities conducted and selected practices, containment equipment, and facilities appropriate to the risk, irrespective of the volume or concentration of bacteria involved.

[9] NCCLS Document M7-A2; NCCLS Document M2-A4; NCCLS Document M11-T2.

Occasions will arise when a biosafety level higher than that recommended in Table III should be selected. For example, a higher biosafety level may be indicated by the unique nature of the proposed activity (e.g., the need for specific containment for experimentally generated aerosols for inhalation studies). It should be noted that many pathogenic bacteria are not listed in Table III. However, many of those agents, which are not listed, have been classified on the basis of hazard.[6] The director of laboratories working with the agents is responsible for appropriate risk assessment and for employing appropriate practices, containment equipment, and facilities for the agent used.

TABLE I

BIOSAFETY LEVELS FOR WORKING WITH PATHOGENIC BACTERIA[a]

Microbiological practices	Containment equipment	Laboratory facilities

Biosafety Level 2
A. Standard
  1. Access to the laboratory is limited or restricted by the laboratory director when work with infectious agents is in progress
  2. Work surfaces are decontaminated at least once a day and after any spill of viable material
  3. All infectious liquid or solid wastes are decontaminated before disposal
  4. Mechanical pipetting devices are used; mouth pipetting is prohibited
  5. Eating, drinking, smoking, and applying cosmetics are not permitted in the work area. Food may be stored in cabinets or refrigerators designated and used for this purpose only. Food storage cabinets or refrigerators should be located outside of the work area
  6. Persons wash their hands after handling either infectious materials or animals and when they leave the laboratory
  7. All procedures are performed carefully to minimize the creation of aerosols
B. Special
  1. Contaminated materials that are to be decontaminated at a site away from the laboratory are placed in durable leak-proof containers which are closed before being removed from the laboratory
	Biological safety cabinets (Class I or II) or other appropriate personal protective or physical containment devices are used whenever
	  1. Procedures with a high potential for creating infectious aerosols are conducted. These may include centrifuging, grinding, blending, vigorous shaking or mixing, sonic disruption, opening containers of infectious materials whose internal pressures may be different from ambient pressures, inoculating animals intranasally, and harvesting infected tissues from animals or eggs
	  2. High concentrations or large volumes of infectious agents are used. Such materials may be centrifuged in the open laboratory if sealed heads or centrifuge safety cups are used and if they are opened only in a biological safety cabinet
		1. The laboratory is designed so that it can be easily cleaned
		2. Bench tops are impervious to water and resistant to acids, alkalis, organic solvents, and moderate heat
		3. Laboratory furniture is sturdy, and spaces between benches, cabinets, and equipment are accessible for cleaning
		4. Each laboratory contains a sink for hand washing
		5. If the laboratory has windows that open, they are fitted with fly screens
		6. An autoclave for decontaminating infectious laboratory wastes is available

(continued)

TABLE I (*continued*)

Microbiological practices	Containment equipment	Laboratory facilities
2. The laboratory director limits access to the laboratory. In general, persons who are at increased risk of acquiring infection or for whom infection may be unusually hazardous are not allowed in the laboratory or animal rooms. The director has the final responsibility for assessing each circumstance and determining who may enter or work in the laboratory		
3. The laboratory director establishes policies and procedures whereby only persons who have been advised of the potential hazard and meet any specific entry requirements (e.g., immunization) enter the laboratory or animal rooms		
4. When the infectious agent(s) in use in the laboratory requires special provisions for entry (e.g., vaccination), a hazard warning sign, incorporating the universal biohazard symbol, is posted on the access door to the laboratory work area. The hazard sign identifies the infectious agent, lists the name and telephone number of the laboratory director or other responsible person(s), and indicates the special requirement(s) for entering the laboratory		
5. An insect and rodent control program is in effect		
6. Laboratory coats, gowns, smocks, or uniforms are worn while in the laboratory. Before leaving the laboratory		

(continued)

for nonlaboratory areas (e.g., cafeteria, library, administrative offices), the protective clothing is removed and left in the laboratory or covered with a clean coat not used in the laboratory

7. Animals not involved in the work being performed are not permitted in laboratory

8. Special care is taken to avoid skin contamination with infectious materials; gloves should be worn when handling infected animals and when skin contact with infectious materials is unavoidable

9. All wastes from laboratories and animal rooms are appropriately decontaminated before disposal

10. Hypodermic needles and syringes are used only for parenteral injection and aspiration of fluids from laboratory animals and diaphragm bottles. Only needle-locking syringes or disposable syringe–needle units (i.e., needle is integral to the syringe) are used for injection or aspiration of infectious fluids. Extreme caution should be used when handling needles and syringes to avoid autoinoculation and generation of aerosols during use and disposal. Needles should not be bent, sheared, replaced in the sheath or guard, or removed from the syringe following use. The needle and syringe should be promptly placed in a puncture-resistant container and decontaminated, preferably by autoclaving, before discard or reuse

TABLE I (*continued*)

Microbiological practices	Containment equipment	Laboratory facilities
11. Spills and accidents that result in overt exposures to infectious materials are immediately reported to the laboratory director. Medical evaluation, surveillance, and treatment are provided as appropriate, and written records are maintained	Biological safety cabinets (Class I, II, or III) or other appropriate combinations of protective or physical containment devices (e.g., special protective clothing, masks, gloves, respirators, centrifuge safety cups, sealed centrifuge rotors, and containment caging for animals) are used for all activities with infectious materials which pose a threat of aerosol exposure. These include manipulation of cultures and of those	Same as for Biosafety Level 2 plus 1. The laboratory is separated from areas open to unrestricted traffic flow within the building. Passage through two sets of doors is the basic requirement for entry into the laboratory access corridors or other contiguous areas. Physical separation of the high containment laboratory from access corridors or other laboratories or activities may also be a double-doored clothes change room (showers may be
12. When appropriate, considering the agents handled, baseline serum samples for laboratory and other at-risk personnel are collected and stored. Additional serum specimens may be collected periodically, depending on the agents handled or the function of the facility		
13. A biosafety manual is prepared or adopted. Personnel are advised of special hazards and are required to read instructions on practices and procedures and to follow them		

Biosafety Level 3

A. Standard

Same as for Biosafety Level 2

B. Special

Same as for Biosafety Level 2 plus

1. Laboratory doors are kept closed when experiments are in progress

2. All activities involving infectious materials are conducted in biological safety cabinets or other physical containment devices within the containment module. No work in open vessels is conducted on open benches

clinical and environmental materials which may be a source of infectious aerosols; the aerosol challenge of experimental animals; harvesting of tissues or fluids from infected animals and embryonated eggs; and necropsy of infected animals

3. The work surfaces of biological safety cabinets and other containment equipment are decontaminated when work with infectious materials is finished. Plastic-backed paper toweling used on nonperforated work surfaces within biological safety cabinets facilitates clean up

4. Laboratory clothing that protects street clothing (e.g., solid front or wraparound gowns, scrub suits, coveralls) is worn in the laboratory. Laboratory clothing is not worn outside the laboratory, and it is decontaminated before being laundered

5. Molded surgical masks or respirators are worn in rooms containing infected animals

6. Animals and plants not related to the work being conducted are not permitted in the laboratory

7. Vacuum lines are protected with high-efficiency particulate air (HEPA) filters and liquid disinfectant traps

included), air lock, or other access facility which requires passage through two sets of doors before entering the laboratory

2. The interior surfaces of walls, floors, and ceilings are water resistant so that they can be easily cleaned. Penetrations in these surfaces are sealed or capable of being sealed to facilitate decontaminating the area

3. Each laboratory contains a sink for hand washing. The sink is foot, elbow, or automatically operated and is located near laboratory exit door

4. Windows in the laboratory are closed and sealed

5. Access doors to the laboratory or containment module are self-closing

6. An autoclave for decontaminating laboratory wastes is available, preferably within the laboratory

7. A ducted exhaust air ventilation system is provided. This system creates directional airflow that draws air into the laboratory through entry areas. The exhaust air is not recirculated to any other area of the building, is discharged to the outside, and is dispersed away from occupied areas and air intakes. Personnel must verify that the direction of airflow (into the laboratory) is proper. The exhaust air from the

(continued)

TABLE I (*continued*)

Microbiological practices	Containment equipment	Laboratory facilities
		laboratory room can be discharged to the outside without being filtered or otherwise treated
		8. The HEPA-filtered exhaust air from Class II or Class III biological safety cabinets is discharged directly to the outside or through the building exhaust system. Exhaust air from Class II biological safety cabinets may be recirculated within the laboratory if the cabinet is tested and certified at least every 12 months. If the HEPA-filtered exhaust air from Class II or III biological safety cabinets is to be discharged to the outside through the building exhaust air system, it is connected to this system in a manner (e.g., thimble unit connection) that avoids any interference with the air balance of the cabinets or building exhaust system

[a] The Biosafety Levels described are those recommended in HHS Pub. No. 86-23.[5]

TABLE II

ANIMAL BIOSAFETY LEVELS FOR WORKING WITH PATHOGENIC BACTERIA[a]

Practices	Containment equipment	Facilities
Animal Biosafety Level 2	Biological safety cabinets, other physical containment devices, and/or personal protective devices (e.g., respirators, face shields) are used whenever procedures with a high potential for creating aerosols are conducted. These include necropsy of infected animals, harvesting of infected tissues or fluids from animals or eggs, intranasal inoculation of animals, and manipulations of high concentrations or large volumes of infectious materials	1. The animal facility is designed and constructed to facilitate cleaning and housekeeping
A. Standard		2. A handwashing sink is available in the room where infected animals are housed
1. Doors to animal rooms open inward, are self-closing, and are kept closed when infected animals are present		3. If the animal facility has windows that open, they are fitted with fly screens
2. Work surfaces are decontaminated after use or spills of viable materials		4. It is recommended, but not required, that the direction of airflow in the animal facility is inward and that exhaust air is discharged to the outside without being recirculated to other rooms
3. Eating, drinking, smoking, and storing of food for human use are not permitted in animal rooms		5. An autoclave that can be used for decontaminating infectious laboratory waste is available in the building with the animal facility
4. Personnel wash their hands after handling cultures or animals and before leaving the animal room		
5. All procedures are carefully performed to minimize the creation of aerosols		
6. An insect and rodent control program is in effect		
B. Special		
1. Cages are decontaminated, preferably by autoclaving, before they are cleaned and washed		
2. Surgical-type masks are worn by all personnel entering animal rooms housing nonhuman primates		
3. Laboratory coats, gowns, or uniforms are worn while in the animal room. The protective clothing is removed before leaving the animal facility		

(continued)

TABLE II (*continued*)

Practices	Containment equipment	Facilities
4. The laboratory or animal facility director limits access to the animal room to personnel who have been advised of the potential hazard and who need to enter the room for program or service purposes when work is in progress. In general, persons who may be at increased risk of acquiring infection or for whom infection might be unusually hazardous are not allowed in the animal room		
5. The laboratory or animal facility director establishes policies and procedures whereby only persons who have been advised of the potential hazard and meet any specific requirements (e.g., immunization) may enter the animal room		
6. When the infectious agent(s) in use in the animal room requires special entry provisions (e.g., immunization), a hazard warning sign, incorporating the universal biohazard symbol, is posted on the access door to the animal room. The hazard warning sign identifies the infectious agent, lists the name and telephone number of the animal facility supervisor or other responsible person(s), and indicates the special requirement(s) for entering the animal room		
7. Special care is taken to avoid skin contamination with infectious materials;		

(continued)

gloves should be worn when handling infected animals and when skin contact with infectious materials is unavoidable

8. All wastes from the animal room are appropriately decontaminated, preferably by autoclaving, before disposal. Infected animal carcasses are incinerated after being transported from the animal room in leak-proof, covered containers

9. Hypodermic needles and syringes are used only for the parenteral injection or aspiration of infectious fluids. Needles should not be bent, sheared, replaced in the sheath or guard, or removed from the syringe following use. The needle and syringe should be promptly placed in a puncture-resistant container and decontaminated, preferably by autoclaving, before discard or reuse

10. If floor drains are provided, the drain traps are always filled with water or a suitable disinfectant

11. When appropriate, considering the agents handled, baseline serum samples from animal care and other at-risk personnel are collected and stored. Additional serum samples may be collected periodically, depending on the agents handled or the function of the facility

Animal Biosafety Level 3

A. Standard

Same as for Animal Biosafety Level 2

TABLE II (continued)

Practices	Containment equipment	Facilities
B. Special 1. Cages are autoclaved before bedding is removed and before they are cleaned and washed 2. Surgical-type masks or other respiratory protection devices (e.g., respirators) are worn by personnel entering rooms housing animals infected with agents assigned to Biosafety Level 3 3. Wraparound or solid-front gowns or uniforms are worn by personnel entering the animal room. Front-button laboratory coats are unsuitable. Protective gowns must remain in the animal room and must be decontaminated before being laundered 4. The laboratory director or other responsible person restricts access to the animal room to personnel who have been advised of the potential hazard and who need to enter the room for program or service purposes when infected animals are present. In general, persons who may be at increased risk of acquiring infection or for whom infection might be unusually hazardous are not allowed in the animal room 5. The laboratory director or other responsible person establishes policies and procedures whereby only persons who have been advised of the potential hazard and meet any specific	1. Personal protective clothing and equipment and/or other physical containment devices are used for all procedures and manipulations of infectious animals 2. The risk of infectious aerosols from infected animals or their bedding can be reduced if animals are housed in partial containment caging systems, such as open cages placed in ventilated enclosures (e.g., laminar flow cabinets), solid wall and bottom cages covered by filter bonnets, or other equivalent primary containment systems	1. The animal facility is designed and constructed to facilitate cleaning and housekeeping and is separated from areas which are open to unrestricted personnel traffic within the building. Passage through two sets of doors is the basic requirement for entry into the animal room from access corridors or other contiguous areas. Physical separation of the animal room from access corridors or other activities may also be provided by a double-doored clothes change room (shower may be included), air lock, or other access facility which requires passage through two sets of doors before entering the animal room 2. The interior surfaces of walls, floors, and ceilings are water resistant so that they may be easily cleaned. Penetrations in these surfaces are sealed or capable of being sealed to facilitate fumigation or space decontamination 3. A foot, elbow, or automatically operated hand washing sink is provided near each animal room exit door 4. Windows in the animal room are closed and sealed 5. Animal room doors are self-closing and are kept closed when infected animals are present

requirements (e.g., immunization) may enter the animal room

6. Hazard warning signs, incorporating the universal biohazard warning symbol, are posted on access doors to animal rooms containing animals infected with agents assigned to Biosafety Level 3. The hazard warning sign should identify the agent(s) in use, list the name and telephone number of the animal room supervisor or other responsible person(s), and indicate any special conditions of entry into the animal room (e.g., the need for immunizations or respirators)

7. Personnel wear gloves when handling infected animals. Gloves are removed aseptically and autoclaved with other animal room wastes before being disposed of or reused

8. All wastes from the animal room are autoclaved before disposal. All animal carcasses are incinerated. Dead animals are transported from the animal room to the incinerator in leak-proof, covered containers

9. Hypodermic needles and syringes are used only for gavage or for parenteral injection or aspiration of fluids from laboratory animals and diaphragm bottles. Only needle-locking syringes or disposable needle syringe units (i.e., the needle is integral to the syringe) are used. Needles should not be bent, sheared, replaced in the sheath or guard, or removed from the syringe following use.

6. An autoclave for decontaminating wastes is available, preferably within the animal room. Materials to be autoclaved outside the animal room are transported in a covered, leak-proof container

7. An exhaust air ventilation system is provided. This system creates directional airflow that draws air into the animal room through the entry area. The building exhaust can be used for this purpose if the exhaust air is not recirculated to any other area of the building, is discharged to the outside, and is dispersed away from occupied areas and air intakes. Personnel must verify that the direction of the airflow (into the animal room) is proper. The exhaust air from the animal room that does not pass through biological safety cabinets or other primary containment equipment can be discharged to the outside without being filtered or otherwise treated

8. The HEPA-filtered exhaust air from Class I or Class II biological safety cabinets or other primary containment devices is discharged directly to the outside or through the building exhaust system. Exhaust air from these primary containment devices may be recirculated within the animal room if the cabinet is tested and certified at least every 12 months. If HEPA-filtered exhaust air from Class I or Class II

(continued)

TABLE II (*continued*)

Practices	Containment equipment	Facilities
The needle and syringe should be promptly placed in a puncture-resistant container and decontaminated, preferably by autoclaving, before discard or reuse. Whenever possible, cannulas should be used instead of sharp needles (e.g., gavage)		biological safety cabinets is discharged to the outside through the building exhaust system, it is connected to this system in a manner (e.g., thimble unit connection) that avoids any interference with the air balance of cabinets or building exhaust system
10. If floor drains are provided, the drain traps are always filled with water or a suitable disinfectant		
11. If vacuum lines are provided, they are protected with HEPA filters and liquid disinfectant traps		
12. Boots, shoe covers, or other protective footwear and disinfectant footbaths are available and used when indicated		

[a] The Animal Biosafety Levels described are those recommended in HHS Pub. No. 86-23.[5]

TABLE III

LABORATORY HAZARDS AND RECOMMENDATIONS FOR WORKING WITH SELECTED BACTERIAL AND RICKETTSIAL AGENTS[a]

Agent	Laboratory hazards	Recommended precautions
Bacterial agents		
Bacillus anthracis	Direct and indirect contact of intact and broken skin with cultures and contaminated laboratory surfaces, accidental parenteral inoculation, and exposure to infectious aerosols. Naturally and experimentally infected animals also pose potential risk to laboratory and animal care personnel	Biosafety Level 2: footnote *b* — Animal Biosafety Level 2: footnote *c* — Biosafety Level 3: footnotes *d* and *e* — Immunoprophylaxis: Licensed vaccine is available through Centers for Disease Control and Prevention (CDC); however, vaccination of laboratory workers is not recommended unless frequent work anticipated. Vaccination is recommended for all persons working in Biosafety Level 3 areas where cultures are handled, and for working with infected animals
Bordetella pertussis	Exposure to infectious aerosols is the primary hazard	Biosafety Level 2: footnote *b* — Biosafety Level 3: footnotes *d* and *e* — Immunoprophylaxis: Administration of pertussis vaccine in combination with diphtheria and tetanus toxoids at 10-year intervals to reduce risk to laboratory personnel
Borrelia burgdorferi	Accidental parenteral inoculation and bites from infected ticks are the primary hazards to laboratory personnel	Biosafety Level 2: footnote *b* — Animal Biosafety Level 2: footnote *c* — Biosafety Level 3: footnotes *d* and *e* — Immunoprophylaxis: Not available for use in humans
Brucella (B. abortus, B. canis, B. melitensis, B. suis)	Direct skin contact with cultures or with infectious clinical specimens from animals, and aerosols are significant hazards. Hypersensitivity to *Brucella* antigens may also be a hazard to laboratory personnel	Biosafety Level 2: footnote *f* — Biosafety Level 3: footnote *g* — Animal Biosafety Level 3: footnote *c* — Immunoprophylaxis: Not available for use in humans

(continued)

TABLE III (continued)

Agent	Laboratory hazards	Recommended precautions
Campylobacter (C. jejuni/C. coli, C. fetus subsp. fetus)	Primary laboratory hazards are ingestion or parenteral inoculation of organisms; importance of aerosol exposure not known. Experimentally infected animals are potential source of infection	Biosafety Level 2: footnotes b and g Animal Biosafety Level 2: footnote c Immunoprophylaxis: Not available for use in humans
Chlamydia psittaci, C. pneumoniae, C. trachomatis	Contact with and exposure to infectious aerosols in handling, care, or necropsy of naturally or experimentally infected birds are major sources of laboratory-associated C. psittaci infections. Infected mice and eggs are less important sources of C. psittaci. Laboratory animals not a reported source of human infection with C. trachomatis. Primary laboratory hazards of C. trachomatis (and probably C. pneumoniae) are accidental parenteral inoculation and direct and indirect exposure of mucous membranes of eyes, nose, and mouth to genital, bubo, or conjunctival fluids, cell culture materials, fluids from infected eggs, and infectious aerosols	Biosafety Level 2: footnotes f and h Gloves are recommended for necropsy of birds and mice, opening of inoculated eggs, and when there is likelihood of direct skin contact with infected tissues, bubo fluids, and other clinical material. Wetting feathers or skin with a detergent–disinfectant prior to necropsy can appreciably reduce risk of aerosols from infected feces and nasal secretions on external surface of animal Animal Biosafety Level 2: footnote i Biosafety Level 3: footnotes d and e Immunoprophylaxis: Not available for use in humans
Clostridium botulinum	Exposure to toxin is primary laboratory hazard. Toxin may be absorbed after ingestion or following contact with skin, eyes, or mucous membranes, including respiratory tract. Accidental parenteral inoculation also represents significant hazard	Biosafety Level 2: footnote j Solutions of sodium hypochlorite (0.1%) or sodium hydroxide (0.1 N) readily inactivate toxin and are recommended for decontaminating work surfaces and spills of cultures or toxin Animal Biosafety Level 2: footnote j Biosafety Level 3: footnotes d and e Immunoprophylaxis: Pentavalent (ABCDE) botulism toxoid available through Centers for Disease Control and Prevention (CDC), as an investigational new drug (IND). This toxoid is recommended for personnel working with cultures of C. botulinum or its toxins

Agent	Hazard	Recommendations
Clostridium tetani	Accidental parenteral inoculation and ingestion of toxin are primary hazards. It is uncertain if tetanus toxin can be absorbed through mucous membranes	Biosafety Level 2: footnote *g* Immunoprophylaxis: Administration of adult diphtheria–tetanus toxoid at 10-year intervals to reduce risk to laboratory and animal care personnel of toxin exposures and wound contamination
Corynebacterium diphtheriae	Inhalation, accidental parenteral inoculation, and ingestion are primary laboratory hazards	Biosafety Level 2: footnotes *b* and *g* Animal Biosafety Level 2: footnote *c* Immunoprophylaxis: Administration of adult diphtheria–tetanus toxoid at 10-year intervals to reduce risk to laboratory and animal care personnel of toxin exposures and work with infectious materials
Escherichia coli (all enteropathogenic, enterotoxigenic, and enteroinvasive strains and strains bearing K1 antigen)	Ingestion or parenteral inoculation of the organism represent primary laboratory hazards; importance of aerosol exposure is not known	Biosafety Level 2: footnotes *b* and *g* Animal Biosafety Level 2: footnote *c* Immunoprophylaxis: Not available for use in humans
Francisella tularensis	Very hazardous. Infection can result from direct contact of skin or mucous membranes with infectious materials, accidental parenteral inoculation, ingestion, and exposure to aerosols and infectious droplets. Human ID_{25-50} is ~10 organisms via respiratory route	Biosafety Level 2: footnote *f* Biosafety Level 3: footnote *g* Animal Biosafety Level 3: footnote *c* Immunoprophylaxis: Investigational live attenuated vaccine is available and recommended for persons working with the agent or infected animals and for persons working in or entering the laboratory or animal room where cultures or infected animals are maintained
Haemophilus ducreyi, *H. influenzae*	Accidental parenteral inoculation and aerosols represent primary laboratory hazards	Biosafety Level 2: footnotes *b* and *g* Animal Biosafety Level 2: footnote *c* Immunoprophylaxis: Not available for use in humans

(continued)

TABLE III (*continued*)

Agent	Laboratory hazards	Recommended precautions
Legionella pneumophila; other *Legionella*-like agents	Generation of aerosols during manipulation of cultures or of other materials containing high concentrations of infectious microorganisms (e.g., infected yolk sacs and tissues) is primary hazard	Biosafety Level 2: footnotes *b* and *g* Animal Biosafety Level 2: footnote *c* Biosafety Level 3 with primary containment devices (e.g., biological safety cabinets, centrifuge safety cups): footnotes *d* and *e* Immunoprophylaxis: Not available for use in humans
Leptospira interrogans (all serovars)	Ingestion, accidental parenteral inoculation, and direct and indirect contact of skin or mucous membranes with cultures, infected tissues, or body fluids (especially urine) are primary laboratory hazards; importance of aerosol exposure not known	Biosafety Level 2: footnotes *b*, *f*, and *g* Animal Biosafety Level 2: footnotes *c* and *h* Gloves are recommended for handling and necropsy of infected animals and when there is likelihood of direct skin contact with infectious materials Immunoprophylaxis: Not available for use in humans
Listeria (all species)	Ingestion and accidental parenteral inoculation represent primary laboratory hazards. Pregnant women may be at increased risk as listeriosis can be transmitted across the placenta and cause fetal death or congenital infection	Biosafety Level 2: footnotes *b* and *g* Animal Biosafety Level 2: footnote *c* Immunoprophylaxis: Not available for use in humans
Mycobacterium leprae	Direct contact of skin or mucous membranes with infectious materials and accidental parenteral inoculation are primary laboratory hazards	Biosafety Level 2: footnote *f* Extraordinary care should be taken to avoid accidental parenteral inoculation with contaminated sharp instruments Animal Biosafety Level 2: footnote *c*
Mycobacterium spp. other than *M. tuberculosis*, *M. bovis*, or *M. leprae*	Direct contact of skin or mucous membranes with infectious materials, ingestion, and accidental parenteral inoculation are primary laboratory hazards associated with clinical materials and cultures; infectious agents in aerosols created during manipulation of broth cultures or tissue homogenates of species associated with pulmonary disease also pose risk to laboratory personnel	Biosafety Level 2: footnotes *b* and *g* Animal Biosafety Level 2: footnote *c* Immunoprophylaxis: Not available for use in humans

Mycobacterium tuberculosis, M. bovis	A proven hazard to laboratory personnel as well as others who may be exposed to infectious aerosols in the laboratory (ID_{50} <10 bacilli). Tubercle bacilli may survive in heat-fixed smears and may be aerosolized during preparation of frozen sections and during manipulation of liquid cultures	Biosafety Level 2 for preparation of acid-fast smears and culturing of sputa or other clinical specimens provided that aerosol-generating manipulations are conducted in a Class I or II biological safety cabinet. Liquefaction and concentration of sputa for acid-fast staining may be conducted safely on the open bench by first treating specimen (in a Class I or II biological safety cabinet) with an equal volume of 5% sodium hypochlorite solution (undiluted household bleach) and waiting 15 min before centrifugation)
		Animal Biosafety Level 2: footnote *c* (guinea pigs or mice)
		Biosafety Level 3: footnote *g*
		Animal Biosafety Level 3: footnote *c* (nonhuman primates experimentally or naturally infected)
		Immunoprophylaxis: Licensed attenuated live vaccine (BCG, *Bacillus* Calmetti–Guerin) is available but not routinely used in laboratory personnel
Neisseria gonorrhoeae	Accidental parenteral inoculation and direct or indirect contact of mucous membranes with infectious clinical materials or cultures are known primary laboratory hazards; importance of aerosols has not been determined	Biosafety Level 2: footnotes *b* and *g*
		Animal Biosafety Level 2: footnote *c*
		Gloves should be worn when handling infected animals and when there is likelihood of direct skin contact with infectious materials
		Biosafety Level 3: footnotes *d* and *e*
		Immunoprophylaxis: Not available for use in humans

(continued)

TABLE III (continued)

Agent	Laboratory hazards	Recommended precautions
Neisseria meningitidis	Parenteral inoculation, droplet exposure of mucous membranes and infectious aerosol and ingestion are the primary hazards to laboratory personnel	Biosafety Level 2: footnotes *b* and *g* Animal Biosafety Level 2: footnote *c* Biosafety Level 3: footnotes *d* and *e* Immunoprophylaxis: Licensed polysaccharide vaccines are recommended for personnel regularly working with large volumes or high concentrations of infectious materials
Pasteurella multocida	Aerosols and accidental parenteral inoculation represent primary laboratory hazards	Biosafety Level 2: footnotes *b* and *g* Gloves should be worn when handling if there is likelihood of direct skin contact with infectious materials. Necropsy of laboratory animals should be conducted in a biological safety cabinet Biosafety Level 3: footnotes *d* and *e* Animal Biosafety Level 3: footnote *c* Immunoprophylaxis: Not available for human use
Pseudomonas pseudomallei	Direct contact with cultures and infectious materials from humans, animals, or the environment; ingestion, autoinoculation, and exposure to infectious droplets and aerosols are primary laboratory hazards	Biosafety Level 2: footnotes *b* and *g* Gloves should be worn when there is likelihood of direct skin contact with infectious materials Biosafety Level 3: footnotes *d* and *e*
Salmonella (all serotypes except *S. typhi*)	Ingestion or parenteral inoculation are primary laboratory hazards; importance of aerosol exposure is not known. Naturally or experimentally infected animals are potential source of infection for laboratory and animal care personnel	Biosafety Level 2: footnotes *b* and *g* Animal Biosafety Level 2: footnote *c*
Salmonella typhi	Ingestion or parenteral inoculation of organism represent primary laboratory hazards; importance of aerosol exposure is not known	Biosafety Level 2: footnotes *b* and *g* Biosafety Level 3: footnotes *d* and *e* Immunoprophylaxis: Licensed vaccines available for personnel regularly working with cultures or clinical materials which may contain *S. typhi*

Organism	Hazards	Biosafety
Shigella spp.	Ingestion or parenteral inoculation of organism are primary laboratory hazards; importance of aerosol exposure is not known. Oral ID_{25-50} for humans ~200 organisms	Biosafety Level 2: footnotes *b* and *g* Animal Biosafety Level 2: footnote *g* Immunoprophylaxis: Not available for use in humans
Staphylococcus aureus	Accidental parenteral inoculation and exposure of mucosal surfaces to aerosols or droplets represent primary laboratory hazards	Biosafety Level 2: footnotes *b* and *g* Animal Biosafety Level 2: footnote *c* Immunoprophylaxis: Not available for use in humans
Streptococcus pyogenes, *S. pneumoniae*	Accidental parenteral inoculation and exposure of mucosal surfaces to aerosols or droplets represent primary laboratory hazards	Biosafety Level 2: footnotes *b* and *g* Animal Biosafety Level 2: footnote *c* Immunoprophylaxis: Licensed pneumococcal vaccine, containing 23 serotypes, is available. Vaccines for other streptococci are not available for use in humans
Treponema pallidum	Accidental parenteral inoculation, contact of mucous membranes or broken skin with infectious clinical materials or organisms, and possibly infectious aerosols are primary hazards to laboratory personnel. ID_{50} by subcutaneous route is ~23 organisms	Biosafety Level 2: For all activities involving use or manipulation of blood or lesion materials from humans or infected rabbits and for manipulation of cell cultures of *T. pallidum*. Gloves should be worn when there is a likelihood of direct skin contact with organisms or lesion materials. Periodic serological monitoring should be considered in personnel regularly working with infectious materials Animal Biosafety Level 2: footnote *c* Immunoprophylaxis: Not available for use in humans

(continued)

TABLE III (continued)

Agent	Laboratory hazards	Recommended precautions
Vibrio cholerae, V. parahaemolyticus	Ingestion of *V. cholerae* and ingestion or parenteral inoculation of other vibrios constitute primary laboratory hazards; importance of aerosol exposure is not known. Human oral ID_{100} of *V. cholerae* in healthy non-achlorhydric individuals is $\sim 10^6$ organisms; risk of infection following oral exposure increased in achlorhydric individuals	Biosafety Level 2: footnotes *b* and *g* Animal Biosafety Level 2: footrote *c* Immunoprophylaxis: Available. but routine use not recommended for laboratory personnel
Yersinia pestis	Direct contact with cultures and infectious materials from humans or rodents, infectious aerosols or droplets generated during the manipulation of cultures and infected tissues and in the necropsy of rodents, accidental autoinoculation, ingestion and bites from infected fleas collected from rodents are primary hazards to laboratory personnel	Biosafety Level 2: footnotes *b* and *g* Gloves should be worn when handling field-collected or infected laboratory rodents and when there is likelihood of direct skin contact with infectious materials. Necropsy of rodents should be conducted in a biological safety cabinet Biosafety Level 3: footnotes *d* and *e* and for work with antibiotic-resistant strains Immunoprophylaxis: Licensed inactivated vaccine available. Immunization recommended for personnel working regularly with cultures or infected rodents
Rickettsial agents *Coxiella burnetii*	Very hazardous. Parenteral inoculation and exposure to infectious aerosols and droplets are most likely sources of infection to laboratory and animal care personnel. Estimated ID_{25-50} via inhalation is 10 organisms	Biosafety Level 2: footnote *k* Biosafety Level 3: footnotes *h* and *m* Animal Biosafety 3: footnote *c* Immunoprophylaxis: Investigational new vaccine available from U.S. Army Medical Research Institute for Infectious Diseases (Fort Detrick, MD). Use of the vaccine should be limited to those at high risk of exposure who have no demonstrated sensitivity to Q-fever antigen

Rochalimaea quintana, R. vinsonii, Rickettsia akari; Ehrlichia chaffeensis	Exposure to naturally or experimentally infected vectors and accidental parenteral inoculation most likely sources of laboratory associated infections. *Rochalimaea henselae* is a recently isolated species of uncertain hazard,[n] but, based on observations with closely related species, it is provisionally recommended that it be handled like *R. quintana* and *R. vinsonii*.	Biosafety Level 2: footnote *l* Animal Biosafety Level 2: footnote *c*
Rickettsia prowazekii, R. typhi (R. mooseri), R. tsutsugamushi, R. canada, and spotted fever group agents of human disease other than *R. rickettsii* and *R. akari*	Accidental parenteral inoculation and exposure to infectious aerosols are primary laboratory hazards	Biosafety Level 2: footnote *k* Animal Biosafety Level 2: footnote *c* (other than flying squirrels or arthropods) Biosafety Level 3: footnotes *b, m,* and *o* Animal Biosafety Level 3: footnote *c* (flying squirrels) Immunoprophylaxis: Not currently available for use in humans
Rickettsia rickettsii	Accidental parenteral inoculation and exposure to infectious aerosols are primary laboratory hazards	Biosafety Level 2: footnote *k* Animal Biosafety Level 2: for holding of experimentally infected rodents Biosafety Level 3: footnotes *b, e, h,* and *m* Immunoprophylaxis: Not currently available for use in humans Other: It is important to have an effective system for reporting and evaluating febrile illnesses in laboratory personnel working with *R. rickettsii*

[a] The recommendations for working with bacterial and rickettsial agents are taken, in part, from HHS Pub. No. 86-23.[5] Additional recommendations were made based on the class of the agent.

[b] For activities using clinical materials and diagnostic quantities of infectious material.

(continued)

TABLE III (continued)

[c] For activities with naturally or experimentally infected laboratory animals (other than necropsy).

[d] For work involving production volumes or high concentrations of cultures or toxins.

[e] For activities which have high potential for aerosol or droplet production (e.g., centrifuging, grinding, blending, vigorous shaking or mixing, sonication, homogenization of tissues).

[f] For activities with clinical specimens from or of animal origin.

[g] For all manipulations of cultures and/or toxin.

[h] For activities involving necropsy of infected animals.

[i] Respiratory protection recommended for personnel working with naturally or experimentally infected caged birds.

[j] For all activities with materials known or potentially containing toxin.

[k] For nonpropagative laboratory procedures, including serological examinations and staining or impression smears.

[l] For propagation.

[m] For activities involving the inoculation, incubation, and harvesting of embryonated eggs or tissue cultures.

[n] R. L. Regnery, B. E. Anderson, J. E. Clarridge III, M. C. Rodriguez-Barradas, D. C. Jones, and J. H. Carr, J. Clin. Microbiol. 30, 265 (1992).

[o] For homogenization of tissue.

Section I

Evaluations of Virulence in Animal Models

[2] Determination of Median Lethal and Infectious Doses in Animal Model Systems

By SUSAN WELKOS and ALISON O'BRIEN

Introduction

Dose–response experiments in animal models are often used to compare the virulence or colonizing capacity of a bacterial mutant to the wild-type parent or to assess the potency of a bacterial toxin. In such studies, the investigator usually calculates the median response dose, namely, the dose that produces a response in 50% of the subjects in a treatment group. The response to be measured may be death, infectivity, or some other quantal (yes-or-no, all-or-none) effect, and the corresponding values to be determined are the 50% lethal dose (LD$_{50}$) of the microbe or toxin, the 50% infectious dose (ID$_{50}$), or, more generally, the 50% effective dose (ED$_{50}$). The 50% end point is usually preferred because it tends to be a more accurate determination than the end point at the extremes of a dose–response curve, particularly when such curves are flat.[1]

To generate data for a 50% end point determination, the usual format with small animals such as mice is to give 3–6 groups of 5–10 animals each different doses of the organism/toxin. A geometric sequence of doses should be administered, such as 2-fold or 10-fold dilutions of the material. Because virulence, toxicity, and infectivity are often relative to the host, reproducible median lethal dose studies may require the use of inbred strains from the same source and of the same age, sex, and housing conditions. Also, precautions should be taken to maintain the genetic stability of the infecting microbe and to culture and administer it identically in every experiment for which one wishes to compare the results with previous studies. The range of doses selected for ED$_{50}$ studies should yield a response range of at least 15 to 85%.[2] Most importantly, an incremental relationship between the dose and the response should be evident from examination of the data.[2] If the foregoing conditions can be satisfied, similar results will be obtained using any of the several methods for estimating end point described below.

---

[1] F. Sheffield, in "Topley and Wilson's Principles of Bacteriology, Virology, and Immunity" (G. Wilson and H. M. Dick, eds.), Vol. 1, p. 429. Williams & Wilkins, Baltimore, Maryland, 1983.

[2] H. C. Batson, "An Introduction to Statistics in the Medical Sciences," p. 62. Chicago Medical Book Company, Chicago, 1956.

The purpose of this chapter is to describe four general approaches for estimating the 50% end point. The most appropriate method for the individual researcher depends on the animal model used, the importance of determining other parameters such as the confidence interval (*CI*) of the LD_{50}, and the complexity of the calculations and/or availability of certain computer programs. Each procedure has inherent advantages and disadvantages and requires that certain conditions be met to use it correctly. Finally, because we are not statisticians, we recommend that an investigator consult a statistician when in doubt about the mathematical appropriateness of an experimental design, interpretation of complex automated analyses, or the best approach for conducting multiple comparisons of 50% end points.

Procedures Used to Estimate Median Lethal Dose (or Median Infectious Dose)

The LD_{50} can be estimated by direct observation of the 50% lethal dose from a plot of the dose–response data or by a graphical method, using log probit[3] or log probability[4] graph paper. However, manual and computerized arithmetic methods are now more commonly used.

Reed–Muench Method

The Reed–Muench procedure[3] is probably the most commonly used method for determining the LD_{50} when the number of animals or amount of reagents is not limiting. In this cumulative approach, the number of survivors and deaths for the various dosage levels are added sequentially and the accumulated values used to calculate an LD_{50}. The likelihood that the result obtained will be in close proximity to the value determined by the more statistically rigorous probit analysis (see next section) is good if on examination of the experimental data the percentage of responders in each group incrementally increases with dose. The problems with the Reed–Muench approach are 2-fold: (1) the estimated LD_{50} can be misleading if the range of doses tested is not symmetrically distributed around the log LD_{50}, and (2) confidence intervals cannot be determined. The procedure for calculating an LD_{50} is illustrated in Example 1 (see Examples section) and Table I.

---

[3] L. C. Miller and M. L. Tainter, *Proc. Soc. Exp. Biol. Med.* **57**, 261 (1944).
[4] J. T. Litchfield, Jr., and F. Wilcoxon, *J. Pharmacol. Exp. Ther.* **96**, 99 (1949).

Probit Method

Probit analysis is probably the most precise statistical approach for determining 50% end points, for data which fulfill its requirements.[2] The probit method depends on the assumption that the observed responses are sampled from a population where the responses are normally distributed. As a rule of thumb, probit analysis requires that at least 3 partial responses (>0%, <100%) are obtained, and at least 6 animals per group and 5–6 dosage levels used. The formal probit analysis involves a weighted linear regression on transformed data which estimates the parameters describing the entire dose–response curve. From these, the ED_{50} and its variance, doses and *CI* values for percentiles other than the median (LD_{10}, LD_{90}, etc.), and the goodness of fit of the model may be calculated. Programs, such as those of the Statistical Analysis System (SAS, Cary, NC), provide this complete analysis. These programs are relatively expensive and difficult to operate, but they can now be run on a personal computer, produce results accepted by the U.S. Food and Drug Administration (FDA), and are used by major pharmaceutical firms. Other probit programs using a small computer or programmable hand calculator have been developed, which, if used appropriately, might be easier to use, less costly, and more convenient than the former.[5-11] For the purposes of this chapter, an example of the Miller–Tainter graphical estimate[3] of the 50% point by probits is presented in Example 1 and Table 1 below. In this and the more formal probit procedures, the median dose is determined after transforming the mortality percentages to probits and the doses to logarithmic values.

If the conditions for probit analysis cannot met, as is the case in certain host–pathogen model systems, much of the statistical information derived will be misleading (i.e., parameters defining the shape of the curve and estimating the probabilities at the extremes, and comparisons of the LD_{50} values of several strains). For example, the probit method might be inap-

[5] S. C. Mehta, D. K. Jain, and C. K. Gupta, *Comput. Biol. Med.* **21,** 167 (1991).

[6] H. R. Lieberman, *Drug Chem. Toxicol.* **6,** 111 (1983).

[7] G. M. Schoofs and C. C. Willhite, *J. Appl. Toxicol.* **4,** 141 (1984).

[8] K. P. Fung, *Comput. Biol. Med.* **19,** 131 (1989).

[9] C. E. Stephan, *in* "Aquatic Toxicology and Hazard Evaluation. ASTM STP 634" (F. L. Mayer and J. L. Hamelink, eds.), p. 65. Amer. Soc. Test. Materials, Philadelphia, Pennsylvania, 1977.

[10] M. M. Abou-setta, R. W. Sorrell, and C. C. Childers, *Bull. Environ. Contam. Toxicol.* **36,** 242 (1986).

[11] Commercially available programs include the SAS version (SAS, Cary, NC) as well as the following available from Walonick Associates, Inc., Minneapolis, MN: Goodness-of-Fit regression modeling package, 1985. Probit Regression, in Stat-Packets Statistical Analysis Package for Lotus Worksheets, 1987.

propriate when the data are not normally distributed, the number of intermediate responses is too few, or the supply of animals or reagents is
limiting. Other tests are available that may require fewer animals, have
more precision than methods such as Reed–Muench, and provide the
information of most interest, namely, the LD_{50} and its confidence interval.

Moving Average Interpolation

Thompson and Weil developed a method using moving averages that
makes no assumptions about the exact nature of the dose–response curve
involved.[12–15] It is helpful, for instance, in tests with too few animals to
accurately know the shape of the curve, yet uses more of the data than
methods such as Reed–Muench that use only the values on both sides of
the 50% response level. The median dose is calculated by interpolation
using the arithmetic means of the doses (in log values) and of the fraction
responding for successive points. The method involves simple calculations
and the use of published tables, and it does not require plotting. The LD_{50}
and its 95% *CI* and estimated standard deviation can be calculated,[13–15]
as can a value for the slope of the curve, if required.[16] The *CI* and slope
may help in estimating the shape of the dose–response curve within the
central region; otherwise, however, the dose–response curve is not well
characterized. The results of the analysis are usually in close agreement
with those calculated by the more rigorous and complex probit analysis,
which requires more animals. A program has been written to automate
and simplify determinations of LD_{50} by moving average interpolation
(MAI), and it is used by several investigators at U.S. Army Medical
Research Institute of Infectious Diseases. It is performed on an IBM-
compatible personal computer and is available on request.[17]

There are only a few requirements to use the MAI method: (1) the
same number of animals for each dose level (tables based on 2–6 or 10
animals per level are available)[14,15]; (2) equal spacing between successive
doses (e.g., 2- or 10-fold); and (3) at least 2 dose levels. However, for
best precision, at least 4 dose levels with 4–5 animals per dose should be

[12] W. R. Thompson, *Bacteriol. Rev.* **11**, 115 (1947).
[13] W. R. Thompson and C. S. Weil, *Biometrics* **8**, 51 (1952).
[14] C. S. Weil, *Biometrics* **8**, 249 (1952).
[15] S. C. Gad and C. S. Weil, *in* 'Principles and Methods of Toxicology'' (A. W. Hayes,
ed.), p. 435. Raven, New York, 1989.
[16] C. S. Weil, *Drug Chem. Toxicol.* **6**, 595 (1983).
[17] H. Dewey and P. Gibbs, Moving Average Interpolation program for the PC, unpublished,
1990. To receive a copy of the program, send the request to Paul Gibbs, BIMD Division,
USAMRIID, Ft. Detrick, Frederick, MD 21702-5011. Include a 1.4 MB/3.5 inch diskette
formatted for an IBM PC-compatible computer and a self-addressed, stamped return envelope.

used. Only 4 points are analyzed although more doses can be used in the experiment, to ensure that the 50% region is bracketed. Determinations of LD$_{50}$ values calculated manually or using the computer program are illustrated in Examples 1 and 2 below.

Staircase

Most methods for performing ED$_{50}$/LD$_{50}$ experiments use a prescribed number of animals tested simultaneously at each of several doses. In the staircase (or up-and-down) method, the doses are tested one at a time with one animal each.[18] This procedure is best used when the supply of animals or test materials is limited, animal purchase or care is expensive, or other factors restricting animal use are present. Such situations occur most often when large animals such as nonhuman primates are used. It is also useful in a pilot experiment to estimate the LD$_{50}$ where few prior data are available, before doing a more detailed analysis. The LD$_{50}$ and standard deviation can be estimated by the staircase method, but information is not provided on the slope of the dose–response curve or lethal dose values at other points on the curve. One drawback of the procedure is the increased length of time required to complete the entire experiment. If time is a limiting factor, but animals are not, a method such as MAI might be more appropriate.

To use the staircase method, a series of equally spaced doses, usually logarithmic dilutions, is selected. Ideally, the spacing should approximate the standard deviation of the responses. For a good estimate of LD$_{50}$ the series should include 2–3 doses for which there are both responding and nonresponding animals. The first dose tested should be at the best estimate of the actual LD$_{50}$ (based on historical data or literature values). Each succeeding dose depends on the outcome of the previous repetition; that is, the dose is decreased if the dose was lethal and increased if the animal survived. Testing is continued until the calculated LD$_{50}$ value, or standard deviation, stabilizes. The configuration of responses is used to obtain a value (from the table of maximum likelihood estimates of LD$_{50}$),[18] which is then used to calculate the LD$_{50}$, as shown in Example 3 below.

Examples of Median Lethal Dose Calculations

Example 1

The following example will be used to show calculation of LD$_{50}$ values by the Reed–Muench, Probit, and MAI methods, namely, determination

---

[18] W. J. Dixon, *J. Am. Stat. Assoc.* **60**, 967 (1965).

TABLE I

CALCULATION OF ORAL MEDIAN LETHAL DOSE FOR EXAMPLE BY REED–MUENCH AND PROBIT METHODS

Dose (log)	Number dead (*D*)	Number survived (*S*)	% dead	Probit[a]	Accumulated[b] *D*	Accumulated[b] *S*	Accumulated[b] Total	Accumulated % dead[c]
7	10	0	100	8.09	30	0	30	100.0
6	9	1	90	6.28	20	1	21	95.2
5	7	3	70	5.52	11	4	15	73.3
4	4	6	40	4.74	4	10	14	28.6
3	0	10	0	2.67	0	20	20	0.0

[a] Value is derived from the % dead (column 4) and obtained from a table.[3]

[b] Accumulated dead are obtained by adding successive entries in column 2 from the bottom to the top; accumulated survivals are calculated by adding successive entries in column 3 from the top to the bottom. Totals (column 8) are obtained by adding the values in columns 6 and 7 at each dose.

[c] Accumulated percent dead are calculated by dividing the accumulated total dead (column 6) by the total (column 8) and multiplying times 100.

of the LD_{50} of a derivative of virulent *Salmonella* strain TML after oral injection of BALB/c mice. A preliminary experiment showed that one dose of 10^7 colony-forming units (cfu) killed all injected mice, and thus the derivative was still virulent. (Note: A provisional LD_{50} could have been determined in this experiment using MAI, by testing 2–3 doses with 4–5 animals each.) In the subsequent experiment, 5 groups of 10 animals each were challenged with 10-fold dilutions and the following mortalities obtained: 10^7 cfu, 10; 10^6 cfu, 9; 10^5 cfu, 7; 10^4 cfu, 4; 10^3 cfu, 0.

Reed–Muench Method

The calculation of LD_{50} values is shown in Table I. The experimental data are entered into the first three columns (see Table I). Accumulated deaths are obtained by adding successive entries in column 2 (number dead) from the bottom to the top; accumulated survivals are obtained from adding successive entries in column 3 (number survived) from the top to the bottom. Accumulated totals in column 8 are obtained by adding the accumulated deaths and accumulated survivors at each dose. Accumulated percent dead (column 9) is obtained by dividing cumulative dead per total times 100 for each dose. The 50% end point is determined by calculating the proportionate distance between the doses that bracket the 50% point, log 5 and 4, respectively[19]:

[19] L. J. Reed and H. Muench, *Am. J. Hyg.* **27,** 493 (1938).

$$\text{Proportionate distance} = \frac{50\% - \text{next lowest }\%}{\text{next highest }\% - \text{next lowest }\%}$$

$$= \frac{50 - 28.6}{73.3 - 28.6} = \frac{21.4}{44.7} = 0.4787$$

To obtain the dose corresponding to the 50% point, the proportionate distance is multipled by the logarithm of the dose increment, and this value is added to the log of the dose of the next lowest percent. The antilog of this sum is the 50% value. For Example 1

$$\log \text{LD}_{50} = (0.4787 \times \log 10) + 4 = 4.4787$$
$$\text{LD}_{50} = 3.0 \times 10^4 \text{ cfu}$$

Alternatively, one can graphically estimate the median dose by plotting the bracketing percentages as linear ordinate values on semilog paper and the corresponding doses on the logarithmic abscissa. The dose corresponding to the 50% point is then extrapolated from the line drawn between these two points.

Probit Method

The LD$_{50}$ and its parameters are estimated by the Miller–Tainter graphical method.[3] The percent deaths are first transformed to probits using a probit table,[6,20,21] as shown in Table I. The probit values are not weighted equally; those furthest from 5.0 are weighted less than those closest to 5.0.[2] The probit values are then plotted as linear ordinate values on semilog paper and the corresponding doses plotted on the logarithmic abscissa. A straight line is drawn that best fits the data. The dose corresponding to probit 5 on the line is the estimated 50% point. The 95% *CI* is estimated by the following formula[3]:

$$95\% \; CI = \text{LD}_{50} \pm (1.96 \times \text{standard error of the LD}_{50})$$

The standard error (SE) of the LD$_{50}$ is calculated as

$$\text{SE} = 2s/(2N)^{1/2}$$

where $2s$ (2 standard deviations) is the difference between the doses corresponding to probits 4 and 6, and N is the total number of animals used in groups spanning the range of probits 3.5 to 6.5. For the data in Table I,

[20] D. J. Finney, "Probit Analysis," p. 1. Cambridge Univ. Press, London, 1971.
[21] R. A. Fisher and F. Yates, "Statistical Tables for Biological, Agricultural and Medical Research," p. 68. Longman Group, Essex, U.K., 1974.

$$\log \mathrm{LD}_{50} \text{ and } 95\% \; CI = 4.40 \pm \left[1.96 \times \frac{(5.66 - 3.08)}{(2 \times 30)^{1/2}} \right]$$

$$= 4.40 \pm (1.96 \times 0.333) = 4.40 \pm 0.653$$

Thus, the 95% CI of the log LD_{50} spans 5.05 (upper) to 3.75 (lower). Results of the probit analysis using the SAS computer program are as follows:

$$\log \mathrm{LD}_{50} \text{ and } 95\% \; CI - 4.52 \pm 5.02 \text{ (upper) and } 3.99 \text{ (lower)}$$
$$\log \mathrm{SE} \text{ of the } \mathrm{LD}_{50} = 1.16$$

Moving Average Interpolation Method

The results using the MAI computer program are

$$\mathrm{LD}_{50} = 3.16 \times 10^4 \text{ cfu}$$
$$95\% \; CI = 1.05 \times 10^4, 9.55 \times 10^4$$

The slope, if required, can be calculated manually,[16] and for this example it was 1.0. The input required by the program and the computer output are given below:

```
C:/mai.exe
Enter the no. of dose levels: 5
Enter the no. of subjects at each dose level: 10
*****Enter dose as either a fraction or decimal value************
[Program accepts arithmetic values or values in scientific notation]
Enter the highest dose: 1.0e7
Enter the next highest dose: 1.0e6
Enter the no. of critical events at level 1 (7.000E+00): 10
Enter the no. of critical events at level 2 (6.000E+00): 9
Enter the no. of critical events at level 3 (5.000E+00): 7
Enter the no. of critical events at level 4 (4.000E+00): 4
Enter the no. of critical events at level 5 (3.000E+00): 0
LD50 = 3.162E +04     95 percent C.I.: (1.047E+04, 9.552E+04)
```

Example 2

Example 2 illustrates manual calculation of the LD_{50} by MAI. The virulence of the TML derivative was to be tested in a mouse strain, such as a congenic or recombinant inbred, that was expensive and limited in availability. Thus, it was desired to reduce the number of mice used in the test. Twenty mice were divided into 4 groups of 5 each and given doses of 10^3 to 10^6 cfu. Mortalities were 0, 2, 3, and 5, respectively. The following are the equations used to derive the LD_{50}, CI, and standard deviation.[12–15]

Median Lethal Dose

$$\log m = \log D + d(K - 1)/2 + df$$

where m is the median effective dose (LD$_{50}$); D, the lowest dose tested; d, the log of the geometric factor between dose levels (0.30 for 2-fold and 1.0 for 10-fold doses); f, the value read from the table in Weil[14] or Gad and Weil[15]; K, number of dose levels used in the calculation $- 1$ ($K = 3$, for Example 2). Values needed to read the tables include n or N, number of animals per dose ($N = 5$, for Example 2); r, number of mortalities in each dose group.

95% Confidence Interval

$$95\% \; CI = \text{antilog of } \log m \pm 2 \times \partial_{\log m}$$

where $\partial_{\log m}$ (standard deviation) $= d \times \partial_f$ (from table in Ref. 14 or 15).
Slope. The value of the slope is rarely required, but it can be calculated as described in Ref. 1.

In Example 2, the table specifying $K = 3$, $N = 5$ is used. It gives values of 0.5000 for f and 0.34641 for ∂_f. Thus,

$$\log \text{LD}_{50} \; (\log m) = 3 + 1(3 - 1)/2 + 1(0.50) = 4.5$$
$$\text{LD}_{50} = 3.2 \times 10^4, \text{ with } 95\% \; CI = 6.4 \times 10^3, 1.6 \times 10^5$$

where 95% $CI = 4.5 \pm 2 \times (1 \times 0.3464)$. An estimate of the standard error is

$$\text{SE of LD}_{50} = 0.3463 \; (\log) = 2.2$$

Note the similarity of the LD$_{50}$ value to that obtained in Example 1, as would be expected from the mortalities; the large CI in Example 2 reflects the reduced size of the experiment.

Example 3

A defined component vaccine for respiratory pathogen X was developed, and its protective efficacy for mice against intraperitoneal challenge was determined. To validate use of this vaccine in humans and as a prerequisite for FDA approval, the efficacy of the vaccine against aerosol challenge was tested in primates. Prior to doing this study, an accurate estimation of the inhalation LD$_{50}$ of the challenge strain was needed. The value was obtained by using the staircase method and a minimum number of monkeys. A 2-fold dilution series of an inoculum containing 200 mouse LD$_{50}$ values was made, and LD$_{50}$ tests were begun at the 1/4 dilution, based on an estimated lethal dose of 50 mouse LD$_{50}$ values from the

Dilution	1/16	1/8	1/4	1/2	1
Log dose	-1.20	-0.90	-0.60	-0.30	0

Animal					
1			O		
2					X
3				X	
4		O			
5				X	
6		O			

FIG. 1. Results of the test series described in Example 3. For the OXXOXO series (O for alive and X for dead), the estimate of the LD_{50} as calculated by the staircase method is 44.6 mouse LD_{50} values. Calculations are shown in Table II and are based on the formula LD_{50} = (final test dose) + (value from table)(difference between dose levels).[18]

literature. Tests were continued until the calculated LD_{50} and standard deviation stabilized. The LD_{50} is estimated based on the formula (final test dose) = (value from table)(difference between dose levels) = LD_{50}, and the results for Example 3 shown in Fig. 1 and Table II. In this example,

TABLE II
CALCULATION OF INHALATION MEDIAN LETHAL DOSE FOR EXAMPLE 3
BY STAIRCASE METHOD

Series[a]	N^b	Equation for log dilution	Log dilution	$LD_{50}{}^c$	SE of LD_{50}
OX	2	$-0.30 + (-0.50)(0.30)$	-0.45	71.0	0.26
OXX	3	$-0.60 + (-0.178)(0.30)$	-0.65	44.8	0.23
OXXO	4	$-0.90 + (1.0)(0.30)$	-0.60	50.2	0.20
OXXOX	5	$-0.60 + (-0.305)(0.30)$	-0.69	40.6	0.18
OXXOXO	6	$-0.90 + (0.831)(0.30)$	-0.65	44.6	0.17

[a] Test results (O, alive; X, dead).
[b] N, Total number of trials reduced by one less than the number of like responses at the beginning of the series.
[c] For example, for the OXX series, where $N = 3 - (1 - 1) = 3$:

1. log LD_{50} dilution $= -0.60 + (-0.178)(0.30) = -0.65$
2. Antilog$(-0.65) = 0.224$
3. Reciprocal $= 4.46$
4. LD_{50} (no. of mouse LD_{50} values) $= 200/4.46 = 44.8$ mouse LD_{50} values
5. SE (of log LD_{50}) $= 0.88 (0.301) = 0.26$, where $0.301 = $ log 2

an LD_{50} value was obtained using only 6 animals, instead of the minimum 30 needed for a probit analysis.

Summary and Recommendations

In most cases, where the LD_{50}/ID_{50} is the major statistic of interest and an approximate value is known, any of several arithmetic or graphical methods could be used and will usually give comparable estimates. When the LD_{50} is unknown or resources are limiting, and sequential experimentation is feasible, the staircase method can first be used to define the region bracketing the LD_{50} value; then a test such as MAI is used. For situations requiring greater precision or more detailed statistical analysis of the data (i.e., curve fitting of the dose–response; estimates of variance of the LD_{50} and of percentiles other than the median; strain comparisons of LD_{50} values), and where the experimental design meets the requirements of the analysis, the initial determinations of LD_{50} by staircase or MAI can be followed by probit analysis. The SAS Probit package is recommended for those with access to it, and the information obtained should be interpreted with the help of a statistician.

Acknowledgments

We thank P. Gibbs, G. Nelson, and C. Stephan for expert advice and statistical consultation and H. Hodgkins for the calculation of LD_{50} using the SAS Probit program. Research in this area done in the laboratory of Dr. O'Brien was supported by National Institutes of Health Grant AL-20148-9 and Uniformed Services University of the Health Sciences Protocol No. RO7313-12.

[3] Experimental Keratoconjunctivitis (Sereny) Assay

By Dennis J. Kopecko

Introduction

Artificial conjunctival infection of animals was first introduced by Calmette in France in the 1920s for the study of pathogenesis.[1] At that time the eyes of experimental animals were first treated with a concentrated bile solution for sensitization and then specific bacterial inocula were introduced into the damaged conjunctival sac. Using this approach Zoeller

[1] B. Sereny, in "Pathogenesis of Intestinal Infections" (M. V. Voino-Yasenetsky and T. Bakacs, eds.), p. 76. Akademiai Kiado, Budapest, 1977.

and co-workers[2,3] were able to induce experimental keratoconjunctivitis in guinea pigs with *Shigella dysenteriae, Salmonella typhi, Salmonella paratyphi*-A, and *S. paratyphi*-B. However, owing to the doubtful efficacy of this bile-pretreatment method and problems with reproducibility, this model never gained widespread usage. Sereny[1,4,5] developed a more practical method for inducing experimental keratoconjunctivitis in guinea pigs and rabbits in 1954 which has since received widespread acceptance. This method of experimental infection has been applied to analyzing the mucosal invasion abilities of *Shigella* spp. and other enteropathogenic bacteria, the attenuation of the virulence state, and the development of immunity to bacterial infection. The Sereny assay continues to be useful for epidemiological studies, for analyses of bacterial virulence, and for vaccine efficacy determinations.

Sereny Assay

Experimental Animals

Induced conjunctival infection with *Shigella* inocula has been obtained in guinea pigs, rabbits, goats, horses, monkeys, mice, and rats,[1,6] but guinea pigs and rabbits are most susceptible. Availability, ease of handling, large eye size for inoculation and visualization, the development of reproducible, pronounced pathological responses to infection, and relatively low cost have led to the predominant use of guinea pigs for the keratoconjunctivitis assay. Sereny[1] has reported that guinea pig sex, age, and weight do not affect experimental symptoms, but albino animals showed more rapid and complete recovery from disease. There still exists considerable variability in the sex, weight, and strains of guinea pigs employed for the Sereny assay. However, homogeneous groups of animals (e.g, female, Hartley strain, 250–300 g)[7] should be employed whenever possible and certainly for experiments involving statistical analyses.

Each experimental animal should be examined carefully for normal eye clarity before use; weight and temperature can also be monitored. Culture of the normal conjunctival flora may be desirable in order to

[2] C. Zoeller, Bastouil, *C. R. Seances Soc. Biol. Ses Fil.* **90**, 1154 (1924).
[3] C. Zoeller, Manoussakis, *C. R. Seances Soc. Biol. Ses Fil.* **91**, 257 (1924).
[4] B. Sereny, *Acta Microbiol. Acad. Sci. Hung.* **2**, 293 (1955).
[5] B. Sereny, *Acta Microbiol. Acad. Sci. Hung.* **4**, 367 (1957).
[6] S. Murayama, T. Sakai, S. Makino, T. Kurata, C. Saskawa, and M. Yoshikawa, *Infect. Immun.* **51**, 696 (1986).
[7] S. Formal, P. Gemski, L. Baron, and E. LaBrec, *Infect. Immun.* **3**, 73 (1971).

establish unequivocally the behavior of a new organism in the experimental conjunctival infection model.

Preparation of Bacterial Inoculum

The bacterial strain to be examined may be cultured in liquid or solid nutrient medium, preferably in a nutritionally rich medium such as brain–heart infusion.[7] As discussed elsewhere,[8] the optimal growth conditions for the best expression of virulence traits vary among bacterial species and have not been empirically established for overall virulence in any bacterial pathogen. Also, simple agar passage or storage of a bacterial pathogen (e.g., *Shigella, Salmonella,* or *Campylobacter*) can lead to virulence attenuation. In fact, the Sereny assay offers an *in vivo* animal model of infection for maintaining shigellae in a highly virulent state.

Although bacteria grown on agar and broth (both fresh overnight and mid-log phase cultures) have been employed for the inocula, experience with *Shigella* indicates that mid-log phase broth cultures incubated at 37° work well. The bacteria are concentrated by centrifugation, are resuspended at ambient temperature in physiological saline or in fresh nutrient broth, and are used immediately. An inoculum of 5×10^8 or 10^9 bacteria should be administered in a volume of 10–20 μl in a plastic-tipped automatic micropipettor or similar device. Care should be taken so as not to damage the eye during the inoculation. A larger inoculum volume cannot be easily accommodated by the conjunctival sac. Appropriate invasive control bacteria should be employed, whereas saline (or sterile broth) alone can serve as a negative control.

Inoculation of Conjunctiva

The guinea pig needs to be grasped firmly at the nape of the neck with one hand while the other hand is gently used to pull out the lower eyelid. The bacterial inoculum is then instilled into the bottom of the conjunctival sac. The upper and lower eyelids of the inoculated eye are held closed for about 1 min while the eye is massaged to disseminate the inoculum. Sereny showed that the inoculated bacteria bind quickly, as washing of the conjunctival sac 1 min after the inoculation does not prevent the development of shigella keratoconjunctivitis.[1] Infection does not typically spread from one eye to the other in the same animal, so that with proper precautions (e.g., small inocula, uncrowded cages) both eyes can be used in separate assays. *Note:* Appropriate safety measures should be taken

[8] E. A. Elsinghorst, this series, Vol. 236.

when working with shigellae, which cause disease in humans at very low doses (see [1], this volume).

Bacterial Dose and Developmental Course of Shigella Keratoconjunctivitis

Because this model has been most successfully employed for the study of *Shigella* invasion of mucosal tissues, the following section specifically summarizes our understanding of *Shigella*-induced, experimental keratoconjunctivitis. Application of this technique to other enteropathogens will be briefly reviewed in a subsequent section.

Ultimate conjunctival infection with *Shigella* is highly dependent on the size of the infective dose and the relative virulence of the strain employed. Using graded doses and groups of animals, Sereny[1] reported median infective doses (ID_{50}) for virulent *Shigella* of between 10^2 and 10^7 cells. However, most assays today are carried out with 10^8–10^9 virulent shigellae as an inoculum, which generally results in a severe keratoconjunctivitis in 24–48 hr.

Conjunctival infection of previously uninfected, healthy guinea pigs with a large dose of virulent *Shigella* triggers the appearance of conjunctival edema and hyperemia generally within 5–10 hr, whereas smaller doses (or less virulent organisms) require up to 5 days to initiate visible infection. Inflammation of the conjunctiva and eyelids develops into a severe, purulent blepharoconjunctivitis within 24 hr, with a highly infective dose. Acute conjunctivitis may be produced by other bacteria (e.g., *Salmonella* spp.) but is, by itself, not characteristic of *Shigella*.

Next, the visible development of keratitis begins in the 24–48 hr time frame and is accompanied by fever. Histopathological analyses[1,9] have revealed that shigellae penetrate the epithelial cells of the conjunctiva and cornea, and multiply in their cytoplasm. The initial conjunctival infection triggers a large polymorphonuclear leukocyte (PMN) emigration from the subepithelial tissues, leading to a purulent exudate in the conjunctival sac. The mucopurulent exudate volume gradually increases to a peak at 2–3 days postinfection and, oftentimes, exudes from the eye. *Shigella* can be cultured from this exudate from infection until clinical healing occurs at 2–3 weeks.

Keratitis develops in characteristic stages (with the cornea changing from smooth brilliance to mildly opaque to a totally opaque, turbid, and undulated surface) beginning with the formation of isolated leukocyte infiltrates which become diffuse within 24 hr.[1,9] In this early stage, the

[9] M. Voino-Yasenetsky and T. Bakacs, *in* "Pathogenesis of Intestinal Infections" (M. Voino-Yasenetsky and T. Bakacs, eds.), p. 98. Akademiai Kiado, Budapest, 1977.

corneal surface becomes heavily infiltrated and ulcerated, and is covered with a whitish gray coat. The whitish gray coat, consisting of cellular debris and leukocytes, then detaches and denudes the swollen red cornea during the next few days. During the second week of infection, the edema and hyperemia disappear, and the cornea regains a more normal smooth surface, starting from the periphery and progressing toward the center. Dark-furred guinea pigs are slower in recovery or never recover completely.[1] For a more complete macroscopic and histopathological characterization of the development of keratoconjunctivitis shigellosa, readers are directed to the original descriptions[4,5] or to more recent reviews.[1,9]

In addition to the above macroscopic evidence of keratoconjunctivitis, shigellae induce a rise in guinea pig body temperature which lasts up to 3 days. In some animals, a loss in body weight occurs during the first few days postinfection. Also, a low level *Shigella* bacteremia has been observed in some conjunctivally infected animals, but the bacteremia is short-lived and does not affect the development of keratoconjunctivitis.[1]

There are differences in the development and severity of experimental keratoconjunctivitis infections. As opposed to the fulminant infection described above, relatively small doses of virulent organisms oftentimes lead to a longer latent period or evoke only moderate symptoms of keratoconjunctivitis. Although the majority of infections heal spontaneously within 2–3 weeks in albino guinea pigs, some last longer. Interestingly, Sereny[1] reports that 2–13% of the cases cause chronic keratoconjunctivitis which can be eliminated with antibiotic therapy. Very exceptionally, shigella sepsis may occur and lead to death.

Development of Immunity

Anti-*Shigella* serum antibodies are not present in previously unexposed, healthy animals. Conjunctival application of living avirulent or killed *Shigella* cells, which do not induce a gross pathological reaction in the eye, may nevertheless stimulate specific anti-*Shigella* antibody production. During experimental keratoconjunctivitis shigellosa, *Shigella* type-specific serum antibodies appear in the second week, rapidly reach a maximum titer of 1:80 to 1:640, and then generally decrease to zero in several months. However, antibody production may be entirely absent,[1] and there is no exact correlation between the presence of specific serum antibodies and protection from infection. As one might expect, specific antibodies can be found in the conjunctival secretions.

The accepted concept that limited acquired immunity develops in guinea pigs following conjunctival infection with virulent shigellae is based on the following observations.[1] First, following recovery from infection the level of guinea pig anti-*Shigella* antibodies that can passively protect mice from experimental *Shigella* sepsis is increased. Second, on reinfection of guinea pigs, *Shigella* bacteremia and hyperpyrexia occur with reduced frequency and duration, relative to primary infection. Third, shigellae are killed rapidly in the conjunctival sac of reinfected, recovered animals (at least within a several month period of the initial infection). Finally, when graded doses of bacteria are employed to reinfect a recovered animal, disease is generally expressed as a mild, abortive form with reduced intensity and duration.

Several important points need to be emphasized concerning the limited immunity acquired through conjunctival infection with *Shigella*. First, this immunity appears to exhibit no serotype specificity within the same eye on reinfection (i.e., there appears to be a localized resistance against infection by different *Shigella* serotypes). Second, in contrast to the serum antibody response which declines rapidly within 3 months, this localized resistance to reinfection in the eye lasts up to 9 months. Third, the major effect of this limited immunity is not to prevent reinfection, but to help arrest the development of full pathologic symptoms. Finally, this acquired limited immunity increases on repeated reinfections, but even serial reinfections cannot induce 100% immunity against a massive *Shigella* challenge.[1]

Current Understanding of Sereny Assay

Cumulative evidence from many studies (reviewed in Refs. 10 and 11) using tissue culture assays and *in vivo* animal models suggests that in humans orally ingested shigellae invade the epithelial mucosa of the terminal ileum and colon. On endocytosis into gut enterocytes, these organisms dissolve the endosomal vacuole, multiply intracellularly, and direct their movement on polymerized actin tails which allow them to spread laterally into adjacent epithelial cells. This intercellular spread plus intracellular multiplication lead to epithelial cell death, trigger an influx of PMNs, and eventually result in focal microulcerations of the mucosa. Dysentery ensues with fever, tenesmus, abdominal cramps, and stools with blood and mucus.

Sereny modified existing conjunctival infection assays to create a simple *in vivo* mucous membrane model in which bacteria, pathogenic for

[10] D. Kopecko, M. Venkatesan, and J. Buysse, *in* "Enteric Infection—Mechanisms, Manifestations, and Management" (M. Farthing and G. Keusch, eds.), p. 41. Chapman & Hall, London, 1989.
[11] P. Sansonetti, *Rev. Infect. Dis.* **13**(4), S285 (1991).

mucosal surfaces, could be assessed for their ability to provoke an orga-
nized pathological process.[4] Histopathologic studies of conjunctivally infected guinea pigs indicate that *Shigella* enter epithelial cells, multiply
intracellularly, provoke an acute inflammatory response, and lead to mucosal ulceration.[9] Thus, this simple animal model measures many, but not
all, of the virulence attributes utilized by shigellae to trigger dysentery.

Full virulence of shigellae (e.g., the abilities to resist the low pH of
the stomach, to resist the many host defense mechanisms of the gastrointestinal tract, and to provoke dysentery) has commonly been studied by
oral administration of organisms to monkeys[12] or starved, opiated guinea
pigs.[13] These *in vivo* assays are impractical for common use to test the
general virulence of a known or suspected *Shigella* isolate. Similarly,
tissue culture invasion assays have significantly advanced our knowledge
of bacterial invasion. Although very useful for studying invasion mechanisms, the standard tissue culture invasion assay does not require a smooth
lipopolysaccharide bacterial cell surface and does not measure other essential virulence traits (e.g., intracellular bacterial multiplication ability).
Oaks *et al.*[14] have developed a tissue monolayer plaque assay which
measures ability to invade, to disseminate to adjacent cells, and to result
in visual dead cell plaques. Although these latter assays are very useful
for specific purposes, the Sereny assay still provides the best single,
cost-effective means of measuring the overall virulence potential of a
Shigella isolate.

Other Applications of Experimental Conjunctival Infection Model

Although the modified conjunctival infection model was developed by
Sereny to analyze the virulence potential of *Shigella* strains, the simplicity
of the model has made it practical for a variety of uses. It became apparent
during the 1960s that certain *Escherichia coli* strains (now termed enteroinvasive *E. coli;* EIEC) cause a dysenteric syndrome in humans and trigger
a keratoconjunctivitis in guinea pigs.[1] For many years these EIEC strains
could only be distinguished from nonpathogenic *E. coli* on the basis of
the Sereny assay. Improved tissue culture invasion assays, serological
tests, and nucleic acid probe hybridization techniques have now been
substituted to identify EIEC on a large scale.[15] However, none of these

---

[12] W. Rout, S. Formal, R. Gianella, and G. Dammin, *Gastroenterology* **68,** 270 (1975).
[13] S. Formal, G. Dammin, E. Labrec, and H. Schneider, *J. Bacteriol.* **75,** 604 (1958).
[14] E. Oaks, T. Hale, and S. Formal, *Infect. Immun.* **53,** 57 (1986).
[15] P. Sansonetti, *in* "Enteric Infection—Mechanisms, Manifestations, and Management"
(M. Farthing and G. Keusch, eds.), p. 283. Chapman & Hall, London, 1989.

assays is equivalent to the Sereny test, which remains the "gold standard" in measurement of *Shigella* and EIEC virulence.

Environmental sources of potential pathogen contamination (e.g., water) can be membrane-filtered and the conjunctival infection carried out on total cells grown on the agar-cultured membrane. The resulting keratoconjunctivitis exudate can be analyzed bacteriologically after several days for the specific causative pathogen. This method has revealed the presence of shigellae which could not be detected by standard bacteriological methods.[1] Further use of this technique might provide important insight into potential environmental sources of *Shigella*/EIEC contamination.

The conjunctival infection model has also been used to examine a variety of other mucosal pathogens for their ability to induce keratoconjunctivitis. Some organisms, like *Corynebacterium diphtheria*,[1] have been found to induce a keratoconjunctivitis, but most enteropathogens stimulate only a conjunctivitis, at best; further studies of their pathogenesis or virulence have not been pursued in this model. However, the following example illustrates that, perhaps, we have not yet utilized the Sereny assay to its full potential. Most salmonellae tested in the Sereny assay are capable of evoking a conjunctivitis reaction but do not attack the cornea. However, there are several reports of strains of *Salmonella typhimurium* and *S. enteritidis* which have provoked a keratoconjunctivitis in the Sereny model (reviewed in Ref. 16). Furthermore, histopathological analyses revealed that these salmonellae penetrate epithelial cells and may multiply intracellularly without spreading to adjacent cells like *Shigella*, and they do not cause such a severe lesion of the epithelial lining. Instead, these salmonellae penetrate into the deeper tissues, lodging in the macrophages of the connective tissues and conjunctival lymphoid follicles where characteristic granulomas develop. Thus, this conjunctival infection model might yet provide new insight into salmonella pathogenesis. Most importantly, however, since many pathogens lose virulence quickly on subculture, the Sereny assay should be carried out with extremely fresh pathogenic isolates to ensure success. Perhaps more rigorous examination will reveal Sereny-positive *Campylobacter* or other enteropathogens, which can be maintained in a heightened virulence state, using this model, for pathogenicity analyses.

Finally, the conjunctival infection model has been shown to be useful in defining *Shigella* virulence properties that affect the Sereny reaction and in maintaining the bacterium in a highly virulent state for analyses of virulence.[8,10] Likewise, the development of live, attenuated bacterial

[16] M. Voino-Yasenetsky and T. Bakacs, *in* "Pathogenesis of Intestinal Infections" (M. Voino-Yasenetsky and T. Bakacs, eds.), p. 215. Akademiai Kiado, Budapest, 1977.

vaccine strains can be monitored for virulence level in the Sereny assay. More recently, guinea pigs have been immunized subcutaneously with various outer membrane protein preparations from shigellae and, on conjunctival challenge with virulent shigellae, have shown complete immunity which is also cross-serotype protective.[17] These studies again echo the tremendous value of the Sereny assay and suggest that it will play an important role in the development of anti-*Shigella* vaccines.

Pros and Cons of Sereny Assay

On the positive side, the assay is simple to perform, reading results is convenient, and results are very reproducible. The assay measures many of the virulence attributes of shigellae, and the model can be used to examine immunity and develop antidysentery vaccines. Moreover, this conjunctival infection model has other applications as described above.

On the negative side, although the pathologic responses to *Shigella* conjunctival infection share important similarities to human dysentery, the induced disease is different from dysentery. The guinea pig conjunctival or corneal epithelium may not express the necessary surface receptors for attachment/invasion by some mucosal pathogens, so that other pathogens may have to be tested in different animal species for conjunctival invasion ability.

Although not exhaustive by any means, these are the key advantages and disadvantages of the Sereny assay. As it has for 40 years, the Sereny guinea pig conjunctival infection model continues to be useful in virulence analyses and may be an important key in the development of antidysentery vaccines.

[17] G. Adamus, M. Mulczyk, D. Witkowska, and E. Romanowska, *Infect. Immun.* **30,** 321 (1980).

[4] Mouse Respiratory Infection Models for Pertussis

By ROBERTA D. SHAHIN and JAMES L. COWELL

Introduction

Systemic infection of laboratory animals has often been used to evaluate the virulence of respiratory pathogens, especially those that invade from the lungs to cause bacteremia, such as encapsulated *Streptococcus pneumoniae* and *Haemophilus influenzae*. Respiratory infection, however, allows for the evaluation of the earliest stages of respiratory bacterial pathogenesis, for example, attachment, colonization, and escape of host defenses. Several models of respiratory bacterial infection have been described, such as models of *Legionella pneumophila, Pseudomonas aeruginosa,* nontypable *Haemophilis influenzae,* and *Bordetella pertussis* infection.

Bordetella pertussis is a highly infectious gram negative coccobacillus that is spread by transmission of aerosol droplets, having a household attack rate of 90%. *Bordetella pertussis* preferentially associates with the cilia of the respiratory epithelium which lines the upper respiratory tract as well as the bronchial tree of the lower respiratory tract. In humans, colonization of the respiratory epithelial cell surface leads to local and systemic disease.

Bordetella pertussis, however, does not disseminate from the lungs to cause bacteremia or meningitis, and therefore it is truly a local respiratory pathogen. Therefore, an animal model of respiratory infection is critical to an understanding of *B. pertussis* pathogenesis. In addition, an animal model of respiratory infection is necessary to evaluate the ability of purified *B. pertussis* antigens in protection from infection and disease, as well as to delineate the mechanism(s) of protective immunity against disease and infection.

Although several laboratory animals have been used in respiratory models of *B. pertussis* infection,[1–3] the most common model has been respiratory infection of mice. Intranasal infection of 2- to 3-day-old mice has been described as one model of pathogenesis,[4] and it was used to demonstrate that gene products regulated by *vir* (also known as *bvg,* a major genetic locus that regulates gene expression in response to environmental stimuli) were virulence determinants.[5] In this model, the mice die within 2–3 days following infection with a virulent strain of *B. pertussis,* which likely reflects death due to the elaboration of toxins but does not reflect the initial stages of infection and outgrowth of bacteria. In addition, the ability to evaluate protective immunity in this model is limited to neutralization of bacterial toxins by passively administered antibody.

Intranasal infection of older mice has also been used in the characteriza-

[1] L. A. E. Ashworth, R. B. Fitzgeorge, L. I. Irons, C. P. Morgan, and A. Robinson, *J. Hyg.* **88,** 475 (1982).
[2] J. W. Hornibrook and L. L. Ashburn, *Public Health Rep.* **54,** 439 (1939).
[3] D. E. Woods, R. Franklin, S. J. Cryz, Jr., M. Ganss, M. Peppler, and C. Ewanowich, *Infect. Immun.* **57,** 1017 (1989).
[4] M. Pittman, B. L. Furman, and A. C. Wardlaw, *J. Infect. Dis.* **142,** 56 (1980).
[5] A. A. Weiss, E. L. Hewlett, G. A. Myers, and S. Falkow, *Infect. Immun.* **42,** 33 (1983).

tion of *B. pertussis* respiratory infection.[6] In this model, a small volume ($\leq$50 μl) of a suspension of bacteria is deposited on the nares, and the mouse is held upright until the inoculum is inhaled. While this method of inoculation has the advantage of convenience (i.e., it is easy to perform with minimal equipment needs), there are important disadvantages in the variability of infection from animal to animal and in the reproducibility of the type of infection obtained.[7] In addition, the localization of the infection has been shown in an intranasal model of influenza virus infection to depend on whether the mice are anesthetized or awake at the time of infection. Intranasal instillation of an inoculum in the nares of anesthetized mice exposes the entire respiratory tract to the agent, whereas instillation in the nares of unanesthetized mice limits the exposure to the nasal epithelium.[8]

Aerosol infection of mice has also been used in the characterization of *B. pertussis* infection.[9,10] In this model, mice inhale aerosol droplets of a bacterial suspension within an exposure box. In a series of careful experiments using [^{35}S]methionine-labeled bacteria, Halperin and colleagues have compared intranasal inoculation to aerosol infection.[7] Aerosol infection was found to produce an infection that was more reproducible in bacterial recovery from mouse to mouse, and the infection was more evenly distributed in the lobes of the lungs compared to intranasal infection. The consistent infection produced with aerosol infection allows for the analysis of the pathogenesis and immunology of *B. pertussis* infection with a minimum of biological variability. Aerosol infection requires a greater initial cost to establish than does intranasal infection, but once the exposure chamber and negative-pressure glove box are set up the procedure is relatively easy to perform. We describe below the apparatus and protocol used in the aerosol infection of mice in our laboratories. Modifications of the aerosol apparatus we describe have also been published.[7,10]

Apparatus

Bordetella pertussis is considered a Class 3 pathogen as an aerosol and thus requires containment under negative pressure during aerosol

[6] B. R. Burnet and C. Timmins, *J. Exp. Pathol.* **18,** 83 (1937).
[7] S. A. Halperin, S. A. Heifetz, and A. Kasina, *Clin. Invest. Med.* **11,** 297 (1988).
[8] R. A. Yetter, S. Lehrer, R. Ramphal, and P. A. Small, *Infect. Immun.* **29,** 654 (1980).
[9] M. Oda, J. L. Cowell, D. G. Burstyn, and C. R. Manclark, *J. Infect. Dis.* **150,** 823 (1984).
[10] Y. Sato, K. Izumiya, H. Sato, J. L. Cowell, and C. R. Manclark, *Infect. Immun.* **29,** 261 (1980).

FIG. 1. Photograph of aerosol chamber.

generation. The challenge inoculum is administered to mice as an aerosol using a Henderson-type aerosol mixing and transit tube[11] mounted in a clear Lucite exposure box (Fig. 1). The exposure box measures 32.5 × 38.5 × 50 cm and contains four removable wire net animal cages (each 10 × 22 × 9 cm) mounted on a mechanical rotation device. The wire net cages also have a removable divider which can be used to divide the cages into two compartments of equal size. Blueprints for this chamber may be obtained on written request to either of the authors.

Because *B. pertussis* is highly infectious as an aerosol, the exposure box is contained in a Biosafety Level 3 glove box (Model NU 704-400, or equivalent, NuAire Corp., Plymouth, MN) in a negative pressure room; high-efficiency particulate air (HEPA)-filtered exhaust from the cabinet is vented with the room air. The glove box has a modified enlarged door on the side that can be fully opened to allow entry or removal of the exposure box. Centered in the door is a smaller port, with an inner and outer door, that allow items to be placed in or removed from the chamber without direct exposure of the operator to the chamber. The glove box contains petcocks on both sides that allow access to compressed breathing air at the inlet as well as to a vacuum pump at the outlet.

The aerosol is generated by passing filtered compressed breathing air, at a pressure of 10 psi, through a nebulizer containing the bacterial suspension at approximately 0.4 ml/min. We have used the Vaponefrin nebulizer (Fisons Corp.), which is no longer made, but have found that

[11] R. F. Berendt, H. W. Young, R. G. Allen, and G. L. Knutsen, *J. Infect. Dis.* **141,** 186 (1980).

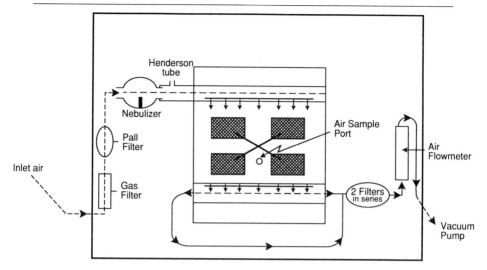

FIG. 2. Schematic of aerosol challenge apparatus, illustrating the airflow and aerosol generation within the Biosafety Level 3 glovebox.

the DeVillbis Model 645 Nebulizer (DeVillbis Corp., Somerset, PA) gives the same result. The Fison nebulizer generates aerosol particles, 90% of which are less than 5 μm and therefore would be deposited in the lung.[12] During the aerosol challenge, the aerosol generated from the nebulizer enters the exposure box through the inlet tube at the top of the chamber, passes over the mice, is removed via an outlet tube at the bottom of the chamber, passes through a series of two disposable 0.2-μm filters (Pall Ultipor Pharmaceutical Filter Assemblies, VWR Scientific, Bridgeport, NJ), and is evacuated at a vacuum pressure of approximately 20 inches of Hg, by a small pump connected to the outlet petcock of the glove box (Fig. 2). An air pressure regulator (Air Products, Washington, DC) on the compressed air tank is used to set and monitor the pressure of the inlet compressed air; a shielded-type calibrated flow meter (Gilmont, Baxter Scientific Products, McGaw Park, IL) is connected to the outlet petcock in order to set and monitor the airflow through the chamber at 26 liters/min.

Aerosol sampling with glass impingers connected to the air sampling portal (see Figs. 1 and 2) has demonstrated that during exposure of an aerosol generated by a 10^9 colony-forming unit (cfu)/ml bacterial suspension the chamber contains approximately 10^3 cfu/ml.[9] It has been shown that, 1 hr after termination of the aerosol, no viable *B. pertussis* cells can

[12] E. W. Larson, H. W. Young, and J. S. Walker, *Appl. Environ. Microbiol.* **31,** 150 (1976).

be detected on the surface of the mice.[9] When the air has been evacuated from the chamber for 30 min after termination of the aerosol, no viable *B. pertussis* cells can be detected in the air of the chamber.[10]

Procedure

The bacterial suspension is prepared from a 21 to 24 hr culture of *B. pertussis* that is screened for contamination by gram stain, harvested from slants, and suspended in sterile phosphate-buffered saline (PBS). The suspension is adjusted to approximately 10^9 cfu/ml by dilution to either the same percent transmission at A_{350} as a 10 opacity unit latex bead standard or to predetermined Klett units which correspond to approximately 10^9 cfu/ml. The bacterial suspension is kept at room temperature until it is used 0.5 to 2 hr later for aerosol challenge. An aliquot of the suspension is diluted and plated in order to establish the viability of the culture.

Mice are placed in the wire mesh cages, which are then mounted within the aerosol chamber. About 80 to 120 neonatal mice or 50 to 60 adult mice can be placed in the cages per aerosol exposure. The nebulizer is connected to the Henderson tube, and 4–5 ml of bacterial suspension is placed in the nebulizer using a syringe with a needle bent such that the droplets of suspension are deposited in the bottom of the nebulizer. Aerosol generation is begun by opening the inlet petcock, and the mice are exposed to the aerosol for 30 min. At approximately 10-min intervals it is necessary to refill the nebulizer by closing the inlet petcock and adding more *B. pertussis* suspension with the syringe. The cages are rotated one quarter turn every 5 min, to assure a uniform exposure, although recent experiments show that uniform exposure is achieved without rotating the cages. Temperature and relative humidity are also measured every 5 min. We have found that the temperature does not fluctuate during the challenge, but the relative humidity reaches between 65 and 85%, depending on the relative humidity of the room air that enters the chamber. On termination of the aerosol, the evacuation pump is left on, or the door of the Lucite chamber containing the mice is left ajar, in order to decrease the humidity.

One hour after termination of the aerosol, the mice are removed from the chamber and transferred without special precautions to conventional animal cages with or without filter tops. These cages are then kept in a conventional animal housing room. Immediately after challenge, two to five animals are sacrificed, their lungs aseptically removed and homogenized in sterile saline, and dilutions of homogenate plated on Bordet–Gengou agar in order to determine the number of viable *B. pertussis* in the lungs. The bacterial recovery immediately following aerosol challenge is approximately 5×10^4 cfu from the lungs of 10- to 18-day-old neonatal mice and approximately 1×10^5 cfu from the lungs of mice greater than 5 weeks of age. The bacteria have been demonstrated to be associated with the cilia of the respiratory epithelium lining the bronchial tree of the infected mice.[10]

One of the authors has evidence that there is no transmission of *B. pertussis* from mouse to mouse.[13] In one experiment, 4 pups from each of two litters of 8 BALB/c mice (20 days of age) were exposed to an aerosol challenge (10^9 cfu/ml inoculum) with *B. pertussis* 18323. After challenge, the infected pups were returned to their mothers, mixing 4 infected weanlings with 4 noninfected littermates. The infected animals began to die 10 days after challenge and had leukocyte counts per microliter of peripheral blood of about 100,000, which is indicative of mice with 10^7 to 10^8 cfu per lung. By day 15 postinfection, all of the infected animals were dead. The noninfected litter mates were followed for 5 weeks after being housed with the infected animals. No deaths or increase in leukocyte levels above normal were seen in the noninfected mice. Five weeks after mixing of the animals the noninfected mice had no detectable (<50 cfu/ lung) *B. pertussis* organisms in their lungs.

Parameters of Infection and Disease

Exposure of the mice by aerosol results in an infectious process with an increase in the number of bacteria recovered from both the tracheas and lungs of infected mice.[14] Whereas infection of the lungs reflects bacteria associated with the bronchial tree, as well as with alveolar spaces, bacterial recovery from the trachea can be monitored to analyze infection of only the respiratory epithelium. This is an important distinction and has resulted in the finding that the filamentous hemagglutinin (FHA) protein of *B. pertussis* is important in the initial colonization of the mouse trachea but is not a factor in colonization of the lungs.[14]

Disease is most often monitored at different times following infection by determining the level of leukocytes in the peripheral blood, the rate of gain in body weight, and number of deaths.[9,10] These disease parameters are useful only when neonatal mice are infected since adult mice, while infected (Fig. 3), do not die, exhibit transient leukocytosis, and do not show weight loss. Disease symptoms in neonatal mice are very age dependent. For example, aerosol infection of mice younger than 18 days postpar-

[13] J. L. Cowell, unpublished observation (1986).

[14] A. Kimura, K. T. Mountzouros, D. A. Relman, S. Falkow, and J. L. Cowell, *Infect. Immun.* **58,** 7 (1990).

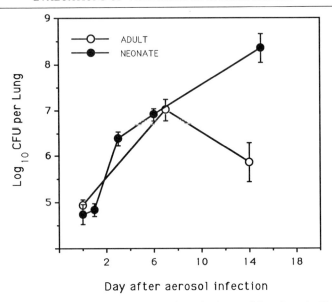

FIG. 3. Time course of bacterial recoveries from the lungs of 5- to 8-week old adult and 18-day neonatal BALB/cAnNcR mice exposed via aerosol to *B. pertussis* 18323 using a challenge inoculum of 10^9 cfu/ml. The geometric mean and standard deviation (bars) of the bacterial recoveries from 5–7 adult mice per point and 8 neonatal mice per point are shown.

tum results in severe leukocytosis, weight loss, and death; however, infected mice 21–22 days postpartum exhibit less pronounced leukocytosis that returns to normal, and these mice do not die.[10]

We have used two strains of *B. pertussis* in the aerosol infection model, namely, strain 18323, which is the same strain used in the intracerebral challenge test used for the assessment of potency of whole cell pertussis vaccines, and Tohama I. Initial infection of 10-day-old mice with 10^4 to 10^5 cfu per lung of 18323 and Tohama I results in 100% death. However, Tohama I kills about 20% of 19-day-old mice, whereas strain 18323 kills 70 to 100% of infected 19-day-old mice.[9,10] The basis for this age-related difference in virulence between the two strains is currently not known. In 17- to 18-day-old mice infected with 18323, the bacterial recovery from the lungs increases from 5×10^4 cfu immediately after infection to over 10^7 cfu per mouse 14 days after infection (Fig. 3). These mice show the greatest leukocytosis and weight loss after 16–17 days postinfection and die shortly thereafter. It has been noted by several authors that the highest leukocytosis and death rates in neonatal mice coincide with the bacterial load growing out to greater than 10^7 cfu per mouse.[15]

[15] Y. Sato and H. Sato, *in* "Pathogenesis and Immunity in Pertussis" (A. C. Wardlaw and R. Parton, eds.), p. 309. Wiley, Chichester and New York, 1988.

Several strains of neonatal mice have been used in aerosol challenge experiments with *B. pertussis,* including inbred and outbred strains. Neonatal challenge experiments, using *B. pertussis* strain 18323, have been performed with outbred DDY, ICR, Swiss Webster, and inbred BALB/c strains, with similar observed leukocytosis and death rates among these strains following aerosol infection.[9,10,13–16] However, it has been noted that certain strains of mice, for example, B10.D2/OSnJ and B10.D2/NSnJ, have little if any leukocytosis and do not die when challenged at 17 days of age with *B. pertussis* 18323, although log 5.95 cfu and log 5.5 cfu were recovered from the lungs of these mice 7 days after aerosol infection.[16] Cowell and Oda have noted a decreased mortality of 12-day-old C57BL/6 and HSFS/N strains following infection with Tohama I, relative to the 100% mortality observed in 12-day-old BALB/c, DBA/2, and C3H mice.[17] Therefore different strains of mice should be screened to establish their susceptibility to infection prior to experimental use.

In 5- to 8-week-old adult mice infected with Tohama I[14] or 18323 (Fig. 3), the bacterial recovery from lungs increases from about 10^5 cfu per mouse immediately after infection to approximately 10^7 cfu per mouse lung 7 to 9 days postinfection; thereafter the number of bacteria in the lungs decline. As stated previously, infected adult mice do not die and show transient leukocytosis that returns to normal. Adult mice that have been exposed to an aerosol of Tohama I to obtain about 10^5 cfu per mouse lung have about 10^3 cfu *B. pertussis* per trachea immediately after exposure. This tracheal colonization increases to about 10^5 cfu per trachea 7 to 9 days after challenge and then declines.[14] Similar bacterial recoveries from aerosol-infected adult mice have been observed in BALB/c and (C57Bl/6 × C3H/HeN) F_1 mice.[16]

Variation of Initial Infective Dose

In one of the author's laboratories[18] conditions have been established to obtain different initial infective doses of approximately 10^2, 10^4, or 10^5 cfu per mouse lung immediately after aerosol challenge (Fig. 4). This was evaluated using 10-day-old outbred Swiss Webster mice and *B. pertussis* strain 18323 and accomplished by using the aerosol exposure procedure as described above except that the number of bacteria in the suspension placed in the nebulizer was varied. A challenge inoculum of 10^7, 10^8, or 10^9 cfu/ml gave viable bacterial lung counts of about 4×10^2, 7×10^3, or 8×10^4 cfu per mouse immediately after aerosol exposure. All of these

[16] M. F. Leef and R. D. Shahin, unpublished observation (1992).
[17] J. L. Cowell and M. Oda, unpublished observation (1984).
[18] J. L. Cowell and K. T. Mountzouros, unpublished observation (1992).

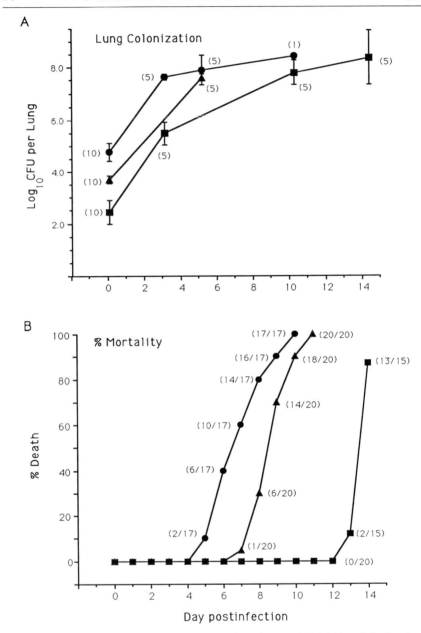

FIG. 4. Effect of varying the aerosol challenge inoculum on the initial lung infective dose and lung colonization (A) and on mortality (B). Ten-day-old Swiss Webster mice were exposed via aerosol to *B. pertussis* 18323 using a challenge inoculum of 10^9 (●), 10^8 (▲), or

doses resulted in colonization of the lung increasing to approximately 10^7–10^8 cfu/mouse 3 to 14 days after challenge. The time to reach these levels was directly related to the initial infective dose level (Fig. 4A). All of the infective doses resulted in lethal infections; however, the time period that elapsed before mortality occurred in the group that received an initial infective dose of 4×10^2 cfu/mouse was substantially delayed compared to the groups of mice that received the higher infective dose (Fig. 4B).

Older Swiss Webster mice (4–6 weeks old) challenged with an inoculum of 2×10^9, 9×10^6, or 6×10^5 cfu/ml of Tohama I also had initial infective doses in the lung of 10^5, 10^4, or 10^2 cfu/ml.[18] Although the mice receiving the highest initial inoculum began to clear the infection in the lungs 10 days after challenge, mice receiving the two lower doses of *B. pertussis* showed an increase in bacterial growth over 5–10 days following challenge, with this growth plateauing by day 14 (data not shown). Similar kinetics of growth were observed in the tracheas of the same animals.

Examples of Applications

The aerosol challenge model has been useful in characterizing purified antigens for the ability to elicit protective immunity from *B. pertussis* infection as well as the ability to mitigate clinical symptoms of infection. Highly purified filamentous hemagglutinin, pertactin, intact fimbriae, and chemically or genetically inactivated pertussis toxin have all been shown to protect against *B. pertussis* aerosol challenge.[19] However, not all purified *B. pertussis* antigens protect neonatal mice against leukocytosis and death. Active immunization with purified *B. pertussis* fimbrial subunits,[20] the major porin of *B. pertussis*,[20] or a purified GroEL-like protein[21] failed

[19] M. J. Brennan, D. L. Burns, B. D. Meade, R. D. Shahin, and C. R. Manclark, *in* "Vaccines: New Approaches to Immunological Problems" (R. Ellis, ed.), p. 23. Butterworth, Shoreham, Massachusetts, 1991.

[20] R. D. Shahin and M. J. Brennan, unpublished observations (1989).

[21] D. L. Burns, J. L. Gould-Kostka, M. Kessel, and J. L. Arciniega, *Infect. Immun.* **59,** 1417 (1991).

10^7 (■) organisms/ml. For each challenge dose, 4 litters of mice (40 to 45 animals) were exposed and then returned to their mothers. For each challenge dose, two litters of mice were used to record the number of deaths, and two litters were used to determine the number of viable organisms per lung at the times shown. In (A) the plots show the geometric mean and standard deviation (bars) for 5 to 10 mice, except at one time point where only one mouse was used because of deaths. In (B), the numbers in parentheses show the number of deaths over the total exposed.

to protect neonates against leukocytosis and death following aerosol challenge, although all of the antigens elicited antibody in 5-week-old mice. Protection experiments with passively administered monoclonal antibodies have allowed for the mapping of protective epitopes on antigens that are candidates for inclusion in acellular pertussis vaccines, and they also offer insights into possible mechanisms of protection afforded by vaccine candidates.

A second application of the aerosol challenge model has been in defining gene products of *B. pertussis* that are involved in pathogenesis. Transposon insertion mutants of *B. pertussis* that affect the production of secreted gene products controlled by the coordinate regulatory locus *vir* (*bvg*) have been used to infect neonatal mice. Using this strategy, two new virulence determinants have been described, namely, a 95-kDa outer membrane protein that is regulated in parallel with pertussis toxin and other known virulence factors[22] and a 12-kDa protein that is regulated reciprocally to pertussis toxin and other known virulence factors.[23]

Points to Consider in Application of Model

Although there is no animal model of pertussis that reflects all aspects of the human disease, aerosol infection of mice with *B. pertussis* results in the association of the bacteria with the cilia of the respiratory epithelium, as in humans, and therefore allows for the analysis of the early stages of *B. pertussis* colonization and the establishment of infection. However, mixed droplet sizes are generated in the aerosol, such that larger aerosol droplets remain in the upper airways, where *B. pertussis* associates with the ciliated epithelium, but small aerosol droplets are deposited into the alveolar spaces of the lung.[12] Therefore, the simultaneous establishment of infection at the respiratory epithelium and in the alveolar spaces should be kept in mind when interpreting data from aerosol infection experiments.

One of the authors (R.D.S.) has found that susceptibility of neonates to lethal *B. pertussis* infection is increased in mouse colonies that test positive for the presence of mouse hepatitis virus. Therefore, the use of specific pathogen-free mice in this infection model, if at all possible, is recommended.[20]

---

[22] T. M. Finn, R. Shahin, and J. J. Mekalanos, *Infect. Immun.* **59,** 3273 (1991).
[23] D. T. Beattie, R. Shahin, and J. J. Mekalanos, *Infect. Immun.* **60,** 571 (1992).

[5] Chinchilla Model of Experimental Otitis Media for Study of Nontypable *Haemophilus influenzae* Vaccine Efficacy

By BRUCE A. GREEN, WILLIAM J. DOYLE, and JAMES L. COWELL

Introduction

Otitis media is one of the most frequent diseases of children from ages 0 to 7 years. Most bacterial otitis is caused by one of three organisms: *Streptococcus pneumoniae,* nontypable *Haemophilus influenzae* (NTHi), or *Moraxella catarrhalis.* Chinchilla, gerbil, and rat animal models have been developed for otitis caused by *S. pneumoniae* and *H. influenzae,* but as of this date no animal model exists for otitis caused by *M. catarrhalis.* The animal models used for *S. pneumoniae* and *H. influenzae* otitis all have limitations which make them somewhat less than perfect, but they are currently used as part of an evaluation scheme of candidates for both prophylaxis and vaccines. The chinchilla model was described first for *S. pneumoniae* by Giebink *et al.*[1] and later adapted for NTHi. It has been used with nontypable *H. influenzae* for evaluation of both antimicrobial agents[2-4] and vaccine candidates.[5,6] Current protocols for use of the model and design of the experiments vary depending on the budget for animals and the pathogenicity of the NTHi challenge strain. This chapter describes use of the chinchilla model to determine the efficacy of vaccine candidates directed against NTHi by either passive or active immunization.

Two modes of challenge are described: direct challenge into the middle ear and nasopharyngeal challenge followed by negative pressure to aspirate NTHi up into the middle ear through the Eustachian tube, the natural route of infection in humans. Each of the challenge modes has its advantages and disadvantages. Direct challenge allows assessment of vaccine

[1] G. S. Giebink, E. E. Payne, E. L. Mills, S. K. Juhn, and P. G. Quie, *J. Infect. Dis.* **134,** 595 (1976).

[2] W. J. Doyle, J. S. Suppance, G. Marshak, E. Y. Cantekin, C. D. Bluestone, and D. D. Rohn, *Otolaryngol. Head Neck Surg.* **90,** 831 (1982).

[3] J. S. Reilly, W. J. Doyle, E. I. Cantekin, J. S. Supance, C. K. Ha-Kyung, D. D. Rohn, and C. D. Bluestone, *Arch. Otolaryngol.* **109,** 533 (1983).

[4] R. Mills and J. Gilsdorf, *J. Laryngol. Otol.* **100,** 255 (1986).

[5] S. J. Barenkamp, *Infect. Immun.* **52,** 572 (1986).

[6] R. Karasic, D. J. Beste, S. C.-M. To, W. J. Doyle, S. J. Wood, M. J. Carter, A. C. C. To, K. Tanpowpong, C. D. Bluestone, and C. C. Brinton, Jr., *Pediatr. Infect. Dis. J.* **8**(Suppl), S62 (1988).

candidates in the middle ear, the site of infection, and also allows use of fewer animals as there is a 100% infection rate in controls using this challenge method. However, vaccines which work at the level of coloniza- tion may not be effective in direct challenge in the middle ear, and it is thus a more stringent test of a vaccine candidate than the nasopharyngeal challenge. In nasopharyngeal challenge, the organisms are first allowed to colonize the nasopharynx for a period of time. This mode of challenge has the advantage of allowing colonization factors of NTHi to be ex- pressed, and vaccine candidates which act on these colonization factors can be tested. The most well-studied example of this is the anti-LKP pili vaccines tested by Karasic et al.[6] The disadvantages of the nasopharyngeal challenge mode are the lower infection rates in negative control groups (i.e., ~50–75%), requiring the use of greater numbers of animals per group for statistical purposes, and the additional manipulation of the negative pressure, which places additional stress on the animal. These and other factors must be taken into account when designing a chinchilla otitis media experiment as described below.

Animals

Chinchillas are outbred animals and are normally obtained from local chinchilla ranchers. The animals obtained usually are young adults be- tween 400 and 500 g whose coats have been deemed unsuitable for fur production. The average cost is approximately $40 per animal.

Active Immunization of Chinchillas

After arrival, chinchillas are kept in quarantine for 2 weeks to allow acclimation to the environment. Animals are then anesthetized with keta- mine hydrochloride at 20 mg/kg by intramuscular injection. Middle ear status is determined by tympanometry and otoscopy (see below) to rule out any preexisting middle ear inflammation or infection. Animals with normal middle ears are enrolled in the study. A 3-ml blood sample is taken via cardiac puncture to serve as baseline for analysis of immune sera. If desired, a preliminary study using small groups of 5 animals may be immunized with candidate vaccines using varying dosage schedules and doses of immunogens to optimize the immune response.

Chinchillas are randomly divided into the required number of groups to be immunized with the vaccines to be tested. For preliminary dose–re- sponse studies, groups are required only for the test vaccines and one saline control group. In larger protection experiments, a positive control group(s) immunized with either outer membranes from the NTHi challenge strain or formalin-fixed whole cells of the challenge strain is also used. All published results using active immunization, both from our laboratory and others[5,7,8] have used powerful adjuvants such as Freund's incomplete adjuvant or monophosphoryl lipid A, or muramyl dipeptide mixed with cell wall ghosts and squalene to elicit a good protective immune response. After the baseline bleed, the animals are left to recover for 5 days, and then the initial immunization is given, usually intramuscularly. The test vaccines are given in a volume of 200 μl containing 15–50 μg of the immunogens emulsified in the appropriate adjuvant. For positive controls, 25 μg of outer membranes are used per injection, whereas whole cells are used at a concentration of 10^7 cells per injection. We have found that an immunization schedule of 3 injections, 30 days apart, works well for the protein immunogens used to date. Fourteen days after the last injection, a 3-ml blood sample is taken by cardiac puncture, and the animals are allowed to recover for 5–7 days.

Passive Immunization of Chinchillas

Chinchillas may be immunized passively with either chinchilla antisera or rabbit antisera. Sera from other animal species have not been used for passive transfer in chinchillas to our knowledge. Hyperimmune rabbit antisera have been used at dilutions ranging from 1 : 1 to 1 : 100,[6] while chinchilla antisera have been used at 1 : 5 to 1 : 10. All serum dilutions are done in phosphate-buffered saline (PBS). The antisera may be administered by intracardiac or intraperitoneal injection. Experiments in our laboratory have shown that rabbit antisera is best transported from the peritoneum into the blood at lower dilutions, so to titer antisera for protective activity it is best to start with undiluted antisera or with 1 : 5 dilutions. However, care should be taken to achieve the highest titers possible as transport of rabbit immunoglobulins (Igs) from the peritoneum to the blood appears to be highly variable from animal to animal. Sera are administered at approximately 10 ml/kg body weight. Passive administration of antisera via the intracardiac route is limited by the volume which can be administered safely. It is best to start with antisera diluted at least 1 : 10 to avoid any problems with untoward reactions to serum components which may lead to death for the recipient animal. In our hands, intracardiac administration of up to 5 ml of diluted antiserum has been performed without

[7] C. C. Brinton, Jr., M. J. Carter, D. B. Derber, S. Kar, J. A. Kramarik, A. C. C. To, and S. W. Wood, *Pediatr. Infect. Dis.* **8,** 54 (1988).
[8] S. I. Pelton, G. Bolduc, S. Guiati, Y. Liu, and P. A. Rice, *Program Abstr. 30th Intersci. Conf. Antimicrob. Agents Chemother.,* Abstr. 610 (1990).

mishap. No problem with transport of antibody exists for intracardiac administration. Although previous experiments seems to indicate that protection can correlate with concentration of antibodies in the blood,[6,8] no end point has been established for protection.

After passive immunization of chinchillas, the animals are challenged between 18 and 24 hr with the appropriate dose of NTHi. Our studies show that the levels of antibodies in the blood will remain fairly high for 2–4 days after administration with either method of delivery, and they peak at 18–24 hr for the intraperitoneal route.

Preparation of Inoculum for Challenge

Strains of NTHi to be used for direct or nasopharyngeal challenge are selected from clinical isolates obtained from patient specimens. Usually, strains isolated from the middle ear of patients with acute otitis media are selected and are kept frozen in brain–heart infusion (BHI)–Levinthal broth containing 20% (v/v) glycerol at $-70°$. To prepare the inoculum for direct challenge, the organism is first passaged in the chinchilla bulla. An overnight broth culture in BHI containing 2 μg/ml NAD and 10 μg/ml hemin (BHI-XV) is diluted into fresh BHI-XV and grown to mid-log phase at 37° with aeration. When the OD_{690} is 0.9, approximately 10^9 colony-forming units (cfu)/ml are present. Approximately 100 μl of a 1 : 100 dilution of the log phase culture are inoculated into the bulla (see below) of a chinchilla using a 22-gauge needle attached to a 1-ml tuberculin syringe. The animal is kept in isolation for 48 hr, and then the bulla is opened at the mucosa, swabbed, and the swab is streaked for isolation of organisms on chocolate agar plates. A high-titer broth (10^7 cfu/ml) is prepared by inoculating a tube of BHI-XV with a few colonies picked from the overnight growth on chocolate agar plates. The broth is incubated for 3 hr at 35° and then diluted in trypticase soy broth containing 20% glycerol, and aliquots are frozen at $-70°$. This results in a fairly stable level of viability being maintained for at least 3 weeks. Two aliquots are thawed and counted daily to monitor colony count. The frozen aliquots provide a pool of known concentration which are used to prepare the inoculum by dilution with 20 mM potassium phosphate buffer, pH 7.2, to the desired concentration. The prepared inoculum is counted via serial dilution on BHI-XV plates at the time of challenge in order to determine the actual number of colony-forming units introduced into the ear.

For nasopharyngeal challenge, the inoculum is prepared as above, and the stock solution containing approximately 10^7 cfu/ml is used to administer the appropriate challenge dose.

Direct and Nasopharyngeal Inoculation

For middle ear inoculation, the chinchilla is sedated with ketamine hydrochloride (20 mg/kg) and the superior bulla is shaved and prepared with Betadine. A 1-cm incision is made through the soft tissue, and the bulla is exposed. A 2-mm bullar perforation is created to expose the epitympanum of the middle ear. A tuberculin syringe equipped with a 22-gauge needle is inserted, and 0.1 ml of inoculum is instilled into the middle ear. The skin is closed with a single surgical staple. For vaccine studies, both ears are inoculated with identical doses of the challenge strain at the same time. One ear is used for culture on every second day, and the other ear is evaluated using noninvasive techniques until the last day of the study. For nasal inoculation, a tuberculin syringe containing the desired concentration (e.g., 10^7 cfu) of inoculum is introduced in the nasal cavity of the sedated chinchilla via the nares, and 0.1 ml of inoculum is instilled into the nose using a 1-ml tuberculin syringe.

The actual concentration of bacteria used for inoculation varies depending on the pathogenicity of the strain. For direct inoculation, concentrations as low as 10 organisms per ear have been used with some strains,[6] whereas other investigators have used concentrations of bacteria as high as 10^3 organisms per ear[5] in order to obtain reproducible infections in the chinchilla middle ear. Before initiating challenge studies for vaccine research, the minimum infectious dose of the challenge strain to be used must be titered by injecting 5-fold dilutions of the strain in question into the middle ears of chinchillas and monitoring the disease state of the ears following challenge. Organism dosages should range from 10 (the lowest concentration which can be statistically put into ears consistently) to approximately 10^4 organisms per ear. Usually, the lowest concentration of organisms which produces consistent acute otitis media with effusion is used for subsequent vaccine studies. If higher number of organisms are desired to achieve more stringent challenges, then this can be done in subsequent experiments.

For nasal challenge, a similar process is performed whereby a 0.1-ml inoculum is used and instilled into the nasopharynx of an anesthetized animal. Test dosages of the strain to be used should start at approximately 10^3 organisms/ml and range up to 10^7 organisms/ml. Approximately 18 hr after administration of the nasal inoculum, the animal is anesthetized and the skin over the superior bulla shaved and prepared with Betadine. An 18-gauge needle connected via tubing to a large volume (e.g., 3 liter) capacitance flask maintained at a pressure of -250 to -300 daPa (dekapascals, a measure of atmospheric pressure) is introduced through the bullar skin and bone into the middle ear space. This results in a middle ear pressure equal to that of the capacitance flask. The induced negative pressure is maintained for approximately 15–20 min, after which the needle is removed from the bulla. The procedure results in the aspiration of bacteria from the nasopharynx to the middle ear for subsequent colonization. Experience using this technique shows that acute otitis media occurs in 50–75% of the nonimmune control animals. Because of the variability in rates of infection, it is important to assume the lowest infection rate for estimates of sample sizes. Thus, the use of this challenge method requires a greater sample size than that needed for direct challenge to achieve the statistical power required to determine if results between different experimental groups are significant. For example, if 20 animals are used per group for direct challenge, then 30–35 animals should be used for nasopharyngeal challenge. This must be taken into account when immunizations are planned and group size determined.

Analysis of Ears of Challenged Animals

Beginning 2 days postchallenge and then on every other day up to day 14 postchallenge, the animals are anesthetized with ketamine, the area over the surgical staple cleaned with Betadine, and the staple removed. The left bullar defect is exposed using a hemostat. A calcium alginate swab moistened in trypticase soy broth or BHI and mounted on a flexible aluminum shaft is introduced into the left middle ear space to contact the mucosa of the hypotympanum. The swab is removed and then streaked directly onto a chocolate agar plate, which is incubated in 5% (v/v) CO_2 for 24–48 hr at 37°. Plates are examined at 24 and 48 hr for growth typical of NTHi colonies, and X and V factor growth requirements are ascertained using the paper strip technique (available from BBL, Cockeysville, MD). Levels of growth on plates may be qualitatively assessed on a scale of 0 to 4 using 0 to represent no growth of NTHi and 4 to represent confluent growth of NTHi. Quantitative assessment of growth requires the initial swab to be placed in BHI broth to allow the bacteria to be suspended and then dilution plates made on either chocolate agar or BHI-XV. Plates are incubated as above and the NTHi colonies counted to obtain the colony-forming units per milliliter of original culture.

When the animals are anesthetized, diagnosis of acute otitis media with effusion is made in the right ears using both otomicroscopy and tympanometry. For otomicroscopy, the chinchilla is positioned on an operating table with the ipsilateral ear up and a speculum introduced into

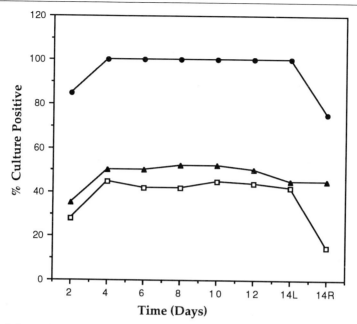

Fig. 1. Protection from infection with NTHi in middle ears in actively immunized chinchillas after direct challenge with NTHi. The percentage of animals in each treatment group with culture positive ears on each day postchallenge is shown. The last time point represents culture of the previously untouched right ear in each animal. (●) Saline; (▲) membranes; (□) whole cells.

the ear canal. The speculum is positioned to allow for visualization of the tympanic membrane under low power (6×) using a Zeiss otomicroscope. A positive diagnosis of acute otitis media is made if the tympanic membrane is either red and inflamed, has observable air fluid levels, is bulging with obvious yellow fluid, or is perforated with drainage to the external canal. Otomicroscopic data are recorded on the following scale: 0 = effusion, 0.5 = mild inflammation, and 1 = normal middle ear.

Tympanometry is then performed using a Teledyne automatic impedance meter (Model TA-7A, Charlottesville, VA) with standard adult probe tips. For each test, volume, compliance, and middle ear pressure are recorded. A measured volume of greater than 1.5 ml and a compliance of 0.0 ml (no change in volume when pressure is applied to the tympanic membrane) is interpreted as a tympanic membrane perforation. Middle ear fluid is diagnosed if the compliance is less than 0.7 ml. Middle ear inflammation is diagnosed if the recorded pressure is less than − 100 daPa or if an effusion is diagnosed to be present.

TABLE I
CULTURE POSITIVE OTITIS MEDIA IN ACTIVELY IMMUNIZED CHINCHILLAS CHALLENGED
WITH NONTYPABLE *Haemophilus influenzae*

Immunogen[a]	No. of animals challenged[b]	No. with culture positive otitis postchallenge[c]	p value[d] versus saline control
Saline	18	18	–
Whole NTHi cells	20	8	0.0073
Outer membranes	18	9	0.0253

[a] Animals immunized with three doses of vaccines in Freund's incomplete adjuvant.
[b] Total number of animals in each group at the completion of the study.
[c] Total number of animals in each group which had any positive middle ear cultures.
[d] The p value from χ^2 analysis of number of animals with culture positive otitis/number of animals in each group versus saline control group.

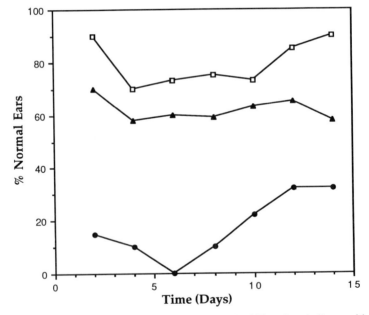

FIG. 2. Middle ear disease in actively immunized chinchillas after challenge with NTHi as diagnosed by tympanometry. A compliance of >1.0 = a normal middle ear was used as the measurement of disease. The percentage of normal middle ears in each treatment group is shown versus day postchallenge. (●), Saline; (▲) membranes; (□) whole cells.

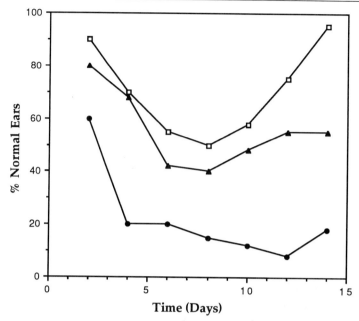

FIG. 3. Otitis media in actively immunized chinchillas directly challenged with NTHi as measured by otoscopy. The percentage of normal ears in each treatment group diagnosed by otoscopy (see text) is shown versus day postchallenge. (●) Saline; (▲) membranes; (□) whole cells.

Data points are collected every second day for culture of the left ear and otomicroscopy and tympanometry of the right ear. On day 14 postchallenge, tympanometry and otoscopy are performed on the right ear as usual and then the right ear is cultured instead of the left ear, to control for persistence of the infection in the left ear due to the continual invasion. After the last time point, the animals are bled once more for a final postchallenge serum sample and then sacrificed.

The culture positive rate is the primary variable used to monitor vaccine effectiveness. The longitudinal results regarding culture positive versus negative animals are viewed graphically as percent culture positive versus day postchallenge. Data for all groups are graphed on the same chart. An example of the protection from infection in actively immunized animals after direct challenge with a homologous NTHi strain is shown in Fig. 1. Note that with this strain, the positive controls, outer membranes and whole formalinized cells, produce a 50% level of protection from infection. The level of protection observed may vary, depending on the virulence of the challenge strain. It is recommended that one of these controls is included to monitor the level of protection which can be obtained when using a particular challenge strain. The level of infection, whether qualitatively or quantitatively assessed, may also be represented graphically, with the y axis showing the level of infection (e.g., 0 to 4 on the qualitative scale) versus day postchallenge.

For analysis, data for otoscopy, tympanometry, and culture are considered to be dichotomous (disease/no disease). Consequently, between-group comparisons can be performed on each evaluation day or as a summary across days using the chi-squared (χ^2) statistic with appropriate adjustments for multiple comparisons. In experiments where mortality censors the extent of observations for a significant number of animals, the Kaplan–Meier survival analysis can be performed with Bernofetti correction for multiple comparisons.[9] An example of simple χ^2 analysis for the experiment in Fig. 1 is shown in Table I. Here, the differences between the saline group and the two positive control groups are statistically significant.

The data for tympanometry may be divided into the two different parameters measured in the experiment. The compliance data may be graphed using a criterion of compliance such that greater than 1.0 equals normal ear and then graphing the percent normal ears versus the day postchallenge. A typical graphic representation of this kind of tympanometry data from the above experiment is shown in Fig. 2. It can clearly be seen that the ears of the animals immunized with the positive controls are much more frequently normal ears as measured by tympanometry than are the ears of sham-immunized animals. Tympanometry data for pressure may be similarly graphed using percent normal ears (normal ear = pressure < − 100 daPa) for the y axis and day postchallenge for the x axis. Data for otoscopy may be viewed and analyzed by graphing the percentage of animals with normal ears (score of 1) on the y axis versus the day postchallenge as the x axis. Data points for all of the groups tested are graphed on the same chart.

A graph of the otoscopy data from the experiment used as an example is shown in Fig. 3. Here, otitis as determined by otoscopy gradually increased in all groups, reaching a peak by day 7 postchallenge. The positive control groups show a much lower level of disease overall and a gradual rise in the percentage of normal ears after day 7. The sham-immunized group shows much more disease and a slower return of ears to normal status.

[9] R. M. Rosenfeld, W. J. Doyle, J. D. Swarts, J. Seroky, and B. Perez-Pinero, *Arch. Otolaryngol. Head Neck Surg.* **118,** 49 (1992).

[6] Animal Models for Ocular Infections

By ROGER G. RANK and JUDITH A. WHITTUM-HUDSON

Introduction

A number of bacterial organisms are capable of causing infections of the eye, but *Chlamydia trachomatis* remains the major cause of preventable blindness in the world and thus represents one of the most important bacterial ocular pathogens. Regrettably, this organism is also one which defies genetic manipulation and is largely undefined with regard to virulence factors. Nevertheless, it has been well documented that the disease is, to a great extent, immune-mediated[1] and thereby presents an excellent opportunity to investigate the host–parasite interactions resulting in clinical disease.

Two separate animal models for chlamydial ocular disease have been described. In the first, varying species of primates, including Taiwan,[2] owl,[3] and cynomolgus monkeys,[4] have been utilized in conjunction with serovars of *C. trachomatis* involved in human disease. The last primate model, in particular, has been extensively described, and the disease has been found to bear a remarkable similarity to human trachoma. The remaining model is that of the guinea pig infected in the eye with the guinea pig agent of inclusion conjunctivitis (GPIC), a natural parasite of the guinea pig first described by Murray.[5] Although it is a *Chlamydia psittaci,* it produces an acute conjunctivitis reminiscent of human inclusion conjunctivitis and has been shown to be capable of eliciting a trachoma-like disease on multiple infections.[6] Certainly, the primate model is the model of choice but is limited somewhat by the expense of primates. In contrast, the guinea pig model has the advantage of being more economical and more easily manipulated. An important caveat of the guinea pig model, however, is that the disease process is considerably more acute than the either human or primate chlamydial ocular disease.

---

[1] J. T. Grayston, S.-P. Wang, L. J. Yeh, and C. Kuo, *Ref. Infect. Dis.* **7**, 717 (1985).

[2] S.-P. Wang and J. T. Grayston, *Am. J. Ophthalmol.* **63**, 1133 (1967).

[3] C. E. O. Fraser, D. E. McComb, E. S. Murray, and A. B. MacDonald, *Arch. Ophthalmol.* **93**, 518 (1975).

[4] H. R. Taylor, S. L. Johnson, R. A. Prendergast, J. Schachter, C. R. Dawson, and A. M. Silverstein, *Invest. Ophthalmol. Visual Sci.* **23**, 507 (1982).

[5] E. S. Murray, *J. Infect. Dis.* **114**, 1 (1964).

[6] M. A. Monnickendam, S. Darougar, J. D. Treharne, and A. M. Tilbury, *Br. J. Ophthalmol.* **64**, 279 (1980).

An example of the potential use of these models in the understanding of virulence concerns the observation that the 57-kDa GroEL homolog of chlamydiae is able to elicit a pathologic response by topical application of the protein to the conjunctiva of immune animals but not normal uninfected animals.[7–9] Indeed, when recombinant strains of chlamydiae become available, both of these models will be extremely valuable in the determination of factors which lead to virulence.

Infection of Subhuman Primate with *Chlamydia trachomatis*

Induction of Ocular Infection

The cynomolgus monkey (*Macaca fascicularis*) has been used as a model for human ocular chlamydial infections for over 10 years.[4,10–14] In addition to developing disease similar to that observed in humans with chronic trachoma, the cynomolgus monkey can be used to test various immunizing regimes with vaccine candidates. Naive, wild-caught monkeys can be purchased from several primate importers. The cost of experimentally naive colony-bred cynomolgus monkeys is 2- to 5-fold higher than for wild-caught animals. However, the Primate Supply Clearing House (University of Washington, Seattle, WA) provides biweekly information on primates for sale. The major consideration with monkeys previously used for other purposes is that the conjunctivae (including proper lid closure) must be normal, and they should not have been previously exposed to antigens which might cross-react with chlamydiae or to drugs which might alter general immune responsiveness, particularly at mucosal surfaces. Adult cynomolgus monkeys (3 years or older) of either sex can be used for ocular infection. Depending on the institution, new monkeys are held in quarantine and tuberculin (TB) tested prior to use in experiments. The usual practice is to TB test on the outer eyelid surface; it is important to arrange for an alternative skin site during the experiment and to never use the lid site if animals have ever received Freund's complete adjuvant.

---

[7] R. P. Morrison, K. Lyng, and H. D. Caldwell, *J. Exp. Med.* **169**, 663 (1989).

[8] R. P. Morrison, R. J. Belland, K. Lyng, and H. D. Caldwell, *J. Exp. Med.* **170**, 1271 (1989).

[9] S. Campbell, S. J. Richmond, and P. S. Yates, *J. Gen. Microbiol.* **135**, 2379 (1989).

[10] J. A. Whittum-Hudson, R. A. Prendergast, and H. R. Taylor, *Curr. Eye Res.* **5**, 973 (1986).

[11] H. R. Taylor, I. W. Maclean, R. C. Brunham, S. Pal, and J. Whittum-Hudson, *Infect. Immun.* **58**, 3061 (1990).

[12] H. R. Taylor, *Am. J. Trop. Med. Hyg.* **42**, 358 (1990).

[13] E. Young and H. R. Taylor, *J. Infect. Dis.* **150**, 745 (1984).

[14] J. A. Whittum-Hudson and H. R. Taylor, *Infect. Immun.* **57**, 2977 (1989).

The major advantage of the monkey model for ocular infection is that human biovars of *Chlamydia trachomatis* may be used as infectious inoculum.[4] This has not been possible with other lower vertebrate models to date. Purified viable chlamydial elementary bodies (EB) are obtained by infection of low passage HeLa cells or McCoy cells and density gradient purification on Percoll or Renografin. *Chlamydia trachomatis* serovars A–C and E have been used to infect eyes of cynomolgus monkeys.[12] Although serovar E (BOUR strain) is generally considered to be a genital strain of *C. trachomatis*, it has been isolated from both the eyes and genital tract of human patients. There are numerous isolate strains available for each chlamydial serovar, and published information regarding the intensity of ocular infection should be used as a guide when choosing a dosage for infection. Some ocular strains are more pathogenic than others even under similar purification protocols: A and B strains are more benign at a similar dosage compared to C and E serovars.[12] Following a topical application of viable chlamydiae [1000–5000 inclusion-forming units (ifu)] in 20 μl, ocular disease develops over a 1- to 2-week period, peaking between 2 and 6 weeks after which inflammation and follicles gradually diminish. Dosages required to induce disease will vary depending on the chlamydial serovar and strain. There are three major criteria for infection which have been used in the cynomolgus monkey model for trachoma: clinical disease scores, culture isolation, and the direct fluorescence antibody (DFA) assay.

Assessment of Clinical Disease

A single topical ocular infection is self-limited, and eyes become culture and DFA negative within 6–8 weeks postinfection. Weekly reinfection was shown to more closely resemble the chronic disease (trachoma) seen in humans.[4] A scoring system based on the human trachoma scoring system[15] has been developed and used successfully in the monkey model to score disease after infection or after induction of ocular delayed hypersensitivity.[16] This score is based on 10 parameters per eye relating to inflammation and follicle formation on both the upper and lower conjunctiva and yields the total clinical disease score (TCDS) when the scores for all signs in both eyes are combined.[16] Ocular disease includes the period in which follicles appear and then decrease in size and number, with inflammation persisting longer and gradually declining over 12 weeks.

[15] B. Thylefors, C. R. Dawson, B. R. Jones, S. K. West, and H. R. Taylor, *Bull. WHO* **65**, 477 (1987).

[16] H. R. Taylor, S. L. Johnson, J. Schachter, H. D. Caldwell, and R. A. Prendergast, *J. Immunol.* **138**, 3023 (1987).

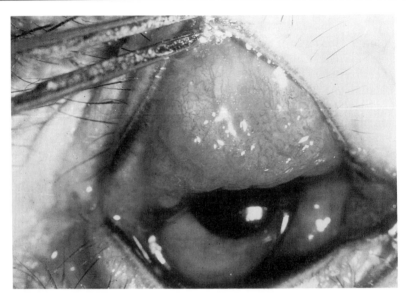

FIG. 1. Infected conjunctiva from a cynomolgus monkey demonstrating the presence of follicles, edema, and hyperemia on the upper conjunctival surface during the peak clinical response to ocular infection.

Eyes should be examined under ketamine anesthesia in a masked fashion with respect to treatment and previous scores. A reasonable examination schedule is weekly for 3–4 weeks following ocular infection, biweekly through 8 weeks, and then once more at 12 weeks. Plans for reinfection or other treatment would obviously influence the examination schedule.

Eyes are examined before any specimens are collected to avoid nonspecific inflammation which swabbing or tear collection could induce. Lids must be inverted nontraumatically by using nontoothed fine ophthalmic forceps. The examiner uses a hand-held slip lamp to enumerate clinical signs in a masked fashion without knowledge of immunization status. Care must be taken to examine all tarsal conjunctival surfaces since occasional nonspecific follicle formation may be observed on the lower conjunctiva. The clinical disease score (CDS) is compiled by grading 10 clinical features on a 0–3 + scale: follicle formation in the bulbar, limbal, superior tarsal, superior fornix, and inferior fornix conjunctiva, the degree of hyperemia (redness) or injection (vessel growth) in the bulbar, superior tarsal, superior fornix, and inferior fornix conjunctiva, and ocular discharge (Fig. 1). Scores for both eyes are added to give a total CDS (TCDS) with the maximum TCDS possible being 60 (generally scores ≤40 are the maxi-

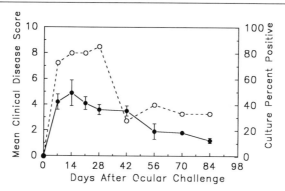

FIG. 2. Representative time course of clinical disease scores (●) and culture positive status (○) after a single infectious challenge. The data points are the means ± the standard error of the mean from four experiments.

mum). The cumulative disease score can also be calculated by summing the scores for each of the 10 signs over the course of the experiment. Data from multiple experiments obtained by this type of analysis are more easily compared if similar examination schedules are used. A typical disease course following a single infection with serovar C (5000 ifu) is shown in Fig. 2.

Biopsies of conjunctiva can be taken, but this should be done by either an ophthalmologist, an ophthalmic veterinarian, or someone surgically trained by the latter. While most of the tarsal conjunctival surface can be removed, usually one-third of a surface is sufficient to yield adequate tissue for immunohistochemistry and other histologic studies. Prior to the start of a longitudinal experiment, conjunctival surfaces which will be biopsied should be randomized so that tissue is not arbitrarily selected based on the severity of disease.

Microbiological Assessment of Infection

The DFA and culture assays closely parallel one another as microbiological measures of infection, although the former detects organisms regardless of viability. Sterile Dacron swabs (Spectrum Laboratories, Inc., Houston, TX) are used to sample the conjunctival surfaces of one or both eyes; the same eye should be used for both culture and DFA if only one eye is to be sampled. For culture, swabs are placed in 1 ml of a standard transport medium for chlamydiae. Isolation is performed in cultures of McCoy cells using standard techniques.

Smears for the DFA assay are made on slides with same Dacron swab by rolling the swab across the sample area, taking care to collect adequate specimens (>1000 cells). Slides are air-dried, fixed in room temperature absolute methanol for 5 min and stored at $-20°$ until use. Prior to staining with the DFA reagent, slides are brought to room temperature, washed in methanol, dried, and stained as directed by the vendor. Control slides should be included in each staining run. Slides are read in a masked fashion under a magnification of $\times 100$ on a fluorescence microscope. There are advantages to both methods: DFA is performed on smears and does not require sterile tissue culture facilities, whereas microscopy for culture assays requires less training.

The presence of the organism may also be assessed by the determination of chlamydial RNA or detection of DNA by the polymerase chain reaction method. Sterile swabs for molecular analyses should be placed either into phenol buffered with sterile 50 mM Tris-HCl, pH 7.4, 2 mM EDTA, 0.15 M NaCl, and 0.5% (w/v) sodium dodecyl sulfate for RNA analysis[17] or into Dulbecco's phosphate-buffered saline with 2% fetal calf serum (FCS), gentamicin (5 μg/ml), vancomycin (12.5 μg/ml), and nystatin (12.5 μg/ml) for polymerase chain reaction analysis.[18]

Histological Evaluation of Disease

Biopsies may be processed for paraffin embedding after fixation in 10% buffered formalin, or they can be snap-frozen in Tissue-Tek O.C.T. compound (Miles Diagnostics, Inc., Elkhart, IN) for frozen sectioning. Frozen sections are preferable for most immunohistochemical studies because monoclonal antibodies are used and formalin or other harsh fixatives destroy the target epitopes. Sections 8–10 μm thick are applied to slides coated with subbed gelatin, fixed for 10 sec, and stored at $-20°$ until fixation in cold acetone for 10 min and staining (alternatively, unfixed slides are stored at $4°$ for 1–14 days before staining). The avidin–biotin immunoperoxidase complex (ABC) (Vector Labs, Burlingame, CA) and horseradish peroxidase (Kirkegaard and Perry, Gaithersburg, MD) staining method works well with this tissue. Many anti-human reagents will stain tissues from the cynomolgus monkey, but not all T cell subset reagents will.

Each antibody of interest should be pretested. We use either cynomolgus spleen or lymph node sections or peripheral blood mononuclear cells as positive controls for conjunctival sections or cytospins, respectively. Submandibular lymph nodes, which are the regional draining lymph nodes

[17] A. P. Hudson, C. M. McEntee, M. Reacher, J. A. Whittum-Hudson, and H. R. Taylor, *Curr. Eye Res.* **11**, 279 (1992).
[18] S. M. Holland, A. P. Hudson, L. Bobo, J. A. Whittum-Hudson, R. P. Viscidi, T. C. Quinn, and H. R. Taylor, *Infect. Immun.* **60**, 2040 (1992).

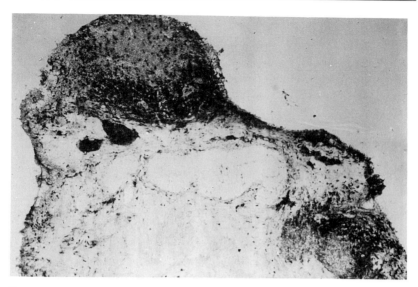

FIG. 3. Immunohistochemical staining for CD8[+] T cells in a monkey in which chronic infection was induced by repeated inoculations of *C. trachomatis* serovar B. The frozen section is immunoperoxidase stained for the CD8 T cell marker and counterstained with methyl green. CD8[+] T cells are found in a perifollicular and subepithelial location. (CD4[+] T cells are also localized perifollicularly.) Original magnification: ×40.

for ocular infection, are also a good source of *Chlamydia*-specific T and B lymphocytes 4–10 weeks after infection.[19] Peripheral (inguinal) nodes do not contain significant numbers of antigen-specific cells during ocular infection.[19] Analyses can include the number of cells per section or 10 high-power fields, distribution and ratios of T cell subsets, size and makeup of B cell-containing follicles, and numbers of macrophage/monocytes. Differing results for T cell subsets can be expected after a single versus repeated infection protocol.[20] Approximately 50% of the conjunctival infiltrate will be T lymphocytes, and as B cell-containing follicles decrease in size and number the T cell infiltrate will also decrease proportionately.

A typical perifollicular staining for CD8[+] T cells is shown in Fig. 3. Follicles are comprised almost exclusively of B lymphocytes; the immunoglobulin (Ig) isotype distribution is IgG > IgM > IgA throughout infection

[19] S. Pal, Z. Pu, R. B. Huneke, H. R. Taylor, and J. A. Whittum-Hudson, *Reg. Immunol.* **3,** 171 (1990).

[20] J. A. Whittum-Hudson, H. R. Taylor, M. Farazdaghi, and R. A. Prendergast, *Invest. Ophthalmol. Visual Sci.* **27,** 64 (1986).

with B serovar,[20] and similar results have been shown in human biopsies.[21] Anti-human immunoglobulin isotype-specific antibodies (Kirkegaard and Perry) which are peroxidase-conjugated for avidin–biotin complex immunoperoxidase staining are very useful for investigation of B cells in frozen sections. In such staining experiments, duplicate slides receive normal IgG at the same dilution and from the same species used for each primary antibody. Nonspecific staining of the conjunctival epithelial cells may be observed even with normal IgG presumably because of Fc receptors on these cells.

Assessment of Cell-Mediated Immunity to Ocular Chlamydial Infection

Lymphoproliferation induced by chlamydial organisms or antigens is used as an *in vitro* measure of cell-mediated immunity to *Chlamydia trachomatis*. Similar to studies in humans, peripheral blood monocytes are obtained by Ficoll–Hypaque density gradient separation.[13,14] Buffering with HEPES should not be used when freezing cynomolgus peripheral blood lymphocytes as it reduces viability of cells. Because cynomolgus monkeys are outbred, cells from different animals cannot be pooled unless the major histocompatibility antigens are identical. This is a major limiting factor for assays which require large cell numbers, such as the limiting dilution assay.

To test local immunity induced by ocular chlamydial infection, larger biopsies of conjunctiva may be obtained. Although clinical examinations depend on retention of sufficient conjunctival tissue, a pair of conjunctival specimens may be collected, one from the upper fornix of one eye and the other from the inferior fornix of the opposite eye. Normal conjunctiva contains insufficient lymphoid cells to warrant this procedure, and attempts should be made to obtain biopsies at the peak of the inflammatory response (2 to 6 weeks) to maximize cell yields. The conjunctival surface will usually heal quickly without complications, and clinical disease can be scored after small biopsy; however, animals should be followed for signs of infection after biopsy. Tissues are placed immediately in cold sterile Hanks' balanced salt solution (HBSS) containing Fungizone (GIBCO, Grand Island, NY; 0.25–1.0 μg/ml), streptomycin (Sigma, St. Louis, MO; 125 μg/ml), and gentamicin (M. A. Bioproducts, Walkersville, MD; 10 μg/ml). Conjunctiva are dispersed to obtain mononuclear cells by digestion in 6 ml collagenase (or 0.02% for 3 hr; No. 4194, Worthington, Freehold, NJ) at 37°. The enzymatic reaction is stopped by dilution in cold HBSS containing 10% FCS. Tissue is triturated to free lymphocytes

[21] M. H. Reacher, J. Pe'er, P. A. Rapoza, J. A. Whittum-Hudson, and H. R. Taylor, *Ophthalmology* **98**, 334 (1991).

from collagen fibrils. Because cell yields are so low (2–10 × 10^6 cells/ two conjunctival surfaces), we resuspend the dispersed tissue, let the collagen clumps settle, and remove the mononuclear suspension from the supernatant.

For the lymphoproliferative assay, mononuclear cells are counted and adjusted to the desired concentration, then cultured according to standard procedures.[13,19,22] Usually sufficient cells from two conjunctival surfaces will be obtained to perform a complete lymphoproliferative assay with cells at a concentration of 1–5 × 10^4/well; 10-fold more cells per well can be used for peripheral blood or lymph node cells. Test wells are run in triplicate for each antigen or dilution of antigen. One hundred microliters of lymphocytes is added to round-bottomed microplate wells followed by 100 μl of purified elementary bodies (EB) (e.g., 10^4–10^7 ifu/ml), chlamydial antigens, or culture medium alone (negative control). Nonspecific stimulation by the mitogens concanavalin A (5–10 μg/ml) or pokeweed mitogen (10 μg/ml or 1 : 100 dilution) can also be included to verify viability. For peripheral blood an S.I. of at least 4.0 is considered positive, because it exceeds the nonspecific responses to chlamydiae observed with peripheral blood mononuclear cells from normal monkeys.[14] Because fewer conjunctival cells are cultured in each well, an S.I. of at least 2.0 is generally considered positive because background counts per minute (cpm) from the lower number of cells are also lower.

To gain information on the *in vitro* antibody-producing potential of cells from immune animals, cells obtained in a similar fashion as above can be cultured with pokeweed mitogen (1 μg/well) for 5–7 days to test for secretion of antichlamydial antibody.[14] At the end of the culture period supernatants are collected and tested for antibody in an appropriate enzyme-linked immunosorbent assay (ELISA). Cells and pokeweed mitogen can be cultured either in replicate (4–12) microwells to pool supernatants or in bulk culture at a similar cell concentration.

Assessment of Antibody Response

Sera obtained from monkeys can be assessed for IgG antibodies to chlamydiae by various assays including microimmunofluorescence, ELISA, and immunoblot analysis. The assessment of both IgG and IgA in tears can also be made. Tears are collected on 1 × 3 mm Weck-Cel (Edward Weck & Co., Inc., Research Triangle Park, NC) sponges (cut and autoclaved in microcentrifuge tubes) that are placed at the interior fornix of the eye. The tears are eluted from the sponges with 220 μl of

---

[22] M. Tuffrey, P. Falder, and D. Taylor-Robinson, *Br. J. Exp. Pathol.* **66**, 427 (1985).

phosphate-buffered saline per sponge by incubation on ice for 30 min before further use. A 1:10 starting concentration is assumed, but this depends on adequate specimen collection. Specific IgA and IgG antibodies will appear and persist in tears after ocular infection, whereas IgG titers in tears are generally lower than those in serum.

Infection of Guinea Pig with Agent of Guinea Pig Inclusion Conjunctivitis

Induction of Ocular Infection

Guinea pigs weighing at least 450–500 g should be used in the experiments. No sex-related variations in ocular infection have been reported, so either sex can be used. Both inbred strains (strain 2 and strain 13) and outbred Hartley guinea pigs from a variety of sources are susceptible to infection. It is extremely critical in choosing a source of guinea pigs for use in ocular infection studies that the colony be checked for endemic GPIC infection. Most suppliers will not be aware of whether their colony is infected, since the infection, if present, occurs primarily in newborns and may go unnoticed. Serology for GPIC is not part of the standard serological survey panel. If the colony has been cesarean-derived and has remained closed, it is likely to be GPIC-free. Nevertheless, it is useful to evaluate sera for antibody to GPIC by a standard microimmunofluorescence assay[23] or ELISA.[24] Guinea pigs should be housed individually to prevent spread of infection to uninfected animals.

Culture of GPIC may be conducted in either McCoy or HeLa cell lines using standard technology, although HeLa cells generally produce greater yields. In addition, GPIC tends to grow more aggressively when compared to the oculogenital strains of *C. trachomatis*. Renograffin or Percoll gradient-purified chlamydiae may be used for infection, but chlamydiae prepared by disrupting an infected monolayer followed by sonication, low-speed centrifugation (500 g) to pellet host cell debris, and a final high-speed centrifugation (30,000 g) to concentrate the organisms may be used with equal effect. Suspensions prepared in either manner are routinely aliquoted and titered for the number of inclusion-forming units (ifu) in 1 ml of the suspension. There has been no recorded loss of virulence in the animals by repeated *in vitro* culture.

Guinea pigs may be infected without anesthesia as long as they are somewhat restrained, usually by having one person hold the animal firmly.

[23] R. G. Rank, H. J. White, and A. L. Barron, *Infect. Immun.* **26,** 573 (1979).
[24] R. G. Rank and A. L. Barron, *Infect. Immun.* **55,** 2317 (1987).

To infect the conjunctiva, the upper and lower palpebral conjunctiva are lifted off the eye by the individual holding the animal. Using an automatic micropipette, 25 μl of a dilution of GPIC in sucrose–phosphate–glutamate buffer (pH 7.4) is deposited directly onto the surface of the eye. In general, the inoculum should contain a total of 10^5 to 10^7 ifu. Lowering the dose will lengthen the time until peak infection occurs. The animals are replaced in the cage gently to prevent spillage of the material from the eye. If it is desired to only infect a single eye, the incidence of crossover infection to the contralateral eye is quite low, in contrast to the monkey.

Assessment of Clinical Disease

Evaluation of clinical disease in the guinea pig GPIC model is not as detailed as in the primate, primarily because of the size of the animal. However, a grading system has been developed by Watkins *et al.*[25] which allows for consistency and ease in scoring without ophthalmologic training. An advantage of this scoring system is also that the animals may be examined without anesthesia. Basically both the bulbar and the palpebral conjunctiva are observed for hyperemia and edema. The scores are reported as follows:

0   Normal conjunctiva
1   Slight hyperemia and edema of either palpebral or bulbar conjunctiva
2   Overt hyperemia and edema of either palpebral or bulbar conjunctiva
3   Overt hyperemia and edema of both palpebral and bulbar conjunctiva
4   Overt hyperemia and edema of both palpebral and bulbar conjunctiva accompanied by purulent exudate

Typically, after a primary GPIC infection, a marked clinical response is observed concomitant with the increase in inclusion score (Fig. 4). In contrast to the primate model, this reaction is primarily composed of an acute inflammatory reaction.[6] However, this initial inclusion conjunctivitis is not unlike the typical conjunctivitis that one sees in humans on the first experience with the organism. On repeated challenge of a previously infected animal, the reactions have a greater mononuclear component, indicative of an immunopathological reaction. In fact, Monnickendam *et*

---

[25] N. G. Watkins, W. J. Hadlow, A. B. Moos, and H. D. Caldwell, *Proc. Natl. Acad. Sci. U.S.A.* **83,** 7480 (1986).

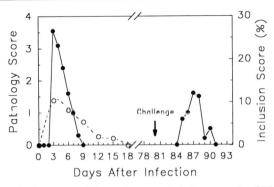

Fig. 4. Pathological response (●) to primary and challenge ocular GPIC infections and course of infection (○) as determined by inclusion scores. Challenge infection did not result in inclusions detectable by the inclusion score method. Such animals were, however, isolation positive.

al.[26] have demonstrated that pannus can be elicited in the guinea pig model after multiple challenges. The intensity of repeated infections increases with the time interval between the primary and secondary infections. The greater the temporal difference between the two infections, the greater the degree of pathological change resulting from the challenge reaction. This observation also applies for the level of infection as measured by inclusion score.[26,27] Certainly, confirmation of the cellular makeup of the response at the various times requires histopathological examination. Unfortunately, at this point, there are no published reports of thorough histopathological analysis of conjunctival infection in the guinea pig.

Microbiological Assessment of Infection

The microbiological course of the infection may be monitored in two ways. The GPIC produces a rather productive infection so that quantification may be accomplished on a scraping from the conjunctiva without the use of anesthesia. The conjunctival scraping is performed with the use of a dental stainless steel plastic filling instrument (No. D-2). The L-shaped end is scraped along the inner surface of either the upper or lower palpebral conjunctiva using sufficient force to obtain epithelial cells. The material is then placed on a glass slide and allowed to dry, after which it is fixed with methanol for 15 min and stained with Giemsa [1 : 50 (v/v) of stock

[26] M. A. Monnickendam, S. Darougar, J. D. Treharne, and A. M. Tilbury, *Br. J. Ophthalmol.* **64**, 284 (1980).
[27] R. Malaty, C. R. Dawson, I. Wong, C. Lyon, and J. Schachter, *Invest. Ophthalmol. Visual Sci.* **21**, 833 (1981).

Giemsa, 7.415 g/liter in methanol in distilled water with 5% phosphate-buffered saline]. Quantification is accomplished by counting at least 100 epithelial cells spread over different areas of the smear and determining the percentage of the cells that have inclusions (inclusion score). Several hundred cells should be observed before a specimen is reported to be negative. The margin of error on repeated counts from a single smear is roughly ±5 units. A sample infection curve is shown in Fig. 4. Groups of animals may be compared statistically using a two-factor (days, group) analysis of variance with repeated measures on one factor (days). Generally, to reduce the trauma to the conjunctiva, specimens are collected for assessment of the inclusion score every third day beginning on day 3 of the infection. This method also has the advantage that the inflammatory exudate can also be evaluated for granulocytic and mononuclear cell components.

The course of the infection may also be determined by quantitative isolation from conjunctival swabs. As above, specimens should be taken no more frequently than every third day. Dacron swabs are rubbed along the inner surface of either the upper or lower palpebral conjunctiva so that epithelial cells can be obtained. The swabs are placed immediately into 1 ml of cold sucrose–phosphate transport (2-SP)[28] and frozen at −70°. The GPIC is isolated from the swabs in shell vials with McCoy cell monolayers according to standard culture procedures. However, because GPIC is relatively fast growing, coverslips should be fixed with acetone about 24 h after initiation of culture, so that primary intact inclusions may be counted. Inclusions may be visualized by incubation with either a genus-specific monoclonal antibody or hyperimmune guinea pig serum from animals which have recovered from the infection, followed by an appropriate fluorescein-labeled antibody. The number of inclusion-forming units per swab can be determined by counting the number of inclusions on the coverslip and relating this to the diluted volume of transport medium added to the shell vial.

Culture of GPIC from the conjunctiva possesses the advantage of being more sensitive than the inclusion score method, with detection being possible for several days longer. The isolation method is particularly valuable in assessing the response to reinfection in previously infected guinea pigs because the immune response is often, depending on the time after the primary infection, sufficient to lower the number of organisms beneath the sensitivity of detection of the inclusion score method. Because of variability in collection of the specimen, however, the range of inclusion-

[28] J. Schachter, in "Manual of Clinical Microbiology" (E. H. Lennette, ed.), p. 357. American Society for Microbiology, Washington, D.C., 1980.

forming units determined among guinea pigs in a single group may be quite broad.

Evaluation of Immune Response

A major advantage of the guinea pig model is the relative ease with which one can evaluate a number of immune parameters, although a major disadvantage is the unavailability of specific reagents, particularly for lymphocyte markers. Serum samples for antibody determinations can be collected essentially as often as one requires by using the method of Lopez and Navia.[29] Briefly, guinea pigs are injected intramuscularly with 0.08 ml/kg of Innovar-vet (Pittman-Moore, Mundelein, IL) which calms the animal but also causes dilatation of peripheral blood vessels. Blood is then collected by puncturing the saphenous vein on the outer aspect of the foot with a blood lancet. As much as 1–2 ml of blood can then be collected with 0.25-ml heparinized capillary tubes. Obviously, if larger quantities of blood are required, cardiac puncture may be performed, but the former method provides virtually no risk of accidently killing the animal.

Ocular secretions for the assessment of IgG or IgA antibody to GPIC can be obtained by first injecting the animal with Innovar-vet as above and then placing a 2 × 7 mm section of Weck-Cel surgical sponge under the conjunctiva or in the corner of the eye for about 5 min until visible swelling of the sponge occurs. Care must be taken, particularly in infected animals, not to abrade the tissues, since any bleeding would obviously contaminate the sponge with serum antibody. The sponge is frozen in a plastic tube without additional fluid. When the antibody assay is to be performed, the sponge is eluted with 0.1 ml of phosphate-buffered saline (pH 7.4). The eluate is centrifuged at 13,000 g in a microcentrifuge to sediment debris and is then included in an appropriate antibody assay. Enough specimen can be obtained to perform both IgG and IgA assays. Both ELISAs and microimmunofluorescence assays can be used for the measurement of either serum or secretion antibodies to GPIC. In addition, immunoblot assays for the antigen-specific response can be performed on either sample, although it may be necessary to pool secretion samples to obtain a sufficient concentration of antibodies for detection by this method.

For cell-mediated immunity, peripheral blood mononuclear cells (PBM) can be readily obtained and used in blast transformation assays. To obtain peripheral blood, the guinea pig is anesthetized by inhalation of methoxyflurane, and blood is collected by a cardiac puncture using a

---

[29] M. Lopez and J. M. Navia, *Lab. Anim. Sci.* **27**, 522 (1977).

1 1/2-inch 22-gauge needle into a syringe containing a solution of trisodium citrate (44 g/liter), citric acid (16 g/liter), and glucose (50 g/liter) (0.1 ml per milliliter of blood). Approximately 4 ml of blood may be safely collected from a 500-g animal. The PBM are isolated on a Histopaque gradient (specific weight 1.077 g/ml; Sigma), washed, and counted.[30] Usually, about 2–6 × 10⁶ cells can be obtained, which is sufficient for assays of multiple antigens or mitogens. If more cells are required or an increased purity of a T cell population is needed, spleens may be used. The T cells can be enriched by filtration of spleen cells over a nylon wool column using standard procedures.[31] For the response to GPIC, 1 μg per well (96-well microculture plate) of purified elementary bodies has been found to produce an optimal response.

[30] R. G. Rank, L. S. F. Soderberg, M. M. Sanders, and B. E. Batteiger, *Infect. Immun.* **57**, 706 (1989).
[31] J. U. Igietseme and R. G. Rank, *Infect. Immun.* **59**, 1346 (1991).

[7] Animal Models for Urogenital Infections

By ROGER G. RANK

Introduction

There are obviously a multitude of bacterial organisms which are infectious for the urogenital tract of humans; however, the major bacterial pathogens which specifically cause infections of the genital tract in humans include *Neisseria gonorrhoeae, Treponema pallidum,* and *Chlamydia trachomatis*. No animal model has been described for *N. gonorrhoeae* genital infections. Although animal models are available for *T. pallidum,* it is only acquired via the genital route, its pathogenesis being more associated with a systemic infection rather than a local genital mucosal infection. In contrast, *C. trachomatis* is restricted primarily to the genital or ocular mucosa and can thus be considered a true genital pathogen. With this in mind, this chapter focuses on animal models of *Chlamydia* which can be used to assess virulence as well as to understand the immune response.

Several animal models have been described for chlamydial genital infections of the female, including *C. trachomatis* infections of rhesus[1]

[1] D. L. Patton, C.-C. Kuo, S.-P. Wang, R. M. Brenner, M. D. Sternfeld, S. A. Morse, and R. C. Barnes, *J. Infect. Dis.* **155**, 229 (1987).

and grivet monkeys,[2] pig-tailed macaques,[3,4] marmosets,[5] and mice[6-9] as well as *C. psittaci* infections of the guinea pig,[10] cat,[11] and koala.[12]. Infections have also been characterized in male guinea pigs[13] and grivet monkeys.[14] While, without a doubt, the primate most closely approximates the human infection, it is also expensive and impractical to use for large studies. Patton has overcome this disadvantage to some degree by implanting oviduct tissue subcutaneously along the abdomen of monkeys so that multiple "pockets" are available for inoculation on a single animal.[1] While this model is extremely valuable for many types of studies, it still requires a major expense. On the other hand, the mouse and guinea pig models lend themselves well to studies involving larger numbers of animals. For this reason, the emphasis in this chapter is on these models. It should be emphasized that no one model is ideal for all studies, so that care should be taken in choosing the animal in order to address the particular research question in mind.

Infection of Guinea Pig with Agent of Guinea Pig Inclusion Conjunctivitis

Although the agent of guinea pig inclusion conjunctivitis (GPIC) is a member of the *C. psittaci,* it produces an infection in the genital tract of guinea pigs which is remarkably similar to human *C. trachomatis* genital infection with regard to pathogenesis, pathology, immunity, and ability

[2] B. R. Moller, E. A. Freundt, and P. A. Mardh, *Am. J. Obstet. Gynecol.* **138**, 990 (1980).
[3] D. L. Patton, S. A. Halbert, C. C. Kuo, S. P. Wang, and K. K. Holmes, *Fertil. Steril.* **40**, 829 (1983).
[4] P. Wolner-Hanssen, D. L. Patton, W. E. Stamm, and K. K. Holmes, *in* "*Chlamydia* Infections" (D. Oriel, G. Ridgway, J. Schachter, D. Taylor-Robinson, and M. Ward, eds.), p. 371. Cambridge Univ. Press, New York, 1986.
[5] A. P. Johnson, C. M. Hetherington, M. F. Osborn, B. J. Thomas, and D. Taylor-Robinson, *Br. J. Exp. Pathol.* **61**, 291 (1980).
[6] A. L. Barron, H. J. White, R. G. Rank, B. L. Soloff, and E. B. Moses, *J. Infect. Dis.* **143**, 63 (1981).
[7] M. Tuffrey and D. Taylor-Robinson, *FEMS Microbiol. Lett.* **12**, 111 (1981).
[8] M. Tuffrey, P. Falder, J. Gale, and D. Taylor-Robinson, *Br. J. Exp. Pathol.* **67**, 605 (1986).
[9] J. I. Ito, Jr., J. M. Lyons, and L. P. Airo-Brown, *Infect. Immun.* **58**, 2021 (1990).
[10] D. T. Mount, P. E. Bigazzi, and A. L. Barron, *Infect. Immun.* **8**, 925 (1973).
[11] J. L. Kane, R. M. Woodland, M. G. Elder, and S. Darougar, *Genitourin. Med.* **61**, 311 (1985).
[12] K. A. McColl, R. W. Martin, L. J. Gleeson, K. A. Handasyde, and A. K. Lee, *Vet. Rec.* **115**, 655 (1984).
[13] D. T. Mount, P. E. Bigazzi, and A. L. Barron, *Infect. Immun.* **5**, 921 (1972).
[14] B. R. Moller and P. A. Mardh, *Fertil. Steril.* **34**, 275 (1980).

to be transmitted sexually.[15,16] The organism is strictly an obligate intracellular parasite of superficial epithelial cells and has as its principal target tissue the exocervix.[17] Of importance is the fact that GPIC is a natural parasite of the guinea pig, so any responses studied can be assumed to be the result of natural host–parasite interactions. One problem with animal studies of human agents is that often the pathogens are not natural parasites of the animal in question, so that one has to consider an entirely different set of variables associated with host specificity. In addition, a major advantage of the guinea pig over other rodent models is that the female reproductive system is more closely related to the human than other rodents with regard to both histology and physiology. The female guinea pig has a 15- to 17-day estrous cycle with spontaneous ovulation and actively secreting corpus lutea. Thus, it can serve as an excellent model to assess hormonal effects on the infection. A variety of considerations and techniques in the use of this model will not be discussed here since they were already described in [6] on the GPIC ocular model. Nevertheless, the majority of those issues are also valid in the use of the genital tract model.

Infection and Assessment of Infection Course in Lower Genital Tract

As with ocular infections, guinea pigs from a GPIC-free colony must be used. Both male and female guinea pigs may be used in this model. However, the majority of studies have been performed with female animals because the major morbidity associated with chlamydial disease is the development of infertility resulting from salpingitis. Sexually mature guinea pigs, weighing 450–500 g, should be used for the studies.

The GPIC for infection purposes is cultured in McCoy or HeLa cells and is purified as described elsewhere in this volume (see [6]). Either a crude or gradient-purified elementary body preparation may be used. Female guinea pigs are infected intravaginally by placing the animal on its back and inserting a blunt micropipette approximately 2–3 cm into the vagina. Then 50 μl of sucrose–phosphate–glutamate (SPG) buffer (pH 7.4)[18] containing from 10^4 to 10^7 inclusion-forming units (ifu) of GPIC elementary bodies is inoculated directly into the vagina. Anesthesia is not required, but the animals should be replaced in the cage carefully in an

[15] D. L. Patton and R. G. Rank, in "Sexually Transmitted Diseases" (T. C. Quinn, ed.), p. 85. Raven, New York, 1992.

[16] R. G. Rank and M. M. Sanders, Am. J. Pathol. **140,** 927 (1992).

[17] A. L. Barron, H. J. White, R. G. Rank, and B. L. Soloff, J. Infect. Dis. **139,** 60 (1979).

[18] J. Schachter, in "Manual of Clinical Microbiology" (E. H. Lennette, ed.), p. 357. American Society for Microbiology, Washington, D.C., 1980.

attempt to prevent them from becoming excited and expelling a portion of the inoculum. A unique aspect of the guinea pig genital tract anatomy is the appearance of a vaginal membrane when the animal is not in estrus. The membrane totally prevents entry into the vagina. Therefore, when animals are to be inoculated, the membrane may be simply broken by pushing gently with the pipette tip. This procedure produces only momentary mild discomfort to the animal with no bleeding or apparent trauma.

Urethral infections in male guinea pigs may be established by extruding the penis and merely "dropping" 50 μl of inoculum onto the external meatus. An alternative method is the direct intraurethral inoculation of the same volume with a 0.76 mm (inside) by 1.22 mm (outside) diameter piece of Tygon tubing attached to a 23-gauge needle with syringe.

As in ocular infections, the course of the infection may be monitored by either the inclusion score method (see [6] in this volume) or by isolation of chlamydiae from cervical or urethral swabs. To obtain material from the female for inclusion scores, a dental stainless steel plastic filling instrument (No. D-2) is inserted about 2 cm into the vagina and the spatula end rotated against the vaginal wall. Care should be taken that epithelial cells are obtained by scraping the wall and not only by obtaining exudative material from the lumen. Most of the exudate will be inflammatory cells with few epithelial cells. The material obtained is spread onto a glass slide and stained with Giemsa after methanol fixation. The inclusion score is determined as described previously. Giemsa staining also allows examination and quantification of the exudate for inflammatory cells. Smears taken at 3-day intervals provide adequate and informative monitoring of the infection course. In general, the infection as assessed by inclusion scores reaches peak levels 6–9 days after vaginal inoculation and lasts about 18–21 days.

Urethral scrapings from male guinea pigs are obtained using the same instrument by inserting the blunt end about 0.5 cm into the urethra of the extruded penis and then scraping the urethral wall. A smear is prepared as above. Because no exudate is routinely seen in infections of male guinea pigs, less material will be apparent on the stained smear.

Cervical swabs for the isolation of GPIC are collected by inserting a Dacron swab about 3 cm into the vagina until it can go no further. The swab is then rotated against the cervix, removed, and placed into transport medium such as sucrose–phosphate buffer (2-SP) (pH 7.4).[18] Urethral swabs are collected by inserting the swab about 1 cm into the urethra, rotating, and then placing in transport medium. The isolation of chlamydiae is performed by standard procedures. Swabs may also be taken every 3 days for the mapping of the infection. Both scraping and swab procedures may be performed without anesthetizing the animal. It has been our observation that, during a primary infection, inclusion scores are sufficient to document the course of the infection, since we have never observed animals to be isolation negative in the presence of positive inclusion scores. Nevertheless, chlamydial isolation is still more sensitive, and, at the end of the infection, isolation attempts may reveal organisms for about 3 days after inclusion scores have become negative. Moreover, when assessing immunity to reinfection, it is common to find that animals are isolation positive in the absence of detectable inclusions on vaginal scrapings. Animals should be monitored until at least two consecutive negative results have been obtained to rule out sampling error.

Assessment of Infection and Disease in Upper Genital Tract of Females

Another advantage of the female guinea pig model is the inclination of GPIC to ascend the genital tract to the uterus and oviducts. Approximately 80% of uterine horns and oviducts become isolation positive for GPIC by day 7 after lower tract inoculation (Fig. 1).[16] Organisms can be isolated up to day 12 from the oviducts. The pathological response is analogous to human salpingitis with acute and chronic inflammatory components as well as plasma cell infiltration in the early stage of infection (days 5–12). Long-term (days 75–85) oviductal damage is characterized by fibrosis and chronic inflammatory response in the mesosalpinx with dilatation of the oviducts, presumably from blockage of the oviduct. Assessment of upper tract disease can readily be made by histopathological analysis of the tissues on sections stained with hematoxylin and eosin.

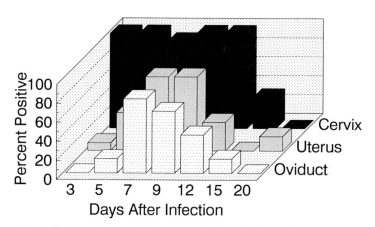

FIG. 1. Percentage of genital tract tissues positive for isolation of GPIC at varying times following intravaginal inoculation with 10^6 ifu of a chlamydial suspension.

Chlamydial inclusions and antigen can also be visualized by immunoperoxidase or immunofluorescent staining. If desired, organisms may be isolated
by mincing portions of the tissue over a 60-mesh wire screen and culturing
the supernatant after low-speed centrifugation to sediment tissue debris.

Assessment of histopathological reactions in the various tissues of the
genital tract can be easily quantified so that data can be obtained for
statistical analysis by nonparametric tests such as the Mann–Whitney U
test. In the guinea pig, we have found that it is convenient to determine
the pathological changes in the exocervix, endocervix, uterine fundus,
both uterine horns, both mesosalpingeal tissues, and both oviducts. In
summarizing results, one should report the data in terms of the total tissues
examined, since it is variable as to whether pathological changes are
unilateral or bilateral. The following parameters are assessed in each
tissue: (1) acute inflammatory response (infiltration with polymorphonuclear leukocytes), (2) chronic inflammatory response (infiltration with
mononuclear leukocytes including lymphocytes, monocytes, and macrophages), (3) plasma cell infiltration, (4) fibrosis, and (5) epithelial erosion
(if appropriate for a given tissue). Each reaction is graded according to
the following scheme.

0.5+   Trace of parameter
1+   Presence of parameter in one tissue site
2+   Presence of parameter at 1–4 foci
3+   Presence of parameter at more than 4 foci
4+   Confluent presence of parameter

Dilatation of the oviduct can also be quantitated using a similar scheme
as follows:

1+   Mild dilatation of single cross section of oviduct
2+   1–3 dilated cross sections of oviduct
3+   >3 dilated cross sections of oviduct
4+   Confluent pronounced dilatation of oviduct

The data can be presented as the percentage of tissues having the parameter and/or the mean score for a given tissue.

An example of the pathological changes occurring in the mesosalpinx
at varying times after infection is presented in Fig. 2. In using this system,
one has a detailed descriptive and quantitative analysis of the effects of
the chlamydial infection which can be used as a baseline for the study of
factors associated with virulence. This is particularly important because
it appears that the host response to the organism via both immunological
and nonimmunological mechanisms is a significant contributor to the disease process. For instance, we have observed that repeated infection with

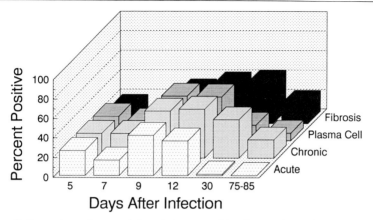

FIG. 2. Percentage of mesosalpingeal tissues positive for varying pathological changes on different days after intravaginal inoculation with 10^6 ifu of GPIC.

GPIC results in an increased percentage of guinea pigs developing tubal dilatation which is associated with infertility (R. G. Rank, unpublished data, 1993). Although the lower genital pathological response is less intense, it is more characteristic of a delayed-type hypersensitivity response than the primary infection, which has both acute and chronic inflammatory components. Moreover, chronic inflammatory and fibrotic changes in the mesosalpinx and oviduct occur and precede the development of increased tubal dilatation.

An important consideration in the evaluation of pathological changes in tissues of the female genital tract is the effect of the estrous cycle. Whereas there does not seem to be any effect on the course of genital infection as determined by inclusion scores on vaginal scrapings or on isolation of chlamydiae from cervical swabs, the production of pathological changes in the oviduct does seem to be influenced. We have noted that animals infected on day 11 of the estrous cycle (day 1 is the day of estrus) have a significantly higher percentage of oviducts with tubal dilatation and mesosalpinxes with chronic inflammation and fibrosis than animals infected on day 6 or day 16 of the cycle.[19a] Thus, in the evaluation of chlamydial disease of the upper genital tract, the stage of the estrous cycle when the animals are infected must be taken into consideration and controlled for by infecting all animals on approximately the same day of the cycle. In fact, it would be best to monitor the cycles of each animal to be used through at least 3 complete cycles to confirm the length of the

[19a] R. G. Rank, M. M. Sanders, and A. J. Kidd, *Am. J. Pathol.* **142,** 1291 (1993).

cycle for the individual animals and to verify that animals are indeed cycling. If the development of upper genital tract pathology is the desirable end point, it would then be advisable to infect all animals on day 11 of the cycle so that the highest percentage of animals develop upper tract disease.

Evaluation of Immune Response

Sera are obtained for determination of antibody by using the method of Lopez and Navia.[19] To evaluate the local antibody response in the genital tract, genital secretions can be collected by the use of a tamponlike arrangement. A 5 cm thread is tied around the center of a 2 × 10 mm preweighed section of Weck-Cel surgical sponge (Edward Weck & Co., Inc., Research Triangle Park, NC). The sponge is folded in half and is inserted into a 5 mm diameter polished glass tube, 3–4 cm in length. To collect the secretions, the guinea pig is lightly anesthetized with Innovar-vet (Pittman-Moore, Mundelein, IL, 0.08 ml/kg body weight), and the glass tube–sponge assembly is inserted approximately 2 cm into the vagina. The sponge is then pushed into the vagina with a wooden applicator stick and the glass tube removed. The thread is cut within a few millimeters of the vagina and left in place for about 1–2 hr. The sponge is removed by use of the string and, after discarding the string, is weighed again so that the weight of the secretions can be determined. The sponge may be frozen at − 20° until needed. The sponge is eluted by adding 0.2 ml of phosphate-buffered saline per 0.1 g of secretion to the sponge, squeezing the sponge repeatedly to express the secretion, and finally centrifuging at high-speed in a microcentrifuge to sediment mucus and debris.

Cell-mediated immunity can be assessed in the guinea pig by the blast transformation response on peripheral blood lymphocytes as described in [6], this volume. Spleen and lymph nodes may be used as well if the animal is to be terminated. *In vivo* delayed-type hypersensitivity may also be measured by injecting 50 μl of antigen into the pinna of the ear and measuring the increase in ear thickness at 24 and 48 hr after injection. This method provides a more quantitative assessment of the response than does injection of antigen intradermally into the flank.

Infection of Mice with Agent of Mouse Pneumonitis

The agent of mouse pneumonitis (MoPn) is a biovar of *Chlamydia trachomatis,* and while its origins were most likely from the human,[20] it

---

[19] M. Lopez and J. M. Navia, *Lab. Anim. Sci.* **27,** 522 (1977).
[20] C. Nigg, *Science* **95,** 49 (1942).

has become adapted to the mouse and is considered to be a parasite of the mouse. Like the human oculogenital strains of *C. trachomatis,* on intravaginal infection, MoPn, remains restricted to the superficial epithelium of the cervix.[6] The major advantage of this system is related to the availability of immunological reagents for the mouse, the variety of strains of mice available including congenitally immunodeficient mice, and the relative inexpensiveness of the mouse as an experimental animal. Thus far, only studies with female mice have been performed. Studies of infection of male mice via the urethra have not been published.

Infection and Assessment of Infection Course in Lower Genital Tract

Mice may be infected intravaginally with 10^5-10^7 ifu of MoPn suspended in 2-SP or SPG buffer. Prior to inoculation, mice are injected intraperitoneally with sodium pentabarbitol (5 mg/100 g body weight). This serves to anesthetize the mice for about 1 hr so that there is less chance for loss of the inoculum. The mice are inoculated in the vagina with 30 μl of the MoPn suspension and are laid on their backs in the cage to reduce the loss into the bedding by capillary action.

A critical factor also in the mouse is the time during the estrous cycle that the animals are inoculated. We have shown that mice inoculated during estrus generally do not become infected, presumably because of the thick mucus which is commonly present. Because it would be impractical to monitor the estrous cycle in all mice, it is suggested that the mice be inoculated for 2–3 consecutive days using a fresh inoculum each day. In this way, mice will be exposed to MoPn on at least one day other than estrus. With this procedure, a high proportion of the mice become infected.

The course of infection is monitored by taking cervical swabs beginning 4–5 days after the last day of inoculation. Dacron swabs are inserted into the vagina of the mouse until they are against the cervix. The swab is then rotated with moderate pressure so that cells may be obtained and is then placed into 2-SP buffer for freezing at $-70°$. No anesthesia is required for collection of the cervical swab. The specimen is processed for isolation of MoPn in tissue culture according to standard techniques and the number of inclusion-forming units per swab determined. Obviously, the quantitation has a high variation based on the collection procedures, but the approximate inclusion-forming units as measured in a number of mice generally fall in the same range, which may be 10^4 to 10^5 at the time of peak infection.[21] The infection lasts 15–21 days, and the mice show a

[21] K. H. Ramsey, W. J. Newhall, and R. G. Rank, *Infect. Immun.* **57,** 2441 (1989).

strong degree of immunity to reinfection for as long as 100–150 days after infection. Thus far, BALB/c, C57B1/6, C3H, C3H/HeJ, and outbred Swiss Webster mice have been infected in our laboratory, and no significant differences have been noted in the course of infection in any individual strain.

Interestingly, the MoPn infection of the lower genital tract appears to relatively avirulent. When congenitally athymic nude mice are inoculated, they remain infected virtually indefinitely and do not appear to suffer any ill effects.[22] We have continuously isolated MoPn from the cervix of nude mice for as long as 300 days. The organisms apparently retain their virulence because an occasional nude mouse will develop a chlamydial pneumonia, presumably as a result of aspiration of organisms during grooming. Thus, the ability of the nude mouse to remain infected provides an excellent system to assess the function of various lymphocyte populations by adoptive transfer techniques.

Infection and Assessment of Infection Course in Upper Genital Tract

The mouse–MoPn model may also be used to evaluate the effect of chlamydial infection of the oviducts on fertility and to study the basic mechanisms of salpingitis. Swenson *et al.*[23] found that infection of the oviducts could be initiated by direct injection of MoPn into the ovarian bursa. This is performed by anesthetizing the mouse with Nembutal and making a 1 cm incision on the retroperitoneal flank after clearing the area of hair with the aid of electric clippers. The ovary is immobilized, and 5 μl of an MoPn suspension (10^6 ifu/ml) in SPG buffer is injected through the fat pad directly into the ovarian bursa using a 30-gauge needle attached to a microdispensing syringe. Care should be taken as best as possible to prevent spillage so that organisms are not introduced by accident into the peritoneum. The wound may be closed simply with a wound clip without suturing the peritoneum. The procedure can be repeated for the contralateral side, or that side may be used as a control.

Generally, an acute salpingitis develops within 6–14 days. The extent of the salpingitis can be assessed grossly by observation of distention of the bursal sac, opacity due to inflammatory exudate, as well as hydrosalpinx. The tissues can also be evaluted microscopically by fixing the oviduct in formalin and staining with hematoxylin and eosin. Additionally, this model is well suited to determining the effect of the infection on fertility. Infected female mice may be housed with males beginning 5 days after

[22] R. G. Rank, L. S. F. Soderberg, and A. L. Barron, *Infect. Immun.* **48,** 847 (1985).
[23] C. E. Swenson, E. Donegan, and J. Schachter, *J. Infect. Dis.* **148,** 1101 (1983).

infection and then sacrificed 2 weeks later.[24] The animals are necropsied and the number of implants per horn of each mouse determined. One may also determine the patency of the oviduct by injecting Evans blue or methylene blue into the end of the uterine horn so that the dye may flow into the oviduct. Tubal obstruction will interfere with the ability of the dye to flow through the entire oviduct and can be equated with infertility of that particular oviduct.

Infection and Assessment of Infection Course in Mice Infected with Human Oculogenital Strains of Chlamydia

In addition to the MoPn–mouse model, mice can be infected in the genital tract with human oculogenital isolates including *Chlamydia trachomatis* serovars D, E, F, G, H, I, K, and L2.[7–9] To accomplish genital tract infection, since these agents do not have apparent biological specificity for the murine host, mice must be treated with progesterone to stabilize the epithelium in a state on anestrus. Mice are injected subcutaneously with 2.5 mg of medroxyprogesterone (Depo-Provera; Upjohn Company, Kalamazoo, MI) 10 and 3 days prior to infection. Infection is accomplished by inoculating via the vaginal route with 10^7–10^8 ifu of the chlamydiae-containing suspension. Depending on the serovar, organisms can be isolated from the cervix 7–8 weeks after infection. In general, the recovery of organisms by isolation is lower than with MoPn at comparable times. Salpingitis can also be studied with the human oculogenital serovars by using the same methodology as described for MoPn, except that progesterone should be given to facilitate infection.

[24] C. E. Swenson and J. Schachter, *Sex. Transm. Dis.* **11,** 64 (1984).

[8] Animal Models for Meningitis

By MARTIN G. TÄUBER and ANDRÉ ZWAHLEN

Introduction

Approximately 15,000 cases of bacterial meningitis occur each year in the United States, while other parts of the world have a substantially higher incidence. Even when treated with highly active antibiotics, the disease is fatal in 5–40% of the patients and causes neurologic sequelae

TABLE I
PATHOGENS OF BACTERIAL MENINGITIS AND AGE-RELATED RELATIVE FREQUENCY[a]

Organism	Overall frequency (%)	Age-related frequency (%)				
		0–1 months	1 month– 4 years	5–29 years	30–59 years	60+ years
Haemophilus influenzae	45	5	70	10	5	3
Streptococcus pneumoniae	18	2	15	20	40	50
Neisseria meningitidis	14	1	10	40	10	2
GB streptococci	6	48	2	1	3	3
Others[b]	17	44	3	29	42	42

[a] Data from United States (1986). Adapted from J. D. Wenger, A. W. Hightower, R. R. Facklam, S. Gaventa, C. V. Broome, and B. M. S. Group, *J. Infect. Dis.* **162,** 1316 (1990).
[b] Includes *Listeria monocytogenes, Escherichia coli,* other enterobacteriaceae, *Pseudomonas aeruginosa,* and *Staphylococcus aureus.*

in up to 30% of survivors.[1–4] The serious prognosis of the disease has generated a strong incentive to understand critical aspects of the pathogenesis and pathophysiology of meningitis. Several informative animal models have been developed, and the design of clinical studies has been stimulated by findings obtained in the animal models.[5,6] Table I summarizes the most common pathogens leading to bacterial meningitis and their relative importance in patients of different age groups.[2]

At least six distinct pathogenic events occur from the acquisition of a microorganism to the development of symptomatic meningitis (Table II). Colonization, invasion, and intravascular multiplication (steps 1 to 3, Table II) are common to all mucosal pathogens causing bacteremia, while steps 4 to 6 (Table II) are specific for bacterial meningitis.

Among the many published animal models of meningitis, the infant rat

[1] W. F. Schlech, J. I. Ward, J. D. Band, A. Hightower, D. W. Fraser, and C. V. Broome, *J. Am. Med. Assoc.* **253,** 1749 (1985).
[2] J. D. Wenger, A. W. Hightower, R. R. Facklam, S. Gaventa, C. V. Broome, and B. M. S. Group, *J. Infect. Dis.* **162,** 1316 (1990).
[3] P. R. Dodge, D. Hallowell, R. D. Feigin, S. J. Holmes, S. L. Kaplan, D. P. Jubelirer, B. Stechberg, and S. K. Hirsh, *N. Engl. J. Med.* **311,** 869 (1984).
[4] J. O. Klein, R. D. Feigin, and G. H. J. McCracken, *Pediatrics* **78,** 959 (1986).
[5] C. M. Odio, I. Faingezicht, M. Paris, M. Nassar, A. Baltodano, J. Rodgers, X. Saez-Llorens, K. D. Olsen, and G. H. J. McCracken, *N. Engl. J. Med.* **324,** 1525 (1991).
[6] M. H. Lebel, B. J. Freij, G. A. Syrogiannopoulos, D. F. Chrane, M. J. Hoyt, S. M. Stewart, B. D. Kennard, K. D. Olson, and G. H. McCracken, *N. Engl. J. Med.* **319,** 364 (1988).

TABLE II
STEPS IN PATHOGENESIS OF BACTERIAL MENINGITIS

1. Colonization of nasopharyngeal mucosa by pathogen
2. Local multiplication and mucosal invasion by pathogen
3. Development of bacteremia with intravascular survival of pathogen
4. Access of pathogen to central nervous system
5. Survival and multiplication of pathogen in cerebrospinal fluid
6. Signaling of presence of multiplying pathogen with development of inflammatory reaction which leads to symptoms and sequelae

model of *Haemophilus influenzae* meningitis[7] and the adult rabbit model,[8] in which pneumococci and *H. influenzae* are the most commonly used pathogens, have stood out with regard to relevance for the human disease and the wealth of information they have generated. The two models are complementary with regard to the steps involved in the pathogenesis of meningitis (Table II). In the infant rat model, the pathogen is inoculated intranasally or intraperitoneally, and the development of meningitis can then be studied (steps 2–5, Table II), thus allowing focus on bacterial and host factors that determine the development of bacteremia and the potential of a pathogen to gain access to the central nervous system (CNS). In contrast, rabbits are infected by direct inoculation of the pathogen into the cerebrospinal fluid (CSF), thereby bypassing steps 1–4 of the pathogenesis of meningitis (Table II). The strength of the rabbit model rests in the ease with which repeated CSF samples can be obtained, providing the possibility of measuring multiple CSF parameters over time. The model is therefore particularly suited to study various aspects of the later steps involved in the pathogenesis of meningitis (steps 5 and 6, Table II). In this chapter, we focus on the two models mentioned. For a summary discussion of other models, the reader is referred to a more extensive review of animal models of meningitis.[9]

Infant Rat Model of *Haemophilus influenzae* Meningitis

General Considerations

Haemophilus influenzae is either nontypable (unencapsulated) or encapsulated, elaborating one of six polysaccharide capsules (designated a

[7] E. R. Moxon, A. L. Smith, D. R. Averill, and D. H. Smith, *J. Infect. Dis.* **129,** 154 (1974).
[8] R. G. Dacey and M. A. Sande, *Antimicrob. Agents Chemother.* **6,** 437 (1974).
[9] W. M. Scheld, *in* "Experimental Models in Antimicrobial Chemotherapy" (O. Zak and M. A. Sande, eds.), p. 139. Academic Press, London, 1986.

through f). Meningitis in humans is almost exclusively caused by strains with the type b capsular polysaccharide, a linear polymer of ribosyl–ribitol phosphate (PRP).[10] Experimental *H. influenzae* meningitis has been produced in mice, rabbits, and primates, but the most commonly used model has been described in infant rats.[7,11] This model reflects several important similarities with the human disease: (a) infant rats, as humans, show preferential susceptibility to type b strains of *H. influenzae;* (b) experimental infection is obtained by the same route as the human disease (intranasal inoculation); and (c) bacteremia is a requirement for the development of meningitis both in human disease and in the model. The most obvious limitation of the model relates to the size of the animals, allowing only small samples of CSF or blood.

Description of Model

Animals. The model uses COBS/CD Sprague-Dawley suckling rats, obtained in litters of approximately 10 with their mother. The litters are kept in standard rat cages throughout the experimentation, and the mothers have free access to regular diet and water. The infant rats are examined for signs of illness, and their weight and respiratory rate are recorded on a daily basis. Infant rats are infected by the nasal or peritoneal route at the age of 5 days, and meningitis is assessed 2 to 3 days later when the posterior fontanella is still open. Although COBS/CD Sprague-Dawley rats do not naturally develop anti-PRP antibodies, their susceptibility to *H. influenzae* type b is age-related, and infant rats develop bacteremia of greater magnitude than older animals. However, 21-day-old weaning rats are well suited for studying nasopharyngeal colonization because they can be kept in single cages.[12]

Bacteria/Inoculum. Strains of *H. influenzae* type b (e.g., strains Eagan[11] or Rd/b+, the latter being a laboratory transformant carrying the type b capsule from strain Eagan[13]) are grown to midlogarithmic phase in brain–heart infusion (BHI) broth (Difco Laboratories, Detroit, MI) supplemented with 10 µg/ml of hemin (Sigma, St. Louis, MO) and 2 µg/ml of diphosphopyridine nucleotide (Sigma). The inoculum is then washed in chilled 10 mM phosphate-buffered saline (PBS) containing 0.1% gelatin

[10] D. C. Turck, in "*Haemophilus influenzae:* Epidemiology, Immunology and Prevention of Disease" (S. H. Sell and P. F. Wright, eds.), p. 3. Elsevier Science Publ., New York, 1982.

[11] A. L. Smith, D. H. Smith, D. R. Averill, J. Marino, and E. R. Moxon, *Infect. Immun.* **8,** 278 (1973).

[12] L. G. Rubin and E. R. Moxon, *J. Infect. Dis.* **143,** 517 (1981).

[13] A. Zwahlen, J. A. Winkelstein, and E. R. Moxon, *J. Infect. Dis.* **148,** 385 (1983).

and diluted to the desired density for inoculation. The titer of the inoculum is confirmed by quantitative cultures on BHI agar containing 10% Levinthal base. For most type b strains, the dose that infects 50% of the animals (ID_{50}; see [2] in this volume) by the intranasal route is 10^3 to 10^4 colony-forming units (cfu), but ID_{50} values have to be established for all strains in initial experiments.[12,14] It may also be advantageous to generate spontaneous one-step streptomycin-resistant mutants of the infecting strain and use these in the experiments to facilitate identification of the infecting strain after recovery from the nasopharynx, CSF, or blood.[11] Although *H. influenzae* is the organism that has been studied most extensively in this model, other meningeal pathogens of the newborn, such as *Escherichia coli* (with a K1 capsule) and group B streptococci also cause hematogenous meningitis in infant rats.[15,16]

Induction of Meningitis. Because the rat is an obligate sniffer, the atraumatic deposition of a 10–25 µl drop of a bacterial suspension (using a 200 µl pipette) onto one nare of the unanesthetized infant rat results in nasal colonization. Nasopharyngeal colonization is assessed 24 to 72 hr after inoculation by means of a nasal wash. Thirty microliters of 10 mM PBS is applied atraumatically to one nare. Sniffing is followed by fluid reflux through the contralateral nare or the oral cavity. The fluid is aspirated and cultured on BHI agar supplemented with 10% Levinthal base. Infant rats may sometimes acquire *H. influenzae* by exposure to colonized littermates, but this is a rare event. For an individual experiment, infant rats from several litters are pooled, animals are randomly assigned to one of the experimental or control groups by writing a corresponding number on the back, and 5 infected and 5 control rats are assigned to a nursing mother. Depending on the inoculum size and the virulence of the bacterial strain, up to 100% of infected animals will develop meningitis within 48 to 72 hr. Clinical signs of severe meningitis are evident when animals appear sick, cyanotic, have an increased respiratory rate, and fail to thrive and nurse.

As an alternative to intranasal inoculation, infant rats can be infected intraperitoneally.[11] One hundred microliters of a bacterial suspension is injected in the left paraumbilical area, using a 30-gauge 1/2-inch needle. This procedure is associated with minimal pain or discomfort and does not require anesthesia. Because bacteria inoculated in the peritoneum are recovered in the blood as early as 20 min later, this route of infection

[14] A. Zwahlen, J. S. Kroll, L. G. Rubin, and E. R. Moxon, *Microb. Pathog.* **7**, 225 (1989).
[15] R. Bortolussi, P. Ferrieri, and L. W. Wannamaker, *Infect. Immun.* **22**, 480 (1978).
[16] P. Ferrieri, B. Burke, and J. Nelson, *Infect. Immun.* **27**, 1023 (1980).

FIG. 1. *Haemophilus influenzae* experimental meningitis in infant rats. Cerebrospinal fluid is obtained by inserting a 30-gauge needle into the cisterna magna through the open fontanella in a 7-day-old suckling rat. A drop of clear CSF appears in the hub of the needle during a successful tap.

is appropriate for studying bacterial survival *in vivo* and hematogenous meningitis.[11] The ID_{50} after intraperitoneal challenge of infant rats with *H. influenzae* type b is virtually one organism.[13]

Older rats (21 days old) can be infected intravenously using a small butterfly and catheterizing one of the tail veins while the rat is temporarily immobilized. One hundred fifty microliters of a bacterial suspension in PBS is injected over 30 sec. Replication of *H. influenzae* type b organisms in the bloodstream begins 30 min after intravenous inoculation of 10 to 100 bacteria.[17]

Blood and Cerebrospinal Fluid. Quantitative blood cultures are performed after 24 to 48 hr by aspirating 10 μl of blood from a small stylet incision of the ventral tail vein and plating the blood (undiluted or in 10-fold dilutions) onto BHI agar supplemented with 10% Levinthal base. The CSF is obtained after 48 to 72 hr (depending on the route of infection and the virulence of the infecting organism) by puncture of the cisterna magna through the lower part of the fontanella, using a 30-gauge 1/2-inch needle (Fig. 1). The unanesthetized infant rat is held with its head flexed in one hand, and the needle is slowly placed through the fontanella with the other hand while the person performing the tap carefully watches for the appearance of a small drop of CSF in the hub of the needle. The needle is then removed, and the CSF collected in the hub (5–10 μl) is allowed to fill a small glass capillary tube and is then processed for quantitative cultures (see above), leukocyte counts, or other analyses. When meningitis

[17] L. G. Rubin, A. Zwahlen, and E. R. Moxon, *J. Infect. Dis.* **152**, 307 (1985).

is present, CSF cultures invariably yield 10^7 to 10^8 cfu/ml, whereas leukocyte counts show more variability, ranging from less than 10 to 10,000/ μl.[7,11] Once the end points of interest have been examined, the animals are euthanized by intraperitoneal injection of pentobarbital sodium.

Lessons Learned from Model

Aspects of both the host and the microorganism relevant for nasopharyngeal colonization, bloodstream invasion, and invasion of the CNS have been studied in this model. Nasopharyngeal colonization results from the survival of a single organism: inoculation with a mixed culture of two *H. influenzae* mutants with different antibiotic resistance is followed by the recovery of only one of them unless the inoculum largely exceeds the ID_{50}.[12] This observation may be due to *in vivo* selection of a few bacteria with acquired resistance to specific host defense mechanisms. In fact, *H. influenzae* cultured from the nasopharynx have greater serum resistance than the same organisms cultured in broth.[18] This phenotypic shift is associated with a change in the lipooligosaccharide composition of the strain.[19] Elaboration of a polysaccharide capsule and of pili is not required for colonization.[14]

Nasopharyngeal colonization of infant rats by *H. influenzae* type b is followed 12 to 48 hr after inoculation by sustained bacteremia.[20] Inoculation of a high number of organisms increases the incidence but not the magnitude of bacteremia. Type b strains penetrate the subepithelial tissue of the nasopharynx within minutes after inoculation.[21] Bloodstream invasion probably occurs by direct invasion of the mucosal blood vessels, since blood cultures become positive within 30 min of nasal inoculation. Different capsular types of *H. influenzae* can cause bacteremia, but only the PRP capsule confers the unique propensity to produce sustained bacteremia and meningitis.[22] The initial rate of intravascular multiplication of *H. influenzae* type b in rats inoculated intravenously achieves values similar to those observed in broth cultures under optimal conditions.[17]

Changes in lipooligosaccharide composition produced in genetically related strains of *H. influenzae* type b also affect the virulence.[23] On the host side, the maturity of the reticuloendothelial system and complement activity determine the efficiency of bacterial clearance from the blood-

[18] L. G. Rubin and E. R. Moxon, *J. Gen. Microbiol.* **131,** 515 (1985).
[19] M. Kuratana and P. Anderson, *J. Infect. Dis.* **163,** 1073 (1991).
[20] E. R. Moxon and P. T. Ostrow, *J. Infect. Dis.* **135,** 303 (1977).
[21] P. T. Ostrow, E. R. Moxon, N. Vernon, and R. Kapka, *Lab. Invest.* **40,** 678 (1979).
[22] L. G. Rubin and E. R. Moxon, *Infect. Immun.* **41,** 280 (1983).
[23] A. Zwahlen, L. G. Rubin, and E. R. Moxon, *Microbial. Pathog.* **1,** 465 (1986).

stream, both in the presence and in the absence of anti-PRP antibodies.[24] Splenectomy increases the incidence and magnitude of bacteremia after intranasal inoculation and results in enhanced susceptibility of rats to encapsulated *H. influenzae* other than type b.[14] Depletion of C3 activity by treating rats with cobra venom factor also favors bacteremia and meningitis by both encapsulated and unencapsulated strains.[13]

Invasion of the subarachnoid space and development of meningitis results from hematogenous spread rather than contiguous infection from the nasopharynx.[20] Sustained bacteremia of at least 10^3 cfu/ml lasting for at least 6 hr is a prerequisite for subarachnoid space invasion by *Haemophilus*. The first histopathological changes of the meninges are detected at sites distant from the nasopharynx.[21] Studies in chimpanzees have suggested that *H. influenzae* may enter the CSF through the choroid plexus.[25] However, the mechanisms of translocation of the organism from the bloodstream to the CNS and the determinants of meningeal tropism remain largely unclear.

Adult Rabbit Model of Meningitis

General Considerations

The adult rabbit model of meningitis, originally described by Dacey and Sande in 1974,[8] has proved to be extremely valuable for the study of events occuring after the infecting organism has reached the subarachnoid space. Studies performed in this model include examination of factors affecting bacterial multiplication and host defenses in the subarachnoid space; identification of bacterial products responsible for triggering CSF inflammation; analysis of the components of meningeal inflammation; studies of the pathophysiology of meningitis; and studies of antibiotic therapy of meningitis. Advantages of the model include the relatively large size of the animals that allows repetitive sampling of CSF and blood samples sufficiently large for multiple studies; easy access to the subarachnoid space that can be utilized, in addition to CSF sampling, for measuring intracisternal pressure; and the possibility of using different pathogens. The most obvious limitation of the model relates to the "unphysiologic" method of infection, in which relatively large numbers of bacteria are directly instilled into the CSF.

[24] P. F. Weller, A. L. Smith, D. H. Smith, and P. Anderson, *J. Infect. Dis.* **138,** 427 (1978).
[25] R. S. Daum, D. W. Scheifele, V. P. Syriopoulou, D. Averill, and A. L. Smith, *J. Pediatr.* **93,** 927 (1978).

FIG. 2. Model of experimental meningitis in adult rabbits. Anesthetized rabbits are secured in a frame via a dental acrylic helmet attached to the skull that contains a turnbuckle fitting a bolt. The cisterna magna can then be punctured with a 25-gauge spinal needle held in a geared electrode introducer mounted horizontally in the frame, allowing repetitive access to CSF. In addition, rabbits have a butterfly needle placed in an ear vein for venous access and a catheter inserted into one of the femoral arteries for arterial access.

Description of Model

Animals/Setup. Most commonly, New Zealand White rabbits, weighing between 2.5 and 3 kg, are used, although other species such as chinchilla rabbits can also be used.[26] Animals are housed in individual cages and have free access to regular diet and water. The first experimental step involves attachment of a dental acrylic helmet to the skull of the animal by four stainless steel screws. The purpose of the helmet is to hold a half-turnbuckle that allows rigid immobilization of the head of the animal in a stereotactic frame via a bolt screw. A second arm of the frame holds a geared electrode introducer in horizontal position which is used to move a spinal needle (25-gauge, 3 1/2 inches) back and forth, thereby facilitating the introduction of the needle into the cisterna magna, where it can be secured in place (Fig. 2). To install the apparatus, animals are anesthetized [we administer a mixture of ketamine (35–50 mg/kg), xylazine 2–5 mg/kg), and azepromazine (0.5–1 mg/kg) intramuscularly]. The head of the animal is then shaved, and local anesthesia (1% lidocaine) is applied to the area of the coronal and saggital suture. The skull is exposed with a longitudinal cut, and four small drill holes are placed in the four quadrants determined by the sutures. The screws are drilled through the outer table

[26] J. L. Kadurugamuwa, B. Hengstler, and O. Zak, *J. Infect. Dis.* **159**, 26 (1988).

of the skull bone, an aluminum half-turnbuckle is placed between the screws, and soft dental acrylic is molded around the screws to hold the turnbuckle firmly in place without obstructing its receptacle. While the animal is still anesthetized, a bolt is screwed into the turnbuckle, and the head of the animal is firmly secured in the frame via the bolt (Fig. 2). The spinal needle is inserted into the cisterna magna, a CSF sample of 0.3 ml is removed through the needle for baseline studies, and a bacterial suspension is injected in the same volume of normal saline or 10 mM PBS to establish infection.

Bacteria/Inoculum. Multiple bacterial species, including *Streptococcus pneumoniae,*[8] *H. influenzae,*[27] group B streptococci,[28] *E. coli,*[28] *Klebsiella pneumoniae,*[29] *Proteus mirabilis,*[30] *Pseudomonas aeruginosa,*[31] and *Listeria monocytogenes,*[32] have been used to induce meningitis in this model. As a rule, only encapsulated bacteria are capable of establishing a productive infection in the CSF, while unencapsulated bacteria are efficiently cleared from the subarachnoid space.[33] In addition to live organisms, heat-killed bacteria or purified bacterial products (lipopolysaccharide, cell wall fragments) can be directly instilled, and the inflammatory response to the stimuli can be measured in CSF samples obtained over time. If meningitis is induced by live organisms, use of bacteria in the midlogarithmic growth phase yields the most reproducible results. Bacteria are washed in saline or PBS, resuspended to the desired concentration, and injected into the cisterna magna. For most pathogens, an inoculum of 10^5–10^6 cfu will induce meningitis in 100% of the animals, even though, at least for pneumococci, a single organism is probably sufficient to induce meningitis.[34] If untreated, the disease is fatal when encapsulated meningeal pathogens are used, but the time between infection and death can greatly vary from less than 24 hr to several days, depending on virulence proper-

[27] G. H. J. McCracken, J. D. Nelson, and L. Grimm, *Antimicrob. Agents Chemother.* **21,** 262 (1982).
[28] U. B. Schaad, G. H. McCracken, Jr., C. A. Loock, and M. L. Thomas, *J. Infect. Dis.* **143,** 156 (1981).
[29] U. B. Schaad, G. H. McCracken, C. A. Loock, and M. L. Thomas, *Antimicrob. Agents Chemother.* **17,** 406 (1980).
[30] L. J. Strausbaugh and M. A. Sande, *J. Infect. Dis.* **137,** 251 (1978).
[31] W. M. Scheld, W. J. Kelly, and M. A. Sande, *in* "Current Chemotherapy and Infectious Disease" (J. D. Nelson and C. Crassi, eds.), p. 1036. American Society for Microbiology, Washington, D.C., 1980.
[32] W. M. Scheld, D. D. Fletcher, F. N. Fink, and M. A. Sande, *J. Infect. Dis.* **140,** 287 (1979).
[33] E. Tuomanen, A. Tomasz, B. Hengstler, and O. Zak, *J. Infect. Dis.* **151,** 535 (1985).
[34] M. A. Sande, E. R. Sande, J. D. Woolwine, C. J. Hackbarth, and P. M. Small, *J. Infect. Dis.* **156,** 849 (1987).

ties of the infecting pathogen. The titer of the inoculum should be routinely confirmed by quantitative cultures.

Cerebrospinal Fluid Sampling and Other Experimental Parameters. Timing of CSF sampling and other experimental measurements obviously depends on the questions addressed in the model. When injecting live organisms ($\sim 10^5$ cfu), CSF pleocytosis and chemical changes become apparent between 6 and 10 hr after infection. If inflammatory bacterial products (e.g., endotoxin) are used to induce meningitis, inflammation in the CSF becomes apparent after about 4 hr.[35,36] Depending on the design of the studies, animals can be allowed to recover from the initial anesthesia after infection and can be returned to the cages until CSF inflammation is fully developed. Alternatively, they can be studied in the early phase after infection and can be left in the frame following injection of the bacterial inoculum. If periods during which the animals are secured in the frame exceed approximately 1 hr, rabbits should be anesthetized with a long-acting anesthetic. Urethane (2 g/kg i.v. over 30 min) has been satisfactory for this purpose. In addition, adequate amounts of fluid should be supplied in all experiments which last for more than a few hours. Once the animals develop clinical signs of meningitis (fever, lethargy, ataxia), they usually stop drinking, even when conscious in the cage. We routinely administer approximately 50 ml/kg of saline subcutaneously every 12 hr to prevent dehydration.

Cerebrospinal fluid can be sampled repetitively while the animal is secured in the frame, either by leaving the spinal needle in place in the cisterna magna or by reintroducing the needle for each sample (the continued presence of a foreign body in the cisterna magna can potentially interfere with some studies). It is possible, for example, to obtain a sample of 0.3 ml of CSF every 2 hr during a 6- to 8-hr experiment in rabbits with pneumococcal meningitis. Bacteria in CSF reach titers of 10^6 to 10^9 cfu/ml in fully developed meningitis in this model, whereas CSF white blood cell counts are typically in the range of 1000–10,000/μl.[37] Chemical parameters of the CSF (lactate, glucose, protein concentration) can be measured and show changes similar to those observed in humans with bacterial meningitis.[37] In addition, CSF samples can be used to measure concentrations of drugs, such as antibiotics, inflammatory products (cytokines, prostaglandins), or any other substance of interest. Blood samples are

[35] E. Tuomanen, H. Liu, B. Hengstler, O. Zak, and A. Tomasz, *J. Infect. Dis.* **152,** 859 (1985).
[36] M. M. Mustafa, O. Ramilo, K. D. Olsen, P. S. Franklin, E. J. Hansen, B. Beutler, and G. H. J. McCracken, *J. Clin. Invest.* **84,** 1253 (1989).
[37] M. G. Täuber, M. Burroughs, U. M. Niemöller, H. Kuster, U. Borschberg, and E. Tuomanen, *J. Infect. Dis.* **163,** 806 (1991).

best obtained through an arterial catheter (PE 90) placed into one of the femoral arteries, while the peripheral ear veins provide easy intravenous access by inserting a 22-gauge butterfly needle. The model can also be used to measure physiological parameters, such as intracranial pressure (by connecting the spinal needle to a pressure transducer),[38] cerebral blood flow (e.g., using radioactively labeled microspheres[39]), or brain edema.[38] The latter is determined by removing the brain from the skull immediately after euthanizing the animal with a pentobarbital overdose (150 mg/kg i.v.) and determining the wet to dry weight ratio. When an experiment is completed, animals are immediately euthanized.

Lessons Learned from Model

As indicated above, multiple aspects of meningitis have been studied using this model, and it is beyond the scope of this chapter to discuss any of them in detail. Relevant to the topic of this volume, the model has elucidated important aspects of the pathogenesis of meningitis. The advantages of the model have been utilized to document a suppressive effect of fever on the growth rate of pneumococci in CSF.[34,40] Studies combining *in vitro* biochemical techniques with the rabbit model have identified bacterial products that are responsible for triggering the inflammatory response in the subarachnoid space. In the case of pneumococci, injection of fragments of the peptidoglycan layers of the cell wall induce inflammation similar to that observed with live organisms.[35,41] Lytic antiobitic therapy causes the release of such bacterial components during experimental pneumococcal meningitis, as suggested by a pronounced increase in CSF inflammation after administration of ampicillin in rabbits with pneumococcal meningitis.[42] Different pneumococcal strains vary in the extent of CSF changes and pathophysiologic alterations that they cause after intracisternal injection, and it is conceivable that these differences are related, at least in part, to the extent to which different strains release biologically active products.[37] In the case of gram-negative bacteria, endotoxin is the major product that triggers inflammation, and lytic antibiotic therapy of experimental *E. coli* meningitis with cefotaxime and of *H. influenzae*

[38] M. G. Täuber, H. Khayam-Bashi, and M. A. Sande, *J. Infect. Dis.* **151**, 528 (1985).
[39] J. H. Tureen, M. G. Täuber, and M. A. Sande, *J. Clin. Invest.* **89**, 947 (1992).
[40] P. M. Small, M. G. Täuber, C. J. Hackbarth, and M. A. Sande, *Infect. Immun.* **52**, 484 (1986).
[41] A. Tomasz and K. Saukkonen, *Pediatr. Infect. Dis. J.* **8**, 902 (1989).
[42] E. Tuomanen, B. Hengstler, R. Rich, M. A. Bray, O. Zak, and A. Tomasz, *J. Infect. Dis.* **155**, 985 (1987).

meningitis with ceftriaxone or chloramphenicol produces a marked accentuation of CSF endotoxin levels.[43,44]

Following the presence of bacterial products in the subarachnoid space, markedly increased concentrations of cytokines can be measured in the CSF, among them tumor necrosis factor α (TNF-α), interleukin 1 (IL-1), and IL-6.[45] After intracisternal inoculation of *H. influenzae* or its lipooligosaccharide in rabbits, TNF-α appears in the CSF within 30 min.[36] The administration of ceftriaxone or chloramphenicol to rabbits with *H. influenzae* meningitis induces a large increase in TNF-α concentrations, most likely in response to the marked release of endotoxin.[44]

The cytokines TNF-α and IL-1β appear to be instrumental for the development of meningeal inflammation as they induce a brisk inflammatory response that can be blocked by antibodies directed against the cytokines, when administered intracisternally to rabbits.[46] The simultaneous inoculation of TNF-α and IL-1β can be synergistic.[47] Other inflammatory mediators also appear to play a role. Prostaglandin E_2 is elevated in the CSF of rabbits with pneumococcal meningitis.[48] Chemotactic activity derived from complement component C5a is associated with early influx of granulocytes into the CSF in pneumococcal meningitis in rabbits,[49] whereas complement depletion in rabbits causes a delay in the onset of pleocytosis.[50] Platelet-activating factor (PAF) is an important autocoid mediator implicated in a variety of CNS diseases. An antagonist to PAF is able to partially block the CSF inflammation due to *S. pneumoniae* in rabbits.[51]

One important consequence of the generation of cytokines and other inflammatory products is the development of a polymorphonuclear CSF pleocytosis. Interestingly, granulocytes appear unable to restrict bacterial

[43] M. G. Täuber, A. M. Shibl, C. J. Hackbarth, J. W. Larrick, and M. A. Sande, *J. Infect. Dis.* **156**, 456 (1987).

[44] M. M. Mustafa, O. Ramilo, J. Mertsola, R. C. Risser, B. Beutler, E. J. Hansen, and G. H. J. McCracken, *J. Infect. Dis.* **160**, 818 (1989).

[45] A. Waage, A. Halstensen, R. Shalaby, P. Brandtz, P. Kierule, and T. Espevik, *J. Exp. Med.* **170**, 1859 (1989).

[46] K. Saukkonen, S. Sande, C. Cioffe, S. Wolpe, B. Sherry, A. Cerami, and E. Tuomanen, *J. Exp. Med.* **171**, 439 (1990).

[47] V. J. Quagliarello, B. Wispelwey, W. J. Long, and W. M. Scheld, *J. Clin. Invest.* **87**, 1360 (1991).

[48] J. H. Tureen, M. G. Täuber, and M. A. Sande, *J. Infect. Dis.* **163**, 647 (1991).

[49] J. D. Ernst, K. Hartiala, I. M. Goldstein, and M. A. Sande, *Infect. Immun.* **46**, 81 (1984).

[50] E. Tuomanen, B. Hengstler, O. Zak, and A. Tomasz, *Microbial Pathog.* **1**, 15 (1986).

[51] C. Cabellos, D. MacIntyre, M. Forrest, M. Burroughs, F. Prasad, and E. Tuomanen, *J. Clin. Invest.* **90**, 612 (1992).

growth of pneumococci in the CSF significantly in the rabbit model.[52] This seems to be the result of insufficient opsonization of the encapsulated pathogen because of the low levels of antibodies and complement present in the CSF.[53] On the other hand, granulocytes play a potentially harmful role in bacterial meningitis by mediating many of the pathophysiologic effects of the disease.[54]

[52] J. D. Ernst, J. M. Decazes, and M. A. Sande, *Infect. Immun.* **41**, 275 (1983).
[53] A. Zwahlen, U. E. Nydegger, P. Vaudaux, P. H. Lambert, and F. A. Waldvogel, *J. Infect. Dis.* **145**, 635 (1982).
[54] M. A. Sande, M. G. Täuber, W. M. Scheld, and G. H. J. McCracken, *Pediatr. Infect. Dis. J.* **8**, 929 (1989).

[9] Animal Models for Periodontal Disease

By Theresa E. Madden and Jack G. Caton

Introduction

Periodontal research requires the use of laboratory animals to study the etiology and pathogenesis of periodontitis, to test therapeutic agents, and to evaluate the safety and efficacy of techniques, drugs, and devices that enhance wound healing. Because most periodontal disease results in irreversible destruction of connective tissue and alveolar bone, it is unethical in humans to induce experimental periodontal disease other than gingivitis.

Animal and human studies have proved that microbial dental plaque is the primary etiologic agent in periodontitis. A number of suspected virulence factors from several plaque bacteria are presently under study.[1] *Actinobacillus actinomycetemcomitans* leukotoxin kills leukocytes *in vitro,* and products from *Capnocytophaga* inhibit polymorphonuclear leukocyte chemotaxis. *Porphyromonas gingivalis, Prevotella intermedia,* and *Capnocytophaga* species express immunoglobulin A (IgA) and IgG proteases. Other proteases from *P. gingivalis, P. intermedia, A. actinomycetemcomitans,* and *Treponema denticola* degrade fibrinogen, fibrin, collagen, fibronectin, albumin, hemopexin, and complement components. This suggests that these putative pathogens are equipped to escape host defenses. As molecular analyses of periodontal bacteria progresses and virulence determinants are cloned and mutated, labora-

[1] J. Slots and R. J. Genco, *J. Dent. Res.* **63**, 412 (1984).

tory animals will remain indispensable in studying *in vivo* pathogenicity.

General Considerations

Criteria for Pathogens

Periodontal disease results from polymicrobial infection of the tissues that surround and support the teeth. Relatively few of the hundreds of diverse oral microbes are thought to be primary pathogens in periodontitis. Gram-positive aerobic cocci, which are colonizers of oral soft tissues, saliva, and immature dental plaque, are associated with periodontal health. In contrast, gram-negative anaerobes and spirochetes, late colonizers of dental plaque, are consistently cultured from inflammatory periodontal lesions. *Porphyromonas gingivalis, A. actinomycetemcomitans, P. intermedia, Fusobacterium nucleatum, Wolinella recta,* and *Eikenella corrodens* are associated with periodontitis, as are species of the genera *Treponema, Bacteroides,* and *Capnocytophaga.*[2] Identification of these organisms as pathogens is based on one or more of the following five criteria: (1) high numbers are cultured from periodontal lesions, (2) serum antibody responses are elicited, (3) potential virulence factors are identified, (4) eradication of the organism accompanies remission of the lesion, and (5) periodontitis results following implantation in an animal model.[2] Although identification of candidate pathogens based on these criteria is generally accepted, more stringent approaches to studying *in vivo* pathogenicity may result in modifications to the above list of periodontal pathogens.

Choice of Animal Model

The primate, beagle dog, and rat models are widely used in periodontal research and are described in detail below. Each species resembles humans in periodontal anatomy, microbiology, and pathophysiology.[3-6] If alternatives to these three are sought, the reader is referred to authoritative reviews of the advantages and disadvantages of the mouse, gerbil, ham-

[2] J. J. Zambon, *in* "Perspectives of Oral Manifestations of AIDS" (P. B. Robertson and J. S. Greenspan, ed.), p. 98. PSG, Littleton, Massachusetts, 1988.
[3] L. R. Brown, S. Hansler, S. S. Allen, C. Shea, M. G. Wheatcroft, and W. J. Frome, *J. Dent. Res.* **52,** 815 (1973).
[4] L. Heijl, B. R. Rifkin, and H. A. Zander, *J. Periodontol.* **47,** 710 (1976).
[5] K. C. Loftin, L. R. Brown, and B. M. Levy, *J. Dent. Res.* **59,** 1606 (1980).
[6] D. M. Simpson and B. E. Avery, *J. Periodontol.* **45,** 500 (1974).

ster, miniature swine, sheep, cat, ferret, mink, wolf, horse, cattle, hedge-hog, lemur, raccoon, coati, suricate mongoose, and shrew.[7-10]

Maintaining Gingival Health in Laboratory Animals

Microorganisms are necessary for the development of gingivitis and periodontitis, which occur naturally in some animals.[9,11] The oral cavity of an animal housed under standard laboratory conditions is colonized soon after birth, as is the dentition immediately after eruption. Consequently, if periodontal health is required in an experimental design, steps must be taken to prevent plaque accumulation and resultant periodontal inflammation. Although unnecessary in most studies, the most stringent approach is to maintain germfree colonies. Alternatively, gingival health in dogs and non-human primates is predictably maintained with plaque removal three times a week.[12,13] Tooth brushing is followed by a 3-min application of 2% (w/v) chlorhexidine gluconate. The use of interproximal cleaning devices, such as dental floss or toothpicks, is optional. When heavy plaque and calculus are present, thorough scaling and polishing are also required.[14] For oral hygiene procedures to be performed safely, animals require sedation (0.1 ml/kg ketamine, i.m.) or behavioral training. Cessation of oral hygiene in most species, as in humans, results in gingivitis within 3 weeks.[15,16]

Diet and Housing

Diet has a profound influence on periodontal health in animals. Coarse chow minimizes plaque accumulation, whereas soft foods and foods with high sucrose content promote plaque formation and periodontitis.[4,17-19] Beagle dogs fed a soft diet can be expected to lose 0.3 mm of periodontal

[7] J. M. Navia, "Animal Models in Dental Research," p. 312. Univ. of Alabama Press, Tuscaloosa, 1977.
[8] B. Klausen, J. Periodontol. 62, 59 (1991).
[9] R. C. Page and H. E. Schroeder, "Periodontitis in Man and Other Animals: A Comparative Review." Karger, Basel, 1982.
[10] K. L. Kalkwarf, R. F. Krejci, and W. C. Berry, Jr., J. Periodontol. 54, 81 (1982).
[11] S. Rovin, E. R. Costich, and H. A. Gordon, J. Periodontal Res. 1, 193 (1966).
[12] J. Caton, J. Clin. Periodontol. 6, 260 (1979).
[13] L. J. van Dijk and W. H. Wright, J. Periodontol. 54, 291 (1983).
[14] N. W. Johnson and E. B. Kenny, J. Periodontal Res. 7, 180 (1972).
[15] M. A. Listgarten and B. Ellegaard, J. Periodontal Res. 8, 199 (1973).
[16] E. Theilade, W. H. Wright, S. B. Jensen, and H. Loe, J. Peiodontal Res. 1, 1 (1966).
[17] F. P. Ashley and M. N. Naylor, J. Periodontal Res. 6, 56 (1970).
[18] E. Kryshtalskyj, J. Sodek, and J. M. Ferrier, Arch. Oral Biol. 31, 21 (1986).
[19] P. R. Garant and M. I. Cho, J. Periodontal Res. 14, 297 (1979).

support per year.[20] If a soft diet is desired, chows and canned food may be moistened with water, or soft pellets may be purchased (Ziegler Brothers, Gardner, PA).

Bedding and animal hair may become impacted between teeth, resulting in localized periodontal bone loss, and is of particular concern in the rat model. Bone loss following inoculation with periodontal pathogens increases proportionally with the volume of impacted material at that site.[20a] Whether hair impaction is a cause or an effect of bone loss (or a combination of both) has not fully been determined. To minimize this complicating factor, rats may be housed in suspended stainless steel mesh cages. Prior to weaning, when pups and dams are housed together, bedding should be changed frequently.

Germ State and Culturing Methods

The oral germ state of the laboratory animal should be known prior to initiating microbial inoculation and experimental periodontitis. Germfree rodents may be utilized to assess response to a specific microorganism. However, anaerobic bacteria do not easily colonize sterile oral tissues, and germfree rodents are thought to differ immunologically from conventional rodents as they are relatively deficient in lymphocytes and in antibody.[8] Gnotobiotic rats, on the other hand, may be obtained free from one or more specific pathogens and are immunologically normal.

Dental plaque from rats less than 1 months old is composed predominantly of streptococci and lacks periodontal pathogens. Therefore, very young conventional rats may be carefully housed to minimize colonization with pathogens. This is accomplished primarily by minimizing contact with infected animals. Rodents are coprophagic and quickly become colonized with the oral flora of cage mates.

The germ state of the experimental animal may be monitored at any point in the protocol. Accurate culturing of the diverse microflora found in dental plaque requires fastidious technique. Plaque is collected with paper points, a curette, or a barbed broach that is flushed with nitrogen gas, and it requires reduced transport conditions, one of several selective growth media, and anaerobic culturing facilities.[21,22] Protocols for sampling, diluting, culturing, and quantifying plaque bacteria, characterizing

[20] S. R. Saxe, J. C. Greene, H. M. Bohannan, and J. R. Vermillion, *Periodontics* 5, 217 (1967).
[20a] B. Klausen, R. T. Evans, N. S. Ramamurthy, L. M. Golub, C. Sfintescu, J.-Y. Lee, G. Bedi, J. J. Zambon, and R. J. Genco, *Oral Microbiol. Immunol.* 6, 193 (1991).
[21] K. W. Frisken, J. R. Tagg, A. J. Laws, and M. B. Orr, *J. Periodontal Res.* 22, 156 (1987).
[22] L. V. Holdeman, E. P. Cato, and W. E. C. Moore (eds.) "Anaerobic Laboratory Manual." Virginia Polytechnic Inst. and State Univ., Blacksburg, Virginia, 1977.

isolates, and correlating these data with clinical measures of periodontitis are found in the literature.[23,24] DNA probe, enzyme-linked immunosorbent assay (ELISA), *in situ* hybridization, latex agglutination, and immunofluorescent techniques are alternative detection methods.[25]

Plaque Toxicity Tests

Toxicity, infectivity, and transmissibility of pure or mixed cultures of plaque bacteria may be tested using the modified methods of Socransky and Gibbons.[26] A heavy suspension of cultured cells (0.5 to 1.0 ml) is injected subcutaneously in the groin of the guinea pig (150–250 g). Adjacent subcutaneous injections of macerated 2% (w/v) agar supplemented with sodium succinate (0.5 M) or hemin (10 μg/ml) are made.[27] Daily examination of the animals for evidence of abscesses and necrotic areas continues for 4 weeks. Aspiration of fluid from the abscesses with sterile syringes requires light ether anesthesia. The aspirate is examined under the light microscope for contamination and injected into a second animal to evaluate transmissibility. Positive infectivity is defined as a spreading necrotic lesion, a localized transmissible abscess, or death of the animal. Negative infectivity is defined as a hard, nodular nontransmissible abscess or mild inflammation. Injection of bacteria into the subcutaneous chamber apparatus (see [10] in this volume) is an alternative technique for testing plaque toxicity.

Clinical Indices

During the course of an animal study, it is usually necessary to monitor plaque accumulation and clinical indicators of periodontal disease. The plaque index (PlI) of Silness and Loe[28] is widely used, although the hygiene index (HI) is a more precise method for measuring plaque.[29] Visual redness and swelling of the gingiva, along with gingival bleeding after stimulation, are reliable signs of periodontal inflammation.[25] Methods for determining the papillary bleeding index (PBI) and the gingival index (GI) are illustrated

[23] R. W. Ali, T. Lie, and N. Skaug, *J. Periodontol.* **63**, 540 (1992).
[24] M. G. Newman and T. N. Sims, *J. Periodontol.* **50**, 350 (1979).
[25] J. Caton, *in* "Proceedings of the World Workshop in Clinical Periodontics" (M. Nevins, W. Becker, and K. Kornman, eds.), p. I-9. American Academy of Periodontology, Chicago, 1989.
[26] S. S. Socransky and R. J. Gibbons, *J. Infect. Dis.* **115**, 247 (1965).
[27] D. Mayrand and B. C. McBride, *Infect. Immun.* **27**, 44 (1980).
[28] P. Silness and H. Loe, *Acta Odontol. Scand.* **22**, 121 (1964).
[29] H. Klaus, E. M. Rateitschak, H. F. Wolf, and T. M. Hassell, "Color Atlas of Dental Medicine 1, Periodontology." Theime Medical Publishers, Stuttgart, 1989.

in the *Color Atlas of Dental Medicine 1, Periodontology.*[29] The Eastman interdental bleeding index (EIBI), more sensitive in detecting interdental inflammation than the PBI,[30] utilizes a soft wooden toothpick which is gently inserted interproximally four times. If bleeding occurs within 15 sec, that site is scored as positive. The number of positive sites is divided by the total sites tested to determine percent bleeding score.[31]

Gingival crevicular fluid can be analyzed for cells or biochemical markers which indicate disease activity. These include collagenase, prostaglandin E_2, and β-glucuronidase.[18,32,33] Pocket depth and clinical attachment level, the "gold standard" for detecting periodontal disease activity, may be measured with manual or automated probes. Clinical attachment level is the distance between a fixed landmark, such as the cementoenamel junction, and the base of the pocket. A comprehensive, cross-sectional study of periodontal disease in beagle dogs details methods for scoring plaque, calculus, gingivitis, loss of attachment, pocket depth, and width of gingiva.[34] These methods may be employed in primates as well.

Periodontal Disease Activity

Periodontitis, the chronic inflammation of gingiva, periodontal ligament, and alveolar bone, is characterized by periods of progressive destruction of these tissues and periods of quiescence. Tissue degradation results from a complex interaction of host and microbial lytic enzymes, cytokine activity, and localized altered bone metabolism. Periodontal disease activity refers only to the destructive phase and is detected by longitudinal clinical attachment level measurements.[25] Because there is no way of knowing exactly when during the interval between measurements disease activity occurred, frequent measurements are advisable in all animal studies.

Specific Models

Primates, dogs, and rats provide great versatility for assessing periodontal tissue invasion and damage by suspected pathogens, mutants, and cloned virulence factors. When large numbers of animals are required and/or when sacrifice is necessary for histologic purposes, the rat and possibly the dog are models of choice.

[30] J. Caton, A. Polson, O. Bouwsma, T. Blieden, B. Frantz, and M. Espeland, *J. Periodontol.* **59,** 722 (1988).

[31] J. Caton and A. Polson, *Compend. Contin. Dent. Ed.* **6,** 89 (1985).

[32] R. Attstrom and J. Egelberg, *J. Periodontal Res.* **5,** 48 (1970).

[33] S. Mukherjee, A. K. Das, and M. K. Patel, *J. Periodontal Res.* **18,** 501 (1983).

[34] W. P. Sorensen, H. Loe, and S. P. Ramijford, *J. Periodontal Res.* **15,** 380 (1980).

In addition to the methods described below, *in situ* hybridization techniques may be used to localize bacteria and/or their gene products within histologic sections taken at periodic intervals. For this type of temporal experiment, a repeated measurement analysis of variance should be used, with time of measurement being the grouping factor and the bacterial parameter being the repeated measurements factor. Significant differences can be further analyzed using the Newman–Keuls procedure, making pairwise comparisons between bacterial parameters.[35]

Primates

Primates in common use are the cynomolgus monkey (*Macaca fasicularis*), squirrel monkey (*Saimiri sciureus*), rhesus monkey (*Macaca mulatta*), baboon (*Papio anubis*), and marmoset. Adult rhesus monkeys are ideal because of their susceptibility to naturally occurring periodontal disease. Plaque-infected primates may be purchased with varying degrees of existing disease. However, if advanced disease is not present at time of purchase, the length of time required for its development generally is prohibitive. To hasten this process in a controlled manner, plaque accumulating devices may be placed and/or periodontal defects may be created surgically.

The primate model of Caton and Zander[36] was the first to induce predictably irreversible alterations in the alveolar bone level and the level of connective tissue attachment. Originally developed to evaluate treatment procedures, the model is easily adapted to microbial studies. Elastic ligatures are placed around selected teeth of the monkeys to encourage plaque accumulation. Orthodontic elastic bands are doubled and placed apical to the interproximal contact points. Ligatures are changed at 2-week intervals until periodontal pocket formation is confirmed by probing. Pockets of 5 to 8 mm can be expected to develop from 66 to 176 days, depending on tooth type.[37] Incisor pocketing generally takes 2 months, premolar 4 months, and molar 5 or 6 months.

In the original protocol, extent of disease was determined histologically. Immediately after sacrifice, the mandible and maxilla are dissected free, fixed for 72 h in Lavdowsky's solution (300 ml formaldehyde, 105 ml glacial acetic acid, 1500 ml of 95% ethanol, 1200 ml distilled water), washed for 24 hr in filtered running tap water, decalcified in 4% nitric acid, cut into blocks, and embedded in celloidin. Mesiodistal sections of

[35] B. A. Smith, J. S. Smith, R. G. Caffesse, C. E. Nasjleti, D. E. Lopatin, and C. J. Kowalski, *J. Periodontal* **58,** 669 (1987).
[36] J. G. Caton and H. A. Zander, *J. Periodontol.* **46,** 71 (1975).
[37] J. G. Caton and C. J. Kowalski, *J. Periodontol.* **47,** 506 (1976).

12 μm thickness are cut parallel to the long axis of the tooth, starting from the facial surface, and step serial sections representing intervals of 48 μm are stained with hematoxylin and eosin.

A calibrated grid in the eyepiece of the light microscope is used to make measurements on the sections at a magnification of $\times$ 35–40. Linear distances are measured between the cementoenamel junction (CEJ) and the crestal bone, apical end of the junctional epithelium (JE), and the apical extent of the angular bone defect (AAD).[36] Attachment loss is the distance from the CEJ to the JE. Bone loss in infrabony pockets is the distance from the crest of bone to the apical end of the defect. When crestal bone is not present, the distance from CEJ to AAD is used to calculate bony loss. Examiner error is determined from the mean of triplicate measurements on 10 randomly chosen sections, done 72 hr apart, and is expressed as average percentage difference.

To compare contralateral defects in an animal, individual pocket measurements from the right-hand side are compared with the corresponding pockets on the left using the paired-sample Student's t-test. Pearson product-moment correlations are determined on the pocket pairs for each parameter. Two animals provide a sufficient number of surfaces for a statistically sensitive comparison of periodontal treatment efficacy. This design controls for "between-animal" variation.[36] In microbial studies, however, implant and control sites in the same animal may become cross-contaminated. Therefore, two is an insufficient number of experimental animals for microbial studies.

Holt et al.[38] modified this model to demonstrate pathogenicity of P. gingivalis, using 4 teeth in each of 8 animals. Plaque-infected cynomolgus monkeys with no previously detectible P. gingivalis were inoculated with the organism. Twenty weeks after ligature placement, two ligated teeth per animal were inoculated three times per week. Anaerobically grown cultures were transported in Brewer jars to the vivarium. Using a Hamilton syringe with a blunted needle tip, 8-μl samples were delivered to the subgingival pocket. The animals were inoculated for 2 weeks, rested for 7 weeks, and inoculated 1 additional week. The sites were monitored for 12 weeks, and plaque and gingival indices were determined. Infection was assessed by selective culturing of subgingival plaque and calculating the percentage of P. gingivalis of the total cultivable microbiota. Detectible levels ranged from 3 to 27%. Serum IgG and IgM antibody titers to P. gingivalis were monitored by a modified ELISA, as additional evidence of infection. Periodontal destruction was determined by radiographic evi-

---

[38] S. C. Holt, J. Ebersole, J. Felten, M. Brunsvold, and K. S. Kornman, *Science* **239**, 55 (1988).

dence of alveolar bone loss. Standardized radiographic images pre- and postinfection were analyzed by computer-assisted densitometric image analysis (CADIA). Change in bone density was measured in gray scale units ranging from 0 to 255. Significant bone loss was represented by a decrease of 6.6 units or greater.

Half of the animals were given rifampin (10 mg/kg, i.m., three times a week) from 8 weeks before implantation through the end of the study at 20 weeks. Half of the ligated sites in each animal were inoculated with a rifampin-resistant *P. gingivalis* strain. Compared with controls, the rifampin-treated animals showed less bone loss at ligated sites at the end of the 20-week ligation period but more bone loss after implantation. It is unclear whether the expression of *P. gingivalis* virulence factors was altered by the multistep process used to generate the rifampin-resistant strain. In addition, it was not reported how composition of dental plaque is altered by long-term rifampin treatment. Despite these two points, 4 of the 5 animals with significant bone loss had rifampin-resistant *P. gingivalis* cultured at greater than 10% of the total cultivable plaque microbiota.

Although periodontitis in primates most closely resembles the human disease, the expense of primate maintenance may limit the number of animals included in a study. The above use of radiographic and clinical measures of periodontitis obviated the need for animal sacrifice. However, reliance on radiographic assessment alone is problematic because bone degradation can occur independent of collagen fiber attachment loss and apical migration of the junctional epithelium.[39] If sacrifice is not planned, then radiographic assessment should be accompanied by probing attachment measurements.

A final limitation of the primate model is that restraints and sedation are usually required to perform intraoral procedures safely. The pole and collar system for training non-human primates[40] is presently being tested at the Eastman Dental Center (Rochester, NY) and may prove to be a useful alternative to sedation for oral hygiene and other nonpainful procedures.

Dogs

Periodontal disease in dogs (*Canis familiaris*) is nearly as well characterized as that in humans.[41,42] Canine periodontium, including the dentogingival attachment, resembles the human histologic arrangement, except

[39] H. E. Schroeder and J. Lindhe, *J. Periodontol.* **51**, 5 (1980).
[40] J. A. Anderson and P. Houghton, *Lab. Anim.* **12**, 49 (1983).
[41] H. E. Schroeder and J. Lindhe, *Arch. Oral Biol.* **20**, 775 (1975).
[42] H. E. Schroeder and J. Lindhe, *J. Periodontol.* **51**, 6 (1980).

for differences in tooth shape and interproximal contacts. When plaque is allowed to accumulate, naturally occurring periodontitis generally develops by 4 to 7 years of age.[34,43] Hence, older dogs with preexisting disease are frequently used to test therapeutic agents.[44,45] Younger and periodontally healthy dogs may be utilized for implantation studies in a similar fashion to the primate model described above. Diagrams of beagle dentition, the typical pattern of plaque and calculus formation,[46] and standard reference points used in microscopic studies are described in detail elsewhere.[47]

At baseline, the teeth of the dogs are scaled and polished, then brushed twice daily for 2 months. To confirm a state of gingival health, a clinical and radiographic evaluation is performed. Histological sections of a small gingival biopsy may be examined for inflammatory cells in the connective tissue. Absence of these cells is consistent with gingival health.[33] Amalgam reference markers are placed in the buccal tooth surface 1 mm coronal to the gingival margin. A soft diet is instituted and dental plaque allowed to accumulate by cessation of oral hygiene procedures. Cotton floss ligatures or silk sutures are secured around premolars in one or more quadrants.[33,40] Because adjacent dog teeth lack tight contacts, it may be necessary to stabilize the ligatures by creating subgingival grooves in the mesial and distal surfaces. Alternative plaque accumulating devices include arch bars and continuous wire splints.[48]

Using these methods, destructive and progressive lesions will develop within 5 months. To accelerate the process or to create more severe defects, ligatures may be mechanically forced deep into the gingival sulcus and advanced biweekly.[18] Alternatively, the gingival sulcus may be deepened surgically and copper bands placed directly on the alveolar bone. After three weeks, the copper bands are removed and replaced with cotton ligatures for an additional 11 weeks. In the first month following ligature removal, a slight gain in clinical probing attachment may occur. However, the defects persist, and the gingival and plaque indices increase.[49,50]

[43] J. Lindhe, S. E. Hamp, and H. Loe, *J. Periodontal Res.* **10**, 243 (1975).
[44] M. J. Jeffcoat, R. C. Williams, H. G. Johnson, W. J. Wechter, and P. Goldhaber, *J. Periodontal Res.* **20**, 532 (1985).
[45] R. C. Williams, M. K. Jeffcoat, M. K. Howell, M. S. Reddy, M. S. Johnson, C. M. Hall, and P. Goldhaber, *J. Periodontal Res.* **23**, 224 (1988).
[46] H. M. Rosenberg, C. E. Rehfeld, and T. E. Emmering, *J. Periodontol.* **37**, 208 (1966).
[47] S. Nyman, G. Sarhed, I. Ericsson, J. Gottlow, and T. Karring, *J. Periodontal Res.* **21**, 496 (1986).
[48] F. H. M. Mikx, D. N. B. Ngassapa, F. M. J. Reijntjens, and J. C. Malta, *J. Dent. Res.* **63**, 1284 (1984).
[49] L. J. Van Dijk, J. Jansen, T. Pilot, and L. T. Van Der Weele, *J. Periodontol.* **53**, 449 (1982).
[50] S. A. Ralls, M. E. Cohen, and E. G. A. Hey, *J. Periodontal Res.* **21**, 264 (1986).

Methods for assessing infection and disease progression in dogs are similar to those in primates. Tooth mobility and gingival inflammation may each be graded from 0 to 3 as described in previous beagle studies.[43] Although it is seldom necessary, gingival crevicular fluid may be used to differentiate between gingivitis and periodontitis. Characteristic changes in the pH, K^+ concentration, and protein composition of crevicular fluid accompany ligature-induced periodontitis.[33]

Most beagle studies that do not rely on histological criteria for loss of connective tissue attachment utilize conventional standardized radiography. The traditional technique for standardizing radiographs uses modified film holders for reproducible angulation and tooth-source–film distances. Identical exposure settings (80 Kilovolt peak, 5 mA, 5/6 sec) and automatic processing result in uniform densities in films taken at different experimental time periods. Magnification ($\times 10$) and a computer digitizer facilitate quantitation of bone around each tooth.[43] Readers unfamiliar with quantification of bone from dental radiographs are referred to a schematic representation of the measurement of canine alveolar bone.[51] Software has been developed to accurately compare nonstandardized radiographs by digital subtraction.[52] Pseudocolor contrast enhancement of digitally subtracted radiographs significantly aids the nonradiologist in detecting minute changes in bone density.[53]

The calculated percent bone loss[54] and rates of bone loss for the experimental periods are subjected to an analysis of variance to determine statistical significance. Baseline measurements allow each tooth surface to serve as its own control. Mean bone loss and mean rate of bone loss per group may be subjected to within-group analysis, paired and unpaired Student's t-tests, and the z-test.[44]

Alveolar bone loss may also be measured by uptake of the bone-seeking radiopharmaceutical technetium-99m-tin–diphosphonate (99^{m}Tc–Sn–MDP).[55] An increase in bone-seeking radiopharmaceutical uptake (BSRU) indicates active bony loss, correlates well with radiographic findings, and yields very low error rates.[56] Intravenous injection of 1.5

[51] R. C. Williams, M. B. Sandler, P. H. Aschaffenburg, and P. Goldhaber, *J. Periodontal Res.* **14**, 342 (1979).

[52] M. K. Jeffcoat, R. L. Jeffcoat, and R. C. Williams, *J. Periodontal Res.* **19**, 434 (1984).

[53] M. S. Reddy, J. M. Bruch, M. K. Jeffcoat, and R. C. Williams, *Oral Surg., Oral Med., Oral Pathol.* **71**, 763 (1991).

[54] H. Bjorn, A. Halling, and H. Thyberg, *Odontol. Revy* **20**, 165 (1969).

[55] M. K. Jeffcoat, R. C. Williams, M. L. Kaplan, and P. Goldhaber, *J. Periodontal Res.* **20**, 301 (1985).

[56] M. K. Jeffcoat, R. C. Williams, W. J. Weichter, H. G. Johnson, M. L. Kaplan, J. S. Gandrup, and P. Goldhaber, *J. Periodontal Res.* **21**, 624 (1986).

mCi/kg is followed by a 4-hr clearance period. Uptake is measured with a miniaturized semiconductor radiation detector probe, which is placed firmly on the gingiva overlying the alveolar crest at the experimental site. Control measurements are made at the nuchal crest and are used to normalize the experimental measurements.[54] Mean BSRU counts per minute are calculated per group and compared using an unpaired Student's t-test.

The number of dogs used in the referenced studies ranges from 5 to 22. When using this number of beagles, pooling of sites based on group, without regard to animal, is generally inappropriate for statistical analysis because high interclass correlations before and after intervention shows that the animal contributes significantly to variability among sites. The mean measurement per animal is the unit of observation for most statistical purposes; however, in certain instances, sites within dogs may be analyzed.[49] Multivariate analysis is recommended for large sample size, and mixed model analysis is best for smaller sample size. When more than one examiner is collecting clinical data, a nonparametric correlation and repeatability scores should always be calculated to determine interexaminer consistency.

Although the beagle is known for its docile nature, general anesthesia is required for examination, scaling, radiography, ligature placement, surgery, and bacterial inoculation and sampling. Acepromazine (10 mg, i.m.) is followed by intravenous sodium pentobarbital (25 mg/kg) for anesthesia.

Rats

Originally developed to evaluate therapeutic modalities, the primate and dog models have been adapted for microbial studies. The rat (*Rattus norvegicus*) model, on the other hand, has been used primarily for investigations of the oral flora. In rat, the periodontal pathogenicity of *P. gingivalis, A. actinomycetemcomitans, F. nucleatum, Capnocytophaga sputigena, Eikenella corrodens, Streptococcus sobrinus, Actinomyces viscosus,* and *A. naeslundii* has been demonstrated.[8] With few exceptions, the pathophysiology of periodontitis in rats resembles that of humans. Keratinization of rat sulcular epithelium differs from that of the human but does not prevent colonization of the sulcus or diffusion of bacterial products into the connective tissue.[8] The immune component of rat periodontitis mimics that of the human disease except that, in rats, polymorphonuclear leukocytes are found in the gingival crevice in the absence of inflammation, and there is a lack of subepithelial inflammatory response.[7]

Ease of handling, low cost, short study time, and variation among strains and germ states make the rat model extremely versatile. Genetically and immunologically altered rats, including T or B cell-deficient, diabetic, generally immunosuppressed, or selectively reconstituted and irradiated strains, facilitate study of the diverse features of periodontitis. A number of conventional strains, most commonly the Sprague-Dawley rat, are used in periodontal studies and include the ODU, a strain which is particularly susceptible to plaque formation.[57] Germfree rats, rats free of specific pathogens (gnotobiotic), and rats infected with indigenous plaque are available. Existing plaque flora may be quite variable between rats of different ages and different colonies. Procedures for maintenance and verification of germfree and monoinfected rat colonies are outlined by Chang et al.[58]

Experimental inoculation with bacteria may begin in animals as young as 5 weeks of age, when eruption of the molars is complete. Periodontal destruction progresses rapidly in the absence of ligatures, obviating the challenge of placing ligatures around the cervices of tiny rat teeth. The oral cavity may be inoculated using contaminated drinking water and food, by swabbing the oral cavity and skin with organisms, or with multiple oral injections of 10^9 bacteria.[7,8,59] It may be necessary to deliver anaerobes by gastric gavage of 7.5×10^{11} cells in 0.5 ml of a 2 to 5% carboxymethylcellulose bolus repeated 3 times at 48-hr intervals.[20a] Because rats are coprophagic, exposure to monocontaminated feces or monoinfected cage-mates may be sufficient to establish a microorganism permanently in the oral cavity. Most human oral pathogens, with the exception of spirochetes, are easily established by any of the above methods. If difficulty with inoculation is encountered, a combination of methods is recommended. Once periodontal pathogens are established, considerable periodontal destruction will occur in 6 to 12 weeks.

To verify that inoculations are successful, cultures of the oral flora may be obtained by cotton swab or by culturing feces.[60] Subgingival plaque samples may be recovered with sterile dental paper points inserted into the periodontal pockets of selected molars. If general anesthesia is required for sampling, Innovar (0.1 ml/100 g, i.v.) containing 20 μg/ml atropine is adequate.[61]

A widely accepted morphometric method for assessing experimental periodontitis in the rat is the measurement of horizontal bone loss on stained, defleshed jaws using a dissecting microscope equipped with a grid eyepiece. Under magnification of $\times 30$ each jaw is stabilized so that

[57] N. Ito, Y. Azuma, and M. Mori, J. Dent. Res. 54, (1975).
[58] K. M. Chang, N. S. Ramamurthy, T. F. McNamara, R. J. Genco, and L. M. Golub, J. Periodontal Res. 23, 240 (1988).
[59] R. J. Gibbons, S. S. Socransky, and B. Kapsimalis, J. Bacteriol. 88, 1317 (1964).
[60] L. Heijl, J. Wennstrom, J. Lindhe, and S. S. Socransky, J. Periodontal Res. 15, 406 (1980).
[61] A. S. Fine and R. W. Egnor, J. Oral Pathol. 15, 138 (1986).

the buccal and lingual cusp tips are aligned. One buccal measurement per root is made on all 12 molars. To avoid concealing significant bone loss in rats, four critical recommendations should be followed[62]: (1) analysis of maxillary and mandibular measurements are separate; (2) inclusion of an equal number of sites from the right- and left-hand sides; (3) exclusion of the mesial surface of the first molar and distal surface of the third molar; and (4) measurement of buccal surfaces only.

Because of asymmetric bone loss from lateral bias, and the likelihood of cross-contamination, a split mouth design is not recommended in rats.[61] Rat incisor teeth should not be included in periodontal studies because they have no roots and are continually erupting. In addition, the generalized passive eruption of the molars results in an age-related, nonpathologic increase in distance between the CEJ and alveolar bone crest. Using age-matched animals overcomes this complication.

Some interproximal lesions are missed by the morphometric method and are better evaluated radiographically[63] or histologically.[20a] If ultrastructural examination of histological sections done, fixation by whole-body perfusion is recommended.[64] The animal is anesthetized with Nembutal (Abbott Labs., Chicago, IL) (5 mg/100 g), and a cannula is inserted into the ascending aorta through an incision in the left ventricle. Fixative (1% v/v glutaraldehyde in a Tyrode's solution at 350 mOsm) is delivered slowly under pressure. Jaws are then dissected free and decalcified in 100 mM EDTA, 0.4% glutaraldehyde, pH 7.4. Block sections are postfixed in 2% (w/v) osmium tetroxide, pH 7.4, buffered with collidine, and dehydrated with ethanol and propylene oxide. Embedding the blocks in Epon 812 is followed by sectioning. Sections for light microscopy are 1 μm thick and are stained with toluidine blue. Sections for electron microscopy are 600 to 1000 Å thick and are stained with uranyl acetate and lead citrate.

Acknowledgments

We thank Dr. Susan J. Hayflick for reading the manuscript.

[62] B. Klausen, C. Sfintescu, and R. T. Evans, *Arch. Oral Biol.* **36,** 685 (1991).
[63] B. Klausen, C. Sfintescu, and R. T. Evans, *Scand. J. Dent. Res.* **97,** 494 (1989).
[64] P. R. Garant, *J. Periodontol.* **47,** 132 (1976).

[10] Animal Chamber Models for Study of Host–Parasite Interactions

By Caroline Attardo Genco and Robert J. Arko

Introduction

The use of laboratory animals in research is becoming more tenuous as tougher guidelines are being implemented in facilities regulated by state, federal, and local laws concerning the humane use and care of research animals. Investigators anticipating studies involving animals must be aware of these regulations and take steps to ensure full compliance to avoid the potential misuse of animals, embarrassment to their organization, and/ or the loss of funding. Future investigators will utilize alternatives to animal models whenever possible; however, there exist situations where the complex interactions of the living organism with the intact host need to be evaluated through animal studies.

An important feature of any animal model for infection is that it simulates an infectious process in humans while mimicking the natural pathogenesis to the greatest extent possible. The validity of animal models for human pathogens may be difficult to define, especially if the infectious agent is species specific. In situations where species barriers prevent establishment of active infections, some uncommon steps may be necessary to develop models that will allow growth of test microorganisms and access of investigators to appropriate research specimens.

The question regarding what constitutes a valid model depends largely on the types of experiments being performed. Certain animals whose immune systems function in many respects like humans can be used to study the complex interaction of multiple cellular and humoral factors that are involved in the inflammatory response, even though these animals may lack the species-dependent receptors found in tissues of human origin. The assessment of host responses to localized infections can be performed quite easily with the use of subcutaneous chambers implanted in small animal species.[1] The subcutaneous chamber models have been developed primarily to study bacterial virulence factors produced by human pathogens and the host response induced by infection. Chambers provide a readily obtainable sample of transudate which contains bacterial cells, host cells, and inflammatory products and allows one to examine host–bacteria

[1] R. J. Arko, *Science* **177**, 1200 (1972).

interactions during the progression of the infection.[2] One can easily sample *in vivo* grown bacteria and assess the modulation of potential virulence factors during infection.

The development and use of animal chamber models have added significantly to the study of microbial pathogenesis. A number of different microorganisms have been successfully cultivated in chambers implanted in animals, including *Neisseria gonorrhoeae, Porphyromonas gingivalis, Bacteroides fragilis, Pseudomonas aeruginosa, Treponema pallidum, Escherichia coli, Staphylococcus aureus, Bordetella pertussis, Streptococcus pyogenes,* and *Haemophilus ducreyi.*[3–16] This chapter addresses the techniques involved in the implantation of chambers and analysis of the composition of fluid obtained from chambers and summarizes the work of many different investigators who have employed these models to study host–parasite interactions.

Techniques Used in Implant Models

Types of Implants

Although a number of different chamber models have been used, all have in common the use of tissue-encapsulated chambers to study bacterial

[2] C. A. Genco, C. W. Cutler, D. Kapczynski, H. Maloney, and R. R. Arnold, *Infect. Immun.* **59,** 1255 (1991).
[3] G. K. Sundqvist, J. Carlsson, B. F. Herrmann, J. F. Hofling, and A. Vaatainen, *Scand. J. Dent. Res.* **92,** 14 (1984).
[4] S. J. Hultgren, T. N. Porter, A. J. Schaeffer, and J. L. Duncan, *Infect. Immun.* **50,** 370 (1985).
[5] I. Maciver, S. H. Silverman, M. R. W. Brown, and T. O'Reilly, *J. Med. Microbiol.* **33,** 139 (1991).
[6] C. Chuard, J.-C. Lucet, P. Rohner, M. Herrmann, R. Auckenthaler, F. A. Waldvogel, and D. P. Lew, *J. Infect. Dis.* **163,** 1369 (1991).
[7] K. D. Coleman and L. H. Wetterlow, *J. Infect. Dis.* **154,** 33 (1986).
[8] J. L. Duncan, *Infect. Immun.* **40,** 501 (1983).
[9] N. M. Kelly, A. Bell, and R. E. Hancock, *Infect. Immun.* **57,** 344 (1989).
[10] A. B. Onderdonk, R. L. Cisneros, R. Finberg, J. H. Crabb, and D. L. Kasper, *Rev. Infect. Dis.* **12,** S169 (1990).
[11] S. E. J. Day, K. K. Vasli, R. J. Russell, and J. P. Arbuthnott, *J. Infect.* **2,** 39 (1980).
[12] R. J. Arko, J. K. Rasheed, C. V. Broome, F. W. Chandler, and A. L. Paris, *J. Infect.* **8,** 204 (1984).
[13] D. L. Trees, R. J. Arko, and S. A. Morse, *Microb. Pathog.* **11,** 387 (1991).
[14] A. B. Onderdonk, R. L. Cisneros, J. H. Crabb, R. W. Finberg, and D. L. Kasper, *Infect. Immun.* **57,** 3030 (1989).
[15] I. Horvath, R. J. Arko, and J. C. Bullard, *Br. J. Vener. Dis.* **51,** 301 (1975).
[16] W. Zimmerli, F. A. Waldvogel, P. Vaudaux, and U. E. Nydegger, *J. Infect. Dis.* **146,** 487 (1982).

growth *in vivo* and the immune response of the host. Differences in implant models stem primarily from the type of materials used and whether the chambers are open or close ended. When considering the material to be used, one must take into account the immunologic effects produced by placing foreign materials in living organisms. High-grade stainless steel is one of the earliest materials found to be compatible with most tissues and has been used extensively in surgical devices to be implanted within the body.[17] Various synthetics, namely, polyethylene, Teflon, and silicones, are also found in implantable devices; however, our knowledge concerning their immunologic effects is less certain.

We typically employ open-ended chambers made of either stainless steel or polyethylene for implant studies.[18] In addition to the stainless steel and polyethylene open-ended chambers, a wide variety of implantable closed chambers have also been used. These include implantable collodion sacs, cellophane dialysis tubing, and plastic syringe barrels having both ends sealed with membrane filters.[7,9,11,14,19] These closed chambers vary considerably from the open-ended implants in that the bacteria, although in close proximity to phagocytes and other host cells, are protected from engulfment. Therefore one cannot monitor natural host–bacteria interactions.

Different chamber materials have pronounced effects on the susceptibility of animals to infection with different microorganisms. For example, although *N. gonorrhoeae* can typically infect stainless steel chambers implanted in mice after injection of as few as 50 colony-forming units (cfu), it cannot be established in polyethylene implants.[20] In contrast to results observed with *N. gonorrhoeae,* Trees *et al.*[13] found that *H. ducreyi* can infect polyethylene but not stainless steel chambers.

The differences observed in the ability of bacteria to grow within different chambers may be due to the host cell response induced by different implant materials. An analysis of chamber fluid obtained from uninoculated stainless steel implants revealed significantly fewer leukocytes when compared to chamber fluid obtained from polyethylene chambers (Table I). Fluid from polyethylene chambers contained a relatively greater number of polymorphonuclear cells (PMNs) and macrophages, whereas fluid from stainless steel implants contained more lymphocytic cells.

[17] R. J. Arko, *Lab. Anim. Sci.* **23,** 105 (1973).

[18] R. J. Arko and A. Balows, *in* "Experimental Models in Antimicrobial Chemotherapy" (O. Zak and M. A. Sande, eds.), Vol. I, p. 355. Academic Press, New York, 1986.

[19] J. A. Moody, C. E. Fasching, L. M. Sinn, D. N. Gerding, and L. R. Peterson, *J. Lab. Clin. Med.* **115,** 190 (1990).

[20] C. A. Genco, R. J. Arko, C. Y. Chen, and S. A. Morse, unpublished data (1989).

TABLE I

MEAN NUMBER OF LEUKOCYTES IN MOUSE CHAMBER
FLUID FROM DIFFERENT CHAMBER TYPES

	Mean number or percent	
Sample	Stainless steel chambers[a]	Polyethylene chambers
Total WBC	658 ± 223	2600 ± 1157
PMNs	38 ± 14 %	44 ± 14 %
Macrophages	15 ± 10 %	38 ± 13 %
Lymphocytes	47 ± 10 %	18 ± 14 %

[a] Chamber fluid samples were obtained with a hypodermic needle and syringe 2 weeks postimplantation into mice. Chamber fluid was stained by Wright stain and the number of white blood cells (WBC), polymorphonuclear cells (PMNs), macrophages, and lymphocytes determined by light microscopy. Percent PMNs, macrophages, and lymphocytes were based on the number of cells relative to the total WBC count in nine separate fields.

Implantation of Chambers and Sampling Technique

The implantation of subcutaneous chambers in laboratory animals involves relatively simple surgical procedures performed with the host animal under general anesthesia. We typically employ coil-shaped subcutaneous culture chambers prepared from 0.5 mm stainless steel wire. These are surgically implanted in the subcutaneous tissue of the dorsolumbar region of each animal. At least 10 days elapses before chambers are inoculated with live bacteria; during this period the outer incision heals completely, with no visible signs or irritation or discomfort. The chambers become encapsulated by a thin layer of fibrous connective tissue containing small blood vessels. As the chamber becomes encapsulated within the subcutaneous tissues, it gradually fills with a light-colored transudate, which is easily sampled with a hypodermic needle and syringe.

The closed-ended chambers employed by many other investigators are typically inoculated with live microorganisms prior to implantation. Sampling from *in vivo* chambers is performed only after the animals are sacrificed, after which the implants are surgically removed.

Characteristics of Chambers and Chamber Fluid

The amount of fluid available from implants varies with the material used, as well as the chamber size and age. Uninfected stainless steel coil implants produce larger amounts of chamber fluid in the first 1–3 weeks after implantation. However, they eventually become filled with connective tissue, which restricts their later use. The tube-type polyethylene implants remain usable over a longer period (>6 weeks) but yield smaller volumes of fluid in mice, hamsters, and guinea pigs.[18]

The encapsulation of chambers with dense connective tissue can alter the rate and type of serum components that gain access to the chamber fluid. The younger, less encapsulated implants generally permit greater exchange of serum components with the chamber fluid. In animals where more than one implant is used, infections in less encapsulated chambers often disseminate to uninoculated implants. Complement activity in guinea pig chamber fluid gradually declines with the age of the implant; however, immunoglobulin M (IgM) and IgG levels remain constant.[18]

Implants with active infections usually produce larger amounts of tissue fluid and cellular debris during the first 2–5 days after inoculation. The inflammatory process may, however, enhance the rate at which the chamber fills with tissue and debris, thus decreasing the quantity of fluid recovered from chambers. We have observed rapid rejection of chambers in mice which were immunized with heat-killed *P. gingivalis* followed by challenge with live organisms.[21] This response is apparently due to the recruitment of host inflammatory cells and their products in response to infection with this organism.[2,21] Thus, the optimal chamber conditions for various studies may be influenced by the selection of an appropriate implant as well as its utilization during the period of desired permeability.

Animal Species

Rabbits

Various types of implants have been used in rabbits including polyethylene "whiffel" balls, stainless steel coils, membrane-enclosed plastic rings, and peristaltic plastic pumps.[1,22,23] The whiffle or practice golf ball has seen extensive use as a porous implant that encapsulates with connective tissue and fills with a tissue fluid transudate, usually by 1 month after implantation. Fluid obtained from whiffle ball chambers has been used to monitor the host response to bacterial challenge as well as the tissue penetration and interstitial drug levels of parenterally administered antibiotics.[23] Because rabbits have well-developed immune and comple-

[21] C. A. Genco, D. R. Kapczynski, C. W. Cutler, R. J. Arko, and R. R. Arnold, *Infect. Immun.* **60**, 1447 (1992).

[22] J. Parsonnet, Z. A. Gillis, A. G. Richter, and G. B. Pier, *Infect. Immun.* **55**, 1070 (1987).

[23] R. Tight, R. B. Prior, R. L. Perkins, and A. Rotilie, *Antimicrob. Agents Chemother.* **8**, 495 (1975).

ment systems, they produce a full range of responses to different antigenic stimuli. A number of different microorganisms have been successfully cultivated in the tissue-encapsulated chambers, including *N. gonorrhoeae, P. gingivalis, Treponema pallidum, Escherichia coli,* as well as toxic shock-producing strains of *S. aureus.*[1,12,14,24,25]

Guinea Pigs

Guinea pigs possess a number of humoral and cellular traits making them ideally suited for immunological experiments. Well-characterized strains are available that demonstrate different complement activities or deficiencies.[26] Both polyethylene and stainless steel chamber implants have been used successfully in guinea pigs; however, polyethylene implants remain intact much longer than the stainless steel implants.[18] Although the extended life span of polyethylene implants allows for sampling of chamber fluid for greater than 1 year following implantation, we have found that both hemolytic and bactericidal complement levels in these chambers decline over time (Table II). Guinea pigs are also extremely sensitive to many commonly used antibiotics including penicillin, lincomycin, clindamycin, and others that can induce a rapidly fatal colitis,[27] and thus they are not the species of choice for testing antimicrobials.

Mice

Successful murine chamber models to examine host–parasite interactions have certain advantages over models employing other animal species. Mice are relatively inexpensive, and they are commercially available in a number of genetically defined, well-characterized strains that demonstrate a variety of immunologic features or genetic defects that can be exploited by scientists.[26] In comparison with other species, however, mice generally show a much lower level of hemolytic complement activities.[26] This defect can be overcome in the mouse chamber model by the injection of fresh guinea pig or human serum at appropriate intervals.[28,29]

[24] G. Dahlen and J. Slots, *Oral Microbiol. Immunol.* **4**, 6 (1989).
[25] T. M. Finn, J. P. Arbuthnott, and G. Dougan, *J. Gen. Microbiol.* **128**, 3083 (1982).
[26] P. L. Altman and D. D. Katz (eds.), *in* "Inbred and Genetically Defined Strains of Laboratory Animals, Mouse and Rat Part 1, Guinea Pig and Hamster Part 2." Federation of American Societies for Experimental Biology, Bethesda, Maryland, 1979.
[27] W. E. Farrar, Jr., and T. H. Kent, *Am. J. Pathol.* **47**, 629 (1965).
[28] R. J. Arko, K. H. Wong, S. E. Thompson, and F. J. Steurer, in "Immunobiology of *Neisseria gonorrhoeae.*" (G. F. Brooks, E. C. Gotschlich, K. K. Holmes, W. D. Sawyer, and F. E. Young, eds.) p. 303. American Society for Microbiology, Washington, D.C., 1978.
[29] R. J. Arko, K. H. Wong, F. J. Steurer, and W. O. Schalla, *J. Infect. Dis.* **139**, 569 (1979).

TABLE II
EFFECT OF IMPLANTATION TIME ON HEMOLYTIC AND
BACTERICIDAL ACTIVITY OF CHAMBER FLUID

Test specimen[a]	Activity	
	50% Hemolytic[b]	50% Bactericidal[c]
Fresh serum	357	40
Chamber fluid		
1 week	92	20
2 weeks	67	20
4 weeks	50	10
8 weeks	45	5
12 weeks	30	2.5
24 weeks	21	<2.5

[a] Chambers were implanted subcutaneously into Hartley guinea pigs and fluid samples obtained with a hypodermic needle and syringe at the designated times. Chamber fluid samples were pooled from four or more guinea pigs and assayed for hemolytic and bactericidal activity.
[b] Hemolytic activity is expressed as units/ml and defined by a standard red blood cell complement fixation assay.[28]
[c] Bactericidal activity is expressed as the dilution factor required for bactericidal activity as determined using *N. gonorrhoeae* (C6 type 1) and a C6 antiserum developed in guinea pigs.

Mouse immunoglobulins have been well characterized and consist of IgA, IgG (1, 2a, 2b, and 3), IgM, and IgE.[30] In mice, IgG antibodies to most protein antigens are primarily of the IgG_1 and IgG_2 subclass, whereas anticarbohydrate antibodies generally belong to the IgG_3 subclass. Monomeric IgG_1 and IgG_3 do not bind complement (C1) while IgG_{2a} and IgG_{2b} do. All mouse IgG subclasses cross placental membranes, but IgG_3 appears to be transported more effectively; IgA and IgM do not cross placental membranes. These characteristics may have some relevance in subcutaneous chambers where blood components must also transverse an endothelial wall and connective tissues to gain access to the chamber fluid. The degree of tissue encapsulation as it relates to the age and type of implant can also affect the distribution of immunoglobulins present in chamber fluid.[18] The typical life expectancy of stainless steel coil implants in mice is 1 month, and that of polyethylene tubing is 3 months. The recoverable specimen size varies from 0.05 to 0.2 ml per day.

[30] R. E. Callard and M. W. Turner, *Immunol. Today* **11**, 200 (1990).

Parameters Examined in Chamber Implant Studies

Bacterial Virulence Factors

It is generally accepted that bacteria are capable of modulating their virulence factors when grown under specific conditions found in the host.[31] These include changes in growth rate, envelope composition, and sensitivity to host-mediated mechanisms of killing. Knowledge of the properties of pathogenic organisms when growing in the human host is derived largely from observations made on freshly isolated organisms or from extrapolations from experiments on *in vitro* grown organisms. The use of chamber models for the study of bacterial virulence factors is unique in that it allows for the continuous growth of bacteria *in vivo*. *In vivo* grown organisms can then be characterized with regard to the production and induction of virulence factors. Characteristics of *in vivo* grown bacteria that have been specifically examined include growth, cell division, outer membrane composition, encapsulation, cell wall content, and toxin production.[3,7,9,11,25,32,33] Chambers have also been used to examine the antimicrobial activity of various agents *in vivo* as well as the emergence of antibiotic-resistant bacteria.[19,34]

In most animal models, virulence of a bacterial strain is based on the gross pathology induced following inoculation into animals. Animals are typically inoculated with graded doses of different strains, and doses that produce a defined disease effect are compared among strains by statistical methods. Because the subcutaneous chamber model allows for continuous growth of bacteria, virulence is typically defined as the relative ability of different strains to grow and become established within chambers.

Host Response

The host response to bacterial infection can be monitored very easily within subcutaneous chambers. This includes both cell-mediated and humoral components of the immune response. Parameters of the host response examined to date include the induction of PMNs, complement activation, and local antibody response. Local antibody is typically detected in chamber fluid within 18 days postchallenge, and levels are comparable to those observed in serum.[2] Chamber models have been used to

[31] H. Smith, *J. Gen. Microbiol.* **136**, 377 (1990).
[32] C. W. Penn, D. R. Veale, and H. Smith, *J. Gen. Microbiol.* **100**, 147 (1977).
[33] C. W. Penn, D. Sen, D. R. Veale, N. J. Parsons, and H. Smith, *J. Gen. Microbiol.* **97**, 35 (1976).
[34] J.-C. Lucet, M. Herrmann, P. Rohner, R. Auckenthaler, F. A. Waldvogel, and D. P. Lew, *Antimicrob. Agents Chemother.* **34**, 2312 (1990).

study the protective role of antibody induced by immunization with *N. gonorrhoeae* and *P. gingivalis* directly into chambers.[21,35]

Host–Parasite Interactions

Perhaps the most unique of parameters that has been examined in the chamber model is the interaction between host and bacterial cells during infection. The ability of bacteria to degrade complement *in vivo*, the susceptibility of bacteria to the bactericidal activity of serum, the binding of host serum proteins to bacterial cells, and the phagocytosis of bacteria by PMNs are just a few examples of the host–parasite events that have been monitored *in vivo* in subcutaneous chambers.[2,3,25,36]

In the remainder of this chapter we summarize the work of many investigators who have examined the *in vivo* growth of various organisms using a subcutaneous implant. We have highlighted the observations made with regard to the detection of specific virulence factors *in vivo*, the host response induced in implants, and observations of specific host–parasite interactions.

Bacterial Pathogens Grown in Chamber Implant Models

Neisseria gonorrhoeae

Growth of *N. gonorrhoeae* in chamber implants represents an active gonococcal infection; however, the model is not entirely analogous to nondisseminated gonorrhea in humans since chamber infections are not mucosal surface infections. Experimental *N. gonorrhoeae* infections within subcutaneous chambers have contributed to our understanding of the *in vivo* characteristics and immunology of this obligate human pathogen. Several studies have also focused on the growth environment *in vivo* as it relates to the acquisition of specific nutrients. Previous reports have suggested that the availability of the essential nutrient iron is important in gonococcal pathogenesis.[37–39] Corbeil *et al.*[37] reported that addition of 15% gastric mucin and hemoglobin to the *N. gonorrhoeae* inoculum increased the lethality of gonococci for the mouse peritoneal model by

[35] C. A. Genco, R. J. Arko, C. Y. Chen, and S. A. Morse, unpublished data, 1989.

[36] R. J. Arko, J. C. Bullard, and W. P. Duncan, *Br. J. Vener. Dis.* **52**, 316 (1976).

[37] L. B. Corbeil, A. C. Wunderlich, R. R. Corbeil, J. A. McCutchan, J. A. Ito, Jr., and A. I. Braude, *Infect. Immun.* **26**, 984 (1979).

[38] C. W. Keevil, D. B. Davies, B. J. Spillane, and E. Mahenthiralingham, *J. Gen. Microbiol.* **135**, 851 (1989).

[39] C. A. Genco, C.-Y. Chen, R. J. Arko, D. R. Ka czynski, and S. A. Morse, *J. Gen. Microbiol.* **137**, 1313 (1991).

more than 100-fold. Keevil *et al.*[38] observed that iron-limited gonococci were highly virulent in the guinea pig chamber model. We have found that iron-uptake mutants of *N. gonorrhoeae* which were unable to grow with human transferrin or hemoglobin as the sole source of iron were also unable to be established *in vivo* in mouse subcutaneous chambers.[39] The median infectious dose (ID_{50}) of one such mutant, Fud14, was greater by a factor of 10^7 than that needed for the infection of chambers by the wild-type strain. Supplementation of chambers with either normal human serum or hemin (iron sources which support the growth of Fud14 *in vitro*), resulted in the establishment of Fud14 *in vivo* for at least 10 days postinoculation. The specific iron source utilized by *N. gonorrhoeae* in this model is unknown; however, an analysis of uninoculated chamber fluid indicated that a substantial amount of hemin was present in the chamber fluid, with the majority present as hemoglobin.[39]

Analysis of *N. gonorrhoeae* grown *in vivo* in chamber implants has defined potential virulence factors produced by this pathogen. For example, *in vivo* grown organisms are more resistant to the effects of fresh guinea pig serum and human PMNs.[19,33,36] In contrast, laboratory-grown organisms when exposed to fresh serum are killed more rapidly and show extensive blebbing (Fig. 1).[36] This resistance is apparently due to surface cytidine 5'-monophospho-*N*-acetylneuraminic acid and other host factors that are acquired during *in vivo* growth but are rapidly lost on subculture.[40] *Neisseria gonorrhoeae* grown in mouse chambers have a well-defined outer membrane and peptidoglycan in comparison to laboratory-adapted cells (Fig. 2) and have been reported to be more pilated.[32,33] *In vivo* grown cells also appear to be more effective as immunogens.[36]

Cross-protection studies employing guinea pigs parenterally immunized with formalin-fixed whole cell vaccines demonstrated that high levels of strain-related protective immunity could be induced.[41] Immunotypic resistance to *N. gonorrhoeae* in immunized mice was significantly increased by injection of exogenous guinea pig complement into chambers before challenge with gonococci.[42] The immunity in mice not treated with complement developed more slowly, was less effective, and waned earlier than that which was complement-dependent. Similar experiments in chimpanzees demonstrated that parenteral immunization with *N. gonorrhoeae* antigens can induce protection against urethral infection.[43]

[40] C. A. Nairn, J. A. Cole, P. V. Patel, N. J. Parsons, J. E. Fox, and H. Smith, *J. Gen. Microbiol.* **134**, 3295 (1988).
[41] R. J. Arko, K. H. Wong, J. C. Bullard, and L. C. Logan, *Infect. Immun.* **14**, 1293 (1976).
[42] K. H. Wong, R. J. Arko, W. O. Schalla, and F. J. Steurer, *Infect. Immun.* **23**, 717 (1979).
[43] R. J. Arko, W. P. Duncan, W. J. Brown, W. L. Peacock, and T. Tomozawa, *J. Infect. Dis.* **133**, 441 (1976).

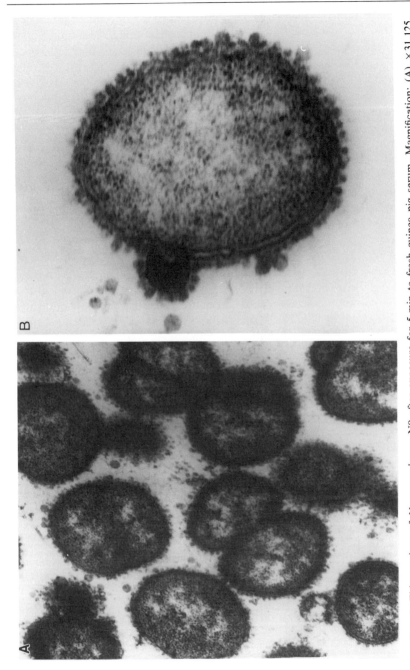

Fig. 1. Thin sections of *N. gonorrhoeae* N9 after exposure for 5 min to fresh guinea pig serum. Magnification: (A) ×31,125, (B) ×85,000.

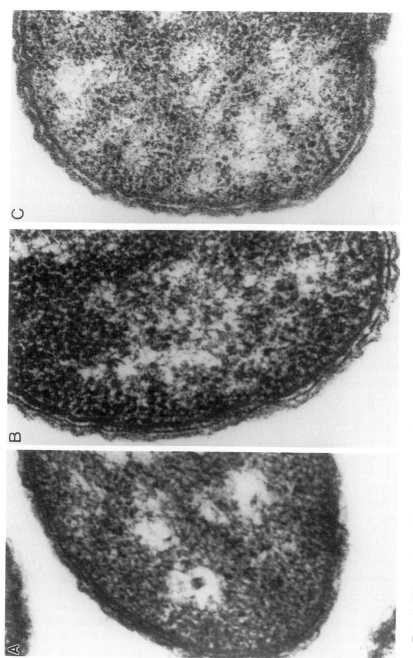

Fig. 2. Thin sections of *N. gonorrhoeae* N9 grown *in vitro* and *in vivo*. (A) *N. gonorrhoeae* N9 cultured directly from mouse chamber fluid; (B) subculture of *N. gonorrhoeae* N9 mouse isolate; (C) *N. gonorrhoeae* mouse avirulent, laboratory-adapted strain of N9. magnification: ×128,000.

The host response induced by infection with *N. gonorrhoeae* in guinea pig chambers is primarily an inflammatory response consisting mainly of PMNs.[41,44] In addition, infection with *N. gonorrhoeae* in chambers induces tissue necrosis, hemorrhaging in chambers, and a perivascular leukocytic response in adjacent tissue. This response is similar to the response found in disseminated gonococcal infection in humans.[45]

Staphylococcus aureus and Staphylococcus epidermidis

A chamber implant model employing rigid polymethacrylate and polytetrafluoroethylene tubes perforated by regularly spaced holes has been used to examine experimental foreign body infections arising from *S. aureus*.[16] As in the gonococcal model, tissue cages can be repeatedly aspirated by percutaneous aspiration. This model allows for the establishment of a persistent and chronic *S. aureus* infection and has been used to evaluate the efficacy of prolonged antibiotic treatment. The combined therapy of fleroxacin with rifampin has been found to be effective in the treatment of methicillin-resistant *S. aureus* with concentrations of antibiotics in cage fluids in the range of those encountered in clinical conditions.[34] In a recent study, *S. aureus* was shown to exhibit a decreased susceptibility to the killing effects of antimicrobials during chronic infections *in vivo*.[6]

Several years ago, we described a rabbit model for toxic shock syndrome using subcutaneous chambers implanted in rabbits.[12] Inoculation of *S. aureus* strains isolated from toxic shock patients in chamber implants resulted in infection characterized by multisystem involvement, as indicated by elevated renal function values, elevated hepatocellular enzymes, and elevated creatine phosphokinase. Periportal inflammation of the liver, erythrophagocytosis in the spleen and lymph nodes, as well as extreme vascular dilation and epithelial lesions similar to those described in patients with toxic shock syndrome were also observed. Both toxic shock syndrome and nontoxic shock syndrome strains produced fever and diarrhea, but toxic shock syndrome strains were significantly more lethal and more likely to produce respiratory distress and lowered blood pressure. This variation in response to different strains was suggested to be related to the many different extracellular products elaborated by staphylococci and to toxic shock toxin 1 (TSS).

A chamber implant model for the growth of *S. aureus* which differs from the model used for growth of *N. gonorrhoeae* has also been described by Day *et al.*[11] The chamber implant in this model is constructed from

[44] D. R. Veale, H. Smith, K. A. Witt, and R. B. Marshall, *J. Med. Microbiol.* **8**, 325 (1975).
[45] F. W. Chandler, S. J. Kraus, and J. C. Watts, *Infect. Immun.* **13**, 909 (1976).

plastic syringe barrels, having both ends sealed with Micropore (Millipore, Bedford, MA) membrane filters. Although one cannot examine bacterial–host cell interactions in this model, it has been used extensively to examine characteristics of *in vivo* grown organisms.[11,44] The exclusion of host cells was perceived as an advantage, in that it decreased potential complications caused by bacteria–host cell interactions, including selective removal of bacterial cells by adherence and phagocytosis.

Growth of *S. aureus* in chamber implants was found to be much slower when compared to growth *in vitro*. *In vivo* grown cultures of *S. aureus* exhibited a prolonged stationary phase, with a final density lower than that achieved *in vitro*. However, higher levels of TSS were produced by *S. aureus* grown *in vivo* as opposed to cells grown *in vitro*. Transmission and scanning electron microscopy of staphylococci did not reveal structural differences between the *in vivo* and *in vitro* grown organisms, with staphylococci growing in characteristic clusters. Following the implantation of chambers containing *S. aureus,* these investigators also observed a rapid influx of PMNs. Both light and scanning electron microscopy revealed PMNs adhering to the outer surface of the Micropore membrane filters.[46]

Polytetrafluoroethylene and titanium chambers containing *S. epidermidis* have been implanted intraperitoneally in rats and have been used to examine the growth of this pathogen *in vivo*.[46,47] Growth *in vivo* resulted in the repression of many cell wall proteins as well as the induction of two iron-repressible cytoplasmic membrane proteins.[47] Immunoblotting experiments revealed that the predominant antibody response to cell envelope proteins was directed against these iron-repressible membrane proteins.

Pseudomonas aeruginosa

Pseudomonas aeruginosa has been grown in implants similar to those described by Day *et al.*[11] In the *P. aeruginosa* model, chambers are made from 1-cm polypropylene syringe barrels sealed at both ends with porous Millipore filters so that host cells are excluded from the bacterial chambers. When examined in this model, mucoid and nonmucoid strains of *P. aeruginosa* isolated from a patient with cystic fibrosis largely retained their phenotypes when grown for up to 1 year *in vivo*.[48] Unwashed *in vivo*

[46] W. J. Pike, A. Cockayne, C. A. Webster, R. C. B. Slack, A. P. Shelton, and J. P. Arbuthnott, *Microb. Pathog.* **10,** 443 (1991).
[47] B. Modun, P. Williams, W. J. Pike, A. Cockayne, J. P. Arbuthnott, R. Finch, and S. P. Denyer, *Infect. Immun.* **60,** 251 (1992).
[48] N. M. Kelly, J. L. Battershil, S. Kuo, J. P. Arbuthnott, and R. E. W. Hancock, *Infect. Immun.* **55,** 2841 (1987).

grown bacteria were also more readily phagocytized when compared to unwashed *in vitro* grown bacteria. This higher degree of phagocytosis was postulated to be due to loosely bound naturally occurring opsonins.

In a later study,[9] the surface characteristics of *P. aeruginosa* grown *in vivo* in the chamber implant was examined. Although the outer membrane protein profile of *in vivo* grown *P. aeruginosa* resembled that of *P. aeruginosa* grown *in vitro*, the lipopolysaccharide (LPS) profile was different. *In vivo* grown cells lacked a series of high molecular weight O-antigen-containing lipopolysaccharide bands, but they did possess a new series of high molecular weight saccharide bands. This alteration in the lipopolysaccharide after growth *in vivo* did not affect the O-antigen serotype or the resistance of the bacteria to serum.

Other chamber models employing agar beads coated with *P. aeruginosa* have demonstrated that there are differences between strains with respect to pulmonary persistence.[49] Strains producing exotoxin A and elastase appear to have increased persistence, which correlates with increased pulmonary pathology. Exotoxin S does not appear to be critical for colonization, but its presence is associated with increased pulmonary damage.

Bacteroides fragilis

Studies with *B. fragilis* have been primarily directed at developing an animal model that simulates the events associated with intraabdominal sepsis in humans. A rat model incorporating gelatin capsules containing cultures of *B. fragilis* has been used to define the virulence of *B. fragilis*. This model was initially designed to avoid the technical problems associated with bacterial contamination of peritoneal fluid samples during evaluation of the host cellular response. As in the model described by Day *et al.*,[11] bacteria cannot leave the chamber. In a study designed to examine the pathogenic potential of encapsulated and unencapsulated *B. fragilis* strains, Onderdonk *et al.*[50] found that implantation of encapsulated *B. fragilis* resulted in abscess formation in the majority of animals. In contrast, unencapsulated strains seldom produced this effect unless they were combined with another organism. Interestingly, implantation of heat-killed, encapsulated *B. fragilis* also resulted in abscess formation. The abscess-potentiating factor in heat-killed preparations was associated with the capsule of *B. fragilis* in that implantation of capsular material alone

[49] M. A. Cash, D. E. Woods, B. McCullough, W. G. Johnson, Jr., and J. A. Bass, *Am. Rev. Respir. Dis.* **119**, 453 (1979).

[50] A. B. Onderdonk, D. L. Kasper, R. L. Cisneros, and J. G. Bartlett, *J. Infect. Dis.* **136**, 82 (1977).

caused abscess formation. These studies thus identified the capsular poly-saccharide as a virulence factor of *B. fragilis*. Studies of the opsonophago-cytic killing of *B. fragilis in vitro* have demonstrated the resistance of encapsulated organisms to this mechanism of killing.[51]

Later studies documented protection against intraabdominal abscess formation in mice by immunization with the capsular polysaccharide of *B. fragilis* before intraperitioneal challenge with viable *B. fragilis*.[52] Immunity to abscess development cannot be passively transferred with hyperimmune antiserum to *B. fragilis* but can be adoptively transferred with CD8$^+$ IJ T cells from actively immunized mice.[53]

The containment chamber has been adapted to use dialysis membranes with an exclusion size of 50 kDa.[10] This modified bacterial containment chamber was used to determine the peritoneal cellular response to *B. fragilis* as well as the fate of bacteria within peritoneal cavities of animal immunized, either actively or through adoptive transfer of cells or cell lysates with the capsular polysaccharide of *B. fragilis*. The dominant cell type produced in response to infection in the peritoneal cavities were neutrophils and macrophages. In immunized animals, a dramatic increase in the lymphocyte population occurred 4 to 6 days following challenge, which coincides with a decrease in viable bacterial cell counts. The factor responsible for killing was not antibody-mediated, as bacteria contained within dialysis sacs were still killed in this model. Killing thus occurred in the absence of cell contact, phagocytosis, or other cell contact-dependent mechanisms. It was concluded that immunization with capsular polysaccharide provides a T cell-dependent immunity supporting a T cell-dependent mechanism of bacterial killing.

Moody *et al.*[19] have also described the use of semipermeable chambers to examine the efficacy of various antimicrobials against *B. fragilis* infection. Results obtained *in vivo* confirmed *in vitro* susceptibility of cefoperazone plus sulbactam, clindamycin, metronidazole, and penicillin G against *B. fragilis*, as well as *B. melaninogenicus, Clostridium perfringens,* and *Peptostreptococcus anaerobius*.

Porphyromonas gingivalis

Tissue cages prepared from Teflon and implanted in guinea pigs have been used to examine the pathogenicity of different *P. gingivalis* strains.[3]

[51] G. L. Simon, M. S. Klempner, D. L. Kasper, and S. L. Gorbach, *J. Infect. Dis.* **145,** 72 (1982).

[52] M. E. Shapiro, A. B. Onderdonk, D. L. Kasper, and R. W. Finberg, *J. Exp. Med.* **154,** 1188 (1982).

[53] A. B. Onderdonk, R. B. Markham, D. F. Zaleznik, R. L. Cisneros, and D. L. Kasper, *J. Clin. Invest.* **69,** 9 (1982).

In this model, *P. gingivalis* virulent strain W83 caused breakdown of tissues, and death of guinea pigs; in contrast, avirulent strains caused only mild pathology. The cellular response against virulent or avirulent strains in the tissue cages was dominated by PMNs. Chamber fluid obtained from mice infected with *P. gingivalis* W83 was found to have a high proteolytic activity, no C3 protein of complement, and only traces of immunoglobulins. Differences in the pathogenicity of *P. gingivalis* strains was attributed to the degradation of C3 by the virulent strain and was suggested to be the crucial event in the perturbation of the host defense. More recent studies have shown that *P. gingivalis* strain W83 fails to accumulate C3 during opsonization by serum owing to its proteolytic capacity.[54,55]

A similar model has been described by Dahlen and Slots,[24] employing tissue cages implanted in the backs of New Zealand White rabbits. In this study, rabbits immunized with *P. gingivalis* whole cells and then challenged with pure cultures of *P. gingivalis* revealed complete elimination or markedly lower postinoculation bacterial counts and considerably weaker tissue reactions than nonimmunized animals. The authors evaluated infectivity according to clinical signs and bacterial counts within cages, but they did not examine cage fluid for local host inflammatory cells.

We have developed a mouse subcutaneous chamber model for the growth of *P. gingivalis* which employs stainless steel chambers implanted in BALB/c mice.[2] When inoculated into chambers, *P. gingivalis* virulent strains caused cachexia, ruffling, general erythema and phlegmonous, ulcerated, necrotic lesions, and death. In contrast *P. gingivalis* avirulent strains, produced localized abscesses at the chamber site, with chamber rejection by day 10. Analysis of chamber fluid from mice infected with avirulent strains revealed inflammatory cell debris, PMNs, and high numbers of dead bacteria. Fluid from mice infected with virulent *P. gingivalis* strains revealed infiltration predominantly of PMNs and live bacteria. Bacteria in chamber fluid samples from *P. gingivalis* invasive strain A7436-infected mice appeared to be primarily extracellular and not associated with PMNs. We have also observed this strain to be highly resistant to phagocytosis in an *in vitro* assay system.[56] The persistence of strain A7436 *in vivo*, its apparent evasion of the PMN *in vivo*, and its resistance to phagocytosis *in vitro* thus appear to be related to its virulence and subsequent pathology observed in mice.

[54] H. A. Schenkein, *J. Periodontal Res.* **24**, 20 (1988).
[55] H. A. Schenkein, *J. Periodontal Res.* **23**, 187 (1988).
[56] C. W. Cutler, J. R. Kalmar, and R. R. Arnold, *Infect. Immun.* **59**, 2097 (1991).

We have examined[21] the effects of immunization with invasive or noninvasive *P. gingivalis* strains on the pathogenesis of infection in the mouse chamber model. We found that immunization with *P. gingivalis* heat-killed cells directly into subcutaneous chambers did not prevent the local survival and colonization of *P. gingivalis* within chambers. Immunization with invasive or noninvasive strains followed by challenge with homologous or heterologous invasive strains protect mice against secondary abscess formation and death; however, *P. gingivalis* persisted in chambers for up to 14 days postchallenge. All strains induced an immunoglobulin G response, and Western blot analysis indicated that sera from mice immunized with different invasive and noninvasive strains recognized common *P. gingivalis* antigens.

Haemophilus influenzae and Haemophilus ducreyi

Trees *et al.*[13] described a mouse subcutaneous chamber model which allows for the long-term growth of *H. ducreyi*. Important *in vivo* alterations in the outer membrane proteins of *H. ducreyi* were noted in this model.[13] The relevance of these alterations to human disease was recently shown when the strain used to produce the mouse infection was accidentally inoculated into the finger of an investigator by a mouse bite. The *H. ducreyi* culture recovered 10 days later from the infected finger demonstrated protein alterations similar to those recovered from mouse chambers.[57] This accident also demonstrated the virulence of a laboratory reference strain that was isolated over 30 years earlier and that had been shown in other animal models to be avirulent.[58]

An agar bead model has been described for the *in vivo* growth of *H. influenzae*.[5] In this model, agar beads are implanted in the lungs of rats, protecting the inoculum from rapid clearance. This infection lasted at least 42 days and induced a specific antibody response to LPS and outer membrane proteins of *H. influenzae*. Despite this encasement in agar beads, pulmonary *H. influenzae* remained susceptible to amoxycillin.

Escherichia coli

Escherichia coli has been grown in close-ended diffusion chambers implanted into the peritoneal cavities of mice and rabbits.[25] The chambers

---

[57] D. L. Trees, R. J. Arko, G. D. Hill, and S. A. Morse, *Med. Microbiol. Lett.* **1,** 330 (1992).
[58] G. W. Hammond, C. J. Lian, J. C. Wilt, and A. R. Ronald, *Antimicrob. Agents Chemother.* **13,** 608 (1978).

were made from syringe barrels, and the ends were covered with 0.22-μm pore filters. In rabbit chambers, *E. coli* K12 strains grew poorly, whereas isogenic strains harboring ColV plasmids and wild-type isolates from extraintestinal infection grew well. Acquisition of either ColVI-K94 (encoding a serum resistance determinant) or ColV-H247 (encoding an iron uptake system) enabled the K12 strains to grow *in vivo*. Differences in sensitivity to the bactericidal action of human or rabbit serum have also been found between organisms grown *in vivo* and *in vitro*. A comparison of the polypeptide composition of bacterial cell envelope preparations on sodium dodecyl sulfate–polyacrylamide gel electrophoresis (SDS–PAGE) revealed differences between *in vivo* and *in vitro* grown *E. coli*, with host immunoglobulins being bound to the surface of *in vivo* grown bacteria. In a more recent study,[4] *in vivo* grown *E. coli* cells were shown to undergo phase variation of type 1 pili. Phase variation has been suggested to be a virulence factor of *E. coli*, and thus the demonstration of this phenomenon *in vivo* is noteworthy.

Bordetella pertussis

In a study by Coleman and Wetterlow,[7] small chambers bounded by 0.22-μm filter membranes implanted into the peritoneal cavity of mice have been used to grow virulent strains of *B. pertussis*. *Bordetella pertussis* were grown *in vivo* in the peritoneal chambers for greater than 100 days. The production of *B. pertussis* toxin was detected *in vivo* as evidenced by the presence of antibody specific for toxin 21–37 days after implantation of chambers. An animal model of lung infection by *B. pertussis* encased in agar beads in rats has also been described.[59] Histopathological examination of infected lungs revealed findings similar to those seen in human disease. A necrotizing inflammation of the tracheobronchial mucous membranes was seen and was characterized by both mononuclear and polymorphonuclear cells, which spread to involvement of the terminal bronchioles and alveoli.

Streptococcus pyogenes

Toxin production by *in vivo* grown *S. pyogenes* has been monitored in chamber fluid obtained from Micropore filter chambers which were

[59] D. E. Woods, R. Franklin, S. J. Cryz, Jr., M. Ganss, M. Peppler, and C. Ewanowich, *Infect. Immun.* **57**, 1018 (1989).

implanted into the peritoneal cavities of mice. Both streptolysin S and streptolysin O were harvested from chamber fluid. In addition, antistreptolysin O antibody was detected 2 weeks after chamber implantation.[8]

Conclusions

Animal models that use subcutaneous chambers allow continual access to the chamber contents throughout the course of infection for microbiological, immunological, and cytological examination. This allows for the longitudinal examination of specific host factors, both cellular and soluble, that are produced locally in response to bacterial challenge during the course of infection, as well as an assessment of specific host cell–bacterium interactions. It should be stressed, however, that the chamber implant models differ in many respects from infection models in that one is not dealing with the natural host or the natural route of infection. However, the potential use of chamber implant models to examine host–parasite interactions and virulence mechanisms is significant.

These models have been an important research tool in our study of the basic pathogenic process associated with bacterial infections. However, the use of these models for the growth of other human pathogens has not been exploited. For example, it should be relatively straightforward to employ these chamber implant models to examine the virulence of and host response to a number of human parasitic pathogens. The lack of suitable small animal models has limited the rapid development and testing of treatment modalities as well as better understanding of the immunologic response to parasites. For example, *Cryptosporidium* infection, a well-recognized cause of diarrhea in immunologically compromised humans, is clearly related to the specific immune defects of the host.[60] The ability to monitor host-mediated responses to this organism clearly would help to elucidate the immune requirements for the control of infection. In addition, like bacterial pathogens, many parasites such as *Borrelia, Trichomonas,* and *Trypanosoma* can undergo surface antigenic changes, making it possible for them to evade the host immune system.[61–63] Using chamber implant models one can easily monitor these antigenic changes *in vivo*. These types of studies would add significantly to our knowledge of the pathogenic mechanisms employed by these parasites.

[60] F. G. Crawford and S. H. Vermund, *Crit. Rev. Microbiol.* **16,** 113 (1988).
[61] K. S. C. Kehl, S. G. Farmer, R. A. Komorowski, and K. K. Knox, *Infect. Immun.* **54,** 899 (1986).
[62] J. F. Alderete, L. Suprun-Brown, L. Kasmala, J. Smith, and M. Spence, *Infect. Immun.* **49,** 463 (1985).
[63] P. J. Myler, A. L. Allen, N. Agabian, and K. Stuart, *Infect. Immun.* **47,** 684 (1985).

Clearly the ability to study host–parasite interactions *in vivo* will help to define disease mechanisms employed by a wide array of human pathogens. The delineation of potential virulence factors may ultimately prove useful for immunological and/or pharmotherapeutic intervention in human infectious disease.

Acknowledgments

The authors acknowledge Tonya Savage for computer support, J. C. Bullard for electron microscopy sections, and Dr. Ramchandra Navalkar for critical review of the manuscript.

[11] Animal Models for Immunoglobulin A Secretion

By DAVID F. KEREN and MARK A. SUCKOW

Introduction

Our laboratory has established experimental methods in both mice and rabbits to elicit a strong secretory immunoglobulin A (IgA) memory response in intestinal secretions against antigens present on or secreted by *Shigella*. The chronically isolated ileal (Thiry–Vella) loop model in rabbits was used to follow the secretory IgA response.[1] We demonstrated previously the existence of a secretory IgA memory response against *Shigella* lipopolysaccharide antigens (LPS) in intestinal secretions following appropriate priming.[2] By using the variety of strains of *Shigella flexneri* produced in the laboratory of Dr. Samuel B. Formal at the Walter Reed Army Institute of Research (Washington, DC), our studies established that secretory IgA memory responses could be elicited by many different types of *Shigella*.[3,4]

Preparation of Chronically Isolated Ileal Loops in Rabbits

The surgical creation of chronically isolated ileal loops in rabbits has been described in detail previously.[1] Briefly, 3 kg New Zealand White rabbits (specific pathogen free) are anesthetized with xylazine and keta-

---

[1] D. F. Keren, H. L. Elliott, G. D. Brown, and J. H. Yardley, *Gastroenterology* **68**, 83 (1975).

[2] D. F. Keren, S. E. Kern, D. Bauer, P. J. Scott, and P. Porter, *J. Immunol.* **128**, 475 (1982).

[3] D. F. Keren, R. A. McDonald, P. J. Scott, A. M. Rosner, and E. Strubel, *Infect. Immun.* **47**, 123 (1985).

[4] D. F. Keren, R. A. McDonald, and S. B. Formal, *Infect. Immun.* **54**, 920 (1986).

mine. A midline abdominal incision is made, and the terminal ileum is identified. Approximately 20 cm of ileum containing a Peyer's patch is isolated with its vascular supply intact. Silastic tubing (Dow-Corning, Midland, MI) is sewn into each end of the isolated segment. The free ends of the tubing are brought out through the midline incision and are tunneled subcutaneously to the nape of the neck where they are exteriorized and secured. Intestinal continuity is restored by an end-to-end anastomosis. The midline incision is closed in two layers. Each day about 2–4 ml of secretions and mucus that collect in the ileal loops are expelled by injecting 20 ml of air into one of the silastic tubes. Mucus is separated from the clear supernatant by centrifugation. The supernatant is stored at −20° and assayed for specific immunoglobulin content. Care of the isolated loops requires a subsequent flush with 20 ml of sterile saline to remove adherent mucus. This saline is removed by repeated gentle flushes of air. With proper daily care, more than 90% of our rabbits have completed experiments lasting for 2 months.

The early studies with the Thiry–Vella loop model used the hybrid strain of *S. flexneri* and *Escherichia coli* (*Shigella* X16) to establish that an IgA response could be elicited to nonpathogenic antigens and to determine the different methods of immunization required to stimulate the mucosal versus the systemic immune response. The *Shigella* X16 strain invades the surface epithelium but does not reproduce once it is within host tissues; therefore, no ulceration is produced. By administering the bacterium directly into isolated intestinal loops, a strong local IgA response was produced.[5] Similar responses were found with the invasive strain M4243 (which does cause ulceration) and with a noninvasive strain, 2457-0, when applied directly into the isolated ileal loops. The presence of a Peyer's patch locally within the isolated loop and the dosage schedule were other important factors influencing development of the mucosal immune response.[5] These early studies proved that when Thiry–Vella loops were stimulated directly with various *Shigella* preparations, secretions collected from those loops would contain considerable antigen-specific secretory IgA but little or no IgG directed against *Shigella*. This was not due to rapid degradation of IgG (which is normally destroyed quickly in intact intestine) since the isolated ileal loops were separated from the proteolytic effects of gastric acid, bile, and digestive enzymes including trypsin, pepsin, and chymotrypsin. Direct stimulation of the isolated loops by *Shigella* antigens resulted in little or no systemic IgG against *Shigella* unless the systemic immune response had been previously primed by a parenteral dose of *Shigella*. To elicit IgG in the serum, it was necessary

---

[5] J. S. Wassef, D. F. Keren, and J. L. Mailloux, *Infect. Immun.* **57,** 858 (1989).

to give the bacteria parenterally; that route resulted in virtually no secretory IgA anti-*Shigella* LPS.

To evaluate the secretory IgA memory response, we immunized rabbits orally with three doses of live or killed *Shigella flexneri* before the Thiry–Vella loops were created. This more natural route of immunization was used to approximate the situation in vaccinating humans. After the animals rested for 2 months, a chronically isolated ileal loop was created. An oral challenge dose of live *Shigella* was given, and secretions from the chronically isolated ileal loops were assayed for the secretory IgA anti-*Shigella* LPS response using an enzyme-linked immunosorbent assay (ELISA). The use of the Thiry–Vella loops as a probe was based on background information about lymphocyte trafficking after intraluminal antigen stimulation for a secretory IgA response.[6,7] The initial step in stimulation of the secretory IgA response involves phagocytosis of antigens by specialized surface epithelial cells, namely, M cells, which are present within the epithelium overlying lymphoid structures throughout the gastrointestinal tract. Through the M cells, antigenic material, microorganisms, and soluble proteins are brought into the underlying gut-associated lymphoid tissues (GALT). Once within GALT, the antigens come into contact with precursor B lymphoblasts which are genetically predisposed to develop secretory IgA response.[8] The presence of M cells overlying isolated follicles and Peyer's patches in the small intestine and colon led us to study the role of M cells in the pathogenesis of the focal and patchy ulcerations of the colon and small bowel which are characteristic of dysentery.

The B lymphocytes that are in GALT come under the influence of specific regulatory cells, originally described as "switch T cells" by Kawanishi *et al.*[9] The latter cells encourage B lymphocytes to alter the phenotype of their surface heavy chain from μ chain to α chain. In addition, there are other helper T cells in GALT which help B lymphocytes mature to IgA-secreting plasma cells.[10-12] Although the precise mechanism of this

[6] D. F. Keren, R. A. McDonald, J. S. Wassef, L. R. Armstrong, and J. E. Brown, *Curr. Top. Microbiol. Immunol.* **146,** 213 (1989).

[7] D. F. Keren, J. E. Brown, R. A. McDonald, and J. S. Wassef, *Infect. Immun.* **57,** 1885 (1989).

[8] J. S. Wassef and D. F. Keren, *Adv. Res. Cholera Relat. Dis.* **7,** 199 (1990).

[9] H. Kawanishi, L. E. Slatzman, and W. Strober, *J. Exp. Med.* **157,** 433 (1983).

[10] D. Campbell and B. M. Vose, *Immunology* **56,** 81 (1985).

[11] H. Kiyono, L. Mosteller-Barnum, A. Pitts, S. Williamson, S. Michalek, and J. McGhee, *J. Exp. Med.* **161,** 731 (1984).

[12] J. Viney, T. T. MacDonald, and J. Spencer, *Gut* **31,** 841 (1990).

switch is unclear, it is known that interleukins 4 and 5 play a role in the process.[13,14] Following their stimulation in GALT, the antigen-specific B-lymphoblasts migrate in turn to the mesenteric lymph nodes, the thoracic duct, and eventually lodge in the spleen where they undergo some degree of maturation.[15,16] The final site of lodging by the B lymphocytes may be somewhat influenced by the location of initial antigen stimulation.[17] In other words, when antigens are applied directly to the respiratory tract, the antigen-specific B-lymphocytes are more likely to return to that site than to the gastrointestinal tract or to mammary secretions. Conversely, when antigen is applied to sites in the gastrointestinal tract, antigen-specific B lymphocytes are more likely to return to the gastrointestinal tract than to other mucosal surfaces.[18,19] In our studies of the mucosal memory response, we took advantage of this lymphocyte trafficking to use the chronically isolated ileal loops as a probe for the mucosal memory response. We found that live, noninvasive *Shigella* were as effective as virulent invasive strains of *Shigella* at eliciting a secretory IgA memory response against *Shigella* LPS.[2-4] These strains did not produce any pathologic lesions in the rabbit intestine.

Mouse Lavage Model for Mucosal Immunity

The mouse model system we use is an adaptation of one used by Elson for studies of the secretory IgA response to cholera toxin.[20] For these studies, specific pathogen-free mice are given four oral doses of lavage solution 15 min apart. Thirty minutes after the final lavage dose, a single intraperitoneal dose of 0.1 mg pilocarpine is given. Pilocarpine encourages the mouse intestine to secrete large volumes of fluid. The mice are placed on wire mesh over beakers containing 3 ml of protease inhibitor solution (soybean trypsin inhibitor in 50 mM EDTA). Intestinal fluid and feces, and saliva, are collected over 30 min. The fluid collected is brought up

[13] D. A. Lebman and R. L. Coffman, *J. Immunol.* **141**, 2050 (1988).
[14] K. W. Beagley, J. H. Eldridge, H. Kiyono, M. P. Everson, W. J. Doopman, T. Honjo, and J. R. McGhee, *J. Immunol.* **141**, 2035 (1988).
[15] M. McWilliams, J. M. Phillips-Quagliata, and M. E. Lamm, *J. Exp. Med.* **145**, 866 (1977).
[16] J. Tseng, *J. Immunol.* **127**, 2039 (1981).
[17] N. F. Pierce, *J. Exp. Med.* **148**, 195 (1978).
[18] R. M. Goldblum, *J. Clin. Immunol.* **10**, 64S (1990).
[19] J. H. Yardley, D. F. Keren, S. R. Hamilton, and G. D. Brown, *Infect. Immun.* **19**, 589 (1978).
[20] C. Elson, W. Ealding, and J. Lefkowitz, *J. Immunol. Methods* **67**, 101 (1984).

to 5 ml volume with phosphate-buffered saline (PBS) and vortexed. After centrifuging the fluid at 650 g for 10 min, 10 μl of 100 mM phenylmethylsulfonyl fluoride (PMSF) in 95% ethanol is added per milliliter of supernatant. The same is then centrifuged 27,000 g at 4° for 20 min, and the supernatant is saved. Again 10 μl of PMSF in 95% ethanol and 10 μl of 1% NaN$_3$ are added along with 50 μl of fetal calf serum to each milliliter of the final supernatant. Samples are stored at $-20°$ until time of assay. Blood samples are taken from the animals by retroorbital bleeding.

Mouse Model for Mucosal Immunity to *Shigella* Antigens

Although the chronically isolated ileal loop model in rabbits has allowed us to characterize many of the important variables involved in stimulating the mucosal memory response to *S. flexneri*, the outbred nature of the rabbit has limited the depth to which we could study the observed reactions. We are limited in terms of available reagents, genetic details of immune responsiveness, and the heterogeneity of the responses elicited. Furthermore, the lack of histocompatability prevents cross-culture experiments with our available *in vitro* systems. Therefore, we have established a mouse model for examining the secretory IgA response to *Shigella* antigens. We chose Shiga toxin (ST) to study, as little is known about the mucosal immune response to this molecule and our studies in rabbits indicate that it may be a strong mucosal immunogen.

We have examined the role of cholera toxin (CT) and ST to serve both as immunogens for the mucosal immune response and as adjuvants to stimulate a secretory IgA response against relatively poor mucosal immunogens. We sought to determine the dose–response curve using purified ST preparations. For these studies, we have used the mouse lavage model system. Specific pathogen-free C57Bl6/J mice (Jackson Laboratories, Bar Harbor, ME) are given intragastric doses ranging from 0.01 to 50 mg of purified ST (provided by Drs. A. Donohue-Rolfe and G. T. Keusch) on days 0, 7, 14, and 21. Secretions are collected as described above.

Samples are assayed for specific antibody content using the ELISA. For this, microtiter wells are coated with a solution containing purified ST preparation (provided by J. Edward Brown). The kinetics of the enzyme–substrate reaction are extrapolated to 100 min. For the Shiga toxin assays, the ELISA Amplification System (Bethesda Research Laboratories, Gaithersburg, MD) is used, with results being measured as $OD_{490\ nm}/15$ min. Specific IgG and IgA standards are processed on each plate with the unknown fluids as previously described.[7]

As shown in Fig. 1, all of the doses of ST were immunogenic for an IgA anti-ST response in the secretions obtained. A classic dose–response curve was achieved. Although we gave doses of 50 mg to a group of mice, 4 of them died soon after the immunization, and, therefore, these data are not included. The animals receiving 25 μg all survived; however, for the studies on adjuvanticity of ST, the 10-μg dose was used to minimize the potential for cytotoxicity of the immunizing dose (see below). The 10-μg dose consistently gave a vigorous IgA anti-ST response in the secretions. By the second week after immunization all of the mice had achieved a detectable response. The response continued to increase and peaked after the third dose. No increase was seen after the fourth dose. The response remained at the peak level with no significant decline through the end of the study at 5 weeks. A modest decline was seen with the 25-μg group; however, this was not significant at 5 weeks compared to the 3 week (peak) value.

Similar dose–response curves were seen with the IgG and IgM in secretions. The IgG response was as strong or stronger than the secretory IgA response as measured by the ELISA (Fig. 2). This is different from the responses described by us in the rabbit loop model using crude toxin

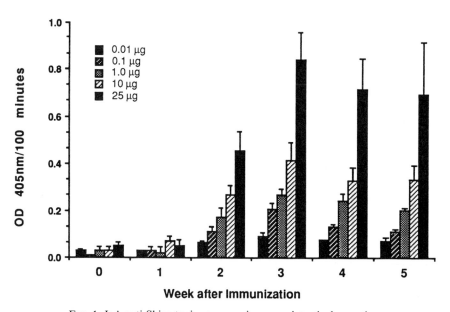

FIG. 1. IgA anti-Shiga toxin response in mouse intestinal secretions.

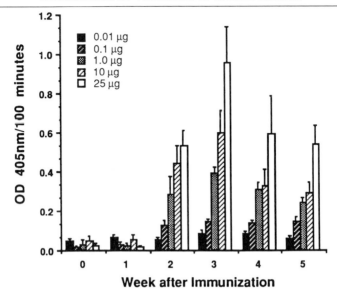

FIG. 2. IgG anti-ST in mouse intestinal secretions.

in previous studies, where relatively trivial amounts of IgG anti-ST were seen in the loop secretions. The difference reflects consistent differences in the mucosal immune systems of mice and rabbits. With virtually all immunogens we have used in the rabbit loop model system [CT, ST, *Shigella* LPS, *Salmonella* LPS, 2-acetylaminofluorene (AAF), keyhole limpet hemocyanin (KLH), and bovine serum albumin], relatively little antigen-specific IgG has been detected in secretions despite the presence of high-titered antigen-specific IgA. With the mouse model, relatively large amounts of both IgA and IgG have been detected.

Because purified ST is an effective mucosal immunogen which gives a classic dose–response curve, we wished to examine whether it was able to provide adjuvant activity like CT, and whether it might even act synergistically with CT to heighten the secretory IgA response to coadministered weak mucosal immunogens. For these studies, we chose to use the standard weak mucosal immunogen KLH. When KLH is given orally, or intraintestinally, only trivial amounts of secretory IgA can be detected against KLH, even after multiple immunizations. In our studies, we give mice 5 mg of KLH orally alone or with various combinations of CT and ST (each used at 10 μg). Our studies indicate that KLH alone produces only a weak IgA anti-KLH response as expected. When CT is added, a strong local IgA and IgG anti-KLH response consistently results.

However, ST is totally ineffective at achieving a secretory IgA anti-KLH response. Indeed, the available information suggests that a slight inhibition of the IgA anti-KLH response may occur (Fig. 3). Whereas ST is one of the strongest mucosal immunogens described thus far, weaker only than CT, ST does not act as a mucosal adjuvant and does not further potentiate the adjuvanticity of CT using the standard KLH system.

Animal Model for Uptake of *Shigella* by Follicle-Associated Epithelium and Villi

To determine the relationship between the virulence of the microorganism and the uptake of the bacteria by the follicle-associated epithelium, an *in vivo* assay procedure was employed. Isolated acute ileal loops 10 cm in length are created in specific pathogen-free New Zealand White rabbits. A single dose containing 2×10^9 *Shigella flexneri* is injected into the acute loop. At 90 min and at 18 hr, the loops are removed and samples are fixed for histological investigation by electron microscopy and light microscopy.

For light microscopy, the sections are fixed in absolute ethanol and stained with Giemsa. For each time period, at least 10 sections of Peyer's

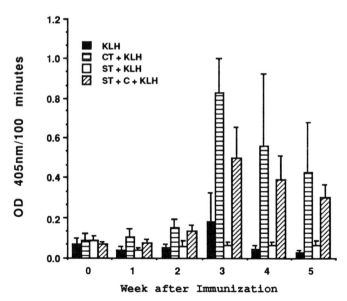

FIG. 3. IgA anti-KLH response in mouse intestinal secretions.

patch and adjacent villi are examined for attachment and uptake of the *Shigella*. Histologically, these sections are divided into two areas: (1) the follicle-associated epithelium overlying the dome areas in Peyer's patches (known to be enriched in M cells) and (2) villi which are outside of the Peyer's patch area. Evaluation is performed using oil immersion light microscopy. Because the normal flora of the rabbit ileum contains less than 10^4 microorganisms, for statistical purposes, less than 0.01% of the flora visualized are from other microorganisms. Further, the *Shigella* have a characteristic size and shape which under the circumstances of the study are readily recognizable. The Bioquant Biometrics Image Analyzer (Nashville, TN) with an IBM computer is used to measure the actual length in millimeters of the lining epithelium over the villi and over the dome regions of the Peyer's patches. The average of 100 fields for dome and villus areas from representative rabbits is calculated. This allows us to express data as bacteria per square millimeter of surface epithelium. Therefore, a direct relationship of villus surface area to follicle-associated epithelium surface area is established, allowing for comparisons. Electron microscopy is performed on some sections, demonstrating the characteristic rod-shaped structure and typical "M" cell location.

Initial Antigen Processing of Different *Shigella* Preparations by Intestine

For studies of initial antigen processing segments of intestine are isolated as outlined above. Each group contains five rabbits (Table I) to ensure reproducibility of the model. The bacteria listed in Table I are injected directly into each loop at time zero and at 30 min, 90 min, and

TABLE I
STRAINS OF *Shigella flexneri* USED FOR M CELL UPTAKE STUDIES IN RABBIT LOOPS

Rabbit group	*S. flexneri* strain	Plasmid[a]	Ulcer[b]	M cell[c]	Sereny[d]
1	M4243	+	+	+	+
2	X16	+	−	+	−
3	2457-O	+	−	+	−
4	M4243A1	−	−	+	−
5	Heat-killed M4243	+	−	+	−

[a] Presence of 140 MDa virulence plasmid.
[b] Mucosal ulceration produced in acute loops.
[c] Uptake by M cells demonstrated ultrastructurally.
[d] Virulence demonstrated by Sereny test in guinea pigs.

18 hr; acute loops are removed and examined by both electron and light microscopy.

For light microscopy, frozen sections are cut at 4 μm and stained with Giemsa. Ten sections are cut from each loop, for each time period. Therefore, 50 slides for each strain of bacteria, for each time period, are examined. All slides are coded so that the observer does not know the time point or the particular animal or strain of *Shigella* being examined. The number of *Shigella* within both the follicle-associated epithelium overlying Peyer's patches and the villus epithelium are counted.

To determine the number of bacteria per unit area, the specimens are standardized using a Bioquant image analyzer. This allows us to express our results as number of bacteria taken up per micron of epithelium. It also allows us to correlate the uptake over isolated follicles with that over surface epithelium. This work has indicated that heat-killed *Shigella* are taken up to a similar extent as nonpathogenic strains of *Shigella*. On average, 1.6 bacteria/mm^2 of surface epithelium are found in the follicle-associated epithelium, whereas 10-fold fewer bacteria are present in the adjacent villus epithelium.[5] Therefore, the follicle-associated epithelium preferentially ingests both live and killed *Shigella*. No evidence of ulceration is seen in any of the acute loops given the heat-killed *Shigella*.[5]

The presence of the bacteria in M cells is confirmed by examining electron microscopic fields of both villus and follicle-associated epithelium. Because light microscopy had shown that the major uptake of *Shigella* had occurred by 90 min, samples from this time period are examined. Denatured (heat-killed) *Shigella* are found within phagocytic vesicles of M cells in the follicle-associated epithelium.[5] The tissue damage at the 18-hr time period is correlated with the degree of uptake by the M cells in the follicle-associated epithelium.[5] All of the rabbits given pathogenic *S. flexneri* M4243 had ulcerations of isolated follicles and Peyer's patch follicles at 18 hr. In marked contrast, none of the animals given nonpathogenic or heat-killed *Shigella* had evidence of ulceration or even microulceration at this time.

In the acute loops given pathogenic *S. flexneri* M4243, evidence of damage to the villus epithelium is observed, but no ulcerations are seen. There is mucus depletion and occasional focal acute inflammation. These findings indicate that *S. flexneri* is initially engulfed by the M cells, wherein they multiply and eventually result in focal ulceration. Both nonpathogenic strains and heat-killed strains are processed by the M cells, which may exlain the ability of the nonpathogenic straims to elicit a mucosal immune response against the *S. flexneri* LPS antigens. It is not clear why the heat-killed strain, which is taken up in similar numbers as the live nonpathogenic strains, has proved to be incapable of priming animals for a mucosal memory response in our previous studies.[2,3] It may be that the heat treatment damages antigens necessary for the appropriate stimulation of mucosal immunity.

Animal Model for Protection by Secretory Immunoglobulin A

New Zealand White rabbits (specific pathogen free) are anesthetized with xylazine and ketamine. A midline abdominal incision is made and the terminal ileum is identified. A series of 5-cm segments from the mid-jejunum to the mid-ileum are created. Double 4.0 silk ligatures are placed between each segment to prevent leakage from one segment to another. Solutions to be tested for toxin or antitoxin activity are injected into the loops in the doses indicated. The midline incision is closed in two layers, and the animals are allowed to rest for 18 hr. At time of sacrifice, the fluid contents of the loops are measured.

Cytotoxicity Assay

Shiga toxin activity is determined by examining the extent of HeLa cell damage by a previously described assay.[21] Briefly, HeLa cell monolayers are grown in 96-well microtiter plates. For the assay, a standard crude toxin lysate of *S. dysenteriae* is incubated with serial dilutions of loop fluids for 30 min at room temperature. The mixture is placed onto the HeLa cell monolayer and allowed to incubate overnight at room temperature. The monolayers are then stained with crystal violet dissolved in 50% ethanol–1% sodium dodecyl sulfate, and the $OD_{620 nm}$ is determined for each well. The dye remaining in each well correlates with the percentage of cells remaining adherent to the microtiter dishes. The $OD_{620 nm}$ readings of wells containing the standard toxin alone are averaged, and that value plus two standard deviations is defined as the end point titer of loop fluids for neutralization of the cytotoxicity of the toxin preparation. All dilutions of loop fluid which give an OD_{620} in the assay greater than this value are scored as positive.

Role of Secretory Immunoglobulin A to Protect against Cytotoxic Effects of Shiga Toxin

Our studies on the mucosal immune response to Shiga toxin and its functional significance were carried out in collaboration with Dr. J. Ed-

[21] M. K. Gentry and J. M. Dalrymple, *J. Clin. Microbiol.* **12**, 361 (1980).

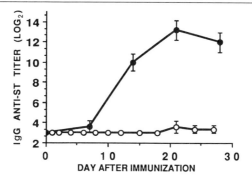

FIG. 4. IgG anti-Shiga toxin in rabbit (○) intestinal secretions and (●) sera of rabbits immunized orally with crude Shiga toxin.

ward Brown. His laboratory provided the Shiga toxin preparations and performed the below listed HeLa cell assay.[22] All the rabbit studies, immunizations, and protection studies were performed in our laboratory. Although the role of Shiga toxin in dysentery is unknown, it is cytotoxic to HeLa cells, causes fluid secretion in rabbit intestine, and is lethal when injected parenterally to rabbits or mice.[21-23]

For the present study, five rabbits are inoculated directly into chronically isolated ileal loops (see above) on the day of surgery (day 0) and on days 7 and 14 postsurgery. They are given 0.5 ml of crude Shiga toxin preparation (see above) in 4 ml of saline. Intestinal secretions are collected daily, and blood samples are collected weekly. A new ELISA for shiga toxin is created for these studies. Although the technical details of the assay are the same as detailed above, a four-point standard curve is assayed on each plate with the unknown samples. The reciprocal of the dilution giving an optical density reading between the two lowest values on the standard curve is defined as the titer.

As shown in Fig. 4, a significant increase in the mean IgG anti-Shiga toxin titer over the day 0 value is detectable in serum by day 7 after the first intraloop immunization. This titer rises after the third dose on day 14 and does not change significantly through the end of the study period on day 30. In contrast to the high titer of IgG anti-Shiga toxin in the serum, only trivial amounts are detected in the loop secretions (Fig. 4). Thus only a small amount of serum IgG anti-Shiga toxin finds its way into the loop secretions (our previous studies have shown good stability of IgG in the chronically isolated ileal loops).

[22] J. E. Brown, D. E. Griffin, S. W. Rothman, and B. P. Doctor, *Infect. Immun.* **36,** 996 (1982).
[23] G. T. Keusch, A. Donohue-Rolfe, and M. Jacewicz, *Ciba Found. Symp.* **112,** 193 (1985).

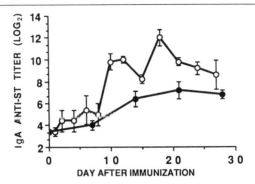

FIG. 5. IgA anti-Shiga toxin in (○) intestinal secretions and (●) sera of rabbits immunized orally with crude Shiga toxin.

The IgA anti-Shiga toxin titer in the serum of the rabbits is lower than the IgA titer in secretions (Fig. 5). A significant ($p < .01$) increase in the IgA anti-Shiga toxin titer of the serum over the day 0 values is seen by day 14. In the loop secretions, as early as day 2, a weak but significant ($p < .05$) increase in the IgA anti-Shiga toxin titer is seen (Fig. 5). The content of IgA anti-Shiga toxin declines on the day after the third intraloop dose (day 14) but has another striking increase 3 days later. After this peak on day 18, the mean IgA anti-Shiga toxin titer slowly declines, although it never drops below the level of activity seen after the second intraloop dose on day 7. It is possible that the slight decline in IgA titer seen the day following each booster immunization reflects the presence of free toxin in the loop which binds to the specific IgA. Alternatively, Shiga toxin may interfere with local antibody synthesis or secretion of IgA into the gut lumen.

To assess the *in vitro* Shiga toxin-neutralizing activity of intestinal loop secretions, a HeLa cell assay is performed. For this assay, HeLa cell monolayers are grown in 96-well microtiter plates and a standard crude toxin lysate of *S. dysenteriae* is incubated with serial dilutions of loop fluids for 30 min at room temperature. This mixture is placed onto the HeLa cells monolayer and allowed to incubate overnight at room temperature. The monolayers are then stained with crystal violet and the $OD_{620\,nm}$ is determined for each well. The dye remaining in each well correlates with the percentage of cells remaining adherent to the microtiter dishes.[21] The $OD_{620\,nm}$ readings of wells containing the standard toxin alone are averaged, and that value plus two standard deviations is defined as the end point titer of loop fluids for neutralization of the cytotoxicity of the toxin preparation. All loop fluids which give an $OD_{620\,nm}$ in the assay greater than this value are scored as positive.

The mean Shiga toxin neutralizing activity in the HeLa cell assay is depicted in Fig. 6. The curve in Fig. 6 shows the same basic triphasic response as the IgA anti-Shiga toxin in loop secretions from Fig. 5. The correlation coefficient of the mean IgA activity in secretions with the mean toxin neutralization titer is .928, whereas the correlation of the IgG level in secretions with the mean toxin neutralization titer is only 0.116.

To assess the *in vivo* Shiga toxin-neutralizing activity of intestinal loop secretions, an acute loop protection model is devised. Pooled loop secretions from rabbits with high-titer IgA anti-Shiga toxin activity as determined by ELISA are diluted $1:2$ in saline and mixed with an equal volume of a $1:256$ dilution of crude toxin. This is injected into 5-cm isolated segments of ileum in unimmunized rabbits. This dose of toxin is chosen as it consistently elicits fluid accumulation when given to acutely ligated loops. As controls, toxin is mixed with secretions from nonimmune animals or saline and injected into other loops in the same rabbit. After 18 hr, the animals are sacrificed, and the volume of fluid in each segment is measured.

Pooled loop secretions from animals immunized with Shiga toxin reduce toxin-induced fluid accumulation in the acutely ligated rabbit intestine. Secretions with no detectable IgA or IgG anti-Shiga toxin by ELISA have no inhibitory effect on the Shiga toxin-induced fluid production by rabbit intestine (Fig. 7). The heterogeneity shown by the standard errors of the means reflects the differential response of the genetically diverse outbred rabbits used in these studies. Even with this degree of heterogeneity, the difference between the fluid production in loops protected with

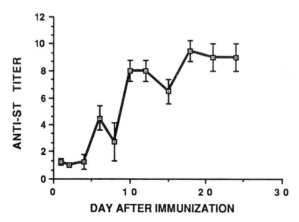

FIG. 6. Anti-Shiga toxin activity *in vitro* (HeLa cell antitoxin assay) of loop secretions from animals given oral immunization with crude Shiga toxin.

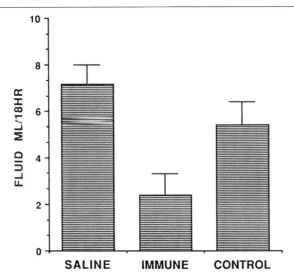

FIG. 7. Anti-Shiga toxin activity *in vivo* of loop secretions from rabbits given oral immunization with crude Shiga toxin.

immune secretions and those given nonimmune secretions is highly significant ($p < .01$).

Conclusions

These studies demonstrate several animal model systems to study the secretory IgA response. Several features are important if major efforts are to be made to use mucosal vaccines to protect humans effectively against enteric diseases. A major route of uptake of the intact *Shigella* is by the M cells which lie in the follicle-associated epithelium overlying lymphoid follicles throughout the gastrointestinal tract. When the *S. flexneri* taken up are virulent, they proliferate within the M cells, eventually producing ulceration at these sites. The presence of many such follicles in the colon and the terminal ileum may explain the frequency of focal ulcerations at these sites in clinical dysentery. Ideally, one would wish to interfere with the initial adherence of the *Shigella* to the surface epithelium to prevent clinical disease.

The mouse model system for evaluating the mucosal immune response to *Shigella* antigens has been expanded to study the role of Shiga toxin. The data from these studies indicate that a strong secretory IgA response to Shiga toxin is elicited and that the level of the response seems to be dose dependent. We strongly encourage the use of the mouse lavage system for future studies because many monoclonal reagents and inbred strains are now available, which allows us to dissect more precisely the cellular basis of the secretory immune response to shigella antigens. This model would be advantageous in trying to determine the subtypes of T cells involved in the mucosal memory process. In the rabbit Thiry–Vella loop studies discussed above, this type of study was not possible.

The inbred mice provide both advantages and disadvantages for future studies. The major advantages relate to the ready availability of biological reagents to specific cell subsets. One disadvantage, which should be remembered, is that the inbred population will not give a good overall indication of the breadth of the response of outbred human populations to a vaccine preparation. For this type of information, subsequent investigations using outbred animals (such as the rabbit Thiry–Vella loop model) are recommended.

Section II

Species and Strain Identification

[12] Serogroup and Serotype Classification of Bacterial Pathogens

By Carl E. Frasch

Introduction

This chapter describes the serological approaches and methods used to characterize strains of several bacterial species of medical importance. These approaches distinguish different strains or clones for the purpose of epidemiological tracing or research into virulence factors. The present discussion is concerned specifically with the serological identification of antigenically distinct types or groups within the following species: *Haemophilus influenzae, Neisseria meningitidis, Pseudomonas aeruginosa, Streptococcus pneumoniae,* and group A and group B β-hemolytic streptococci. It is assumed that the bacterial strains are isolated and in pure culture. Direct detection of bacterial antigens in patient materials is not discussed.

Invasive bacterial pathogens have evolved antigenic diversity in their surface antigens as a means of evading host immune clearance. These organisms have chemically and antigenically variable surface structures in the form of capsular polysaccharides, as in the case of *Escherichia coli, H. influenzae, N. meningitidis,* and *S. pneumoniae,* or in the form of lipopolysaccharides (LPS) as in the case of *E. coli, Salmonella, Shigella, P. aeruginosa, N. meningitidis,* and *Neisseria gonorrhoeae.* A serotyping system should be based on surface antigens that are stably expressed in all strains, yet show clear antigenic differences between strains. Some bacterial pathogens have antigenically variable surface protein antigens identified by serologic techniques. The group A streptococcus is an example where antibodies to the antiphagocytic M protein can confer protective immunity.

Antigens are complex structures often displaying multiple epitopes. In the case of proteins these epitopes are generally unique. In contrast, epitopes found on LPS and other polysaccharides often represent di- or trisaccharides, which are repeated many times in a single molecule, and can be shared among a number of unrelated bacterial species. Serological cross-reactions must therefore be considered in the evaluation of typing results.

Why identify or differentiate below the species level? First, not all strains of a given species have the same potential to cause disease. Al-

METHODS IN ENZYMOLOGY, VOL. 235

though some strains, types, or clones of a species are generally harmless members of the normal flora, other members of the same species are strongly disease associated. For example, there are six capsular types of *H. influenzae*, yet type b causes 95% of all invasive disease; *N. meningitidis* is divided into 12 different serogroups, yet serogroups, A, B, and C cause 90% of meningococcal disease. For some, the nonpathogenic members may have surface polysaccharides that cross-react with those of commensal bacteria, such as *Streptococcus viridans* or *Lactobacillus* spp., which induce cross-protective antibodies. However, in general the reasons as to why only some members of a species are virulent remains unclear.

Second, classification of strains within a species allows identification and tracing of outbreaks. When the normal endemic level of a bacterial disease suddenly increases, it is often due to entry into the population of a more virulent or antigenically variant clone. The isolates must be identified below the species level to identify such a situation, or the appearance of an antibiotic-resistant strain. Thus, it becomes important to identify that a new variant clone is in the population to allow epidemiological tracing or to alert other communities.

Although not considered in this chapter, different strains of some gram-negative species may also be identified by differences in the molecular sizes of their outer membrane proteins on sodium dodecyl sulfate–polyacrylamide gel electrophoresis (SDS-PAGE), as, for example, nontypable *H. influenzae* strains[1] and *N. gonorrhoeae*.[2] The method of enzyme electromorph typing (ET typing), where the different alloenzymes differing in electrophoretic mobility of 10 to 15 critical function enzymes are used to identify ET types, is used to identify clones and genetic relatedness. The ET typing approach has been applied to many different bacterial species.[3]

Most serologic methods rely on antigen–antibody reactions, such as agglutination or precipitation. Antibodies to surface antigens have been used to classify strains of various species into serogroups or serotypes (Table I). The classification methods illustrated for the species discussed below are directly applicable to other species. Classification of the Enterobacteriaceae is very complex and has been reviewed by the Orskovs.[4]

A point that should be considered in cultivation of strains to be typed is possible phenotypic heterogeneity within a single isolate, associated with some surface proteins and LPS. To prevent selection of unknown

---

[1] L. van Alpen and H. A. Bijlmer, *Pediatrics Suppl.*, 636 (1990).
[2] R. C. Noble and S. C. Schell, *Infect. Immun.* **19**, 178 (1978).
[3] R. K. Selander, J. M. Musser, D. A. Caugant, M. N. Gilmore, and T. S. Whittam, *Microbiol. Pathog.* **3**, 1 (1987).
[4] F. Orskov and I. Orskov, *Methods Microbiol.* **10**, 1 (1978).

TABLE I

BACTERIAL SPECIES OF MEDICAL IMPORTANCE THAT HAVE BEEN TYPED FOR PURPOSES OF EPIDEMIOLOGICAL TRACING OR STUDIES ON DISEASE ASSOCIATIONS

Species	Identification	Biochemical basis for type	Number of types	Type designation
Escherichia coli	O type	Lipopolysaccharide	>100	O1, O2
	H type	Flagella	>50	H1, H2
Haemophilus influenzae	Type	Polysaccharide	6	a, b, c
Klebsiella pneumoniae	O type	Lipopolysaccharide	>100	O11, O32
Neisseria meningitidis	Group	Polysaccharide	13	
	Serotype	Outer membrane porin proteins	>20	A, B, C 1, 2, 4
Pseudomonas aeruginosa	Type	Lipopolysaccharide	17	O:1, O:8
Staphylococcus aureus	Type	Polysaccharide	8	1, 2, 5
Streptococcus agalactiae	Type	Polysaccharide	6	Ia, Ib, II, III
Streptococcus pneumoniae	Type	Polysaccharide	83	2, 6F, 19A
Streptococcus pyogenes	Group	Carbohydrate	20	A, B, C, D
	M type	Protein	>70	6, 12, 49
Vibrio cholerae	O type	Lipopolysaccharide	3	A, B, C

variants, 10 or more colonies should be picked for subculture to maintain the dominant phenotype.

Haemophilus influenzae

Haemophilus influenzae is the causative agent of a spectrum of human diseases, the most serious of which are meningitis and epiglottitis. There are six serologic types: a, b, c, d, e, and f. The six types, first described in 1931 by Pittman,[5] have structurally and immunologically distinct capsular polysaccharides.[6] Most invasive *H. influenzae* disease is caused by type b, and antibodies against the type b polysaccharide are protective. *Haemophilus influenzae* type b is the most frequent cause of bacterial meningitis in young children, and before the use of *Haemophilus* b conjugate vaccines became routine there were about 20,000 cases of invasive disease annually in the United States due to this pathogen, with 25 to 40% of the survivors having some postinfection sequelae.

[5] M. Pittman, *J. Exp. Med.* **53,** 471 (1931).
[6] W. M. Egan, F.-P. Tsui, and G. Zon, in "*Haemophilus influenzae*" (S. H. Sell and P. A. Wright, eds.), p. 185. Elsevier Biomedical, New York, 1982.

Capsular Serotyping Methods

A number of methods have been described to identify the different capsular types including capsular swelling (the Quellung reaction), precipitation by agar gel double diffusion, slide agglutination, countercurrent immunoelectrophoresis, latex agglutination, and antiserum agar. There is, however, no clear consensus on the best methods to identify types of *H. influenzae*. Strains of *H. influenzae* type b can become nonencapsulated on subculture. Young colonies that are capsulated appear iridescent and slightly opaque on a transparent medium such as Levinthal agar, whereas nonencapsulated colonies are bluish and translucent.[7]

Two serotyping methods are recommended here, bacterial cell agglutination and latex agglutination.[8] Typing sera and anti-type b latex reagents are commercially available. Prepare dense cell suspensions in phosphate-buffered saline (PBS), pH 7.2 to 7.4, containing 0.5% (v/v) formaldehyde from bacteria grown on plates no more than 6 to 8 hr. Use the cell suspensions for slide agglutination by mixing equal amounts of the bacterial suspension and a suitable dilution of the typing serum. The typing serum is diluted in 2-fold dilutions, and the highest dilution giving rapid agglutination is the optimal dilution. Agglutination should occur within about 1 min with constant rocking of the slide. If latex reagents are used, either cell suspensions or cell extracts may be used, and the agglutination occurs almost immediately.

For some purposes it may be preferable to isolate the capsular polysaccharide for use in immunoprecipitation by agar gel double diffusion (Ouchterlony) or by countercurrent immunoelectrophoresis (CIE).[9] The polysaccharides of the different *H. influenzae* types may be easily recovered from young cultures, either in broth or from plates, by extraction with EDTA.[10] Suspend unwashed cells in 10 mM HEPES, pH 7.5, containing 10 mM EDTA (about 10 ml/g wet weight cells) and incubate with stirring for 1 hr at 37°. The cells are removed by centrifugation at 20,000 g for 30 min. The extract can be used directly for serotyping, or the polysaccharide can be precipitated by addition of $CaCl_2$ to 20 mM and ethanol to 70% final concentration. The precipitate is dissolved in water, dialyzed, then freeze-dried. The resulting polysaccharide will contain about 10% protein, but is sufficiently pure for the above purposes and as a stable reference antigen

---

[7] M. Pittman, H. E. Alexander, W. L. Bradford, G. Eldering, and P. L. Kendrick, *in* "Diagnostic Procedures and Reagents" (A. H. Harris and M. B. Coleman, eds.), p. 414. American Public Health Association, New York, 1963.

[8] M. Leinonen and S. Sivonen, *J. Clin. Microbiol.* **10**, 404 (1979).

[9] T. Omland, *Methods Microbiol.* **10**, 235 (1978).

[10] F. L. A. Buckmire, *Methods Microbiol.* **10**, 171 (1978).

when stored at $-20°$. The protein can be removed by phenol extraction. To do so dissolve the ethanol precipitate in 0.2 M sodium acetate and add an equal volume of 90% phenol (prepared by adding 10 ml of saturated sodium acetate to 100 g of phenol crystals). Shake for 15 min, then let stand to allow phase separation. Centrifuge 5000 g for 30 min in a refrigerated centrifuge and remove the clear upper aqueous layer. Recover the polysaccharide by ethanol precipitation. Polysaccharide can also be prepared as described by Gotschlich et al. for the meningococcal polysaccharides.[11]

To conduct epidemiological studies on the spread of H. influenzae type b and to examine strain specific differences in virulence, type b strains need to be further characterized. A number of methods have been used to subdivide type b strains including ET typing, outer membrane protein patterns on SDS-PAGE, biotyping, and LPS heterogeneity on SDS–polyacrylamide gels. These are reviewed by van Alphen and Bijlmer.[1] Kilian has described 8 biotypes based on production of indole, presence of urease, and activity of orithine decarboxylase, but invasive strains are mostly biotype I.[12] A combination of outer membrane protein subtyping, LPS typing, and biotyping provide sufficient discriminatory power for meaningful epidemiological studies.

Neisseria meningitidis

Neisseria meningitidis is the second leading cause of bacterial meningitidis in the United States and the leading cause in many other countries. Other diseases caused by meningococci include bacteremia and pneumonia. The organism is of epidemiological importance because it may cause outbreaks or epidemics. In the United States there are 3000 to 4000 cases per year, mostly in young children, with an overall incidence of about 1/100,000.

Neisseria meningitidis is divided into 12 serogroups based on the immunological specificity of the capsular polysaccharides; A, B, C, H, I, K, L, X, Y, Z, 29E, and W135. The chemical structures of each of the polysaccharides are known.[13] Serogroup D no longer exists. Studies in our laboratory have shown that the type strain M158, used for preparation of group D typing sera, as well as several other old group D strains are no longer encapsulated. Most meningococcal disease is caused by serogroups A, B, and C. Group A is rare in the United States and Western Europe but is common in Asia and in the so-called meningitis belt of

[11] E. C. Gotschlich, T.-Y. Liu, and M. Artenstein, J. Exp. Med. **129,** 1349 (1969).
[12] M. Kilian, J. Gen. Microbiol. **93,** 9 (1976).
[13] C. E. Frasch, Med. Microbiol. **2,** 115 (1983).

Northern Africa. In the United States in 1986 (most recent surveillance data), strains of groups B and C were isolated from 48 and 46% of patients with meningococcal disease.[14] An effective polysaccharide vaccine is available for serogroups A and C, whereas experimental outer membrane protein vaccines have been used effectively for group B.[15]

Neisseria meningitidis shows marked diversity in many surface-exposed antigens, and over 300 ET types have been described.[16] Considerable phenotypic diversity can at times be found among different colonies on the same culture plate, as evidenced by differences in colonial opacity seen in young cultures (no more than 14 to 16 hr) on clear agar. It is therefore particularly important not to select individual colonies for subculture or preserving a meningococcal strain.

Identification of strains within the serogroups by serotyping and subtyping is important, because there are clear strain differences in ability to cause disease. Some group B serotypes, for example, are common among carriers but are rarely recovered from patients. It is therefore of epidemiological importance to identify strains likely to cause outbreaks.

Serogrouping Methods

Strains of *N. meningitidis* are generally serogrouped by bacterial or latex agglutination. They may also be grouped by whole cell enzyme-linked immunosorbent assay (ELISA) using monoclonal reagents.[17] The later technique is especially useful when large numbers of strains are to be examined. The antisera for serogrouping by agglutination are commercially available, as are latex reagents. For group B, high-titered monoclonal anti-B reagents should be used where available, since the group B polysaccharide is poorly immunogenic. Although grouping antisera are available from a variety of sources, Vedros has described production of grouping sera in rabbits.[18]

Agglutination is the most reliable procedure for routine serologic classification of meningococcal isolates. Although carrier isolates are frequently noncapsulated, essentially all invasive disease is caused by capsulated

[14] R. W. Pinner, B. G. Gellin, W. F. Bibb, C. N. Baker, R. Weaver, S. B. Hunter, S. H. Waterman, L. F. Mocca, C. E. Frasch, and C. V. Broome, *J. Infect. Dis.* **164**, 368 (1991).
[15] G. Bjune, E. A. Hoiby, J. K. Gronnesby, O. Arnensen, J. H. Fredriksen, A. Halstensen, E. Holten, A.-K. Linbak, H. Nokleby, E. Rosenqvist, L. K. Solberg, O. Closs, J. Eng. L. O. Froholm. A. Lystad, L. S. Bakketeig, and B. Hareide, *Lancet* **338**, 1093 (1991).
[16] D. A. Caugant, L. F. Mocca, C. E. Frasch, L. O. Froholm, W. D. Zollinger, and R.K. Selander, *J. Bacteriol.* **169**, 2781 (1987).
[17] T. Abdillahi and J. T. Poolman, *FEMS Microbiol. Lett.* **48**, 367 (1987).
[18] N. A. Vedros, *Methods Microbiol.* **10**, 293 (1978).

strains. Prepare a dense suspension of cells from a fresh culture (no more than 12 to 14 hr old) in PBS, not saline (a low pH can adversely effect the agglutination reaction). If the bacteria are not easily removed from the plate, do not use them for serogrouping, because nonspecific agglutination may occur. Allow any clumps of bacteria to settle to the bottom of the tube. Using a glass slide or a plate with ceramic rings as used for ABO blood typing, mix one drop each of bacteria and diluted grouping sera. As a negative control use diluted normal rabbit serum in PBS. Rock the slide for 1 to 2 min at room temperature and read the agglutination reaction. The reaction is best read against a dark background with indirect lighting. There should be no agglutination in the normal serum, but some strains especially from carriers do agglutinate. These strains are likely noncapsulated. To standardize the grouping sera, they should be diluted in normal rabbit serum and the highest dilution giving a 4+ reaction within 1 min only with the homologous serogroup should be used for routine grouping.

Using of latex reagents offers a highly accurate means of rapid serogroup identification by coagglutination.[19] Latex reagents use very small amounts of antibody, are standardized, and are subject to quality control in a central laboratory. The latex reagents can be used either with cell suspensions, mixing equal volumes of reagents as described above, or with extracted polysaccharide, the latter method avoiding the problem of autoagglutinable strains. The polysaccharide may be extracted by preparing a dense suspension of the bacteria (greater than McFarland standard No. 4) in 1 to 2 ml of PBS containing 2 mM EDTA, then strongly vortexing the sealed tubes for 10 to 20 sec. The cells are removed by centrifugation and the supernatant saved. Latex agglutination is performed by mixing one drop of supernatant with one drop of the latex reagent, and agglutination is noted after 1 min.

Serotyping and Subtyping Methods

Meningococcal serotyping was originally developed based on strain-specific differences in bactericidal killing.[20] Later, the extracted serotype antigens were used for serotyping by agar gel double diffusion. However, at that time there was not clear definition of what proteins or LPS antigens were responsible for serotype specificity. With the advent of monoclonal antibodies it became critical to define which surface antigens were involved in serotype specificity.

[19] F. Nato, J. C. Mazie, J. M. Fournier, B. Slizewicz, N. Sagot, M. Guibourdenche, D. Postic, and J. Y. Riou, *J. Clin. Microbiol.* **29,** 1447 (1991).
[20] C. E. Frasch and S. S. Chapman, *Infect. Immun.* **5,** 98 (1972).

The meningococcal outer membrane contains three to five major proteins between about 46 and 26 kDa, identified as Class 1 through Class 5.[21] Serotypes were defined as epitopes on the Class 2 and Class 3 protein. These proteins have molecular masses between 34 and 40 kDa. Subtypes are defined by antigenic differences among the Class 1 proteins, which are between 43 and 46 kDa.[22] There are over 20 different serotypes and over 12 subtypes within group B, and these are shared among other sero groups. This classification system including the serogroup is used to help identify distinct strains or clones, as, for example, B : 4 : P1.15 (serogroup : serotype : subtype). There are 11 different LPS immunotypes[23]; however, typing reagents are not generally available, and the same strains grown under different conditions may express or not express a given LPS epitope.

Although a number of serotyping methods have been used to identify serotypes within serogroups, only methods using monoclonal antibodies are described here. There are two approaches for preparation of antigens for use with the monoclonal serotyping and subtyping reagents, namely, use of whole cells or extracted outer membranes. The latter is preferred when it is necessary to retain the preparations for reanalysis up to years later, especially as new monoclonal reagents become available. The antigens are analyzed either by ELISA or dot immunoblots. If whole cells are used, they are prepared as described by Abdillahi and Poolman.[17] Briefly, bacteria are removed from young overnight cultures with a swab and suspended in PBS, pH 7.2, heated to 50° for 30 min, then diluted to an absorbance of 0.1 at 620 nm. The cells can be used immediately or stored at 4° for up to several weeks. A rapid extraction method is used to recover outer membranes. Cells are harvested from one plate of a young overnight culture and suspended in 5 ml of 0.2 M LiCl, 0.2 M sodium acetate, pH 6.0 in a small screwcapped tube containing about 1 g of sand. The membranes are extracted with vigorous shaking for 2 hr in a 50° water bath. The cells are removed in a refrigerated centrifuge at 10,000 g for 20 min, and the membranes are pelleted by ultracentrifugation at 100,000 g for 1 hr. Alternatively, the membranes can be precipitated from the cell-free extract by addition of 4 volumes of 95% ethanol. The recovered antigen is suspended in about 0.3 to 0.5 ml of water containing 0.02% sodium azide and stored at 4°. The ELISA and dot immunoblot procedures are performed as described.[17,24]

[21] C.-M. Tsai, C. E. Frasch, and L. F. Mocca, *J. Bacteriol.* **146,** 69 (1981).
[22] C. E. Frasch, W. D. Zollinger, and J. T. Poolman, *Rev. Infect. Dis.* **7,** 504 (1985).
[23] R. E. Mandrell and W. D. Zollinger, *Infect. Immun.* **16,** 471 (1977).
[24] E. Wedege, E. A. Hoiby, E. Rosenqvist, and L. O. Froholm, *J. Med. Microbiol.* **31,** 195 (1990).

Pseudomonas aeruginosa

Pseudomonas aeruginosa accounts for 10% of all nosocomial infections, and 11% of all blood isolates. It is also a major cause of morbidity and mortality among patients with depressed immune systems, with a case fatality rate of around 50%. It is the leading cause of bacterial infections in burn units. Antibiotic treatment is problematical, because the organism is often resistant to many antibiotics, and some of the highly effective antibiotics have toxicity problems. For these reasons it is of particular importance to identify quickly nosocomial outbreaks in the hospital. *Pseudomonas aeruginosa* is an opportunistic pathogen, and community-acquired infections are generally confined to contamination of traumatic wounds or are associated with intravenous drug abuse.

Simple growth requirements allow *P. aeruginosa* to grow anywhere there is moisture and high humidity. There are, thus, many possible sources of hospital outbreaks, from the flowers brought to the patient to the nebulizer fluid. The presence of an outbreak is often not recognized, and therefore not acted on, until demonstrated by serotyping data. Typing is needed to investigate specific outbreaks, to control the source of an outbreak, and to identify the more invasive types such as O : 11.

The most frequently used serotyping methods are based on antigenic differences in the O-polysaccharides of the LPS and the flagellar H antigens.[25] Since four or five H antigen types account for over 90% of *P. aeruginosa* isolates, H antigen typing, alone, is not very useful. Although several typing schemes have been proposed based on LPS immunotypes, the International Antigenic Typing System (IATS) with 17 immunotypes (O : 1–O : 17) is recommended.[26] Approximately 95% of clinical isolates are typable by the IATS system. The IATS types include the 12 Habs types and the 7 Fisher types.[27] Difco Laboratories (Detroit, MI) and the Pasteur Institute (Paris, France) produce standardized sera to the 17 IATS types.

The structures of the different O-polysaccharides have been determined.[28] The O-polysaccharides consist primarily of monoamino and diamino sugars and frequently have acidic functions. Some strains express, in addition to the LPS, the O-polysaccharide antigen as an LPS-free high molecular weight polysaccharide.

[25] T. L. Pitt, *Eur. J. Clin. Microbiol. Infect. Dis.* **7**, 238 (1988).
[26] G. B. Pier and D. M. Thomas, *J. Infect. Dis.* **145**, 217 (1982).
[27] C. D. Brokopp and J. J. Farmer III, in *"Pseudomonas aeruginosa"* (R. G. Doggett, ed.), p. 89. Academic Press, New York, 1978.
[28] Y. A. Knivel, E. V. Vinogradov, N. A. Kocharova, N. A. Paramov, N. K.Kochetkov, B. A. Dmitriev, E. S. Stanislavsky, and B. Lanyi, *Acta Microbiol. Hung.* **35**, 3 (1988).

Serotyping

Slide agglutination using either live bacteria or autoclaved (or boiled) bacteria is generally used for serotyping. For most purposes suspensions of live bacteria are best. Problems of cross-reactions with heat-labile antigens can be alleviated using the heated antigen. Grow the bacteria overnight at 36° on Mueller–Hinton agar or other suitable plating media. Harvest several colonies into thimerosal saline (0.405 g thimerosal and 8.5 g NaCl in 100 ml water) to prepare a cell suspension with an absorbance of 0.25 at 650 nm. As discussed above, it is important not to select single colonies for subculture or for preparing suspensions for immunotyping so as not to select for LPS variants present in the culture. Isolates should also be typed soon after isolation with a minimum of subcultures. The heated antigen is prepared by suspending the growth from one plate in about 10 ml of saline in a screw-capped tube, then autoclaving for 10 min or boiling for 30 min. The heated cells are pelleted, the supernatant discarded, and the cells suspended in 0.8 ml saline.[27] The heated antigen is stable and may be stored at 4° for prolonged period of time, but it has the problem of being more autoagglutinable than live cell suspensions.

The commercial sera are generally diluted 1 : 10 and further if problems of cross-reactions occur with the 17 standard type antigens (available from Difco Laboratories). The agglutination test is done either on a glass plate or in a plastic petri dish, by mixing equal volumes, about 10 μl each, of cell suspension and diluted typing sera. The reactions are read at 1 min. Longer reaction times are not recommended to avoid weak reactions. Only 2+ or greater reactions are considered positive.

Streptococcus pneumoniae

Streptococcus pneumoniae or the pneumococcus is a serious health problem at the extremes of life. The organism is found in the throat and nasopharynx of most people, but not all pneumococci are able to cause disease. The pneumococcus is the most common cause of bacterial pneumonia, with an estimated 400,000 to 500,000 hospitalizations annually in the United States, with 50,000 deaths. The rates are highest in individuals over 60 years of age. The pneumococcus is the second or third most common cause of bacterial meningitis, with an incidence of about 2/ 100,000. There is a high case fatality rate (40%) in young children. The organism is isolated from about one-third of all cases of otitis media, and it is estimated that 70% of children will have at least one episode of otitis media by the age of 3 years. Thus, prevention of pneumococcal disease can have a major public health impact in both children and older adults.

The mechanism by which the pneumococcus is able to cause invasive disease is still not understood, even though the organism has been studied for most of the twentieth century. The ability of pneumococci to invade the blood is highly associated with the state of capsulation. The capsular polysaccharide is antiphagocytic, and antibodies to the polysaccharide are opsonic and protective.

There are 83 known serotypes based on chemically and antigenically distinct capsular polysaccharides.[29] There are two systems of nomenclature, the American and Danish. Most investigators currently use the Danish nomenclature. This system recognizes 27 single types that are antigenically unrelated to any others, for example, types 1, 2, 3, 4, 5, and 14. The remaining 56 types fall into 19 groups of cross-reacting types, such as group 6 (types 6A and 6B) and group 19 (types 19F, 19A, 19B, and 19C). Typing of strains within groups is usually done only for special studies, because it requires an extensive set of "factor" antisera. Each type in the Danish system corresponds to a single type in the American system. To avoid confusion when both systems are cited, it is common to indicate the Danish type followed by the American type in parentheses.[30]

Although there are more than 80 capsular types, over 60% of all pneumococcal disease is caused by approximately 10 types. Some pneumococcal types almost never cause systemic disease, presumably due in part to sharing of capsular antigens with normal commensal bacteria. The observed relative prevalence of the different pneumococcal types led to the formulation of the 14- and then the 23-valent polysaccharide vaccine. It is now important to determine prevalent pneumococcal types in different age groups and geographic areas to determine the expected level of vaccine protection in a given population. It is also important to type pneumococcal isolates recovered from patients with a history of pneumococcal immunization to determine whether the disease was due to a vaccine serotype.

The 23 types included in the vaccine are 1, 2, 3, 4, 5, 6B, 7F, 8, 9N, 9V, 10A, 11A, 12F, 14, 15B, 17F, 18C, 19F, 19A, 20, 22F, 23F, and 33F. The chemical structures of the 23 polysaccharides are shown in a review paper.[31] These are complex polysaccharides and cross-react with a number of other organisms.

The pneumococcus has a number of different surface polysaccharides and proteins, of which only the capsular polysaccharide is commonly used for type identification. The C polysaccharide is found on all pneumococcal types, including rough organisms, and is a teichoic acid, containing phos-

[29] C.-J. Lee, S. D. Banks, and J. P. Li, *Crit. Rev. Microbiol.* **18**, 89 (1991).
[30] E. Lund, *J. Syst. Bacteriol.* **20**, 321 (1970).
[31] H. J. Jennings, *Curr. Top. Microbiol. Immunol.* **150**, 97 (1990).

phocholine as the major antigenic determinant. Typing sera when prepared contains C polysaccharide antibodies, which should be removed by adsorption with a rough nonencapsulated strain.[32] All clinically important pneumococcal types express surface protein A (PspA), a highly variable protein useful for serotyping.[33] Over 30 PspA types have been identified using monoclonal and polyclonal antibodies.

Capsular Serotyping

Before serotyping is attempted it should be verified that the cultures are pure, and that the strains are optochin sensitive or bile soluble. There are a number of methods for serotype determination, including slide agglutination, precipitation, the Quellung reaction, and countercurrent immunoelectrophoresis. The simplest and most accurate method is the Quellung reaction. The Neufeld Quellung test is a test for capsular swelling in presence of the homologous type-specific antibody, and the work Quellung means swelling. In this test the antibody interacts with and stabilizes the capsule, excluding the methylene blue included with the antibody. The antibody also causes a change in the refractive index of the capsule, increasing its visibility under the microscope.

To perform the Quellung reaction the method described by Austrian is generally used.[34] Mix an equal volume of 1% methylene blue in saline with the typing serum. Add a small drop of a light cell suspension to a glass slide and allow to dry. Add the typing serum and dye, cover with a coverslip, and observe under oil immersion (1000 × objective) using oblique illumination, almost "dark field like." There should be no more than 50–100 cells per high-power field. A positive Quellung reaction will be characterized by a sharply defined halo surrounging the pneumococcal cells. The Danish Statens Seruminstitut (Copenhagen, Denmark) produces an Omni serum containing antibody to all 83 types, which can be used first.[32] They also prepare nine serum pools, each containing antibodies to several types. Use of the pools reduces the amount of serum required to determine the exact type. Production and standardization of anticapsular typing sera are described by Lund and Hendrichsen.[32]

Serologic Identification of β-Hemolytic Streptococci

Strain of *Streptococcus pyogenes* are differentiated into groups A through V, based on antigenically and chemically distinct cell wall polysac-

[32] E. Lund and J. Henrichsen, *Methods Microbiol.* **12**, 241 (1978).
[33] M. J. Crain, W. D. Waltman, J. S. Turner, J. Yother, D. H. Talkington, L. S. McDaniel, B. M. Gray, and D. E. Briles, *Infect. Immun.* **58**, 3293 (1990).
[34] R. Austrian, *Mount Sinai J. Med.* **43**, 699 (1976).

charides or teichoic acids. Only groups A, B, C, D, F, and G are generally associated with human disease. Latex reagents for these groups are commercially available. About 95% of group A strains are inhibited by bacitracin. Serotyping systems have been developed for group A and group B streptococci.

Group A Streptococci

Group A streptococci are the most frequent cause of bacterial pharyngitis or strep throat in school-age children, accounting for up to 10% of clinic visits. Streptococcal skin infections are also common. Two serious postinfection sequelae, acute rheumatic fever and acute glomerulonephritis, make identification and prompt treatment of group A streptococcal infections of critical importance.

The work of Rebecca Lancefield in the early 1930s on the β-hemolytic streptococci resulted in a rational classification system. This was a pivotal point in our understanding of the epidemiology of group A streptococcal infection.[35] She first developed the cell wall grouping scheme, then went on to study types within group A using antibodies to the M protein.[36] It is important to determine streptococcal groups, because group A and group B strains are associated with severe infections, whereas strains of other serogroups, with the exception of groups C and G, are much less likely to cause serious disease.

The major virulence factor in group A streptococci is the surface fimbrial M protein, and strains lacking an M protein are avirulent.[37] The M protein is anchored in the cell membrane and extends onto the cell surface as fine fimbrialike structures.[38] The M protein is antiphagocytic, and protective immunity is M protein specific. The M typing of group A streptococci is important, because certain M types are more associated with skin infections and glomerulonephritis and others are associated with throat infections and rheumatic fever. The M typing can help trace virulent strains through a population; for example, M-1 isolates appear to be more virulent than many other types, are associated with scarlet fever, and produce potent pyrogenic exotoxins.

Another surface protein of epidemiological importance is the T protein, but T proteins are not virulence related. However, T typing may be used to characterize clones of group A streptococci further and to suggest which M sera to use for a given isolate.[39]

---

[35] R. C. Lancefield, *J. Exp. Med.* **57,** 571 (1933).
[36] H. F. Swift, A. T. Wilson, and R. C. Lancefield, *J. Exp. Med.* **78,** 127 (1943).
[37] R. C. Lancefield, *J. Immunol.* **89,** 307 (1962).
[38] V. A. Fischetti, *Clin. Microbiol. Rev.* **2,** 285 (1985).
[39] E. Wilson, R. A. Zimmerman, and M. D. Moody, *Health Lab. Sci.* **5,** 199 (1968).

The scheme for serologic identification of group A streptococci includes several steps. (1) Determine serogroup on acid extracts using capillary precipitation. It may be sufficient to determine group A, group B, and non-A or B. (2) Prepare trypsinized cells and determine T type by slide agglutination. (3) Determine opacity factor (OF) positive or negative by the horse serum agar or microtitration plate methods. (4) Do microtitration plate OF typing by inhibition of either typing sera or selected human sera containing OF antibodies. (5) Based on the T typing pattern and the results of the OF typing do selected M protein typing by the agar gel double diffusion method using acid extracts. Sera for M and T typing are available from the Institute of Hygiene and Epidemiology (Prague, Czech Republic).

Determination of Serogroup Using Acid Extracts Prepared by Lancefield Method. Grow strains in 30 ml Todd–Hewitt broth for 18 to 20 hr at 35°–37°.[36] Pellet cells by centrifugation and add 0.3 ml of 0.2 N HCl in saline. Place in a boiling water bath for 10 min, then cool. Pellet cells and discard. Add one drop of phenol red to the extract and neutralize to slight pink color with 0.2 N NaOH. The extract is used for serogroup determination by capillary precipitation. A narrow capillary is filled approximately one-third full with hyperimmune rabbit grouping sera, then wiped clean. Next an equal volume of extract is drawn into the tube. The tube is inverted into a Plasticine-filled capillary holder. The precipitin reaction is read in about 5 min and no longer than 10 min.

Determination of T Type by Method of Moody. The T typing is done on streptococcal cells treated with trypsin to remove the M protein.[39] Each T typing pattern is associated with 1 to 9 specific M types. Grow the strains overnight in 5 ml of Todd–Hewitt broth at 30° (35° is satisfactory). Pellet cells by centrifugation, and remove all but about 0.5 ml of broth. Add two drops of 5% (w/v) trypsin in PBS, pH 7.8. Mix vigorously to break up clumps then incubate for 2 hr at 37°. Mix on a slide a 1 mm loop of typing serum with a small drop of trysinized cell suspension and read the agglutination within 1 min. The method and interpretation of the results are described further by Wilson *et al.*[39]

Opacity Factor Reaction. The opacity factor (OF) is a lipoproteinase that acts on the α_1-lipoprotein in serum, causing the serum to become opaque.[40] There are now 27 antigenically different OF types, each associated with a single M type. Most strains that are OF positive are nontypable or poorly M-typable. The first step in OF typing is to determine whether the strain produces an OF. Horse serum is generally used, but different lots need to be screened with a known positive OF preparation to choose

---

[40] F. H. Top and L. W. Wannamaker, *J. Exp. Med.* **127,** 1013 (1968).

a suitable normal horse serum. The acid extract described above can be used to measure OF. Culture supernatants or 1% (w/v) SDS cell extracts (10 min, at room temperature) may also be used.[41] To detect the presence of OF, spot small amounts of the acid extracts onto 1% ion agar containing 50% horse serum and incubate overnight at 37°. Opaque zones indicate OF positivity. The microtitration plate method of Johnson and Kaplan can also be used.[42] Here 100 μl of screened horse serum is thoroughly mixed with 10 μl of the extract. The plate is sealed and incubated overnight at 37°, then 100 μl of saline is added. A positive reaction is read visually against a dark background with oblique lighting. The plate can also be read on an ELISA reader at 450 or 490 nm. Positive and negative controls need to be included with each assay.

Opacity Factor Inhibition Typing by Micromethod of Johnson and Kaplan. Using microtitration plates combine 10 μl each of extract and either type-specific human serum or animal typing serum and incubate for 1 hr at 35°–37°.[42] Add 100 μl of the screened horse serum and continue as above. This provides a surrogate method for M typing of strains that are OF positive.

M Typing by Method of Rotta. The problem is that M typing sera are not generally available and are in short supply. For this reason agar gel double diffusion is the preferred method, a method conservative of reagents. In addition, unabsorbed hyperimmune rabbit typing sera can be used, prepared as described by Rotta or Lancefield.[36,43] The acid extracts described above are used with selected M typing sera based on the results of the T and OF typing. The gel is formed using 1% Noble agar in PBS, pH 7.4. Place the acid extracts in the outer wells and serum in inner well, then place in moist chamber. The M protein forms precipitin lines distinct from the group A polysaccharide after 24 hr at room temperature.

Group B Streptococci

Group B β-hemolytic streptococci are a serious medical problem in neonates and women. *Streptococcus agalactiae* is a major cause of neonatal meningitis and septicemia within the first 4 months of life. Two distinct forms, early and late onset disease, occur in neonates, occurring up to 3 to 4 days and up to 3 to 4 months after birth. Early onset disease is correlated with a combination of maternal colonization and lack of type-specific antibodies in the neonate. The organism is also one of the most important causes of perinatal and urinary tract infections in women. The

[41] G. A. Sarvani and D. R. Martin, *J. Med. Microbiol.* **33**, 35 (1990).
[42] D. R. Johnson and E. L. Kaplan, *J. Clin. Microbiol.* **26**, 2025 (1988).
[43] J. Rotta, *Methods Microbiol.* **12**, 177 (1978).

gastrointestional tract and the female genital tract are the primary reservoirs of group B streptococci.

Group B streptococci have been subdivided into six serotypes based on immunochemical differences in a surface carbohydrate, but most disease is caused by groups Ia, Ib, II, and III. The typing was originally worked out by Lancefield using the same methods applied earlier to the group A streptococci.[44] The structures of the four major type polysaccharides were elucidated by Jennings.[31] All have terminal N acetyl neuraminic acid (NeuNAc) residues, and all have very similar structures, differing mostly in the side chains and some internal linkages. Group B streptococci also contain a number of surface proteins including two c protein serovars, α and β, and the R protein. The major use of these other antigens is to better define clones, because, for example, about half of early onset disease is caused by type III.

Serotyping Methods. Production of typing sera using formalinized whole cell vaccines is described by Wilkinson and Eagon.[45] The chemical similarity between some types requires that the typing sera be cross-adsorbed. Acid extracts are prepared as described above for group A, except that the cells are extracted for 2 hr at 50°, since the terminal sialic acid is partially removed if extracted at higher temperatures. Colman found that group B streptococci grown in nutrient broth containing 0.5% glucose yielded better carbohydrate extracts than cells grown in Todd–Hewlett broth.[46] Serotyping can be done either by capillary precipitation or by agar gel diffusion. Both methods are as described for group A streptococci. If the capillary method is used, the reactions should be read after 5 to 10 min. The agar gel method gives distinct precipitin bands with acid extracts.

[44] R. C. Lancefield, *J. Exp. Med.* **67**, 25 (1938).
[45] H. W. Wilkinson and R. G. Eagon, *Infect. Immun.* **4**, 596 (1971).
[46] G. Coleman, *Eur. J. Clin. Microbiol. Infect. Dis.* **7**, 226 (1988).

[13] Analysis of Genetic Variation by Polymerase Chain Reaction-Based Nucleotide Sequencing

By KIMBERLYN NELSON and ROBERT K. SELANDER

Introduction

With the advent of polymerase chain reaction (PCR) technology, comparative analysis of the complete nucleotide sequences of multiple genes in large samples of bacterial strains has become feasible, so that, at long last, the full extent of genetic diversity within and among bacterial populations and species can be directly measured. This is a singularly important development for population genetics, because the testing of theories concerning the relative contributions of various deterministic and stochastic factors to the generation and maintenance of genetic polymorphism requires full assessment of allelic diversity at individual loci.[1] Comparative sequence data make it possible to address the question of the role of horizontal transfer and recombination in bacterial evolution. For epidemiological research, sequence data are, of course, required for the design of optimally efficient strain- and species-specific probes and PCR primers. Moreover, technological advances in automated sequencing may soon make it possible for comparative sequence analysis to be routinely employed in a variety of epidemiological applications, including reconstruction of the evolutionary origin and history of new pathogenic strains.

With the objective of understanding the evolutionary mechanisms that generate and maintain genetic diversity and determine genetic population structure in bacteria, we have used PCR methods to examine a representative sample of strains of *Salmonella* and *Escherichia coli* for nucleotide sequence variation in several chromosomal genes encoding proteins that serve a variety of cellular functions.[2,3] Here we have described the general methods used in our laboratory to obtain population samples of nucleotide sequences of various genes.

Preparation of Genomic DNA from Bacteria

Relatively crude minipreparation procedures yield DNA of sufficient quantity and quality for PCR applications in our laboratory. Here we outline two protocols for DNA preparation, the first adapted from Wilson[4] for gram-negative bacteria such as species of *Escherichia, Salmonella, Citrobacter,* and *Klebsiella* and the second developed in our laboratory for the gram-positive species *Streptococcus pyogenes.*[5]

For each bacterial isolate studied, a sample of the liquid bacterial culture is stored as a new stock at −70°C. This is important, for, if

[1] M. Kimura, "The Neutral Theory of Molecular Evolution." Cambridge Univ. Press, Cambridge, 1983.
[2] K. Nelson, T. S. Whittam, and R. K. Selander, *Proc. Natl. Acad. Sci. U.S.A.* **88**, 6667 (1991).
[3] K. Nelson and R. K. Selander, *J. Bacteriol.* **174**, 6886 (1992).
[4] K. Wilson, *in* "Current Protocols in Molecular Biology" (F. M. Ausubel, R. Brent, R. E. Kingston, D. D. Moore, J. G. Seidman, J. A. Smith, and K. Struhl, eds.), p. 2.4.1. Wiley, New York, 1990.
[5] K. Nelson, P. M. Schleivert, R. K. Selander, and J. M. Musser, *J. Exp. Med.* **174**, 1271 (1991).

necessary, one can return to the actual culture that was used to prepare the DNA.

Preparation of DNA from Gram-Negative Species

1. Plate the bacterial strain of interest onto appropriate agar medium and incubate overnight. Check for contamination.
2. Inoculate 5 ml of liquid culture medium with a single colony and grow at 37° until the culture reaches stationary phase, generally overnight.
3. Remove 1.5 ml of the culture and harvest the cells by centrifugation (11,000 g, 5 min). Discard the supernatant.
4. Resuspend pelleted cells in 500 μl of Tris–EDTA buffer (TE) by repeated pipetting. Add 30 μl of 10% sodium dodecyl sulfate (SDS). Incubate for 20 min at 65°.
5. Add 10 μl of RNase A (5 mg/ml) and incubate for 30 min at 37°.
6. Add 5 μl of proteinase K (10 mg/ml) and continue incubation for 1 hr at 37°.
7. Add 100 μl of 5 M NaCl and mix thoroughly.
8. Add 80 μl of CTAB/NaCl solution [10% CTAB (hexadecyltrimethylammonium bromide) in 0.7 M NaCl]. Mix thoroughly and incubate for 10 min at 65°.
9. Add 700 μl of chloroform, mix thoroughly but gently, and centrifuge for 5 min.
10. Remove the aqueous supernatant to a new microcentrifuge tube, but be careful to leave the interface behind. Add 700 μl of phenol/chloroform (50 : 50, v/v), mix thoroughly, and centrifuge for 5 min.
11. Transfer the supernatant to a new tube and fill the tube to the top with 2-propanol. Gently invert the tube several times. The DNA can be pelleted immediately, or the sample can be stored at −20° for several hours or overnight.
12. Wash the pelleted DNA with 70% (v/v) ethanol and centrifuge for 5 min. Remove the supernatant and dry the pellet either at room temperature or, briefly, in a lyophilizer.
13. Dissolve the pellet in 100 μl of TE.

Preparation of DNA from Streptococcus pyogenes

1. Plate isolates onto BHI (brain–heart infusion) agar and grow overnight.
2. Use a sterile wooden stick or inoculating loop to scrape a small ball of cells from the plate, but avoid removing medium.

3. Suspend the cells in 800 μl of TE by repeated pipetting. Incubate for 15 min at 65°.
4. Centrifuge for 5 min to pellet cells. Discard the supernatant. Resuspend the cells in 800 μl of TE. Centrifuge and discard the supernatant.
5. Resuspend the cells in 500 μl of TE containing 10 μg/ml mutanolysin. Incubate for 2 hr at 37°.
6. Add 100 μl of 10% SDS and invert the tube to mix. Incubate for 20 min at 65°C.
7. Centrifuge for 10 min and transfer the supernatant to a clean tube.
8. Add 10 μl of RNase (5 mg/ml) and incubate for 30 min at 37°.
9. Add 5 μl of proteinase K (10 mg/ml) and incubate overnight at 37°C.
10. Extract the lysate once with phenol/chloroform and once with chloroform.
11. Add 250 μl of 7.5 M ammonium acetate and fill the tube to the top with 95% ethanol. Mix gently. The DNA can be pelleted immediately or the sample can be stored at $-20°$ for several hours or overnight.
12. Wash the pelleted DNA with 70% ethanol and centrifuge for 5 min. Remove the supernatant and dry the pellet either at room temperature or in a lyophilizer.
13. Dissolve the pellet in 100 μl of TE.

Amplification of Target Sequence

There are many approaches to the PCR amplification of a target sequence. It is beyond the scope of this chapter to discuss all the variations of PCR technology; the reader is encouraged to consult other sources for additional protocols.[6,7] In our experience, fairly standard procedures work well for a variety of genes.

Polymerase Chain Reaction Amplification

The typical PCR run profile is a 5-min denaturation at 94°, followed by 30 cycles of 1 min at 94°, 2 min at the annealing temperature, and 2.5 min at 72°C. This is followed by a slow cool down to 25° over a 30-min period, and the samples are then held at 4° until the tubes are removed from the thermocycler (Perkin-Elmer Cetus, Norwalk, CT). Annealing

[6] H. A. Erlich, "PCR Technology." Stockton, New York, 1989.
[7] M. A. Innis, G. H. Gelfand, J. J. Sninsky, and T. J. White, "PCR Protocols." Academic Press, San Diego, 1990.

temperatures range from 55° to 65°, but most target sequences amplify well at 60°C.

Polymerase Chain Reaction Conditions

Because we generate single-stranded template for sequencing directly from the double-stranded PCR product, two PCR amplifications are required for each target sequence to make DNA for sequencing both orientations. For target sequences less than 1000 base pairs (bp) in length, one 100-μl reaction per orientation is sufficient to sequence the entire gene. For longer sequences, two 100-μl reactions per orientation are necessary to produce sufficient template.

In each reaction tube, 1 μl of genomic DNA from the strain of interest is mixed with 99 μl of a cocktail. The mixture is covered with a layer of mineral oil before it is placed in the thermocycler. The typical cocktail (per 100-μl reaction) contains 53 μl of sterile distilled water, 16 μl of 1.25 mM deoxynucleoside triphosphates (dNTPs), 10 μl of 10$\times$ reaction buffer (100 mM Tris-HCl, pH 8.3, 500 mM KCl, 15 mM MgCl$_2$, and 0.1% gelatin), 10 μl (2 μM stocks) of each of two primers [one labeled with phosphate (see below)], and 0.5 μl of Taq DNA polymerase (United States Biochemical, Cleveland, OH, or Boehringer Mannheim, Indianapolis, IN).

Primer Design

Primers for PCR are generally 28–30 nucleotides in length and have approximately a 50% G + C composition. If the primer is within a coding region, it is best to place the 3′ end at either the first or second position of a codon. We also avoid placing the 3′ end at a run of consecutive bases, such as GGG of TTT, as this seems to cause some mispriming. Several computer programs for optimizing the design of primers are commercially available, and these can be useful in avoiding primer/primer interactions as well as self-priming due to bases pairing within the same primer.

Generation of Single-Stranded Template DNA for Sequencing

Some investigators directly sequence the double-stranded DNA; others clone the PCR product DNA into M13 or some other vector and then produce single-stranded DNA; and still others use the double-stranded PCR DNA in a second round of PCR with an asymmetric concentration of the two primers to produce single-stranded product. We prefer an approach developed by Higuchi and Ochman[8] that takes advantage of the

[8] R. G. Higuchi and H. Ochman, *Nucleic Acids Res.* **17**, 5865 (1989).

properties of λ-exonuclease. One PCR primer is labeled at the 5' end with phosphate, and λ-exonuclease is subsequently used to digest preferentially the strand of the PCR product that contains the labeled primer, thus producing single-stranded DNA for sequencing.

Phosphorylating Primers

1. Mix 20 μg of primer with 20 units of T4 polynucleotide kinase (BRL, Gaithersburg, MD), in buffer (10× kinase buffer: 500 m*M* Tris-HCl, pH 7.5, 100 m*M* MgCl₂, 50 m*M* dithiothreitol (DTT), 10 m*M* ATP, and 500 μg/ml bovine serum albumin (BSA)].
2. Incubate for 1 hr at 37°.
3. Concentrate the primer with a Centricon-3 microconcentrator (Amicon, Danvers, MA). Fill the tube with 2 ml of sterile distilled H₂O. Spin in a superspeed centrifuge (2–4 hr) until the volume is 40–60 μl.
4. Store the stock solution of the primer at −20°.

Treatment of Product DNA with λ-Exonuclease

1. Put the PCR mixture into microconcentrator (100,000 molecular weight cutoff, Millipore, Bedford, MA) and fill to 400 μl with sterile distilled water. Spin in a microcentrifuge at 4300 rpm until the final volume is 50–60 μl. The DNAs will concentrate at different rates depending on the fragment size and concentration.
2. To each microconcentrator, add 5–6 μl of 10× λ-exonuclease buffer (667 m*M* glycine–KOH, pH 9.4, 25 m*M* MgCl₂, 500 μg/ml BSA) and 4 units of λ-exonuclease (BRL) per 100 μl of the original PCR mixture. Incubate samples in the microconcentrator for 30 min at 37°.
3. Fill the microconcentrator to 400 μl with sterile distilled water and centrifuge at 4300 rpm until the volume is reduced to 60–70 μl.

Sequencing Single-Stranded DNA

The above procedure generates single-stranded DNA suitable for sequencing with commercially available reagents. We have used Sequenase (United States Biochemical) and TAQuence (United States Biochemical). The primers used for the PCR may also be used for sequencing the 5' end of each template. Additional internal primers (generally 18 bp in length) are employed to sequence completely both orientations.

The results of each sequencing reaction are entered into a computer by use of a digitizer and a program written by Stephen W. Schaeffer (Pennsylvania State University, University Park, PA). Individual sequences are compiled and edited with the SEQMAN and SEQMANED options of DNASTAR (DNASTAR, Madison, WI).

Applications of Comparative Sequence Analysis

We have generated a database of nucleotide sequences of several genes in diverse strains of *Salmonella* and *E. coli* that were recovered from natural populations and represent the ranges of genomic diversity in these bacteria previously revealed by DNA–DNA hybridization and multilocus enzyme electrophoresis. Two examples illustrate the types of information comparative sequence analysis has yielded.

The *GapA* gene, encoding glyceraldehyde-3-phosphate dehydrogenase, is highly conserved in both prokaryotes and eukaryotes,[9] whereas activity of the *put* operon is not required for cell survival and growth of enteric bacteria, except when proline is the sole nitrogen source.[10] Because these two genes encode proteins with very different cellular functions, we were interested in comparing the genes with respect to the rate of nucleotide sequence divergence, the degree of constraint on amino acid substitution, the rate of horizontal gene transfer and recombination, and the extent of evolutionary divergence.

Virtually complete sequences of *gapA* (924 bp) and the proline permease gene, *putP* (1467 bp), and complete sequences (416–422 bp) of the control region of the *put* operon were determined for 16 strains of *Salmonella*, representing all 8 subspecies, and 12 strains of *E. coli*. For both genes, the average difference between strains was much larger for *Salmonella* than for *E. coli*, and in both taxa *putP* was more divergent than *gapA* (Table I). For *gapA*, there were differences in the rate of nucleotide substitution among functional domains, with the rate of substitution at synonymous sites being significantly higher for codons specifying the catalytic domain of the enzyme than that for those encoding the NAD^+-binding domain; however, the nonsynonymous substitution rate showed the opposite relationship. In contrast, there was no difference in the distribution of polymorphic amino acid positions between the membrane-spanning and loop regions of the permease molecule, and rates of synonymous nucleotide substitution were virtually the same for the two domains. Statistical analysis of the sequences of the two genes yielded evidence of four probable cases of recombination: the acquisition by strains of *Salmonella*

[9] R. F. Doolittle, D. F. Feng, K. L. Anderson, and M. R. Alberro, *J. Mol. Evol.* **31**, 383 (1990).

[10] S. R. Maloy, in *"Escherichia coli* and *Salmonella typhimurium:* Cellular and Molecular Biology," p. 1513. American Society of Microbiology, Washington, D.C., 1987.

TABLE I
Sequence Variation in *gapA* and *putP* Genes
among Strains of *Salmonella* and *Escherichia coli*

Species and sample	Mean percent difference between strains	
	gapA	*putP*
Nucleotide sites		
Salmonella	3.8	4.6
Within subspecies	0.2	0.5
Between subspecies	4.0	5.1
Escherichia coli	0.2	2.4
Salmonella versus E. coli	6.2	16.3
Amino acid positions		
Salmonella	1.1	1.3
Within subspecies	0.04	0.1
Between subspecies	1.2	1.4
Escherichia coli	0.1	0.3
Salmonella versus *E. coli*	1.7	5.5

subspecies V of a segment of the *gapA* gene from a source related to *Klebsiella pneumoniae;* the presence in strains of *Salmonella* subspecies VII of a large segment of *putP* transferred from an unidentified source; the exchange of a 21-bp segment between two strains of *E. coli;* and the occurrence in one strain of *E. coli* of a cluster of 14 unique polymorphic control region sites derived from an unknown donor. An evolutionary tree for *putP* and the *put* control region sequences was generally concordant with a tree for the *gapA* gene (Fig. 1) and also with a tree based on multilocus enzyme electrophoresis.

General Conclusions

The picture emerging from comparative studies of sequence variation in *Salmonella* and *E. coli* is that, for most genes, intragenic recombination does not occur at rates sufficiently high to obscure phylogenetic relationships dependent on mutational divergence. Consequently, particular multilocus genotypes may persist as clonal lineages for long periods and, in many cases, achieve global distribution. This is true for *gapA* and for *putP*, and the general concordance of individual trees for these genes with a tree for the same strains based on multilocus enzyme electrophoresis[2] suggest that it is true for metabolic enzyme genes in general (see also Ref.

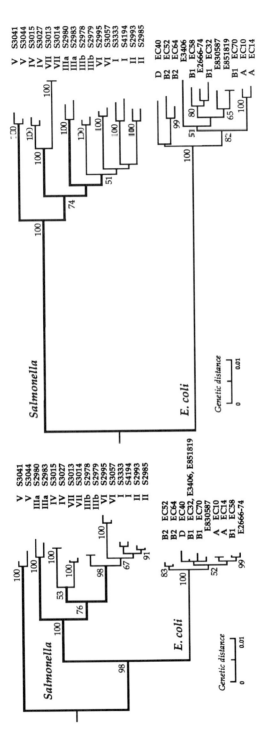

gapA

putP and put control region

FIG. 1. Evolutionary trees for the *gapA* and *put* operon sequences of 16 strains of *Salmonella* and 12 strains of *E. coli*, constructed by the neighbor-joining method [N. Saitou and M. Nei, *Mol. Biol. Evol.* **4**, 406 (1987)] from matrices of pairwire genetic distance. A number adjacent to a node indicates the percentage of bootstrap trees that contained that node.

11). However, it is not the case for flagellin-encoding genes in *Salmonella*[12] or the 6-phosphogluconate dehydrogenase locus (*gnd*) in E. coli,[13,14] for which it has been clearly demonstrated that recombination is a primary proximate source of allelic diversity within populations.

From an adaptive evolutionary standpoint, there is reason to expect that genes encoding highly antigenic proteins, such as flagellins, are especially likely to exhibit a mosaic structure resulting from recombination. For genes of this type, recombination events will occasionally lead to an increase in frequency of the recombinant strain owing to the selective advantage to a cell of presenting altered cell-surface structures to the environment (host defense mechanisms and phages).[15] Recombination in genes encoding subunits of restriction-modification systems may also be favored by frequency-dependent selection exerted by phages.[16] Of course, recombination events that increase resistance to antibiotics may confer a great advantage, as in the case of chromosomal genes encoding penicillin-binding proteins.[17] But, for many genes, such as *putP* and *gapA*, that encode polypeptides for which there may be no adaptive premium on diversity in amino acid sequence per se, recombination is unlikely to confer an adaptive advantage. The hypothesis that the effective recombination rate varies among genes in relation to the functional type of gene product will be tested as additional genes are sequenced.

[11] B. G. Hall and P. M. Sharp, *Mol. Biol. Evol.* **9,** 654 (1992).

[12] N. H. Smith, P. Beltran, and R. K. Selander, *J. Bacteriol.* **172,** 2209 (1990).

[13] M. Bisercic, J. Y. Feutrier, and P. R. Reeves, *J. Bacteriol.* **173,** 3894 (1991).

[14] D. E. Dykhuizen and L. Green, *J. Bacteriol.* **173,** 7257 (1991).

[15] R. K. Selander, P. Beltran, and N. H. Smith, *in* "Evolution at the Molecular Level" (R. K. Selander, A. G. Clark, and T. S. Whittam, eds.), p. 25. Sinauer, Sunderland, Massachusetts, 1990.

[16] P. M. Sharp, J. E. Kelleher, A. S. Daniel, G. M. Cowan, and N. E. Murray, *Proc. Natl. Acad. Sci. U.S.A.* **89,** 9836 (1992).

[17] B. G. Spratt, L. D. Bowler, Q.-Y. Zhang, J. Zhou, and J. Maynard Smith, *J. Mol. Evol.* **34,** 115 (1992).

[14] Determinations of Restriction Fragment Length Polymorphism in Bacteria Using Ribosomal RNA Genes

By Rivka Rudner, Barbara Studamire, and Erich D. Jarvis

Introduction

Analysis of restriction fragment length polymorphism of ribosomal RNA (rRNA) genes is a powerful and reliable tool for, but not limited to, two purposes: (1) as a taxonomic technique and (2) as a pathogen diagnostic technique. For taxonomists, the degree of conservation and variation in restriction patterns provides information about the genetics and evolution of related and distant bacteria. For diagnostic purposes as an epidemiological tool, one can determine with confidence whether clinically isolated species of a given genus represent the pathogenic form. The Southern transfer technique[1] is used in conjunction with rDNA–DNA hybridization to determine the distribution of homologous restriction fragments of the genome under comparison. If cloned DNA is not available, rRNA or transfer RNA (tRNA) can be used instead.

The rRNA and tRNA genes remain the most useful of the chronometers to study restriction site polymorphisms since they are ubiquitous and highly conserved, and different positions in the sequences change at different rates, allowing most phylogenetic relationships (including the most distant) to be measured.[2] In addition, many species have multiple identical *rrn* copies surrounded by variable sequences, allowing for the characterization of numerous forms of nonribosomal polymorphisms not possible with single-copy genes. A restriction pattern of multiple *rrn* bands of varying sizes can serve as a "hallmark" for a given bacterial species. Furthermore, owing to the high degree of conservation, a cloned rDNA from one species, namely, *Escherichia coli,* has been successfully used as the hybridization probe to explore the degree of homology among a group of many closely or distantly related organisms.[3] Thus, the most compelling reason for using rDNA as chronometers is that they can determine the hallmarks of many organisms at various taxonomic levels up to species differences. However, below the species level, the method is limited by the smaller probability of finding polymorphisms adjacent to RNA genes among strains of the same species. In this case it may be

[1] E. Southern, *J. Mol. Biol.* **98,** 503 (1975).
[2] C. L. Woese, *Microbiol. Rev.* **51,** 221 (1987).
[3] F. Grimont and P. A. D. Grimont, *Ann. Inst. Pasteur* **137B,** 165 (1986).

useful to examine other DNA probes from common single-copy genes like *trp, tna, thy,* and *lacZ* used in early studies with the Enterobacteriaceae[4-6] or *leu, thr,* and *trp* used in studies with members of the genus *Bacillus*[7,8] which provided useful information on the mechanisms that led to polymorphism such as small deletions, insertions, and other chromosomal rearrangements.

We have successfully used rRNA and tRNA as probes to determine the multiplicity, arrangement, and location of *rrn* and *trn* loci as well as differences between strains of the same species, namely, *Bacillus subtilis,* and between members of the same genus (*Bacillus*).[7-12] Ribosomal RNA restriction pattern analysis has been applied to the taxonomy of staphylococci,[4] members of *Enterobacteriaceae,*[5,6,13] *Candida* spp,[14] and *Bacillaceae,*[10,11] to cite just a few examples. The essential tools have been restriction enzymes that cut only once or not at all in the *rrn* operons and RNA probes that hybridize to chromosomal segments bounded at one end by non-*rrn* sequences. In the case of the enterobacteria (*E. coli* or *Salmonella typhimurium*),[15] Southern analysis of double-digested genomic DNA with *Bam*HI–*Pst*I revealed seven highly resolved bands. Neither restriction endonuclease cleaved within the seven operons. In the *Bacillaceae* (i.e., *Bacillus subtilis*) we used the enzyme *Bcl*I, which cuts only once in all ten operons, 79 base pairs (bp) before the 5′ end of the 23 S rDNA.[16] Southern hybridizations of *Bcl*I digests of genomic DNA from *B. subtilis* strains NCTC3610 or 168T yielded 10 distinct fragments when probed with any rDNA fragment 5′ to the internal *Bcl*I site.[9,11] Using an enzyme which cuts once in the operon is preferable because it allows the identification of clustered *rrn* operons as was found in *B. subtilis,* where 5 out of

[4] F. M. Thomson-Carter, P. E. Carter, and T. H. Pennington, *J. Gen. Microbiol.* **135,** 2093 (1989).

[5] A. Anilonis and M. Riley, *J. Bacteriol.* **143,** 355 (1980).

[6] M. Riley and A. Anilonis, *J. Bacteriol.* **143,** 366 (1980).

[7] P. Gottlieb and R. Rudner, *J. Syst. Bacteriol.* **35,** 244 (1985).

[8] P. Gottlieb, Ph.D. Dissertation, City Univ. of New York, New York (1984).

[9] G. LaFauci, R. L. Widom, R. L. Eisner, E. D. Jarvis, and R. Rudner, *J. Bacteriol.* **165,** 204 (1986).

[10] R. L. Widom, E. D. Jarvis, G. LaFauci, and R. Rudner, *J. Bacteriol.* **170,** 605 (1988).

[11] E. D. Jarvis, R. L. Widom, G. LaFauci, Y. Setauchi, I. R. Richter, and R. Rudner, *Genetics* **120,** 204 (1988).

[12] R. Rudner, A. Chevrestt, S. R. Buchhotz, B. Studamire, A.-M. White, and E. D. Jarvis, *J. Bacteriol.* **175,** 503 (1993).

[13] L. Harshman and M. Riley, *J. Bacteriol.* **144,** 560 (1980).

[14] B. B. MaGee, T. M. D'Souza, and P. T. MaGee, *J. Bacteriol.* **169,** 1639 (1987).

[15] A. F. Lehner, S. Harvey, and C. W. Hill, *J. Bacteriol.* **160,** 682 (1984).

[16] C. J. Green, G. C. Stewart, M. A. Hollis, B. S. Vold, and K. F. Bott, *Gene* **37,** 261 (1984).

10 are arranged in two sets of tandem repeats (Fig. 1d).[10] The genomic locations of *rrn* fragments can be determined through the integrative mapping technique using transductional crosses as was done in *B. subtilis*[9,11] or through genomic restriction mapping as was done in *H. influenzae.*[17]

Among strains of *E. coli, S. typhimurium,* or *B. subtilis,* restriction site polymorphisms 5' and 3' to ribosomal operons were easily ascertained on either *Bam*HI–*Pst*I[15] or *Bcl*I[10] Southern hybridizations, respectively. In *B. subtilis* we found strains with naturally occurring *rrn* deletions which arose by homologous recombination between adjacent *rrn* gene sets and exhibit losses of unique fragments.[10] We also found strains with heterologous intrachromosomal rearrangements adjacent to *rrn* operons.[18] Restriction enzymes that cut more than once within the *rrn* offer Southern hybridization patterns that are useful for the identification of intragenic restriction polymorphism. They allow for the evaluation of heterogeneities in the arrangement and sequence of the individual genes, namely, 16 S, 23 S, and 5 S, and tRNAs within the *rrn* loci among strains of related species.

For example, the basic physical map of all the *rrn* operons of *B. subtilis* consists of three internal *Eco*RI and *Sma*I sites, two *Pst*I sites, and single *Bcl*I, *Bam*HI, and *Hin*dIII sites[19] (Fig. 1c). Two of the ten operons (*rrnO, rrnA*) contain tRNA genes in the spacer regions between the segments that code for 16 S and 23 S.[20] In that spacer region, one *Eco*RI site is variable in the genus *Bacillus.* It is found in the 5' end of the 23 S gene, is present in strains of *B. subtilis* and *B. licheniformis,* and is absent in strains of *B. globigii, B. pumilus,* and *B. amyloliquefaciens.*[7] This heterogeneity allowed us to construct a dendrogram analysis of the relationships among five species of *Bacillus* and their strains[7] (Fig. 2).

In this chapter, we describe procedures which deal with the most optimal conditions required to obtain high resolution of *rrn* and *trn* fragments on Southern blots of genomic DNA from members of the genus *Bacillus.* Included are methods for the preparation of genomic DNA, restriction digests, isolation and labeling of rDNA and tDNA, gel electrophoresis including pulse-field gel electrophoresis (PFGE), Southern blotting, and hybridization analyses. The experimental approach requires the following considerations.

Choice of Restriction Endonuclease. (1) Enzymes that do not cut in *rrn* operons will generate fragments that are larger than the average size of an operon, namely, 4.8 kilobases (kb). These enzymes (i.e., *Bgl*I, *Sal*I)

[17] J. J. Lee, H. O. Smith, and R. J. Redfield, *J. Bacteriol.* **171,** 3016 (1989).
[18] E. D. Jarvis, S. Cheng, and R. Rudner, *Genetics* **126,** 784 (1990).
[19] G. Stewart, F. Wilson, and K. Bott, *Gene* **19,** 153 (1982).
[20] K. Loughney, E. Lund, and J. E. Dahlberg, *Nucleic Acids Res.* **10,** 1607 (1982).

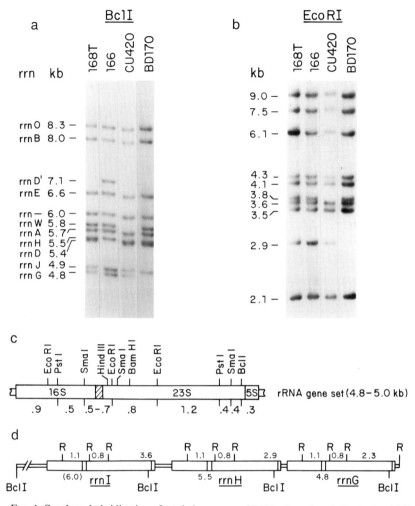

FIG. 1. Southern hybridization of total chromosomal DNAs from *B. subtilis* strains 168T, 166, CU420, and BD170 containing either a chromosomal rearrangement upstream of *rrnD* (166) or naturally occurring deletions of *rrnW* (CU420) or *rrnG* (BD170). (a) *Bcl*I digests were electrophoresed on 0.75% agarose for 5–7 days at 15–20 mA and probed with a labeled *Eco*RI–*Pst*I 23 S fragment; the assignment of the individual *rrn* operons is according to E. D. Jarvis, R. L. Widom, G. LaFauci, Y. Setauchi, I. R. Richter, and R. Rudner, *Genetics* **120**, 204 (1988). (b) *Eco*RI digests were electrophoresed on 0.75% agarose for 48 hr at 15–20 mA and probed with a labeled *Eco*RI–*Hind*III 23 S, 5 S fragment. (c) Generalized restriction map of a *B. subtilis* rRNA gene set as proposed by G. Stewart, F. Wilson, and K. Bott, *Gene* **153**, (1982). The hatched area represents the abutment region between 16 S and 23 S rRNA with or without tRNA genes [K. Loughney, E. Lund, and J. E. Dahlberg, *Nucleic Acids Res.* **10**, 1607 (1982)]. (d) Structure of a cluster of *rrn* operons in *B. subtilis*. The triplet *rrnI–rrnH–rrnG* is located at 14° on the map [E. D. Jarvis, R. L. Widom, G. LaFauci, Y. Setauchi, I. R. Richter, and R. Rudner, *Genetics* **120**, 204 (1988)]. The *Bcl*I and *Eco*RI (R) fragment sizes shown are those identified by restriction analysis.

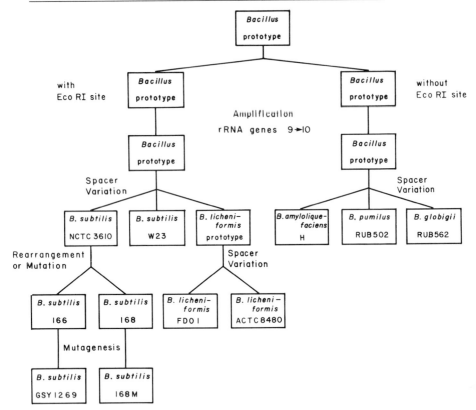

FIG. 2. Dendrogram analysis of the restriction relationships among five species of *Bacillus* and their strains. The bacteria examined were divided into two lines of descent based on the structures of the rRNA gene sets. Type I *Bacillus* species have rRNA genes possessing the 5′ 23 S sequence *Eco*RI site (see Fig. 1c), and type II species have rRNA gene sets without the *Eco*RI site. [By permission from P. Gottlieb and R. Rudner, *J. Syst. Bacteriol.* **35**, 244 (1985).]

are suitable for determining the location of *rrn* operons on a genomic restriction map, but they are not suitable for determining multiplicities, especially when clustering is involved. These enzymes might not cut between adjacent operons in a cluster. (2) Enzymes that cut once in *rrn* operons will ensure the determination of the total number of gene sets and the extent of clustering, whereby a single cutter will release operons from one another (e.g., the *Bcl*I site in the clusters shown in Fig. 1d). These enzymes (i.e., *Bcl*I, *Hin*dIII) are the preferred ones for taxonomic and diagnostic studies since differences between adjacent *rrn* operons are easily deciphered. (3) Enzymes that cut more than once in *rrn* operons are useful for intragenic restriction polymorphism (i.e., *Eco*RI, *Pst*I).[7]

Choice of Probe. (1) All three rRNA genes (16 S, 23 S, 5 S) are suitable as probes when one restricts genomic DNA with an enzyme that does not cut inside the *rrn* operon. (2) When the genomic DNA is restricted with an enzyme that cuts only once it is desirable but not essential to probe either the 5' side to the restriction site in the operon or the 3' side of the restriction site. (3) When genomic DNA is cut more than once, a probe which hybridizes to internal *rrn* sequences overlapping or between the restriction sites is suitable for evaluating the existence of intragenic variations as discussed above.[7,8] (4) In the absence of cloned rDNA and restriction site analysis of a given species, isolated 16 S, 23 S, and 5 S can be synthesized as cDNA by a reverse transcriptase reaction.[21] These cDNAs can then be cloned or tested for restriction sites. In all cases we recommend probing with either 16 S or 23 S alone; otherwise, the patterns are more complex, with excessive numbers of bands.

Bacterial Strains and Isolation of Genomic DNA

Bacillus strains are best maintained on slants of tryptose blood agar base (TBAB; Difco Laboratories, Detroit, MI). Asporogenous strains are stored in vials in 20% (v/v) glycerol at $-70°$. To culture for the purpose of DNA isolation, the following three media are recommended: (1) VY, 2.5% veal infusion (Difco)–0.5% yeast extract (Difco); (2) LB (Luria–Bertani) broth, containing (per liter) Bacto-tryptone 10 g, Bacto-yeast extract 5 g, NaCl 10 g, 1 M NaOH 1.0 ml; or (3) BHIB, 3.7% Bacto-brain–heart infusion broth (Difco).

For large-scale DNA preparation, a 5-ml starter culture is used to inoculate 500 ml of the above media in a 2.0-liter Erlenmeyer flask. Growth is allowed to proceed with vigorous agitation at $37°$ overnight. Genomic DNA is isolated and purified by a modification[22] of the procedure of Marmur.[23] The cells are collected by centrifugation at 5000 rpm, washed with a 0.15 M NaCl–0.1 M EDTA (pH 8.0) solution, and lysed at $37°$ in the presence of lysozyme (50 μg/ml, final concentration). The suspension of crude DNA fibers is then treated with an RNase mixture (50 μg/ml of pancreatic RNase and 50 U/ml of T1) for 30 min at $37°$, followed by digestion with pronase (500 μg/ml, self-digested at $37°$ for 4 hr) at $37°$ for

[21] T. Maniatis, E. F. Fritsch, and J. Sambrook, "Molecular Cloning: A Laboratory Manual." Cold Spring Harbor Laboratory, Cold Spring Harbor, New York, 1982.
[22] R. Rudner, H. Lin, S. Hoffman, and E. Chargaff, *Biochim. Biophys. Acta* **144**, 199 (1967).
[23] J. Marmur, *J. Mol. Biol.* **3**, 208 (1961).

at least 2 hr until the solution becomes clear. This treatment is followed by two deproteinizations with 90% phenol (pH 7.8), ethanol precipitation, and dialysis against TE buffer (10 mM Tris-HCl, 0.1 mM EDTA, pH 7.5). High molecular weight DNA should have an $OD_{280/260}$ ratio of about 0.5 and is stored at 2 mg/ml in the cold over a drop of chloroform. All genomic DNAs used to perform "Zoo Blots" should be uniformly diluted from the stock to 100 μg/ml in TE buffer and subjected to restrictions.

Alternatively, small-scale rapid lysates for processing many bacterial species can be done when small amounts of DNA are needed. Chromosomal DNA is prepared from 5-ml overnight cultures according to the method of Jarvis et al.[18] The cells are centrifuged at 5000 rpm and suspended in 0.2 ml of SET buffer (20% sucrose, 50 mM EDTA, and 50 mM Tris, pH 7.6) containing 5 mg/ml lysozyme and 1 mg/ml RNase, vortexed well, transferred to a 1.5-ml Eppendorf tube, and incubated at 37° for 15 min. To the lysate 0.4 ml of 1% sodium dodecyl sulfate (SDS) is added and mixed well by inversion, followed by a 10-min incubation at 60°. An equal volume (0.6 ml) of a 1 : 1 mixture containing 80% Tris-buffered (pH 8.0) phenol : chloroform/isoamyl alcohol (24/1) is added, and the sample is vortexed and centrifuged for 5 min. The aqueous layer is removed and reextracted with chloroform/isoamyl alcohol. Nucleic acids are then precipitated with an equal volume of 2-propanol after adjustment to 0.3 M sodium acetate and centrifuged for 10 min. The pellet is washed twice with 70% ethanol, dried for 15 min in a Speed Vac (Savant Instruments, Inc., Hicksville, NY), and dissolved in 50–100 μl of sterile water. Typical recoveries are 25–50 μg of chromosomal DNA per sample.

Restriction Patterns of Genomic DNA and Southern Blotting

Chromosomal DNA (5 to 10 μg/ml) is digested with the desired restriction endonuclease at 3 U of enzyme per microgram of DNA for 12 hr at 37° under the conditions for digestion recommended by the supplier (New England Biolabs, Beverly, MA; Boehringer Mannheim, Indianapolis, IN; etc.). The choice of enzyme is dictated by the generalized restriction map of the B. subtilis rRNA gene set as proposed by Stewart et al.[19] (Fig. 1c). The enzymes BclI, BamHI, and EcoRI are ideally suited because they yield a characteristic and reproducible banding pattern of the 10 rrn operons in Bacillus when hybridized to the appropriate probe (Fig. 1a). The HindIII and EcoRI enzymes have also been useful for locating tRNA gene clusters associated downstream of rrn operons.[12]

A large horizontal gel apparatus (21 × 24 cm) containing a cooling device (IBI, New Haven, CT) is used for high resolution of fragments. The comb has 20 teeth and under the conditions used forms wells able to contain 50–60 μl of restricted DNA. Agarose type II (low electroendosmosis, EEO; Sigma, St. Louis, MO) at 0.75–0.85% (w/v) is used to cast the gel, and the running buffer is Tris–borate (TBE; 8.9 mM Tris–HCl, pH 8.0, 89 mM boric acid, 2.5 mM EDTA, containing 0.5 μg/ml ethidium bromide).[23] The key to high resolution is to run the gels at low voltage (5 mA and 10 V) for 2–5 days to obtain maximum separation of closely sized bands (Fig. 1a). A lane of phage λ DNA, cleaved by HindIII, is used as a fragment size marker. Gels loaded with BclI- or BamHI-restricted genomic DNA are run until the fourth band of HindIII-cleaved λ (4.3 kb) reaches the bottom of the gel. The other restrictions (i.e., EcoRI, HindIII) need not run that long so that the internal rrn fragments or tRNA genes with molecular sizes 0.9–4.0 kb are retained.[7,12]

Another useful approach is the separation of large restriction fragments generated from digestions with the rare cutters ApaI, EagI, NaeI, NotI, SmaI, SfiI, and RsrI in pulsed-field gel electrophoresis using one of the two systems: (1) OFAGF (orthogonal field alternation gel electrophoresis) and (2) CHEF (counter-clamped homogeneous field electrophoresis). Studies done in Haemophilus influenzae Rd with these techniques were used to generate a physical restriction map of the entire genome and to map all six rrn.[17] Both PFGE systems use 1% agarose gels in 0.5× TBE buffer at 0.5°–5°. The OFAGE experiments are run at 280 V for 12 hr with pulse times ranging from 1 to 36 sec, depending on the resolution range desired. The CHEF experiments are run at 195 V for 24 hr with pulse times ranging from 6 to 80 sec. The use of PFGE is particularly useful in constructing a physical map of uncharacterized genomes from clinically isolated serotypes using the rrn sequences as the initial probes, as was done for Mycoplasma mobile, Caulobacter crescentus, and Mycoplasma pneumoniae.

The electrophoretically separated DNA fragments are transferred onto either nitrocellulose filters (Schleicher & Schuell, Keene, NH; BA-S NC) or Nytran nylon membranes (Schleicher & Schuell) according to the procedure of Southern[1] following a modified approach of the established protocols[23] and recent advances. After electrophoresis, the gel is soaked for 1 hr in a denaturing solution (1.5 M NaCl–0.5 M NaOH), followed by a 1-hr soak in a neutralizing buffer (1 M Tris-HCl, pH 7.5, 1.5 M NaCl). The gel is soaked in 10× SSC (1× SSC is 150 mM NaCl, 15 mM sodium citrate) for 30 min, followed by blotting on a flat surface onto a membrane which had been presoaked in 10× SSC for 5 min. The DNA bands are fixed to the membrane by UV-mediated cross-linking using a UV 300 Transilluminator (Fotodyne, New Berlin, WI) at 254 nm for 90–120 sec. The membranes are kept moist throughout to allow for rehybridization with a second probe. The blot is prehybridized in a solution containing

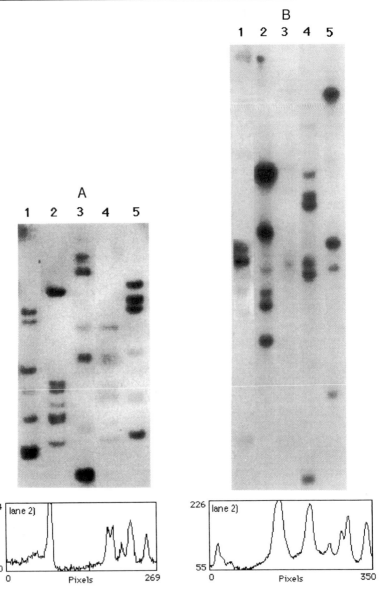

Fɪɢ. 3. Southern hybridization using *B. subtilis* rRNA and tRNA as probes of total chromosomal DNAs from the following five members of the genus *Bacillus:* (1) *B. subtilis* 168T, (2) *B. subtilis* W23, (3) *B. pumilus* RUB502, (4) *B. licheniformis* ACTC8480, and (5) *B. amyloliquefaciens* H. (A) *Bcl*I digests were probed with a labeled 2.0-kb *Hind*III–*Pst*I fragment of the 23 S gene (see Fig. 1c). The gel was electrophoresed as described for Fig. 1a. A density profile from a line plot of lane 2 (*B. subtilis* W23) was done with the NIH

40% formamide, $4 \times$ SSC, 50 mM sodium phosphate, pH 6.5, 250 μg/ml sonicated heat-denatured salmon sperm DNA, 1% glycine, and $5 \times$ Denhardt's solution ($1 \times$ Denhardt's is 0.02% Ficoll, 0.02% polyvinyl-pyrrolidone, 0.02% bovine serum albumin) as described by Ostapchuck *et al.*[24] in a heat-sealable bag at 42°–45° for 3–5 hr. After removal of the prehybridization buffer, hybridization buffer containing 40% formamide, $4 \times$ SSC, 20 mM sodium phosphate, pH 7.5, 100 μg/ml sonicated dena-tured salmon sperm DNA, 1% Denhardt's solution, and 0.1% SDS is added (4 ml per each 100 cm^2 of membrane), followed by the addition of ^{32}P-labeled rRNA, or single-stranded DNA, or polymerase chain reaction (PCR) fragment probes. The filters are incubated overnight at 42°–45° in a water bath. Alternatively, the HB 1100D Red Roller II (Hoeffer Scientific Instruments, San Francisco, CA) designed for incubating and hybridizing membranes from Southern transfers is recommended as a suitable oven that holds up to 6 large or small hybridization tubes; the apparatus can perform hybridizations with minimal volumes of solutions, as little as 5 ml.

The hybridization solution is removed and the membrane is then washed in $2 \times$ SSC–0.5% SDS for 15 min followed by a second wash in $2 \times$ SSC–0.1% SDS at room temperature, and then in $0.1 \times$ SSC–0.5% SDS at 42° for 1 hr with gentle rocking, followed by a stringent wash in $0.1 \times$ SSC–0.5% SDS at 68°–72° for 1 hr. The same procedure is used when the probe is the end-labeled RNA. The membrane is kept moist, sealed in a heat-sealable bag, and exposed to Kodak (Rochester, NY) XAR-5 film at $-70°$ with (or from room temperature, when sharper bands are desired) Du Pont Cronex Lightning Plus intensifying screens for sev-eral days. Fragment sizes are determined from the autoradiogram by measuring the migration distances and applying the relationship of Bear-den.[25] Typical restriction patterns of *Bcl*I and *Eco*RI digests of genomic DNA from strains of *B. subtilis* and from five species of the genus *Bacillus* obtained by the conventional gel electrophoresis–Southern blot analysis are shown in Figs. 1 and 3.

[24] P. Ostapchuk, A. Anilionis, and M. Riley, *Mol. Gen. Genet.* **180,** 475 (1980).
[25] J. Bearden, *Gene* **6,** 221 (1979).

---

image analysis software. (B) *Eco*RI digests were probed with a labeled 1.6-kb PCR fragment containing the amplified 16 tRNA gene cluster of *trnD* [R. Rudner, A. Chevrestt, S. R. Buchhotz, B. Studamire, A.-M. White, and E. D. Jarvis, *J. Bacteriol.* **175,** 503 (1993)]. The gel was electrophoresed as described for Fig. 1b. A density profile of lane 2 (*B. subtilis*) is also presented.

Hybridization Probes and Densitometry

Ribosomal RNA for use as a probe is isolated from *B. subtilis* 168T grown in VY medium and purified by the method of Margulies *et al.*[26] The 5 S, 16 S, and 23 S rRNA species are resolved in low melting point agarose (BRL, Gaithersburg, MD) or Sea-Plaque (FMC, Rockland, ME). The RNA bands are excised from the gel are 5′ end labeled with [γ-^{32}P]adenosine triphosphate and T4 polynucleotide kinase according to the method of Maizels.[27] Moreover, ^{32}P-labeled cDNAs of these rRNAs can be synthesized in agarose by a reverse transcriptase reaction according to published procedures.[21] Alternatively, cloned rDNA, tDNA, or PCR fragments,[7,9–12] all gel-purified and cleaned with GlasPac (National Scientific Supply Co., San Rafael, CA), can be used as hybridization probes. Labeling is done either by nick translation[28] or with the random primer extension kit as directed by the supplier (USB, Cleveland, OH) using [α-^{32}P]dCTP and/or [α-^{32}P]dATP. The labeled DNA is freed from low molecular weight material by passage through a 1-ml spun column of Sephadex G-50 (Pharmacia Fine Chemicals, Piscataway, NJ) equilibrated with TE buffer. The rRNA and the DNA are routinely labeled to a specific activity of 0.5×10^8 to 2×10^8 counts/min (cpm)/μg. For hybridization, 1×10^6 to 4×10^6 cpm of ^{32}P-labeled RNA or DNA probe is added per lane of the hybridization membrane.

Autoradiographed films are scanned for hybridization intensity using various commercial densitometers equipped with recorders such as the Zeineh soft laser scanning densitometer (Model SL-DNA; Biomed Instruments, Fullerton, CA) or Wayne Rasband's NIH image analysis software using a photometric charge-coupled diode (CCD) camera system (Biological Detection Systems). In our initial studies, densitometry measurements[7,8] were done with a linear transport drive containing a holder for X-ray film strips inserted between the spectrometer monochromatic light source (580 nm) and the spectrophotometer (Gilford Instruments). Scan profiles were made by a recorder (Model 6051, Gilford) equipped with a pyroscribe device onto heat-sensitive chart paper (Honeywell). Peak areas are determined with a planimeter (Nobis Instruments) equipped with a mobile cursor to scan the densitometric tracing. The intensities of the hybrid bands on autoradiograms of "Zoo Blots" reflect the relative amounts of probe DNA that is bound and therefore serve as a rough measure of the total amount of DNA present in a particular restriction

---

[26] L. Margulies, V. Remeza, and R. Rudner, *J. Bacteriol.* **107**, 610 (1971).

[27] N. Maizels, *in* "ICN–UCLA Symposium on Molecular and Cellular Biology" (G. Wulcox, J. Abelson, and C. F. Fox, eds.), Vol. 8, p. 247. Academic Press, New York, 1977.

[28] P. W. J. Rigby, M. Dieckmann, C. Rhodes, and P. Berg, *J. Mol. Biol.* **113**, 237 (1977).

fragment of the related species that is well matched with the corresponding DNA sequences in the probe preparation. The sum of the measured intensities for all hybrid bands in a bacterial chromosome then serves as an estimate of the relative homology index to the *B. subtilis* 168T probe DNA.

Analysis of Differentiation and Similarity Coefficients

To determine the extent of conservation of sequences in *rrn* regions and the extent of strain differentiation the following analysis is performed.[5] A "similarity coefficient" (*S*) representing a fractional estimate of the conserved fragment sizes between two chromosome digests is expressed as $S = 2a/(x + y)$, where *a* equals the pairs of size-conserved fragments in paired digests of chromosomal DNA and *x* and *y* are the total number of homolog fragments observed in paired autoradiogram lanes. The fraction of conserved fragments is related to the fraction of base substitutions[29] and to the number of fragments with a specific size range. For example, as seen in Fig. 3A, the *B. subtilis* 168T chromosome has 10 *Bcl*I restriction fragments ($x = 10$) and W23 has 7 ($y = 7$), with four bands being similar ($a = 4$); therefore, the calculated *S* value is 0.47. The same analysis between *B. subtilis* and *B. pumilus* gives an *S* value of 0.11. Data previously compiled on eight members of the genus *Bacillus* revealed variations between two restriction digests (*Eco*RI and *Hin*dIII) at all levels of similarity coefficients from highly conserved ($S = 0.86$) to random ($S = 0.00–0.14$).[8] We note that the *Eco*RI restriction patterns of the *Bacillus* species are considerably more conserved than those obtained with the *Hin*dIII enzyme.[8] Similarly, rRNA gene restriction pattern analysis in seven *Staphylococcal* species and their strains revealed a greater band conservation among the *Eco*RI homologs than among the *Hin*dIII fragments.[4]

Concluding Remarks

The use of *E. coli* rRNA (16 S, 23 S) as a broad-spectrum probe has allowed the application of fingerprinting of genomic DNA from a large number of isolates from diverse gram-positive and gram-negative bacteria.[3] This single RNA probe reacted with the *rrn* genes in the genomes of species as phylogenetically remote from *E. coli* as *Bacillus, Brucella, Acinetobacter, Mycobacterium,* and *Listeria*. Each species generated by *Eco*RI, *Hin*dIII, and *Bam*HI digests gave a novel band pattern of rDNA restriction fragments. When these patterns are combined they constitute the beginning of a database which could easily serve for the identification

---

[29] W. B. Upholt, *Nucleic Acids Res.* **4**, 1257 (1977).

of unknown isolates. Ideally, the comparison of the restriction pattern of an unknown isolate with the corresponding pattern of a known species would require screening the database, assuming that all the techniques have been standardized.

Determination of rDNA provides a novel means for detecting minor genomic differences and can be used to characterize clinical isolates, which apparently show fewer obvious variations in total DNA fingerprints compared to the differences seen between species.[3,4,14,30] As an epidemiology tool, rRNA gene restriction patterns are highly reproducible and easy to compare if one limits the number of bands attainable by the use of highly defined probes. The assignment of strains to a particular group or subgroup is arbitrary unless computer-assisted analyses are used.

[30] R. J. Owen, A. Beck, P. A. Dayal, and C. Dawson, *J. Clin. Microbiol.* **26**, 2161 (1988).

[15] DNA Fingerprinting of *Mycobacterium tuberculosis*

By Dick van Soolingen, Petra E. W. de Haas, Peter W. M. Hermans, and Jan D. A. van Embden

Introduction

The use of genetic markers to type pathogenic mycobacterial strains is of great value for studying the epidemiology of tuberculosis. This is particularly important because multidrug resistant organisms are increasingly prevalent and because the acquired immunodeficiency syndrome (AIDS) pandemic is posing new challenges to tuberculosis control strategies. Until recently, mycobacteriophage and drug susceptibility patterns were the only markers available for *Mycobacterium tuberculosis* strain typing. This method has limited utility because it is technically difficult to perform and only a limited number of types can be distinguished.[1]

Mycobacterium tuberculosis belongs to the *M. tuberculosis* complex group of bacteria which includes *M. tuberculosis, Mycobacterium bovis, M. africanum*, and *Mycobacterium microti*. These species constitute a genetically closely related group of bacteria which are often difficult to differentiate using biochemical markers and growth characteristics. This close taxonomic relatedness is also reflected in DNA–DNA hybridization studies which have shown nearly 100% chromosomal homology between

[1] J. Crawford and J. H. Bates, *in* "The Mycobacteria: A Sourcebook," Part A, (G. P. Kubica and I. G. Wayne, eds.), pp. 123. Dekker, New York, 1984.

METHODS IN ENZYMOLOGY, VOL. 235

these species.[2] Furthermore, fragment patterns of restriction endonuclease-digested total chromosomal DNA have been found to show little polymorphism.[3] The recent discovery of certain repetitive DNA elements in mycobacteria has permitted the elucidation of a high degree of DNA polymorphism among different isolates of the *M. tuberculosis* complex bacteria.

Two classes of DNA have been found to be involved in the genetic diversity: transposable DNA elements or insertion sequences (ISs) and short tandemly repeated DNA sequences. Although the latter type of repetitive DNA is an excellent tool to differentiate *M. tuberculosis* complex strains and also certain nontuberculosis mycobacteria like *M. kansasii* and *M. gordonae*,[4–6] we will restrict the methods described here to IS-associated DNA polymorphism. Transposable elements are discrete DNA segments that can move to new genetic locations without extensive sequence homology. Two unrelated transposable elements have been identified in *M. tuberculosis:* IS*1081* and IS*6110*.[7,8] The host range of both IS elements is limited to species of the *M. tuberculosis* complex and therefore these multicopy sequences are excellent targets for sensitive and specific detection of these pathogenic mycobacteria by polymerase chain reaction (PCR).[9,10] The 1324-base pair (bp) element IS*1081* is invariantly present in five or six copies in *M. tuberculosis* complex strains and the polymorphism of restriction fragments carrying IS*1081* is very limited. Therefore, this element generally is not a useful marker in the epidemiology of tuberculosis, however it can be used for reliable identification of *M. bovis* BCG.[11] The extremely high genetic stability of IS*1081* is in sharp contrast to the other tuberculosis complex-specific insertion sequence, IS*6110*. This 1.4 kb IS element belongs to the IS*3* family of insertion elements

[2] J. E. Clark-Curtiss, *in* "Molecular Biology of the Mycobacteria." (J. J. McFadden, ed.), p. 77. London Univ. Press, Surrey, U.K., 1990.

[3] D. M. Collins and G. W. de Lisle, *J. Clin. Microbiol.* **21**, 562 (1985).

[4] P. W. M. Hermans, D. van Soolingen, and J. D. A. van Embden, *J. Bacteriol.* **174**, 4157 (1992).

[5] P. M. A. Groenen, A. E. Bunschoten, D. van Soolingen, and J. D. A. van Embden, *Mol. Microbiol.* in press (1994).

[6] B. C. Ross, K. Raios, K. Jackson, and B. Dwyer, *J. Clin. Microbiol.* **4**, 942 (1992).

[7] D. M. Collins and D. M. Stephens, *FEMS Lett.* **83**, 11 (1991).

[8] D. Thierry, M. D. Cave, K. D. Eisenach, J. T. Crawford, J. H. Bates, B. Gicquel, and J. L. Guesdon, *Nucleic Acids Res.* **18**, 188 (1990).

[9] P. W. M. Hermans, D. van Soolingen, J. W. Dale, A. R. Schuitema, R. A. McAdam, D. Catty, and J. D. A. van Embden, *J. Clin. Microbiol.* **28**, 2051 (1990).

[10] D. Thierry, A. Brisson-Noël, V. Vincent-Lévy-Frébault, S. Nguyen, J. Guesdon, and B. Gicquel, *J. Clin. Microbiol.* **28**, 2668 (1990).

[11] D. van Soolingen, P. W. M. Hermans, P. E. W. de Haas, and J. D. A. van Embden, *J. Clin. Microbiol.* **30**, 1772 (1992).

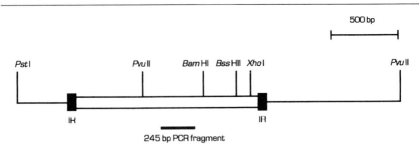

FIG. 1. Physical map of the chromosomal region of *M. bovis* BCG containing the IS*6110*-like element IS*987* of *M. bovis* BCG. IS*987* is depicted as an open bar, flanked by short inverted repeats (closed bar). The position of the 245 bp fragment amplified by the PCR using the INS1 and INS2 primers is depicted.

originally discovered in *Escherichia coli* and *Shigella*. The sequence of three independent copies have been determined and these were virtually identical, differing in only a few base pairs.[10,12,13] The number of IS*6110* copies and the sites of insertion in the chromosome are highly variable from strain to strain, indicating that this is a rather unstable genetic element. It is this very property that makes the IS*6110* element extremely powerful for studying the epidemiology of tuberculosis.

General Principle of Method

The DNA typing of clinical isolates of *M. tuberculosis* complex strains is based on polymorphism generated by variability in both the copy number and the chromosomal positions of IS*6110*[14]. The technique of fingerprinting entails the growth of *M. tuberculosis*, extraction of DNA, restriction endonuclease digestion, Southern blotting, and probing for the IS element. If it is desirable to compare fingerprints with those obtained in other laboratories, a standardized method can be used.[15] Only three parameters are critical for a standardized method of IS*6110*-based DNA fingerprinting: the specificity of the restriction enzyme, the nature of the DNA probe, and the use of appropriate molecular weight standards.

The physical map of the IS*6110* sequence is depicted in Fig. 1. *Pvu*II

[12] R. A. McAdam, P. W. M. Hermans, D. van Soolingen, Z. F. Zainuddin, D. Catty, J. D. A. van Embden, and J. W. Dale, *Mol. Microbiol.* **4**, 1607 (1990).

[13] P. W. M. Hermans, D. van Soolingen, E. M. Bik, P. E. W. de Haas, J. W. Dale, and J. D. A. van Embden, *Infect. Immun.* **59**, 2695 (1991).

[14] D. van Soolingen, P. W. M. Hermans, P. E. W. de Haas, D. R. Soll, and J. D. A. van Embden, *J. Clin. Microbiol.* **29**, 2578 (1991).

[15] J. D. A van Embden, J. T. Crawford, J. W. Dale, B. Gicquel, P. W. M. Hermans, R. McAdam, T. Shinnick, and P. Small, *J. Clin. Microbiol.* **31**, 406 (1993).

is used as restriction enzyme in the standardized method, because this enzyme cleaves the IS6110 sequence once. Since *M. tuberculosis* usually contains up to 25 IS6110 copies, the use of a DNA probe which overlaps both sides of the *Pvu*II site in the IS6110 sequence would result in overcrowded lanes with many overlapping bands. Therefore, we have arbitrarily chosen a DNA probe, which contains sequences only to the right of the *Pvu*II site on the physical map as shown in Fig. 1. This reduces the number of IS6110-containing bands in the fingerprint to half of the maximum number possible and the minimum size of any hybridizing band is 0.9 kb (see Fig. 1).

In exceptional cases, when the differentiation of the patterns is not adequate, the membranes can be reprobed with the "left arm" of IS6110 or with other repetitive DNA. In order to compare fingerprints between *M. tuberculosis* isolates prepared in different laboratories, the size of each IS6110-hybridizing fragment must be determined. This requires the use of molecular weight markers, which span the 0.9-15 kb range of most IS6110-hybridizing fragments. External size markers should be run in two or three lanes of each gel. Furthermore, as a control to check the hybridization conditions it is recommended to include in each gel a lane containing digested DNA from a *M. tuberculosis* reference strain of *M. tuberculosis* resulting in multiple bands of known size and intensity (see Fig. 2A). Although the use of size markers in separate lanes is adequate when comparing small numbers of similar strains (such as outbreak investigations) it will not provide sufficient precision to permit the computerized comparison of hundreds or thousands of strains. For maximal precision, size markers which do not cross-hybridize with IS6110 DNA are mixed with *M. tuberculosis* DNA samples to be fingerprinted prior to loading onto the gel. After hybridization with IS6110, the membrane can be reprobed with labeled molecular weight marker standards. This results in a second autoradiogram with molecular weight standards in each lane which can be superimposed onto the first autoradiograph. This procedure allows very accurate molecular weight determination of the IS-containing bands in each land (see Fig. 2B).

Procedure

Isolation of Genomic DNA from Mycobacteria

1. Inoculate a Loewenstein slant (Bio Merieux) with the mycobacterial strain of interest.[14]
2. Incubate at 37° until a clearly visible culture has formed at the surface, usually for 2 to 3 weeks.

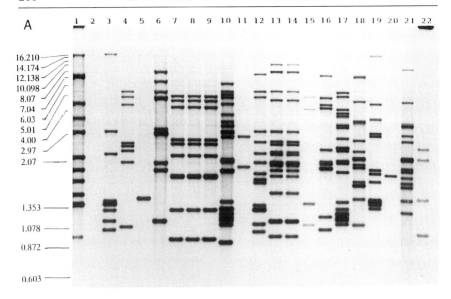

Fig. 2. (A) DNA fingerprints of PvuII-cleaved mycobacterial DNA, after hybridization with the peroxidase-labeled 245-bp PCR fragment of IS6110. All lanes contain DNA from clinical isolates. Lane 1, reference *M. tuberculosis* strain Mt.14323; lane 2, *M. chelonei chelonei*; lanes 7–9, *M. tuberculosis* isolates obtained from a single patient from sputum, feces, and urine, respectively; lanes 13 and 14, *M. tuberculosis* isolates from a single patient, lung tissue and ascites, respectively; lane 20, *M. bovis* BCG human isolate; all other lanes contain *M. tuberculosis* DNA from nonepidemiologically related strains. Note that *M. bovis* BCG contains a single IS-copy. Numbers on the left-hand side indicate sizes of the markers in kilobase pairs. (B) Internal size markers visualized by hybridization with labeled size markers. The filter is the same one used in Fig. 2A. After hybridization with the 245-bp IS6110 PCR fragment the filter was rehybridized with labeled size markers. By superimposing Fig. 2A and 2B, very accurate molecular weight estimations are possible for *M. tuberculosis* with IS6110 containing PvuII fragments in each individual lane. Note that prior to the vacuum blotting procedure the left and the right slot were filled with chromosomal DNA from the reference strain Mt14323 plus size markers to obtain reference points for correct alignment of both autoradiograms. The 2.07-kb fragment of the supercoiled marker represents open circular DNA, and its position relative to the linear fragments might differ depending on the gel conditions. The faint band between the 1.353 and the 1.078 kb represents a 2.07 kb supercoiled fragment. Numbers on the left-hand side indicate sizes of the markers in kilobase pairs.

3. Heat for 20 min at 80° to kill the cells for safety reasons.
4. Suspend two loopfuls of cells in 400 μl of 1× TE buffer (10 mM Tris-HCl, 1 mM EDTA, pH 8.0) by vortexing.
5. Add 50 μl of 10 mg/ml lysozyme, vortex, and incubate for 1 hr at 37°.

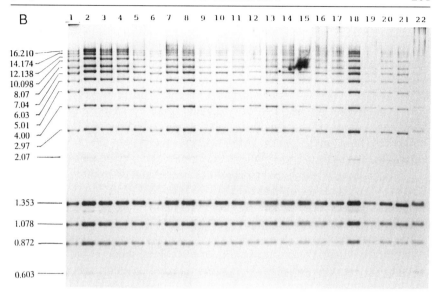

FIG. 2. (*continued*)

6. Add 70 μl of 10% SDS and 6 μl of 10 mg/ml proteinase K. Vortex and incubate for 10 min at 65°.
7. Add 100 μl of 5 M NaCl and vortex.
8. Add 80 μl of CTAB/NaCl solution (10% N-cetyl-N,N,N,-trimethyl-ammonium bromide, 0.73 M NaCl), vortex until the liquid content becomes "milky," and incubate for 10 min at 65°.
9. Add an equal volume of chloroform/isoamyl alcohol (24 : 1, v/v), vortex for 10 sec, and centrifuge for 5 min at 12,000 g.
10. Precipitate the nucleic acid by adding 0.6 volume of 2-propanol to the aqueous supernatant and keep the mixture for 30 min at −20°.
11. Centrifuge for 15 min at room temperature in a microcentrifuge at 12,000 g.
12. Wash the DNA pellet by adding 1 ml of cold 70% ethanol and centrifuge for another 5 min at 12,000 g to remove residual CTAB and NaCl.
13. Carefully remove the supernatant and permit the pellet to dry at room temperature for approximately 5 min.
14. Redissolve the pellet in 20 μl of 0.1× TE buffer. The DNA is stored at 4°. The yield amount is usually 5–20 μg of genomic DNA.

Southern Blotting of Mycobacterial DNA

Digest about 2 μg of genomic mycobacterial DNA with 10 U of the restriction enzyme *Pvu*II in a volume of 20 μl of digestion buffer (10 mM Tris-HCl, 10 mM MgCl$_2$, 50 mM NaCl, 1 mM DTE, pH 7.5). Add to the 20 μl of digested mycobacterial DNA sample 5 μl of sample buffer (50 mM Tris HCl, 5 mM EDTA, 50% glycerol, pH 7.5, 0.05% bromphenol blue), containing 15 ng *Pvu*II digested supercoiled ladder DNA (BRL, Bethesda, ML) and 5 ng of *Hae*III-digested PhiX174 DNA (Clontech, Palo Alto, CA).

This mixture is subjected to electrophoresis in 0.8% agarose (Seakem LE) in 89 mM Tris, 89 mM boric acid, 25 mM EDTA, 0.5 μg/ml ethidium bromide, pH 8.2) at 3.2 V/cm for about 18 hr. It should be noted that best resolution of fragments is obtained at low field strength.

As the mobility of DNA in agarose is slightly dependent of the DNA concentration, most accurate results are obtained when equal quantities of digested *M. tuberculosis* DNAs are subjected to electrophoresis. This can be done by visual estimation of the DNA amount after a short electrophoresis run, or by measuring the OD$_{260}$ with a spectrophotometer (Gene Quant, Pharmacia LKB Biochrom Ltd., Cambridge, UK).

Prior to transfer of DNA to a filter, the gel is exposed to ultraviolet light (5 min at 302 nm, 8000 μW/cm^2, UV Products) and is treated for 10 min with 0.25 M HCl to create breaks in the DNA. After denaturation for 20 min in 0.4 M NaOH (twice) and rinsing in distilled water, the DNA is transferred to a positively charged Nylon DNA membrane (Hybond N+, Amersham) by vacuum blotting (Miniblot-V, Millipore, Bedford, MA). The filters can be stored at $-20°$ for several years, provided that they are protected from dehydration by sealing in plastic.

Preparation of the Peroxidase-Labeled DNA Probe

The IS6110 DNA probe is prepared by *in vitro* amplification of a 245 bp fragment (see Fig. 1), using the polymerase chain reaction (PCR).[14] The reaction mixture consists of 5 μl of 10× PCR buffer (500 mM KCl, 100 mM Tris-HCl, pH 8.3, 30 mM MgCl$_2$, 0.2% gelatin), 4 μl of dNTP mix (0.2 mM of each dNTP), 4 μl of 0.4 μM primer INS-1 (5′ CGTGAGGG-CATCGAGGTGGC), 4 μl of 0.4 μM primer INS-2 (5′ GCGTAGGCGTC-GGTGACAAA), and 0.25 μl of *Taq* polymerase (1.25 unit, Perkin Elmer Cetus). Ten milliliters of target DNA (100 ng from *M. bovis* BCG) is added and the mixture is overlaid with 75 μl mineral oil to prevent evaporation. After mixing the centrifugation for 5 sec in a microcentrifuge the preparation is thermocycled as follows: 3 min at 94°; 20 cycles of 1 min at 94°, 1 min at 65°, and 2 min at 72°; and finally 4 min at 72°. The amplified DNA is examined after agarose electrophoresis and ethidium bromide staining.

If the correct fragment has been obtained, it is purified by Sephadex G50 molecular sieving: Remove a Quick Spin Sephadex G-50 column (Boehringer Mannheim, Germany) from its zip-lock bag and gently invert it several times to resuspend the medium. Remove the top cap from the column and then remove the bottom tip. This sequence is necessary to avoid a vacuum and uneven flow of the buffer. Allow the buffer to drain. Place the column into a tube (10 ml) and centrifuge in a swing out rotor for 2 min at 800 g. Bring the column in an upright position, carefully apply 100 μl of the sample to the surface of the column bed and place the column into a fresh 10-ml tube. Centrifuge for 4 min at 800 g in a swing-out rotor. The eluate of the second tube contains the purified DNA.

Precipitate the DNA by adding 2.5 volumes of ethanol and keep the mixture for 30 min at $-20°$. Spin for 15 min in a microcentrifuge at 12,000 g. Wash the pellet by adding 1 ml of 70% ethanol and repellet for 5 min in a microcentrifuge 12,000 g. Remove the supernatant, air dry the pellet, and dissolve in 50 μl of 0.1× TE. Determine the concentration of the PCR fragment by measurement of the OD_{260} of a 1 : 50 dilution of the sample. Store the purified DNA in small aliquots at $-20°$.

Hybridization and Detection of IS6110 DNA

The IS6110-containing DNA fragments immobilized onto the filter are visualized after hybridization with the peroxidase-labeled DNA probe by enhanced chemiluminescense (ECL). The 245-bp IS6110 DNA fragment is labeled with peroxidase, using glutaraldehyde as a crosslinker according to the instructions of the manufacturer (ECL, Amersham, UK). The labeled probe can be stored for 6 months at $-20°$ in 50% glycerol.

The filter is placed into a roller bottle or a plastic bag and prehybridized in hybridization (ECL) buffer containing 0.5 M NaCl for 15 min at 42°. After prehybridization the hybridization buffer is poured into a clean bottle. The labeled probe is mixed with the hybridization buffer and this mix is applied back into the roller bottle or bag. Hybridization takes place overnight at 42° under gentle shaking the bag or, if a roller bottle is used, rolling at 6 rpm. The filter is washed twice in prewarmed (42°) wash buffer [36% (w/v) urea, 0.4% SDS, 75 mM NaCl, 7.5 mM sodium citrate, pH 7] under gentle shaking or rolling at 6 rpm for 20 min at 42°. Additionally, it is washed twice in 75 mM NaCl, 0.75 mM sodium citrate, pH 7 for 5 min under gentle shaking at room temperature. In a dark room the filter is transferred to a clean box. The chemiluminescence detection mixture (0.125 ml/cm^2) containing hydrogen peroxidase and luminol (ECL) is applied onto the filter for exactly 1 min. The moist filter is wrapped in Saran wrap and autoradiographed in a cassette using Hyperfilm (Amersham,

UK). Depending on the degree of chemiluminescence, the film is exposed from 1 min to 2 hr.

Analysis and Interpretation of DNA Fingerprints

Small numbers of fingerprints can be compared for similarity by eye and this is usually satisfactory for the analysis of isolates from outbreaks of tuberculosis. However the handling of large numbers of polymorphic banding patterns requires computer-assisted data storage and analysis methods. In principle, variants of such methods are presently available,[16] but an evaluation of the usability of these systems for the epidemiology of tuberculosis has yet to be carried out. For computer-assisted analysis the use of internal size standards during electrophoresis is of critical importance for reliable comparison of large numbers of DNA finger-prints.[17] To fully exploit the potential of fingerprint data, comparison of fingerprints between different laboratories is needed. The procedure of fingerprinting described above is consistent with the recent recommendations on a standardized method of RFLP analysis of *M. tuberculosis*.[15] The *M. tuberculosis* genome contains one or more hot spots for integration of IS6110.[13] One of these loci has been identified as a region containing multiple direct repeats (DRs). About 85% of the *M. tuberculosis* strains contain one or more IS6110 copies at this DR region. Therefore, strains containing only one IS6110 element usually carry this element at a similar or identical restriction fragment, although these strains may be epidemiologically unrelated. If further differentiation of such strains is required, the filters could be reprobed with labeled IS1081 DNA, or probes derived from short repetitive DNA like DR or the polymorphic GC-rich repeat can be used.[17] In latter cases the genomic DNA should be cut with enzymes, like *Alu*I, which produce smaller restriction fragments.

Final Comments

The high degree of IS6110-dependent RFLP could suggest that IS6110 is a very unstable genetic element and that the IS6110 fingerprints might change after only a few generations of growth thus invalidating its use for epidemiological investigations. However, a large body of experimental and epidemiological evidence indicates that IS6110 transposition is an extremely rare event. For example, laboratory strains, which have been

[16] J. D. A. van Embden, D. van Soolingen, P. Small, and P. W. M. Hermans, *Res. Microbiol.* **143**, 385 (1992).
[17] D. van Soolingen, P. de Haas, P. Hermans, P. Groenen, and J. D. A. van Embden, *J. Clin. Microbiol.* **31**, 1987 (1993).

propagated for many decades in different laboratories, usually display a very similar IS6110 banding pattern.[17] Single transpositional events have been observed in isolates from microepidemics, but the change in the size of a single IS6110-containing restriction fragment usually does not affect the interpretation, as *M. tuberculosis* usually harbors multiple IS copies on different restriction fragments.

The differentiation of *M. tuberculosis* by DNA fingerprinting has led to many practical applications: monitoring the efficacy of chemotherapy, tracing and control of common sources of infection (outbreak investigations), identification of main routes of TB transmission, impact of immunosuppression on susceptibility to (re)infection, tracing nosocomial and laboratory infection, etc. It is to be expected that fingerprints might help to also answer long-standing questions such as the relative contribution of reinfection versus reactivation, the impact of HIV on the transmission to the non-HIV-infected community; the recognition of strains with high transmissability, virulence or tissue tropism and the protection by BCG against a subclass of *M. tuberculosis*. Finally, the observation that fingerprints of *M. tuberculosis* strains are associated with geographic origin[14] offers in principle the possibility to track the movement of strains over large distances. Such data may provide important insights in the global transmission of tuberculosis and identify strains with particular properties such as high transmissability and virulence.

Acknowledgments

We thank Janetta Top and Betty van Kranen for secretarial assistance during the preparation of the manuscript. Part of this work was supported by the World Health Organization, Programme for Vaccine development, and the European Community "Science and Technology for Development" programme of the Commission of the European Communities, Contract No. TS2-M-CT91-0335.

[16] Phylogenetic Identification of Uncultured Pathogens Using Ribosomal RNA Sequences

By THOMAS M. SCHMIDT and DAVID A. RELMAN

Introduction

Determining the evolutionary relatedness, or phylogeny, of organisms remains a fundamental goal of biology. Constructing phylogenies provides not only a view of the pathways down which evolution has proceeded, but also establishes a logical basis for taxonomic classifications. Microorganisms lack the rich variety of morphologies and behaviors that form the basis for classification in plants and animals. Consequently, microbial taxonomy has been based largely on numerical systems, where organisms that share some combination of metabolic traits are grouped. The traits selected for microbial taxonomy do not necessarily reflect the organism's phylogeny.

By establishing phylogenetic relationships, some properties of an otherwise unknown organism may be predicted on the basis of its known relatives: representatives of particular phylogenetic groups are expected to have properties common to that group. The ability to predict traits would be particularly useful in the description of uncultured microorganisms, because the establishment of pure cultures is necessary for the observation of most characteristics. With perhaps as few as 1% of extant microorganisms maintained in pure culture, the requirement for culturing hinders the description of novel microorganisms. Although the fraction of pathogenic microorganisms that have been cultured may be somewhat higher than 1% because of the importance placed on understanding these microbes, many human pathogens remain uncultured and the number of novel opportunistic pathogens may increase along with the expanding population of immunocompromised hosts. This chapter will focus on methods to obtain and analyze rRNA sequences for the phylogenetic identification of uncultured human pathogens.

Use of Macromolecular Sequences to Infer Organismal Phylogenies

Nucleotide and amino acid sequences provide a multitude of characters useful for inferring phylogenetic relationships and may well be the primary source of systematic data in the near future. The identification and phylogenetic classification of organisms can be based on the comparison of sequences of several macromolecules. The choice of macromolecule on which to base organismal phylogenies is crucial.[1] The ideal macromolecule is (1) present in all organisms under consideration, (2) homologous,[2] (3) sufficiently constrained so that multiple mutations at a single site are rare, (4) not transferred laterally, and (5) sufficient in size to permit a statistically valid comparison. The most commonly used macromolecule to establish

---

[1] C. R. Woese, *Microbial. Rev.* **51,** 221 (1987).

[2] Molecular biologists often equate the terms homology and similarity. This has unfortunately obscured the original definition of homology in describing features that are similar due to common evolutionary ancestry. In this chapter, homology will be used as originally intended and the degree of similarity between sequences will be given as percent similarity or identity.

organismal phylogenies is the major structural RNA of the ribosomal small subunit. The small subunit ribosomal RNA (rRNA) has a sedimentation coefficient of 16 S in most prokaryotes and 18 S in most eukaryotes. Other macromolecules that have been used to construct universal phylogenies include elongation factors Tu and G and the α and β subunits of F_1-ATPase.[3] The general picture that emerges from phylogenetic analysis of these macromolecules is that there are three primary lines, or domains, of evolutionary descent: the Bacteria (formerly eubacteria), the Archaea (formerly archaebacteria), and the Eucarya (formerly eukaryotes).[4]

The use of rRNA sequences to infer evolutionary relationships among organisms is now widely accepted. As a result, many taxonomists now employ a comparative analysis of macromolecular sequences to identify and classify organisms. Ribosomal RNA sequences are particularly attractive for use in the identification and classification of organisms because of the availability of a growing database of rRNA sequences and the ability to amplify rRNA-encoding genes (rDNA) from the limited amount of material that is routinely available for analysis via the polymerase chain reaction (PCR). In addition, the sequences on which these phylogenies are based may be useful in the development of specific diagnostic reagents. These reagents may prove useful in the elucidation of the natural ecology of organisms that are diffcult to study by other means.

Methods for Obtaining Ribosomal RNA Sequences from Uncultured Microorganisms

The demonstration that small subunit rRNA sequences could be used to establish phylogenetic relationships has motivated the development of improved methods for obtaining these sequences. Most of the early rRNA-based phylogenetic studies were based on RNase T1 oligonucleotide sequence catalogs of 16 S rRNA.[5] The creation of these catalogs is cumbersome and requires purification of 16 S rRNA in amounts which would rarely be available from uncultured microorganisms. Technical advances have facilitated direct rRNA sequencing (discussed below), and, with the

[3] N. Iwabe, K. Kuma, M. Hasegawa, S. Osawa, and T. Miyata, *Proc. Natl. Acad. Sci. U.S.A.* **86,** 9355 (1989).
[4] C. R. Woese, O. Kandler, and M. L. Wheelis, *Proc. Natl. Acad. Sci. U.S.A.* **87,** 4576 (1990).
[5] G. E. Fox, E. Stackebrandt, R. B. Hespell, J. Gibson, J. Maniloff, T. A. Dyer, R. S. Wolfe, W. E. Balch, R. S. Tanner, L. J. Magrum, L. B. Zablen, R. Blakemore, R. Gupta, L. Bonen, B. J. Lewis, D. A. Stahl, K. R. Luehrsen, K. N. Chen, and C. R. Woese, *Science* **209,** 457 (1980).

advent of the polymerase chain reaction,[6] it is possible to amplify bacterial rDNA directly from infected human tissue. These techniques permit the phylogenetic identification of an organism without the need to isolate or culture the organism.

Reverse Transcriptase Sequencing of Ribosomal RNAs

A significant breakthrough in 16 S rRNA sequence determination was achieved when universally conserved regions of the 16 S rRNA sequences were identified. From these regions, complementary oligodeoxynucleotide primers could be synthesized for use with reverse transcriptase in dideoxy-nucleotide-terminated sequencing from bulk cellular RNA.[7] Despite the efficiency of this method, there are some limitations. Reverse transcriptase sequencing of bulk RNA results in an error rate of approximately 1%.[7] In addition, and more relevant to the study of rare and uncultured microorganisms, sufficient amounts of rRNA must be available to conduct sequencing reactions. Because samples available for the identification of unknown pathogens are usually small, this approach has not been practical, and so details of the reverse transcriptase method are not presented here.

Direct Cloning of 16 S Ribosomal RNA Genes

Ribosomal RNA sequences have been obtained from environmental microbial communities by direct cloning of mixed-population DNA.[8] This method does not require cultivation of the organisms, nor does it impose the potential bias created by the initial selection of genes that may be preferred targets for broad-range PCR primers. Bulk genomic DNA is purified from a sample, subjected to limited digestion with a restriction endonuclease, and size-fractionated. A recombinant library is created with the size-fractionated DNA and screened with universal or broad-range rRNA probes, and selected clones are sequenced (Fig. 1).

This approach requires that adequate DNA be available for library construction, again a potential limitation when working with clinical samples. Also, with extremely few microorganisms amid a far greater number of host cells, the process of screening a recombinant library may be tedious. A PCR amplification step provides a means for obtaining an

---

[6] R. K. Saiki, S. Scharf, F. Faloona, K. B. Mullis, G. T. Horn, H. A. Erlich, and N. Arnheim, *Science* **230,** 1350 (1985).

[7] D. J. Lane, B. Pace, G. J. Olsen, D. A. Stahl, M. L. Sogin, and N. R. Pace, *Proc. Natl. Acad. Sci. U.S.A.* **82,** 6955 (1985).

[8] T. M. Schmidt, E. F. DeLong, and N. R. Pace, *J. Bacteriol.* **173,** 4371 (1991).

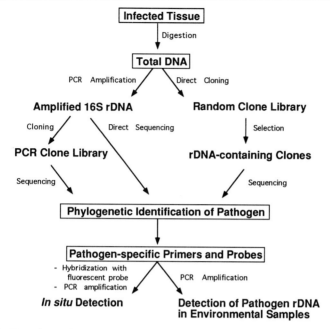

FIG. 1. Experimental approaches to the identification of uncultured human pathogens using 16 S rRNA sequences.

adequate number of microbial 16 S rDNA molecules for subsequent cloning or sequencing.

Polymerase Chain Reaction Amplification of 16 S Ribosomal RNA Genes

Several groups of investigators have demonstrated the utility of the PCR in generating large amounts of 16 S or 16 S-like rDNA.[9-12] The target material in these studies was microbial genomic DNA. In one study, bacterial 16 S rRNA sequences were amplified from a mixture of bacterial and human genomic DNA.[9] The key to these studies was the use of universal or broad-range small subunit rRNA sequences for the design of PCR oligonucleotide primers. There are several conserved sequence

[9] K. Chen, H. Neimark, P. Rumore, and C. R. Steinman, *FEMS Microbiol. Lett.* **48,** 19 (1989).
[10] L. Medlin, H. J. Elwood, S. Stickel, and M. L. Sogin, *Gene* **71,** 491 (1988).
[11] W. G. Weisburg, S. M. Barns, D. A. Pelletier, and D. J. Lane, *J. Bacteriol.* **173,** 697 (1991).
[12] K. H. Wilson, R. B. Blitchington, and R. C. Greene, *J. Clin. Microbiol.* **28,** 1942 (1990).

regions throughout the small subunit rRNA molecule (Fig. 2). Oligonucleotides complementary to any of these regions may be useful as sequencing
or PCR primers or as hybridization probes. Sequences conserved within
the domain of the bacteria are the best characterized and most widely
tested. Several bacterial broad-range sequences are listed in Table I.

We have described previously an approach for the identification of
uncultured bacterial pathogens based on the amplification of bacterial
16 S rDNA directly from infected host tissue.[13,14] This approach is most
successfully applied to fresh-frozen tissue; if this is unavailable, formalin-
fixed, paraffin-embedded tissue is sufficient. The stages of approach are
illustrated in Fig. 1. As soon as a 16 S rRNA sequence that may correspond
to that of the microbial pathogen is obtained, it is crucial to demonstrate
that this sequence is found in most, if not all, affected tissues. At the
same time, it should not be found in tissues uninvolved by this disease.
The design of specific PCR primers from the putative pathogen-associated
16 S rRNA sequence can be extremely useful in analyzing numerous
tissues in this kind of investigation, especially if some of them contain
commensal host microbial flora. A discussion of the criteria by which one
may identify an uncultured microorganism with this approach is presented elsewhere.[15]

From an experimental perspective, it is most important to (1)
minimize the number of manipulations of the tissue during preparation
of the digest[16]; (2) isolate physically the processes of sample digestion,
PCR setup, and PCR product manipulation, as well as adhere to
protocols for prevention of contamination[17]; and (3) intersperse numerous negative control tissue samples among the experimental samples
during the entire procedure. The following is an outline of the experimental steps of this approach.

1. To prepare formalin-fixed, paraffin-embedded tissue for PCR amplification, cut 10 μm thick tissue sections using disposable, sterile
microtome blades and instruments, and place the tissue in sterile
microcentrifuge tubes. Extract paraffin by incubating sections in
octane for 30 min twice, and then by washing with ethanol twice.

[13] D. A. Relman, J. S. Loutit, T. M. Schmidt, S. Falkow, and L. S. Tompkins, N. Engl. J. Med. 323, 1573 (1990).
[14] D. A. Relman, T. M. Schmidt, R. P. MacDermott, and S. Falkow, N. Engl. J. Med. 327, 293 (1992).
[15] D. A. Relman, J. Infect. Dis. 168, 1 (1993).
[16] D. K. Wright and M. M. Manos, in "PCR Protocols: A Guide to Methods and Applications" (M. A. Innis, D. H. Gelfand, J. J. Sninsky, and T. J. White, eds.), Academic Press, San Diego, 1990.
[17] S. Kwok and R. Higuchi, Nature (London) 339, 237 (1989).

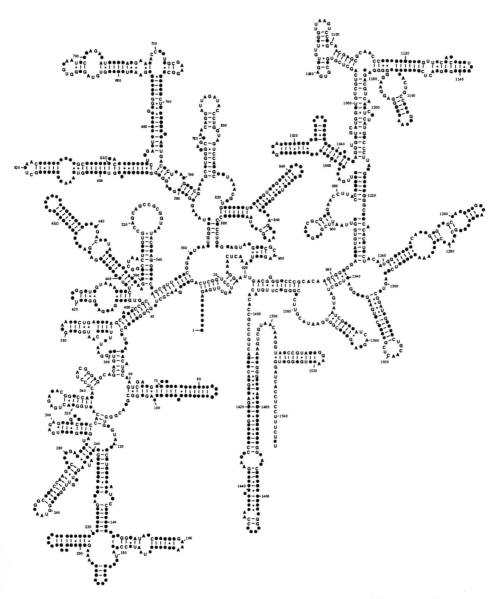

FIG. 2. Secondary structure of the 16 S rRNA from *Escherichia coli* [R. R. Gutell, B. Weiser, C. R. Woese, and H. F. Noller, *Prog. Nucleic Acic Res. Mol. Biol.* **32,** 155 (1985)], with bases that occur at a frequency of less than 90% in bacteria denoted (●). The 90% consensus was determined from bacterial 16 S rRNA sequences in the ribosomal database project (release 2.0) [G. J. Olsen, R. Overbeck, N. Larsen, T. L. Marsh, M. J. McCaughey, M. A. Macinkenas, W. Kuan, T. J. Macke, Y. Xing, and C. R. Woese, *Nucleic Acids Res.* **20,** 2199 (1992)].

TABLE I
PRIMERS FOR BROAD-RANGE POLYMERASE CHAIN REACTION AMPLIFICATION OF
BACTERIAL 16 S RIBOSOMAL DNA SEQUENCES

Name	Primer sequence (5' to 3')	16 S rRNA positions[a]	Refs.
Forward			
8F	AGA GTT TGA TCC TGG CTC AG	8–27	c, d, e
515F	GTG CCA GCA GCC GCG GTA A	515–533	f
911F	TCA AAK[b] GAA TTG ACG GGG GC	911–930	f, g
1175F	GAG GAA GGT GGG GAT GAC GT	1175–1194	g, h
Reverse			
806R	GGA CTA CCA GGG TAT CTA AT	806–787	f, i
1390R	AGG CCC GGG AAC GTA TTC AC	1390–1371	g, h
1492R	GGT TAC CTT GTT ACG ACT T	1510–1492	c, d, e

[a] *Escherichia coli* 16 S rRNA positions that correspond to the 5'–3' ends of each of the primers.

[b] K indicates that a 1 : 1 mixture of dT and dG is incorporated at this position.

[c] P. A. Eden, T. M. Schmidt, R. P. Blakemore, and N. R. Pace, *Int. J. Syst. Bacteriol.* **41,** 324 (1991).

[d] W. G. Weisburg, S. M. Barns, D. A. Pelletier, and D. J. Lane, *J. Bacteriol.* **173,** 697 (1991).

[e] D. A. Relman, P. W. Lepp, K. N. Sadler, and T. M. Schmidt, *Mol. Microbiol.* **6,** 1801 (1992).

[f] D. A. Relman, T. M. Schmidt, R. P. MacDermott, and S. Falkow, *N. Engl. J. Med.* **327,** 293 (1992).

[g] D. A. Relman, J. S. Loutit, T. M. Schmidt, S. Falkow, and L. S. Tompkins, *N. Engl. J. Med.* **323,** 1573 (1990).

[h] K. Chen, H. Neimark, P. Rumore, and C. R. Steinman, *FEMS Microbiol. Lett.* **48,** 19 (1989).

[i] K. H. Wilson, R. B. Blitchington, and R. C. Greene, *J. Clin. Microbiol.* **28,** 1942 (1990).

Dry in a heating block at 55°. For frozen tissue, excise approximately a 2 mm cube of the tissue, wash twice in ethanol, and dry.

2. Digest the dried tissue in 50–300 μl of freshly prepared digestion buffer (200 μg/ml proteinase K, 1% (w/v) Laureth-12 nonionic detergent, 50 mM Tris, pH 8.5, 1 mM EDTA), the volume used depending on the amount of tissue and ease with which it forms a homogeneous suspension. Incubate at 55° for 3–18 hr.

3. Heat-inactivate the proteinase K at 95° for 10 min. Pellet the insoluble debris in a microcentrifuge at 14,000 rpm for 5 min. Use 1–10 μl of the supernatant directly in PCR amplifications. Store unused digest at $-20°$ and, if required, use within 2 weeks.

4. Individual PCR conditions must be optimized by varying concentrations of Mg^{2+} (1–6 mM), deoxyribonucleoside triphosphate (20–200 μM each), and *Taq* polymerase (0.5–5 U/100 μl). Use of the "hot start" technique (adding all reaction components except an essential component, e.g., *Taq* polymerase, denaturing at 95° for 3 min, then adding the essential component while holding temperature at 85°) may improve the sensitivity of the PCR. Use the highest annealing temperature possible, as permitted by primer length and composition. Forty PCR cycles may be necessary with rare targets and fixed tissue.

5. For amplification of 16 S rRNA targets shorter than 800 base pairs (bp) with broad-range primers, contamination of PCR reagents with exogenous DNA may necessitate irradiation of the PCR mixture with UV light before addition of tissue digest. Place the tubes on the surface of a 302 nm wavelength transilluminator. Titrate the duration of irradiation that is tolerated by each set of primer pairs and PCR reagents.

6. The quality of each tissue digest should be tested by amplifying a known host genomic target, for example, β-globin gene sequence, from varying amounts of the digest supernatant.

7. The presence of a PCR product can be determined with agarose gel electrophoresis and ethidium bromide staining. DNA sequence determination of PCR products can be performed directly or after cloning the product using standard recombinant techniques. Nucleotides misincorporated by *Taq* polymerases late in the amplification cycle are less likely to result in erroneous sequence determination if the PCR product is sequenced directly. This is because most current sequencing, techniques will not detect a template that comprises ~ <10% of the total templates available.

Aligning and Comparing Sequences

Once the rRNA sequence has been determined, it can be compared to previously determined sequences. Known sequences can be retrieved from several sequence databases such as GenBank, the European Molecular Biology Laboratory (EMBL) data library, and the DNA Data Bank of Japan (DDBJ). Information on access to these databases is available through electronic mail at the following addresses:

GenBank: Info@NCBI.nlm.nih.gov
EMBL Data Library: DataLib@EMBL-Heidelberg.DE
DDBJ: ddbj@ddbj.nig.ac.jp

Because of the utility of aligned sequence sets, a new database of aligned rRNA sequences has been created, the ribosomal RNA database project (RDP).[18] Information on the database is available by sending the one-word message "help" to "server @ rdp.life.vivc.edu." The aligned sequence can be retrieved with a file transfer protocol (FTP) to "rdp.life.vivc.edu." Information about the database is also available by contacting Dr. Gary J. Olsen (Department of Microbiology, University of Illinois, Urbana, IL 61810).

Another database of aligned rRNA sequences is maintained by Rupert DeWachter and colleagues.[18a] Information about this database can be obtained by sending an electronic mail message to: rrna@ccv.via.ac.be.

The editorial policy of most journals requires submission of sequence data to an accessible database before publication of a paper. This includes sequences that are used to construct phylogenetic trees. When retrieving sequences from the databases for comparison, be aware that some sequences are submitted with the sequence of PCR primers or portions of a cloning vector appended to the rRNA sequence. These extra bases can confuse the alignment process (discussed below) and influence the outcome of a phylogenetic analysis.

To determine relationships between sequences it is not sufficient simply to compare homologous sequences; sequences must be properly aligned so that homologous positions are compared. Alignment is a crucial, and too often overlooked, step in phylogenetic analysis; all subsequent manipulations of data are predicated on the comparison of homologous positions in sequences. Erroneous measures of similarity and phylogeny result from inappropriate alignments.

At present, alignment of rRNA sequences is best done manually. Most computer alignments are designed to optimize the alignment between two sequences by inserting gaps to minimize the number of differences between the sequences. A score is assigned to alignments, with points awarded for the number of positions with identical bases and points subtracted for dissimilar bases or the insertions of gaps. Computer alignments can serve as a good first approximation, but the best computer alignment does not necessarily reflect an alignment of homologous positions. This can best be done by taking into consideration the secondary structure of the molecule.

---

[18] G. J. Olsen, R. Overbeck, N. Larsen, T. L. Marsh, M. J. McCaughey, M. A. Macinkenas, W. Kuan, T. J. Macke, Y. Xing, and C. R. Woese, *Nucleic Acids Res.* **20**, 2199 (1992).
[18a] P. DeRijk, J.-M. Neefs, Y. V. dePeer, and R. DeWachter, *Nucleic Acids Res.* **20**, 2075 (1992).

All 16 S-like rRNAs contain the same core structure.[19] The nucleotides which are conserved in 90% of the bacterial sequences available through the 16 S ribosomal database project are shown in Fig. 2. Fitting a newly determined sequence to this structure aids in the alignment of the new sequence because one can reasonably predict which positions are homologous for much of the molecule. We routinely use an enlarged version of a 16 S rRNA secondary structure and pencil in changes between the template and a new sequence. This is useful in suggesting the correct alignment of nucleotides, and it also is a reasonable measure of the quality of the sequence data, because base-pairing between the majority of the helices shown in Fig. 2 should be preserved by compensatory base changes. Hypervariable regions of the molecule, for example, from positions 1000 to 1040 (Fig. 2) are particularly diffcult to align because of variation in the number of nucleotide positions and in the secondary structure. These regions stand out in Fig. 2, where frequently varied positions are denoted with a filled circle. If alignment in a region is ambiguous, the region is best left out of comparisons of the sequence and phylogenetic analyses because the assumption of positional homology is not fulfilled. These regions can be useful for the comparisons of closely related organisms and for the design of specific oligodeoxynucleotide probes (discussed below).

There will necessarily be gaps inserted into sequences to achieve positional homology because the lengths of 16 S-like ribosomal RNAs vary. Gaps are often omitted from sequence comparisons and phylogenetic analyses because a proper treatment of gaps is not obvious. For instance, in hypervariable regions where a stem and loop structure is present in one group of organisms and absent in another, that difference could be the result of a single ancestral addition or deletion or a series of stepwise additions or deletions. Should all these positions be given equal weight, or should they be treated as a single event? When there is no indication of how questions such as these should be answered, the region in question is often omitted from subsequent analyses.

Selection of Treeing Algorithms

Inferring phylogenetic relationships from sequences of macromolecules requires selection of an appropriate method from the many available. Several reviews present and compare the conceptual frameworks of the

---

[19] R. R. Gutell, B. Weiser, C. R. Woese, and H. F. Noller, *Prog. Nucleic Acid Res. Mol. Biol.* **32**, 155 (1985).

various methods (see Refs. 20 and 21 and references therein) and are useful to anyone anticipating a phylogenetic analysis of traits. An optimal method for inferring phylogenetic relationships should be efficient (fast), powerful enough to handle large data sets, product consistent and statistically significant results, and be falsifiable. Unfortunately, these characteristics are often incompatible, and so one must accept the limitations of the method employed. Three commonly used approaches to infer phylogenies from 16 S rRNA sequences are the distance matrix, parsimony, and maximum likelihood methods.

Distance programs construct a phylogenetic tree from a matrix of pairwise comparisons generated for each pair of sequences selected for analysis. A tree is optimized as the lengths of line segments connecting two sequences are adjusted to fit most closely to the original similarity computations. The sum of the adjusted lengths connecting any two organisms is commonly referred to as evolutionary distance and is frequently represented on the horizontal axis in phylogenetic trees. The evolutionary distance then is based on the number of fixed point mutations that differ between sequences (sequence similarity).

The fundamental goal of parsimony analyses is to minimize the number of mutations required to organize sequences in a phylogenetic tree. The most parsimonious tree is the one that requires the fewest number of fixed mutations to produce the inferred relationships. The algorithms used to search for the optimal tree are varied.[20]

Another frequently used method to infer phylogenies from rRNA sequences is maximum likelihood. The objective of maximum likelihood methods is to evaluate the net likelihood that a given evolutionary model will yield the observed nucleotide sequences. In comparing multiple sequences, this approach considers the probability of each nucleotide occurring at each node in the tree as a function of both branch lengths and branching order.

The methods to infer phylogenies result in a tree that "best" fits the data, as an exact solution is usually not possible. Because the best tree is often only marginally better than other trees that fit the data, it is useful to provide some measure of significance to any given tree. The most frequently used method to achieve this is the bootstrap.

In a bootstrap analysis, the number of data points compared remains fixed, but positions are sampled at random, recorded and then returned (replacement), and may be resampled. The outcome of such a resampling

[20] D. L. Swofford and G. J. Olsen, in "Molecular Systematics" (D. M. Hillis and C. Mortiz, eds.), p. 411 Sinauer, Sunderland, Massachusetts, 1990.
[21] D. Penny, M. D. Hendy, and M. A. Steel, TREE 7, 73 (1992).

is that some positions are selected for analysis multiple times and others not at all. Analyses then proceed with the "new" data set. By repetitive sampling and analysis of the data, the variation among the resulting estimates is taken to indicate the size of the error involved in making the original estimates. If a particular relationship appears in 95% or more of the trees, then it is taken to be statistically significant.[22]

Using several phylogenetic methods to infer the phylogenetic classification of an organism can also increase the confidence in a particular association. Similarly, the analysis of signature sequences[23] offers confirmation of correct placement.

There are two widely distributed programs for phylogenetic analysis. PHYLIP (Phylogeny Inference Package) is a free package of programs for inferring phylogenies and performing certain related tasks. At present it contains 31 programs, which carry out different algorithms on different kinds of data, including nucleotide sequences. The package includes extensive documentation files that provide the information necessary to use and modify the program. PHYLIP is available through network distribution by anonymous FTP from "evolution.genetics.washington.edu". Other information can be obtained from Dr. Joseph Felsenstein [Department of Genetics, University of Washington, Seattle, WA 98195 (Internet: joe@genetics.washington.edu)]. Fast DNAml, a modified version of the Phylip maximum likelihood program, is available through RDP.

PAUP (Phylogenetic Analysis Using Parsimony) performs parsimony analyses using several models and includes bootstrapping routines. PAUP is distributed in precompiled form for IBM-PC and Macintosh computers and as C source code for workstations, minicomputers, and mainframes. Contact Mary Lou Williamson (Center for Biodiversity, Illinois Natural History Survey, 607 E. Peabody Dr., Champaign, IL 61820).

Technical Issues and Problems

The 16 S rRNA-based approach we have described for the identification of uncultured pathogens raises some technical and theoretical issues. These issues focus on the nature of the tissue sample, the choice of primers, the problem of contamination, and the interpretation of multiple, closely related 16 S rRNA sequences within the product of a single PCR amplification.

[22] J. Felsenstein, *Evolution* **39**, 783 (1985).
[23] C. R. Woese, E. Stackebrandt, T. J. Macke, and G. E. Fox, *Syst. Appl. Microbiol.* **6**, 143 (1985).

Quality and Type of Tissue

The quality of the experimental sample is crucial for subsequent broad-range 16 S rDNA PCR amplification. Several factors pertaining to the tissue sample play important roles in determining whether a PCR product can be generated. Tissue fixation, in comparison with quick-freezing, reduces the likelihood of target amplification with the PCR. Storing fixed samples causes degradation of average DNA molecular size, cross-linking, and oxidative damage.[24] In addition, older tissue samples may contain amplification inhibitors. Increased amounts of *Taq* polymerase may overcome this inhibition.[24] The method of tissue fixation is also known to influence PCR results.[25] Other tissue factors of importance are anatomic site and tissue pathology. Sites contaminated with commensal flora are less desirable. Tissue necrosis may lead to degradation of DNA targets, and large amounts of fibrosis may hinder tissue digestion in the laboratory. One should always test tissue digests as reagents in the PCR by attempting to amplify a known host tissue DNA target such as the β-globin gene.[26] The size of this test DNA target should approximate the size of the intended 16 S rDNA target. A tissue sample should not be classified as negative in a bacterial 16 S rDNA amplification reaction unless such a tissue quality control is positive.

Choice of Primers

To begin the process of amplifying 16 S rDNA sequences from a previously uncharacterized microbial pathogen, one must select regions of the 16 S rRNA molecule that are conserved within the relevant phylogenetic grouping. For a suspected eubacterial pathogen the latter grouping would be the bacterial domain. For eukaryotic pathogens (e.g., fungi, protozoa) conserved sequences for various monophyletic groups are less well-defined. In all cases, primer sequences should not be complementary to the corresponding regions of the host (e.g., human) small subunit rRNA genes. Some bacterial broad-range primer sequences are listed in Table I. Endonuclease recognition sites may be added to the 5' end of primer sequences to facilitate the cloning of the reaction products.

Regions of conserved sequence occur throughout the 16 S rRNA-encoding gene; as a result, one may choose 16 S rDNA targets of variable size. Relatively smaller target sizes are more appropriate for fixed or older

[24] S. Paabo, *Proc. Natl. Acad. Sci. U.S.A.* **86**, 1939 (1989).
[25] C. E. Greer, S. L. Peterson, N. B. Kiviat, and M. M. Manos, *Am. J. Clin. Pathol.* **95**, 117 (1991).
[26] H. M. Bauer, Y. Ting, C. E. Greer, J. C. Chambers, C. J. Tashiro, J. Chimera, A. Reingold, and M. M. Manos, *J. Am. Med. Assoc.* **265**, 472 (1991).

tissue specimens, or for tissues with a significant degree of necrosis. At the same time, the choice of a small target size increases the likelihood of amplifying extraneous or contaminating DNA fragments. It also provides one with less sequence information for subsequent phylogenetic analysis. Separate amplifications of small, overlapping 16 S rRNA-encoding gene regions may be an alternative to the amplification of relatively large portions of the gene. Perhaps even more so than with other PCR amplifications, one should select primer pairs with similar lengths and melting temperatures, as well as pairs that do not anneal with one another or with themselves.[27]

Contamination of Tissue or Reagents

What does it mean to find multiple, unrelated 16 S rRNA sequences within the broad-range amplification product from a single tissue? Possibilities include a true polymicrobial disease process, contamination of the tissue sample with an extraneous organism or its DNA during procurement or digestion, and contamination of PCR reagents. The analysis of multiple tissue samples from different anatomic sites and multiple recombinant product clones may help resolve these possibilities. DNA contamination of at least one PCR reagent, namely, *Taq* polymerase, has been reported.[28,29] In one instance, DNA product from a "reagent-only" PCR control reaction was sequenced and found to correspond to a *Pseudomonas* sp.[13] In addition, one would suspect that column purification [e.g., high-performance liquid chromatography (HPLC)] of oligonucleotide PCR primers may contaminate the latter with extraneous DNA. Acrylamide gel purification of oligonucleotides may be less likely to cause this problem. Ultraviolet irradiation of PCR reagents before the addition of digested sample reduces PCR amplification of contaminating DNA within the PCR reagents.[30] One should titrate the UV radiation dose for the intended target DNA size and the specific PCR reagents being used.

Product carryover in the PCR becomes an increasingly important problem as one continues to produce a particular amplified product in a given laboratory. This problem is revealed with the use of numerous negative control tissue samples and the analysis of PCR products. Strict adherence to physical isolation methods may prevent or eliminate carryover.[17] In

[27] M. A. Innis, D. H. Gelfand, J. J. Sninsky, and T. J. White (eds.), "PCR Protocols: A Guide to Methods and Applications." Academic Press, San Diego, 1990.
[28] K. H. Rand and H. Houck, *Mol. Cell. Probes* **4**, 445 (1990).
[29] T. M. Schmidt, B. Pace, and N. R. Pace, *BioTechniques* **11**, 176 (1991).
[30] G. Sarkar and S. S. Sommer, *Nature (London)* **343**, 27 (1990).

certain circumstances, however, enzymatic or photochemical techniques for PCR product inactivation may be useful.[31,32]

Ribosomal RNA Sequence Microheterogeneity

Current taxonomic definitions do not specify the degree to which 16 S rRNA sequence heterogeneity occurs among the members of species or genera.[33] In the future, an expanded sequence database may facilitate the establishment of 16 S rRNA sequence criteria for species assignments. In the meantime, investigators must struggle to interpret the finding of multiple, closely related 16 S rRNA sequences within a single tissue amplification reaction. One possible interpretation is that there are multiple, related strains or species within the tissue. Another possibility is amplification of different rRNA operons from the same organism. In addition, damaged DNA can cause substitutions or mosaic sequence products due to "jumping PCR."[34] We have found a 3.5-fold higher rate of nucleotide substitutions among cloned, amplified products from formalin-fixed samples than from fresh-frozen samples.[14] In comparison, *Taq* polymerase contributes a background, intrinsic error rate of approximately 0.1–0.2%.[35] Direct sequencing of amplified product minimizes the influence of random nucleotide substitutions that may occur late during amplification. Another approach to identifying this problem is to look for paired complementary base changes in regions of known secondary structure. Paired changes suggest true 16 S rNA variation in the original target molecules.

Treeing Artifacts

Any analysis is only as reliable as the original data. In addition to the quality of the sequence data, the quantity of data affects the validity of the analysis. Unless unavailable due to experimental constraints, complete 16 S rRNA sequences should be determined and used.

Several factors may affect the placement of an organism in a phylogenetic tree. For example, if there is a group of sequences in an analysis whose G + C ratio clusters and varies from the other sequences under consideration, this may lead to the erroneous placement of sequences on a tree.[36] Analysis of only those positions where there has been a transversion (transversion analysis) removes the G + C bias that may be introduced.[36]

[31] H. A. Erlich, D. Gelfand, and J. J. Sninsky, *Science* **252**, 1643 (1991).
[32] D. H. Persing, *J. Clin. Microbiol.* **29**, 1281 (1991).
[33] G. E. Fox, J. D. Wisotzkey, and P. Jurtshuk, Jr., *Int. J. Syst. Bacteriol.* **42**, 166 (1992).
[34] S. Paabo, D. M. Irwin, and A. C. Wilson, *J. Biol. Chem.* **265**, 4718 (1990).
[35] P. D. Ennis, J. Zemmour, R. D. Salter, and P. Parham, *Proc. Natl. Acad. Sci. U.S.A.* **87**, 2833 (1990).

The selection of sequences compared can also influence the topology of a phylogenetic tree. One approach to check for this influence is to generate multiple trees, varying the composition of the sequences compared. Consistent placement of a sequence within a particular phylogenetic group, despite tree composition, suggests that the placement is appropriate.

Techniques for Confirming Identification *in Situ*

The evidence that the rRNA sequence belongs to the pathogen of interest can be strengthened by the visual association of the sequence with microorganisms in the context of tissue pathology.

In Situ Probing

Oligodeoxynucleotide probes, complementary to rRNAs, can be synthesized with an amino linkage on the 5' nucleotide. Several fluorescent dyes can be attached to the amino linkage, making the probes suitable for use in the *in situ* detection of pathogens. The fluorescently labeled probe can permeate fixed cells and associate specifically with rRNAs in the ribosome. When viewed using the appropriate excitation wavelength of light, cells fluoresce when the probe hybridizes with the complementary sequence in the rRNA. A major disadvantage of using rRNAs as targets for *in situ* hybridization is the number of ribosomes, which can be as high as 70,000 per cell in *E. coli*. Variable regions of the rRNA are often attractive sites for probe design. A single base pair mismatch can be sufficient to discriminate among closely related organisms. The rationale behind the design of specific probes and procedures for the production of fluorescently labeled probes are presented elsewhere.[37]

In Situ Polymerase Chain Reaction

In situ PCR shares with *in situ* hybridization the advantage of enabling one to localize and correlate specific 16 S rRNA sequences with tissue histology. In addition, by incorporating target amplification, *in situ* PCR may increase the detection sensitivity of rare or low density targets over conventional *in situ* hybridization.[38] This technique has been used to detect

[36] C. R. Woese, L. Achenbach, P. Rouviere, and L. Mandelco, *Syst. Appl. Microbiol.* **14,** 364 (1991).
[37] D. A. Stahl and R. Amann, *in* "Nucleic Acid Techniques in Bacterial Systematics" (E. Stackebrandt and M. Goodfellow, eds.), p. 205. Wiley, New York, 1991.
[38] G. J. Nuovo, F. Gallery, P. MacConnell, J. Becker, and W. Bloch, *Am. J. Pathol.* **139,** 1239 (1991).

lentiviruses in isolated human cells[39] and human papillomavirus in fixed tissue.[38] Modifications in the PCR protocol have facilitated primer–target annealing and improve sensitivity and specificity.[40] With further technical improvements, *in situ* PCR may become an important adjunct to investigations of previously uncharacterized pathogens.

Acknowledgments

We thank P. W. Lepp for assistance in preparing Fig. 2 and B. J. Bratina for her critical reading of this manuscript.

[39] G. J. Nuovo, P. MacConnell, A. Forde, and P. Delvenne, *Am. J. Pathol.* **139,** 847 (1991).
[40] A. T. Haase, E. F. Retzel, and K. A. Staskus, *Proc. Natl. Acad. Sci. U.S.A.* **87,** 4971 (1990).

Section III

Bacterial Cell Fractionation

[17] Isolation of Outer Membranes

By HIROSHI NIKAIDO

Introduction

The outer membrane is the outermost continuous structure on the surface of most gram-negative bacteria, and as such is important in the interaction of bacteria with the host defense machinery. Because of this, it frequently becomes necessary to isolate outer membranes. As a lipid bilayer membrane, the outer membrane in many ways behaves similarly to the cytoplasmic membrane during cell fractionation. However, one can separate the outer membrane from the cytoplasmic membrane by taking advantage of differences in the buoyant densities. Furthermore, because most of the lipids in the outer membrane are lipopolysaccharides (LPS), which have very different properties from those of the glycerophospholipids comprising the bilayer of the cytoplasmic membrane, one can often solubilize cytoplasmic membrane lipids preferentially without solubilizing much of the LPS of the outer membrane. The solubilization of the lipids is accompanied by the solubilization of the cytoplasmic membrane proteins, leaving behind a partially lipid-depleted outer membrane, still containing nearly all of its proteins. Finally, the presence of large amounts of LPS makes the outer membrane an intrinsically unstable structure. Because of this, fragments of the outer membrane are shed off as blebs or vesicles, especially under conditions that further destabilize the membrane architecture. Collecting these vesicles is another convenient way of isolating samples of outer membrane. Reviews are available on the structure and functions of the outer membranes.[1-5]

Isolation by Equilibrium Density Gradient Centrifugation

The outer membrane usually shows a significantly higher buoyant density than the cytoplasmic membrane, because the polysaccharide portion

[1] H. Nikaido and M. Vaara, *Microbiol. Rev.* **49,** 1 (1985).
[2] H. Nikaido and M. Vaara, *in* "*Escherichia coli* and *Salmonella typhimurium*: Cellular and Molecular Biology" (J. L. Ingraham, K. B. Low, B. Magasanik, M. Schaechter, and H. E. Umbarger, eds.), Vol. 1, p. 7. American Society for Microbiology, Washington, D.C., 1987.
[3] H. Nikaido and T. Nakae, *Adv. Microb. Physiol.* **20,** 163 (1979).
[4] R. Benz, *Crit. Rev. Biochem.* **19,** 145 (1985).
[5] H. Nikaido, *Mol. Microbiol.* **6,** 435 (1992).

METHODS IN ENZYMOLOGY, VOL. 235

of the LPS has a much higher density than lipids or proteins. Thus, after the breakage of cells, the outer membrane can be separated from the cytoplasmic membrane by centrifugation through sucrose density gradients.[6–8] The procedure described below is that routinely carried out in our laboratory, and it is based on the original procedures of Osborn *et al.*[6] and of Schnaitman,[7] with some modifications.[8]

Materials

Bacterial strains: The procedure described can be used for any strains of *Escherichia coli* and *Salmonella* spp., as well as other related genera of the Enterobacteriaceae.

L broth: Add 10 g of Bacto yeast extract (Difco, Detroit, MI), 10 g of Bacto-tryptone (Difco), and 5 g of NaCl to 1 liter of distilled water, and sterilize by autoclaving. Adjustment of pH is not necessary.

HEPES buffer: Adjust the pH of a 10 mM solution of HEPES (N-2-hydroxyethylpiperazine-N'-2'-ethanesulfonic acid) to 7.4 with NaOH. Keep the buffer refrigerated, preferably after filter sterilization; otherwise, microbial contamination may occur.

Sucrose solutions: For a 25% (w/w) sucrose solution, dissolve 25 g of sucrose in 75 ml of HEPES buffer. For a 60% (w/w) solution, dissolve 60 g of sucrose in 40 ml of HEPES buffer. More correctly, 73 and 38 ml, respectively, of distilled water can be used as the solvent, and 2 ml of 0.5 M HEPES buffer, pH 7.4, can then be added to make the final buffer concentration exactly 10 mM. However, small differences in buffer concentration usually do not affect separation.

Procedure and Comments

Preparation of Crude Cell Envelope Fraction. Twenty-five milliliters of an overnight culture of the bacterial strain is diluted into 1 liter of prewarmed, sterile L broth in a 6-liter Erlenmeyer flask. The flask is incubated at 37° with aeration by shaking. When the culture has attained a cell density of about 0.3 mg dry weight/ml [this corresponds to a Klett reading of 130 (red filter) or an OD$_{600}$ of about 0.9 measured with spectrophotometers equipped with large sample compartments such as Cary 16 or Zeiss PMQ II (some other spectrophotometers give much lower read-

[6] M. J. Osborn, J. E. Gander, E. Parisi, and J. Carson, *J. Biol. Chem.* **247**, 3962 (1972).
[7] C. A. Schnaitman, *J. Bacteriol.* **104**, 890 (1970).
[8] J. Smit, Y. Kamio, and H. Nikaido, *J. Bacteriol.* **124**, 942 (1975).

ings],[9] the cells are harvested by centrifugation. From this point on, all operations are carried out at 4°.

The cells are washed once with cold HEPES buffer, by resuspension and centrifugation. The pellet can be stored frozen at −70° until needed. The cells are resuspended in 60 ml of 10 mM HEPES buffer containing about 1 mg each of pancreatic deoxyribonuclease and pancreatic ribonuclease, and the suspension is passed through a French pressure cell (SLM Instruments, Urbana, IL) at 14,000 psi. With bacterial species producing proteases of significant activity, it is prudent to add to the bacterial suspension, just before French pressure cell treatment, protease inhibitors such as phenylmethylsulfonyl fluoride (PMSF; this can be kept as a 0.1 M solution in 2-propanol and added to the suspension so that the final concentration would become 0.5–1 mM). The addition of EDTA may also be necessary to inhibit metal proteases; however, the concentration of EDTA should be kept to a minimal value (usually 1–3 mM) in order to avoid the destabilization of the LPS leaflet (see below). The extract is immediately centrifuged at 1000 g for 15 min to remove unbroken cells, and the supernatant fraction is centrifuged at 100,000 g for 2 hr to yield a pellet corresponding to the crude cell envelope fraction, containing both the outer membrane and the cytoplasmic membrane.

In the original procedure of Osborn et al.,[6] the cells were converted to spheroplasts by treatment with EDTA and lysozyme in 0.75 M sucrose, and then the spheroplasts were lysed by osmotic shock. A major disadvantage of that procedure is that the conditions required for the formation of spheroplasts are sensitive to a number of parameters, including the physiological state of the cell, the composition of the growth medium, and the nature of the particular strain used. Thus, considerable fine-tuning is usually necessary to make this step reproducible for any new bacterial strain. Disruption by the French pressure cell has the advantage that it is very reproducible, at least for all the strains of the Enterobacteriaceae. The disadvantage of the use of the French pressure cell is that the cytoplasmic membrane is converted to very small vesicles, which are often difficult to pellet even after 2 hr of centrifugation at 100,000 g; thus, if one also wants to isolate the cytoplasmic membrane fraction, it may be advantageous to skip the pelleting step and to apply the cell lysate directly on top of the sucrose density gradient, described below.

With some species, separation of the membranes is less than ideal in the standard procedure. One modification, used successfully for *Pseudomonas aeruginosa,* is to resuspend cells in a hypertonic solution of sucrose, so that the outer and inner membranes are already maximally

[9] A. L. Koch, *Anal. Biochem.* **38,** 252 (1970).

separated at the time of cell disruption.[10] We need a critical evaluation of this modification in order to see whether the procedure has general applicability.

In principle, sonication can also be used to break cells. However, extended sonication is known to produce chimeric membrane vesicles. Thus, if sonication has to be used, the conditions should be just sufficient to disrupt most of the cells.

It should be noted that the "outer membrane" fraction obtained by fractionation of cells disrupted either by French pressure cell treatment or sonication contains a large amount of the peptidoglycan layer, still associated with the outer membrane. In contrast, much of the peptidoglycan layer should be absent if the envelope is prepared from EDTA–lysozyme spheroplasts.

Sucrose Density Gradient Centrifugation. The crude envelope fraction obtained from 1 liter of culture is carefully resuspended in 4–5 ml of HEPES buffer; repeated drawing in and squeezing out through a syringe needle may be the easiest way of achieving a uniform suspension. The suspension is layered carefully on top of a continuous sucrose density gradient [25–60% (w/w) sucrose in 10 mM HEPES buffer], and then the tube is centrifuged at 4° in the SW28 rotor of a Beckman (Fullerton, CA) centrifuge for 16 hr at 26,000 rpm. The lower, white band with strong turbidity corresponds to outer membranes (with the still associated peptidoglycan layer), whereas cytoplasmic membranes appear as a translucent, brownish upper band. Because of the translucence, the cytoplasmic membrane band is often not recognizable unless the tube is viewed against a white background, such as a sheet of white paper. With the *E. coli* and *S. typhimurium* strains, the outer and cytoplasmic membranes are usually found at positions where the sucrose solutions have densities around 1.22–1.24 and 1.16–1.18 g/ml, and it is useful to determine the densities of the solutions throughout the gradient by using a refractometer. The outer membrane band is recovered, and the sucrose is removed by centrifugation of the sample after dilution in HEPES buffer.

The most often encountered problem in the separation arises from the application of too much envelope material, a situation that encourages aggregation between membrane vesicles because of the high concentrations. The conditions given above represent about the maximum amount of *E. coli* or *S. typhimurium* envelope that can be applied to the 35-ml gradient in a centrifuge tube for the SW28 rotor. Resolution will improve if the sample is divided and added to two or three tubes, and this may be absolutely necessary for certain species.

[10] R. E. W. Hancock and H. Nikaido, *J. Bacteriol.* **136**, 381 (1978).

Sodium dodecyl sulfate (SDS)–polyacrylamide gel analysis of each fraction is useful because the protein pattern of the outer membrane is dominated by a few major proteins (porins, OmpA, and murein lipoprotein), whereas the cytoplasmic membrane shows a much more complex protein composition.[11] In addition, quantitative markers of the outer and cytoplasmic membranes are very useful. For identification of the outer membrane, 2-keto-3-deoxyoctonic acid (KDO) in the LPS can be quantitated colorimetrically as described by Osborn et al.[6] (Note that KDO determination involves a periodate oxidation step and therefore cannot be used until most of the sucrose, which also consumes periodate, has been removed from the membrane sample.) For the cytoplasmic membrane, NADH dehydrogenase, succinate dehydrogenase, or NADH oxidase can be used as a marker.[6] If the separation is successful, the degree of cross-contamination as assessed by the use of these markers will be lower than 10%, typically about 3–5%.

As mentioned earlier, the cytoplasmic membrane vesicles are very small, and some may take a long time to come to the equilibrium position. Thus, a much longer centrifugation time (e.g., 72 hr in a SW41 rotor) is needed for the membranes to come near the final equilibrium position. (That some of the vesicles have still not reached the final position can be revealed by the sucrose density gradient flotation technique, which is useful in the finer fractionation of membranes.[12] Use of metrizamide, which has a lower viscosity, instead of sucrose further accelerates the attainment of equilibrium.[13]) However, for the purpose of obtaining outer membrane fractions, the short centrifugation period described above is sufficient. In fact, with some bacteria, a fraction enriched in the outer membrane can be obtained by simply centrifuging the crude envelope fraction at an intermediate speed (e.g., at 13,000 g for 15 min[14]), totally bypassing the sucrose density step, because of the large differences in size and density between the outer and inner membrane vesicles.

The procedure above uses a continuous sucrose density gradient. When one is dealing with the wild-type strains of enteric bacteria, one can substitute this with an appropriate discontinuous gradient [e.g., a gradient containing 13 ml of 24% (w/w), 17 ml of 42% (w/w), and 5 ml of 55% (w/w) sucrose for a SW28 rotor[8]]. The cytoplasmic membrane will be found at the interface between the 24 and 42% sucrose, and the outer membrane between the 42 and 55% sucrose. However, the buoyant density

[11] G. F.-L. Ames, E. N. Spudich, and H. Nikaido, J. Bacteriol. 117, 406 (1974).
[12] K. Ishidate, E. S. Creeger, J. Zrike, S. Deb, B. Glauner, T. J. MacAlister, and L. I. Rothfield, J. Biol. Chem. 261, 428 (1986).
[13] J. R. Thom and L. L. Randall, J. Bacteriol. 170, 5654 (1988).
[14] H. Nikaido, W. Liu, and E. Y. Rosenberg, Antimicrob. Agents Chemother. 34, 337 (1990).

of the outer membrane approaches more closely that of the cytoplasmic membrane in ''deep rough'' mutants that are lacking most of the carbohydrate side chain of LPS and in such cases the use of discontinuous gradient may produce poor separation. With nonenteric species, the buoyant densities of vesicles may not be easily predictable. In one extreme case, the outer membrane of *Myxococcus xanthus* was shown to have a lower buoyant density than the cytoplasmic membrane.[15]

In the procedure described above, chelating agents such as EDTA are not used either in the lysis step or in the gradient centrifugation step. This is done to avoid the EDTA-induced loss of LPS[16] and possibly some outer membrane proteins. However, with some bacteria, the divalent cation-bridged aggregation of membranes could become a problem, and in such cases the inclusion of very low concentration of EDTA may become necessary. Osborn *et al.*[6] added 5 mM EDTA to sucrose solutions.

Isolation by Differential Extraction with Detergents

In contrast to the lipid bilayer present in the common biological membranes, the lipid bilayer in the outer membrane contains one leaflet composed entirely of LPS. A molecule of LPS contains 6–7 saturated fatty acid chains, and therefore it interacts strongly with its neighbor through hydrophobic and van der Waals interactions. It is therefore difficult for perturbants, such as detergents (or even small hydrophobic solutes) to penetrate into the space in between the LPS.[1] On the other hand, a LPS molecule contains a large number of negative charges on its surface, about 6–7 net negative charges per molecule in *Salmonella*. The density of negative charges on the surface of the LPS leaflet is therefore very high, up to the level of about one negative charge per hydrocarbon chain (in contrast, in usual biological membranes, including the cytoplasmic membrane, the density is of the order of 0.1 net negative charge per hydrocarbon chain). The stability of the outer membrane therefore requires that these charges are bridged by divalent metal cations and by polyamines.

These unusual features in the structure of LPS are reflected in the behavior of the outer membranes to detergent solubilization. Thus, DePamphilis and Adler[17] as well as Schnaitman[18] discovered that the outer membrane of *E. coli* cannot be solubilized by nonionic detergents such as Triton X-100, as long as the LPS leaflet is stabilized by high concentra-

[15] P. E. Orndorff and M. Dworkin, *J. Bacteriol.* **141**, 914 (1980).
[16] L. Leive, *Biochem. Biophys. Res. Commun.* **21**, 290 (1965).
[17] M. L. DePamphilis and J. Adler, *J. Bacteriol.* **105**, 396 (1971).
[18] C. A. Schnaitman, *J. Bacteriol.* **108**, 533 (1971).

tions of divalent cations such as Mg^{2+}. This differential extraction procedure in the presence of Mg^{2+} is still being used to prepare a partially extracted outer membrane that retains most of the proteins and LPS (but has lost most of the phospholipids[18]). A more widely applied procedure is that of Filip et al.,[19] using an anionic detergent, Sarkosyl, in the absence of added Mg^{2+}. Sarkosyl, with one negative charge per hydrocarbon chain, has about the same charge density per surface area as LPS, and it does not disaggregate the LPS-based outer membrane, presumably because it mimics LPS in this aspect. In contrast, it disaggregates and solubilizes effectively the cytoplasmic membrane.

Materials

Sarkosyl (sodium N-lauroylsarcosinate) solution (10%, w/v)

Procedure

Bacterial cells are broken either by the osmotic lysis of spheroplasts,[6] by sonication, or by French pressure cell treatment as described above. In each case, one should be careful to avoid proteolysis by carrying out the preparation procedure quickly at low temperature, and by using appropriate protease inhibitors. The crude cell envelope fraction is obtained by centrifuging the cell lysate at 100,000 g for 1–2 hr. The cell envelope fraction from 1 liter of culture is resuspended evenly in 60 ml of a dilute buffer, such as HEPES, and to this 3 ml of 10% Sarkosyl is added. The mixture is stirred at room temperature for 20 min, and then is centrifuged at 100,000 g for 1 hr. The pellet contains most of the outer membrane proteins and much of the LPS.

Comments

When detergents are used for solubilization of membranes, concentration of the detergent is important because it has to be above the critical micellar concentration. However, a factor that is often neglected is that the mass ratio between detergent and the membrane has to be large, so that the detergent micelles can compete effectively against the lipids within the membrane. Unpredictable results have occasionally been produced due to the lack of appreciation of this factor. As recommended by Helenius and Simons,[20] the total mass of detergent must be at least 10 times that of the total mass of proteins in the membrane. If the envelope fraction is prepared from a larger amount of cells than indicated above, the volume

[19] C. Filip, G. Fletcher, J. L. Wulff, and C. F. Earhart, *J. Bacteriol.* **115**, 717 (1973).
[20] A. Helenius and K. Simons, *Biochim. Biophys. Acta* **415**, 29 (1975).

of the extraction mixture must be increased accordingly in order to fulfill this requirement.

Differential extraction with Sarkosyl is probably the most widely used method for the preparation of outer membrane proteins. However, caution is needed in using this method. First, as pointed out above, the detergent will probably extract a substantial fraction of phospholipids from the outer membrane. Second, although the method has been carefully evaluated for use in enteric bacteria, there is no guarantee that it will work ideally with other organisms. Finally, even with *E. coli,* the procedure is reported to extract some minor outer membrane proteins.[21]

Some bacteria are reported to produce outer membrane with a mixed LPS–glycerophospholipid outer leaflet (e.g., *Vibrio cholerae*[22]). The detergent extraction procedure may not work properly with these bacteria.

Another detergent solubilization procedure, originally developed for the solubilization of *Neisseria gonorrhoeae* porin,[23] uses a zwitterionic detergent, 5% Zwittergent 3.14, in the presence of 0.5 M CaCl$_2$ and a varying concentration of ethanol. Although this interesting protocol undoubtedly solubilizes effectively the porins of such organisms as *Legionella pneumophila*[24] and *Bordetella pertussis,*[25] there is no published evidence to indicate that only the outer membrane proteins are extracted by this scheme. Regrettably the method has been used to obtain what has been uncritically called "outer membrane proteins" by several workers. Similarly, extraction with Triton X-114 does extract outer membrane proteins of some strains into the detergent phase.[26] However, again there appears to be no published evidence that this method does not extract cytoplasmic membrane proteins as well.

Isolation Based on Selective Release of Outer Membranes from Intact Cells

The extremely high negative charge density of the LPS molecule makes the outer membrane an inherently unstable structure. Consequently, in many bacterial species, portions of outer membrane tend to form blebs on the surface of growing cells, and some of these are released into media as outer membrane vesicles from normally growing cells. This

[21] I. Chopra and S. W. Shales, *J. Bacteriol.* **144,** 425 (1980).
[22] S. Paul, K. Chaudhuri, A. N. Chatterjee, and J. Das, *J. Gen. Microbiol.* **138,** 755 (1992).
[23] M. S. Blake and E. C. Gotschlich, *J. Exp. Med.* **159,** 452 (1984).
[24] J. E. Gabay, M. Blake, W. D. Niles, and M. A. Horowitz, *J. Bacteriol.* **162,** 85 (1985).
[25] S. K. Armstrong, T. R. Parr, Jr., C. D. Parker, and R. E. W. Hancock, *J. Bacteriol.* **166,** 212 (1986).
[26] R. N. Thwaits and S. Kadis, *Infect. Immun.* **59,** 544 (1991).

phenomenon occurs even in *E. coli* and *S. typhimurium*, but it appears to be especially prominent in organisms such as *Neisseria, Vibrio, Campylobacter,* and *Brucella melitensis* (see papers cited in Ref. 27). When cells of some species are resuspended in solutions of NaCl, the Na^+ ion presumably competes with the divalent cations in the outer membrane and destabilizes the membrane, resulting in the release of fragments containing LPS, proteins, and some phospholipids. This method has been used in the isolation of outer membrane vesicles from *Rhodobacter capsulatus*.[28] When LiCl or lithium acetate is used,[29] the Li^+ ion with its high surface charge density competes with the Mg^{2+} ions much more effectively than Na^+, causing a more effective release of outer membrane vesicles; this method is described below. Finally, when the cells are converted to spheroplasts, much of the outer membrane fragments are released into the medium, and the outer membrane vesicles can be isolated directly from the supernatant fluid after the removal of spheroplasts (an example is the procedure of Mizuno and Kageyama for the isolation of *Pseudomonas aeruginosa* "outer membranes").[30]

The LiCl/lithium acetate extraction procedure described below is that of McDade and Johnston,[31] and has been widely used in the isolation of outer membrane vesicles from *Neisseria* species as well as *Haemophilus influenzae*.

Materials

Bacterial cells: *Neisseria gonorrhoeae* cells, grown in a liquid medium up to a midexponential phase

LCA buffer: 0.2 *M* LiCl/0.1 *M* lithium acetate, with the pH adjusted to 6.0

TS buffer: 10 m*M* Tris-HCl buffer, pH 8.0, containing 0.2 *M* NaCl and 0.02% NaN_3

Procedure

About 10 g (wet weight) cells are resuspended in 200 ml of LCA buffer and incubated for 2 hr at 45° with vigorous shaking in the presence of 6-mm glass beads. The glass beads are then removed by filtering the bacterial suspension through a metal mesh screen, and the beads are washed once with LCA buffer. The wash and bacterial suspension are

[27] C. Gamazo and I. Moriyon, *Infect. Immun.* **55,** 609 (1987).
[28] J. Weckesser, G. Drews, and R. Ladwig, *J. Bacteriol.* **110,** 346 (1972).
[29] K. H. Johnston, K. K. Holmes, and E. C. Gotschlich, *J. Exp. Med.* **143,** 741 (1976).
[30] T. Mizuno and M. Kageyama, *J. Biochem. (Tokyo)* **84,** 179 (1978).
[31] R. I. McDade, Jr., and K. H. Johnston, *J. Bacteriol.* **141,** 1183 (1980).

combined, then centrifuged at 12,000 g for 15 min at 4° to remove intact cells. The supernatant fluid is centrifuged again at 25,000 g for 15 min to remove large aggregates, and the supernatant fluid is applied to a large gel filtration column (Sepharose 6B-CL, 5.0 by 60 cm) equilibrated with TS buffer. The column is eluted with the TS buffer, and the effluent is monitored at 280 nm. The outer membrane vesicles are eluted at the void volume. These fractions are pooled and concentrated by vacuum dialysis.

Comments

In an earlier paper, Johnston *et al.*[29] compared NaCl with lithium acetate and found that the latter was superior in its ability to effect the release of outer membrane vesicles; this is consistent with the considerations described above. Because procedures of this type involve destabilization of the outer membrane to effect the release of its fragments, there is no guarantee that the composition of the vesicles are quantitatively identical to that of the bulk outer membrane. It is likely that most of the outer membrane proteins are recovered in the vesicles, but they are not suitable for the quantitative analysis of LPS, for example. In the procedure described above, gel filtration is used for the isolation of vesicles. Centrifugation at high speed can also be used to recover vesicles, although the exact conditions have to be worked out for each case because it depends on the density and the size of the outer membrane vesicles released.

[18] Isolation and Purification of Periplasmic Binding Proteins

By GIOVANNA FERRO-LUZZI AMES

Introduction

Periplasmic active transport systems (permeases) in gram-negative bacteria such as *Escherichia coli* and *Salmonella typhimurium* are composed of a substrate-binding protein located in the periplasm and a membrane-bound complex containing four subunits. Two of the membrane-bound subunits are hydrophobic proteins which span the membrane several times, while the other two have a hydrophilic sequence (reviewed in Refs. 1 and 2). The latter (also referred to as the conserved components) display extensive sequence similarity across the many systems that have been characterized, in particular in two conserved motifs that have been implicated in the ability of these proteins to bind ATP.[3-6] Comparison of the various permeases indicates that they are very similar to one another, and it has been suggested that they are all related evolutionarily.[7-9] Hypotheses regarding the mechanism of action of periplasmic permeases usually depict the substrate-binding protein becoming liganded on entry of the substrate into the periplasm and then interacting with the membrane-bound complex; this interaction would trigger a series of conformational changes in the membrane-bound components which result in translocation of the substrate. Once the binding protein has become unliganded it is postulated to have a lower affinity for the membrane-bound complex and therefore to be released from the membrane.

An understanding of the mechanism of action of the periplasmic binding proteins is necessary in order to understand the general mechanism of action of these permeases. Their analysis, of course, starts with their isolation and purification. Many of the general characteristics of binding proteins in this respect have been found to be the same when several systems were compared. Extensive studies have been performed in this laboratory on the periplasmic component of the histidine permease, namely, the histidine-binding protein HisJ. It is probably reasonable to assume that the methodologies used for HisJ are generally applicable. Thus, this chapter is based mainly on the isolation and purification of HisJ, with the addition of those details that are deemed to be of general usefulness.

[1] G. F.-L. Ames, *Annu. Rev. Biochem.* **55**, 397 (1986).
[2] C. E. Furlong, in *"Escherichia coli* and *Salmonella typhimurium:* Cellular and Molecular Biology" (F. C. Neidhardt, ed.), p. 768. American Society for Microbiology, Washington, D.C., 1987.
[3] A. Hobson, R. Weatherwax, and G. F.-L. Ames, *Proc. Natl. Acad. Sci. U.S.A.* **81**, 7333 (1984).
[4] C. F. Higgins, I. D. Hiles, K. Whalley, and D. K. Jamieson, *EMBO J.* **4**, 1033 (1985).
[5] C. S. Mimura, S. R. Holbrook, and G. F.-L. Ames, *Proc. Natl. Acad. Sci. U.S.A.* **88**, 84 (1991).
[6] C. F. Higgins, I. D. Hiles, G. P. C. Salmond, D. R. Gill, J. A. Downie, I. J. Evans, I. B. Holland, L. Gray, S. D. Buckel, A. W. Bell, and M. A. Hermodson, *Nature (London)* **323**, 448 (1986).
[7] G. F.-L. Ames, *Curr. Top. Memb. Transp.* **23**, 103 (1985).
[8] G. F.-L. Ames, C. Mimura, S. Holbrook, and V. Shyamala, *Adv. Enzymol.* **65**, 1 (1992).
[9] G. F.-L. Ames and H. Lecar, *FASEB J.* **6**, 2660 (1992).

Isolation Procedures

The classic method for the isolation of periplasmic proteins is by the osmotic shock procedure.[10] This procedure is based on the plasmolysis of the cells with sucrose in the presence of EDTA, followed by the rapid addition of a low osmotic medium, which causes sudden swelling of the cell volume by uptake of water in an effort to compensate for the sudden drop in external osmotic pressure. The sudden change in the cell volume, together with the modifications of the outer cell membrane caused by EDTA, results in damage to the cell surface (presumably to both the inner and outer membranes and to the cell wall) and in the release of the periplasmic components and some of the cytoplasmic components into the medium.[10,11] It is important to keep in mind that the so-called periplasmic proteins are released to different extents by this procedure. Indeed, the definition of a protein as being periplasmically located is somewhat ambiguous, because it is based only on our ability to release them from the cell surface. Therefore, it should be understood that the definition is entirely operational, and caution should be exercised when defining the nature of proteins on this basis only. Particularly, relevant examples in this respect are the cytoplasmic elongation factor EF-Tu, which is released by osmotic shock when the low osmotic shock medium does not contain any Mg^{2+} ions but is not released in the presence of Mg^{2+},[12] and a periplasmic RNase which was considered for years as being ribosomal because on cell rupture it was always found associated with ribosomes (an artifact arising from its basic nature).[10]

The following procedure (as performed on *Salmonella typhimurium* LT2 and *Escherichia coli* K12) has been reproducibly successful,[13] using volumes from a few milliliters to several liters and was modified from a previous publication.[14] The cells are cultured to saturation in minimal medium (supplemented with whatever nutrient the bacteria require), using glucose as the carbon source. It is most convenient to inoculate the culture in the evening so that the bacteria are fully grown the next morning. In the case of HisJ, the level of production is higher in cells grown to saturation in minimal medium than in cells either grown in rich medium or at the exponential phase of growth. A high level of aeration during growth is also a requirement for good yields. The specific conditions leading to the

---

[10] N. G. Nossal and L. A. Heppel, *J. Biol. Chem* **241**, 3055 (1966).
[11] G. K. Anraku, *J. Biol. Chem.* **243**, 3116 (1968).
[12] G. R. Jacobson, B. J. Takacs, and J. P. Rosenbusch, *Biochemistry* **15**, 2297 (1976).
[13] Nikaido and G. F.-L. Ames, *J. Biol. Chem.* **267**, 20706 (1992).
[14] J. E. Lever, *J. Biol. Chem.* **247**, 4317 (1972).

highest level of production of an individual protein should be established for the particular strain used. To the culture, 0.5 M Tris-Cl, pH 7.8, is added to a final concentration of 50 mM. The cells are immediately and rapidly harvested by centrifugation and resuspended at room temperature in 50 mM Tris-Cl, pH 7.8, containing 40% sucrose (one-tenth to one-twentieth the original culture volume). The process of resuspension has to be performed rapidly and must achieve complete uniformity. The best procedure is to disperse the cell pellet in whatever small amount of residual medium is left in the centrifuge tube(s), prior to the addition of the sucrose-containing buffer; this results in a uniformly "creamy" and very thick cell suspension that is then rapidly dispersed in the relatively large volume of sucrose-containing buffer.

Then 0.5 M NaEDTA is added immediately to a final concentration of 2 mM, and the suspension is gently or occasionally stirred at room temperature for 10 min. The cells, which should not be exposed to the high osmolarity medium for any longer than the prescribed 10 min, are harvested rapidly by centrifugation, paying attention to achieve a "tight" pellet, which should be drained very thoroughly by decantation. The cells are dispersed thoroughly and rapidly as described above, and cold water (one-tenth the volume of the initial culutre volume) is immediately and rapidly added with definite mixing, but without causing frothing. Some proteins may require the presence of Mg^{2+} in the water for full release or for maintaining activity. In general, it seems that in the absence of Mg^{2+} less cytoplasmic material is released and therefore its absence is to be preferred.

The shocked cell suspension is immediately centrifuged at high speed to separate cleanly the supernatant, that is, the shock fluid, from the shocked cells, which are discarded. The centrifugal force and the length of time to be used in this step depend on the size of the sample: the minimum length of time that is compatible with yielding a clear supernatant fraction is to be established individually. The purpose of this step is to remove all intact cells as thoroughly as possible.

The procedure can be adapted to a variety of cell culture volumes, from several liters to 100 ml or less. If the size of the culture exceeds 10 to 15 liters, the volume of the shock fluid becomes cumbersomely large and other procedures, such as French press cell breakage, is to be preferred, even though they may yield a crude protein solution that contains periplasmic components at a much lower level of initial purity than the shock fluid.

An alternate procedure for the isolation of periplasmic binding proteins makes use of chloroform as a membrane-disrupting agent. This approach, referred to as "chloroform shock," was developed for the rapid handling of numerous small cultures for screening purposes.[15] A comparison of this method to the standard osmotic shock procedure was also presented.

Stationary cells grown as described above in small test tubes cultures (2 ml) are harvested at 3000 rpm in a Sorvall SS-34 rotor (Dupont, Wilmington, DE) for 10 min. The supernatant fraction is decanted, or drawn off with a Pasteur pipette, and the tubes are drained carefully, avoiding loss of cells. The cells are resuspended by a brief vortexing step, 10 μl of chloroform is added, and the tubes are vortexed and allowed to stay at room temperature for 15 min. Then 0.2 ml of 10 mM Tris-Cl pH 8.0 is added, and the cells are removed by centrifugation at 7000 rpm in the Sorvall SS-34 or SA600 rotor for 20 min. The entire procedure can be performed in the glass culture tubes, thus avoiding time-consuming manipulations. Care must be taken to use tubes and adapters that withstand the centrifugal force.[15] The supernatant fraction (the chloroform shock fluid) is withdrawn with a Pasteur pipette, sacrificing, if necessary, some of the supernatant material. It is important that the bacterial pellet be firm and that the material withdrawn be clear of any cells. The procedure has been applied to larger volumes of cultures (50 ml)[15] and is conceivably applicable to much larger volumes.

Many binding proteins are peculiarly resistant to inactivation by heat and by protease action, and HisJ was also shown to be resistant to this chloroform treatment. Other binding proteins should be tested for resistance to exposure to chloroform if activity retention is an important factor. In any case the method is excellent for the simultaneous and rapid screening of many strains for the presence and nature of their binding protein complement, such as is necessary in the isolation and characterization of mutants. Such a screening would be best followed by an electrophoretic analysis of the chloroform shock followed by immunoblotting, if the protein is not recognizable by Coomassie blue staining.[15]

Activity Assays

Assay of binding protein activity is always based on the substrate-binding ability, because this is the only measurable property.[16] The most reliable assay method is equilibrium dialysis, which is best performed against a very large volume of radioactively labeled substrate solution. Because the affinity of binding proteins for the substrates is generally very high and many substrates are available at very high specific activities,

[15] G. F.-L. Ames, C. Prody, and S. Kustu, *J. Bacteriol.* **160**, 1181 (1984).
[16] J. E. Lever, *Anal. Biochem.* **50**, 73 (1972).

dialysis even against volumes as large as 1 liter or more are economically feasible. In brief, 0.3 ml of protein solution (diluted depending on the level of activity expected) is placed in a dialysis tubing bag, and sealed by knots at each end. Union Carbide (New York, NY) tubing No. 8 is appropriate and is prepared for the assay by being boiled gently in two changes of distilled water for 10 min each. The tubing is precut in 10 cm lengths to decrease the level of contamination by subsequent handling. When the bags are knotted a very small amount of air is included. The tied bags are hung from the looped end of glass rods which are suspended by bent ends from the edge of a beaker containing 1.5 liters of 10 mM sodium phosphate, pH 7.0 at 4°. After all samples are in the beaker, the labeled substrate is added (for HisJ, this is usually 1 nM [^{3}H]histidine), and the solution is stirred gently with a magnetic stirrer for 16 to 20 hr. The bag is cut open, and 0.2 ml is withdrawn and counted in an appropriate scintillation fluid. Blank levels are measured (and subtracted) by assaying for radioactivity present in the external liquid and inside bags that do not contain protein. All manipulations in setting up the assay and in opening the bags are performed with gloved hands. One dialysis units is 1 pmol of substrate retained (above blank level) at 10 nM substrate under the conditions of the assay. The units would be defined differently for different proteins.

When determining parameters such as the K_D for binding or the protein specificity by competition with various substrate analogs (which require many different dialysis conditions), individual small volume dialysis assays can be set up in several small beakers or, better, in screw-capped bottles. Numerous vessels can be placed on a platform shaker set for gentle agitation. The individual vessels can contain dialysis volumes of 50 ml of less, and the bags can be allowed to float freely in them (the bags should then be coded such as with different numbers of knots or with specially shaped cut ends).

An alternative method makes use of the protein-adsorption properties of nitrocellulose filters[16]: the adsorbed protein retains its substrate-binding activity, thus allowing the rapid assay by filtration through nitrocellulose filters of mixtures containing the binding protein and its substrate. The protein solution (100 μl) is added at room temperature to 20 μl of 62.2 mM sodium phosphate buffer, pH 7.0, containing labeled substrate (6 μM[^{3}H]histidine). Then 100 μl of the mixture is applied to an individual Schleicher and Schuell (Keine, NH) 0.45 μm nitrocellulose filter, prewetted with water and placed on an individual filtration apparatus. The vacuum, which must be applied initially to remove excess water, must be released before application of the sample. As soon as the sample is delivered, vacuum is immediately applied, released, and then followed by the addition of 600 μl of 10 mM sodium phosphate buffer, pH 7.0, which is also immediately followed by vacuum. The filter is carefully removed with the vacuum on, blotted on a paper towel, and dried thoroughly. The radioactivity retained on the filter is then determined by scintillation counting. One filter unit is 1 pmol of substrate bound to the filter above the blank value at 1 μM histidine at approximately 23°.

Details of both assays and their drawbacks and advantages have been described.[16] It should be noted that while the filter binding assay is rapid, it is not useful at the beginning stages of purification because contaminating proteins compete effectively for retention of the binding protein by the filter. In contrast, equilibrium dialysis is independent of the presence of other proteins; in addition, its sensitivity can be increased considerably, for example, in cases when the binding affinity constant is not very good, by increasing the protein concentration while maintaining the substrate concentration low (which also decreases the chances of artifacts by non-specific binding which can occur at high substrate concentrations).

Equilibrium dialysis assays using microdialysis cells, as opposed to dialysis against large volumes of substrate solution as recommended above, has also been described.[17] It has the disadvantage that it results in substrate concentration changes during the assay: this is because the total amount of substrate available is very low and comparable to that of the binding protein. An equilibrium dialysis assay in large volumes, rather than a filter binding assay, should always be used for the final establishment of kinetic constants on the pure binding protein.

Because binding proteins are usually very stable, it is convenient and usually feasible to follow the entire purification process by immunoelectrophoresis or immunoslot-blotting, once antibodies against the protein of interest become available. This would limit activity assays only to the initial and final assessments of the activity and purity of the protein and to the determination of kinetic constants.

Purification

Starting from the shock fluid, several methods for the purification of binding proteins can be followed. An initial heat treatment is often a good purification step since many binding proteins are quite stable. Similarly, a low pH (5.0) treatment was found useful for HisJ[13] because it caused the precipitation of contaminating proteins and lipopolysaccharide. A variety of chromatographic steps can be applied subsequently. Because the shock fluid is usually produced in a relatively large volume, a first chro-

---

[17] R. C. Willis and C. E. Furlong, *J. Biol. Chem.* **250,** 2574 (1975).

matographic step that binds the protein of interest tightly is highly preferable because it concentrates the protein. Such a step would be desirable even though it did not yield a large increase in specific activity. Standard ion-exchange chromatographic procedures (carboxymethylcellulose or DEAE-cellulose) are effective.[14] A combination of two columns of opposite charge is often sufficient to yield protein that is 95% pure or more. A final molecular sieve chromatography step can yield protein of a purity level satisfactory for crystallization purposes. If the molecular weight of the binding protein is known, the appropriate sieving column can be easily deduced since all known binding proteins appear to function as monomers. Availability of an affinity chromatography matrix[17] or a hydrophobic chromatography step[19] can significantly simplify the purification process. High-performance liquid chromatography (HPLC) on columns of the same type as described above are useful for producing small amounts of pure protein very rapidly.[13]

General Notes of Interest

A common characteristic of binding proteins, which is a consequence of their high substrate affinity, is that they often maintain the bound substrate through all the purification procedures.[20] For some studies, in particular for X-ray crystallography studies, it is necessary to deligand the proteins. This has been achieved by their denaturation with guanidine hydrochloride, followed by dialysis and renaturation.[20] A gentler procedure has been developed based on exposure of the purified protein to a high concentration of substrate known to have a poor affinity for the protein, which results in the replacement by competition of the tightly bound substrate with the poor substrate. The protein liganded with the poor substrate can now be deliganded easily by dialysis. Dialysis at a pH value removed from the pH optimum for binding also favors the release of the substrate. This method has yielded protein that is acceptable for crystallization purposes.[13] Another deliganding procedure makes use of removal of the substrate by native gel electrophoresis; however, this method can only be used if the ligand carries a charge and on small amounts of protein.[21]

Acknowledgments

This work was supported by National Institutes of Health Grant DK12121.

[18] T. Ferenci and U. Klotz, *FEBS Lett.* **94**, 213 (1978).
[19] S. Shaltiel, G. F.-L. Ames, and K. D. Noel, *Arch. Biochem. Biophys.* **159**, 174 (1973).
[20] D. M. I. Miller, J. S. Olson, J. W. Pflugrath, and F. A. Quiocho, *J. Biol. Chem.* **258**, 13665 (1983).
[21] A. Joshi and G. F.-L. Ames, unpublished results, 1991.

[19] Isolation and Characterization of Lipopolysaccharides, Lipooligosaccharides, and Lipid A

By Michael A. Apicella, J. McLeod Griffiss,
and Herman Schneider

Introduction

In 1941, Shear determined that the component of the *Serratia marcescens* cell wall responsible for tumor-destroying activity was composed of a polysaccharide and a lipid.[1] He was the first to designate this material lipopolysaccharide (LPS). These are a family of glycolipids which consist of a lipid, which has been designated lipid A, and a carbohydrate component of variable length. The lipid A is embedded in the external layer of the bacterial outer membrane and is linked through 2-keto-3-deoxyoctulosonic acid to a carbohydrate component which extends into the surrounding environment.[2] The carbohydrate portion of LPS consists of a relatively conserved core region linked to a series of repeating four to six sugar units (the O-antigen). These units can form polysaccharides with molecular masses in excess of 65 kDa.[3,4] Lipooligosaccharides (LOS) are analogous to LPS and have carbohydrate components which are smaller (3.2 to 7.2 kDa) than LPS and tend to be more intricately branched than the LPS carbohydrate.[5-9] They can be isolated from *Haemophilus influenzae, Neisseria gonorrhoeae, Neisseria meningitidis, Haemophilus ducreyi, Branhamella catarrhalis,* and *Bordetella pertussis.*[5-13]

[1] M. J. Shear, *Cancer Res.* **1,** 731 (1941).

[2] O. Ludertiz, A. M. Staub, and O. Westphal, *Bacteriol. Rev.* **30,** 192 (1966).

[3] R. C. Goldman and L. Leive, *Eur. J. Biochem.* **107,** 145 (1980).

[4] L. Palva and P. H. Makela, *Eur. J. Biochem.* **107,** 137 (1980).

[5] N. J. Philips, M. A. Apicella, J. M. Griffiss, and B. W. Gibson, *Biochemistry* **31,** 4515 (1992).

[6] B. W. Gibson, J. W. Webb, R. Yamasaki, S. J. Fisher, A. L. Burlingame, R. E. Mandrell, H. Schneider, and J. M. Griffiss, *Proc. Natl. Acad. Sci. U.S.A.* **86,** 17 (1989).

[7] H. Schneider, T. L. Hale, W. D. Zollinger, R. C. Seid, C. Hammack, and J. M. Griffiss, *Infect. Immun.* **45,** 544 (1984).

[8] W. McLaugh, N. J. Philips, A. A. Campagnari, R. Karalus, and B. W. Gibson, *J. Biol. Chem.* **267,** 13434 (1992).

[9] H. J. Jennings, K. G. Johnson, and L. Kenne, *Carbohydr. Res.* **121,** 233 (1983).

[10] C. M. Johns, J. M. Griffiss, M. A. Apicella, R. E. Mandrell, and B. W. Gibson, *J. Biol. Chem.* **266,** 19303 (1991).

[11] A. A. Campagnari, S. M. Spinola, A. J. Lesse, Y. A. Kwaik, and M. A. Apicella, *Microb. Pathog.* **8,** 353 (1990).

The purpose of this chapter is to describe methods for the isolation and analysis of LPS and LOS. A variety of methods for isolation of these glycolipids has evolved since the 1930s. These include extractions with trichloroacetic acid,[14] ether,[15,16] water,[17] EDTA,[18] pyridine,[19] phenol,[20-23] and after solubilization with sodium dodecyl sulfate (SDS).[24]

Two procedures can suffice for the isolation of preparations of high purity. These are the modified phenol–water method[23] and the SDS solubilization method of Darveau and Hancock.[24] The phenol–water technique, which was first decribed by Westphal and co-workers,[21,22] takes advantage of the amphipathic nature of the lipopolysaccharide and the solubility of the majority of bacterial proteins in phenol. Polysaccharides, mucopolysaccharides, lipopolysaccharides with O side chains, and nucleic acids are usually soluble in aqueous solutions and insoluble in phenol. Phenol is a weak acid, its dissociation constant at 18° in water being 1.2×10^{-10}. Mixtures of phenol and water have a high dielectric constant. These facts form the basis of a method of partitioning proteins and polysaccharides and/or nucleic acids between phenol and water. Minor modifications have been made in the basic protocol described by Westphal and Jann[22] and by Johnson and Perry.[23] These changes result in LPS and LOS preparations that have less contamination with nucleic acids and produce somewhat greater yields. One major limitation of the phenol–water extraction method is that LPS with truncated polysaccharide components or the more hydrophobic, shorter chain LOS molecules frequently partition into the phenol phase.[25] Erwin and co-workers[25] have isolated the LOS of *H. influenzae aegyptius* from the phenol phase.

The method of Darveau and Hancock[24] utilizes SDS to solubilize LPS,

[12] R. E. Mandrell, H. Schneider, M. A. Apicella, W. Zollinger, P. Rice, and J. M. Griffiss, *Infect. Immun.* **54,** 63 (1986).
[13] M. S. Peppler, *Infect. Immun.* **43,** 224 (1984).
[14] A. Boivin and L. Mesrobeanu, *C. R. Seances Soc. Biol. Fil.* **112,** 76 (1933).
[15] E. Ribi, W. T. Haskins, M. Landy, and K. C. Milner, *J. Exp. Med.* **114,** 647 (1961).
[16] C. Galanos, O. Luderitz, and O. Westphal, *Eur. J. Biochem.* **9,** 245 (1969).
[17] R. S. Roberts, *Nature (London)* **209,** 80 (1966).
[18] L. Leive, *Biochem. Biophys. Res. Commun.* **21,** 290 (1965).
[19] W. F. Goebel, F. Benkley, and E. Perlman, *J. Exp. Med.* **81,** 315 (1946).
[20] O. Westphal, O. Luderitz, E. Eichenberger, and W. Keiderling, *Z. Naturforsch. B: Anorg. Chem. Org. Chem. Biochem. Biophys. Biol.* **B7,** 536 (1952).
[21] O. Westphal and O. Luderitz, *Angew. Chem.* **66,** 407 (1954).
[22] O. Westphal and K. Jann, *Methods Carbohydr. Chem.* 83V (1965).
[23] K. G. Johnson and M. B. Perry, *Can. J. Microbiol.* **22,** 29 (1975).
[24] R. P. Darveau and R. E. W. Hancock, *J. Bacteriol.* **155,** 831 (1983).
[25] A. Erwin, R. S. Munford, and the Brazilian Purpuric Fever Study Group, *J. Clin. Microbiol.* **27,** 762 (1989).

allowing separation of the insoluble peptidoglycan. Bacterial proteins are removed from the LPS by protease digestion, and the LPS is precipitated with ethanol. RNA and DNA are removed by nuclease digestion prior to the addition of the SDS. This technique is more complex than the phenol–water method. Its major advantage is that more hydrophobic LPS preparations can be isolated. Unless care is taken, the high pH of some of the buffers can cause O-deacylation of the LPS, modifying the biological activity of the LPS or LOS preparation.

Modified Phenol–Water Technique

In the modified phenol–water procedure,[23] 5 g of freeze- or acetone-dried bacteria should be placed in a mortar and pestle and ground until a very fine powder is formed. The powder is suspended in 25 ml of 50 mM sodium phosphate, pH 7.0, containing 5 mM ethylenediaminetetraacetic acid (EDTA). The suspension is allowed to hydrate completely before proceeding. The suspension is then stirred in a shearing mixer such as a Sorvall Omnimixer or a blender at top speed for 1 min. Hen egg lysozyme (100 mg) is added and the suspension stirred overnight at 4°. The suspension is placed at 37° for 20 min and then stirred at top speed in a shearing-type mixer for 3 min. The volume of the suspension is increased to 100 ml with 50 mM sodium phosphate, pH 7.0, containing 20 mM MgCl$_2$. Ribonuclease A and deoxyribonuclease I are added to a final concentration of 1 μg/ml. The suspension is incubated for 60 min at 37° and then for 60 min at 60°. Occasionally, the suspension will become gelatinous at this stage. If this occurs, stir for 3 min in the shearing mixer before phenol extraction.

The bacterial suspension is then placed in a 70° water bath until the temperature equilibrates. An equal volume of 90% (w/v) phenol which has been preheated to 70° is added and throughly mixed. The resulting mixture is rapidly cooled by stirring for 15 min in an ice–water bath. The phenol–bacterial suspension mixture is then centrifuged at 4° at 18,000 g for 15 min. A sharp interface occurs between the aqueous, phenol, and interface layers. Occasionally there is a sediment. The aqueous and phenol layers should be removed by aspiration. The aqueous layer containing the lipopolysaccharide is retained while the phenol layer is discarded. The aqueous layer is dialyzed against frequent changes of distilled water until no detectable phenol odor remains. The dialyzate is lyophilized.

This is still a crude LPS preparation which needs further purification by centrifugation. The preparation is suspended in distilled water at concentrations of 25–35 mg/ml. At times, this crude LPS can be difficult to suspend. It should be vortexed vigorously to obtain a smooth suspension.

Gentle sonication may be required to obtain an even distribution of the LPS suspension. The suspension is sedimented at 1100 g for 5 min. The sediment is discarded, and the supernatant fractions are centrifuged at 105,000 g for 16 hr at 4°. The gel-like pellet is saved and the supernatant discarded. The pellet is resuspended in the original volume of distilled water and the process of low-speed and high-speed centrifugation repeated until the desired purity is obtained. The final pellet can be resuspended in distilled water and lyophilized.

Detergent Solubilization Method of Lipopolysaccharide Preparation

In the SDS solubilization method,[24] dried bacterial cells (500 mg), which have been ground with a mortar and pestle, are suspended in 15 ml of 10 mM Tris-HCl buffer, pH 8.0, 2 mM MgCl$_2$ containing 100 μg/ml pancreatic DNase I and 25 μg/ml pancreatic RNase. The suspension is allowed to hydrate for several hours and is passed twice through a French pressure cell at 15,000 psi. The cell lysate is then sonicated twice at 6 W for 30 sec using a microprobe. DNase and RNase are added again to a final concentration of 200 and 50 μg/ml. The suspension is incubated at 37° for 2 hr. After the incubation period, 5 ml of 0.5 M tetrasodium EDTA in 10 mM Tris-HCl (pH 8), 2.5 ml of 20% SDS dissolved in 10 mM Tris-Hcl, and 2.5 ml of 10 mM Tris-HCl (pH 8) are added. The final volume of the sample is 25 ml and contains 0.1 M EDTA, 2% SDS, and 10 mM Tris-HCl at a pH of approximately 9.5 (according to Darveau and Hancock,[24] the Tris is added to potentiate the effects of the EDTA rather than for its buffering effects). The sample is vortexed to solubilize the components and then subjected to centrifugation at 50,000 g for 30 min at 20°. The supernatant is decanted. The sediment, which contains peptidoglycan and associated proteins, is discarded. Pronase is added to the supernatant to a final concentration of 200 μg/ml. The sample is incubated overnight at 37° with constant shaking. Any precipitate which forms can be removed by centrifugation at 1000 g for 10 min.

The LPS or LOS is next precipitated with ethanol. Two volumes of 0.376 M MgCl$_2$ in 95% (v/v) ethanol is added to the supernatant and cooled to 0° (check with a thermometer) by placing the flask in a −70° freezer or a dry ice bath. After the sample has cooled to 0°, it is centrifuged at 12,000 g for 15 min at 0° to 4°. It is important that the sample be kept as close to 0° while the precipitate is formed and during the centrifugation. The pellet is suspended in 25 ml of 2% SDS–0.1 M tetrasodium EDTA, dissolved in 10 mM Tris-HCl (pH 8), and sonicated as described above. The solution is then incubated at 85° for 10 to 30 min to ensure the denaturation of SDS-resistant proteins. After cooling, the solution is adjusted to pH 9.5. Pronase is added to 25 μg/ml, and the sample is incubated overnight at 37° with constant agitation.

The LPS is again precipitated with 0.376 M MgCl$_2$ in 95% ethanol at 0° and sedimented by centrifugation at 12,000 g for 15 min at 0° to 4°. The pellet is resuspended in 15 ml of 10 mm$M$ Tris-HCl (pH 8), sonicated, and centrifuged at 1000 rpm for 5 min to remove insoluble Mg^{2+}–EDTA crystals which sometimes precipitate at this stage. The supernatant is then centrifuged at 200,000 g for 2 hr at 15° in the presence of 25 mM MgCl$_2$. The pellet, which contains LPS, is suspended in distilled water.

Small-Scale Preparations of Lipopolysaccharide or Lipooligosaccharide

Occasionally, analytical studies such as Western blot analysis of LPS or LOS will require preparation of small quantities of the glycolipids from multiple bacterial samples. Two methods have been described that can be used for such isolations.[26,27] One relies on a modified phenol–water extraction, and the second utilizes SDS solubility of bacteria followed by enzymatic degradation of bacterial proteins. Neither approach gives a highly purified preparation, but both enrich for LPS or LOS and can serve as adequate substitutes in acrylamide gel and Western blot studies. The methods are presented below.

Rapid Isolation Micromethod for Lipopolysaccharide

In modified phenol–chloroform method,[26] a bacterial suspension [10^8 colony-forming units (cfu)/ml] in 2 ml phosphate-buffered saline (PBS) is centrifuged in a 15-ml tube at 10,000 g for 5 min at 4°. The pellet is washed once in PBS (pH 7.2) containing 0.15 mM CaCl$_2$ and 0.5 mM MgCl$_2$. The washed cells are resuspended in 300 μl of water and transferred to a 1-dram vial containing a stir bar. An equal volume of hot (65°–70°) 90% phenol is added. The mixture is stirred vigorously at 65°–70° for 15 min. The suspension is chilled on ice, transferred to a 1.5-ml polypropylene tube, and centrifuged at 8500 g for 15 min at 4°. The supernatant is transferred to a 15-ml conical centrifuge tube. The phenol phase is reextracted with 300 μl of distilled water. The aqueous phases are pooled. Sodium acetate is added to 0.5 M final concentration. Ten volumes of 95% ethanol are added, and the sample is placed at $-20°$ overnight in order to precipitate the LPS. The precipitate is centrifuged at 2000 g and 4° for 10 min. The supernatant is discarded. The pellet is suspended in 100 μl of distilled water and transferred to a 1.5-ml polypropylene tube. The ethanol precipitation is

[26] T. Inzana, *J. Infect. Dis.* **148**, 492 (1983).
[27] P. J. Hitchcock and T. M. Brown, *J. Bacteriol.* **154**, 269 (1983).

repeated. The precipitate is dried and resuspended with 50 μl of distilled water. It is stored at $-20°$.

Lipopolysaccharide Microextraction Using Proteinase K Digestion

Organisms grown on solid medium are harvested with a sterile swab and suspended in 10 ml of cold PBS, pH 7.2, to a turbidity of 0.4 absorbance units at 650 nm.[27] A portion (1.5 ml) of the suspension is centrifuged for 1.5 min in a microcentrifuge at 14,000 g. The pellet is solubilized in 50 μl of lysing buffer containing 2% SDS, 4% 2-mercaptoethanol, 10% glycerol, 1 M Tris, pH 6.8, and bromphenol blue. This is heated at 100° for 10 min. To digest bacterial proteins, 25 μg of proteinase K in 10 μl of lysing buffer is added to each boiled lysate and incubated at 60° for 60 min. The preparation can then be used in acrylamide gel electrophoresis or for Western blots in volumes ranging from 0.5 to 2 μl.

Purity of Lipopolysaccharide and Lipooligosaccharide Preparations

After isolation of the LOS or LPS, the degree of purity should be ascertained. Nucleic acid contamination is the principal concern. Spectral analysis from 245 through 290 nm using 1 mg/ml purified LPS or LOS can be performed to determine nucleic acid and protein contamination. Agarose gel electrophoresis with ethidium bromide staining can also be useful as a highly sensitive means of determining the degree of nucleic acid contamination. Protein contamination can be determined using the Lowry method or a similar colorimetric method.[28] Acrylamide gel electrophoresis of the LOS/LPS preparation will define the physical characteristics of the preparation.[3,4] The methodology for acrylamide gel electrophoresis of LOS/LPS is different from the Laemmli method[29] since neither the running buffer nor the resolving gel contains SDS or urea. This method, utilized in a number of laboratories studing LPS and LOS, is described below. The pore size of the gel can be adjusted according to the preparation under study and can range from 10 to 16% acrylamide.

Lipooligosaccharide and Lipopolysaccharide Acrylamide
   Gel Electrophoresis

Acrylamide gel electrophoresis of LOS and LPS can be an exasperating procedure because of day-to-day variation in resolution. To ensure consis-

[28] O. H. Lowry, N. J. Rosenbrough, A. L. Farr, and R. J. Randall, *J. Biol. Chem.* **193**, 265 (1951).
[29] U. K. Laemmli, *Nature (London)* **227**, 680 (1970).

tency of the gels, all glassware must be acid washed and rinsed well before drying, all reagents must be of the highest quality, and the quality of water should be in the 18 megohm range. Finally, stock solutions should not be used past the time recommended.

Stock Solutions

30% Acrylamide stock: Dissolve 29.2 g acrylamide, 0.8 g bisacrylamide in 40 ml distilled water, and fill to 100 ml. Filter into a dark bottle and store at 4°. This stock should not be used after 2 weeks.

Resolving gel buffer, 1.88 M Tris-HCl, pH 8.8: Dissolve 22.78 g Tris in 70 ml distilled water, adjust to pH 8.8 with concentrated HCl, and fill to 100 ml. Store at 4°. This stock should not be used after 2 weeks.

Spacer gel buffer, 1.25 M Tris-HCl, pH 6.8: Dissolve 15.12 g Tris in 70 ml distilled water, adjust to pH 6.8 with concentated HCl, and fill to 100 ml and store at 4°. This stock should not be used after 2 weeks.

Sample buffer: Add 0.727 g Tris, 0.034 g EDTA, and 2.0 g SDS to 70 ml distilled water, adjust to pH 6.8 with concentrated HCl, fill to 100 ml with distilled water, filter, sterilize, and store at room temperature.

Dye buffer: Combine 2.5 ml sample buffer, 2.0 ml glycerol, 400 μ 2-mercaptoethanol, and 200 μl saturated solution of bromphenol blue. This buffer should be made fresh for each run. A stock solution can be made without adding the 2-mercaptoethanol, which should then be added just prior to use.

Ammonium persulfate: Dissolve 50 mg ammonium persulfate in 1 ml of distilled water. This solution should be made fresh for each run.

Reservoir buffer, Tris–glycine buffer, pH 8.3: Dissolve 115.2 g glycine, 24 g Tris, and 8.0 g SDS in 8 liters of distilled water. No pH adjustment should be necessary if correct reagents and quantities are added.

Procedure. To prepare two 14% resolving gels, combine 18.45 ml of 30% acrylamide, 7.9 ml resolving buffer, and 12.57 distilled water, place in a vacuum flask, and degas for 15 min. Add 0.3 ml ammonium persulfate solution. The resolving gel should be poured between the glass plates as soon as the TEMED is added and the solution gently mixed. Overlay the gel with 2 mm distilled water and allow the gel to polymerize for at least 2 hr. This will make a total of 39.56 ml resolving gel, which is sufficient for two 12 × 14 cm slab gels with a 0.75 mm spacer. Double the volumes if 1.5 mm spacers are used. To make 16% resolving gels, mix 10.55 ml of

30% acrylamide, 4.95 ml distilled water, and 3.95 ml resolving buffer. The quantities of the other reagents are unchanged.

To prepare the spacer gel, combine 2 ml of 30% acrylamide, 2 ml spacer buffer, and 15.6 distilled water. Degas the solution for 15 min and add 0.2 ml ammonium persulfate, 0.2 ml of 10% SDS and 10 μl TEMED. Mix gently. The total volume of the spacer gel is 20.01 ml. The layer of distilled water over the resolving gel should be removed and the spacer gel poured. The comb chosen for the run should be inserted at this time. The spacer gel should be allowed to polymerize for at least 1 hr but preferably overnight.

Solubilize the LPS or LOS in sample buffer to the desired concentration (0.1 to 1 mg/ml should be sufficient). Dilute 1 : 1 in dye buffer, boil for 5 min in a water bath, and allow to cool. Load 5–10 μl per well.

For electrophoresis, remove the comb and rinse the sample wells with reservoir buffer. Fill the wells with reservoir buffer and add samples, allowing them to sink to the bottom. Place the gels into the chamber and electrophorese at constant current under the following conditions: For one 12 × 14 cm slab gel, use 10–12 mA through the spacer gel, then raise to 15 mA through the resolving gel (total run time will be ~5 hr); for two 12 × 14 cm slab gels, use 20 mA through spacer gel, then raise to 30 mA through resolving gel (the total run time will be 5 to 6 hr). The gels can be stained using the silver stain method described by Tsai and Frasch[30] or may be prepared for Western blotting.[31]

Use of Long Gels to Resolve Poorly Separated Low Molecular Weight Lipooligosaccharides

Low molecular weight LOS molecules may be poorly separated through the usually used 12 × 14 cm slab gels.[7] Increased resolution of these molecules can be achieved by using a longer gel. Figure 1 shows the resolution obtained with the use of a 29 cm gel of 13.1% acrylamide to separate meningococcal and gonococcal LOS (kindly provided by B. Brandt, Department of Bacterial Diseases, Walter Reed Army Institute of Research, Washington, DC). The LOS bands of strain 8032 (lane i) have estimated M_r values of 3200–5400.[7] They are clearly separated into seven bands by the long gel but appear as only five and occasionally six bands in a short gel.[7]

Long gels must be run overnight and are even more subject to day-to-day variation, reduced mobility of the components at the sides of gels

[30] C.-M. Tsai and C. S. Frasch, *Anal. Chem.*, 119 (1982).
[31] D. A. Knecht and R. L. Dimond, *Anal. Biochem.* **136**, 180 (1984).

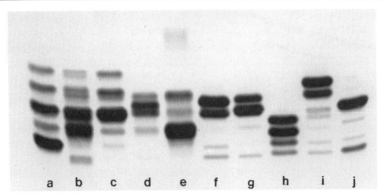

FIG. 1. A "Long" 13.1% acrylamide gel shows increased resolution of LOS from gonococcal strain MS11 mkD (lane a) and the LOS from meningococcal strains 1207-0 (lane b),
1032-6 (lane c), 1255-0 (lane d), 1157-0 (lane e), 138I (lane f), M136 (lane g), 126E (lane h),
8032 (lane i), and 8529 (lane j). The LOS bands have been fixed and stained by the method
of Tsai and Frasch.[30]

("wall effects"), and incomplete resolution than are shorter gels, but the
increased resolution can easily compensate for the increase in frustration.
Although we do not recommend long gels for routine monitoring of LOS
preparations, they can be invaluable for assigning chemical structure to
epitopes that have been identified by monoclonal antibodies in immunoblots of components separated by SDS–polyacrylamide gel electrophoresis.

Tricine–SDS–Acrylamide Gel Electrophoresis

Increased resolution of LPS and LOS can be achieved using Tricine–SDS–polyacrylamide gel electrophoresis.[32] This technique can be
particularly useful in the analysis of lower molecular mass LOS and LPS
mutant preparations, where differences in individual bands as small as
140 Da can be distinguished. The drawbacks of this technique are that
the time for runs can exceed 72 hr in 12 × 14 cm slab gels and, in addition,
Western blot transfers can be quite variable.

The technique is a modification of the method of Schagger and Von
Jager.[33] Acrylamide and bisacrylamide are used to prepare the stock solu-
tions of acrylamide. These solutions should be discarded after two weeks.
Stock solutions of acrylamide are prepared in two concentrations of 49.5%

[32] A. J. Lesse, A. A. Campagnari, W. E. Bittner, and M. A. Apicella, *J. Immunol. Methods*
126, 109 (1990).
[33] H. Schagger and G. Von Jager, *Anal. Biochem.* **166**, 368 (1987).

T, 6% C and 49.5% T, 3% C where T represents the total percentage of acrylamide (both acrylamide and bisacrylamide) and C represents the percentage of cross-linker (bisacrylamide) to the total concentration of acrylamide. Anode buffer (0.2 M Tris-HCl, pH 8.9), cathode buffer (0.1 M Tris-HCl, 0.1 M tricine, 0.1% SDS, pH 8.25), and gel buffer (3.0 M Tris, 0.3% SDS, pH 8.45) are prepared as described by Schagger and Von Jager.[33]

Gels can be prepared in any apparatus chosen. We prepare gels in a multiple casting apparatus (SE-295) from Hoeffer Scientific (San Francisco, CA). Separating gels 0.75 mm thick are prepared five at a time in a final concentration of 16.5% T, 6% C by the addition of 16.6 ml of 49.5% T, 6% C solution, 16.6 ml gel buffer, 5.2 ml glycerol, and 11.5 ml distilled water. Stacking gels of approximately 1 cm are made in a 4% T, 3% C concentration by mixing 1 ml of 49.5% T, 3% C acrylamide stock solution, 3.1 ml gel buffer, and 8.4 ml distilled water and are polymerized individually on top of the previously polymerized separating gels with the addition of 150 μl of 10% ammonium persulfate and 15 μl TEMED. Spacer gels are not necessary. The LOS or LPS are dissolved in sample buffer as described in the previous section. Samples up to 10 μl in size with concentrations from 0.1 to 1 μg/ml LPS or LOS are utilized. The electrophoresis is carried out using the SE 280 Tall Mighty Small gel apparatus (Hoeffer Scientific). Samples are loaded under the cathode buffer and run at 30 V constant voltage until the sample is totally within the stacking gel. The voltage is then increased to 105 V for the remainder of the run. Separating gels averaged about 9 cm in length. Gels are run for about 10 hr, approximately 1 hr after the bromphenol blue dye front has left the gel.

The LPS and LOS banding patterns are best visualized on acrylamide gels using the silver staining procedure of Tsai and Frasch.[30] A modification of this technique is given below.

Silver Staining Technique for Gels

The LPS and LOS bands can be rapidly identified in acrylamide gels using the silver stain method of Tsai and Frasch.[30] The method is very sensitive and requires careful attention to cleanliness of the glassware used. Only a glass tray should be used for the procedure, and it is best to have dedicated trays for each step in the procedure. Glassware should be precleaned before each study with nitric acid and washed thoroughly with distilled water. Gloves should be worn while performing the procedure. Mild rotary agitation (70 rpm) is required for each step. The rinses and washes are crucial and should not be reduced in time or number.[27]

As soon as the electrophoresis is complete, the gel should be placed in a fixing solution consisting of 40% ethanol and 5% acetic acid in distilled water overnight. Add 0.9% periodic acid (1.8 g/200 ml) to fixing solution. Rinse three times with distilled water. Then transfer to a separate dish and wash three additional times with distilled water (500–1000 ml), agitating for 10 min each time. After the last 10-min wash, rinse three times with distilled water. In a separate dish, pour in freshly prepared staining reagent (this should be made during the last distilled water wash). The stain is made by mixing, in the following order, in a hood 28 ml of 0.1 N NaOH, 2.1 ml NH$_4$OH (concentrated), 5 ml of 20% silver nitrate (the silver nitrate should be added dropwise), and 115 ml distilled water. Agitate for 10 min. Transfer the gel to a separate dish and rinse gel three times with distilled water. The gel is transferred to another dish, and fresh formaldehyde developer (25 mg anhydrous citric acid, 150 ml of 37% formaldehyde in 500 ml distilled water) is added. For best results, the gel should be developed in the dark using a Kodak (Rochester, NY) GBX-2 safelight. The bands will develop over the next 10 to 15 min. Background staining will intensify the longer the gel is left in the developer. The reaction can be stopped by rinsing the gel in water, transferring to a separate dish, and adding rapid fix (10 ml in 100 ml distilled water).

Isolation of Lipid A

Lipid A can be isolated by mild acid hydrolysis of the purified LPS.[34] The LPS or LOS is solubilized in aqueous 0.02% triethylamine, and acetic acid is added to a final concentration of 1.5% (v/v). The mixture is heated for 2 hr at 100° and then cooled. The lipid A is quantitatively precipitated by the addition of 1 M HCl to a final pH of 1.5. The insoluble lipid A is centrifuged, washed three times with cold distilled water, and lyophilized.

[34] V. A. Kulshin, U. Zahringer, B. Lindner, C. E. Frasch, C.-M. Tsia, B. A. Dimtriev, and E. T. Rietschel, *J. Bacteriol.* **174**, 1793, (1992).

[20] Isolation of Peptidoglycan and Soluble Peptidoglycan Fragments

By RAOUL S. ROSENTHAL and ROMAN DZIARSKI

Introduction

Peptidoglycan (PG; synonyms: murein, mucopeptide) is a heteropolymer, unique to bacterial cell walls, that consists of a glycan backbone of alternating units of N-acetylglucosamine (GlcNAc) and N-acetylmuramic acid (MurNAc) with short peptides linked to the lactyl groups of the MurNAc moieties. In virtually all bacteria, the sugars in PG are linked by β-1,4-glycosidic bonds, and the general structure of the peptide is L-alanine–D-glutamic acid–a diamino acid–D-alanine–D-alanine. The dibasic amino acid in position 3 is typically lysine in gram-positive cocci and diaminopimelic acid (DAP) in gram-positive bacilli and gram-negative bacteria. Peptide cross-linking bonds between amino acids located on different glycan chains lead to the formation of a complex three-dimensional macromolecule that has been likened to an enormous, covalently closed basket surrounding the cytoplasmic membrane.[1] The rather rigid arrangement of polymeric glycan (up to 100 disaccharide units), cross-linked by peptides, plays a major role in determining cell shape and in maintaining the physical integrity of the bacterium. It is not surprising, therefore, that although there are numerous subtle variations in the structure of PG among different organisms, mostly limited to the cross-linking peptides,[2] the composition (including the unique MurNAc, DAP, and D-amino acid residues) and the supramolecular organization of PG have been well conserved in nature.

Although PG was long regarded as merely a biologically inert corset which served only a protective function, it is now clear that, given access to host tissues and cells, both insoluble PG and soluble PG fragments are versatile and potent biological effectors. Indeed, the large family of PG compounds obtained from various bacteria is, perhaps, rivaled only by endotoxins [the lipopolysaccharide (LPS) component of gram-negative bacteria] in the number and diversity of biological systems that it is capable of modulating. Among the many in vitro activities of naturally occurring PG fragments or synthetic PG derivatives, for example, muramyl dipeptide

---

[1] W. Weidel and H. Pelzer, Adv. Enzymol. 26, 193 (1964).
[2] K. H. Schleifer and O. Kandler, Bacteriol. Rev. 36, 407 (1972).

(MDP), are stimulation of antibody production,[3] activation of the metabolic and killing capacity of leukocytes,[4] stimulation of macrophages to release endogenous mediators such as interleukin 1,[5] toxicity for ciliated epithelial cells in culture,[6] and activation of complement.[7] *In vivo* PG fragments, collectively, mediate a seemingly endless series of responses, that is, they possess adjuvanticity,[8] are pyrogenic,[9] induce autoimmunity,[10] provoke arthritis,[11] enhance nonspecific resistance to microbial infections and tumors,[12] induce nonspecific antibacterial proteins in insects,[13] and modulate neurological functions, such as slow wave sleep[14] and appetite[15] in higher animals. Although long and diverse, the list reveals a common denominator which links many PG-mediated activities, namely, PG fragments share a propensity to modulate the behavior of cells, principally macrophages and lymphocytes, involved in the development of inflammation and immunity. Thus, it is not surprising that some PG derivatives are currently being investigated for the deleterious effects of arthritogenicity and toxicity, while similar (or identical) compounds are being employed in human trials as adjuvants in vaccines and as anticancer agents.

   Given this rich biological potential of PG, there is increasing need for methods to derive and purify the diverse members of this chemical class for biological testing. Considerable work has already dealt with the chemical synthesis of low molecular weight PG fragments modeled around MDP[16]; indeed, MDP and some of its analogs are available from commercial distributors, for example, Calbiochem (La Jolla, CA) and Sigma Chemical Co. (St. Louis, MO). We will not detail methods for chemical synthesis of PG compounds; rather, we concentrate on intact PG and natural PG

[3] C. D. Leclerc, D. Juy, and L. Chedid, *Cell. Immunol.* **42,** 336 (1979).
[4] N. P. Cummings, M. J. Pabst, and R. B. Johnston, Jr., *J. Exp. Med.* **152,** 1659 (1980).
[5] M. R. Gold, C. L. Miller, and R. I. Mischell, *Infect. Immun.* **49,** 731 (1985).
[6] M. A. Melly, Z. A. McGee, and R. S. Rosenthal, *J. Infect. Dis.* **149,** 378 (1984).
[7] J. Greenblatt, R. J. Boackle, and J. H. Schwab, *Infect. Immun.* **19,** 296 (1978).
[8] A. R. Adam, R. Ciorbaru, F. Ellouz, J. F. Petit, and E. Lederer, *Biochem. Biophys. Res. Commun.* **56,** 561 (1974).
[9] B. Heymer, *Z. Immunitaetsforsch.* **149,** 245 (1975).
[10] I. Lowy, C. Leclerc, and L. Chedid, *Immunology* **30,** 441 (1980).
[11] T. J. Fleming, D. E. Wallsmith, and R. S. Rosenthal, *Infect. Immun.* **52,** 600 (1986).
[12] D. Juy and L. Chedid, *Proc. Natl. Acad. Sci. U.S.A.* **72,** 4105 (1975).
[13] P. E. Dunn, W. Dai, M. R. Kanost, and C. Geng, *Dev. Comp. Immunol.* **9,** 559 (1985).
[14] J. M. Krueger, J. R. Pappenheimer, and M. L. Karnovsky, *J. Biol. Chem.* **257,** 1664 (1982).
[15] W. Langhans, R. Harlacher, B. Balkowski, and E. Scharrer, *Physiol. Behav.* **47,** 805 (1990).
[16] P. Lefrancier, M. Derrien, L. Lederman, F. Nief, J. Choay, and E. Lederer, *Int. J. Pept. Protein Res.* **11,** 289 (1978).

derivatives that, to date, have not been completely and successfully synthesized.

Isolation of Insoluble Peptidoglycans

General Considerations

Three major classes of bacteria, grouped according to the complexity of the cell wall structure in which the PG resides, namely, gram-positive, gram-negative, and acid-fast bacteria, present three different challenges. Fundamentally, the optimal methods for isolation and purification depend on what (if any) accessory polymers are covalently bound to PG. Acid-fast bacteria (*Mycobacterium tuberculosis* has been the main choice for PG studies) are clearly the most complex in this regard because the PG is linked to a diverse series of lipophilic constituents, as well as carbohydrates and proteins. Methods are available to isolate mycobacterial PG in stages with various sets of its accessory polymers attached and to obtain purified mycobacterial PG.[17] However, unless one specifically desires to study PG-containing complexes from acid-fast bacteria, the difficulty of the methods needed preclude the use of these organisms for general applications.

Gram-positive PG is embedded in a relatively thick cell wall and is usually covalently attached to other polymers, with teichoic acids and carbohydrates being the most common. As a result, isolation of gram-positive PG generally requires a more elaborate procedure (often involving mechanical cell disruption, multiple enzymatic steps, and/or strong extraction procedures) than does isolation of gram-negative PG, which is commonly attached covalently to a single type of lipoprotein for a given species or, in a few cases, simply exists as the free macromolecule *in situ*. Investigators who choose to employ gram-positive PG should take some consolation in the fact that the typical gram-positive bacterium contains at least 10-fold more PG on a weight basis than does a typical gram-negative counterpart.

Gram-Positive Bacteria

Isolation of PG from cell walls involves the following steps: (i) growing and harvesting the cells; (ii) inactivating autolytic enzymes; (iii) rupturing the cells; (iv) obtaining crude cell walls by differential centrifugation; (v)

---

[17] E. Lederer, *J. Med. Chem.* **23**, 819 (1980).

obtaining purified cell walls by removing RNA, DNA, and proteins; and (vi) obtaining purified PG by removing covalently bound teichoic acid and carbohydrates. There are several procedures that can be used to accomplish each of these tasks. The choice of a specific method primarily depends on the type of bacteria used for PG isolation.

(i) Rich growth media are usually used to allow abundant growth of bacteria. The use of liquid or solid media is a matter of personal preference and availability of equipment. The type of medium and growth conditions, however, may affect the chemical composition of PG, for example, the amino acid composition or extent of O-acetylation.[2] The objective usually is to obtain a large mass of bacterial cells. Liquid cultures in fermenters or in flasks on a shaker can be efficiently used for this purpose. Alternatively, bacteria can be harvested from lawns grown on large agar plates, which, in large-scale experiments, reduces the volume of liquid that has to be centrifuged.

(ii) Inactivation of autolytic enzymes is usually accomplished by boiling the bacteria immediately after the harvesting. This step is important to prevent partial degradation of PG by the enzymes, which can change the structure of PG and reduce its yield. Although some PG hydrolases are not inactivated by boiling (e.g., staphylococcal endo-β-N-acetylglucosaminidase, which is inactivated at $120°$ but not at $100°$[18]), autoclaving the cells is not recommended.

(iii) Several methods are available to rupture the cells, and the choice of a method usually depends on the availability of equipment. French press (X-press, AB Biox, Jarfalla, Sweden, U.S. availability through Tekmar Company, Cincinnati, OH) or one of several instruments that break the cells by shaking or mixing with small glass beads (Bead Beater, Biospec Products, Bartlesville, OK, or MSK Braun Cell Homogenizer, both available from Tekmar) can be used successfully. Some bacteria, for example, *Streptococcus pyogenes,* are more difficult to break than others, and the conditions should be optimized for a given bacterium. A convenient method to monitor the efficiency of cell disintegration is gram staining: disintegrated cells (cell ghosts) should lose the gram-positive staining and stain gram-negative, even though they usually retain the original cell shape.

(iv) The next step involves obtaining crude cell walls by differential centrifugation, that is, removing unbroken cells by low-speed centrifugation and then sedimenting crude cell walls from the supernatant by high-speed centrifugation and washing.

[18] S. Valisena, P. E. Varaldo, and G. Satta, *J. Clin. Invest.* **87,** 1969 (1991).

(v) To obtain purified cell walls, namely, complexes of PG and covalently linked teichoic acid and often also a carbohydrate, such as the C carbohydrate in streptococci, it is necessary to remove noncovalently bound RNA, DNA, and proteins, and often covalently bound proteins. The task is best accomplished by sequential digestion with RNase, DNase, and proteases. Although digestion with the RNase and the DNase can be performed simultaneously, it is important to perform a separate digestion with a protease, to avoid digestion and inactivation of the RNase and the DNase. Trypsin is the most often used protease and yields sufficiently pure protein-free cell walls. If all proteins are not removed by trypsin (as determined by amino acid analysis), digestion with trypsin can be followed by digestions with pepsin (in 0.05 N HCl)[19] and pronase. However, the endopeptidase activity of pronase can digest peptide bridges of some PG molecules. Moreover, some commercial preparations of proteases of bacterial or fungal origin (e.g., proteinase K) can be contaminated with PG-lytic enzymes, and the use of such preparations should be avoided. Some investigators use boiling in sodium dodecyl sulfate (SDS) to remove noncovalently bound molecules, but, in general, this procedure is the least effective in removing cell-wall bound RNA, DNA, and proteins[20] and is not recommended.

(vi) Most gram-positive bacteria have cell wall teichoic acid covalently bound to PG.[21] Some bacteria, for example, streptococci, may also have other carbohydrates covalently bound to PG. Therefore, obtaining purified PG requires removing these teichoic acids and other carbohydrates. Because specific enzymes that can accomplish this task are not available, chemical hydrolysis has to be used. The conditions of hydrolysis have to be harsh enough to remove the carbohydrates but gentle enough to preserve the structure of PG. The best method to remove teichoic acids is extraction with 5–10% trichloroacetic acid (TCA). Extraction with 10% TCA for 24 to 48 hr at 4° is often used, but it does not remove all teichoic acid. Short extraction with 10% TCA at higher temperature (15 min, 90°)[19] can be used, but this treatment may partially degrade PG.[22] Extraction at room temperature with 5% TCA is a reasonable compromise.[22,23] TCA extraction has been successfully used for purification of PG from numerous bacteria, such as *Staphylococcus, Micrococcus, Bacillus, Listeria,* and

[19] E. Janczura, H. R. Perkins, and H. J. Rogers, *Biochem. J.* **80**, 82 (1961).
[20] D. C. Brown, R. J. Doyle, and U. N. Streips, *Prep. Biochem.* **6**, 479 (1976).
[21] R. Dziarski, *Curr. Top. Microbiol. Immunol.* **74**, 113 (1976).
[22] K. H. Schleifer, *Z. Immunitaetsforsch. Exp. Klin. Immunol.* **146**, 104 (1976).
[23] R. Dziarski and K. Kwarecki, *Zentralbl. Bakteriol. Parasitenkd. Infektionskrankh. Hyg. I Abt. Suppl.* **5**, 393 (1976).

Propionibacterium.[22–27] Other extraction procedures using nitrous acid, sulfuric acid, sodium hydroxide, and *N,N*-dimethylhydrazine have also been used, but, in general, they tend to be more destructive for PG than the TCA extraction.[22] In some bacteria, however, such as *Streptococcus pyogenes* or *Corynebacterium,* TCA extraction is not sufficient, because it does not remove the PG-bound carbohydrates. Therefore, PG from these bacteria is most often purified by hot formamide extraction (20 min at 150°–170°).[28,29] This method, however, introduces *O*-formyl[30] and *N*-formyl[29] groups into PG, which changes its chemical structure and may also change the biological activity (e.g., *O*-formylation renders PG resistant to lysozyme[30]).

Below is the TCA extraction procedure for isolation of cell wall PG from *Staphylococcus aureus*[19,23,24] This procedure can be used for isolation of PG from most gram-positive bacteria, except for the species in which formamide extraction has to be used to remove PG-bound carbohydrate.[28] All reagents are sterile, and disposable plasticware or glassware baked at 180° is used, to avoid contamination with endotoxin.

1. Grow lawns of bacteria by spreading 0.5 ml of an overnight culture onto 14-cm nutrient agar plates and incubating them overnight at 37°.

2. Wash the bacteria off the plates with a small amount of cold saline, by scraping with a bent glass rod. Collect cells into a flask and filter through a gauze to remove small pieces of agar.

3. Heat the bacterial cell suspension at 100° for 20 min.

4. Sediment the bacteria by centrifugation (2000 g, 4°), and wash two times with saline, once with water, and three times with acetone. Dry the sediment at 37°. This acetone powder of bacteria can be processed immediately or stored frozen.

5. Suspend 2.5 g of dry bacterial powder in 25 ml of ice-cold water and mix with 25 ml of 0.1–0.15 mm glass beads in a Bead Beater (Biospec Products, also available from Tekmar) until total disruption of cells is achieved (5–15 min for staphylococci), as judged by uniform gram-negative staining of disrupted cells (compare them with control gram-positive staining of cells before disruption).

[24] J. T. Park and R. Hancock, *J. Gen. Microbiol.* **22**, 249 (1960).

[25] H. Shimizu, G. Tamura, and K. Arima, *Agric. Biol. Chem.* **43**, 1486 (1979).

[26] F. Fiedler, J. Seger, A. Schrettenbrunner, and H. P. R. Seeliger, *Syst. Appl. Microbiol.* **5**, 360 (1984).

[27] K. Kamisanago, H. Fujii, Y. Yanagihara, I. Mifuchi, and I. Azuma, *Microbiol. Immunol.* **27**, 635 (1983).

[28] R. M. Krause and M. McCarty, *J. Exp. Med.* **114**, 127 (1961).

[29] H. R. Perkins, *Biochem. J.* **95**, 876 (1965).

[30] H. Heymann, L. D. Zeleznick, and J. A. Manniello, *J. Biol. Chem.* **239**, 2981 (1964).

6. Filter the mixture of broken cells and glass beads through a 10-15 M Pyrex (Corning, NY) sintered glass filter to separate the beads, then wash the beads with a small amount of water.

7. From the filtrate, sediment unbroken cells by a 10-min centrifugation at 1500 g and 4°, and then sediment the cell walls from the supernatant by 30 min centrifugation at 6500 g at 4°. Wash the sedimented cell walls three times with water.

8. Suspend the sedimented cell walls in 200 ml of 50 mM phosphate buffer, pH 7.6, add filtered (0.2 μm pore size) solutions of ribonuclease (RNase A, Sigma R4875) to a final concentration of 100 μg/ml and deoxyribonuclease (DNase I, Sigma D4527) to 50 μg/ml, and add 0.5 ml toluene and phosphate buffer to 250 ml. Incubate for 18 hr at 37° with slow or occasional mixing. Add sterile trypsin to 200 μg/ml and incubate for another 18 hr at 37°. The enzyme digestion should substantially reduce the optical density of the cell wall suspension (e.g., an initial OD_{480} of 0.82 decreases after digestion to 0.21).

9. Sediment the cell walls by centrifugation at 6500 g and 4° for 30 min, wash four times with water, and lyophilize. The purified cell walls can be stored frozen for several years.

10. Mix 1 g of cell walls with 250 ml of 5% TCA for 18 hr at 22°, sediment the insoluble material by centrifugation at 6500 g for 30 min, and re-extract the sediment first with 160 ml and then with 100 ml of 5% TCA. Save the supernatants for recovery of teichoic acid (Step 13).

11. Heat the sediment (which is purified PG) before resuspension at 90° for 15 min, wash three times with water and three times with acetone, and dry the white PG powder. Store frozen.

12. Drying causes aggregation of PG, and these aggregates usually have to be dispersed by ultrasonication. Ultrasonication is difficult to standardize in different laboratories, because of the differences in the power output of different sonicators under different conditions. Aggregated PG can be easily sedimented by centrifugation at 1500 g, whereas dispersed insoluble PG forms a white smooth suspension that does not sediment at 1500 g but can be sedimented by centrifugation at 10,000 g. To obtain insoluble PG dispersed by sonication, suspend PG in 2 ml of saline or phosphate-buffered saline (PBS) at 10 mg/ml and sonicate for 15 to 30 min on ice with a 1/8 inch microprobe (20–30 W output on a W220-F Sonicator, Heat Systems–Ultrasonic, Plainview, NY). Sterilize the PG suspension by heating at 90° for 30 min and store in solution at 4° (do not freeze or lyophilize because PG will aggregate). Sterility of the PG can be checked by plating a drop of PG on two nutrient agar plates and incubating one at 20° and one at 37°.

13. To recover teichoic acid from the TCA extraction supernatant

(Step 10), pool the three supernatants, remove TCA by three extractions with 2 parts (each time) of ethyl ether, and precipitate teichoic acid from the water phase with three volumes of acetone for 24 hr at 4°. Sediment the precipitated teichoic acid by centrifugation at 6500 g, wash several times with acetone, and dry. Store in a desiccator in a freezer. Teichoic acid readily dissolves in water, yielding a clear colorless solution.

Gram-Negative Bacteria

The fundamental principle behind most purification procedures for gram-negative PG is straightforward: dissolve the bacterium in hot detergent to leave PG as the sole insoluble residue, treat with proteinase if the presence of covalently attached lipoprotein necessitates, and wash extensively to remove soluble material. The following method, which we modified from that of Hebeler and Young,[31] was designed for isolation of PG from *Neisseria gonorrhoeae*. However, given sufficient numbers of organisms, the method should be generally applicable to gram-negative bacteria. Our typical recipe for preparation of gonococcal PG, as detailed below, is based on nine 4-liter flasks, each containing 1.5 liters of liquid culture, but we can easily scale down to 1 liter and have scaled up to as much as 45 liters in a small fermenter.

Day 0. Heavily inoculate from a previously frozen stock culture of *N. gonorrhoeae* onto two agar plates of Clear Typing Medium (CTM) with 2.1% glucose[32] for each 1.5 liters of liquid medium to be used. Incubate for 18 hr at 37° in candle extinction jar. We routinely use "transparent" (*tr*) colonial morphologies[33] because clones of *tr* gonococci grow as smoother suspensions in liquid culture than do their "opaque" (*op*) colonial morphology counterparts. The PG of *tr* and *op* gonococci appears identical.[34]

Day 1

1. Harvest each solid culture with 1–2 ml of warm (37°) liquid growth medium[34] (LGCB$^+$) using sterile glass spreaders and combine into a single homogeneous inoculum.

2. Inoculate, drop wise, each 4-liter flask containing 1.5 liters LGCB$^+$ until the OD_{540} reaches 0.05.

3. Incubate on reciprocal shaker at 36.5°.

[31] B. H. Hebeler and F. E. Young, *J. Bacteriol.* **126,** 1180 (1976).
[32] R. S. Rosenthal, R. S. Fulbright, M. E. Eads, and W. D. Sawyer, *Infect. Immun.* **15,** 817 (1977).
[33] J. Swanson, *Infect. Immun.* **19,** 320 (1978).
[34] S. C. Swim, M. A. Gfell, C. E. Wilde III, and R. S. Rosenthal, *Infect. Immun.* **42,** 446 (1983).

4. Monitor growth (OD_{540}) in pilot flask at about 45-min intervals.

5. After one doubling, add 10 μCi of [1-^{14}C]glucosamine (~55 mCi/ mmol) or [6-^{3}H]glucosamine (~35 Ci/mmol) to one flask. Do not add "cold" glucosamine as carrier since glucosamine at high concentrations is toxic to gonococci. The final PG will contain a very small amount of radioactivity (in both the glucosamine and muramic acid moieties) that serves as tracer and facilitates quantitation of PG hydrolase products generated subsequently from the parent PG preparation.

6. Harvest cultures in late exponential phase at an OD_{540} of 0.6 (~6 × 10[8] cells/ml) using Pellicon Cassette System (Millipore Corp., Bedford, MA) equipped with a 100K molecular weight cutoff NMWL filter or other suitable rapid harvest procedure. The tangential flow filtration feature of the Pellicon Cassette can concentrate 13.5 liters of culture to less than 500 ml in approximately 30 min, and rapidity of harvest is essential to preserve the integrity of PG (especially gonococcal PG) in the face of hydrolytic autolysins. Prior to harvest, cultures should be streaked on CTM plates and examined 24 hr later to ensure purity of gonococcal cultures.

7. Centrifuge (8000 g, 5 min, 10°) the residual culture after filtration and resuspend cell pellets homogeneously in total of 150 ml distilled water.

8. Immediately add 150 ml of cold 10% trichloroacetic acid, mix, and refrigerate for at least 30 min.

9. Centrifuge (8000 g, 5 min, 10°) and wash three times in distilled water.

10. Resuspend final pooled cell pellets in 250 ml of 50 mM sodium acetate buffer, pH 5.3, and mix thoroughly. Warning: Use of alkaline buffers will promote de-O-acetylation of PG from strains in which the PG is extensively O-acetylated on the C-6 of muramic acid residues.

11. Add 250 ml of 8% SDS in the sodium acetate buffer and incubate at 95° for 1 hr. This treatment almost immediately clarifies preparations of gonococcal PG, but manual dispersion of clumps with a stir rod will facilitate complete dissolution. Incubate overnight at room temperature.

Day 2

1. Centrifuge (43,000 g, 1 hr, 12°) the SDS-treated preparations. Lower temperatures may cause the SDS to precipitate.

2. Wash three times by centrifugation in distilled water.

3. Resuspend pooled pellets in 68 ml of 50 mM Tris-HCl, pH 6.8, and mix thoroughly.

4. Add 8 ml of 2% SDS in Tris-HCl plus 3.6 ml of Tris-HCl, gently mix to prevent excessive bubbling of SDS solution, and add 400 μl of proteinase K (10 mg/ml in 1% SDS) to make the final proteinase K concen-

tration 50 μg/ml. Incubate for 12 hr at 37°. Although under most conditions gonococci do not possess detectable covalently bound protein, we have found, empirically, that proteinase K treatment reduces the content of non-PG amino acids in the final preparation from 1–2% to less than 0.1% by weight. However, based on our practical experience, proteinase K prepared according to certain methods by certain manufacturers contains additional nonspecified enzymatic activities capable of degrading PG and, thus, can be disastrous for use in PG isolation. The major deleterious constituent appears to be a glycan-splitting activity, probably a glucosaminidase. It is conceivable that this trace activity is detectable only because we employ amounts of proteinase K and incubation times in great excess of that needed to adequately reduce protein content. We have detected PG-degrading activity in proteinase K obtained from EM Science (Cherry Hill, NJ) and Boehringer Mannheim (Indianapolis, IN) and also in pronase from various sources. Particularly interesting is the fact that the undesirable activity is more easily detectable in proteinase K purified using contemporary techniques, such as isoelectric focusing, which typically yield a superior product by most criteria. The activity is not detectable in "less pure" preparations isolated by traditional chromatographic techniques. For *Escherichia coli* and other gram-negative organisms that possess lipoprotein covalently associated with PG, thorough proteolysis is even more important than for neisserial PG. For these applications, the methods of choice seem to employ various other proteases,[35,36] for example, pronase and trypsin; in some applications,[35] α-amylase is also added to degrade high molecular weight glycogen trapped inside the PG matrix.

Day 3

1. Wash the proteinase K-insoluble residue (purified PG) twice in distilled water at 43,000 g for 1 hr at 12°. We typically centrifuge and wash the entire batch in four equal portions. Pellets should be transparent. The small amount of insoluble residue (probably precipitated salts) which may be evident can be removed before subsequent high-speed washes by low-speed centrifugation (1200 g, 10 min, 12°). Alternatively, we typically resuspend the residue in just enough water to cover the pellets and vortex gently. The clear gel containing the PG easily dislodges into suspension, leaving the heavier precipitate stuck to the side of the tube.

2. Wash the PG (43,000 g, 1 hr, 12°) three times in the Tris-HCl buffer and three times in pyrogen-free water.

[35] B. Glauner, J. V. Holtje, and U. Schwarz, *J. Biol. Chem.* **263**, 10088 (1988).
[36] H. D. Heilmann, *Eur. J. Biochem.* **31**, 456 (1972).

3. Resuspend the final combined pellets in around 10 ml pyrogen-free water and lyophilize. The typical yield from a batch of nine 1.5-liter cultures of gonococci is approximately 50 mg of white, hydroscopic cotton-like PG. The specific activity is typically around 2×10^6 disintegrations/min (dpm)/mg for ^{3}H and 2×10^5 dpm/mg for ^{14}C.

Isolation of Soluble Polymeric Peptidoglycan

Sonicated Peptidoglycan

Ultrasonication devices which concentrate shearing forces around the tip of a probe are capable of breaking covalent bonds in insoluble PG and in releasing soluble but high molecular weight PG fragments in the M_r range of 10^6–10^7. This technique has long been used successfully and efficiently. However, the extent of solubilization is very sensitive to variables, for example, viscosity of the suspension, PG concentration, height and width of fluid in the sonication tube, and condition of the probe, which influence effective energy output. Furthermore, different sonicators vary widely in power output. Indeed, in most instances, output is simply defined in arbitrary units. As a result of these factors, ultrasonication procedures are extremely difficult to standardize between laboratories. Accordingly, we recommend that conditions for a given application in a given laboratory be established empirically based on the percent solubilization and on the size range of the fragments desired. In general, percent solubilization is probably best determined by monitoring the relative amount of radioactively labeled PG in the supernatant resulting from high-speed centrifugation, and a variety of chromatographic methods have been employed to fractionate the soluble fragments on the basis of molecular weight. Because of its soluble nature, sonicated PG more closely approximates than does insoluble PG the high molecular weight forms of PG that are likely released during natural infections by, for example, host PG hydrolases.[37] Accordingly, preparations of sonicated PG often have been employed as physiologically reasonable model compounds for biological testing, and they have found particular use in systems to examine the arthritogenicity of PG.

Gram-Positive Bacteria

The following procedure is used to obtain soluble high molecular weight PG from *S. aureus* insoluble PG.[38]

[37] R. S. Rosenthal, W. J. Folkening, D. R. Miller, and S. C. Swim, *Infect. Immun.* **40,** 903 (1983).
[38] R. Dziarski, *Cell. Immunol.* **109,** 231 (1987).

1. Suspend insoluble PG in 3 ml of water at 10 mg/ml in a 5-ml glass tube and sonicate for 180 min (18 10-min cycles with breaks for cooling in between cycles) in ice bath with a 1/8 inch microprobe (20–30 W output on a W220-F Sonicator, Heat Systems–Ultrasonic). Use the maximal output possible with a given probe that does not result in spillage of the fluid.

2. Because not all of the PG is usually solubilized this way, remove the remaining insoluble PG by centrifugation at 10,000 g. Determine the concentration of PG in the supernatant by glucosamine and amino acid content on an amino acid analyzer, or spectrophotometrically at 218 nm. Alternatively, PG can be biosynthetically labeled with ^{14}C, and the concentration can be determined based on the amount of radioactivity left in the supernatant. As defined above, the procedure should result in greater than 70% solubilization, with most soluble fragments in the M_r range of $5-10 \times 10^6$. Other investigators have employed differential centrifugation, ultrafiltration, and column chromatography to further fractionate the soluble PG on the basis of size.[39,40]

3. Sterilize the white opalescent PG solution by heating at 90° for 30 min and store in solution at 4° (do not freeze or lyophilize because PG will aggregate). Sterility of the PG can be checked by plating a drop of PG on two nutrient agar plates and incubating one at 20° and one at 37°.

Gram-Negative Bacteria

The following procedure is used to obtain soluble high molecular weight PG from *Neisseria gonorrhoeae* insoluble PG.

1. Suspend insoluble PG to 4 mg/ml in a 17 × 100 mm pyrogen-free glass or polypropylene tube in an ice–water bath. We typically sonicate batches of approximately 20 mg.

2. Sonicate for three 15-min cycles at 50% duty (with 2-min breaks for cooling in between cycles) using a Branson Sonifier 450 equipped with a 1/8 inch microprobe (Branson Ultrasonics Corp., Danbury, CT). The tip of the probe is fixed about 2 mm from the bottom of the tube in a fashion such that the probe is unable to contact the sides or bottom of the tube. We use the maximal output possible that does not result in spillage or foaming of the suspension (about 35% of maximal output of the instrument as indicated on the relative output scale). Note that we employ probes which are dedicated solely to sonication of purified PG;

[39] A. Fox, R. R. Brown, S. K. Anderle, C. Chetty, W. J. Cromartie, H. Gooder, and J. H. Schwab, *Infect. Immun.* **35,** 1003 (1982).

[40] C. Chetty, D. G. Klapper, and J. H. Schwab, *Infect. Immun.* **38,** 1010 (1982).

these probes are exhaustively cleaned by alcohol and by sonicating and rinsing in pyrogen-free water prior to use to reduce endotoxin contamination.

3. Centrifuge (43,000 g, 1 hr, 8°) the sonicate to remove insoluble PG (and small fragments of glass or plastic that occasionally break off the tube), and carefully pipette off the supernatant. Approximately 85% (range 80–95%) of the original insoluble PG will be rendered soluble, as defined by determining the percentage of radioactive PG in the supernatant. The final soluble product is a heterogeneous mixture consisting almost exclusively of fragments with M_r greater than 10^6, as determined by chromatography on Sepharose 6B. Store sonicated soluble PG at 4°; do not freeze.

Soluble Polymeric Peptidoglycan Secreted by Cells

Bacteria grown in the presence of β-lactam antibiotics typically secrete soluble polymeric PG fragments, owing to continued synthesis of PG chains, whereas transpeptidation and incorporation of PG into cell walls by peptide cross-linking are inhibited.[41–46] Teichoic acid is also synthesized and secreted into the medium, but it is not covalently bound to PG.[45] The secreted soluble, uncross-linked PG has an average chain length of 50 to greater than 100 disaccharide–peptide units and has an M_r of 50,000–200,000. Isolation of this soluble PG involves the following steps: (i) obtaining exponentially growing cells in enriched medium; (ii) harvesting the cells and transferring them into synthetic medium; (iii) growing the cells in the synthetic medium in the presence of penicillin; (iv) obtaining and concentrating the PG-containing supernatant; and (v) isolating and purifying soluble PG from the supernatant.

The yield of soluble PG is relatively small (compared to the yield of insoluble cell wall PG), and several factors are important for optimizing the procedure. The choice of bacteria is important. Although theoretically any β-lactam-sensitive and β-lactamase-negative bacterium could be used, soluble polymeric PG is efficiently secreted only by gram-positive bacteria,[41–46] probably because in gram-negative bacteria soluble polymeric PG molecules are too large to pass through the outer membrane. Gram-positive bacteria vary in the amount of secreted PG; for example, even when

[41] D. Keglevic, B. Ladesic, O. Hadzija, J. Tomasic, Z. Valinger, M. Pokorny, and R. Naumski, *Eur. J. Biochem.* **42**, 389 (1974).

[42] D. Mirelman, R. Bracha, and N. Sharon, *Biochemistry* **13**, 5045 (1974).

[43] Z. Tynecka and J. B. Ward, *Biochem. J.* **146**, 253 (1975).

[44] A. R. Zeiger, W. Wong, A. N. Chatterjee, F. E. Young, and C. U. Tuazon, *Infect. Immun.* **37**, 1112 (1982).

[45] H. Fischer and A. Tomasz, *J. Bacteriol.* **157**, 507 (1984).

[46] J. F. Barrett and G. D. Shockman, *J. Bacteriol.* **159**, 511 (1984).

several strains of the same species (*S. aureus*) were compared, both high and low secretors were observed.[44] Moreover, all bacteria produce autolytic enzymes, which are activated by β-lactam antibiotics and which may partially digest the cell wall and release soluble PG fragments from the cell wall. These fragments may or may not subsequently copurify with the uncross-linked soluble secreted PG. For this reason, some studies may require the use of autolysin-negative mutants, such as *Bacillus licheniformis* 94.[43] The exponential growth of bacteria and their proper concentration in the penicillin-containing synthetic medium are essential.[44] All media should be prewarmed, and the harvesting and washing of the bacteria during transfer into the synthetic medium should be done with warm buffers and in as short a period of time as possible. For large bacteria growing in chains, such as *Bacillus licheniformis* 94,[43] harvesting the cells and washing can be done by filtration through Whatman (Clifton, NJ) No. 1 filter paper, rather than centrifugation. Prolonging the incubation time with penicillin does not increase and often decreases the yield of soluble PG.[44]

Purification of soluble PG can be accomplished by ion-exchange chromatography,[43] Sephadex gel filtration,[44] vancomycin affinity chromatography,[44,45,47] or a combination of these methods. Vancomycin binds to the D-Ala–D-Ala sequence that is uniquely found in uncross-linked PG[44,48,49] and, therefore, can be used to purify soluble uncross-linked PG fragments or biosynthetic PG intermediates. To simplify the detection of soluble PG during the purification procedure, it is convenient to label biosynthetically the newly synthesized PG with N-acetyl[^{14}C]glucosamine[43] or [^{14}C]alanine.[44,47] For large-scale purification of staphylococcal soluble PG, [^{14}C] alanine has been used because it is relatively inexpensive. However, before performing a large-scale purification procedure with other bacteria, it should be determined if a given bacterium effectively incorporates a given radioactive compound into its PG. To restrict the biosynthetic activity of bacteria, some investigators also employ chloramphenicol to inhibit protein synthesis during penicillin treatment.[43]

Below is a detailed procedure for isolation and purification of soluble PG from *S. aureus*.[44,47] Similar procedures can be used for isolation of soluble PG from most gram-positive bacteria, except that for different bacteria the growth media and incubation conditions may have to be modified. The cell wall minimal medium should contain amino acids that are particular to PG of that species. Endotoxin-free water is used for all

[47] R. Dziarski, *J. Biol. Chem.* **266**, 4713 (1991).
[48] H. R. Perkins, *Biochem. J.* **111**, 195 (1969).
[49] M. A. De Pedro and U. Schwarz, *FEMS Microbiol. Lett.* **9**, 215 (1980).

media and buffers, all reagents are sterile, and disposable plasticware or glassware baked at 180° is used, to avoid contamination with endotoxin. Two batches, staggered by approximately 2 hr, are usually done on one day.

1. Inoculate two flasks containing 30 ml each of Todd–Hewitt broth with pure colonies of *S. aureus* from an agar plate; incubate overnight at 37° with shaking.

2. Inoculate 2.5 liters of warm (37°) Todd–Hewitt broth with 25 ml of the above overnight culture. This inoculated culture should have an OD_{660} of 0.08–0.1. Incubate at 37° with vigorous shaking until the OD_{660} reaches 0.8–0.9 (usually 2 hr).

3. Centrifuge at no less than 1500 g at 30°–37° for 15 min. Wash the cells by centrifugation using 2 liters of warm (37°) PBS, pH 6.8, containing 28.5 mM glucose.

4. Suspend the cells in 1 liter of warm (37°) cell wall minimal medium supplemented with vitamins (28.5 mM glucose, 1 mM glycine, 1 mM DL-glutamic acid, 0.75 mM DL-alanine, 0.2 mM L-lysine, 0.17 mM uracil, 0.125 mM glycerol, 1 mM $MgCl_2$, 0.1 mM $MnCl_2$, 3 μM thiamin, and 8.2 μM nicotinamide in 80 mM potassium phosphate buffer, pH 6.8, filter sterilized). The OD_{660} of this cell suspension should be 1.4–1.6.

5. Incubate with abundant aeration by vigorous shaking in a water bath at 37° for 3–5 min. Add 7.5 ml of warm 10 mg/ml benzylpenicillin, and after 3 min add 10 μCi of DL-[^{14}C]alanine (Sigma). Incubate with shaking at 37° for 60 min.

6. Heat at 95°–100° for 30 min, mixing frequently. Chill to 4° in ice–water and centrifuge at no more than 1500 g for 20 min at 4°. Collect the supernatant and centrifuge for 20 min at 6500 g and 4°. Discard both pellets (note that the sample is radioactive and should be treated accordingly).

7. Filter the supernatant through a 0.2- or 0.45-μm membrane filter. Concentrate in a heated (75°) rotary evaporator to approximately 80–100 ml. Dialyze three times at 4° (12,000–14,000 molecular weight cutoff) against water and once against PBS, pH 7.2.

8. Prepare a 50-ml column of unsubstituted Sepharose and a 45 ml column of vancomycin–Sepharose (see below for the preparation of vancomycin–Sepharose), equilibrating both columns with PBS.

9. Pass the concentrated dialyzed supernatant first through the unsubstituted Sepharose column and then through the vancomycin–Sepharose column (20 drops/min, room temperature), and wash extensively with PBS until no radioactivity (^{14}C), no proteins (OD_{280}), and no carbohydrates (OD_{218}) are present in the eluate.

10. Elute the vancomycin-bound soluble PG with a 200-ml gradient of NH$_4$OH in water, pH 7.5 to 10.5, at 14 drops/min. Collect 2-ml fractions and monitor for the presence of ^{14}C. Soluble PG elutes as a single peak or closely spaced double peak at pH 8.4–8.7. The fractions containing PG are pooled and concentrated or lyophilized. The PG concentration is determined on an amino acid analyzer. The yield from 1 liter of cell wall minimal medium is usually 2–2.5 mg soluble PG. Soluble PG can be stored in solution, frozen, or lyophilized.

11. The vancomycin–Sepharose column can be washed with 4 N NaCl and then PBS, pH 7.2, and reused.

The procedure for preparing vancomycin–Sepharose is given below. Activation of Sepharose (steps 1–3) is performed at room temperature. Coupling of vancomycin (steps 4–8) is performed at 4°.

1. Wash 50 ml Sepharose (agarose, Sigma CL-4B-200) in water, then suspend in 50 ml of water. Add 50 ml of 2 M sodium carbonate and mix.

2. Under the hood, add 2.5 ml of 2 g/ml CNBr (from a stock dissolved in acetonitrile and stored frozen) and mix for 2 min.

3. Wash the resin on a sintered glass filter (150 ml, medium) sequentially with 500 ml of 0.1 M bicarbonate, pH 9.5, water, and 0.2 M phosphate buffer, pH 8.3.

4. Dissolve 500 mg vancomycin (Sigma V-2002) in 50 ml of 0.2 M phosphate buffer, pH 8.3, adjust pH to 8.3 with 1 N NaOH (~10 drops), measure the OD$_{280}$ of a 1 : 100 dilution (~0.45), and cool to 4°.

5. Suspend 50 ml CNBr-activated Sepharose (Step 3) in 50 ml of vancomycin solution (Step 4). Mix slowly end-over-end for 20 hr at 4°.

6. Filter (or centrifuge at 400 g); the OD$_{280}$ of a 1 : 100 dilution of the filtrate (or supernatant) should be reduced by 60–70% compared with the preactivation OD$_{280}$ (Step 4).

7. Wash with 500 ml of phosphate buffer, pH 8.3. Suspend in 50 ml of 1 M ethanolamine and mix for 18 hr at 4°.

8. Wash sequentially with 1000 ml of (i) phosphate buffer, pH 8.3, (ii) 0.1 M acetate buffer with 0.5 M NaCl, pH 4.0, (iii) 0.1 M sodium bicarbonate with 0.5 M NaCl, pH 10, and (iv) PBS, pH 7.2 (until the pH is 7.2). Store at 4° in PBS, pH 7.2, with 0.02% NaN$_3$.

Isolation of Low Molecular Weight Peptidoglycan Oligomers and Monomers

Gram-Positive Bacteria

Peptidoglycan oligomers and monomers have been isolated from several gram-positive bacteria. Peptide-free di-, tetra-, hexa-, octa-, and deca-saccharides[50,51] and oligosaccharide peptides[52] have been isolated from lysozyme-digested PG from *Micrococcus lysodeikticus*. Mono- and disaccharide pentapeptide PG fragments have been purified from lysozyme digests of high molecular weight PG secreted by *Brevibacterium divaricatum* grown in the presence of penicillin.[53] Disaccharide hexapeptide PG fragments, produced by digestion with endogenous muramidases, have been isolated from *Streptococcus faecium*.[54]

Gram-Negative Bacteria

Peptidoglycan from gram-negative bacteria seems the most frequent source of purified, low molecular weight, chemically defined fragments for biological testing. Indeed, the single most common classes of low molecular weight PG used to date are probably disaccharide peptide monomers, dimers, and higher oligomers derived from *Neisseria gonorrhoeae* PG by the action of a hexosaminidase, such as lysozyme (a muramidase).

Muramidase-Derived Peptidoglycan Fragments. Muramidases cleave PG to release fragments with a reducing MurNAc end. Unless endopeptidase activity is also present, the resulting products are peptide cross-linked and reflect the native composition of uncross-linked monomers and cross-linked oligomers in the parent PG. The choice of a muramidase depends on the specificity of the enzyme and on the presence of substituents (most notably O-acetyl groups on the C-6 of MurNAc) which mediate resistance to muramidase action. Some muramidases such as the *Chalaropsis* B enzyme[55] and the muramidase from *Streptomyces globisporus* break β-1,4-glycosidic bonds between GlcNAc and MurNAc regardless of the presence of O-acetyl groups; such nonspecific activities are preferred for most applications. Extensively O-acetylated PG is resistant to hydrolysis by many muramidases including hen egg white lysozyme and lysozyme purified from human polymorphonuclear leukocytes.[56] We routinely employ the *Chalaropsis* muramidase and refer to the final PG products as "*Chalaropsis* fragments." The *Chalaropsis* enzyme previously had been available commercially through Miles Laboratories (Elkhart, IN), but production is now discontinued. The original methods of Hash,[55,57]

[50] N. Sharon, T. Osawa, H. M. Flowers, and R. W. Jeanloz, *J. Biol. Chem.* **241,** 223 (1966).
[51] D. M. Chipman, J. J. Pollock, and N. Sharon, *J. Biol. Chem.* **243,** 487 (1968).
[52] D. Mirelman and N. Sharon, *J. Biol. Chem.* **242,** 3414 (1967).
[53] D. Keglevic, B. Ladesic, J. Tomasic, Z. Valinger, and R. Naumski, *Biochim. Biophys. Acta* **585,** 273 (1979).
[54] J. F. Barrett, V. L. Schramm, and G. D. Shockman, *J. Bacteriol.* **159,** 520 (1984).
[55] P. L. Fletcher and J. H. Hash, *Biochemistry* **11,** 4274 (1972).
[56] R. Rest and E. Pretzer, *Infect. Immun.* **34,** 62 (1981).
[57] J. H. Hash and M. V. Rothlauf, *J. Biol. Chem.* **242,** 5586 (1967).

although tedious, are excellent and still recommended. We have not yet used the commercially available muramidase SG (ICN ImmunoBiologicals from Seikasaka Kogyo Co., Ltd.) for large-scale preparations destined for biological use, but all indications are that the following procedure could easily be modified (simple buffer changes, etc.) to accommodate this muramidase. The following details the method for *Chalaropsis* muramidase-derived PG fragments. All reagents are made with pyrogen free water; all glassware or plasticware is pyrogen free or, when this is not achievable, rinsed extensively with pyrogen-free water.

1. Suspend intact PG (typically about 50 mg) in 50 mM sodium acetate buffer, pH 5.2, to 20 mg/ml. The PG contains a small amount of radioactive tracer (see above) to monitor the reaction by determining the loss of TCA-precipitable (insoluble) PG during digestion.

2. Treat with *Chalaropsis* B muramidase (final concentration 0.5–20 units per milligram of PG) at 37° until all of the PG is rendered soluble based on the criterion of TCA solubility. We have used progressively lower concentrations of enzyme as our supply of this precious reagent has decreased, and, even with the lowest concentration indicated, the PG was 100% solubilized within 2 days (i.e., the reaction was complete).

3. Centrifuge at 43,000 g for 60 min.

4. Remove the supernatant, confirm by radioactivity determination that all the labeled PG is in the soluble phase, and lyophilize.

5. Rehydrate in 1–5 ml water depending on the means of gel-permeation chromatography to be employed (see below).

Transglycosylase-Derived Peptidoglycan Fragments. The novel class of bacterial PG hydrolases known as transglycosylases (PG : PG-6-mur-amyltransferases, EC 3.2.1.), like muramidases, causes the cleavage of glycosidic linkages between the C-1 of MurNAc and C-4 of GlcNAc. However, the transglycosylases simultaneously catalyze the intramolecular cyclization of the MurNAc residue between C-1 and C-6 and, thereby, result in the release of PG fragments containing a nonreducing 1,6-anhydro-MurNAc end.[58,59] These compounds have been referred to as anhydro fragments.

To date, the best characterized transglycosylases are those of *E. coli*[59,60] (referred to as lytic transglycosylases to distinguish from biosynthetic transglycosylases), but the activity is also prominent in gonococci[61] and also occurs in *Bordetella pertussis*[62] and probably many other

[58] J. V. Holtje, D. Mirelman, N. Sharon, and U. Schwarz, *J. Bacteriol.* **124**, 1067 (1975).
[59] A. Taylor, B. C. Das, and J. van Heijenoort, *Eur. J. Biochem.* **53**, 47 (1975).
[60] H. Mett, W. Keck, A. Funk, and U. Schwarz, *J. Bacteriol.* **144**, 45 (1980).
[61] R. K. Sinha and R. S. Rosenthal, *Infect. Immun.* **29**, 914 (1980).
[62] B. T. Cookson, A. N. Tyler, and W. E. Goldman, *Biochemistry* **20**, 1744 (1989).

bacteria. In both *E. coli* and neisseriae, transglycosylase is a major contributor to the remodeling and autolytic digestion of the PG layer. Furthermore, the *Neisseria gonorrhoeae* transglycosylase is responsible for an uncommonly high rate of PG turnover (~30% per generation)[61] in exponential gonococci and mediates release of large quantities of anhydro monomers. Thus, anhydro monomers are regarded as physiologically achievable compounds that likely occur as by-products of natural infections. Indeed, their generation may be no mere triviality since anhydro monomers containing various peptide side chains are much more potent as sleep enhancers than are the respective muramidase-derived disaccharide peptide analogs.[63]

Preparation of Lytic Transglycosylase. Escherichia coli, the only enzyme source for generating anhydro fragments to date, actually has at least two distinct transglycosylases, one of which is soluble and the other, membrane-bound. At present, preparation of transglycosylase activity is critical to production of this key family of PG fragments because chemical synthesis of the prototypic DAP-containing anhydro monomer (M_r 921) has been elusive, although important intermediates have been synthesized.[64,65]

Several procedures for purification of transglycosylase activity have been reported, primarily by Schwarz, Keck, and colleagues.[58,60,66] The optimal method for our purposes has been a variation of methods employing CM-Sepharose CL-6B[67,68] modified to extract total (soluble plus membrane-associated) transglycosylase activity and to purify the enzyme only partially; this procedure was originally suggested to us by U. Schwarz (personal communication, 1980). A tremendous advantage of this method is that the final product retains DD-endopeptidase activity, which breaks peptide cross-linking bonds and facilitates the virtual quantitative conversion of insoluble PG to anhydro monomers (the principal anhydro fragment desired for biological testing to date). Because all strains of *E. coli* appear to possess transglycosylase activities, we initially simply chose the well-studied ATCC (Rockville, MD) strain, 9637. The construction of strain 4B101 harboring the plasmid pAB58,[69] which carries the *Slt* (soluble lytic transglycosylase) gene and produces a 30-fold overexpression of the soluble enzyme, would seem an irresistible improvement. However, if purified

[63] J. M. Krueger, R. S. Rosenthal, S. A. Martin, J. Walter, D. Davenne, S. Shoham, S. L. Kubillus, and K. Biemann, *Brain Res.* **403**, 249 (1987).
[64] H. Paulsen, P. Himpkamp, and T. Peters, *Liebigs Ann. Chem.* **1986**, 664 (1986).
[65] D. Kantoci, D. Keglevic, and A. E. Derome, *Carbohydr. Res.* **162**, 227 (1987).
[66] W. Kusser and U. Schwarz, *Eur. J. Biochem.* **103**, 277 (1980).
[67] W. Keck and U. Schwarz, *J. Bacteriol.* **139**, 770 (1979).
[68] H. Engel, B. Kazemier, and W. Keck, *J. Bacteriol.* **173**, 6773 (1991).
[69] A. S. Betzner and W. Keck, *Mol. Gen. Genet.* **219**, 489 (1989).

to homogeneity, the absence of endopeptidase activity in such preparations would, ironically, leave cross-linking peptides intact and reduce the yield of anhydro monomers.

1. To prepare transglycosylase/endopeptidase, sonicate with a microprobe about 24 g (wet weight) washed, late-exponential phase *E. coli* cells in 25 ml of 10 mM sodium phosphate buffer, pH 7.4, until more than 90% of the cells are broken. We employ a chemically defined liquid medium (pH 7.0) containing, per liter, 0.29 g MgSO$_4$ · 7H$_2$O, 10 g K$_2$HPO$_4$, 1.75 g Na$_2$HPO$_4$, 2 g citric acid, 2.3 g (NH$_4$)$_2$SO$_4$, and 5 g dextrose. However, the choice of medium does not seem critical.

2. Add DNase (40 μg/ml, final) plus CaCl$_2$ (4 mM) and MgCl$_2$ (5 mM), and incubate at 37° for 2 hr.

3. Extract the DNase-treated sonicate twice for 90 min with stirring at 4° in 10 mM Tris–maleate buffer (pH 6.8) containing 10 mM EDTA, 0.5 M NaCl, 0.1 mM dithioerythritol, and 2% (v/v) Triton X-100. The final volume is approximately 100 ml.

4. Centrifuge (35,000 g, 50 min, 5°).

5. Dialyze the supernatant three times against 4.5 liters of 10 mM Tris–maleate (pH 5.2) containing 10 mM MgSO$_4$, 0.1 mM dithioerythritol, and 2% Triton X-100. Recentrifuge the retentate to remove precipitated salts.

6. Apply to column of carboxymethyl Sepharose CL-6B (Pharmacia Fine Chemicals, Inc., Piscataway, NJ) eluted, stepwise, first with 10 mM Tris–maleate buffer (pH 6.8) containing 10 mM MgSO$_4$, 0.1 mM dithioerythritol, and 0.2% (v/v) Triton X-100, then with elution buffer plus 60 mM KCl, and finally with elution buffer plus 300 mM KCl. Partially purified transglycosylase/endopeptidase used to prepare anhydro-PG fragments elutes after addition of 300 mM KCl.

7. Fractions are assayed[37] for their ability to solubilize radiolabeled TCA-insoluble gonococcal PG and pooled. Confirmation that the reaction products are mainly the desired anhydro monomers is performed using gel-permeation chromatography and reversed-phase high-performance liquid chromatography (HPLC) (see below).

8. Aliquots of enzyme can be stored frozen at $-20°$ with little or no loss of activity for at least 2 years.

Preparation of Anhydro Monomers

1. Suspend weakly radioactive non-O-acetylated, intact gonococcal PG (typically 25–50 mg) to 20 mg/ml in 10 mM Tris–maleate–NaOH buffer (pH 5.0) containing 10 mM MgSO$_4$ and 0.2% Triton X-100.

2. Add transglycosylase/endopeptidase and incubate the mixture at 37°. The enzymatic potency of each lot of enzyme can vary greatly, and we do not try to standardize different batches; therefore, the amount of enzyme is determined empirically based on what is needed to achieve 90–100% solubilization of PG with concurrent conversion to anhydro fragments. Usually, we start with 100 μl of enzyme (~2% of a typical preparation of transglycosylase/endopeptidase described above from 24 g of E. coli).

3. Monitor the progress of the reaction at daily intervals by determining the loss of TCA-insoluble substrate (by radioactivity measurements).

4. When the rate of the reaction slows (and this may even approach zero in the first 3 or 4 days of the reaction), centrifuge (43,000 g, 1 hr) residual insoluble substrate, remove the soluble fraction (containing the anhydro monomers), and reincubate insoluble PG with fresh enzyme. Repeat until the degree of solubilization desired is achieved. We employ our enzyme preparations very conservatively and have had some reactions which require four or five enzyme additions and take up to 2 weeks.

Because of the presence of the endopeptidase and because, even at intermediate stages of the reaction, the soluble products are almost exclusively monomers (the enzyme seems to be an exoglycosylase), the recovery is excellent. Thus, over 85% of insoluble PG is typically recovered as anhydro monomers, and, even after chromatography (see below), the final yield of anhydro monomers from insoluble PG is about 60%.

Fractionation of Low Molecular Weight Peptidoglycans by Size

1. Any of several systems may be used to isolate the various classes of muramidase- and transglycosylase-derived PG fragments according to size (i.e., monomers, dimers, etc.). We suggest the gel-filtration or molecular-sieve HPLC systems below.

1a. Rehydrate the enzyme-derived fragments in up to 5 ml water and filter the entire batch on connected columns of Sephadex G-50 fine (~400 ml packed volume) and G-25 fine (~300 ml packed volume), eluting with 0.1 M LiCl. This scheme effectively permits isolation of three well-distinguished fractions from gonococcal PG: monomers (~30%), peptide cross-linked dimers (40%), and higher oligomers (30%), the latter consisting mainly of trimers and tetramers.

1b. Rehydrate the enzyme-derived fragments in about 1 ml water and subject 200 μl (10 mg) to molecular-sieve HPLC operating at 0.5 ml/min using three columns connected in series: (i) TSK SW guard column (7.5 cm × 7.5 mm, i.d.), (ii) TSK 3000 SW (60 cm × 7.5 mm, i.d.), and (iii) TSK 2000 SW (60 cm × 7.5 mm, i.d.), all from Varian Instruments (Palo

Alto, CA). PG fragments are eluted under isocratic conditions in 10 mM sodium phosphate buffer, pH 6.6, and the PG is monitored by absorbance at 210 nm or by radioactivity determination. This system can, under optimal analytical conditions, achieve baseline separation between disaccharide peptide pentamers and hexamers and, even in the semipreparative mode above, can obtain the requisite separation between dimers and trimers.

2. Pool the appropriate fractions after gel-filtration or molecular-sieve HPLC, lyophilize, and desalt using a column (~250 ml) of Sephadex G-15 (or G-10) eluted in pyrogen-free water.

3. Relyophilize and hydrate the product (typically 10 mg/ml is convenient for our purposes) for chemical characterization and quality assurance analysis (see below) and biological testing.

Reversed-Phase High-Performance Liquid Chromatography

At the completion of gel-permeation chromatography, each isolated fraction consists of a mixture of PG fragments of the designated size class, the best studied of which is PG monomers (Table I). Thus, in the most complicated class (when extensively O-acetylated PG is the substrate), *Chalaropsis* monomers consist of a family of about nine identifiable monomeric fragments with reducing MurNAc ends.[70] In addition, some fragments (<1%) possess nonreducing anhydro-MurNAc ends, reflecting the ends of some PG chains *in vivo*. Of course, the structural possibilities become even more complex when dimers and higher oligomers are employed.[71] When purification of the individual members of a given family is desired, reversed-phase HPLC is necessary. Such highly purified PG compounds have been particularly useful in elucidating the structural basis for PG-mediated activities such as sleep enhancement. Since the initial report of Glauner,[72] other HPLC systems have been developed, each seeming to offer its own advantage.[73,74] We originally employed and still favor the methods of Martin *et al.*,[70] which enable purification of all requisite fragments without resorting to altering the structure of individual compounds by borohydride reduction. Reduction of muramidase-derived fragments is done to simplify subsequent HPLC analyses because the

[70] S. A. Martin, R. S. Rosenthal, and K. Biemann, *J. Biol. Chem.* **262**, 7514 (1987).

[71] L. Johannsen, R. S. Rosenthal, S. A. Martin, A. B. Cady, F. Obal, Jr., M. Guinand, and J. M. Krueger, *Infect. Immun.* **57**, 2726 (1989).

[72] B. Glauner, *Anal. Biochem.* **172**, 451 (1988).

[73] H. Harz, K. Burgdorf, and J. V. Holtje, *Anal. Biochem.* **190**, 120 (1990).

[74] G. Allmaier, M. Caparros, and E. Pittenauer, *Rapid Commun. Mass Spectrom.* **6**, 294 (1992).

corresponding unreduced fragments exist in various anomeric configurations that complicate elution patterns considerably. Of course, the borohydride-reduced compounds would not constitute physiologically reasonable reagents for biological testing, and, thus, we avoid this technique.

Our standard HPLC system employs Vydac C_{18} columns (46 × 250 mm, The Separations Group, Hesperia, CA) packed with 5 μm beads and operated at a flow rate of 1.0 ml/min. The PG is detected by absorbance at 214 nm. Typical elution gradients are linear and cover the range from 100% water (containing 0.05% CF_3COOH) to 25% CH_3CN in water (containing 0.035% CF_3COOH), but the specific gradient employed depends on the nature and complexity of the sample. For example, separation of all major components of anhydro monomers and *Chalaropsis* monomers can be accomplished using a gradient of 4 to 13% CH_3CN over 30 min (Fig. 1), although isolation of minor components often requires rechromatography using the same or more shallow gradient conditions. Further, to increase the yield of compounds, for example, MH^+ 940 and 869 (Fig. 1a), for which one anomer of each coelutes with an anomer of the other, it is helpful to isolate the mixture and rechromatograph after allowing sufficient time for reanomerization. Using this column system, the maximum load for relatively simple mixtures (e.g., anhydro monomers) that are not complicated by the presence of anomers is around 1 mg. For more complex mixtures (e.g., *Chalaropsis* monomers containing anomeric mixtures of both O-acetylated and non-O-acetylated species), the maximum load with reasonable resolution is approximately 0.5 mg. Use of semipreparative pumps capable of delivering up to 40 ml/min and larger reversed-phase columns, such as Vydac C_{18}, 22 × 250 mm, with 10 μm bead size can easily increase the yield per run by 10- to 20-fold.

Other Classes of Peptidoglycan Fragments

Most studies to date have focused on the biological function of PG fragments consisting of one or more of the disaccharide peptide subunits. However, as elucidation of the structural basis for PG-mediated activities continues, future emphasis may include other PG derivatives, such as peptide-free glycan chains and free peptides. Although we cannot detail the procedures here, others have provided methods to isolate free glycan chains by HPLC following treatment of PG with human serum N-acetylmuramyl-L-alanine amidase[73] and to prepare free peptides or lactyl peptides by amidase action[75] or by treatment with NaOH,[76] respec-

[75] J. Tomasic, L. Sesartic, S. A. Martin, Z. Valinger, and B. Ladesic, *J. Chromatogr.* **440**, 405 (1988).
[76] D. J. Tipper, *Biochemistry* **7**, 1441 (1968).

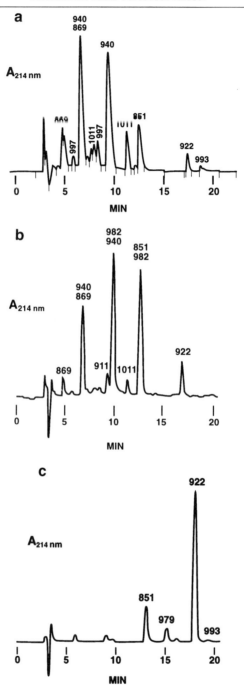

tively. These procedures should prove useful in generating interesting biological reagents from both gram-negative and gram-positive PG.

Elimination/Reduction of Endotoxin Contamination

Uncertainty as to the contributing role of endotoxin contamination is widely considered the biggest obstacle to defining unambiguously a cause and effect relationship between PG and any given biological response (see section on quality assurance below). Accordingly, we recommend that the final step in preparation of soluble PG destined for biological testing be treatment to reduce levels of endotoxin. The logic for this suggestion is obvious when the PG is derived from gram-negative bacteria but, ironically, is equally valid for gram-positive bacteria (lacking endotoxin). This paradox exists because environmental endotoxin contamination may actually be a more legitimate concern than endogenous endotoxin from the bacterial source of PG.

The numerous treatments to reduce endotoxin contamination are based on the selective binding of lipid A by another reagent, for example, polymyxin B[77] or *Limulus* anti-LPS factor,[78] or on the selective partitioning of endotoxin using detergents.[79] Treatment with polymyxin B coupled to agarose (to facilitate separation of the soluble phase from the LPS, which binds to the particulate phase) is probably the most widely used and the one we favor.

For low molecular weight PG derivatives (e.g., MDP or disaccharide peptide monomers or dimers), add equal volumes of 1.1 mg/ml polymyxin B–agarose (Sigma) in sterile pyrogen-free saline and PG (1–5 mg/ml in pyrogen-free water). Incubate the mixture at 25° for 30 min and centrifuge at 110,000 g for 3 hr. The lengthy centrifugation at high speed (in great

---

[77] A. C. Issekutz, *J. Immunol. Methods* **61**, 275 (1983).
[78] A. Biondi, S. Landolfo, D. Fumarola, N. Polentarutti, M. Introna, and A. Mantovani, *J. Immunol. Methods* **66**, 103 (1984).
[79] Y. Aida and M. J. Pabst, *J. Immunol. Methods* **132**, 191 (1990).

FIG. 1. Reversed-phase HPLC on a Vydac C$_{18}$ column (5 μm, 0.46 × 25 cm) in a linear gradient of 4 to 13% CH$_3$CN over 30 min at 1.0 ml/min of (a) *Chalaropsis* non-O-acetylated monomers, 500 μg [the maximum absorbance of the major peak (MH$^+$ 940 and 869) is 0.7]; (b) *Chalaropsis* O-acetylated monomers, 25 μg [maximum absorbance of major peak (MH$^+$ 982 and 940) is 0.07]; and (c) anhydro monomers, 20 μg [maximum absorbance of major peak (MH$^+$ 922) is 0.07]. The numbers above the peaks refer to the mass of the protonated molecular ion(s), as determined by fast atom bombardment mass spectrometry, of the included PG fragments (see Table I).

excess of that needed simply to remove the agarose beads) also helps by removing any unbound endotoxin and freeing soluble PG trapped in the fluid phase of the beads. The supernatant is transferred to a 0.22-μm Spin-x filter unit (Costar, Cambridge, MA) and centrifuged at 16,000 g for 10 sec to sterilize the preparation and to remove any residual beads. Recovery of PG is typically over 85%. Under these conditions, it is assumed that the level of polymyxin B is in great excess of the amount of endotoxin to be removed. This seems reasonable based on our finding that polymyxin B–agarose is able to bind and remove more than 24.7 μg of commercial endotoxin per milliliter from a solution containing 25 μg/ ml, an amount greatly exceeding that likely ever to be encountered in PG preparations.

For high molecular weight PG, e.g., soluble macromolecular PG (S-PG) obtained by sonication (see this volume, p. 263), treatment with polymyxin B–agarose is as for low molecular weight PG except that the initial high-speed centrifugation is omitted.

Structural Analysis of Peptidoglycans

Chemical and General Characterization of Peptidoglycans

Historically, the most routine means to define PG are amino acid/ amino sugar analysis and susceptibility to specific PG-degrading enzymes. Thus, for example, if a preparation happens to contain only the expected PG constituents in an appropriate molar ratio and happens to be fully solubilized by lysozyme, then there is little question that the material is PG. Although somewhat modified by contemporary technology, this simple logic is still fundamental to the characterization of PG.

The amino sugar and amino acid content of the PG should be quantitated after acid hydrolysis using any one of numerous HPLC or standard ion-exchange systems available.[35,80] It should be noted that the amino acid composition in some bacteria may vary depending on the strain and growth conditions and that small amounts of alternative amino acids (0.1–3% by weight) may be present, some of which may be actual constituents and not simply exogenous contaminants. Whereas insoluble or S-PG molecules, especially from gram-positive organisms, frequently retain some contaminating amino acids that are not present in the PG from the test organism, it is reasonable to expect enzyme-generated, soluble, purified PG fragments to contain only the amino sugars and amino acids characteristic of that species. Further, these constituents should be present at the appropriate molar ratios, and the amount of PG estimated from amino acid

[80] R. L. Heinrikson and S. C. Meridith, *Anal. Biochem.* **136**, 65 (1984).

and amino sugar analysis should approximate the mass of the sample determined by other methods, for example, dry weight or spectrophotometric measurements. Intact gonococcal PG contains GlcNAc, MurNAc, Ala, Glu, DAP, and Gly in the molar ratio, $1:1:1.8:1:1:0.1$. The amount of Ala is lower than predicted from the pentapeptide-containing cytoplasmic precursor which contains 1 mol of L-Ala and 2 mol of D-Ala per monomer subunit. This reflects an active gonococcal carboxypeptidase which results in a preponderance of disaccharide tri- and tetrapeptide in the final cell wall PG.

As with some proteinaceous amino acids, the amino sugars of PG are rather sensitive to the extreme conditions normally used for hydrolysis of protein. Accordingly, hydrolysis in 4 N HCl at 100° for 6 hr under vacuum is recommended. Even under these conditions, the MurNAc and GlcNAc are readily de-N-acetylated and ultimately detected as the free amines. In manually performed hydrolysis, further destruction of amino sugars may still occur, resulting in lower than expected molar ratios of muramic acid and/or glucosamine (relative to glutamic acid). Thus, it may be necessary to perform analyses after various times of hydrolysis and to extrapolate the content of the amino sugars back to time zero.

Determination of Percent Peptide Cross-Linking

Determination of percent peptide cross-linking involves quantitation of the percentage of monomer subunits in intact PG (or in PG fragments) in which the ε-amino group on the dibasic amino acid (e.g., DAP) is involved in a peptide cross-linking bond. The procedure is based on quantitation of free versus bound amino groups using 2,4-dinitrofluorobenzene[61] or on the distribution of uncross-linked monomers and various peptide cross-linked oligomers, as determined by sizing analysis following hydrolysis of all muramidase-sensitive glycosidic bonds.[81]

Determination of Glycan Chain Length

An earlier procedure employs derivatization of the free reducing end of glycan chains with borohydride and, after acid hydrolysis, quantitation of the ratio of reduced muramic acid at the ends and unlabeled (internal) muramic acid residues.[82] This method yields only the average chain length and provides an inaccurate estimate of chain length when chains in vivo end in the nonreducing 1,6-anhydromuramic acid. A newer technique to provide a more direct estimate of chain length is based on HPLC analysis

[81] R. S. Rosenthal, R. M. Wright, and R. K. Sinha, Infect. Immun. 28, 867 (1980).
[82] J. M. Ghuysen, D. J. Tipper, and J. L. Strominger, this series, Vol. 8, p. 685.

TABLE I
STRUCTURES OF GONOCOCCAL PEPTIDOGLYCAN MONOMERS
DETERMINED BY FAST ATOM BOMBARDMENT
MASS SPECTROMETRY[a]

MH[+]	Primary sequence
851	GlcNAc-anhydroMurNAc-Ala-Glu-DAP
869	GlcNAc-MurNAc-Ala-Glu-DAP
911	GlcNAc-(O-Ac)MurNAc-Ala-Glu-DAP
922	GlcNAc-anhydroMurNAc-Ala-Glu-DAP-Ala
940	GlcNAc-MurNAc-Ala-Glu-DAP-Ala
979	GlcNAc-anhydroMurNAc-Ala-Glu-DAP-Ala-Gly
982	GlcNAc-(O-Ac)MurNAc-Ala-Glu-DAP-Ala
993	GlcNAc-anhydroMurNAc-Ala-Glu-DAP-Ala-Ala
997	GlcNAc-MurNAc-Ala-Glu-DAP-Ala-Gly
1011	GlcNAc-MurNAc-Ala-Glu-DAP-Ala-Ala

[a] MH[+], Protonated molecular ion (1 mass unit more than M_r); O-Ac, 6-O-acetyl.

of free glycan chains that have been stripped of the peptide side chains by N-acetylmuramyl-L-alanine amidase.[73]

Determination of Percent O-Acetylation

Determination of percent O-acetylation is typically based on the complete conversion of test PG compounds to monomers by combined muramidase and endopeptidase action followed by chromatographic quantitation of O-acetylated monomers and total (non-O-acetylated plus O-acetylated) monomers.[34] Chemical means to estimate O-acetyl content are also available.[83]

Mass Spectrometry

Fast atom bombardment mass spectrometry (MS) and, in particular, tandem MS have revolutionized structural analysis of low molecular weight PG fragments (<~3000). Indeed, these and related technologies, which are rapidly becoming available at major research centers, offer efficient determination of the molecular weight and unambiguous proof of the primary structure of PG fragments[62,70,74] (Table I).

[83] L. Johannsen, H. Labischinski, B. Reinicke, and P. Giesbrecht, FEMS Microbiol. Lett. 16, 313 (1983).

Quality Assurance of Peptidoglycan Preparations

The relative freedom of final PG preparations from chemical contaminants which could cause the results of biological testing to be ambiguous should be customized to the particular bacterial origin of PG and to the biological systems employed. However, there are some widely applied strategies to deal with the most obvious classes of contaminating substances encountered.

Endotoxin

Given the biological potency of endotoxin and the overlap in biological reactions mediated by both endotoxin and PG, it is difficult to overestimate the "endotoxin problem." Indeed, even when there is probably minimal chance for a given activity to be due to LPS contamination, one can almost always count on some serious colleagues still being obsessed with the possibility. Thus, it is not sufficient simply to treat PG reagents to reduce endotoxin. One should also quantitate residual contamination and take measures to assure that this level of endotoxin is unlikely to account for any biological effect observed.

Determination of endotoxin levels using the *Limulus* amebocyte lysate (LAL) assay has evolved into the most widely used quantitative test, based on its sensitivity, specificity, and ease of execution. We employ the Pyrotell test (Associates of Cape Cod, Inc., Woods Hole, MA) performed according to the manufacturer's instructions. The test is offered in lots that range in sensitivity from 0.03 (our choice) to 0.5 EU (endotoxin units) per milliliter. Various endotoxin preparations may differ in LAL potency; at maximum sensitivity, most commercial endotoxins and our gonococcal endotoxin standard from strain 1291 (supplied by M. Apicella, Iowa City, IA) are detected in the range of about 0.5 to 5 pg/ml.

It should be noted that high molecular weight soluble PG (e.g., S-PG), is also LAL positive (albeit less potent than LPS), whereas low molecular weight PG fragments retain little, if any, reactivity. Thus, we routinely digest samples of PG lots with PG-specific muramidase prior to LAL testing to make the assay more specific for endotoxin. Obvious controls include assuring that addition of the muramidase does not alter significantly the LAL reactivity of endotoxin standards. When in doubt, the conservative approach would be to error on the side of "overestimating" the level of endotoxin in PG preparations.

Depending on the potency and the structural basis of PG action in the biological system of interest, various independent means to address the role of endotoxin may be considered. One fundamental strategy is simply to test several endotoxins in the biological assay and to demand that the minimum amount of endotoxin needed to cause a positive biological response be in great excess of the amount of endotoxin (measured by LAL) in the PG preparation. The strongest kind of statement that one is able to offer after following this logic is something like: The amount of purified endotoxin standard needed to produce the reaction is x-fold greater than that actually present in the PG preparation used. The actual number for x that one will accept is rather arbitrary and is subjectively chosen by the investigator. We tend to set a lower acceptability limit of 10^2 for x, although, in most systems, the value for x is greater than 10^3 and sometimes approaches 10^5. This type of statement, while useful and comforting, cannot constitute proof that endotoxin does not contribute to the observed response. In fact, one does not know the source of endotoxin actually present in the PG sample, and one can only assume that its potency in the LAL and biological test of interest falls in the range of known endotoxins. Thus, if the contaminating endotoxin were less potent in the LAL assay or more potent in the relevant biological system than known endotoxins, one might erroneously dismiss the LPS effect.

Moreover, endotoxin and PG fragments may act synergistically to enhance several biological reactions, for example, interleukin 1 (IL-1) production.[84] Such an activity might further reduce the strength of the safest concluding statement to something like: It is unlikely that endotoxin, alone, could account for the observed biological response attributed to PG. Despite the concerns indicated, progress in some biological systems may ultimately demand accepting this strategy as the best available to evaluate the possible role of endotoxin.

Other sensitive indicators of endotoxin presence are lethality tests in galactosamine-treated mice (which are at least 10,000 times more sensitive to LPS than PG)[85,86] or in adrenalectomized mice (which are at least 3000 times more sensitive to LPS than PG).[87] However, these tests are not routinely performed because they are less sensitive and more expensive and labor-intensive than the LAL assay.

A second approach which might come closer to establishing the irrelevancy of endotoxin in a PG-dependent reaction depends on eliminating the activity with PG-specific enzymes (or, conceivably, other PG-specific reagents such as antibodies). Such a strategy is effective in biological systems in which any given PG fragment is much more potent than the products obtained by enzymatic digestion. For example, this approach has

[84] M. W. Vermeulen, J. R. David, and H. G. Remold, *J. Immunol.* **139**, 7 (1987).
[85] C. Galanos, M. A. Freudenberg, and W. Reutter, *Proc. Natl. Acad. Sci. U.S.A.* **76**, 5939 (1979).
[86] J. R. Ulrich and R. Dziarski, unpublished (1992).
[87] R. Dziarski and A. Dziarski, *Infect. Immun.* **23**, 706 (1979).

been used to show that complete muramidase digestion of high molecular weight PG eliminates the ability of the PG to activate complement[88] and to stimulate release of IL-1 from murine P388D1 cells,[89] thus supporting the argument that the activities could not be accounted for simply on the basis of endotoxin contamination. Of course, to ensure the validity of this result, one must simultaneously demonstrate that muramidase treatment does not alter the corresponding biological activity of endotoxin standards.

Other special cases exploit more subtle differences in the structure–function relationships among PG fragments. For example, muramidase-derived monomers from gonococci stimulate IL-1 release from rat macrophages under conditions in which the corresponding peptide cross-linked dimers (obtained from adjacent fractions in the same purification scheme and containing similar levels of endotoxin) are completely inactive.[89] Thus, in this example, it seems highly unlikely that the IL-1 stimulating capacity of monomers could be attributable to endotoxin contamination.

Another approach is to inhibit the observed effects with endotoxin-specific reagents, such as polymyxin B[77] and deacylating enzymes.[90] Such reagents cannot, however, be used to prove that PG is responsible for the biological response. Nevertheless, abolition of activity by these reagents should be taken as a strong warning that contaminating endotoxin may be contributing to the biological effects of PG preparations.

It would seem possible to employ so-called endotoxin-resistant mice strains (e.g., C3H/HeJ) to distinguish functionally between endotoxin and PG, on the basis that such animals (and their cells) would still respond to PG. Indeed, in contrast to endotoxin, PG effectively activates B cells and macrophages from C3H/HeJ mice.[91,92] However, because the precise defect in endotoxin-resistant strains is unknown, we cannot recommend this strategy. Thus, even hyporesponsive C3H/HeJ mice may respond to crude "environmental" endotoxins containing the lipid A-associated proteins or some rough LPS.[93] Contamination by these forms of endotoxin could mislead one into concluding that an endotoxin-independent effect was observed. Furthermore, there is no assurance that the genetic defect leading to endotoxin resistance would not simultaneously mediate resis-

[88] B. H. Petersen and R. S. Rosenthal, *Infect. Immun.* **35,** 442 (1982).

[89] B. D. Reynolds and R. S. Rosenthal, unpublished (1989).

[90] A. L. Erwin and R. S. Munford, *J. Biol. Chem.* **265,** 16444 (1990).

[91] T. Saito-Taki, M. J. Tanabe, H. Mochizuki, T. Matsumoto, M. Nakano, H. Takada, M. Tsujimoto, S. Kotani, S. Kusumoto, T. Shiba, K. Yokogawa, and S. Kowata, *Microbiol. Immunol.* **24,** 209 (1987).

[92] F. Vacheron, M. Guenounou, and C. Nauciel, *Infect. Immun.* **42,** 1049 (1983).

[93] L. Flebbe, S. W. Vukajlovich, and D. C. Morrison, *J. Immunol.* **142,** 642 (1989).

tance to some PG fragments, as if mechanisms for stimulation by each individual substance occurred via partially shared pathways (e.g., a common cellular receptor for endotoxin and PG).[94]

Proteins

The purity of soluble PG fragments should be called into question if the preparation contains higher than background levels of additional amino acids or amino sugars (especially aromatic and sulfur-containing amino acids) that are not found in the PG from a given species. A reasonably sensitive test for non-PG amino acids should be able to detect any individual contaminant at a level of 0.05% by weight.

For soluble high molecular weight PG, no contaminating protein bands should be evident in overloaded ($\geq$100 μg per lane) silver-stained polyacrylamide gels, capable of detecting 0.3 ng of protein per band. Note, however, that high molecular weight soluble PG stains very weakly with silver stain.[47] Functional distinction between protein and PG is straightforward, since, in contrast to proteins, PG-mediated effects would normally be resistant to heat (100°) and proteinase treatment.

Carbohydrates/Teichoic Acids

The major concern with carbohydrates and teichoic acids is endogenous cell wall substances commonly bound covalently to PG (almost exclusively in gram-positive bacteria). Because different bacteria have different carbohydrates bound to PG, the analyses for possible contaminants will depend on the bacterial species. The simplest analysis for teichoic acids, which are present in most gram-positive bacteria, is determination of phosphorus. This can be done qualitatively or semiquantitatively by performing paper or thin-layer chromatography of hydrolyzates of PG preparations, and by visualizing phosphoric acid released from teichoic acid by spraying with freshly prepared spray I (3 g ammonium molybdate dissolved in 255 ml water, plus 30 ml of 1 N HCl and 15 ml of 60% HClO$_4$), drying 3–4 min at 105°, and then spraying with spray II (0.2 g SnCl$_2$ dissolved in 5 ml 36% HCl and diluted to 250 ml with water). The relative amounts of teichoic acid phosphoric acid in PG can be estimated by titration of the lowest amounts of PG and teichoic acid hydrolyzates that give a positive reaction. A quantitative determination of phosphorus can be performed after mineralization.[95]

[94] R. Dziarski, J. Biol. Chem. 266, 4719 (1991).
[95] D. Herbert, P. J. Phipps, and R. E. Strance, Methods Microbiol. 1, 209 (1971).

Typically, PG from *Staphylococcus aureus* or *Streptococcus pyogenes* contains less than 1% contamination with teichoic acid based on phosphorus analysis.[96] If specific possible contaminants of PG can be predicted based on the known composition of the cell wall, the presence of these components should be analyzed. For example, *Streptococcus pneumoniae* teichoic acid contains galactosamine, choline, and ethanolamine, and these compounds can be quantified by nuclear magnetic resonance (NMR) spectroscopy.[97] Typical PG preparations from this bacterium contain around 2% teichoic acid.[97] In *S. pyogenes,* the most difficult to remove contaminant is the C carbohydrate, the presence of which can be detected by the analysis of the major component, rhamnose.[39] Typical preparations of *S. pyogenes* PG contain 0.5–0.9% rhamnose.[39]

Detergents

The possibility that residual detergents, such as SDS and Triton X-100, used in the preparation of intact PG and anhydro monomers, respectively, might contribute to the biological effects of PG preparations seems largely overlooked. Yet, in light of the potency of such agents as cellular toxins, their binding to PG fragments and availability during subsequent biological testing would be a most unwelcome confounding variable. To approach this problem, we prepare certain lots of intact PG or PG fragments which employ addition of [³H]SDS or [³H]Triton X-100 during the appropriate step in the purification procedure. After preparation of the appropriate PG reagent, we then estimate by radioactivity measurements the amount of residual SDS or Triton X-100. Assuming a worse case scenario the maximal level of contamination of PG by SDS and Triton-X-100, respectively, is about 0.06 and 0.1% (by weight). As a criterion to argue that the detergent is not responsible for any given PG effect, we demand that the maximum level of detergent present in the biological assay be at least 10-fold lower than the minimal amount necessary to produce the given biological response. Such controls seem particularly appropriate in bioassays to determine the toxicity of PG fragments.[6]

[96] R. Caravano and J. Oberti, *Ann. Immunol. (Paris)* **132C**, 257 (1981).
[97] J. F. Garcia-Bustos and A. Tomasz, *J. Bacteriol.* **169**, 447 (1987).

[21] Purification of Streptococcal M Protein

By VINCENT A. FISCHETTI

Introduction

M protein is one of the major virulence determinants for the group A streptococcus, the causative agent for serious human diseases such as rheumatic fever, acute glomerulonephritis, and streptococcal toxic shock. The capacity of the M molecule to limit the deposition of complement component C3 on the streptococcal surface is attributed to the antiphagocytic property of the molecule.[1] Based on the DNA sequence of the M protein gene,[2] the molecule is composed of three regions: a short N-terminal nonhelical region, a central helical region exhibiting a seven-residue repeat pattern of nonpolar amino acids, and a C-terminal cell wall and membrane-spanning region responsible for anchoring the molecule to the cell (Fig. 1) (for a review of M protein structure, see Fischetti[3]). The seven-residue pattern of apolar amino acids in the central helical region, which is made up of tandem repeat blocks (A, B, and C), is responsible for the formation of the coiled-coil structure of the M molecule. The N-terminal nonhelical region and adjacent A repeats are hypervariable and responsible for the type specificity of the M proteins. The B repeat contains variable epitopes, while the C-repeat region is conserved among 31 different serotypes. The wall- and membrane-associated regions are highly conserved among all known M molecules as well as other surface proteins on gram-positive cocci.[4] To date, over 80 serologically distinct M proteins have been identified.

M protein is functionally defined (by Lancefield[5]) as a molecule with the capacity to enable streptococci to resist phagocytosis by human granulocytes, and antibodies to the M molecule can reverse this effect. Because the structure of M protein resembles that of other surface molecules from streptococci,[6-9] one cannot assume that the molecule that has been isolated

[1] R. D. Horstmann, H. J. Sievertsen, J. Knobloch, and V. A. Fischetti, *Proc. Natl. Acad. Sci. U.S.A.* **85,** 1657 (1988).
[2] S. K. Hollingshead, V. A. Fischetti, and J. R. Scott, *J. Biol. Chem.* **261,** 1677 (1986).
[3] V. A. Fischetti, *Clin. Microbiol. Rev.* **2,** 285 (1989).
[4] V. A. Fischetti, V. Pancholi, and O. Schneewind, *Mol. Microbiol.* **4,** 1603 (1990).
[5] R. C. Lancefield, *J. Immunol.* **89,** 307 (1962).
[6] E. Frithz, L.-O. Heden, and G. Lindahl, *Mol. Microbiol.* **3,** 1111 (1989).
[7] D. G. Heath and P. P. Cleary, *Proc. Natl. Acad. Sci. U.S.A.* **86,** 4741 (1989).
[8] H. Gomi, T. Hozumi, S. Hattori, C. Tagawa, F. Kishimoto, and L. Bjorck, *J. Immunol.* **144,** 4046 (1990).

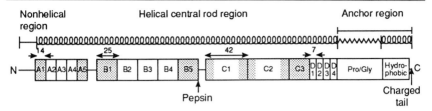

FIG. 1. Characteristics of the complete M6 protein sequence [S.K. Hollingshead, V. A. Fischetti, and J. R. Scott, *J. Biol. Chem.* **261**, 1677 (1986); V. A. Fischetti, D. A. D. Parry, B. L. Trus, S. K. Hollingshead, J. R. Scott, and B. N. Manjula, *Proteins: Struct. Funct. Genet.* **3**, 60 (1988)]. Blocks A, B, C, and D designate the location of the sequence repeats. Numbers above the block indicate the number of amino acids per repeat. Shadowed blocks indicate those in which the sequence diverges from the central consensus sequence. Pro/Gly denotes the region rich in proline and glycine likely located in the peptidoglycan. "Hydrophobic" is a 19 hydrophobic amino acid region adjacent to a 6 amino acid charged tail. "Pepsin" identifies the position of the pepsin-sensitive site after amino acid 228. The positions of the nonhelical region, helical central rod region, and cell anchor region are noted. Reproduced from Ref. 3 with permission from the American Society for Microbiology.

is in fact M protein based solely on structural or serological data. Antibodies must be raised to the suspected M protein and used in an opsonization assay[10] to test for the phagocytosis of homologous streptococcal types. Opsonic antibodies prepared against whole streptococci or a known purified M protein should be able to be absorbed by the purified M protein molecule, rendering the serum nonopsonic.[10] Without these functional tests, the molecule should be termed M-like based on similarities found by sequence and structural data.

Lancefield first extracted M protein from group A streptococci over 60 years ago to obtain a preparation to identify serologically different streptococcal types. The extraction procedure of suspending the streptococci in dilute hydrochloric acid and heating to 100° for 10 min was a quick and effective method of releasing the group A-specific carbohydrate and M protein from the streptococci as well as a number of other antigens to be used in serological assays.[11]

Although never intended as a method for purification of the M protein, the acid extraction procedure has been used by several groups for the preparation of M protein (for review, see Fox[12]). Although not known for several years, it was apparent that this method fragmented the native M

[9] U. Sjobring, *Infect. Immun.* **60**, 3601 (1992).
[10] R. C. Lancefield, *J. Exp. Med.* **110**, 271 (1959).
[11] R. Lancefield and Perlmann, *J. Exp. Med.* **96**, 71 (1952).
[12] E. N. Fox, *Bacteriol. Rev.* **38**, 57 (1974).

protein into many polypeptides. Because of the heterogeneity of the M protein fragments and the extent of contamination with a large number of other molecules, only Fox and co-workers were successful in purifying M protein from an acid extract for use in human vaccine trials.[13,14]

Other methods were devised for extracting M protein from whole streptococci or isolated cell walls. They include treating the cells or walls with alkali,[15,16] sonic oscillation,[17,18] group C phage-associated lysin,[19,20] nonionic detergent,[21,22] nitrous acid,[23] cyanogen bromide,[24] and guanidine hydrochloride.[25] These procedures usually result in a heterogeneous preparation requiring an elaborate purification scheme. However, despite these elaborate procedures, in nearly all cases the final product exhibited extensive heterogeneity of the M protein. Described here are two methods for the isolation and purification of M protein from group A streptococci: the pepsin extraction procedure yields the N-terminal half of the molecule, whereas the recombinant procedure results in the whole M molecule.

Pepsin Extraction

Probably the first important method for the preparation of M protein that offered milligram quantities of purified protein was the pepsin digestion method published by Cunningham and Beachey[26] and further described by Beachey et al.[27] The authors found that treating streptococci with pepsin at the suboptimal pH of 5.8 results in the release of a biologically active fragment of the M protein (termed PepM) in solution.[26,28] By

[13] E. N. Fox and L. O. Krampitz, Fed. Proc. 12, 442 (1953).
[14] E. N. Fox, R. H. Waldman, M. K. Wittner, A. A. Mauceri, and A. Dorfman, J. Clin. Invest. 52, 1885 (1973).
[15] E. N. Fox and M. K. Wittner, Immunochemistry 6, 11 (1969).
[16] M. W. Cunningham and E. H. Beachey, J. Immunol. 115, 1002 (1975).
[17] I. Ofek, S. Berger-Rabinowitz, and A. M. Davies, Isr. J. Med. Sci. 5, 293 (1969).
[18] R. W. Besdine and L. Pine, J. Bacteriol. 96, 1953 (1968).
[19] V. A. Fischetti, J. B. Zabriskie, and E. C. Gotschlich, in "Streptococcal Disease and the Community" (M. J. Haverkorn, ed.), p. 26. Excerpta Medica, Amsterdam, 1974.
[20] J. O. Cohen, H. Gross, and W. K. Harrell, J. Med. Microbiol. 10, 179 (1977).
[21] V. A. Fischetti, J. Exp. Med. 146, 1108 (1977).
[22] V. A. Fischetti, J. Exp. Med. 147, 1771 (1978).
[23] K. Hafez, A. M. ElKholy, and R. R. Facklam, J. Clin. Microbiol. 14, 530 (1981).
[24] K. L. Vosti and W. K. Williams, Infect. Immun. 21, 546 (1978).
[25] H. Russell and R. R. Facklam, Infect. Immun. 12, 679 (1975).
[26] M. W. Cunningham and E. H. Beachey, Infect. Immun. 9, 244 (1974).
[27] E. H. Beachey, G. Stollerman, E. Y. Chiang, T. M. Chiang, J. M. Seyer, and A. H. Kang, J. Exp. Med. 145, 1469 (1977).
[28] E. H. Beachey, G. L. Campbell, and I. Ofek, Infect. Immun. 9, 891 (1974).

removing the streptococci by centrifugation and filtration, one is left with a starting material where the M protein is in high concentration compared with other contaminants. After purification, the fragment is found to be homogeneous by sodium dodecyl sulfate–polyacrylamide gel electrophoresis (SDS–PAGE), strongly suggesting that the M molecule on the cell surface was also likely to be homogeneous, contrary to earlier findings. Depending on the serotype, the size of the extracted PepM molecule varies from 20 to 40 kDa.[29–34] When injected into rabbits, the pepsin-derived fragment elicits antibodies capable of initiating phagocytosis of streptococci of the homologous serotype. The PepM fragment can also absorb type-specific opsonic antibodies from serum.[28,35] Thus, the pepsin-derived M protein fragment retains some of the biologically important determinants of the native M molecule present on the streptococcal surface. Pepsin extraction, therefore, provided the starting material in a variety of studies.[29,32,34–40]

Based on the sequence of the M protein gene,[2] M protein is a molecule of about 400 amino acids, the majority of which is composed of sequence repeats (as described above, Fig. 1). Pepsin at pH 5.8 cleaves the molecule at a hinge region located near the center of the molecule between the B and C repeats (Fig. 1). M proteins exhibit size variation[41] resulting from differences in the number of repeat blocks[42] among different strains, explaining the difference in the size of the pepsin fragment.

[29] B. N. Manjula, A. S. Acharya, S. M. Mische, T. Fairwell, and V. A. Fischetti, *J. Biol. Chem.* **259,** 3686 (1984).

[30] W. Kraus, E. Haanes-Fritz, P. P. Cleary, J. M. Seyer, J. B. Dale, and E. H. Beachey, *J. Immunol.* **139,** 3084 (1987).

[31] K. M. Khandke, T. Fairwell, A. S. Acharya, B. L. Trus, and B. N. Manjula, *J. Biol. Chem.* **263,** 5075 (1988).

[32] L. Moravek, O. Kuhnemund, J. Havlicek, P. Kopecky, and M. Pavlik, *FEBS Lett.* **208,** 435 (1986).

[33] M. S. Bronze, E. H. Beachey, and J. B. Dale, *J. Exp. Med.* **167,** 1849 (1988).

[34] G. N. Phillips, P. F. Flicker, C. Cohen, B. N. Manjula, and V. A. Fischetti, *Proc. Natl. Acad. Sci. U.S.A.* **78,** 4689 (1981).

[35] B. N. Manjula and V. A. Fischetti, *J. Immunol.* **124,** 261 (1980).

[36] E. H. Beachey, J. M. Seyer, and A. H. Kang, *Proc. Natl. Acad. Sci. U.S.A.* **75,** 3163 (1978).

[37] E. H. Beachey and J. Seyer, *Semin. Infect. Dis.* **4,** 401 (1982).

[38] E. H. Beachey, G. H. Stollerman, R. H. Johnson, I. Ofek, and A. L. Bisno, *J. Exp. Med.* **150,** 862 (1979).

[39] I. Ofek, W. A. Simpson, and E. H. Beachey, *J. Bacteriol.* **149,** 426 (1982).

[40] K. M. Khandke, T. Fairwell, and B. N. Manjula, *J. Exp. Med.* **166,** 151 (1987).

[41] V. A. Fischetti, K. F. Jones, and J. R. Scott, *J. Exp. Med.* **161,** 1384 (1985).

[42] S. K. Hollingshead, V. A. Fischetti, and J. R. Scott, *Mol. Gen. Genet.* **207,** 196 (1987).

Pepsin Extraction Method

The pepsin extraction procedure works well for many serotypes of M protein, primarily those classified as class I.[43] In a class II M protein (M49) the cleavage was found to be localized after a C repeat,[44] suggesting that the hinge region in class II M proteins may differ.

Streptococcal Strains

Well-characterized group A streptococcal strains may be obtained from research groups working on specific types. Group A streptococcal strains may also be obtained from hospital clinical laboratories; however, the serotype is usually unknown. Before growth for M protein extraction, the cells should be rotated in normal human blood to select for M protein-rich organisms.[45]

Culture Conditions

Milligram quantities of purified M protein may be obtained if the cells are grown in large quantities of broth. For the method described here 60 liters of Todd–Hewitt broth (Difco Laboratories, Detroit, MI) supplemented with 0.2% yeast extract (Difco) results in the best yields. The packed cell paste is collected by centrifugation and washed three times in saline. The yield of bacteria can vary from 200 to 300 g (wet weight).

Isolation of Crude M Protein with Pepsin

The method described here is modified from previous methods by Beachey et al.[27] and Manjula and Fischetti.[35] The streptococci are washed twice with 3 to 4 volumes of 67 mM phosphate buffer, pH 5.8, and the pellet is resuspended in the same buffer to a final concentration of 1.0 g bacteria in 2 ml of buffer. The cell suspension is warmed to 37°, and pepsin (Worthington, Freehold, NJ) is added to a concentration of 1.0 mg of enzyme to 10 g bacteria. Digestion is carried out at 37° for 45 min, with slow stirring. At the end of the digestion period, the flask containing the suspension is transferred to an ice bath, solid sodium bicarbonate is added to raise the pH to about 7.4 to stop the digestion process, and the bacteria are sedimented by centrifugation. The pellet is resuspended in phosphate buffer, pH 5.8, and the pepsin digestion procedure is repeated. The super-

[43] D. Bessen, K. F. Jones, and V. A. Fischetti, J. Exp. Med. 169, 269 (1989).
[44] K. M. Khandke, T. Fairwell, A. S. Acharya, and B. N. Manjula, J. Protein Chem. 9, 511 (1990).
[45] C. G. Becker, Am. J. Pathol. 44, 51 (1964).

natants from the two digestions are then filtered separately through a 0.2-μm Millipore (Bedford, MA) filter, concentrated to about one-fifth the original volume by pressure filtration through a PM10 membrane (Amicon Corp., Lexington, MA), dialyzed against 50 mM ammonium bicarbonate, and lyophilized. From previous experience,[35] one may expect about 1.0 g of crude protein from 300 g (wet weight) of bacteria. Analysis by SDS–PAGE will reveal a major protein band at about 20 to 40 kDa (depending on the strain[41]), representing the PepM molecule. This should be verified by Western blotting using type-specific antibody to the PepM protein being extracted.

Purification Procedure

Because the pepsin-derived region of the M protein represents the hypervariable and variable regions of the molecule, a different purification scheme will need to be devised for each M protein type. In most instances the initial procedure would be ion-exchange chromatography, usually DEAE or CM, followed by molecular sieve or hydroxylapatite columns for final purification. The PepM24 and PepM5 proteins have been partially purified from crude digests by ion-exchange chromatography on QAE[27] and DEAE,[35] respectively. The purification of the PepM5 protein is described here in detail.

Chromatography on DEAE-Sephadex A-25 (0.9 × 26 cm column) is carried out in 10 mM phosphate buffer, pH 8.0. After applying about 100 mg of the crude M5 extract to the column, the nonadherent proteins are washed out with the equilibration buffer until the optical density (OD) at 280 nm returns to baseline. Adherent proteins may then be eluted with stepwise NaCl at 100, 200, and 300 mM concentrations in the phosphate buffer. Analysis by SDS–PAGE indicates that the major protein band representing the complete PepM5 molecule is found in the fall-through and wash fractions of the DEAE column along with some small molecular weight contaminants (this is in contrast to the PepM24 protein chromatographed on QAE, which remained in the column and is eluted after 300–400 mM NaCl).[27] The pooled fall-through and wash fractions are concentrated by dialysis against ammonium bicarbonate and lyophilized.

Gel filtration is used to remove the small molecular weight contaminants from the PepM5 protein in the DEAE-purified fractions. The lyophilized pool from the DEAE column (~50 mg of protein) is suspended in about 300 μl of ammonium bicarbonate and applied to a 2.5 × 95 cm column of Sephadex G-200, equilibrated and eluted with 100 mM ammonium bicarbonate, pH 8.2. Column fractions are monitored for protein by measuring the absorbance at 280 nm. Fractions containing the purified PepM5 protein without the small molecular weight contaminants, as revealed by SDS–PAGE, are pooled and lyophilized. From previous results on the purification of the PepM5 protein[35] one may expect about 150 mg of purified PepM5 protein from the total crude pepsin digest.

Recombinant M Protein

Gram-positive surface molecules are synthesized with N-terminal leader sequences and a C-terminal anchor motif[46] to allow them to be translocated through the cytoplasmic membrane to the surface of the bacterial cell to be anchored. Because these surface proteins are unable to penetrate the outer membrane of gram-negative bacteria, gram-positive surface molecules that are cloned in *Escherichia coli* are found to be concentrated in the periplasmic space of the *E. coli*.[47] One can exploit this characteristic to purify large quantities of these proteins. Because the *E. coli* cytoplasm is known to contain proteolytic enzymes, the placement of cloned molecules into the periplasm and isolation therefrom without release of the cytoplasmic contents will usually result in larger and more homogeneous yields of protein.

In this section a procedure is outlined to determine if the cloned M protein (or in fact any protein) is concentrated in the periplasm or found in the cytoplasmic compartment of the *E. coli*. The M6 protein will be used as an example for the isolation and purification process. In most instances, the procedure used here for the isolation of the M6 protein from the *E. coli* periplasm may be followed for other gram-positive surface proteins as well as other proteins containing leader sequences. The final purification method described here for the M6 protein may not apply for M protein from other M serotypes because of the difference in the hypervariable region of these molecules. For these molecules, a different purification scheme may need to be devised. However, a combination of ion-exchange and molecular sieve chromatography steps will usually be successful.

Cell Fractionation of Escherichia coli

To determine the cellular compartment in which the M protein is found, the *E. coli* strain containing the M protein clone is grown in YT broth (50 ml) at 37° to an OD of 0.7 at 650 nm. Cells are collected by centrifugation

[46] O. Schneewind, P. Model, and V. A. Fischetti, *Cell (Cambridge, Mass.)* **70**, 267 (1992).
[47] V. A. Fischetti, K. F. Jones, B. N. Manjula, and J. R. Scott, *J. Exp. Med.* **159**, 1083 (1984).

at 7000 g (at 4°) and resuspended in 2.4 ml of ice-cold TSE buffer (100 mM Tris, pH 8.0, containing 20% (w/v) sucrose and 5 mM EDTA). Lysozyme is added to a final concentration of 0.5 mg/ml, and, following gentle mixing, the suspension is placed on ice for 20 min. For the whole cell control, 0.5 ml of the suspension is removed and processed as described below. MgCl$_2$ is added to the remaining 2.0 ml (50 mM final concentration) to stabilize the resultant spheroplasts, and these are then sedimented at 7000 g (at 4°) for 15 min. The supernatant is filtered through a 0.4-μm Millipore membrane to yield the periplasmic fraction. The spheroplasts are washed once in 2.0 ml TSE buffer and sedimented at 7000 g for 15 min.

To lyse the spheroplasts, 0.5 ml of a solution containing 10 mM MgCl$_2$ and 100 μg/ml DNase is added to the pellet along with 1.5 ml of water. The spheroplasts are then aspirated vigorously several times through a Pasteur pipette and frozen (in dry ice/ethanol) and thawed twice. The lysate generated by this treatment is centrifuged at 50,000 rpm for 1 hr in a Beckman type 65 rotor and the supernatant (cytoplasmic fraction) filtered as above.

Samples of the periplasmic and cytoplasmic fractions (100 μl) are immediately added to an equal amount of SDS loading buffer, boiled for 3 min, and applied to an SDS–polyacrylamide gel for analysis. The remainder of each fraction is maintained in aliquots at −70°. Analysis by SDS–PAGE and Western blots of the fractions using type-specific antibodies to the M protein reveals that nearly all of the reactive protein is found in the periplasmic space.

Large-Scale Isolation of Crude M Protein from Escherichia coli

Sixty liters of a YT culture of *E. coli* containing the M6 clone is grown to an OD of 0.7–1.5 at 650 nm. The cells are harvested in a Sharples centrifuge (Sharples-Stokes, Warminster, PA) and washed once in 100 mM Tris, pH 8.0. The pellet is then suspended in 1.0 liter of ice-cold TSE buffer containing 500 mg of lysozyme and incubated on ice for 20 min. MgCl$_2$ is added to a final concentration of 50 mM, and the spheroplasts sedimented at 7000 g (at 4°) for 30 min. The supernatant (crude periplasm) is immediately aspirated from the pellet and filtered through a 0.45-μm Millipore membrane. The filtered supernatant is divided into two equal aliquots, one of which is frozen at −70° and purified separately. The 500-ml supernatant is brought to 65% saturation with ammonium sulfate and placed at 4° for 18 hr. The resulting fine precipitate is sedimented at 7000 g for 1 hr, suspended in 50 ml of 5 mM sodium acetate, pH 5.5, dialyzed extensively against this buffer, and frozen at −70° until used for purification.

Purification of M Protein Isolated from Escherichia coli Periplasm

All purification steps should be monitored by SDS–PAGE and Western blotting using type-specific antibodies to the M protein. The crude periplasmic extract (after ammonium sulfate fractionation) is applied to a 1.5×12 cm column of CM cellulose (Whatman, Clifton, NJ) equilibrated in 5 mM sodium acetate, pH 5.5. The column is washed with this buffer until the absorption at 280 nm reaches baseline. The adsorbed M protein is eluted with 100 mM sodium phosphate, pH 7.0. The eluted M protein fractions (containing some contaminating proteins) are pooled and applied directly to a 1.0×17 cm column of hydroxylapatite (Bio-Rad Laboratories, Richmond, CA) equilibrated in 25 mM sodium phosphate, pH 7.0. The column is washed with 200 mM sodium phosphate, pH 7.0, until there is no detectable absorbance at 280 nm. The M protein is then eluted with 400 mM sodium phosphate, pH 7.0. Appropriate fractions as judged by SDS–PAGE and Western blotting are pooled, dialyzed extensively against 50 mM ammonium bicarbonate, and stored in the lyophilized state.

Yields of purified M6 protein range from 50 to 120 mg per 60 liters of culture, depending on the final optical density. *Escherichia coli* grown to late log to early stationary phase (an average OD of 1.5) yield larger quantities of M protein. Apparently the longer the cells are allowed to grow the more protein is exported into the periplasm. The recombinant M protein exhibits a multiple banded structure on SDS–PAGE of about four closely spaced proteins, an as yet undefined characteristic of M proteins.[3] As of this writing seven M protein molecules have been cloned and sequenced, namely, M6,[2] M5,[48] M12,[49] M24,[50] M49,[51] M57,[52] and M2[53]; however, only the M6 protein molecule has been fully characterized.[3]

[48] L. Miller, L. Gray, E. H. Beachey, and M. A. Kehoe, *J. Biol. Chem.* **263,** 5668 (1988).
[49] J. C. Robbins, J. G. Spanier, S. J. Jones, W. J. Simpson, and P. P. Cleary, *J. Bacteriol.* **169,** 5633 (1987).
[50] A. R. Mouw, E. H. Beachey, and V. Burdett, *J. Bacteriol.* **170,** 676 (1988).
[51] E. J. Haanes and P. P. Cleary, *J. Bacteriol.* **171,** 6397 (1989).
[52] B. N. Manjula, K.M. Khandke, T. Fairwell, W. A. Relf, and K. S. Sripakash, *J. Protein Chem.* **10,** 369 (1991).
[53] D. E. Bessen and V. A. Fischetti, *Infect. Immun.* **60,** 124 (1992).

[22] Isolation and Assay of *Pseudomonas aeruginosa* Alginate

By THOMAS B. MAY and A. M. CHAKRABARTY

Introduction

Pseudomonas aeruginosa that are isolated from sputum of patients having cystic fibrosis often have a mucoid appearance owing to the synthesis and secretion of an exopolysaccharide. Linker and Jones[1,2] first reported that this polysaccharide was similar to alginate, a linear copolymer composed of β-1,4-linked D-mannuronic acid and its C-5 epimer L-guluronic acid (Fig. 1). Alginate is a commercially important polymer obtained from marine algae.[3] Alginate has also been identified as a component of the slime layer of the bacterium *Azotobacter vinelandii*.[4] Alginate produced by *P. aeruginosa* has an uncharacteristically high ratio of mannuronate to guluronate residues and principally contains poly(D-mannuronate) or random block structures.[5] Bacterial alginates have also been shown to be *O*-acetylated.[2,5] *O*-Acetyl modification has been localized primarily to the O-2 position of mannuronic acid, but also to the O-3 position or to both the O-2 and O-3 positions.[6]

The secretion of alginate into the growth medium greatly simplifies purification of this polysaccharide from mucoid strains of *P. aeruginosa*. In fact, removal of the cell material and precipitation of the alginate from solution is a selective purification method that suffices for routine monitoring of alginate production by mucoid strains of *P. aeruginosa*. However, additional purification steps will likely be necessary for positive identification of uncharacterized polysaccharides from mucoid strains of *P. aeruginosa*. Contaminating nucleic acids, proteins, and polysaccharides also affect the rheological properties of the polymer.[7] In addition, further purification is required when raising antialginate antibodies.[8] The aim of this chapter is to detail the methods for isolating and measuring the quantity of alginate produced by mucoid strains of *P. aeruginosa* as well as

[1] A. Linker and R. S. Jones, *Nature* (*London*) **204**, 187 (1964).
[2] A. Linker and R. S. Jones, *J. Biol. Chem.* **241**, 3845 (1966).
[3] T.-Y. Lin and W. Z. Hassid, *J. Biol. Chem.* **241**, 3283 (1966).
[4] P. A. J. Gorin and J. F. T. Spencer, *Can. J. Chem.* **44**, 993 (1966).
[5] L. R. Evans and A. Linker, *J. Bacteriol.* **116**, 915 (1973).
[6] G. Skjak-Braek, H. Grasdalen, and B. Larsen, *Carbohydr. Res.* **154**, 239 (1986).
[7] N. J. Russell and P. Gacesa, *Mol. Aspects Med.* **10**, 1 (1988).
[8] G. B. Pier, J. M. Saunders, P. Ames, M. S. Edwards, H. Auerbach, J. Goldfarb, D. P. Speert, and S. Hurwitch, *N. Engl. J. Med.* **317**, 793 (1987).

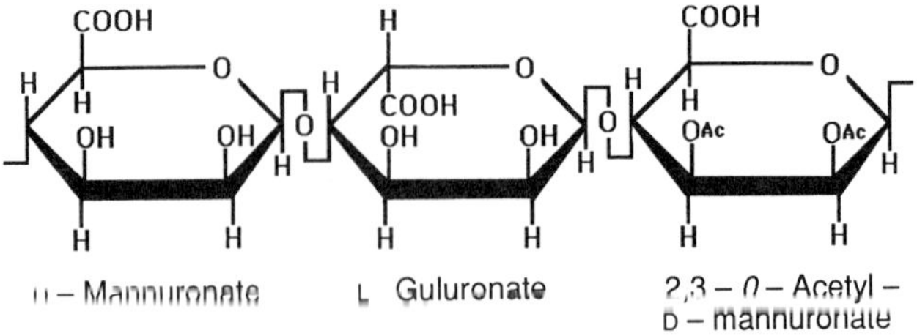

FIG. 1. Structure of alginate. Alginate from *P. aeruginosa* consists largely of poly(D-mannuronate) interspersed with variable amounts of the C-5 epimer L-guluronate. Poly(L-guluronate) block structure is characteristically absent from pseudomonal alginate. Mannuronate residues of the bacterial polymer may be O-acetylated, as indicated by OAc, at the O-2, O-3, or O-2 and O-3 positions.

to provide additional steps for purification and characterization of this exopolysaccharide.

Isolation of Alginate

Bacterial Strains and Growth Conditions

The mucoid alginate-producing *P. aeruginosa* 8821 was isolated from the sputum of a patient having cystic fibrosis. Like other mucoid strains, 8821 spontaneously loses the capacity to produce alginate when propagated on standard laboratory growth media. Darzins and Chakrabarty[9] isolated the stable alginate-producing strain 8830 after chemical mutagenesis of one of the nonmucoid revertants of strain 8821. A number of Alg⁻ mutants were obtained by further mutagenesis of the stable mucoid strain 8830.

Pseudomonas aeruginosa strains are first grown overnight in Luria broth (Difco, Detroit, MI) at 37°. The overnight cultures (100 μl each) are then plated on *Pseudomonas* isolation agar (PIA; Difco) containing 1% glycerol as a carbon source and incubated for 24 hr at 37°. A single plate of mucoid *P. aeruginosa* yields enough alginate for quantitation purposes, but additional plates will likely be required for purification and extensive analysis. Mucoid strains of *P. aeruginosa* (such as strain 8821) will also produce alginate in Luria broth and in minimal salts media containing 1% glucose as a carbon source.[9] However, we usually grow cells on solid media for direct confirmation of mucoidy because of the unstable nature of the alginate phenotype.

[9] A. Darzins and A. M. Chakrabarty, *J. Bacteriol.* **159**, 9 (1984).

Removal of Cell Material

The major problem in alginate isolation is separating the highly viscous polymer away from the bacterial cells. The mucoid growth is washed from the PIA plate(s) with 0.9% NaCl and then further diluted with saline (the volume depending on the viscosity). The cell suspension is thoroughly mixed and centrifuged for 30 min at 13,700 g. The cell pellet is washed once with the saline solution for quantitative recovery of alginate. The supernatants are then combined and recentrifuged to remove the remaining cell material. The cell pellet is kept for determination of total cell protein or dry weight (see section on uronic acid assay).

Pier *et al.*[8] heat the culture supernatant to render any remaining bacteria nonviable when using the purified alginate to raise antibodies. In addition, heat treatment appears to precipitate contaminating proteins.[10] Some methods recommend addition of EDTA to the culture supernatant to chelate divalent cations that cause alginate to adhere to the cell surface. This treatment is best avoided since EDTA is known to solubilize outer membrane components and the improvement in cell removal is questionable.[7,11]

Alginate Isolation from Culture Supernatant

Alginate isolation procedures are generally permutations of the initial methods used for obtaining alginate from cultures of *P. aeruginosa*.[2,5] Briefly, 3 volumes of ice-cold 95% ethanol is added to the culture supernatant (with stirring) to precipitate the alginate. We store the ethanol solution at $-70°$ for several hours to increase alginate recovery. Alternatively, sodium acetate (1% final concentration) can be added to the culture supernatant to aid ethanol precipitation.[2] The alginate precipitate is then recovered by centrifugation at 13,700 g for 15 min. The alginate pellet is washed twice with 95% ethanol and once with absolute ethanol, dried under vacuum, and resuspended at 1 mg dry weight per milliliter of distilled water or of an appropriate buffer unless indicated otherwise. Alginate obtained from liquid grown cultures are next dialyzed for 24 hr against three changes (4 liters each) of distilled water to remove small molecules that interfere with the uronic acid colorimetric assay.

Cetylpyridinium chloride precipitation has been used as a second purification step since it is selective for acidic polysaccharides like alginate.[12,13]

[10] S. S. Pedersen, F. Espersen, N. Høiby, and G. H. Shand, *J. Clin. Microbiol.* **27**, 691 (1989).
[11] A. J. Anderson, A. J. Hacking, and E. A. Dawes, *J. Gen. Microbiol.* **133**, 1045 (1987).
[12] J. R. W. Govan, P. Sarasola, D. J. Taylor, P. J. Tatnell, N. J. Russell, and P. Gacesa, *J. Clin. Microbiol.* **30**, 595 (1992).
[13] A. M. Carlson and L. W. Matthews, *Biochemistry* **5**, 2817 (1966).

However, we do not use the cetylpyridinium chloride precipitation step because it has been reported that the alginate yield is reduced and that the benefits are limited compared to ethanol precipitation alone.[2] In general, methods based on alkaline extraction should also be avoided since alkali deacetylates bacterial alginates.[2]

Assay for Alginate

There are three basic methods of assaying uronic acids: the carbazole assay,[14] the orcinol–FeCl$_3$ assay,[15] and the decarboxylation method measuring CO$_2$ values.[16] The modified carbazole assay of Knutson and Jeanes[14] has been well tested in our laboratory, and it is easy and reproducible if the methodology is consistent. The modified carbazole assay uses borate and heat to increase the assay sensitivity toward mannuronate and guluronate by 10-fold over the original carbazole assay.[14] We use a microassay version of the method.

Materials and Reagents

A borate stock solution is prepared by dissolving 24.74 g of H$_3$BO$_3$ in 45 ml of 4 M KOH and diluting to 100 ml with distilled water. The stock solution can be warmed gently to redissolve any precipitate that forms on storage. A borate working solution is prepared fresh daily by diluting the borate stock solution 1 : 40 (v/v) with concentrated H$_2$SO$_4$. A 0.1% (w/v) carbazole solution is prepared in absolute ethanol.

Uronic Acid Assay

The borate working solution is first equilibrated in an ice–water bath. A 70-μl aliquot of the uronic acid sample is then carefully layered on top of 600 μl of ice-cold borate working solution. The sample mixture is cooled, vortexed for 4 sec, and immediately returned to the ice–water bath. Twenty microliters of the carbazole solution is then layered on top of the mixture. The reaction mixture is cooled, vortexed for 4 sec, and returned to the ice–water bath until all the samples are processed. The reaction mixture is heated for 30 min at 55° to develop the color reaction (stable for 2 hr). Absorbance at 530 nm is indicative of a positive uronic acid test. Uronic acid concentration is determined from a standard curve of D-mannuronic acid or seaweed alginate (Sigma, St. Louis, MO) ranging

---

[14] C. A. Knutson and A. Jeanes, *Anal. Biochem.* **24,** 470 (1968).
[15] A. H. Brown, *Arch. Biochem.* **11,** 269 (1946).
[16] M. V. Tracey, *Biochem. J.* **43,** 185 (1948).

from 10 to 1000 μg/ml. Uronic acid content is usually expressed per milligram cell dry weight or per milligram cell protein. Total cellular protein is measured by the Bio-Rad protein assay (Bio-Rad Laboratories, Richmond, CA) after first lysing the cells with 5% trichloroacetic acid, centrifuging, and resuspending the precipitate in 50 mM potassium phosphate (pH 7.0) buffer.

Preventing uncontrolled heating during mixing of the carbazole reagents is particularly important for accurate measurement of uronic acids. The modified carbazole assay is still sensitive to interference by neutral sugars, so we dialyze alginate samples that were obtained from liquid grown cultures. This extra step removes impurities that interfere with the carbazole assay, particularly glucose (used as a carbon source) which inevitably caramelizes in the borate–H_2SO_4 solution.

Quantitation of Alginate Produced by Mucoid Pseudomonas aeruginosa

The mucoid strain 8821 produces 1.6 mg alginate per milliliter of culture supernatant after 24 hr of growth in minimal media broth, whereas strain 8830 appears Alg$^-$ under these conditions.[9] *Pseudomonas aeruginosa* 8830 does produce low amounts of alginate in liquid culture, provided the cells are grown in a rich medium such as Luria broth with 0.1% glucose. It is not clear why the stable mucoid strain 8830 is unable to produce substantial quantities of alginate in liquid culture within 24 hr, although the alginate yield from a 48 hr culture of strain 8830 exceeds that of strain 8821. We usually grow *P. aeruginosa* 8830 and the isogenic Alg$^-$ mutants on solid growth media because of the variability in alginate production by strain 8830 in liquid culture. In addition, growth of mucoid *P. aeruginosa* (e.g., strain 8821) on solid media provides direct visual evidence that the majority of the cell population has maintained a mucoid phenotype and has not reverted to the nonmucoid form.

Table I shows that strain 8830 produces copious amounts of alginate when grown on solid media for 24 hr. Alginate levels double when 8830 is incubated for an additional 24 hr.[17] In general, extending the incubation period may be beneficial in cases where the yield of alginate is important. The *algC* mutant strain 8858, which is defective for phosphomannomutase activity, produces extremely low levels of alginate (Table I). This result suggests that cell removal and ethanol precipitation reduce impurities below levels that interfere with the uronic acid assay. Because strain 8858 is a derivative of the Alg$^+$ strain 8830, and thus also of strain 8821, these

[17] N. A. Zielinski, A. M. Chakrabarty, and A. Berry, *J. Biol. Chem.* **266,** 9754 (1991).

TABLE I
ALGINATE PRODUCTION BY *Pseudomonas
aeruginosa* STRAINS[a]

Strain[b]	Alginate concentration[c] (mg/mg cell protein)
8830	1.17
8858	ND[d]
8858[pCP13]	0.04
8858[pAB8]	3.43

[a] Data from N. Zielinski, A. M. Chakrabarty, and A. Berry, *J. Biol. Chem.* **266**, 9754 (1991).
[b] Strain 8858 is an *algC* mutant (defective for phosphomannomutase) derived from the stable Alg+ strain 8830. The plasmid pAB8 contains a wild-type copy of the *algC* gene that restores alginate synthesis by strain 8858. Plasmid pCP13 is the vector used in the construction of pAB8.
[c] Alginate was isolated after 24 hr of growth on *Pseudomonas* isolation agar containing 1% glycerol.
[d] ND, Not detectable.

strains must not produce any other uronic acid-containing materials that coprecipitate with alginate. Furthermore, the mucoid phenotype is restored (complemented) by the plasmid pAB8 which contains a wild-type copy of the *algC* gene.

Colorimetric Tests

A few simple colorimetric assays can be used to assure that an uncharacterized polysaccharide from *P. aeruginosa* is alginate and not another polymer containing uronic acids. In addition to a positive uronic acid test, bacterial alginate should be tested for the presence of *O*-acetyl groups as determined by the method of Hestrin[18] with β-D-glucose pentaacetate as the standard. The polysaccharide should lack neutral, hexoses, amino sugars, 3-keto-D-mannooctulosonic acid (a component of lipopolysaccharides), and phosphorus.[19–22]

[18] S. Hestrin, *J. Biol. Chem.* **180**, 249 (1949).
[19] S. Seifter, S. Dayton, B. Novic, and E. Muntwyler, *Arch. Biochem.* **25**, 191 (1950).
[20] R. Belcher, A. J. Nutten, and C. M. Sambrook, *Analyst* (*London*) **79**, 201 (1954).
[21] M. J. Osborn, *Proc. Natl. Acad. Sci. U.S.A.* **50**, 499 (1963).
[22] P. S. Chen, T. Y. Toribara, and H. Warner, *Anal. Chem.* **28**, 1756 (1956).

A positive reaction with any of the colorimetric tests, except those for uronic acid and O-acetylation, is an indication that the alginate sample is contaminated with another polysaccharide or that the polysaccharide is not alginate. Thus, the polysaccharide will need to be purified further. Increased purity may also be desired depending on the intended use of the alginate sample. We have provided additional suggestions for purification and identification of alginate based on methods in the literature that have been well tested.

Purification of Alginate

Removal of Nucleic Acids

Alginate prepared as described above may contain as much as 20% nucleic acids.[2] Nucleic acid contamination is monitored by absorbance at 260 nm. Removal of nucleic acids has been accomplished by incubating the polysaccharide samples with nucleases.[2,5,10] Alginate samples are resuspended in 100 mM potassium phosphate (pH 7.0) buffer containing 0.9% NaCl, 10 mM MgCl$_2$, and 1 mM CaCl$_2$ and incubated with RNase A and DNase I (100 μg/ml each) for 4 hr at 37°. The nucleases are heat-inactivated at 80° for 30 min and removed by centrifugation at 20,000 g for 20 min. Alginate is then ethanol-precipitated as described above. Nucleic acids are not detectable following this treatment.[10]

Removal of Protein

Protein contamination is monitored by absorbance at 280 nm. The alginate samples are dissolved in 100 mM Tris-HCl (pH 7.4) buffer at 5 mg/ml and incubated with 1 mg/ml pronase (Calbiochem, LaJolla, CA) for 48 hr at 37° to remove contaminating protein.[5] The solution is dialyzed for 17 hr against 2 changes (40 volumes) of distilled water. Any precipitate that forms is removed by centrifuging at 20,000 g for 20 min. Alginate is then precipitated as before, washed twice with 95% ethanol, washed once with absolute ethanol, washed with ether, and dried under vacuum. Chromatographic purification of alginate (see below) precludes the need for protease treatment.

Purification by Column Chromatography

Alginate purification by column chromatography allows easy monitoring of the separation of proteins, nucleic acids, and other cell components (such as lipopolysaccharide) from uronic acid-containing fractions. The Dowex 1-X2 anion-exchange chromatography method of Sherbrock-Cox *et al.*[23] has been extensively used to purify alginate. However, a method based on the DEAE anion-exchange resin[8,10] appears to be more effective in separating alginate from nonalginate polysaccharides. Briefly, an alginate sample is dissolved in a minimum volume of 25 mM ammonium carbonate and loaded onto a DEAE-Sephacel column that has been equilibrated with the same buffer (1 ml resin/ml of alginate). It should be noted that ammonium carbonate treatment does not deacetylate the alginate sample.[10] Alginate is then eluted from the column with a linear gradient of 25 mM to 1 M ammonium carbonate (10 times bed volume). The uronic acid-containing fractions are dialyzed 24 hr against three changes of distilled water. Pier *et al.*[8] include an additional gel-filtration step (Sephacel S-300) when preparing antibody against alginate.

Positive Identification of Alginate

Spectroscopy Methods

Infrared spectroscopy provides a semiquantitative measurement of the ratio of D-mannuronic acid to L-guluronic acid (M : G ratio). The M : G ratio, characteristic of a given alginate from a given source, provides useful information about the nature of the alginate gel; high M : G alginates are elastic and low M : G alginates are brittle.[7] Bacterial alginates also have IR absorbance bands at 1250 and 1730 cm^{-1} owing to presence of O-acetyl groups.[2,10,23] These bands are absent from the spectra of algal alginates as well as alkali-treated bacterial alginates. In addition, the absorption peak at 893 cm^{-1} suggests a strong negative optical rotation indicative of β linkages.[2]

High-field Fourier transform ^{1}H nuclear magnetic resonance (NMR) provides information not only about the fraction of D-mannuronic acid (F_M) and L-guluronic acid (F_G) but also about the fraction of alginate composed of homopolymeric (F_{MM} and F_{GG}) and heteropolymeric (F_{MG} and F_{GM}) block structures.[23,24] The work of Grasdalen *et al.*[24] provides an excellent description for assigning ^{1}H NMR peaks to alginate block structures. Alginate from *P. aeruginosa* typically has a large M : G ratio and no GG block structure compared to alginate from other sources.[23] The lack of extensive GG block structure indicates that *P. aeruginosa* produces an elastic alginate gel.[7]

Infrared spectroscopy and high-field Fourier transform ^{1}H NMR are the most convenient methods to identify alginate positively, provided that

[23] V. Sherbrock-Cox, N. J. Russell, and P. Gacesa, *Carbohydr. Res.* **135**, 147 (1984).
[24] H. Grasdalen, B. Larsen, and O. Smidsrød, *Carbohydr. Res.* **68**, 23 (1979).

the equipment is available.[7] In addition, ^{13}C NMR spectroscopy provides similar information about newly synthesized alginate.[25] The major drawback to this method is the low sensitivity.[24] However, uronic acids can also be identified by conventional chromatographic methods (described below).

Chromatography Methods

Alginate samples (1 mg/ml) are first hydrolyzed in 88% formic acid at 100° for 5 hr.[2] The hydrolyzate is dried under vacuum over NaOH pellets and redissolved at 10 mg/ml. We use sodium dodecyl sulfate (SDS)–polyacrylamide gel electrophoresis (15% resolving gels[26]) to monitor the hydrolysis reaction. Alginate is stained with 0.1% toluidine blue O in 30% methanol for 1 hr and destained with 1% acetic acid. Although hydrolysis is incomplete under these conditions, stronger mineral acids should be avoided because they tend to decarboxylate uronic acids, particularly guluronic acid.[13]

Mannuronate and guluronate can be identified by descending paper (Whatman, Clifton, NJ, No. 1 paper) chromatography in ethyl acetate–acetic acid–pyridine–water (5 : 1 : 5 : 3, v/v), acetone–95% ethanol–2-propanol–20 mM borate buffer, pH 10 (3 : 1 : 2 : 2, v/v), butyl acetate–acetic acid–95% ethanol–water (3 : 2 : 1 : 1, v/v), or acetone–2-propanol–95% ethanol–water (3 : 1 : 1 : 1, v/v).[2,23,27] Because mannuronolactone and guluronolactone are formed during acid hydrolysis, D-mannuronic acid and seaweed alginate (Sigma) standards should be treated under the same conditions of hydrolysis. Uronic acid spots are specifically stained by anisidine red reagent[28] and also by silver nitrate.[29]

Dowex 1-X8 column chromatography can also be used to separate uronic acids,[30] but this method can be prohibitive since it requires substantially more sample than thin-layer techniques. The uronic acids of *P. aeruginosa* alginate are identified by comparison with the Dowex 1-X8 profile of seaweed hydrolyzate.[2] Furthermore, the M : G ratio of the alginate sample can be estimated by quantitation of the uronic acid fractions. We have used similar thin-layer and Dowex 1-X8 chromatography methods to identify GDP-D-mannuronic acid, an intermediate of the alginate biosynthetic pathway.[27]

[25] M. J. E. Narbad, J. E. Hewlins, P. Gacesa, and N. J. Russell, *Biochem. J.* **267,** 579 (1990).
[26] U. K. Laemmli, *Nature (London)* **227,** 680 (1970).
[27] S. Roychoudhury, T. B. May, J. F. Gill, S. K. Singh, D. S. Feingold, and A. M. Chakrabarty, *J. Biol. Chem.* **264,** 9380 (1989).
[28] L. Hough, J. K. N. Jones, and W. H. Wadman, *J. Chem. Soc.,* 1702 (1950).
[29] W. E. Trevelyan, D. P. Proctor, and J. S. Harrison, *Nature (London)* **166,** 444 (1950).
[30] A. Haug and B. Larsen, *Acta Chem. Scand.* **16,** 1908 (1962).

Alginase Treatment

Degradation by an alginase enzyme is another means of confirming the identification of alginate. We are unaware of any commercial sources of alginase enzymes. In general, the enzyme is obtained from bacterial soil organisms that were isolated on the basis of using commercially available algal alginates as a sole source of carbon.[31] Alginase (alginate lyase) has also been shown to be produced by mucoid *P. aeruginosa* and is encoded by the *algL* gene mapping in the *alg* gene cluster at 34 min.[32,33] The action of bacterial alginases can be followed by the increase of α,β-unsaturated uronides as measured by absorbance at 235 nm.[5] Enzyme products can also be checked by paper chromatography [acetone–ethanol–2-propanol–5 m*M* borate buffer, pH 12.0 (3 : 1 : 1 : 2, v/v)] using alginase-treated algal alginate as a standard. Degradation by alginase suggests that the polysaccharide contains β-1,4 linkages.[5]

Acknowledgments

This work was supported by U.S. Public Health Service Grant AI16790-13 from the National Institutes of Health and in part by Grant P-455 from the Cystic Fibrosis Foundation.

[31] J. Preiss and G. Ashwell, *J. Biol. Chem.* **237,** 309 (1962).
[32] N. L. Schiller, S. R. Monday, C. M. Boyd, N. T. Keen, and D. E. Ohman, *J. Bacteriol.* **175,** 4780 (1993).
[33] A. Boyd, M. Ghosh, T. B. May, D. Shinabarger, R. Keogh, and A. M. Chakrabarty, *Gene* **131,** 1 (1993).

[23] Purification of *Escherichia coli* K Antigens

By WILLIE F. VANN and STEPHEN J. FREESE

Introduction

Bacterial capsular polysaccharides have been shown to assist bacteria in the evasion of host defense in several ways. These include (1) providing a barrier to opsonization, (2) presentation of self-antigens, and (3) defeat of the complement system.[1] The similarity of some capsular polysaccharides to structures found in mammalian tissues[2,3] has stimulated an interest in their use as reagents or probes in the investigation of cell interactions and eukaryotic polysaccharide function. The *Escherichia coli* K1 polysac-

[1] K. Jann and B. Jann (eds.), *in* "Current Topics in Microbiology," Vol. 150, p. 1. Springer-Verlag, New York, 1990.
[2] F. A. Troy, *Glycobiology* **2,** 5 (1992).
[3] W. F. Vann, A. Schmidt, B. Jann, and K. Jann, *Eur. J. Biochem.* **116,** 359 (1981).

charide is poly[α(2-8)-N-acetylneuraminic acid] and is the same polysialic acid found on embryonic neural cell adhesion molecules (N-CAM).[2] Antibodies and a neuraminidase specific for the K1 polysaccharide are used routinely to study N-CAM function.[4] Similarly, the *E. coli* K5 polysaccharide bears homology with N-acetylheparosan.[3] This bacterial polysaccharide has been used to study heparin biosynthesis and function.[5]

Bacterial capsular polysaccharides are usually repeat structures and vary in charge. This subject has been reviewed in a previous volume in this series.[6] The polysaccharides of pathogenic gram-negative bacteria most frequently have negatively charged repeat units, which facilitates their fractionation from lipopolysaccharide and neutral O-antigenic polysaccharides. Capsular polysaccharides are often loosely associated extracellular structures which are shed into the culture media. Thus, the capsule of a strain which excretes polysaccharide can often be purified by fractionation of the culture medium. Capsular polysaccharides of *Haemophilus influenzae*,[7] *Klebsiella*,[8] and *Neisseria meningitidis*[9] have been isolated using such an approach. Where possible, bacteria should be grown on a low molecular weight culture medium to obviate the problem of contamination by high molecular weight media components.

Escherichia coli K antigens, like *H. influenzae*[7] and *N. meningitidis*[9] capsules, are negatively charged[6] and can therefore be readily purified by precipitation with positively charged detergents. Use of such detergents allows for precipitation of polysaccharide shed into the culture medium. When Cetavlon (hexadecyltrimethylammonium bromide) is added to the whole culture, soluble polysaccharide is precipitated along with polysaccharide remaining with the bacteria. Because many *E. coli* capsular polysaccharides are acid or alkali labile or possess labile substituents,[6] care must be taken to prevent degradation by using buffered reagents and purifying the polysaccharide at 4°. The procedure described below is a modification of a method developed in the laboratories of Dr. E. C. Gotschlich at the Rockefeller University[10] and Dr. John Robbins[11,12] at the

[4] E. R. Vimr, R. D. McCoy, H. F. Vollger, N. C. Wilkinson, and F. A. Troy, *Proc. Natl. Acad. Sci. U.S.A.* **81,** 1971 (1984).

[5] M. Kusche, H. H. Hanesson, and U. Lindahl, *Biochem. J.* **275,** 151 (1991).

[6] K. Jann and B. Jann, this series, Vol. 50 [25].

[7] P. Anderson and D. H. Smith, *Infect. Immun.* **15,** 472 (1977).

[8] K. Okutani and G. G. S. Dutton, *Carbohydr. Res.* **86,** 259 (1980).

[9] D. R. Bundle, H. J. Jennings, and C. P. Kenny, *J. Biol. Chem.* **249,** 4797 (1974).

[10] E. C. Gotschlich, M. Rey, C. Etienne, W. R. Sanborn, R. Triau, and B. Cventanovic, *Prog. Immunobiol. Stand.* **5,** 458 (1972).

[11] F. Orskov, I. Orskov, A. Sutton, R. Schneerson, W. Lin, W. Egan, G. E. Hoff, and J. Robbins, *J. Exp. Med.* **149,** 669 (1979).

[12] W. F. Vann, T.-Y. Liu, and J. B. Robbins, *Infect. Immun.* **13,** 1654 (1976).

National Institutes of Health. The method begins with detergent precipitation and removes protein by phenol extraction. Contaminating lipopolysaccharide forms micelles and is removed by ultracentrifugation. Variations of these procedures have been used to purify *Haemophilus* and meningococcal polysaccharides.[7,9] Also described is a method for the isolation of ^{13}C-labeled polysaccharide in a 1-liter shake flask.

Purification of *Escherichia coli* K5 Capsular Polysaccharide

Reagents

Growth medium: Each liter of medium should contain the following ingredients: Sodium phosphate (dibasic) (2 g), casamino acids (Difco, Detroit, MI) (12 g), KCl (0.5 g), NaCl (1 g), L-cysteine hydrochloride (15 mg), $MgSO_4 \cdot 6H_2O$ (0.5 g), citric acid (0.1 g), and dextrose (10 g). The medium is supplemented with 100 ml of the low molecular weight fraction (LMW) of yeast extract (10×) and adjusted to pH 7.5 with ammonium hydroxide.

LMW yeast extract (10×): The LMW fraction is prepared by filling dry dialysis bags with yeast extract powder. The bags are then dialyzed against 10% of the recommended reconstitution volume of water for 48 hr at 4°. The low molecular fraction outside of the dialysis bag is then sterile filtered and stored frozen until use.

Buffered phenol: Buffered phenol is prepared by dissolving 500 g of crystalline phenol in 140 ml of 10% saturated sodium acetate, pH 7.4.

Procedure. Bacteria may be grown until stationary phase in a fermenter for large quantities or, for smaller quantities, overnight in a Fernbach flask. Figure 1 presents a typical scheme for purification of polysaccharide from a 50-liter fermentation run.

A culture of *E. coli* O10 : K5 is inoculated into 50 liters of the above medium and allowed to grow until stationary phase at pH 7.0. The culture is then diluted 2-fold with 0.2% Cetavlon (hexadecyltrimethylammonium bromide) and is allowed to stand for at least 2 hr. The Cetavlon complex containing polysaccharide and cells is harvested by centrifugation (yield 282 g) and stored frozen. Polysaccharide is dissociated from the Cetavlon complex as follows. The frozen paste is resuspended in 500 ml water, and the suspension is adjusted to 1 M CaCl$_2$ with 82.5 g CaCl$_2$ and 750 ml water. The paste is then extracted by homogenizing vigorously in an Omnimixer (Omni International, Waterbury, CT) and centrifuging at 16,000 g (at 4°) for 30 min. The 750 ml of viscous supernatant mixture is slowly adjusted to 25% ethanol stirred for 1 hr, and centrifuged to remove

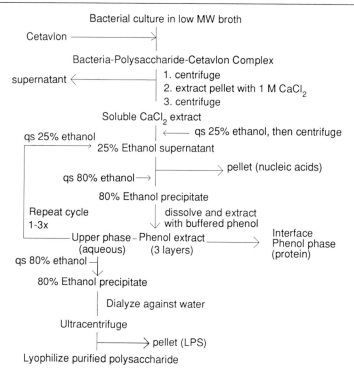

FIG. 1. Outline for the purification of K antigens. Isolation of *E. coli* capsular polysaccharides.

cell debris and precipitated nucleic acid and protein. The supernatant is adjusted to 80% ethanol, and the crude polysaccharide fraction is recovered by centrifugation (16,000 *g* at 4°).

The paste is dissolved in 400 ml of 10% saturated sodium acetate and stirred vigorously with an equal volume of cold buffered phenol for 30 min in an ice bath. The phases are separated by centrifugation at 10,000 rpm. The upper aqueous phase (360 ml) is carefully removed. Care should be taken not to include the interface. The aqueous layer is adjusted to 25% ethanol and allowed to stand in the cold for 1 hr and then centrifuged at 10,000 rpm. Polysaccharide is recovered from the supernatant by adjusting to 80% ethanol and centrifuging after standing overnight at 4°. If the polysaccharide solution does not become turbid on addition of ethanol, a precipitate may be induced by the dropwise addition of 1 *M* CaCl$_2$. The 80% ethanol precipitate is redissolved in 10% saturated sodium acetate, pH 7.4, and then processed through at least one additional cycle of phenol extraction and ethanol precipitation until the interface is minimal.

The final ethanol precipitate is dissolved in 10% saturated sodium acetate and dialyzed extensively against water. The dialyzate is then centrifuged at 100,000 g for 2 hr to remove lipopolysaccharide. The supernatant is then lyophilized. This process yields 1–3 g of polysaccharide from 300 g of Cetavlon cell paste, depending on strain and polysaccharide.

Preparation of ^{13}C-Labeled *Escherichia coli* Capsular Polysaccharides

Bacterial polysaccharides can be conveniently analyzed by ^{13}C nuclear magnetic resonance (NMR) spectroscopy using conventional one-dimensional (1D) techniques[9,13] and by ^{1}H NMR using modern two-dimensional (2D) methods.[14] Procedures dependent on the 1% natural abundance of ^{13}C can require long periods of time for acquisition of data. The availability of ^{13}C-enriched glucose (33 and 99%) has made the use of ^{13}C-labeled polysaccharides for NMR studies feasible. A 1D ^{13}C NMR spectrum of 33% ^{13}C-enriched K1 polysialic acid may be obtained in a few minutes rather than overnight. The 99% [13]C enrichment allows one to take advantage of the interaction of carbon nuclei[15] and thus obtain more structural information about a polysaccharide using more sophisticated 2D and three-dimensional (3D) ^{13}C experiments.[16] Data for the [13]C–[13]C correlation spectroscopy (COSY) experiment[14,17] on the K14 polysaccharide in Fig. 2 were obtained overnight on a 5 mg sample. An equivalent experiment using natural abundance would require an acquisition time of about 6 months.

The procedures for isolation of ^{13}C-labeled polysaccharides are very similar to those for the unlabeled polysaccharide. However, to maintain a high degree of labeling in the polysaccharide, bacteria are grown in M9 minimal media[18] supplemented with enough nutrients to facilitate reasonable growth with [^{13}C]glucose (33 or 99% enriched) as the major carbon source.

Reagents

Growth medium: M9 minimal medium[18] is supplemented with 250 mg casamino acids and 1 ml 10× yeast extract dialyzate per liter of medium. To this is added 2 g/liter sterile-filtered [^{13}C]glucose (33 or 99% enrichment) in 10 ml water.

[13] H. J. Jennings and I. C. P. Smith, *Methods Carbohydr. Chem.* **8,** 97 (1980).
[14] J. Dabrowski, this series, Vol. 179 [12].
[15] R. Barker, H. A. Nunez, P. Rosevear, and A. S. Serianni, this series, Vol. 83 [3].
[16] S. W. Fesik and E. R. Zuider, *Q. Rev. Biophys.* **23,** 97 (1990).
[17] L. Mueller, *J. Magn. Reson.* **72,** 191 (1987).
[18] J. Miller, "Experiments in Molecular Genetics." Cold Spring Harbor Laboratory, Cold Spring Harbor, New York, 1972.

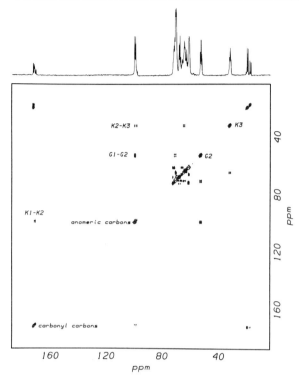

FIG. 2. ^{13}C–^{13}C COSY spectrum of 99% ^{13}C-labeled K14 polysaccharide obtained using the P.E.COSY experiment.[17] Data were acquired using a General Electric (Palo Alto, CA) GN-300 spectrometer operating at a ^{13}C frequency of 75.47 MHz. The sample, composed of 5 mg of polysaccharide in 0.5 ml of D_2O, was maintained at 32° and was not spun. A sweep width of 13.9 kHz was used in both dimensions. In the T2 dimension 280 free induction decays (FIDs) were summed into 1024 data points. Two dummy scans were discarded for each T1 point. The recycle time between FIDs was 450 msec. In the T1 dimension 296 increments were acquired and zero-filled to a final size of 1024 data points. In both dimensions data were processed using a squared sine bell shifted by 30°. All processing was performed using the FTIRIS2 software obtained from Hare Research Corporation (Bothel, WA).

Procedure. Polysaccharide is purified in a manner similar to the procedure described above. Volumes are kept to a minimum to improve yields. Cetavlon paste from 1 liter of culture grown on 33% enriched glucose (4.33 g) is resuspend in 60 ml of 1 M $CaCl_2$ and stirred vigorously. To the stirring cell suspension is added 12 ml of 95% ethanol. After 15 min the cells are removed by centrifugation. The polysaccharide is precipitated from the supernatant by the addition of 228 ml of 95% ethanol. The turbid suspension is stored in the cold for 30 min and the precipitate recovered by centrifugation. The crude polysaccharide is redissolved in 40 ml of 10% saturated sodium acetate and extracted with an equal volume of buffered phenol. The polysaccharide is recovered from the aqueous phase after adjusting to 80% ethanol and redissolved in 25 ml of 10% saturated sodium acetate. The preparation is processed once more through a cycle of phenol extraction and ethanol precipitation and is then dialyzed. The dialyzate is centrifuged for 1–2 hr at 150,000 g and lyophilized to yield 55 mg of 33% ^{13}C-enriched polysaccharide.

Analysis of Purified Polysaccharides

After completion and at some stage during purification the identity of the polysaccharide in question needs to be established. Because the structure and composition of many capsular polysaccharides of *E. coli* are known, a chromatographic or colorimetric assay to detect an intrinsic component of the polysaccharide repeat unit such as a monosaccharide or phosphate is used. Most of the methods needed to analyze the carbohydrate components of these polysaccharides have been described in several contributions in this series.[19–21] Whitfield and co-workers have described a sensitive chromatographic method for compositional analysis of bacterial polysaccharide based on high pH anion-exchange chromatography.[22] When quantitating capsular polysaccharides care must be taken to assay intrinsic components of the repeat unit since substituents such as *O*-acetyl groups are variable and are often labile, leading to inaccurate estimates of polysaccharide content.

Antibodies are useful reagents for detecting antigens during the purification by simple immunoprecipitation.[19] Because K-antigens are negatively charged, they can be detected by immunoelectrophoresis using Cetavlon or antibody precipitation.[19,23]

Nuclear magnetic resonance spectroscopy is a very useful method for the analysis of purified polysaccharide even in the absence of a known structure or composition. A natural abundance ^{13}C NMR spectrum can be readily used to check the polysaccharide in question and give some idea of the degree of contamination above 10%. Because NMR is not a destructive technique it is appropriate for polysaccharides not conve-

---

[19] I. C. Hancock and I. R. Poxton (eds.), "Bacterial Cell Surface Techniques." Wiley, New York, 1988.
[20] B. Lindberg and J. Lönngren, this series Vol. 50 [1].
[21] M. R. Hardy, this series, Vol. 179 [7].
[22] A. Clarke, V. Sarabia, W. Keenleyside, P. R. MacLachlan, and C. Whitfield, *Anal. Biochem.* **199,** 68 (1991).
[23] F. Orskov, *Acta Pathol. Microbiol. Scand. Sect. B* **84,** 319 (1976).

niently hydrolyzed by acid for compositional analysis. The reader is referred to other articles in this series for a more detailed discussion of NMR analysis of carbohydrates.[13,14]

Figure 2 shows an example of a $^{13}C-^{13}C$ P.E.COSY spectrum[17] obtained from 5 mg of 99% ^{13}C-labeled K14 polysaccharide [repeat unit ...→ 6)-β-D-Gal$p$NAc-(1 → 5)-β-D-KDOp-(2 → ..., where KDO is 2-keto-3-deoxyoctulosonic acid.][24] Even a cursory examination of the spectrum provides the following information. There are three cross-peaks with the carbonyl carbon signals, one of which correlates with the anomeric carbon signals (K1–K2). These indicate the presence of two acyl groups and one uronic acid in the repeat unit. The presence of only two upfield cross-peaks with the anomeric carbon signals (G1–G2, K2–K3) indicates that the repeat unit is a disaccharide. By tracing out the connectivities of the carbon signals, they all can be assigned. In the case of an unknown polysaccharide these data can provide a wealth of information about saccharide identity, anomeric configuration, and location of substituents.

Because ^{13}C enrichment greatly improves the sensitivity of the NMR experiment, structural analysis of small amounts of bacterial polysaccharide becomes feasible. This could be applied to the analysis of polysaccharide products of biosynthesis mutants or the analysis of structural variations on polysaccharide isolated from pathogens.

[24] B. Jann, P. Hofmann, and K. Jann, *Carbohydr. Res.* **120**, 131 (1983).

Section IV

Microbial Acquisition of Iron

[24] Deferration of Laboratory Media and Assays for Ferric and Ferrous Ions

By CHARLES D. COX

Introduction

Iron is important in human diseases. Its importance ranges from direct toxicity for human tissue to a controlling effect on the production of bacterial virulence factors. The measurement of iron and the ability to control iron concentrations are vital components to experimental strategies. Instead of measuring iron concentrations by atomic absorption spectroscopy, as chemists would be inclined to do, most investigators of pathogenesis use colorimetric assays. Therefore, this discussion will combine details of colorimetric assays with brief, simple reviews of the advantages of some alternative analytical methods. Finally, we review methods for media construction.

This chapter focuses on a mixed ecosystem: prokaryotic parasites infecting mammalian hosts. Any pathogenic interaction contains mixtures of host-associated and parasite-associated iron in possible ferric and ferrous forms. A complete understanding of pathogenesis demands accurate measurements of the components of these mixtures. During the analysis of the equilibrium between host- and parasite-associated iron, it is important to consider the history of iron acquisition systems. We begin with a theory that explains the predominant appearance of iron, out of all other metals, in reactions important to biology and pathogenesis. Early prokaryotic life and the initial events in prokaryotic parasitism of eukaryotic cells took place under iron-rich conditions. The abundance of certain transition metals, iron in particular, in prebiotic times probably determined their present predominance in enzymatic reactions of biological importance.[1] However, following the advent of oxygen-generating photosynthesis, this common element in the earth's crust[2] was oxidized to the ferric ion, which is sequestered in crystallized salts and in high molecular weight hydroxide precipitates.[3] The prokaryotic cells that survived this critical epoch appear today with elaborate mechanisms for solubilizing and acquiring extracellular iron.

---

[1] F. Egami, *J. Biochem.* (*Tokyo*) **77**, 1165 (1975).
[2] J. E. Zajic, "Microbial Biogeochemistry." Academic Press, New York, 1969.
[3] T. G. Spiro and P. Saltman, *Struct. Bonding* (*Berlin*) **6**, 116 (1969).

METHODS IN ENZYMOLOGY, VOL. 235

Extensive studies have described aerobic iron metabolism, in particular the iron chelators, called siderophores, produced by bacteria for the extracellular solubilization and membrane transport of iron.[4,5] In contrast, mammalian biology survived the same selective pressures of oxygen with an exquisite sensitivity to the toxicity of free iron, maintaining iron in complexes with the proteins transferrin and lactoferrin. These proteins, together with conalbumin from eggs, are similar bacteriostatic[6] agents and constitute the theory of nutritional immunity.[7] Additionally, serum transferrin appears to be only 30% saturated with iron in healthy humans. More quantitative research is needed on the effects of iron on interbacterial and bacteria–host relationships. An example of such a study revealed an enteric bacterium producing two siderophores, mobilizing iron from intracellular stores with one of its siderophores, aerobactin, and acquiring iron from extracellular protein complexes with the other, enterobactin.[8] This is a controversial finding, and only further research, perhaps involving new methods of iron detection, will result in a definition of the actual activities of and selective pressures for these siderophores.

Measurements of Iron Concentration and Iron Status

Biological Assays for Iron

Several new sensitive assays for iron are under development. One assay depends on the toxicity of Fe(II) in the presence of bleomycin, a microbial metabolite that demonstrates antitumor activity.[9] The Fe(II)–bleomycin complex cleaves DNA in a reaction that has been developed into an assay to detect minute amounts of iron in samples such as cerebrospinal fluid.[10] Current research concerns improving the assay and understanding the activity of bleomycin and analogs.

There are several bacterial regulatory genes that respond to environmental iron concentrations. A luxAB gene fusion in Escherichia coli employed the bacterial luminescence reaction to report the bacterial detection of Fe, Ni, Cu, and Al.[11] Additional iron-responsive regulatory genes which

[4] J. H. Crosa, Microbiol. Rev. 53, 517 (1989).
[5] W. S. Waring and C. H. Werkman, Arch. Biochem. 1, 425 (1942).
[6] A. L. Schade and L. Caroline, Science 100, 14 (1944).
[7] E. D. Weinberg, Physiol. Rev. 64, 65 (1984).
[8] J. H. Brock, P. H. Williams, J. Riceaga, and K. G. Wooldridge, Infect. Immun. 59, 3185 (1991).
[9] Y. Sugiura, T. Takita, and H. Umezawa, Met. Ions Biol. Syst. 19, 81 (1985).
[10] J. M. Gutteridge, Clin. Sci. 82, 315 (1982).
[11] A. Guzzo, C. Diorio, and M. S. DuBow, Appl. Environ. Microbiol. 57, 2255 (1991).

could yield reporter systems for environmental iron levels are *fur* in *E. coli*[12] and *Pseudomonas aeruginosa*[13] and *regA* in *P. aeruginosa*.[14]

Spectroscopy of Naturally Occurring Iron Complexes

The unpaired electrons along with bonding structures and angles determine the electronic spectra and paramagnetic behavior of iron complexes. Iron, through its versatile reactivity, forms different oxygen, nitrogen, and sulfur complexes that demonstrate characteristic visible and ultraviolet absorption spectra. In these complexes, iron is in close contact with bonding atoms called ligands. For example, ferric ion tends to react with oxygen ligands in catechols and hydroxamates. Ferrous ion tends to react with nitrogen ligands, found in pyridine or phenanthroline derivatives. The ligands or donor atoms are spatially arranged according to the electronic states of the metal, and the number of ligands yields a coordination number. Examples of the unfilled *d* orbitals and prominent ligand structures are displayed in Table I. These considerations are vital for detection, quantitation, and characterization of iron and iron complexes. In addition, iron reactivity, for example, the reactions involved in iron toxicity to human tissue, varies between the complexes with citrate, siderophores, and transferrin, because of the variations in the ligand fields in each structure.

Many naturally occurring iron complexes may be measured directly by spectrophotometry. The assay depends on electronic transitions, namely, the *d–d* transitions for iron and the transitions between ligand and metal. A good example of these transitions can be found in the studies of the iron complexes ferritin and phosvitin.[15] Table I displays the relative energies associated with the *d–d* transitions and their relationships to ligand structure. The octahedral complex of ferritin possesses a much smaller energy change in the $^6A_1$–$^4T_1$ electronic transition compared to the same transition in the tetrahedral complex common to phosvitin.[15] Spectrally, these electronic transitions can be observed in the red-brown color of ferritin, associated with a 900 nm absorbance maximum, and the green color of phosvitin, associated with shorter wavelength (i.e., 447, 426, and 400 nm) absorption maxima. This is a very simplistic view of these relationships, and other ligand structures are much more complex. These energetic concerns also reveal information about metal–ligand tran-

[12] S. Silver and M. Walderhaug, *Microbiol. Rev.* **56**, 195 (1992).
[13] R. W. Prince and M. L. Vasil, *Abstr. Am. Soc. Microbiol.* **92**, 103 (1992).
[14] D. W. Frank and B. H. Iglewski, *J. Bacteriol.* **170**, 4477 (1988).
[15] H. B. Gray, *in* "Proteins in Iron Storage and Transport in Biochemistry and Medicine" (R. R. Crichton, ed.), p. 3. North-Holland, Amsterdam, 1975.

TABLE I

ELECTRONIC AND LIGAND STRUCTURES OF IRON[a]

Parameter	Ferrous	Ferric
Designation of oxidation state	Fe(II)	Fe(III)
Valence state	2+	3+
$3d$ Orbital designation	d^6	d^5
Electron arrangement		
Low spin	(filled orbitals)	(filled orbitals)
High spin	(filled orbitals)	(filled orbitals)
Coordination number	4 — 5 — 6	4 — 5 — 7
Stereochemistry	Tetrahedral — Trigonal bipyramidal — Octahedral	Tetrahedral — Octahedral — Pentagonal bipyramidal
Example		Phosvitin — Ferritin EDTA
Lowest electronic energy levels of high-spin orbitals		Phosvitin: $^4E,\ ^4A_1$ —, $^4T_2$ —, $^4T_1$ —, $^6A_1$ — Ferritin: $^4E,\ ^4A_1$ —, $^4T_2$ —, $^4T_1$ —, $^6A_1$ —

[a] The formal designations of the oxidation states used in this chapter are associated with the valence states of iron. The electronic statuses of the 3d orbitals are associated with the spin arrangements. Finally, these considerations are related to ligand structures, stereochemistry, and spectroscopic behaviors. Two examples are noted for the relative energies of the electronic transitions between the $^6A_1$ and $^4T_1$ levels which have been associated with the different spectroscopic absorbances of phosvitin and ferritin.

sitions in both oxidized and reduced forms. For example, in cytochrome studies, the Sôret region (~420 nm) of the visible light spectrum can be used quantitatively. However, the α and β absorbance bands, from approximately 500 to 650 nm in reduced minus oxidized spectra, provide descriptions of the type of cytochrome and its oxidized or reduced state.

Colorimetric Reagents for Determining Ferric and Ferrous Ions

Most analyses involve addition of a variety of chelators to form chromogenic complexes with Fe(II). There are several benefits of the colorimetric analyses, particularly the ease of methodology and the selective measurements of Fe(II) and Fe(III) in the same environment. 1,10-Phenanthroline,[16] an Fe(II) chelator, has been largely replaced in iron assays by Ferrozine.[17] Ferrozine reagent [disodium 3-(2-pyridyl)-5,6-bis(4-phenylsulfonate)-1,2,4-triazine], used alone, reacts with the Fe(II) present in the environment, yielding a complex which has a molar extinction coefficient of 27,900 at 562 nm. Total iron can be obtained by reducing all of the iron in the sample with hydroxylamine,[17] and measuring the Ferrozine reaction. The difference between the measurements should yield the Fe(III) concentration.

The Ferrozine assay usually involves samples of 100 to 500 μl that can contain mixtures of siderophores, serum, or transferrin. Iron is released from the protein complexes by reduction with 500 μl of 0.2% (w/v) ascorbic acid in HCl.[18] The proteins are precipitated with 500 μl of 11.3% (w/v) trichloroacetic acid. It is most convenient to conduct the reactions in 2.0-ml microcentrifuge tubes so that the precipitates can be removed easily by centrifugation. The supernatants are removed to borosilicate tubes for mixing with 500 μl of 0.5 M ammonium acetate buffer and 100 μl Ferrozine–neocuproine reagent[18] (75 mg of each in 25 ml water). Neocuproine (2,9-dimethyl-1,10-phenanthroline) binds any copper also released from serum proteins. Copper would interfere with the chromogenic detection of ferrous ion, but copper–neocuproine absorbs at 457 nm. This assay has been sensitive to 1 to 2 nmol iron with linearity to 100 nmol through absorption measurements of both concentrated and diluted reactions.

Dry- and wet-ashing techniques[19] along with an acid–permanganate treatment have been described for releasing iron[20] from all organic complexes and insoluble compounds. A reexamination of the activities of

[16] A. Jensen and S. E. Jorgensen, *Met. Ions Biol. Syst.* **18**, 5 (1984).
[17] L. L. Stookey, *Anal. Chem.* **42**, 779 (1970).
[18] P. Carter, *Anal. Biochem.* **40**, 450 (1971).
[19] H. Beinert, this series, Vol. 54, p. 435.
[20] W. W. Fish, this series, Vol. 158, p. 357.

Ferrozine, bathophenanthroline disulfonate (BPDS), and ferricyanide reduction showed that BPDS was a more rapid reactant in enzymatic assays.[21] Another study[22] described a modification of the Ferrozine assay and compared it with atomic absorption spectrometry for iron values in human plasma samples containing EDTA.

The Ferrozine reagent has been widely used because of its sensitivity and its solubility in buffered biological systems. These factors also resulted in its use to assay "iron reductase," an enzymatic activity that reduces Fe(III) added in specific chelates to Fe(II).[23] The Fe–Ferrozine complex has been monitored continuously during enzymatic reactions.

Another common reagent is tripyridyl-s-triazine[24] (TPTZ), which can be used in water-soluble systems or with nitrobenzene extraction. The ferrous chelate has a molar extinction coefficient of 22,600 at 593 nm. Bathophenanthroline disulfonate, namely, 4,7-diphenyl-1,10-phenanthroline disulfonate (BPDS), is similarly used in aqueous reactions to form ferrous complexes with a molar extinction coefficient of 22,140 at 534 nm.[25] In both assays, modifications have been made for total iron-binding capacity measurements in complex systems, such as serum. The reactions have also been modified for measurements in a single tube.[26] The use of TPTZ and bathophenanthroline allow pH-dependent assessment of transferrin iron binding because the protein is active in binding only at pH 7.0 and above. Below pH 7.0, transferrin loses iron to the colored Fe(II) complexes. In some cases, particularly when turbid, colloidal, or precipitated samples are used, extraction of the Fe(II) complex is desired. Bathophenanthroline without the sulfonate group forms a ferrous complex with a molar extinction coefficient of 22,150 at 534 nm that can be extracted into an isoamyl alcohol layer.[27]

Although these are sensitive assays for ferrous ion, it must be remembered that there are problems with positive drift in ferrous assays. In addition, the predominance of ferrous iron is some tissues, such as plant tissue,[28] means that careful controls must be conducted to assure that

[21] A. Berczi, J. A. Sizensky, F. L. Crane, and W. P. Faul, *Biochim. Biophys. Acta* **1073**, 562 (1991).

[22] T. A. Walmsley, P. M. George, and R. T. Fowler, *J. Clin. Pathol.* **45**, 151 (1992).

[23] H. A. Dailey and J. Lascelles, *J. Bacteriol.* **129**, 815 (1977).

[24] J. A. O'Malley, A. Hassan, J. Shiley, and H. Traynor, *Clin. Chem.* **16**, 92 (1970).

[25] J. F. Goodwin, *Clin. Biochem.* **3**, 307 (1970).

[26] H. L. Willims and M. E. Conrad, *J. Lab. Clin. Med.* **67**, 171 (1966).

[27] K. A. Doeg and D. M. Ziegler, *Arch. Biochem. Biophys.* **97**, 37 (1962).

[28] S. C. Mehrotra and P. Gupta, *Plant Physiol.* **93**, 1017 (1990).

reduced iron is not a reaction artifact. Increasing the levels of ferrous chelators pulls the equilibrium to Fe(II) at ever greater rates.

High-Performance Liquid Chromatography

High-performance liquid chromatography (HPLC) has been applied to metal analysis. Metal chelates of dibenzyl dithiocarbamate (DBDC) have been made in water samples for separation on a C_{18} reversed-phase column using a solvent composed of 64% (v/v) methanol, 12% (v/v) acetonitrile, 5% (v/v) tetrahydrofuran, and 19% (v/v) acetate buffer, pH 5.[29] An ultraviolet (254 nm) monitor is used to monitor the effluent for metal chelates and unreacted chelator. Ten metals have been successfully analyzed: Cd, Cr, Cu, Hg, Ni, Pb, Sb, Se, Tl, and Zn. Although iron was not considered a toxic metal in this study, it should form a chelate and successfully elute from the column. Of the ions studied, only silver and arsenic acid fail to form detectable chelates.

In addition, stereospecific reactions between trivalent metals and the chelator N,N'-ethylenebis(2-o-hydroxyphenylglycine) were separated and detected on reversed-phase HPLC columns.[30] Because this area of investigation is in its infancy, many different chelators, including siderophores, have yet to be tried. A study of 16 fungal siderophores revealed separation on XAD-2 columns combined with successful detection.[31] Catechol siderophores from enteric bacteria have also been separated as iron-free compounds and were detected by ion-spray mass spectrometry.[32] Presumably, iron chelates of these siderophores could also be detected with different retention times. To this point, it is important to mention that a large metal chelator, metallothionein, has been isolated and detected by HPLC. Detection by atomic absorption spectrometry allowed metallothionein analysis together with metal analysis and quantitation.[33]

Although conventional stainless steel HPLC columns and silicic acid supports are notorious for metal contamination, mobile phase modifiers, such as nitrilotriacetic acid (NTA) and ethyleneglycol bis(β-aminoethyl ether)-N,N,N',N'-tetraacetic acid (EGTA), have been used in our laboratory to clean columns successfully and maintain low-iron effluents.

[29] J. H. Shofstahl and J. K. Hardy, *J. Chromatogr. Sci.* **28**, 225 (1990).
[30] S. L. Madsen, C. J. Bannochie, A. E. Martell, C. J. Mathias, and M. J. Welch, *J. Nucl. Med.* **31**, 1662 (1990).
[31] S. Konetschny-Rapp, H. G. Huschka, G. Winkelmann, and G. Jung, *Biol. Metals* **1**, 9 (1988).
[32] I. Berner, M. Greiner, J. Metzger, G. Jung, and G. Winkelmann, *Biol. Metals* **4**, 113 (1991).
[33] M. P. Richards, *J. Chromatogr.* **482**, 87 (1989).

Atomic Absorption Spectroscopy

The most accepted technique for metal analyses is atomic absorption spectroscopy (AAS). Flame AAS yields total iron values down to the range of 0.2 ng/ml. The standard flame spectrometer has evolved to the more sensitive furnace AAS[34] with quantitation possible to the 10 pg level. The AAS technique has been used on many biological samples including ashed or lyophilized needle aspirates of human liver biopsy material. Methods have been developed for the specific use of flame and furnace AAS to determine specific metals. The sample is burned in a flame or heated in a graphite boat to yield an atomic vapor through which the specific emission wavelength from the excited metal ion is passed. Because exquisite sensitivity is not required for most iron determinations, flame AAS has been used most frequently. Modifications of these methods allow determinations of iron overload, total iron binding capacity, and the levels of iron bound to transferrin. Unfortunately, iron contamination is always a problem owing to the abundance of the metal in the environment.

A new development of this technology is the simultaneous multielement atomic absorption continuous source spectrometer (SIMAAC).[34] In this case a high-intensity xenon light source is chromatically refined so that changing elements is not necessary. The name of the instrument indicates the ease of use for multielement analysis.

Flame Atomic Emission Spectrometry

Flame emission or flame photometry, used to determine the identity and concentration of metals,[35] requires high flame temperatures to produce excited state ions in the sample. A spectrometer measures light emitted at 372 nm to determine iron quantitatively at the 10 ng/ml level.

Inductively Coupled Plasma-Optical Emission Spectroscopy

Inductively coupled plasma-optical emission spectroscopy (ICP-OE) is based on transformation of the metal into a plasma (charged ion cloud). The plasma is formed by radiofrequency generated high-frequency currents flowing through an induction coil.[36] The plasmas reach temperatures of 10,000 K and are kept moving by argon, flowing at approximately 1 liter/min. Coolant argon also flows around the coil at 10 to 20 liters/min. Plasma emissions are detected by scanning monochrometers or photodiode array detectors. The range of sensitivity for iron, depending on the

[34] W. Slavin, this series, Vol. 158, p. 117.
[35] T. H. Risby, this series, Vol. 158, p. 180.
[36] K. A. Wolnik, this series, Vol. 158, p. 190.

method of sample administration, is between 0.09 and 0.2 ng/ml. Another variation of this instrument is the inductively coupled plasma-mass spectrometer (ICP-MS). The plasma formed is injected into a mass spectrometer for analysis. Future developments will reduce analytical costs and combine these methods for detection and quantitation of samples separated by HPLC.

Atomic Fluorescence Spectroscopy

As in the previous case, the atomic fluorescence spectrometer also creates a plasma, but detection of ions is achieved by irradiating the plasma with light characteristic of the spectral line of the metal under investigation. A spectrophotometer, at a right angle to the light source, measures the fluorescence emission of the excited atoms. The sensitivity for iron using this technique would be in the range of 2 to 5 ng/ml.[37]

Isotopes of Iron

Two forms of radioactive iron can be used to trace iron through biological systems. Iron-55 decays by electron capture with a half-life of 2.94 years. It can be counted by scintillation counters using the preset tritium channel at approximately 40% efficiency. Many different methods of analysis can be envisioned. For example, this radionuclide has been used to trace the active accumulation of both ferric and ferrous forms of iron by *Bacteriodes* species.[38] On the other hand, iron-59 decays by γ emission, has a half-life of 45 days, can be counted with γ or scintillation counters, and is more suitable for autoradiography. Iron-59 is often used for systems requiring greater sensitivity. In all of these iron tracer systems, the background levels of iron-56 must be appreciated so that total iron accumulation can be accurately determined. Tracing iron at picomole levels is typical. Because the energies of the two decay processes are so different, the two isotopes can be easily employed for dual-label studies.

Neutron Activation Analysis

Neutron bombardment of a complex sample results in the generation of radioactive elements. In the case of iron, iron-59 is formed, and a sensitive Geiger counter detects the γ emission.[39] Multielement determinations require high purity germanium detectors. These data are quantitative, can be used on native materials without denaturation, and have diminished

[37] R. G. Michel, this series, Vol. 158, p. 222.
[38] E. R. Rocha, M. deUzeda, and J. H. Brock, *FEMS Microbiol. Lett.* **68,** 45 (1991).
[39] J. Versieck, this series, Vol. 158, p. 267.

problems with contamination from introduced iron. Although the sensitivity for iron is low, approximately 3 μg, the technique is very useful for checking concentrations of metals which have been determined by more destructive analytical methods.

Electronic Paramagnetic Resonance Spectroscopy

Electron paramagnetic resonance (EPR) spectroscopy is valuable for revealing structural, physical, and quantitative information about iron.[40] The single unpaired electron in the $^4T_2$ configuration of low-spin Fe(III) displays a spectroscopic splitting factor (*g* factor) when placed in a spectrometer under a magnetic field (Table I). Obviously, the high-spin d^5 Fe(III) electrons reveal additional complexities when placed in a magnetic field. However, EPR can be used to describe the proteinaceous environments of reactive iron, polynuclear iron complexes, and coordination structures.

In addition to the study of iron–ligand interactions, EPR is being used to study the involvement of iron in free radical generation and the resultant toxicity to mammalian tissues.[41] Features of this research involve the study of iron in these reactions and the spin trapping of the radicals.[42] Free radical-generating reactions depend on available iron, and, in one instance, siderophore-bound iron has been found active in these toxic reactions.[43] Additional iron assays may be developed from these investigations.

Mössbauer Spectroscopy

Mössbauer spectroscopy measures the electronic resonances of ^{57}Fe irradiated by the γ rays of ^{57}Co decay.[44] Because ^{57}Fe is 2.2% in natural abundance, a high concentration of sample or enrichment of the sample with ^{57}Fe is required. This technique has been used to determine the oxidation states of iron in proteins and to determine the number of iron atoms per active site and the number of iron clusters per protein. For example, Mössbauer spectroscopy was combined with other techniques to describe the Fe(II) activities in isopenicillin N synthase.[45]

[40] G. Palmer, *Adv. Inorg. Biochem.* **2,** 153 (1980).
[41] J. T. Flaherty, *Am. J. Med.* **91,** 795 (1991).
[42] C. L. Ramos, S. Pou, B. E. Britigan, M. S. Cohen, and G. M. Rosen, *J. Biol. Chem.* **267,** 8307 (1992).
[43] T. J. Coffman, C. D. Cox, B. L. Edeker, and B. E. Britigan, *J. Clin. Invest.* **86,** 1030 (1990).
[44] B. H. Huynh and T. A. Kent, *Adv. Inorg. Biochem.* **6,** 163 (1984).
[45] L. J. Ming, L. Que, A. Kriauciunas, C. A. Frolik, and V. J. Chen, *Biochemistry* **30,** 11653 (1991).

Magnetic Resonance Imaging

Magnetic resonance imaging (MRI) is being combined with iron stains to detect neoplasms[46] and bacterial infections.[47] Iron stains such as super-paramagnetic iron oxide nanoparticles have been used to explore iron in Kupffer cells of spleen and liver[48] in techniques which may have applications to studies of pathogenesis.

Ion Microscopy

Electron microscopy has been developed into a powerful analytical tool for elemental analysis, utilizing the primary ion beam to sample photons, electrons, neutral atoms, molecules, and ions sputtered from bombarded samples.[49] Improved resolution has resulted in typical sampling cross-sections in the 0.01 μm range. Scanning transmission electron microscopy (STEM)[50] has been combined with X-ray diffraction analysis of biopsy material containing suspected fungal hyphae. The analysis revealed the filaments to be composed mainly of iron, phosphorus, and calcium.[51] Similarly, scanning electron microscopy has been coupled with energy-loss spectroscopy (EELS)[50] at high beam energies to achieve single atom sampling. A beam energy of 100 keV was used to determine four Fe atoms in a single 65-kDa hemoglobin molecule.[52]

Deferration of Laboratory Media

Iron Removal from Labware

Glass is a notorious ion-exchange surface. Long-term problems are posed by the ion exchange from the deeper layers, continually contaminating the surface with iron. Therefore, concentrated nitric acid treatment for 4 hr followed by extensive washing in distilled, deionized water must be conducted periodically even if the glassware is unused. Other acid washing solutions may be used, but nitric acid poses the least interference with subsequent fluorometric analyses for aromatic compounds.

[46] C. Borgna-Pignatti and A. Castriota-Scanderberg, *Haematologica* **76**, 409 (1991).
[47] J. L. Stone, G. R. Cybulski, M. E. Gryfinski, and R. Kant, *J. Neurosurg.* **70**, 879 (1989).
[48] R. Weissleder, A. S. Lee, B. A. Khaw, T. Shen, and T. J. Brady, *Radiology* **182**, 381 (1991).
[49] S. Chandra and G. H. Morrison, this series, Vol. 158, p. 157.
[50] J. J. Bozzola and L. D. Russell, "Electron Microscopy, Principles and Techniques for Biologists," p. 331. Jones and Bartlett, Boston, 1992.
[51] J. Connelly, J. Y. Ro, and J. Cartwright, *Arch. Pathol. Lab. Med.* **115**, 1166 (1991).
[52] R. D. Leapman and S. B. Andrews, *J. Microsc.* **165**, 225 (1992).

Dialysis tubing is another source of metal contamination. Tubing should be soaked in 1% acetic acid for 1 hr, followed by washing in distilled, deionized water. Metals are removed by treatment in three changes of 1% (w/v) sodium carbonate and 1 mM ethylenediaminetetraacetic acid (EDTA), heating each change of the solution to 75°. Finally, the EDTA is washed out by three changes of heated 1% sodium carbonate. The sodium carbonate is removed by extensive water washes. Complaints about EDTA contamination of dialysis tubing usually results from the inadequate attempts to remove EDTA in water. The tubing is best stored in a plastic bottle with a few drops of chloroform added to prevent bacterial growth.

Iron Removal from Water

Controlling iron concentrations in media starts with removing available iron from the water source. The most common approach is distillation. Distilled and double-distilled water can be stored in plastic containers for a period of 1 week. It is common for bacterial growth to reach levels of 10^4 colony-forming units (cfu)/ml within 3 weeks of storage in glass or plastic containers. Most modern laboratories have "house-distilled" water that magically appears at the tap and tragically causes problems. Common problems, occurring in the still, the pipes, and the storage reservoir, can be easily monitored by observation. Contaminated water often involves easily observable storage problems, from corroding metal surfaces to dead rodents in the storage reservoir. Most laboratories using water for cell culture, HPLC, or metal analyses subject distilled water to an ion-exchange system. The Milli-Q (Millipore, Bedford, MA) and NANOpure (Barnstead Thermolyne, Dubuque, IA) systems very conveniently yield water of high quality. Both systems have conductivity meters that indicate to the operator the conductive quality of the water. Water and media for low- or controlled-iron studies should be sterilized by filtration rather than by autoclaving.

Iron Removal from Media

Addition of iron during the construction of bacteriological media is unavoidable, especially with the addition of phosphate and carbonate anions. Once added, these contaminants must either be removed or quantitatively realized by assay. The trivial term, deferration, has been coined to describe these processes. Waring and Werkman[5] originally used an excess of 8-hydroxyquinoline to bind Fe(III) in media with a binding coefficient of 10^{38}. The excess reagent and chelate were then removed by exhaustive chloroform extraction. This technique is truly exhausting if done properly, that is, if the 8-hydroxyquinoline level is assayed to determine complete removal. Residual quinoline may cause problems in subsequent iron assays, bacteriological reactions, and fluorescence assays. Diethylenetriaminepentaacetic acid (DTPA) has been recommended in a similar method to remove iron by extraction or for iron assay.[53]

A similar procedure for the removal of as many as 23 elements employs dithizone (diphenylthiocarbazone).[54] This reagent turns red when forming coordination complexes with metals. The compound and the chelates are insoluble in water and can be extracted into chloroform or carbon tetrachloride. Oxygen sensitivity is a problem, and recrystallization of the reagent is recommended.

Conversely, a straightforward mechanism for iron removal involves adding magnesium carbonate (5 g/liter), stirring for 5 minutes, and removing the insoluble magnesium carbonate by centrifugation and filtration. This technique reduces iron levels from 2.3 to 0.4 μM in a relatively complex medium. A problem may occur with bacteria unable to tolerate the magnesium or carbonate ions that saturate the solution from the low solubility of the magnesium carbonate.

Chelex 100 is a styrene–divinylbenzene copolymer sold by Bio-Rad (Richmond, CA) and is commonly employed for removing iron. Most investigators employ the resin directly from the bottle in batch treatments of media just prior to autoclaving. However, there are unsubstantiated rumors that the imidodiacetate groups, active in metal binding, may exist as free iron-binding compounds in some older bottles of resin. The best procedure is a washing of the resin through an acid cycle immediately before use. This process can be rapidly accomplished on a Büchner funnel, mixing 1 volume of resin with 2 volumes of 1 M HCl. The slurry is then washed with 5 volumes of deionized water. The sodium form is generated by washing with 2 volumes of 1 M NaOH. Finally, the sodium hydroxide is washed from the resin using 5 volumes of clean water. The product is typically sold in the sodium form and should be very active.

Intrinsic Iron Removal

It may be impossible to use these procedures for some complex media used for growing fastidious bacteria. A variety of chelating agents can be added to media to sequester the iron away from bacteria. Most pertinent to this discussion is the addition of purified transferrin as an iron-chelating, growth-inhibiting agent.[55] Transferrin has been used to inhibit the growth

[53] P. E. Powell, G. R. Cline, C. P. P. Reid, and P. J. Szanezlo, *Nature (London)* **287**, 833 (1980).

[54] B. Holmquist, this series, Vol. 205, p. 7.

[55] J. J. Bullen, H. J. Rogers, and J. E. Lewin, *Immunology* **20**, 391 (1971).

of a variety of bacteria,[6] including *Salmonella,*[56] *E. coli,*[57] *Serratia marcescens,*[58] and *Pseudomonas aeruginosa.*[59] Although more expensive, lactoferrin has also been shown to be inhibitory for *E. coli.*[60,61] Ovotransferrin or conalbumin also has a broad range of antibacterial activity.[62,63] A theoretical problem with using these antibacterial proteins is the bactericidal activity that has also been reported for transferrin and lactoferrin against both gram-positive and gram-negative bacteria.[64] Therefore, these proteins may have other activities, also inhibitable by iron, that contribute to antibacterial phenomena.

One of the most widely used bacteriostatic agents is ethylenediamine-di(*o*-hydroxyphenylacetic acid) (EDDA, EDDHA, or EHPG). EDDA is preferred to EDTA because it interferes with bacterial iron metabolism with less bacteriolytic activity at levels up to 200 μM. EDDA, contrary to transferrin and lactoferrin, binds Fe(II) with a binding coefficient of 10^{38}. Fe(III) is also bound tenaciously. The toxicity of these chelates is of some concern, and recrystallization of the compound is recommended before use.[65] EDDA has been used as a bacteriostatic agent against iron metabolism in *E. coli,*[65] *Pseudomonas* species,[66,67] and *Staphylococcus aureus.*[68,69] A paper disk assay was also developed to display the abrogation of EDDA-mediated growth inhibition by crude and purified bacterial siderophores.[70]

Another Fe(II)-binding bacteriostatic agent, namely 2,2'-dipyridyl, has been used against *E. coli.*[56,71,72] Evidence has suggested that 2,2'-dipyridyl acts mainly against the activity of enterochelin, whereas transferrin inhibi-

[56] R. J. Yancey, S. A. L. Breeding, and C. E. Lankford, *Infect. Immun.* **24**, 174 (1979).
[57] M. DerVartanian, *Infect. Immun.* **56**, 413 (1988).
[58] W. H. Traub, *Exp. Cell Biol.* **45**, 184 (1977).
[59] R. Ankenbauer, S. Sriyosachati, and C. D. Cox, *Infect. Immun.* **49**, 132 (1985).
[60] M. D. Drew, I. M. Bevandick, and B. D. Owen, *Can. J. Anim. Sci.* **70**, 647 (1990).
[61] P. Rainard, *Vet. Microbiol.* **11**, 103 (1986).
[62] S. M. Payne and R. A. Finkelstein, *J. Clin. Invest.* **61**, 1428 (1978).
[63] L. M. Simpson and J. D. Oliver, *Infect. Immun.* **41**, 644 (1983).
[64] R. T. Ellison and T. J. Giehl, *J. Clin. Invest.* **88**, 1080 (1988).
[65] H. J. Rogers, *Infect. Immun.* **7**, 445 (1973).
[66] P. A. Vandenberg, C. F. Gonzalez, A. M. Wright, and B. S. Kunka, *Appl. Environ. Microbiol.* **46**, 128 (1983).
[67] D. Hohnadel, D. Haas, and J.-M. Meyer, *FEMS Microbiol. Lett.* **36**, 195 (1986).
[68] J. P. Maskell, *Antonie van Leeuwenhoek* **46**, 343 (1980).
[69] J. H. Marcelis, H. J. Den Daas-Slagt, and J. A. A. Hoogkamp-Korstainje, *Antonie van Leeuwenhoek* **44**, 257 (1978).
[70] M. Luckey, J. R. Pollack, R. Wayne, B. N. Ames, and J. B. Neilands, *J. Bacteriol.* **111**, 731 (1972).
[71] B. A. Ozenberger, M. S. Nahlik, and M. A. McIntosh, *J. Bacteriol.* **169**, 3638 (1987).
[72] M. A. McIntosh and C. F. Earhart, *J. Bacteriol.* **131**, 331 (1977).

tion is exerted against the activity of aerobactin.[73] In addition, nitrilotri-acetic acid is another chelator binding predominantly Fe(II) and has been utilized against siderophore-defective mutants of *E. coli*.[74] The activities of many of these compounds remain unresolved, and minimal controls, such as the reversal of inhibition by iron overload, are essential for confirming the basis of iron deprivation.

In all iron experiments, the concentration of iron must be controlled. The problem with this dictum is that this abundant metal must first be removed from the growth medium and then be supplied in the proper, controlled form. In addition, the metal must be analyzed for concentration and valence state. Therefore, analysis must always be used to confirm the activity of the deferration methods employed to construct culture media.

[73] M. derVartanian, *Infect. Immun.* **56**, 413 (1988).
[74] L. Langman, I. G. Young, G. E. Frost, H. Rosenberg, and F. Gibson, *J. Bacteriol.* **112**, 1142 (1972).

[25] Detection, Isolation, and Characterization of Siderophores

By SHELLEY M. PAYNE

Introduction

Siderophores, low molecular weight high-affinity ferric iron chelators, are synthesized and secreted by many microorganisms in response to iron deprivation.[1] These compounds solubilize and bind iron and transport it back into the microbial cell, usually through specific membrane receptors.[2] Some siderophores have the ability to remove iron from mammalian iron-binding proteins, and the ability to express these iron transport systems has been associated with bacterial pathogenicity.[3]

Most siderophores may be classified as either hydroxamates or phenolates–catecholates, although a few have been isolated which do not fall into these classes. A variety of techniques have been developed to detect, to isolate, and to characterize siderophores. This chapter is a summary

[1] C. E. Lankford, *Crit. Rev. Microbiol.* **2**, 273 (1973).
[2] J. B. Neilands, *Annu. Rev. Microbiol.* **36**, 285 (1982).
[3] S. M. Payne, *Crit. Rev. Microbiol.* **16**, 81 (1988).

of those techniques commonly used in working with siderophores and is primarily an updating of two excellent reviews of methods by Neilands.[4,5]

Growth Conditions

Because siderophore synthesis is regulated by iron and is often influenced by other environmental factors, growth conditions are important for maximizing siderophore production.[6] The medium should be deferrated to remove excess iron and prevent repression of siderophore synthesis. Techniques for deferration of growth media are discussed in [24] of this volume. Conditions which increase the requirement of the cell for iron or further reduce the availability of iron will enhance siderophore yields. These include use of carbon sources such as succinate or lactate instead of glucose, addition of manganese to the medium, and increased aeration of the culture during growth. Growth temperature is also important, and siderophore production may decrease with increasing temperature. In *Salmonella typhimurium*, for example, it was found that enterobactin production was greater at 31° than at 37° and was not detected at 40.3°.[7] Cultures should be grown to stationary phase, since maximum siderophore production often occurs during late log and early stationary phase. Monitoring and adjustment of culture pH during growth may be necessary; acidic conditions increase the solubility of iron and reduce siderophore production, while high pH may cause breakdown of catechols.

In general, cultures are passaged in defined medium containing a suboptimal concentration of iron. The cells are subcultured in this iron-deficient medium and grown with aeration at or below the optimal growth temperature until the cells reach stationary phase. Sufficient iron starvation to induce siderophore synthesis is confirmed by monitoring absorbance of the culture during growth; the iron-starved culture should show a slower growth rate and/or a lower final cell density than a duplicate culture containing excess iron.

Detection of Siderophores

With the exception of the cell wall-associated mycobactins, siderophores are generally found in the culture supernatant. Cells are removed

[4] J. B. Neilands, *in* "Development of Iron Chelators for Clinical Use" (A. E. Martell, W. F. Anderson, and D. G. Badman, eds.), p. 13. Elsevier/North-Holland, New York, 1981.
[5] J. B. Neilands, *Struct. Bonding* **58**, 1 (1984).
[6] A. Bagg and J. B. Neilands, *Microbiol. Rev.* **51**, 509 (1987).
[7] J. A. Garibaldi, *J. Bacteriol.* **110**, 262 (1972).

by centrifugation, and the supernatant is assayed for the presence of the compounds. The assays are based on chemical or biological properties of the siderophores. A positive assay often provides not only a means of detecting the siderophore during purification but also considerable information about the structure of the compound.

Siderophore Assays Based on Chemical Properties

Addition of Iron

Many of the siderophores form distinctive colored complexes with iron. Their presence, therefore, can often be detected by the formation of a colored compound when iron is added to the culture supernatant.[5] Iron may be added either as ferric or ferrous salts, since the ferrous iron is rapidly converted to the ferric form on complexing with the siderophore. Although many siderophores can be detected by this method, it is the least sensitive means of detection.

Ferric Perchlorate Assay

Hydroxamates or other siderophores capable of forming stable iron complexes at low pH can be detected with ferric perchlorate in perchloric acid.[8] Catechols typically do not react in this assay since the iron is dissociated from these compounds at low pH.

Reagent

5 mM Fe(ClO$_4$)$_3$ in 0.1 M HClO$_4$

Procedure. Mix 0.5 ml culture supernatant with 2.5 ml ferric perchlorate reagent. Complexes are orange to purple in color. The absorbance is measured at a wavelength appropriate for the color formed by a particular siderophore.

Csáky Assay

The Csáky assay is also used to detect hydroxamates.[9] It is more sensitive than the ferric perchlorate assay but is more qualitative than quantitative. The assay detects the presence of secondary hydroxamates and depends on oxidation to nitrite and formation of a colored complex via diazonium coupling.

[8] C. L. Atkin, J. B. Neilands, and H. J. Phaff, *J. Bacteriol.* **103**, 722 (1970).
[9] T. Csáky, *Acta Chem. Scand.* **2**, 450 (1948).

Reagents

Sulfanilic acid: 1 g sulfanilic acid dissolved by heating in 100 ml 30% acetic acid (v/v)

Iodine solution: 1.3 g iodine in 100 ml glacial acetic acid

Sodium arsenite: 2 g Na_3AsO_4 in 100 ml water

Sodium acetate: 35 g sodium acetate in 100 ml water

α-Naphthylamine solution: 3 g α-naphthylamine dissolved in 1000 ml of 30% acetic acid

Procedure. The siderophore solution or supernatant (1 ml) is hydrolyzed with 1 ml of 6 N H_2SO_4 in a boiling water bath for 6 hr or at 130° for 30 min. The solution is then buffered by adding 3 ml sodium acetate solution. Add 1 ml sulfanilic acid solution followed by 0.5 ml iodine solution. After 3–5 min, excess iodine is destroyed with 1 ml sodium arsenite solution. Add 1 ml α-naphthylamine solution. Add water to 10 ml, allow color to develop for 20–30 min, and measure the absorbance at 526 nm.

Arnow Assay

The colorimetric Arnow assay is useful for the detection of catechol siderophores in culture supernatants.[10] It is preferable to grow the bacteria in a low-iron minimal medium as some broth media will produce a high background in this test.

Reagents

0.5 N HCl

Nitrite–molybdate reagent: 10 g sodium nitrite and 10 g sodium molybdate dissolved in 100 ml water

1 N NaOH

Procedure. To 1 ml of supernatant or siderophore solution, add the following in order, mixing after each: 1 ml HCl, 1 ml nitrite–molybdate (catechols produce a yellow color at this point), and 1 ml NaOH (color should change to red). The color is stable for at least 1 hr, and the solution has an absorption maximum at 510 nm.

Functional Assays of Siderophores

Because the Csáky and Arnow assays are based on chemical properties of the siderophores, compounds which did not fall into the hydroxamate or catechol class may not be detected. Assays which rely on functional or biological properties provide a more general method for detection of

[10] L. E. Arnow, *J. Biol. Chem.* **118,** 531 (1937).

siderophores. Two that are widely used are the Chrome azurol S assay and the bioassay.

Chrome Azurol S Assay

The chrome azurol S (CAS) assay developed by Schwyn and Neilands[11] uses a iron–dye complex which changes color on loss of iron. Siderophores, which have higher affinity for the iron than does the dye, can remove the iron, resulting in a change in color of the dye from blue to orange. This is the most universal assay developed thus far for siderophores; it depends only on the ability of the compound to bind iron with relatively high affinity. The iron–dye complex used to detect the siderophore can be incorporated into agar media or can be used in a liquid assay.

Medium Components

10× MM9

KH_2PO_4	3 g
NaCl	5 g
NH_4Cl	10 g
Water	1 liter

Autoclave.

CAS–HDTMA

Chrome azurol S	605 mg
Water	500 ml

Dissolve. Add 100 ml 1 mM $FeCl_3 \cdot 6H_2O$ in 10 mM HCl.
Add slowly to hexadecyltrimethylammonium bromide (HDTMA) solution (729 mg HDTMA in 400 ml water).
Autoclave.

Deferrated casamino acids: Dissolve 10 g casamino acids in 100 ml water. Extract the casamino acids solution with an equal volume of 3% 8-hydroxyquinoline in chloroform to remove contaminating iron. Extract with an equal volume chloroform to remove traces of 8-hydroxyquinoline.

CAS agar

Water	750.0 ml
NaOH	6.0 g (dissolved in the water)
PIPES	30.24 g
10× MM9	100.0 ml
Bacto-agar (Difco, Detroit, MI)	15.0 g

Autoclave and cool to 50°.

[11] B. Schwyn and J. B. Neilands, *Anal. Biochem.* **160,** 47 (1987).

Add the following:

Deferrated casamino acids	30 ml
20% Glucose (or other carbon source)	10 ml
1 M MgCl$_2$	1 ml
100 mM CaCl$_2$	1 ml

Vitamin and amino acid supplements if required

Mix. Add 100 ml CAS–HDTMA solution. Mix gently to avoid foaming.

Pour plates. Colonies of siderophore-producing bacteria grown on this medium will be surrounded by a yellow or orange halo.

Chrome Azurol S Liquid Assay

Reagents

CAS assay solution

2 mM CAS stock solution: 0.121 g CAS in 100 ml water

1 mM Fe stock solution: 1 mM FeCl$_3$ · 6H$_2$O in 10 mM HCl

Piperazine buffer: dissolve 4.307 g piperazine in 30 ml water. Add 6.75 ml concentrated HCl to bring the pH to 5.6

HDTMA: dissolve 0.0219 g HDTMA in 50 ml water in a 100-ml mixing cylinder.

Mix 1.5 ml Fe solution with 7.5 ml CAS solution and add to the HDTMA in the mixing cylinder.

Add piperazine solution to the mixing cylinder and bring volume up to 100 ml with water.

Shuttle solution: 0.2 M 5-Sulfosalicylic acid. Store in the dark.

Procedure. Grow culture to stationary phase in minimal medium with reduced phosphate, for example, MM9 (above, buffer with 100 mM PIPES or Tris) or T medium (below). Excess phosphate or other weak iron chelators should be avoided as they interfere with the reaction. Add 0.5 ml CAS assay solution to 0.5 ml culture supernatant and mix. Then add 10 μl of shuttle solution and mix. Let stand a few minutes. Siderophores, if present, will remove iron from the dye complex, resulting in a reduction in blue color of the solution. Measure the absorbance (A_{630}) for loss of blue color. At higher siderophore concentrations, the color of the ferric siderophore may interfere with absorbance measurements. In this case, dilute the supernatant and repeat to get a more accurate reading. For A_{630} measurements, use the minimal medium as a blank, and use the minimal medium plus CAS assay solution plus shuttle as a reference (r). The sample (s) should have a lower reading than the reference. Siderophore units are defined as $[(A_r - A_s)/A_r]100 = \%$ siderophore units.

Bioassays

Bioassays are the most sensitive assays for siderophores. The specificity of the assays is determined by the choice of indicator strain. In general, the indicator is seeded into a low-iron agar medium, and the compound or supernatant to be tested is placed on a sterile filter disk or into a well cut in the agar. The size of the zone of growth of the indicator strain around the siderophore will be proportional to the siderophore concentration.

Aureobacterium (Arthrobacter) flavescens Bioassay

The following assay is primarily used for the detection of hydroxamates. *Aureobacterium flavescens* is relatively promiscuous with regard to transport of hydroxamates but does not appear to respond to catechol siderophores. The procedure detailed below was developed in the laboratory of Szaniszlo[12] and is similar to that published by Antoine *et al.*[13]

Strain

Aureobacterium flavescens JG9 (ATCC, Rockville, MD)

Media

Assay medium, per liter

K_2HPO_4	2.0 g
$(NH_4)_2HPO_4$	0.5 g
$MgSO_4 \cdot 7H_2O$	0.1 g
Yeast extract	1.0 g
Vitamin-free casamino acids	1.0 g
Sucrose	10.0 g

For solid medium, add 10 g Gelrite (Kelco, San Diego, CA) or 20 g agar (Gelrite produces a clearer background, making it easier to measure zones of growth).

Maintenance medium: Assay medium plus 0.2 μM desferrioxamine B. Add 20 ml of a 10 $\mu g/ml$ stock solution to 1 liter of maintenance medium.

Procedure. Streak *A. flavescens* onto maintenance medium and incubate for 2 days at 25°. Pick colonies which are medium size, smooth edged, yellow, and glistening and transfer to 100 ml of liquid maintenance medium. Grow at 25° on a rotary shaker (250 rpm) for 48 hr. Pellet cells at 10,000 rpm for 15 min at 4° and resuspend in 25 ml saline. Wash three

[12] C. B. Frederick, P. J. Szaniszlo, P. E. Vickrey, M. D. Bentley, and W. Shive, *Biochemistry* **20**, 2432 (1981).
[13] A. D. Antoine, N. E. Morrison, and J. H. Hanks, *J. Bacteriol.* **88**, 1672 (1964).

times, resuspending in 100 ml assay medium to a density of 100 Klett units after the last centrifugation. Incubate at 25°, 250 rpm, for 48 hr. Centrifuge and wash cells four times, then resuspend to 200 Klett units in assay medium. Keep the cell suspension at 4° until used. Add cells to solid assay agar medium (1%, v/v, inoculum), mix thoroughly, and dispense 20 ml per plate. Allow plates to harden and place siderophores or culture supernatants on sterile filter paper disks on the surface of the plates. Incubate plates at 25° for 48–72 hr and examine for growth of *A. flavescens* around the disk.

A more quantitative liquid assay also may be used.[13] The siderophore or culture supernatant to be tested is filter sterilized and added in a volume of 100 μl to 10 ml of liquid assay medium. Add 50 μl of the washed cell suspension and incubate at 27° with shaking in a tube with a slightly loosened cap. Measure the optical density (A_{500}) over a 72-hr period. Without added siderophore, the A_{500} should remain in the 0.04–0.05 range.

General Bioassay

Aureobacterium flavescens will only detect siderophores for which it has the appropriate receptor. Thus many siderophores, especially catechols, will not be recognized in this assay. A more general method is to use as an indicator the same strain from which the siderophore is to be isolated. In most cases, this will involve using a wild-type strain which synthesizes and transports the siderophore. It is necessary to perform the assay under conditions where the strain is severely iron limited and cannot grow in the absence of exogenously added siderophore. The following procedure,[14] based on that developed by Miles and Kjimji,[15] is used for this purpose.

Media

Deferrated ethylenediaminedi(*o*-hydroxyphenylacetic acid) (EDDA): Excess iron is removed from the EDDA by the procedure of Rogers.[16] Four grams of EDDA is added to 75 ml of 1 *N* HCl which has been brought to a boil. The solution is stirred with a nonmetal stirrer until the EDDA is completely dissolved, then the solution is returned to the heat until it boils. The cooled solution is filtered through a Whatman (Clifton, NJ) No. 1 filter to remove precipitated iron, and 600 ml of reagent grade acetone is added to the filtrate.

[14] S. M. Payne, *J. Bacteriol.* **143,** 1420 (1980).
[15] A. A. Miles and P. L. Kjimji, *J. Med. Microbiol.* **8,** 477 (1975).
[16] H. J. Rogers, *Infect. Immun.* **7,** 445 (1973).

The pH is adjusted to 6.0 with 1 N NaOH. The EDDA is allowed to precipitate overnight in the cold. The precipitated EDDA is collected by filtration through a Whatman No. 1 filter, and the EDDA on the filter is washed with cold actone. The EDDA is dried completely and stored in a plastic container. It should be very pale in color. A working stock (50 mg/ml) is prepared by adding 1 g of the deferrated EDDA to 15 ml 1 N NaOH. The pH is adjusted to 9.0 with concentrated HCl. Water is added to 20 ml, and the solution is filter sterilized and stored in sterile plastic tubes.

EDDA agar: Deferrated EDDA (10 to 1000 μg/ml) is added to molten L agar (per liter: 10 g Bacto-tryptone, 5 g yeast extract, 10 g NaCl, and 15 g agar or 10 g Gelrite). The EDDA agar is allowed to cool and is held at 4° for 24 hr prior to use.

Alternative medium, dipyridyl agar: The iron chelator 2,2′-dipyridyl is added to agar media at a final concentration of 100–400 μM.

Procedure. The EDDA or dipyridyl agar is melted, seeded with the indicator strain at low density (10^3 to 10^4 cells/ml), and poured into dishes. Cultures are tested for siderophore production by spotting 10 μl of a fully grown broth culture directly on the surface of the hardened agar. Siderophore solutions or culture supernatants are tested by placing 10 μl on a sterile filter paper disk on the agar surface or by filling wells cut into the agar. If the siderophore is dissolved in an organic solvent, the solvent is allowed to evaporate from the disk before placing it on the agar. The plates are incubated at the normal growth temperature for the organism and then examined for growth of the indicator strain in the agar surrounding the producer strain or culture supernatant. The size of the zone of growth of the indicator will be proportional to the concentration of the siderophore.

Comments. The amount of EDDA and number of organisms seeded in the agar is determined empirically for each strain. The concentration of EDDA must be sufficiently high to prevent growth of the indicator but not so high that growth of the producer on the surface of the agar is inhibited. Under optimal conditions, the bacteria in the lawn are unable to overcome the iron limitation created by EDDA and are present at too low a density to cross-feed in the agar medium. A culture on the surface will be able to overcome the effects of EDDA because it is plated at very high density; it will grow and secrete siderophore into the surrounding agar. If the bacteria in the lawn are able to use the secreted siderophore, a halo of colonies in the agar will appear around the producer. Similarly, purified siderophores or culture supernatants containing usable siderophores will allow growth of colonies surrounding the disk or well.

A number of variations can be made in the bioassay procedure. It is only necessary that the indicator plates and strains be prepared in a consistent manner to provide reproducible results. For example, it is not essential that the EDDA be added to the agar 24 hr in advance. It may be added immediately prior to use, although this may necessitate increasing the concentration of EDDA. Minimal media can be used for less fastidious organisms, less EDDA will be required than when L agar is used.

Isolation and Purification of Siderophores

Once siderophore production is detected, the assay which detects the siderophore can be used to follow the purification of the compound. Siderophores may be isolated as the iron complex or the ligand, depending on the nature of the particular compound. If the iron complex is not charged and multinuclear complexes are not formed, it may be easier to isolate the siderophore in this form; the color of the complex makes it easy to follow the compound during purification.

Extraction with organic solvents is an efficient first step in the purification of siderophores.[4] Extraction separates the siderophore from most of the salts and high molecular weight compounds in the medium and provides a highly effective enrichment. Catechol-type compounds are usually extracted into ethyl acetate, whereas benzyl alcohol or chloroform–phenol (1 : 1 v/v), is used for hydroxamates. Some siderophores, for example, aerobactin, are charged and do not readily partition into organic solvents. Ion-exchange chromatography can be used to isolate charged siderophores. Purification schemes for several of the well-characterized siderophores are detailed below. These provide examples of the types of procedures used to isolate siderophores.

Enterobactin and Other Catechols: Procedure 1

The following purification scheme was developed in the laboratory of C. F. Earhart[17] and is a modification of the procedure of Young.[18]

Strain

AN102, a *fepC* mutant of *Escherichia coli* defective in transport of the siderophore is used for overproduction of enterobactin

[17] J. R. Pierce, Dissertation, Univ. of Texas at Austin (1986).
[18] I. G. Young, *Prep. Biochem.* **6,** 123 (1976).

Growth medium

M9

Na_2HPO_4	6 g
KH_2PO_4	3 g
NaCl	0.5 g
NH_4Cl	1 g
Water	1 liter

Adjust pH to 7.4. Autoclave, then add the following to the cooled medium:

$1 M$ $MgSO_4$	2 ml
20% Glucose	10 ml
$1 M$ $CaCl_2$	0.1 ml

For AN102, supplement with 40 μg/ml each of proline, leucine, and tryptophan, 20 μg/ml thiamin, and 20 μM $FeSO_4$.

Enterobactin Isolation Medium

M9 with precipitated iron

Because enterobactin breaks down rapidly in cultures, iron is added to the medium to stabilize the enterobactin. The iron must be in a relatively insoluble form, however, to prevent it from being taken up by the low-affinity iron transport system of the cells and repressing siderophore synthesis. To accomplish this, $FeSO_4$ to give a final concentration of 200 μM is dissolved in the minimum amount of HCl and added to the M9 salts medium prior to autoclaving. After autoclaving, the medium will contain a visible greenish precipitate. The medium is supplemented with glucose, amino acids, and thiamine as for the growth medium.

Isolation. AN102 is inoculated into 100 ml of M9 growth medium. The culture is incubated overnight at 37° with aeration. This 100-ml culture is used to inoculate 10 1-liter portions (10 ml inoculum per liter of medium) of enterobactin isolation medium. The cultures are grown to stationary phase at 37° with vigorous aeration (18–24 hr). The accumulation of enterobactin in the medium results in a dark purplish color. The cells are removed by centrifugation, the pH of the medium is adjusted to 3 with concentrated HCl (which results in a loss of color as the iron is dissociated from the siderophore), and the supernatant is extracted three times with 1/5 volume of ethyl acetate. The pooled organic phases are concentrated to approximately 200 ml by rotary evaporation and washed with 100 ml of 0.1 M sodium phosphate, pH 5, containing 25 mM $FeSO_4$, transferring the enterobactin to the aqueous phase as the iron complex. Traces of ethyl acetate are removed from the aqueous phase by evaporation. The pH is raised to 7 with 2 N NaOH and the solution washed three times with ethyl acetate. The pH is then reduced to pH 3 with concentrated HCl, and the enterobactin is returned to the organic phase by extracting three times with equal volumes of ethyl acetate. The volume of the pooled ethyl acetate extracts is reduced to 150 ml by flash evaporation. The extract is washed with an equal volume of 0.1 M sodium phosphate, pH 2, and then washed three times with equal volumes of deferrated 0.1 M sodium phosphate, ph 7.4. The organic phase is dried over anhydrous $MgSO_4$. At this point, the enterobactin can be precipitated with hexane, or the iron complex can be made. If enterobactin is to be stored, the more stable iron complex should be formed by adding $FeSO_4$ in sodium phosphate, pH 7, to the ethyl acetate. Traces of ethyl acetate are removed from the aqueous phase by evaporation, and the ferric enterobactin is stored at $-20°$.

Enterobactin and Other Catechols: Procedure 2

A simplified procedure was developed by Mark McIntosh in the laboratory of Neilands.[4] Cells are grown as described above, but a modified Tris-buffered medium (T medium described below) with the addition of 0.4% glucose, 40 μg/ml each of proline, leucine, and tryptophan, 20 μg/ml thiamin, 100 μg/ml casamino acids, and 17 mg of $MnSO_4 \cdot H_2O$ is substituted. The cells are removed by centrifugation from 10 liters of fully grown culture and the supernatant extracted with ethyl acetate. However, in this procedure the pH is not lowered before extraction. The pooled ethyl acetate extracts are reduced to 500 ml, washed with 0.1 M sodium citrate, pH 5.5, and dried over anhydrous $MgSO_4$. The extract is further concentrated to approximately 5 ml, and the enterobactin is precipitated with hexane. The residue is dissolved in a small amount of ethyl acetate and the maximum amount of benzene added which will not cause precipitation. This is chromatographed on a short column of Mallinckrodt (St. Louis, MO) 100-mesh silicic acid to remove oxidation products. The eluate is concentrated and crystallized by addition of hexane yielding several hundred mg of white enterobactin crystals.

This second procedure is applicable to other catechols such as vibriobactin, the siderophore produced by *Vibrio cholerae*. For isolation of vibriobactin, an El Tor strain such as Lou15 is grown in T medium containing sodium lactate and sodium succinate as carbon sources. The pH of the supernatant is adjusted to 6, and the vibriobactin extracted as described above, omitting the chromatography step. Following the initial hexane precipitation, vibriobactin is dissolved in methanol and reprecipitated with hexane.[19]

[19] G. Griffiths, S. Sigel, S. M. Payne, and J. B. Neilands, *J. Biol. Chem.* **259**, 383 (1984).

Enterobactin and Other Catechols: Procedure 3

An alternative procedure to isolate catechols using chromatography rather than extraction into organic solvents was developed in the laboratory of Winkelmann.[20] The strain is grown in 10 liters of medium as described above. The cells are removed by centrifugation, and the supernatant is acidified to pH 2 by the addition of H_2SO_4. The supernatant is passed through an XAD-2 (Mallinckrodt, St. Louis, MO) column. The column is washed with two volumes of distilled water and the catechols eluted with one column volume of methanol. The methanol is evaporated and the residue dissolved in a small volume of methanol. Enterobactin is further purified by chromatography on Sephadex LH-20 with methanol as the eluting solvent. The purity of the preparation can be confirmed by thin-layer chromatography (TLC) or by high-performance liquid chromatography (HPLC: C_{18} reversed-phase column, isocratic elution with methanol–0.1% phosphoric acid 1 : 1, 1 ml/min).

Aerobactin

Aerobactin is isolated by a modification[5,14] of the procedures of Gibson and Magrath[21] and Neilands.[5]

Strain

Shigella flexneri SA101

Medium

T Medium[22]

NaCl	5.8 g
KCl	3.7 g
$CaCl_2$	0.113 g
$MgCl_2 \cdot 6H_2O$	0.1 g
NH_4Cl	1.1 g
KH_2PO_4	0.272 g
Na_2SO_4	0.142 g
Tris (trizma base)	12.1 g
Water	to 1000 ml

Adjust to pH 7.4 using concentrated HCl. Autoclave.

Supplement with 0.4% glucose and 5 μg/ml nicotinic acid.

Procedure. Inoculate 100 ml of T medium with 1 ml of a fully grown broth culture of *S. flexneri*. The culture is incubated at 37° with aeration

[20] I. Berner, M. Greiner, J. Metzger, G. Jung, and G. Winkelmann, *Biol. Metals* **4**, 113 (1991).
[21] F. Gibson and D. I. Magrath, *Biochim. Biophys. Acta* **192**, 175 (1969).
[22] E. H. Simon and I. Tessman, *Proc. Natl. Acad. Sci. U.S.A.* **50**, 526 (1963).

until fully grown, then diluted 1 : 50 into five 1-liter portions of T medium. The cultures are grown at 37° with vigorous aeration to stationary phase (~24 hr). The cells are removed by centrifugation and the supernatant evaporated to dryness in a rotary evaporator. The residue is taken up in a small volume of water (~6 ml) and applied to a column (1.5 × 40 cm) of BioGel (Bio-Rad, Richmond, CA) P-10 equilibrated with water. The column is eluted with water, and the fractions giving a positive reaction with the ferric perchlorate reagent are pooled and evaporated to dryness. the residue is dissolved in water and applied to a column (3 × 10 cm) of Dowex 1 (chloride form) and the siderophore eluted with a gradient of 0.4 to 1 M NH_4Cl. The ferric perchlorate-reactive fractions are pooled, applied to a column of Dowex 50W-X8 (H^+ form), and eluted with water. The elute is adjusted to pH 2, saturated with $(NH_4)_2SO_4$, and extracted with benzyl alcohol (approximately three times) to remove all the aerobactin from the aqueous phase. The benzyl alcohol extracts are pooled and filtered through a double layer of filter paper. The filtrate is diluted with 10 volumes of diethyl ether and the hydroxamate extracted into a small volume of water. Traces of benzyl alcohol are removed by ether extraction, and the aqueous phase is lyophilized, yielding approximately 60 mg of aerobactin.

Ferrichrome

Ferrichrome is isolated from cultures of *Ustilago sphaerogena* by the procedure of Neilands.[4]

Medium

Growth medium, per liter

K_2SO_4	1 g
K_2HPO_4	3 g
$(NH_4)_2SO_4$	3 g
Sucrose	20 g
Citric acid	1 g
Thiamin	2 mg
Cu^{2+}	5 μg
Mn^{2+}	350 μg
Zn^{2+}	2 mg
Mg^{2+}	80 mg

The pH is adjusted to 6.8 to 7.0 with concentrated NH_4OH.

Procedure. An isolated colony of *U. sphaerogena* on nutrient agar is used to inoculate 100 ml of growth medium. The culture is incubated at

30° until turbid. Ten Fernbach flasks, each containing 1 liter of the same medium, are each inoculated with 10 ml of the culture, and cultures are grown with vigorous aeration for 1 week.

The cells are removed from the culture by centrifugation, and ferrous sulfate is added to the supernatant until maximum color is reached. The pH is adjusted to pH 3 by addition of H_2SO_4, and ammonium sulfate is added to 50% saturation. The orange-colored ferrichrome is extracted into benzyl alcohol and filtered through two layers of filter paper. The benzyl alcohol is diluted with several volumes of diethyl ether and the ferrichrome extracted with several small aliquots of water. Remaining traces of benzyl alcohol are removed by extraction with a small volume of diethyl ether. The aqueous phase is concentrated under vacuum and stored in the cold to allow crystallization of the ferrichrome A, which is removed by filtration. The filtrate is neutralized and reduced to dryness. The residue is extracted into dry, hot methanol. The methanol phase is reduced in volume until precipitation is noted. The solution is allowed to stand overnight to permit the ferrichrome to crystallize. The crystals are collected by filtration and recrystallized from hot methanol. The yield of ferrichrome is approximately 10 mg per liter of culture.

Characterization

The assays used for detection give considerable information about the nature of the siderophore. Based on the results of the Arnow and Csáky assays, the compound can be classified as a catechol or hydroxamate, respectively, and some siderophores have been found to contain both functional groups. Chemical, electrophoretic, and spectral analyses are used to determine the precise structure of the compound.

The ferric hydroxamates exhibit absorption maxima in the 425 to 520 nm range. By measuring the absorbance as a function of pH, it is possible to determine the number of hydroxamate moieties in the molecule.[5] Monohydroxamates are reddish-orange at neutral pH, shifting to purple as the solution is acidified to pH 5 or lower. Thus the maximum shifts from 420–450 nm at neutral pH to 500–520 nm at acidic pH. Trihydroxamates, however, do not show this color transition, retaining the orange color at pH 2–3. Rhodotorulic acid, which has two hydroxamate groups, displays an intermediate behavior.

Catechol siderophores display fluorescence characteristic of the 2,3-dihydroxybenzoyl group. These compounds show three absorbance peaks in the ultraviolet, at 320, 250, and 210 nm.[5] The ferric catechols are reddish-purple colored at neutral pH, but the iron is dissociated and the color lost at acidic pH.

Acid hydrolysis can be used to cleave the compound, and the components can be separated and characterized by chromatography.[4,5] Paper electrophoresis and thin-layer chromatography in comparison with known standards are often used to characterize the intact siderophore as well as hydrolysis products. Iron should be removed prior to hydrolysis. The catechols are freed of iron by reducing the pH to 2 and extracting the ligand into ethyl acetate. Hydroxamates can be treated with 0.5 N NaOH in the cold, which will cause the iron to form insoluble ferric hydroxide. Alternatively, the iron can be removed by extraction with 3% 8-hydroxyquinoline in methanol. The ligand is then hydrolyzed with 6 N HCl.

Complete structural analysis is aided by the use of nuclear magnetic resonance (NMR) and mass spectrometry. If the iron complex of the siderophore can be crystallized, X-ray diffraction can be used. This not only reveals information on structure, but also shows the sites of attachment of the iron.

Acknowledgments

Work on iron transport in the author's laboratory has been supported by Grants AI16935 from the National Institutes of Health and DMB 8819169 from the National Science Foundation. The assistance of Janice Stoebner in the preparation of the manuscript is gratefully acknowledged.

[26] Effects of Iron Deprivation on Outer Membrane Protein Expression

By J. B. Neilands

Introduction

Information on iron-regulated outer membrane proteins of bacteria can be found in books[1,2] dealing with microbial iron assimilation or in particular publications such as the *Journal of Bacteriology*. The correlation between iron assimilation and virulence has been treated authoritatively by Weinberg,[3] and there is also a book on the topic.[4] The scope of this chapter is

[1] G. Winkelmann (ed.), "Handbook of Microbial Iron Chelates." CRC Press, Boca Raton, Florida, 1991.
[2] G. Winkelmann, D. van der Helm, and J. B. Neilands (eds.), "Iron Transport in Microbes, Plants and Animals," VCH, Weinheim, 1987.
[3] E. D. Weinberg, *Drug Metab. Rev.* **22,** 531 (1990).
[4] J. J. Bulen and E. Griffiths (eds.), "Iron and Infection." Wiley, New York, 1987.

restricted to a few microbial species for which substantial knowledge is available.

The primary function of the outer membrane of gram-negative bacteria is to offer protection against a variety of noxious agents which may be encountered in the environment, such as the detergents which occur in the gastrointestinal (GI) tract of animals. On the other hand, some provision must be made for uptake of the nutrients that the microorganism needs for survival. Small molecules such as monosaccharides and amino acids, where the molecular mass is about 600 Da or less, permeate the outer membrane through small, water-filled pores created by the porins. These structures are not effective for absorption of iron, a mineral believed to be essential for all forms of life with the possible exception of the lactobacilli. Since the advent of an oxidizing atmosphere some billions of years ago, the surface iron has existed in the Fe(III) state and, as such, is highly polymerized and quantitatively insoluble at biological pH.

A number of bacteria and fungi respond to iron deficiency by elaboration of low molecular weight, virtually ferric-specific agents generically designated siderophores (Greek: iron bearer). The siderophore spontaneously and rapidly associates with ferric ion. To surround and chelate the six-coordinate ferric ion, a molecular mass in excess of 600 Da is frequently required. In the case of gram-negative bacteria this has mandated the synthesis and incorporation in the outer membrane of specific structures, receptors, designed to recognize and transport the ferric siderophores. In the course of evolution the siderophore receptors, intended for uptake of the nutritious ferric ion, have in turn become "parasitized" by lethal agents such as bacteriophages, bacteriocins, and antibiotics. Siderophores are multidentate, typically oxygen-containing ligands with affinity constants for ferric ion ranging from 10^{20} to 10^{30}. The complexed iron is transported back into the cell via the receptor, and the iron is released internally by reduction to Fe(II), which is a substantially "softer acid" than Fe(III), and as a consequence it has a much lower complexation constant for the siderophore. The reduction, sometimes accompanied by dismemberment of the ligand, completes the process of high affinity, siderophore-dependent solubilization and internalization of ferric ion.

Although most aerobic and facultative anaerobic bacteria as well as fungi synthesize at least one siderophore, some species make no detectable levels of a high-affinity iron-scavenging system. Some have devised a receptor for heme,[5] whereas others have imitated mammalian cells in synthesis of an uptake device, a receptor, for ferric transferrin.[5] Still other microbes, such as *Saccharomyces cerevisiae,* reduce Fe(III) to Fe(II) at

---

[5] S. M. Payne, *Crit. Rev. Microbiol.* **16,** 81 (1988).

the cell surface.[6] The relatively high solubility of Fe(II) has obviated synthesis of siderophores by strict anaerobes.

In regard to supply of iron it is necessary to distinguish between total and available iron. If the total amount of iron is high but is present in some unavailable form, then a siderophore uptake system may be expressed. No living system, animal or vegetable, will have an abundant supply of "loose" and microbiologically available iron. This would expose the system to infection and/or oxidative damage. For example, enteric bacteria incubated in the peritoneal cavity of experimental animals are iron-stressed to the extent that siderophores are produced.[7] Common laboratory media, if synthetic, may contain 0.1 to 10 μM total iron. The former is regarded as "low" and the latter as "high" in iron. Complex media, such as LB, may be considered to be high in iron, and when growing under such conditions microorganisms will acquire iron by a low-affinity pathway. It may be possible to bind iron *in situ* and thus make it unavailable. Reagents that have been used for this purpose include α,α'-dipyridyl, ethylenediaminedi(*o*-hydroxyphenylacetic acid), deferriferrichrome A, citrate, and conalbumin. If the culture is truly iron stressed, then the addition of a pure iron salt to a level of 10 μM in minimal media or to a stoichiometric excess in complex media containing a specific iron-binding agent should result in superior growth.

A comparison of the outer membrane protein profiles by sodium dodecyl sulfate–polyacrylamide gel electrophoresis (SDS–PAGE) analysis of high- and low-iron growth cells should reveal which components are iron regulated. The next step is to acquire a mutant which fails to grow on low iron or in the presence of one of the complexed forms of iron. A Tn5–phoA fusion is particularly useful for collection of membrane mutants since the product, alkaline phosphatase, must be external to the cell to be active. The mutant should be restored to growth competency following transformation with a vector carrying an insert bearing the gene for the desired receptor. It then remains to reduce the insert to the smallest possible size, sequence it, and examine its mode of regulation. This level of analysis is most advanced in the case of the aerobactin determinants carried on the virulence plasmid pColV-K30 of *Escherichia coli*.[8] If desired, the protein(s) can be expressed in mini- or maxicells and a partial sequence determined in order to assure correspondence between the gene and its product. There is a limit to the flexibility of the outer membrane

[6] A. Dancis, D. G. Roman, G. J. Anderson, A. G. Hinnebusch, and R. D. Klausner, *Proc. Natl. Acad. Sci. U.S.A.* **89,** 3869 (1992).

[7] E. Griffiths, *Biol. Metals* **4,** 7 (1991).

[8] J. B. Neilands, *Can. J. Microbiol.* **38,** 728 (1992).

as regards its protein composition, and cells carrying expression vectors for these genes may not grow well.

Iron receptors display relative specificity for the ligands. Individual siderophores usually have separate receptors, although the ferrichrome-binding protein in the outer membrane of *E. coli* interacts with a variety of lethal agents including various bacteriophages, colicin M, and albomycin. Some attempts have been made to chemically modify standard antibiotics in such a way as to make them resemble siderophores. Rifampicin and cephalosporin have been thus modified with the result that the minimum inhibitory concentrations have been reduced substantially.

Iron-regulated membrane proteins have been found in the envelope of the following gram-negative bacteria: *Aeromonas, Anacystis, Aquaspirillum, Azomonas, Azospirillum, Azotobacter, Campylobacter, Escherichia, Haemophilus, Klebsiella, Neisseria, Paracoccus, Pasteurella, Proteus, Pseudomonas, Salmonella, Shigella, Synechococcus, Vibrio,* and *Yersinia.* The list includes the phytopathogen *Erwinia*[9] and the anaerobic periodontiopathic *Porphyromonas* (formerly *Bacteriodes*) *gingivalis.*[10] Clearly, the phenomenon is a general one.

Escherichia coli

The main iron-regulated outer membrane proteins of *E. coli* K12 are listed in Table I. These are generally referred to in terms of the appropriate mass, for example, the band at 81 kDa is the ferric enterobactin receptor (FepA), which is also a binding site for colicins B and D. These are the bands that are enhanced 10- to 20-fold in outer membrane preparations from iron-stressed cells. They are all in the range of 80 kDa and are thus about twice as large as the porins. The 80.5-kDa ferric dicitrate receptor is an exception in that it is induced by the presence of the ligand. These receptors are not all derepressed to the same extent, and it is characteristic of the 78-kDa ferrichrome receptor that it is only weakly induced. The 74-kDa receptor for ferric aerobactin has been most extensively studied as the IutA protein of pColV-K30. This receptor is also encoded on the chromosome of *E. coli*, and it is present in several enteric bacterial species, although the molecular mass may depart from 74 kDa. Williams[11] showed that production of a siderophore system, subsequently found to be aerobactin, could account for the capacity of *E. coli* to cause disseminating infections in animals.

[9] D. Expert and A. Toussaint, *J. Bacteriol.* **163,** 221 (1985).
[10] S. C. Holt and T. E. Bramanti, *Crit. Rev. Oral Biol. Med.* **2,** 177 (1991).
[11] P. H. Williams, *Infect. Immun.* **26,** 925 (1979).

TABLE I
IRON-REGULATED OUTER MEMBRANE PROTEINS
OF *Escherichia coli* K12

Gene	Gene product[a] (kDa)	Ligands[b]
fiu	83	Ferric monocatechols?
fepA	81	Ferric enterobactin
fecA[c]	80.5	Ferric dicitrate
fhuA (*tonA*)	78	Ferrichromes
fhuE	76	Coprogen, ferric rhodotorulate, ferrioxamines
cir	74[d]	Ferric monocatechols?
iutA	74	Ferric aerobactin

[a] Mass determined by analysis.
[b] Presumed biochemical function, but may also serve as receptors for specific lethal agents.
[c] Induced by ferric dicitrate.
[d] Mass for *E. coli* (pColV-K30) but differs for other sources.

All outer membrane low-iron induced proteins that have been sequenced thus far have been shown to carry a "TonB box" array near the N terminus, illustrated in Table II. It is taken as evidence for a direct interaction between the TonB protein, which is located in the inner membrane, and the iron-regulated receptors on the outer membrane. This concept needs further work, however, since these large proteins may by mere happenstance display regions of homology. Biochemically and genetically, the transport of ferric siderophores and vitamin B_{12} have much in common.

TABLE II
TonB BOX OF OUTER MEMBRANE FERRIC
SIDEROPHORE RECEPTORS OF *Escherichia coli*

Protein	Sequence[a]
FepA	D T I V V
FecA	F T L S V
FhuA	D T I T V
FhuE	E T V I V
Cir	E T M V V
IutA	E T F V V
Consensus	Acidic T V

[a] Near N terminus.

TABLE III

IRON BOX IRON(II)-Fur BINDING OPERATOR SITES IN IRON-REGULATED PROMOTERS

Gene	Sequence[a]						
fepA	TAT	TAT	GAT	A	ACT	ATT	TGC
fecA	GAA	AAT	AAT	T	CTT	ATT	TCG
	TGT	AAG	GAA	A	ATA	ATT	CTT
fhuA (tonA)	CTT	TAT	AAT	A	ATC	ATT	CTC
fhuE	TAC	AAA	CAA	A	ATT	ATT	CGC
	GCG	TAT	ATT	T	CTC	ATT	TGC
cir	TGG	ATT	GAT	A	ATT	GTT	ATC
	GAT	AAT	TGT	T	ATC	GTT	TGC
iucA (aerA)[b]	GAT	AAT	GAG	A	ATC	ATT	ATT
iucA (aerA)[b]	CAT	AAT	TGT	T	ATT	ATT	TTA
fur	TAT	AAT	GAT	A	GCG	ATT	ATC

[a] Usually located between or near -10 and -35 sites.
[b] Confirmed by footprinting analysis with isolated Fur protein.

The ferric aerobactin receptor encoded by pColV-K30 was cloned by taking advantage of the fact that cells forming the receptor become sensitive to cloacin produced by *Enterobacter cloacae* DF-13. A double screen, namely, resistance to the antibiotic specified by the vector and sensitivity to cloacin, gave about 16 kilobases (kb) of DNA carrying the gene for *iutA*. This turned out to be the ultimate gene in an operon coding for synthesis of aerobactin as well as for transport of the ferric siderophore across the outer membrane.

As may be seen from Table III, the genes for all iron-regulated outer membrane receptors of *E. coli* carry a common "iron box" sequence located near the -10 and -35 sites of the promoter. Iron by itself is, however, not sufficiently "semantic" to bind and regulate the DNA without participation of an adapter protein called Fur (ferric uptake regulation). The aerobactin promoter has two such sites, labeled primary and secondary. The consensus of the iron box sequence is 5' GATAATGATAAT-CATTATC, which is a symmetrical, interrupted palindrome. The Fur repressor is a 17-kDa protein containing 147 amino acids. It is activated by ferrous ion to bind most likely as a dimer at the iron box of regulated promoters. According to an analysis,[12] the activating divalent metal ion binds at the C-terminal domain to cause a conformational change that increases the affinity of the N-terminal region for the operator. Gel shift and footprint analyses with both DNase and hydroxyl radical ($\cdot$OH) indicate that the affinity of the repressor for the operator is in the low nanomo-

[12] M. Coy and J. B. Neilands, *Biochemistry* **30**, 8201 (1991).

lar range.[13] Thus the Fur and MerR proteins are somewhat analogous except that the latter[14] is always bound to the operator and acts as an activator in the presence of high levels of mercury. Because *fur* mutants are constitutive and overexpress the surface receptors and siderophore(s), it can be concluded that the promoters of these genes are "strong" and hence require no activation. The *fur* gene itself is both negatively autoregulated, which requires a metal, and activated by an upstream CAP system.[15]

The isolation and properties of the ferric enterobactin receptor (FepA) of *E. coli* K12 may be considered typical for preparation of the outer membrane iron-regulated proteins.[16] In this case strain BN3040 was selected as the source since it is colicin Ia resistant and lacks the 74-kDa protein, which interferes with the isolation and assay of FepA. The strain is also defective in synthesis of enterobactin since it is blocked between 2,3-dihydroxybenzoic acid and enterobactin (*entA*). The organism is thus chronically starved for iron when grown on the minimal salts medium of Simon and Tessman.[17] The cells are broken by sonication and the inner and outer membranes separated in the usual way. Because FepA is a slightly acidic protein, it could be subjected to chromatography using the anion-exchanger DE-52, to which it adheres at pH 7.2. After a second pass through the DE-52 column, the receptor is about 86% pure, as judged by SDS–PAGE. The preparation is then applied to an affinity column containing molybdenum(VI) bis(2,3-dihydroxybenzoyl)lysine covalently bonded to the free amino groups of Affi-Gel 102. The sample is introduced in 2% Triton X-100–10 mM Tris (pH 7.2)–2 mM Na$_2$MoO$_4$ and eluted with a gradient of 0–0.2 M NaCl in the same buffer. The yield is about 19 mg from 150 ml of culture with A_{650} of 0.8. The protein, which gives a single band on SDS–PAGE, is insoluble in water and requires at least 0.5% Triton X-100 or 0.1% SDS for solution. It is still capable of binding [^{55}Fe]enterobactin, for which the K_D is approximately 10 nM. The pI was found to be 5.5; stains for carbohydrate were negative. The purification scheme is summarized in Table IV.

The gene sequence suggests a mature protein of 723 amino acids, preceded by a signal peptide.[18] There are only 2 cysteines in the molecule.

[13] V. de Lorenzo, F. Giovannini, M. Herrero, and J. B. Neilands, *J. Mol. Biol.* **203**, 875 (1988).
[14] D. M. Ralston, B. Frantz, M. Shin, J. G. Wright, and T. V. O'Halloran, in "Metal Ion Homeostasis" (D. H. Hamer and D. R. Winge, eds.), p. 407. Alan R. Liss, New York, 1989.
[15] V. de Lorenzo, M. Herrero, F. Giovannini, and J. B. Neilands, *Eur. J. Biochem.* **173**, 537 (1988).
[16] E. H. Fiss, P. Stanley-Samuelson, and J. B. Neilands, *Biochemistry* **21**, 4517 (1982).
[17] E. H. Simon and I. Tessman, *Proc. Natl. Acad. Sci. U.S.A.* **50**, 536 (1963).
[18] M. D. Lundrigan and R. J. Kadner, *J. Biol. Chem.* **261**, 10797 (1986).

TABLE IV
PURIFICATION OF FERRIC ENTEROBACTIN OUTER MEMBRANE RECEPTOR (FepA)[a]

Fraction	Units[b]/ml	Total units ($\times 10^{-4}$)	Protein (mg/ml)	Units/mg
Whole membranes	1730	69.2	5.1	339
Soluble outer membranes	1102	24.0	2.45	450
DE-52 column 1	2090	14.6	2.46	839
DE-52 column 2	2819	6.8	2.1	1432
Mo(VI)-diDHB-Lys	250	1.8	0.1	2500

[a] Adapted with permission from E. H. Fiss, P. Stanley-Samuelson, and J. B. Neilands, *Biochemistry* **21,** 4517 (1982). Copyright 1982 American Chemical Society.
[b] One unit equals 1 pmol bound [^{55}Fe]enterobactin.

The net charge is -14. An iron box contained within a bidirectional promoter is the putative Fur regulatory site.[19]

Erwinia Species

Erwinia chrysanthemi, the bacterium responsible for soft rot of vegetables, produces an iron-regulated siderophore, N-[N^2-(2,3-dihydroxybenzoyl)-D-lysyl]-L-serine, and its outer membrane receptor, a polypeptide with molecular mass of 80 kDa. The synthesis and transport of the siderophore have been correlated with the virulence of the bacterium, the first and as yet only plant pathogen for which this relationship has been shown.[20]

Pseudomonas Species

Various members of the genus *Pseudomonas* are important in plant nutrition and human pathology. All produce a plethora of fluorescent, pigmented siderophores variously known as pseudobactins, pyoverdins, or azotobactins. All contain a derivative of 2,3-diamino-6,7-dihydroxy-quinoline in amide linkage to a peptide of 6 to 10 amino acids. In addition to the sidephores, outer membrane proteins are formed in iron deficiency. The receptors are generally specific for a particular siderophore, although some will apparently accept a variety of pseudobactins.[21] In *Pseudomonas*

[19] G. S. Pettis, T. J. Brickman, and M. A. McIntosh, *J. Biol. Chem.* **263,** 18857 (1988).
[20] T. Franza and D. Expert, *J. Bacteriol.* **173,** 6874 (1991).
[21] W. Bittner, J. D. Marugg, L. A. de Weger, J. Tommassen, and P. J. Weisbeek, *Mol. Microbiol.* **5,** 647 (1991).

aeruginosa an 85-kDa protein acts as receptor for pyocin Sa and ferripyoverdin, but the cell contains a second uptake system for the siderophore with lower affinity and wider specificity.[22]

Expression of the high-affinity iron absorption system in *Pseudomonas* spp. appears to be transcriptionally regulated as in *E. coli*, with some modifications. A *fur*-type repressor is present, and, in addition, an activator protein may be required.[23]

Vibrio Species

Vibrio cholerae, like *E. coli*, is equipped with a variety of iron uptake systems. When stressed for iron it makes the catechol-type siderophore vibriobactin, although this does not seem to be a virulence factor. The organism also uses heme and iron citrate as iron sources, and it is significant that its hemolysin is regulated by iron and a Fur-like protein. The iron-regulated proteins of cells grown *in vitro* include a component with molecular mass of 220 kDa and several others ranging in mass from 62 to 77 kDa. The ferric vibriobactin receptor has an SDS-PAGE molecular mass of 74 kDa.[24]

Virulence is clearly associated with iron assimilation in *Vibrio anguillarum*, the causative agent of a lethal hemorrhagic septicemia of salmonid fishes. The high-affinity iron uptake system is plasmid borne and involves a structurally unique siderophore, anguibactin, and its cognate receptor, an 86-kDa outer membrane receptor (OM2).[25]

[22] A. J. Smith, P. N. Hirst, K. Hughes, K. Gensberg, and J. R. W. Govan, *J. Bacteriol.* **174**, 4847 (1992).
[23] D. J. O'Sullivan and F. O'Gara, *Mol. Gen. Genet.* **228**, 1 (1991).
[24] J. A. Stoebner, J. R. Butterton, S. B. Calderwood, and S. M. Payne, *J. Bacteriol.* **174**, 3270 (1992).
[25] J. H. Crosa, *Microbiol. Rev.* **53**, 517 (1989).

[27] Identification and Isolation of Mutants Defective in Iron Acquisition

By J. B. NEILANDS

Introduction

Iron is an essential element for all microbial species, with the possible exception of certain members of the lactobacilli. The mineral must be acquired from the immediate environment, and it is apparent that a number of different mechanisms have evolved for performance of this task. Most aerobic and facultative anaerobic bacteria, and the majority of fungal species, form siderophores and transport the ferric complexes of these high-affinity carriers. Iron is abundant on the surface of the earth, where it ranks second to aluminum among the metals and fourth among all elements, but is rendered unavailable owing to the very small solubility product constant of ferric hydroxide. In a neutral, aqueous, aerobic environment the K_{sp} of 10^{-38} limits the concentration of free ferric ion to approximately 10^{-18} M. The affinity of siderophores for Fe(III) is sufficiently great to allow them to compete effectively with $Fe(OH)_3$ in soil and water, or with ferric transferrin–ferric lactoferrin *in vivo*. Some microbes make no detectable siderophore and, provided they require iron, must obtain the element via an alternative pathway such as reduction, utilization of heme, or extraction from the specific iron-containing metalloproteins.

Because most microorganisms form at least one siderophore, we focus on the chrome azurol S (CAS)[1] technique as a nearly universally applicable method for detection of mutants defective in biosynthesis, transport, or regulation in this pathway. If an initial screen by CAS is negative, then, assuming the cell has a need for iron, one of the several alternative acquisition systems must be operational.

Siderophores generally belong to one of two chemical types, namely, hydroxamates or catechols. The former are present in both bacteria and fungi, whereas the latter appear to be restricted to bacterial. The Csáky[2] and Arnow[3] reactions may be used, respectively, to detect these siderophores (see [25] in this volume). A battery of indicator strains is also available.[4,5] Thus *Escherichia coli* RW193 ATCC (Rockville, MD) 33475, an *entA* mutant, will respond to the catechol-type siderophore, enterobactin, whereas *Aureobacterium* (*Arthrobacter*) *flavescens* Jg-9 ATCC 25091, a soil isolate, is stimulated by all known hydroxamate but not by any catechol-type siderophores. Hence the Csáky and Arnow reactions, used in conjunction with the indicator strains, provide information on the chemical nature of the CAS-positive agent.

Because of the rather narrow range of response by the indicator strains and the difficulty in maintaining these cultures in stock, a universal chemi-

[1] B. Schwyn and J. B. Neilands, *Anal. Biochem.* **160**, 47 (1987).
[2] A. H. Gillam, A. G. Lewis, and R. J. Anderson, *Anal. Chem.* **53**, 841 (1981).
[3] L. E. Arnow, *J. Biol. Chem.* **118**, 531 (1937).
[4] G. Winkelmann (ed.), "Handbood of Microbial Iron Chelates." CRC Press, Boca Raton, Florida, 1991.
[5.] J. B. Neilands, *Struct. Bonding* (*Berlin*) **58**, 1 (1984).

cal test for siderophores has several advantages. Furthermore, among the more than 100 individual siderophores described to date, some recent additions to the roster are lacking both hydroxamate and catechol functions. Rhizobactin,[6] staphyloferrin A,[7] rhizoferrin,[8] and the phytosiderophores[4] are examples. Obviously, the most useful chemical detector would be one that could be added directly to agar medium and thus facilitate the isolation of mutants. The CAS assay approaches the ideal in this regard, but it suffers from the fact that the detergent used to disperse the dye may be toxic to gram-positive bacteria and fungi. The detergent must be present to enable the dye to reach full color potency, with or without iron, otherwise an extinction of only a few thousand instead of approximately 105,000 is realized. The toxicity may be mitigated in various ways as, for example, by inclusion of a polystyrene resin to adsorb excess detergent[9] or by substitution of the anionic by a zwitterionic detergnet.[1] In the latter instance the color change will be from green to yellow rather than from blue to orange. The sensitivity of the CAS assay depends on the inordinately high extinctions of both the ferric and metal-free forms of the dye, and its effectiveness can be attributed to the fact that most siderophores have the power to extract iron from the triphenylmethane dye complex. Thus colonies of siderophore-producing strains will display an orange halo against a blue background.

When a culture is mutagenized, for example, via a Tn5 insert, the size of the halo at low and high iron as well as the ability of the mutant to grow in the presence of nonutilized ferric chelators afford information on the nature of the mutation in a siderophore pathway. This is illustrated in Table I. An analysis of this type has been applied to the nitrogen-fixing symbiont *Rhizobium meliloti* 1021 for isolation of biosynthetic, transport, and regulatory mutants affected in the rhizobactin 1021 pathway.[10]

The CAS assay to be described is essentially the same as the published procedure,[1] with only minor modifications. The method was shown to be applicable to siderophore biosynthetic transport and regulatory mutants of *E. coli*.

Chrome Azurol S Assay

Reagents. Chrome azurol sulfonate (CAS), hexadecyltrimethylammonium bromide (HDTMA), and piperazine-N,N'-bis(2-ethanesulfonic

[6] M. J. Smith, J. N. Shoolery, B. Schwyn, I. Holden, and J. B. Neilands, *J. Am. Chem. Soc.* **107**, 1739 (1985).
[7] S. Konetschny-Rapp, G. Jung, J. Meiwes, and H. Zähner, *Eur. J. Biochem.* **191**, 65 (1991).
[8] H. Drechsel, J. Metzger, S. Freund, G. Jung, J. R. Boelaert, and G. Winkelmann, *Biol. Metals* **4**, 238 (1991).
[9] F. A. Fekete, V. Chandhoke, and J. Jellison, *Appl. Environ. Microbiol.* **55**, 2720 (1989).
[10] P. R. Gill and J. B. Neilands, *Mol. Microbiol.* **3**, 1183 (1989).

TABLE I
PHENOTYPES OF WILD-TYPE AND SIDEROPHORE MUTANTS

Strain	CAS halo		Growth in presence of chelators
	High iron	Low iron	
Wild type	Small	Large	Good
Biosynthesis mutant	Small	Small	Poor
Uptake mutant	Small	Large	Poor
Regulation mutant[a]	Large	Large	Good

[a] These mutations could be in a gene for synthesis of a repressor, such as Fur, or in an operator.

acid) (PIPES) are obtained from a supply house, such as Sigma Chemical Company (St. Louis, MO). Casamino acids, an acid hydrolyzate of casein, is available from Difco Laboratories (Detroit, MI) and is deferrated by shaking a 10% solution of casamino acids with a solution of 3% 8-hydroxy-quinoline in chloroform until the blue-black color of the ferric complex of the chelating agent disappears from the chloroform layer. Modified M9[11] growth medium (MM9) is obtained by reduction of the total phosphate of M9 to 0.03% KH_2PO_4. All reagents are prepared in doubly distilled water.

Chrome Azurol Sulfonate Agar Plates. To prepare a liter of blue agar, 60.5 mg CAS powder is dissolved in 50 ml of water and mixed with 10 mM of 1 ml $FeCl_3 \cdot 6H_2O$ in 10 mM HCl. This solution is added slowly, with stirring, to 72.9 mg HDTMA dissolved in 40 ml water. The dark blue liquid is autoclaved. A mixture of 750 ml water, 100 ml of 10× MM9 salts, 15 g of agar, 30.24 g (0.1 mol) PIPES (free acid), and 12 g of a 50% (w/w) NaOH solution (to bring the pH to the pK_{a_1} of PIPES, 6.8) is also autoclaved. When the solution has cooled to 50°, 30 ml of 10% casamino acids, a carbon source, and any required supplements such as vitamins of antibiotics are added from sterile solutions. Finally, the ferric dye solution is poured down the wall of the vessel with just enough agitation to achieve thorough mixing without generation of foam. A 30 ml volume is poured into each petri plate. After the medium has solidified the culture to be tested is scratched out on the surface in the usual way.

Discussion

Obviously, the composition of the medium will have to be adapted to the nutrient requirements of the microbial strain under study. Tryptophan

[11] T. Maniatis, E. F. Fritsch, and J. Sambrook, "Molecular Cloning: A Laboratory Manual." Cold Spring Harbor Laboratory, Cold Spring Harbor, New York, 1986.

is destroyed in acid hydrolysis and, if required, must be added to a level of 30 μg/ml. Relatively weak ferric chelators, such as 2,3-dihydroxybenzoic acid and citrate, give smaller haloes than do genuine siderophores. Phosphate is remarkably effective in removing the iron of ferric–CAS, and for this reason the total phosphate of the medium must be limited and PIPES used as a neutral buffer. MOPS [3-(N-morpholino)propanesulfonic acid] may be used as a substitute for PIPES. Complex media, such as LB, also interfere. Exchange rates for transfer of the iron from CAS to the siderophore are markedly dependent on the structure of the latter, catechols being generally much faster than hydroxamates. The shuttle chelator sulfosalicylic acid speeds the transfer when the CAS analysis is performed in solution or by paper electrophoresis. The problems most commonly encountered are an improper pH or an excess of phosphate.

[28] Identification of Receptor-Mediated Transferrin-Iron Uptake Mechanism in *Neisseria gonorrhoeae*

By Cynthia Nau Cornelissen and P. Frederick Sparling

Introduction

Iron (Fe) is required for the growth of virtually all microorganisms. In biological systems, Fe is either insoluble or complexed with Fe-binding proteins, so bacteria have developed mechanisms by which they can solubilize and utilize this essential element. Most aerobic bacteria, in response to growth in Fe-deficient environments, synthesize and secrete low molecular weight high-affinity Fe-binding molecules called siderophores. Expression of a system by which the siderophore–Fe complex is transported through the cell envelope is coregulated with expression of siderophore biosynthesis genes. Several species of pathogenic bacteria, however, do not synthesize soluble siderophores but, instead, utilize host Fe-binding proteins directly,[1-3] often by means of specific receptors for binding the host Fe proteins. The purpose of this chapter is to outline an approach for characterization of a nonsiderophore, receptor-mediated iron-uptake system. Utilization of transferrin (Tf)-bound Fe by *Neisseria gonorrhoeae* has been described and will be used as a prototype. The approach, how-

[1] P. A. Mickelsen and P. F. Sparling, *Infect. Immun.* **33**, 555 (1981).

[2] B. R. Otto, A. M. J. J. Verweij-van Vught, J. van Doorn, and D. M. MacLaren, *Microb. Pathogen.* **4**, 279 (1988).

[3] D. J. Morton and P. Williams, *FEMS Microbiol. Lett.* **53**, 123 (1989).

ever, is broadly applicable, and portions have been used to successfully characterize other iron acquisition systems by various organisms.[4]

Identifying Iron Uptake Systems

The first step in characterization of an Fe uptake system is to determine if the organism makes its own siderophore. Siderophores are generally categorized as either phenolate (catecholate) or hydroxamate, based on the chemical structure. Bioassays have been described which can detect excreted siderophores of both types.[5-7] Chemical assays are available that detect phenolates,[8] hydroxamates,[9] or any Fe-binding compound.[10] Growth with inorganic Fe in the presence of a stoichiometric concentration of an Fe chelator such as desferrioxamine mesylate (Desferal), α,α'-dipyridyl, or ethylenediaminedi(o-hydroxyphenylacetic acid) (EDDA) implies that the organism is capable of synthesis, secretion, and utilization of a high affinity, Fe-binding compound.[1] If sequestration of a high molecular weight Fe-binding protein, such as transferrin or hemoglobin, in a dialysis bag in the growth medium supports growth, secretion and utilization of a low molecular weight siderophore is inferred.[11]

The pathogenic *Neisseria* species have been reported to secrete soluble hydroxamate-type siderophores[7]; however, West and Sparling found that culture medium alone contains small amounts of a siderophore-like activity.[12] Using the experimental procedures described above a consensus has been reached that the pathogenic *Neisseria* species do not synthesize and secrete siderophores. They do, however, utilize the Fe from *Escherichia coli*-derived siderophores, namely, aerobactin[13] and enterochelin.[14]

If siderophore synthesis cannot be detected, the next step is to determine what Fe-containing compounds can support growth of the bacterium.

[4] B. R. Otto, A. M. J. J. Verweij-van Vught, and D. M. MacLaren, *Crit. Rev. Microbiol.* **18**, 217 (1992).
[5] M. Luckey, J. R. Pollack, R. Wayne, B. N. Ames, and J. B. Neilands, *J. Bacteriol.* **111**, 731 (1972).
[6] C. V. Reich and J. H. Hanks, *J. Bacteriol.* **87**, 1317 (1964).
[7] R. J. Yancey and R. A. Finkelstein, *Infect. Immun.* **32**, 600 (1981).
[8] L. E. Arnow, *J. Biol. Chem.* **118**, 531 (1937).
[9] T. Z. Csaky, *Acta Chem. Scand.* **2**, 450 (1948).
[10] B. Schwyn and J. B. Neilands, *Anal. Biochem.* **160**, 47 (1987).
[11] W. R. McKenna, P. A. Mickelsen, P. F. Sparling, and D. W. Dyer, *Infect. Immun.* **56**, 785 (1988).
[12] S. E. West and P. F. Sparling, *Infec. Immun.* **477**, 388 (1985).
[13] S. E. West and P. F. Sparling, *J. Bacteriol.* **169**, 3414 (1987).
[14] J. M. Rutz, T. Abdullah, S. P. Singh, V. I. Kalve, and P. E. Klebba, *J. Bacteriol.* **173**, 5964 (1991).

Iron sources can be screened in a chemically defined, low-Fe medium or in complex medium to which an iron chelator has been added. *Neisseria gonorrhoeae* has been shown to utilize the Fe from human glycoproteins Tf and lactoferrin (Lf) as well as from heme, hemin, and hemoglobin.[1] The presence of weak Fe-binding acids such as citrate and nitriloacetate also promote growth of the gonococcus in low Fe medium.[1] Other metabolic intermediates that bind Fe and thereby provide a usable Fe source include isocitrate, pyruvate, malate, and pyrophosphate.[1]

Characterization of Iron Uptake Systems

The Tf-Fe uptake mechanism of *N. gonorrhoeae* has been characterized in some detail. Utilization of Tf-Fe by the gonococcus requires direct contact between the protein and the cell surface. Isolation of Tf in a dialysis bag in low-Fe medium does not support growth[11]; in fact, growth of the gonococcus is inhibited by the presence of Tf in a dialysis bag, due to the Fe scavenging ability of the sequestered protein. When $^{125}I-^{59}Fe$ double-labeled Tf is supplied to Fe-starved cells, only the ^{59}Fe is incorporated into the cell. The ^{125}I-labeled Tf binds to whole cells but is not accumulated intracellularly.[11] Iron uptake from Tf is KCN sensitive and thus energy dependent, whereas binding alone is not.[11] Iron uptake from Tf is only detectable when cells are previously grown in Fe-depleted medium.[11] These results taken together indicate the presence of an Fe-repressed, energy-dependent receptor for Tf on the gonococcal cell surface.

Two assays have been used to demonstrate direct binding of Tf to whole cells of *N. gonorrhoeae*. Lee and Schryvers used a solid-phase binding assay and horseradish peroxidase (HRP)-labeled Tf as the ligand to demonstrate the specificity of the receptor for human Tf.[15] The solid-phase binding assay measures the amount of labeled Tf that binds to dried, whole bacteria on a nitrocellulose membrane. Although rapid and useful, the assay is difficult to quantitate. Other investigators have attempted to use a similar solid-phase assay employing dead, dried bacteria to perform Scatchard analysis of a meningococcal hemoglobin receptor but have acknowledged the serious technical limitations of this approach.[16] Blanton *et al.*[17] used a liquid-phase equilibrium binding assay and

[15] B. C. Lee and A. Schryvers, *Mol. Microbiol.* **2**, 827 (1988).
[16] B. C. Lee and P. Hill, *J. Gen. Microbiol.* **138**, 2647 (1992).
[17] K. J. Blanton, G. D. Biswas, J. Tsai, J. Adams, D. W. Dyer, S. M. Davis, G. G. Koch, P. K. Sen, and P. F. Sparling, *J. Bacteriol.* **172**, 5225 (1990).

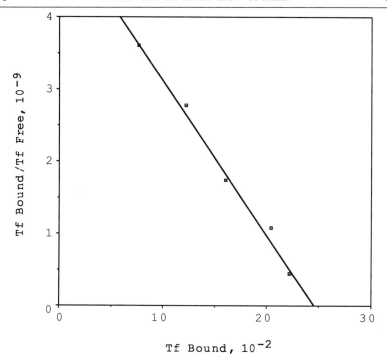

FIG. 1. Scatchard plot of ^{125}I-labeled Tf specific binding to FA19. The line was fitted by linear regression. Data generated by J. Tsai and reported in modified form in Ref. 17.

Scatchard analysis (Fig. 1) to estimate the copy number of the gonococcal receptor to be about 2500 per colony-forming unit (cfu) and the affinity to be approximately 5×10^{-7} M. To prevent cell lysis it is necessary to perform the binding assay at room temperature and to use centrifugation to separate the unbound ligand. The cells and bound ligand are thus maintained at infinite dilution for significant periods of time. Consequently, some ^{125}I-labeled Tf is presumably shed from the surface, and, due to these technical considerations, the affinity and copy number of the receptor may have been underestimated. Nonetheless, this analysis demonstrates that the classical criteria for a receptor, saturability and specificity, are satisfied for the gonococcal Tf receptor. Whole cell binding of human Tf in both solid- and liquid-phase assays has been shown to be Fe repressed, consistent with the observation that Fe uptake from Tf is induced in low-Fe medium (see above).

A critical step in the characterization of the gonococcal Tf-Fe uptake system is isolation of mutants specifically deficient in the ability to utilize and bind human Tf.[17] The mutants are generated by ethyl methanesulfonate mutagenesis, followed by selection for Tf-nonutilizing mutants on streptonigrin, a mutagen that selectively kills cells with high internal pools of Fe. Thus growth of a mutagenized culture on minimal medium containing Tf and streptonigrin enriches for those mutagenized cells that cannot efficiently utilize Tf as an Fe source. Representative mutations are transformed into the parental strain FA19 by congression so that unlinked mutations could be eliminated. Mutants of two phenotypic classes have been isolated and characterized; *trf* mutants (Tf receptor function) cannot bind or utilize human Tf, whereas *tlu* mutants (Tf/Lf utilization) cannot internalize Tf- or Lf-bound Fe, although binding of both ligands is unaffected. This mutation indicates the presence of a common, nonreceptor component in Tf and Lf utilization pathways. The Lf receptor function (*lrf*) mutants, isolated in an analogous manner, have also been characterized and indicate that the Tf and Lf receptors are separate proteins since *lrf* mutants bind wild-type levels of Tf. The *trf*, *tlu*, and *lrf* mutations are unlinked in the gonococcal chromosome, but the gene products have not been identified.[17]

To identify potential components of a Tf receptor, proteins to consider are those that are Fe repressed, able to bind Tf, and absent in receptor mutants. Two Fe-repressed, Tf-binding proteins have been identified in gonococcal strain FA19.[18] A 100-kDa membrane protein (TBP1) binds to Tf[18] in the affinity purification procedure developed by Schryvers and Morris.[19] An 85-kDa protein (TBP2) binds HRP–Tf following sodium dodecyl sulfate–polyacrylamide gel electrophoresis (SDS–PAGE) and electroblotting, although it is not readily isolated by Tf affinity procedures.[18] The *trf* mutants are specifically deficient in the expression of TBP1, the presence of which is detected using polyclonal antiserum raised against affinity-purified TBP1.[18] The *tlu* and *lrf* mutants, however, express wild-type levels of TBP1.[18] The expression of TBP2 in all the mutants is unaffected.[18]

The gonococcal genes encoding TBP1 (*tbpA*)[18] and TBP2 (*tbpB*)[20] have been cloned, sequenced, and mutagenized. The predicted protein sequence of TBP1 is homologous to that of a class of outer membrane receptors found in gram-negative bacteria.[18] Transposon mutagenesis of the *tbpA* gene generates mutants that cannot utilize Tf as an Fe source, although reduced but significant Tf binding remains, as assayed by the solid-phase HRP–Tf binding assay.[18] All the characterized *trf* mutations

[18] C. N. Cornelissen, G. D. Biswas, J. Tsai, D. K. Paruchuri, S. A. Thompson, and P. F. Sparling, *J. Bacteriol.* **174,** 5788 (1992).
[19] A. B. Schryvers and L. J. Morris, *Infect. Immun.* **56,** 1144 (1988).
[20] J. E. Anderson, P. F. Sparling, and C. N. Cornelissen, manuscript in preparation (1993).

A   B

[Competitor] µM   [Competitor] µM

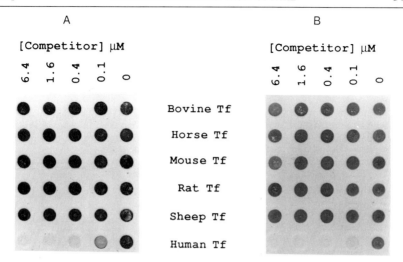

FIG. 2. Specificity of Tf binding in *E. coli* expressing gonococcal TBP1. Blots were probed with a mixture of horseradish peroxidase-labeled Tf (0.4 µg/ml, 2.8 n*M*) and unlabeled mammalian Tf at the concentrations indicated above the blots. (A) Gonococcal strain FA19 grown under iron-deficient conditions; (B) *E. coli* FS1033, expressing TBP1, grown with IPTG (isopropylthiogalactoside). (Reprinted from Ref. 21 with permission from the American Society for Microbiology.)

are located in the TBP1 structural gene since they are all repaired by reconstructed *tbpA*.[18] To date, we have not been able to isolate *tbpB* mutants by their ability to grow on Tf and streptonigrin.

The predicted protein sequence of TBP2 is characteristic of a lipid-modified protein.[20] The *tbpB* gene is located directly upstream of *tbpA* such that transposon mutagenesis of *tbpB* causes a polar effect on *tbpA* expression. Transposon mutants, which express low levels of TBP1, are capable of significant growth on Tf, although they express no detectable TBP2. Eliminating expression of TBP2 and diminishing expression of TBP1 decreases HRP–Tf binding to below the limit of detection in the solid-phase binding assay. These results, as well as analysis of the DNA sequence of the upstream and intergenic regions, suggest that *tbpB* and *tbpA* are part of a single transcriptional unit.[20] Although the function and sequence of TBP1 are consistent with its role as a Tf receptor, the precise contribution of TBP2 to Tf receptor function remains unclear. We presume that TBP2 does play a role either in Tf binding or in Fe release and transport since both TBP1 and TBP2 are Fe repressed, bind human Tf specifically, and appear to be cotranscribed.

TBP1 conditionally expressed in *E. coli* causes the organism to bind

TABLE I
CHARACTERISTICS OF BACTERIAL AND EUKARYOTIC TRANSFERRIN RECEPTORS

Characteristic	Bacterial receptor	Eukaryotic receptor[a]
Copy number	2500/cfu[b]	90,000–300,000/cell
Affinity	$5 \times 10^{-7} M$[b]	$1 \times 10^{-9} M$
Recognition	Apo and ferrated	Ferrated only
Specificity	Human	Various
Iron unloading	No internalization	Internalizes Tf and receptor
Structure	Heterodimer?	Homodimer

[a] Characteristics shown for eukaryotic receptor are found in Ref. 22.
[b] These numbers are likely to be underestimated (see text).

human Tf with the same specificity as seen in the gonococcus (Fig. 2).[21] This observation indicates that TBP1 alone is sufficient for Tf receptor activity and argues that no other gonococcal factors are required to impose specificity on the receptor. Expression of TBP1, however, is not sufficient to confer on *E. coli* the ability to grow on human Tf as a sole Fe source.[21]

Conclusions

Use of a receptor-mediated, nonsiderophore uptake system is conceivably advantageous to the microorganism. Siderophore biosynthesis requires energy; thus, scavenging an Fe-binding protein from the host would constitute a considerable energy savings. Additionally, if the neisserial and eukaryotic Tf receptors recognize different domains of Tf, it is possible that the receptor–Tf complex could serve as an adhesin, promoting interaction with the host cell membrane. Different Tf-binding domains on these receptors are suggested by the observation that the gonococcal receptor recognizes only human Tf,[15,17] whereas the eukaryotic receptor recognizes Tf molecules from various mammalian sources.[22]

Table I compares the characteristics of the neisserial Tf receptor with those of the eukaryotic counterpart. These comparisons must be taken into account when formulating a model of how the neisserial Tf receptor might function *in vivo*. Because the serum concentration of Tf is about 30-fold higher than that required to saturate the neisserial receptor,[17,23]

[21] C. N. Cornelissen, G. D. Biswas, and P. F. Sparling, *J. Bacteriol.* **175**, 2448 (1993).
[22] R. Newman, C. Schneider, R. Sutherland, L. Vodinelich, and M. Greaves, *Trends Biochem. Sci.* **7**, 397 (1982).
[23] J. Tsai, D. W. Dyer, and P. F. Sparling, *Infect. Immun.* **56**, 3132 (1988).

direct competition with the high-affinity eukaryotic receptor may not be necessary. Also, because the bacterial receptor does not effectively distinguish between apo and ferrated Tf,[17] a receptor with low affinity may be necessary to ensure rapid turnover of the unloaded ligand. Although the structure of the eukaryotic receptor is known to consist of two polypeptides, each of which binds a Tf molecule,[22] the structure of the neisserial receptor is presently unclear. The contribution of TBP2 to gonococcal Tf receptor function and the topology of the receptor are the subjects of ongoing research.

[29] Isolation of Genes Involved in Iron Acquisition by Cloning and Complementation of *Escherichia coli* Mutants

By Susan E. H. West

Introduction

Cloning of genes by complementation is a powerful method for gene isolation in prokaryotic systems. Microbial iron acquisition systems are highly conserved. Thus, interspecies complementation of *Escherichia coli* mutants is a plausible strategy for cloning iron acquisition genes from most pathogenic microorganisms. Briefly, the strategy consists of (1) detecting an analog of the *E. coli* gene by DNA hybridization, (2) choosing an appropriate *E. coli* mutant to complement, (3) constructing a library, and (4) identifying a complementing clone. This approach has proved useful for studying iron acquisition in pathogens that are difficult to manipulate genetically. For example, West and Sparling[1] identified and cloned from *Neisseria gonorrhoeae* an analog of the *E. coli fhuB* gene, and Barghouthi *et al.*[2] cloned the *Aeromonas hydrophila amoA* gene, an analog of the *E. coli entC* gene. Additionally, numerous investigators have used interspecies complementation to demonstrate that a cloned gene was analogous to a specific *E. coli* iron acquisition gene.

[1] S. E. H. West and P. F. Sparling, *J. Bacteriol.* **169**, 3414 (1987).
[2] S. M. Barghouthi, S. M. Payne, J. E. L. Arceneaux, and B. R. Byers, *J. Bacteriol.* **173**, 5121 (1991).

Escherichia coli Acquisition Systems

Escherichia coli possesses multiple distinct, but interconnected mechanisms for iron acquisition.[3-5] These systems consist of an iron chelator which solubilizes environmental iron and specific proteins which aid in internalization of the iron–siderophore complex. The iron chelators may be either citrate[6] or a siderophore,[7] an iron-chelating compound of low molecular weight produced by various microorganisms and fungi. All *E. coli* strains produce the phenolate siderophore enterobactin, and some strains produce the hydroxamate siderophore aerobactin.[8] *Escherichia coli* can also use several exogenously applied hydroxamate siderophores as iron sources.[9] The expression of the enterobactin system, as well as the other *E. coli* iron transport mechanisms, is controlled by intracellular iron levels. The controlling element is the protein product of the *fur* gene.[10] Additionally, the protein products of the *tonB* and *exbB* genes are required to couple cytoplasmic membrane energy to high-affinity active transport of the iron–siderophore complex across the outer membrane.[11,12] These systems have all been characterized genetically; thus, numerous iron acquisition mutants are available for interspecies complementation studies. Table I is a compilation of the genes which comprise the various *E. coli* iron acquisition systems.

Detection of Escherichia coli Iron Acquisition Gene Analog by DNA Hybridization

The decision to clone an *E. coli* iron acquisition gene analog from a particular microorganism is often based on physiological observations of the ability of that microorganism to produce a specific siderophore or to use a specific iron source. These observations, by themselves, do not prove that a microorganism possesses an analog of an *E. coli* iron acquisition gene. Thus, it is often helpful to confirm that the microorganism possesses the homologous gene(s) by hybridization of the cloned *E. coli* gene(s) to genomic DNA from the pathogen. Detection of homologous sequences can identify which genes are possessed and can provide infor-

[3] S. Silver and M. Walderhaug, *Microbiol. Rev.* **56**, 195 (1992).
[4] J. H. Crosa, *Microbiol. Rev.* **53**, 517 (1989).
[5] A. Bagg and J. B. Neilands, *Microbiol. Rev.* **51**, 130 (1987).
[6] G. E. Frost and H. Rosenberg, *Biochim. Biophys. Acta* **330**, 90 (1973).
[7] J. B. Neilands, *Microb. Sci.* **1**, 9 (1984).
[8] P. J. Warner, P. H. Williams, A. Bindereif, and J. B. Neilands, *Infect. Immun.* **33**, 540 (1981).
[9] J. B. Neilands, *Annu. Rev. Nutr.* **1**, 27 (1981).
[10] K. Hantke, *Mol. Gen. Genet.* **197**, 337 (1984).
[11] K. Postle, *Mol. Microbiol.* **4**, 2019 (1990).
[12] K. Hantke and L. Zimmerman, *FEMS Microbiol. Lett.* **12**, 31 (1981).

mation, such as the size of the hybridizing fragment and restriction endonuclease sites, that may aid in the design of the cloning strategy.

Standard procedures have been described for performing DNA hybridizations and for labeling of the DNA probe.[13,14] Because most *E. coli* iron acquisition genes (Table 1) have been cloned and sequenced, they should be readily available from the investigators who originally cloned the gene. The probe should be specific for the *E. coli* gene and should not include any flanking or vector sequences. It is important that the hybridization be performed under reduced stringency conditions, as well as the standard high-stringency conditions, to detect genes which are not fully homologous. The variables that affect the hybridization stringency include the salt concentration of the hybridization solution, the presence of formamide in the hybridization reaction, and the temperature at which the hybridization is performed.[15]

Even though a microorganism may synthesize or use a particular iron source, it may not possess analogs of all the genes that comprise a specific *E. coli* iron acquisition system. Thus, it is important to confirm which genes the microorganism possesses so that an appropriate *E. coli* mutant can be chosen for interspecies complementation. A simplified procedure to determine whether a microorganism contains one or more genes comprising an operon is to probe a Southern blot containing restriction fragments specific for each gene with labeled genomic DNA from the pathogen. With this procedure, it is not necessary to purify specific probes for each gene within the operon. This strategy is feasible for bacterial pathogens because of the low degree of complexity of their genomes. Genomic DNA can be labeled with $[\alpha\text{-}^{32}\text{P}]\text{dCTP}$ by nick translation[13,14] to a specific activity of at least 10^8 counts (cpm)/μg of DNA. As mentioned above, hybridizations should be carried out under both reduced and high-stringency conditions. With this approach, West and Sparling[1] demonstrated that *N. gonorrhoeae* possesses a *fhuB* analog but not the other genes which comprise the *E. coli fhuACDB* operon.

Choice of Specific Escherichia coli Iron Acquisition Mutant

Even though they are all regulated by the *fur* gene, each of the *E. coli* iron acquisition systems functions independently. For example, an *E. coli*

[13] J. Sambrook, E. F. Fritsch, and T. Maniatis, "Molecular Cloning: A Laboratory Manual," 2nd Ed. Cold Spring Harbor Laboratory, Cold Spring Harbor, New York, 1989.

[14] F. M. Ausubel, R. Brent, R. E. Kingston, D. D. Moore, J. G. Seidman, J. A. Smith, and K. Struhl, "Current Protocols in Molecular Biology." Greene Publ. and Wiley (Interscience), New York, 1989.

[15] G. A. Beltz, K. A. Jacobs, T. H. Eickbush, P. T. Cherbas, and F. C. Kafatos, this series, Vol. 100, p. 266.

TABLE I
Escherichia coli IRON ACQUISITION GENES

Gene symbol	Function	Refs.
entA	2,3-Dihydro-2,3-dihydroxybenzoate dehydrogenase (Ec 1.3.1.28)	*a, b, c*
entB	2,3-Dihydro-2,3-dihydroxybenzoate synthase (isochorismatase, Ec 3.3.2.1)	*c, d*
entC	Isochorismate synthase (EC 5.4.99.6)	*e*
entD	Enterobactin synthase, component D	*f*
entE	Enterobactin synthase, component E	*g*
entF	Enterobactin synthase, component F	*h, i*
entG	Enterobactin synthase, component G	*c, d*
fepA	Ferric enterobactin receptor	29, *j, k, l*
fepB	Periplasmic ferric enterobactin-binding protein	*j, m*
fepC	Ferric enterobactin uptake, membrane-associated ATP-binding protein	*j*
fepD	Ferric enterobactin uptake, inner membrane protein	*j, n, o*
fepE	Ferric enterobactin uptake	*j*
fepG	Ferric enterobactin uptake, inner membrane protein	*n, o*
fes	Ferric enterobactin esterase	*h, k*
iucA	Aerobactin biosynthesis, synthase	*p, q*
iucB	Aerobactin biosynthesis, acetylase	*p, q*
iucC	Aerobactin biosynthesis, synthase	*p, q*
iucD	Aerobactin biosynthesis, oxygenase	*p, q, r*
iutA	Aerobactin outer membrane receptor	30, *p, q*
fhuA	Ferrichrome receptor	31, *s, t, u*
fhuE	Rhodotorulate/coprogen receptor	*v, w, x*
fhuB	Ferric hydroxamate uptake, cytoplasmic membrane component	*s, y, z, aa*
fhuC	Ferric hydroxamate uptake, cytoplasmic membrane component	*s, y, bb, cc*
fhuD	Ferric hydroxamate uptake, cytoplasmic membrane component	*s, y, bb, cc*
fecA	Citrate-dependent iron transport outer membrane receptor	*dd, ee, ff*
fecB	Periplasmic ferric dicitrate-binding protein	*dd, ee, ff, gg*
fecC	Cytoplasmic membrane ferric dicitrate transport protein	*gg*
fecD	Cytoplasmic membrane ferric dicitrate transport protein	*ff, gg*
fecE	ATP-binding ferric dicitrate transport protein	*gg*
fecR	Citrate-dependent iron transport regulation	*hh*
fecI	Citrate-dependent iron transport regulation	*hh*
tonB	Required for ferric siderophore transport	*ii*
exbB	Required for ferric siderophore transport	12, *jj, kk*
fur	Regulation of all iron acquisition systems	10, *ll, mm, nn*
feo	Ferrous iron transport	*oo*

[a] M. S. Nahlik, T. P. Fleming, and M. A. McIntosh, *J. Bacteriol.* **169,** 4163 (1987).
[b] J. Liu, K. Duncan, and C. T. Walsh, *J. Bacteriol.* **171,** 791 (1989).

c M. S. Nahlik, T. J. Brickman, B. A. Ozenberger, and M. A. McIntosh, *J. Bacteriol.* **171**, 784 (1989).

d J. F. Staab and C. R. Earhart, *J. Bacteriol.* **172**, 6403 (1990).

e B. A. Ozenberger, T. J. Brickman, and M. A. McIntosh, *J. Bacteriol.* **171**, 775 (1989).

f S. K. Armstrong, G. S. Pettis, L. J. Forrester, and M. A. McIntosh, *Mol. Microbiol.* **3**, 757 (1989).

g J. F. Staab, M. F. Elkins, and C. R. Earhart, *FEMS Microbiol. Lett.* **59**, 15 (1989).

h G. S. Pettis and M. A. McIntosh, *J. Bacteriol.* **169**, 4154 (1987).

i F. Rusnak, M. Sakaitani, D. Drueckhammer, J. Reichert, and C. T. Walsh, *Biochemistry* **30**, 2916 (1991).

j B. A. Ozenberger, M. S. Nahlik, and M. A. McIntosh, *J. Bacteriol.* **169**, 3638 (1987).

k G. S. Pettis, T. J. Brickman, and M. A. McIntosh, *J. Biol. Chem.* **263**, 18857 (1988).

l M. D. Lundrigan and R. J. Kadner, *J. Biol. Chem.* **261**, 10797 (1986).

m M. F. Elkins and C. F. Earhart, *J. Bacteriol.* **171**, 5443 (1989).

n C. M. Shea and M. A. McIntosh, *Mol. Microbiol.* **5**, 1415 (1991).

o S. S. Chenault and C. F. Earhart, *Mol. Microbiol.* **5**, 1405 (1991).

p N. H. Carbonetti and P. H. Williams, *Infect. Immun.* **46**, 7 (1984).

q V. de Lorenzo, A. Bindereif, B. H. Paw, and J. B. Neilands, *J. Bacteriol.* **165**, 570 (1986).

r M. Herreo, V. de Lorenzo, and J. B. Neilands, *J. Bacteriol.* **170**, 56 (1988).

s L. Fecker and V. Braun, *J. Bacteriol.* **156**, 1301 (1983).

t J. W. Coulton, P. Mason, and M. S. Dubow, *J. Bacteriol.* **156**, 1315 (1983).

u J. W. Coulton, P. Mason, D. R. Cameron, G. Carmel, R. Jean, and H. N. Rode, *J. Bacteriol.* **165**, 181 (1986).

v K. Hantke, *Mol. Gen. Genet.* **191**, 301 (1983).

w M. Sauer, K. Hantke, and V. Braun, *J. Bacteriol.* **169**, 2044 (1987).

x M. Sauer, K. Hantke, and V. Braun, *Mol. Microbiol.* **4**, 427 (1990).

y V. Braun, R. Gross, W. Koster, and L. Zimmerman, *Mol. Gen. Genet.* **192**, 131 (1983).

z C. A. Prody and J. B. Neilands, *J. Bacteriol.* **157**, 874 (1984).

aa W. Koster and V. Braun, *Mol. Gen. Genet.* **204**, 435 (1986).

bb R. Burkhardt and V. Braun, *Mol. Gen. Genet.* **209**, 49 (1987).

cc J. W. Coulton, P. Mason, and D. D. Alatt, *J. Bacteriol.* **169**, 3844 (1987).

dd S. Hussein, K. Hantke, and V. Braun, *J. Biochem.* (*Tokyo*) **117**, 431 (1981).

ee W. Wagegg and V. Braun, *J. Bacteriol.* **145**, 156 (1981).

ff U. Pressler, H. Staudenmaier, L. Zimmerman, and V. Braun, *J. Bacteriol.* **170**, 2716 (1988).

gg H. Staudenmaier, B. Van Hove, Z. Yaraghi, and V. Braun, *J. Bacteriol.* **171**, 2626 (1989).

hh B. Van Hove, H. Staudenmaier, and V. Braun, *J. Bacteriol.* **172**, 6749 (1990).

ii K. Postle and R. F. Good, *Proc. Natl. Acad. Sci. U.S.A.* **80**, 5235 (1983).

jj K. Eick-Helmerich and V. Braun, *J. Bacteriol.* **171**, 5117 (1989).

kk K. Eick-Helmerich and V. Braun, *Mol. Gen. Genet.* **206**, 246 (1987).

ll K. Hantke, *Mol. Gen. Genet.* **182**, 288 (1981).

mm A. Bagg and J. B. Neilands, *J. Bacteriol.* **161**, 450 (1985).

nn S. Schaffer, K. Hantke, and V. Braun, *Mol. Gen. Genet.* **200**, 110 (1985).

oo K. Hantke, *FEMS Microbiol. Lett.* **44**, 53 (1987).

mutant that contains a mutation blocking enterobactin synthesis or uptake can grow on iron-limiting medium if citrate or a hydroxamate siderophore is supplied as the iron source. Likewise, mutants unable to utilize citrate or exogenously supplied hydroxamate siderophores or to synthesize aerobactin can grow on iron-limiting medium if the enterobactin system is functional. Thus, the mutant chosen for interspecies complementation must be unable to use the other *E. coli* iron acquisition systems because of either a mutation in a specific system or unavailability of a particular iron source.

To clone an analog of the enterobactin system, the *E. coli* mutant must be blocked in the ability to utilize hydroxamate siderophores and citrate. Thus, *E. coli* strains which do not synthesize or transport aerobactin or contain a *fhuB, fhuC,* or *fhuD* mutation could be used. Citrate induces the citrate-dependent system when *E. coli* is grown under iron-limiting conditions.[16] Thus, citrate should be eliminated from the growth medium to prevent expression of this system. Exogenous siderophores should not be included in the growth medium unless a *fhuB, fhuC,* or *fhuD* mutant is used.

To clone an analog of the aerobactin biosynthesis genes, the *E. coli* mutant should be unable to synthesize or transport enterobactin. In addition to *ent* and *fep* mutants, several *E. coli aro* mutants, including *aroA* or *aroB* mutants, meet this requirement. The *aro* mutants are unable to produce enterobactin because they are blocked in the production of chorismate, a precursor required for synthesis of aromatic amino acids and enterobactin. Additionally, to prevent induction of the citrate-dependent system, the medium should not contain citrate.

To clone one of the *fhuBCD* genes, the *E. coli* mutant must be deficient in production and/or utilization of enterobactin. Additionally, a hydroxamate siderophore must either be supplied exogenously to or be synthesized by the *fhuB, fhuC,* or *fhuD* mutant. The aerobactin receptor and biosynthesis genes are often encoded on a plasmid such as pColV.[8] Thus, to construct a strain which synthesizes aerobactin, the ColV plasmid can be conjugatively transferred to the appropriate mutant as described by Miller.[17] Again, the medium must be free of citrate to prevent induction of the citrate-dependent system.

To clone an anlog of the *E. coli fur* gene, the *E. coli fur* mutant should contain a fusion of a *fur*-regulated promoter to a reporter gene such as β-galactosidase. Techniques for constructing gene or operon fusions have

[16] B. Van Hove, H. Staudenmaier, and V. Braun, *J. Bacteriol.* **172**, 6749 (1990).
[17] J. H. Miller, "Experiments in Molecular Genetics," 2nd Ed. Cold Spring Harbor Laboratory, Cold Spring Harbor, New York, 1992.

recently been described by Slauch and Silhavy.[18] This strategy does not provide a direct selection for complementation but does provide an easily visible assay to screen for complementation of a *fur* mutant. For example, colonies expressing β-galactosidase are blue as a result of hydrolysis of the chromogenic substrate 5-bromo-4-chloro-3-indolyl-β-D-galactoside (X-Gal) when grown on iron-limiting medium plus X-Gal and are white or pale blue when grown on iron-replete medium plus X-Gal.

Construction of Genomic Library for Complementation of Specific *Escherichia coli* Mutant

Detailed procedures for construction of genomic libraries are presented in several standard cloning manuals such as earlier volumes in this series, Sambrook *et al.*,[13] and Ausubel *et al.*[14] However, the vector for cloning an *E. coli* iron acquisition analog by interspecies complementation should be chosen carefully. Many of the *E. coli* iron acquisition genes encode membrane-associated proteins. Overexpression of membrane proteins, as a result of increased gene dosage because a high-copy vector was used, is often lethal for *E. coli* or causes the recombinant plasmid to be unstable. Therefore, the use of a single-copy vector, or at least a low-copy-number vector, is required for the successful cloning of iron acquisition genes which are membrane associated. Additionally, overexpression of a gene, especially regulatory genes, can often result in expression of an altered phenotype.

Several vectors, which are derivatives of phage λ and are termed phasmids, have been constructed specifically for complementation of *E. coli* mutants.[19] These vectors contain the origin of replication from the single-copy plasmid NR1. Thus, in *E. coli* strains that carry the λ cI gene when grown at a permissive temperature, the vector replicates as a single-copy plasmid. Use of the plasmid λSE4 has resulted in the successful cloning of the *N. gonorrhoeae fhuB* analog[1]; the gene could not be subcloned into a multicopy vector, presumably because it was lethal for *E. coli* (S.E.H. West, S. Thompson, and P. F. Sparling, unpublished observations). Alternatively, low-copy-number broad-host-range cosmid vectors such as pLAFR[20] and derivatives[21,22] can also be used. Approximately 1–3 copies of these vectors are present per bacterial cell. Using

[18] J. M. Slauch and T. J. Silhavy, this series, Vol. 204, p. 213.
[19] S. J. Elledge and G. C. Walker, *J. Bacteriol.* **162,** 777 (1985).
[20] A. M. Friedman, S. R. Long, S. E. Brown, W. J. Buikema, and F. M. Ausubel, *Gene* **18,** 289 (1982).
[21] B. Staskawicz, D. Dahlbeck, N. Keen, and C. Napoli, *J. Bacteriol.* **169,** 5789 (1987).
[22] L. N. Allen and R. S. Hanson, *J. Bacteriol.* **161,** 955 (1985).

a library constructed in pLAFR3 which conferred on *E. coli* DH5α the ability to synthesize and utilize aerobactin, Ishimaru and Loper[23] have successfully cloned the aerobactin biosynthesis and receptor genes from *Erwinia carotovora* subsp. *carotovora*.

In contrast to genes encoding membrane proteins and regulatory proteins, biosynthetic genes can be successfully cloned on medium-copy number vectors. For example, Barghouthi *et al.*[2] used a library of *Aeromonas hydrophila* DNA in an RSF1010-derived vector to complement an *E. coli entC* mutant.

It may not be feasible or desirable to construct the library for interspecies complementation directly in the *E. coli* iron acquisition mutant. For example, many of the available *E. coli* iron acquisition mutants do not contain the appropriate mutations for propagation of λ or do not express LamB, the λ receptor and, therefore, cannot be used as hosts for a λ- or cosmid-based library. Alternatively, to maximize the efficiency of library construction, it may be desirable to use an *E. coli* strain which is recombination deficient, which lacks specific restriction systems, and, thus, will not degrade methylated DNA, or which can be made highly competent for transformation. Therefore, it may be necessary either to transfer the library from a standard *E. coli* cloning host to the *E. coli* iron acquisition mutant or to introduce the required mutations or genes into the *E. coli* iron acquisition mutant. Alternatively, a specific iron acquisition mutation can be introduced into one of the standard *E. coli* strains used for library construction by mutagenesis. For example, to obtain a recombination-deficient enterobactin-deficient mutant, Barghouthi *et al.*[2] have mutagenized *E. coli* HB101, a *recA*-deficient strain, and screened for mutants deficient in enterobactin synthesis. The HB101 *ent* mutant was then used as the recipient for complementation with the *A. hydrophila* library. Standard techniques for the genetic manipulation of *E. coli* have been described by Miller[17] and can be used to construct the required strains.

Identification of Complementing Clone

Restoration of the ability to grow on iron-limiting medium provides a direct selection for identification of a clone which complements the *E. coli* iron acquisition mutant. M9 minimal medium[17] supplemented with the appropriate nutrients for the particular strain being studied is a good medium to use for this direct selection. *Escherichia coli* iron acquisition mutants are unable to grow on M9 minimal medium unless the mutation is complemented or the medium is supplemented with an iron source that

[23] C. A. Ishimaru and J. E. Loper, *J. Bacteriol.* **174,** 2993 (1992).

can be utilized. With the availability of high-quality deionized water and reagents, the M9 medium may be iron limiting even for wild-type E. coli strains. Therefore, the addition of trace amounts of iron to the medium may be necessary; the quantity to be added must be determined empirically.

A good screen for complementation of a siderophore biosynthetic mutation is the universal siderophore assay developed by Schwyn and Neilands.[24] This colorimetric assay is based on the affinity of siderophores for iron(III) and therefore is independent of siderophore structure. The chromogenic substrate can be incorporated into agar plates. This assay has been used to distinguish between biosynthetic, transport, and regulatory mutations in E. coli.

Once a complementing clone has been identified, it is often necessary to confirm that the ability to grow on an iron-limiting medium was not due to complementation of another mutation. For example, the E. coli strains used for cloning a gonococcal fhuB analog also contained mutations which blocked biosynthesis of enterobactin.[1] Therefore, it was necessary to confirm that the complementing clone had not restored the ability to synthesize enterobactin. To test for production of a specific siderophore, "cross-feeding" of the appropriate biosynthetic mutant can be used. Escherichia coli LG1419 which cannot synthesize enterobactin[25] and E. coli LG1522 which cannot synthesize aerobactin[26] are two mutants that have been used as indicator strains for this purpose. For the "cross-feeding" assay, culture supernatants of the complemented strain, grown in iron-limiting medium to induce siderophore production, are spotted onto a lawn of the indicator strain.

If a siderophore biosynthesis gene was cloned, the ability of the complemented mutant to synthesize a specific siderophore should be confirmed. Chemical assays which can be used are the Arnow method for catechols[27] and the Csáky test for hydroxamates (see [25] in this volume).[28] The "cross-feeding" assays described above for enterobactin and aerobactin are specific for each siderophore and may also be used.

Many of the E. coli outer membrane proteins which are receptors for specific siderophores also serve as receptors for colicins and bacteriophages. The enterobactin receptor FepA is the receptor for colicins B and D,[29] the aerobactin receptor Iut is also the receptor for cloacin produced

[24] B. Schwyn and J. B. Neilands, Anal. Biochem. **160**, 47 (1987).
[25] P. H. Williams and P. J. Warner, Infect. Immun. **29**, 411 (1980).
[26] N. H. Carbonetti and P. H. Williams, Infect. Immun. **46**, 7 (1984).
[27] L. E. Arnow, J. Biol. Chem. **118**, 531 (1937).
[28] A. H. Gilliam, A. G. Lewis, and R. J. Anderson, Anal. Chem. **53**, 841 (1981).
[29] S. K. Armstrong, C. L. Francis, and M. A. McIntosh, J. Biol. Chem. **265**, 14536 (1990).

by *Enterobacter aerogenes* DF13,[30] and FhuA is the receptor for colicin M and bacteriophages T1, T5, and φ80.[31] If a siderophore receptor gene was cloned, the complemented mutant could be screened for susceptibility to the appropriate colicin or bacteriophage. However, the failure to detect sensitivity to a particular colicin or bacteriophage should be interpreted with caution as the phage and colicin binding domains of the receptor proteins may not be conserved.

Conclusions

It is not possible to predict the success of this strategy for cloning an iron acquisition gene by complementation of an *E. coli* mutant. However, the approach was proved useful for several genes from pathogenic microorganisms that are difficult to manipulate genetically.[1,2,23]

[30] A. Bindereif, V. Braun, and K. Hantke, *J. Bacteriol.* **150,** 1472 (1982).
[31] G. Carmel, D. Hellstern, D. Henning, and J. W. Coulton, *J. Bacteriol.* **172,** 1861 (1990).

Section V

Genetics and Regulation

[30] Bacterial Transformation by Electroporation

By JEFF F. MILLER

Introduction

Electroporation is a versatile method for the introduction of DNA into bacteria.[1-3] The technique involves the application of a brief high-voltage pulse to a sample of cells and DNA. The result is transient membrane permeability and DNA uptake by a subpopulation of the surviving organisms. Electroporation (or electrotransformation) is broadly applicable. It has also been successfully used to introduce DNA into animal,[4] plant,[5] fungal,[6] and protozoan cells.[7]

The mechanism by which electroporation induces reversible membrane permeability is unclear, especially in the case of bacteria. Nevertheless, the technique has been successfully applied to over 100 bacterial species, including gram-positive, gram-negative, and acid-fast organisms, as well as pathogens for humans and/or other animals.[8] Efficiencies of electrotransformation vary widely. An optimized technique capable of generating 10^9 to 10^{10} transformants/μg of plasmid DNA for several *Escherichia coli* strains is in common use.[3] In contrast, some bacterial species respond with much lower efficiencies, and others appear to be intractable. In many cases, however, low electrotransformation efficiencies can be increased by systematic evaluation of several key variables.

This chapter is intended as a practical guide for the use, establishment, and optimization of electroporation protocols for bacterial transformation. Detailed discussions of the principles and practice of the technique can be found in several reviews.[8-11]

[1] B. M. Chassy and J. L. Flickinger, *FEMS Microbiol. Lett.* **44,** 173 (1987).
[2] J. F. Miller, W. J. Dower, and L. S. Tompkins, *Proc. Natl. Acad. Sci. U.S.A.* **85,** 856 (1988).
[3] W. J. Dower, J. F. Miller, and C. W. Ragsdale, *Nucleic Acids Res.* **16,** 6127 (1988).
[4] E. Neumann, M. Schaefer-Ridder, Y. Wang, and P. H. Hofschneider, *EMBO J.* **1,** 841 (1982).
[5] M. L. Fromm, P. Taylor, and V. Walbot, *Nature (London* **319,** 791 (1986).
[6] D. M. Becker and L. Guarente, This series, Vol. 194, p. 182.
[7] V. Bellofatto and G. A. M. Cross, *Science* **244,** 1167 (1989).
[8] J. T. Trevors, B. M. Chassy, W. J. Dower, and H. P. Blaschek, *in* "Guide to Electroporation and Electrofusion" (D. C. Chang, B. M. Chassy, J. A. Saunders, and A. E. Sowers, eds.), p. 265. Academic Press, San Diego, 1992.
[9] W. J. Dower, *Genet. Eng. (N.Y.)* **12,** 275 (1990).

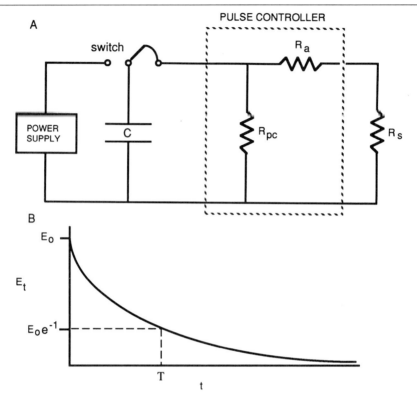

FIG. 1. (A) Capacitor discharge circuit and (B) exponential decay curve with time constant T. See text for details.

Electroporation Devices and Exponential Decay Waveform

A basic understanding of the electrical parameters is necessary for efficient use and optimization of electrotransformation. Capacitor discharge devices are commonly employed to deliver pulses of intensity and duration suitable for electroporation.[3] The pulse generator shown in Fig. 1 generates an electric pulse characterized by an exponential decay waveform. Figure 1A and the specific methods presented below are based primarily on the Gene Pulser apparatus manufactured by Bio-Rad Labora-

[10] W. J. Dower and S. E. Cwirla, in "Guide to Electroporation and Electrofusion" (D. C. Chang, B. M. Chassy, J. A. Saunders, and A. E. Sowers, eds.), p. 291. Academic Press, San Diego, 1992.

[11] W. J. Dower, B. M. Chassy, J. T. Trevors, and H. P. Blaschek, in "Guide to Electroporation and Electrofusion" (D. C. Chang, B. M. Chassy, J. A. Saunders, and A. E. Sowers, eds.), p. 485. Academic Press, San Diego, 1992.

tories (Richmond, CA). Suitable instruments with a variety of features are also available from other sources.[12]

The electric field (E) experienced by a sample held between parallel electrodes during electroporation is given by

$$E = V/d \tag{1}$$

where V is the voltage applied across electrodes separated by a distance d. Although several sample chamber designs are in current use, the most common configuration involves a spectrophotometer-type cuvette with parallel aluminum electrodes. The electric field strength may therefore be adjusted in two ways, by setting the voltage output of the power supply and by using cuvettes with the desired gap distance (usually 0.1 to 0.4 cm).

Two parameters of central importance for bacterial electroporation are the initial electric field strength (E_0) and the duration of the discharge (characterized by the time constant, T). On delivery of the pulse, the electric field rises quickly to a peak value (E_0) and decays over time (t) according to the relationship

$$E_{(t)} = E_0 \, e^{-(t/T)} \tag{2}$$

At time $t = T$, $E_{(t)} = E_0 \, e^{-1}$. T is therefore equal to the time required for the initial field to decay to approximately 37% of its original value, as illustrated in Fig. 1B. The rate of decay is a function of the total resistance and capacitance of the circuit, and

$$T = RC \tag{3}$$

where T is in seconds, R is resistance in ohms (Ω), and C is capacitance in farads (F). Manipulation of the time constant involves adjusting the resistance of the system as well as the size of the capacitor, and most electroporation devices offer a range of capacitors to choose from. In the system shown in Fig. 1A, one of several resistors (R_{pc}) can be switched into the circuit in parallel with the sample (R_s), and a small resistor (R_a) is placed in series with the sample to limit the current and protect the instrument should arcing occur (R_{pc} and R_s comprise the pulse-controller portion of the Bio-Rad instrument). The total resistance of the circuit in Fig. 1A is given by

$$1/R_{total} = 1/(R_s + R_a) + 1/R_{pc} \tag{4}$$

---

[12] B. M. Chassy, J. A. Saunders, and A. E. Sowers, *in* "Guide to Electroporation and Electrofusion" (D. C. Chang, B. M. Chassy, J. A. Saunders, and A. E. Sowers, eds.), p. 555. Academic Press, San Diego, 1992.

If (and only if) the resistance of the sample is very high in comparison to the parallel resistor, the total resistance is essentially determined by the choice of R_{pc} (i.e., when $R_s + R_a \gg R_{pc}$, then $R_{total} \approx R_{pc}$). For this reason the general method described below uses a sample of very low conductivity, consisting of cells suspended in distilled water plus 10% glycerol (~5000 Ω). This allows convenient adjustment of the total resistance and the resulting time constant. For example, with a sample of 5000 Ω in series with a 20-Ω protective resistor (R_a), selecting a 200-Ω parallel resistor (R_{pc}) will result in a total resistance of about 192 Ω [i.e., close to that of R_{pc}, Eq. (4)]. If a 25-μF capacitor is then selected, the resulting RC time constant is given by Eq. (3): 192 $\Omega \times (25 \times 10^{-6}$ F) = 4.8×10^{-3} sec = 4.8 msec. Because the resistance of the sample is also affected by the electric field strength, R_s will vary throughout the pulse, and this is an estimate rather than an exact calculation. Sample resistance is also dependent on temperature, volume, and the distance between the electrodes. Because R_a is small in relation to R_s, it has a negligible effect on the voltage drop across the sample. In addition, most of the current passes through R_{pc}, and this decreases the chance of arcing in the sample chamber during delivery of the pulse.

General Method for Bacterial Electroporation: High Efficiency Transformation of Escherichia coli

The highest efficiencies of electrotransformation reported thus far have been obtained with *E. coli* strains used for recombinant DNA manipulations.[3,13,14] The protocol described below, adapted from Dower *et al.*,[3] is capable of giving 10^9 to 10^{10} transformants/μg of plasmid DNA with several *E. coli* strains including LE392, DH5α, MC1061, and WM1100.[3,9] It should be noted that considerable variability may be observed between strains, and between different preparations of the same strain. In some cases, especially with gram-negative bacteria, application of the basic *E. coli* method with a few modifications will allow transformation of new species, and optimization will mainly involve adjustment of the electrical parameters. Other organisms may require special attention to variables in addition to those associated with the electrical pulse, and some bacteria may be recalcitrant. In general, gram-positive bacteria appear to be more difficult to transform by electroporation than are gram-negative bacteria, although the technique is widely used for both classes of organisms. Variables affecting the technique and methods for optimizing electrotransformation

[13] N. M. Calvin and P. C. Hanawalt, *J. Bacteriol.* **170**, 2796 (1988).
[14] M. Smith, J. Jessee, T. Landers, and J. Jordan, *Focus (BRL)* **12**, 38 (1990).

of gram-negative, gram-positive, and acid-fast species are discussed in a following section.

Growth and Preparation of Cells

1. Inoculate rich medium (e.g., L broth or "superbroth," 3.2% tryptone, 2.0% yeast extract, 0.5% NaCl) with a 1/100 dilution of an overnight culture. Use a flask that can accommodate at least four times the culture volume. Grow cells at 37° with vigorous shaking to aerate the culture.

2. Harvest cells in early to mid-log phase of growth (for many *E. coli* strains OD_{600nm} ~0.5). Cultures in later growth stages may yield significantly lower efficiencies. Chill the growth flask on wet ice (15–30 min) and pellet cells by centrifugation at 4° in a prechilled rotor at 4000 *g* for 15 min. Remove as much of the growth medium as possible.

3. Electroporation at high voltage requires a cell suspension of very low conductivity, and thorough washing of the cells is therefore required. During the washing procedure cells and solutions should be kept ice cold. Resuspend the pellet in 1 culture volume of ice-cold distilled or deionized water and centrifuge as above.

4. Resuspend the pellet in 1/2 volume of cold deionized water and repeat centrifugation.

5. Resuspend pellet in 1/50 volume of either ice-cold deionized water or 10% glycerol (as a cryoprotectant) if cells are to be frozen for storage. Centrifuge as above.

6. Resuspend in 1/500 volume of deionized water or 10% glycerol. The final concentration should be in the range of $2–5 \times 10^{10}$ cells/ml.

7. If desired, aliquots in 10% glycerol can be frozen by incubation on powdered dry ice and stored at −70° (for at least 1 year).

8. The procedure can be scaled as desired. A 1-liter culture yields enough cells for about 50 standard electrotransformations.

Electrotransformation

1. Chill the shocking chamber and electroporation cuvette on ice. If using frozen cells, thaw an aliquot on ice.

2. In a microcentrifuge tube on ice mix 40 μl of cold cells with 1–4 μl of DNA in a low ionic strength medium such as deionized water or TE (10 mM Tris-HCl, 1 mM EDTA, pH 8.0).

3. Transfer the mixture to a chilled electroporation cuvette (0.2 cm gap) and tap the solution to the bottom of the chamber. Chill on ice for 1–2 min.

4. Deliver a pulse of 12.5 kV/cm with a time constant of approximately 5 msec. On the Bio-Rad Gene Pulser the settings are 2.5 kV (for a 0.2-cm

cuvette) with a 25-μF capacitor and a parallel resistance of 200 Ω. Analogous conditions for other instruments can be obtained from the manufacturer's instructions.

5. *Immediately* after the pulse, add 1 ml of outgrowth medium (SOC: 2% tryptone, 0.5% yeast extract, 10 mM NaCl, 2.5 mM KCl, 10 mM MgCl$_2$, 10 mM MgSO$_4$, 20 mM glucose) at room temperature to the cuvette and gently resuspend the sample.

6. Transfer to a culture tube and outgrow for 45 min (1–2 generations) with aeration at 37°. This allows establishment of plasmid replication and expression of antibiotic resistance.

7. Dilute and plate on selective media. When optimizing protocols it is also useful to determine the number of viable cells before and after the pulse by plating dilutions on nonselective media.

8. A typical 0.2-cm cuvette will accommodate up to 400 μl of sample, and low ionic strength DNA solutions can be added to 10% of the final volume. Cuvettes of 0.1 cm holding 20–80 μl are also available; however, the applied voltage should be adjusted for the smaller gap distance.

Reagents

During electroporation, components of the sample in addition to DNA are free to enter cells. Reagents of the highest quality should therefore be used throughout the procedure.

Precautions

High-voltage, high-current discharges such as those used for electroporation are potentially lethal, and manufacturer's recommendations for avoiding safety hazards should be followed. Explosive arcing may also occur, especially with samples of lower resistance, and all or part of the sample may be aerosolized as a result. Appropriate measures should be taken with pathogenic bacteria to avoid danger to personnel.

Establishing and Optimizing Electroporation Protocols

Electroporation is an extremely adaptable method, and many variations of the technique are reported.[8,9] As noted above, differences between strains and species can result in large fluctuations in transformation efficiency, and repeated attempts with some bacteria have been unsuccessful. Nonetheless, electroporation should be considered as a potential method for transforming any bacterial species. Modifications in the technique are often required, and the best guide is usually an understanding of the anatomy and physiology of the organism. The cell wall of gram-positive bacteria is a notable example. Although evidence is fragmentary, the density and thickness of the peptidoglycan layer appear to interfere with electrotransformation, and agents or growth conditions that negatively influence cell wall synthesis and structure have proved beneficial.[11,15–17] Some factors that influence electrotransformation efficiency (transformants per microgram or picomole of DNA) and/or frequency (fraction of surviving cells transformed) are discussed below. Further information can be found in references noted at the beginning of the chapter.

Cell Growth and Preparation

A major factor in cell preparation is the point in the growth curve at which the cells are harvested, and this is an important variable that is usually easily optimized. *Escherichia coli*[9] and *Actinobacillus pleuropneumoniae*[18] give highest efficiencies when harvested in early to mid-exponential growth phase. In contrast, late log or early stationary phase cells perform best for some gram-positive organisms including *Clostridium perfringens*[19,20] and *Lactococcus lactis*.[21] The growth medium can also be chosen to fill particular needs, and the use of rich broths is recommended when possible since higher cell yields result. The concentration of cells also has a profound effect on the efficiency of transformation. With *E. coli,* maximal efficiencies are obtained around $3–5 \times 10^{10}$ cells/ml.[2,9] Alternatively, for bacterial strains that secrete nucleases, lower cell concentrations may be beneficial. The effect of DNases may be decreased by adding DNA immediately before the pulse,[22] by including carrier DNA, or possibly by using DNase inhibitors.

Wash solutions and electroporation buffers can be supplemented with sucrose or other nonionic agents to increase osmolarity. Sucrose at 300 mM has been used to prevent lysis of *Pseudomonas aeruginosa* during electroporation.[23] Gram-positive organisms subjected to conditions that destabilize the cell wall are usually stabilized by 0.5–0.625 M sucrose.[15–17]

[15] G. M. Dunny, L. N. Lee, and D. J. LeBlanc, *Appl. Environ. Microbiol.* **57,** 1194 (1991).
[16] S. F. Park and G. S. A. B. Stewart, *Gene* **94,** 129 (1990).
[17] I. B. Powell, M. G. Achen, A. J. Hillier, and B. E. Davidson, *Appl. Environ. Microbiol.* **54,** 655 (1988).
[18] G. Lalonde, J. F. Miller, L. S. Tompkins, and P. O'Hanley, *Am. J. Vet. Res.* **50,** 1957 (1989).
[19] M. K. Phillips-Jones, *FEMS Microbiol. Lett.* **66,** 221 (1990).
[20] S. P. Allen and H. P. Blaschek, *FEMS Microbiol. Lett.* **70,** 217 (1990).
[21] D. A. McIntyre and S. K. Harlander, *Appl. Environ. Microbiol.* **55,** 2621 (1989).
[22] J. E. Alexander, P. W. Andrew, D. Jones, and I. S. Roberts, *Lett. Appl. Microbiol.* **10,** 179 (1990).
[23] A. W. Smith and B. H. Iglewski, *Nucleic Acids Res.* **17,** 10509 (1989).

Salts (e.g., 1 mM MgCl$_2$ for *Bacteroides* spp.[24]) should only be added when necessary, and the concentration kept as low as possible. The same holds true for pH-buffering agents. Manufacturer's recommendations should be consulted when electroporating samples of appreciable ionic strength. Temperatures used during cell preparation and electroporation can be increased if necessary for cold-intolerant organisms, although temperature is an important variable during pulse delivery (see below). The effect and specific manner of freezing should also be empirically determined for new species, and initial trials are best conducted with freshly prepared cells.

Gram-Positive Cell Walls

Several methods have been used to partially overcome the apparent barrier to transformation posed by the gram-positive cell wall. Growth of *Enterococcus faecalis* in inhibitory concentrations of glycine (1.5–6%) to interfere with cell wall synthesis results in significant improvements in transformation efficiency.[15,25] Under these conditions sucrose is usually added for osmotic stabilization. Pretreatment of *Listeria monocytogenes* by growth in media containing penicillin G (10 μg/ml) and sucrose (0.5 M) significantly increases transformation.[16] Finally, mild lysozyme treatment (2 kU/ml for 20 min at 37°) followed by electroporation and growth in media containing sucrose increases the efficiency of electroporation of *Streptococcus lactis* by 300- to 1000-fold.[17] Under these conditions, 90% of the recovered transformants were derived from osmotically fragile cells. The optimal concentration and treatment time should be empirically determined when using agents that affect cell wall integrity.

Electrical Parameters: E_0, T

The small size of bacterial cells necessitates high initial electric field strengths for transformation, typically in the range of 5–20 kV/cm. The electric field strength and time constant are primary determinants of electrotransformation efficiency, and these parameters must be optimized for specific species, strains, and methods of cell preparation. Within certain boundaries, the relative effects of pulse amplitude and duration are compensatory,[2,3] and the relationship between efficiency and either parameter can manifest as a narrow window[18] or as a line with positive slope.[2] Most bacteria seem to display unique response curves. It should also be noted that high values of E_0 or T often result in unacceptable levels of cell killing and increase the occurrence of arcing during the pulse.

[24] C. J. Smith, A. Parker, and M. B. Rogers, *Plasmid* **24**, 100 (1990).
[25] H. Holo and I. F. Nes, *Appl. Environ. Microbiol.* **55**, 3119 (1989).

A reasonable starting point for optimization is to set the time constant at approximately 5 msec by choosing an appropriate capacitor and parallel resistor. Then E_0 is varied from 5 to 20 kV/cm in increments of 1–2 kV (0.1-cm cuvettes are used to achieve high values for E_0), and both the efficiency of transformation and cell survival are measured. The effect of changing the time constant can then be assessed at an appropriate value for E_0. Significant cell death may[3] or may not[2] accompany electrotransformation. When cell death does occur, optimal transformation conditions may represent a compromise between killing and permeabilization. Alterations in electric field strength, total resistance, and capacitance also change the current, power, charge, and energy applied to the sample. The possibility of significant heating and hydrolysis should also be considered. A detailed discussion of electrical variables can be found in Ref. 3.

Sample and Pulse Delivery

The volume, temperature, and ionic strength of the sample during pulse delivery often have a noticeable effect on the outcome. Decreased resistance increases current and sample heating, and at high voltages media that are too conductive may cause explosive arcing and loss of the sample.

In most cases the temperature of the sample, cuvette, and shocking chamber are kept as low as possible during pulse delivery.[3] Low temperature may have a beneficial effect by reducing ohmic heating, by affecting physical properties of the cell membrane, or by some other mechanism. Low temperature also reduces the chance of explosive arcing.

In the procedure described above, the sample is immediately resuspended in outgrowth media after the pulse. Delaying this step by as little as 10 min can result in a nearly 5-fold decrease in efficiency with *E. coli*.[3] The time of outgrowth may also need adjustment depending on the particular antibiotic resistance determinant or phenotypic property used for selection.[24] For plasmids encoding inducible antibiotic resistance, outgrowth in the presence of subinhibitory concentrations of antibiotic may improve recovery of transformants.

DNA

The number of transformants obtained during electroporation is highly dependent on DNA concentration. For *E. coli* the linear relationship between transformation frequency and DNA concentration holds for over 6 orders of magnitude,[3] and similar results with many other bacterial species have been reported. At high DNA concentrations (7.5 μg/ml) using the *E. coli* procedure described above, almost 80% of the surviving cells can be transformed,[3] and the need for plasmid selection can be circumvented if desired. In contrast to chemical treatments for inducing DNA uptake by *E. coli,* electroporation shows a high capacity for DNA and is not easily saturated.[2,3,9] Carrier DNA (salmon sperm) can also be used at concentrations up to 10 μg/ml with no decrease in transformation efficiency.[10] In contrast, other species may have a lower DNA capacity. The subpopulation of cells that are electrocompetent in a given situation appears to vary and may be as low as 1–10% in the case of *Lactobacillus casei.*[26]

The ionic strength of the DNA sample should be low to prevent arcing. Ethanol precipitation followed by extensive washing can be used to change buffers. Alternatively, microdialysis of small volumes can be performed by spotting the sample onto a microdialysis membrane (Millipore, Bedford, MA, type VS, pore size 0.025 μm) floating in an appropriate buffer. Microdialysis of ligation mixtures against 10% glycerol/1 mM EDTA, pH 8, for 45 min can significantly increase transformation efficiencies.[27] Plasmid DNA prepared by a variety of methods can be used for electrotransformation, although preparations of high purity usually give the best results.

The effect of plasmid size and topology on transformation efficiency have also been examined. For *E. coli,* plasmids ranging from 2.7 to 21 kilobases (kb) produce the same molar efficiency of transformation, and converting plasmids from supercoiled to relaxed circular form does not affect transforming activity.[9] The latter observation highlights the utility of electroporation for transforming ligation mixtures. In some cases topology does appear to be important (e.g., *Listeria monocytogenes*[16]), and linear molecules usually have low transformation efficiencies. A report describes the introduction of a 136-kb plasmid into *E. coli* at an efficiency of 1.7 × 10^6 transformants/μg,[28] and a 250-kb megaplasmid was transformed into *Agrobacterium,* although at a much lower efficiency.[29] Although bacteriophage DNA can be used for electrotransformation, little information on the use of linear chromosomal DNA has appeared. Deletions or rearrangements do not usually accompany plasmid electrotransformation.

Genetic Factors

Replication and the expression of antibiotic resistance are obvious requirements for successful plasmid transformation. In many cases, re-

[26] B. M. Chassy, A. Mercenier, and J. Flickinger, *Trends Biotechnol.* **6,** 303 (1988).

[27] M. Jacobs, S. Wnendt, and U. Stahl, *Nucleic Acids Res.* **18,** 1653 (1990).

[28] E. D. Leonardo and J. M. Sedivy, *Bio/Technology* **8,** 841 (1990).

[29] T. Mozo and P. J. J. Hooykaas, *Plant Mol. Biol.* **16,** 917 (1991).

striction–modification systems can present formidable barriers to transformation. For example, the efficiency of electrotransformation of *Campylobacter jejuni* with plasmid DNA increases by at least 4 orders of magnitude when the DNA is prepared from *Campylobacter* as opposed to *E. coli* HB101.[2] During optimization it may therefore be advantageous to use plasmid DNA isolated from the same strain or species being examined, and endogenous plasmids,[18] or conjugative transfer systems,[2] may be useful for this purpose. Although general methods for overcoming restriction barriers to electrotransformation have not been reported, simply using large amounts of DNA may offer a partial solution. Alternatively, it may be practical to select or construct mutations that eliminate a particular restriction system. In a similar vein, a recent report describes the isolation and characterization of efficient plasmid transformation (Ept) mutants of the acid-fast bacterium *Mycobacterium smegmatis*.[30] The Ept mutations increase electrotransformation efficiencies by 4 to 5 orders of magnitude over that of the parent strain, giving efficiencies of greater than 10^5 transformants/μg of plasmid DNA.[30] The Ept phenotype does not appear to result from mutations in restriction–modification systems or mutations affecting DNA uptake, leaving plasmid replication and/or maintenance among the likely possibilities. For *E. coli, recA* mutations usually decrease transformation efficiencies when compared to isogenic strains with a wild-type allele, probably resulting from the reduced viability of *recA* strains.[9]

Although many bacterial species are naturally competent, natural uptake systems are often associated with deletions and/or rearrangements of transforming plasmid DNA, and the relative attributes of electroporation as opposed to natural transformation remain to be determined. A study with *Haemophilus influenzae* reported that naturally competent cells were several orders of magnitude less efficient than noncompetent cells for electrotransformation with plasmid DNA.[31]

[30] S. B. Snapper, R. E. Melton, S. Mustafa, T. Keiser, and W. R. Jacobs, Jr., *Mol. Microbiol.* **4,** 1911 (1990).
[31] M. A. Mitchell, K. Skowronek, L. Kauc, and S. Goodgal, *Nucleic Acids Res.* **19,** 3625 (1991).

[31] Analysis and Construction of Stable Phenotypes in Gram-Negative Bacteria with Tn5- and Tn10-Derived Minitransposons

By Víctor de Lorenzo and Kenneth N. Timmis

Introduction

Molecular genetics is one of the most powerful tools available to microbiologists to investigate the basis of bacterial phenotypes. Even where biochemical or structural data are scarce, genetic analysis (particularly when combined with DNA sequencing) often reveals considerable insight into the mechanisms involved in microbial processes of medical, industrial, and environmental interest. Current genetic assets available for studying the properties of different microorganisms have been reviewed in Volume 204 of this series and also this volume and elsewhere.[1,2] In this chapter, we discuss the use of minitransposons, developed in our laboratory, for the analysis, construction, and manipulation of complex phenotypes in a wide range of gram-negative bacteria.[3-5] The only two conditions limiting the use of these minitransposons is that a target strain should (1) be an effective recipient in RP4-mediated conjugal transfer and (2) support transposition of Tn5 and/or Tn10. These two conditions are generally satisfied by most genera of interest.

Transposons and Minitransposons

Transposons are the most versatile tools for the genetic analysis of bacteria.[6] Unlike the situation with chemical mutagenesis, transposons produce complete disruption of the mutated gene, resulting in nonleaky phenotypes; moreover, multiple insertions or retranspositions are rare events, and, most important, the resulting phenotype is genetically linked

[1] K. Chater and D. Hopwood, in "Genetics of Bacterial Diversity" (D. Hopwood and K. Chater, eds.), p. 53. Academic Press, London, 1989.
[2] V. de Lorenzo and K. N. Timmis, in "Pseudomonas, Molecular Biology and Biotechnology" (E. Galli, S. Silver, and B. Witholt, eds.), p. 415. American Society for Microbiology, Washington, D.C., 1992.
[3] M. Herrero, V. de Lorenzo, and K. N. Timmis, J. Bacteriol. 172, 6557 (1990).
[4] V. de Lorenzo, M. Herrero, U. Jacubzik, and K. N. Timmis, J. Bacteriol. 172, 6568 (1990).
[5] V. de Lorenzo, L. Eltis, B. Kessler, and K.N. Timmis, Gene 123, 17 (1993).
[6] C. M. Berg, D. E. Berg, and E. A. Groisman, in "Mobile DNA" (D. E. Berg and M. M. Howe, eds.), p. 879. American Society for Microbiology, Washington, D.C., 1989.

to a selectable marker such as resistance to an antibiotic. The spectrum of different types of transposons, their mechanisms of transposition, and their applications in genetic engineering have been the subject of excellent reviews.[6] We discuss here only those properties of the Tn*10* and Tn*5* transposons which are relevant to the subject of this chapter.

Tn*10* and Tn*5* belong to the class of the so-called composite transposons, namely, those whose mobility is determined by two insertion sequences (IS) flanking the DNA region determining the selectable phenotype. Although their mechanisms of transposition differ substantially, both Tn*5*[7] and Tn*10*[8] are able to transfer from one replicon to another as a consequence of the action of the transposase encoded by one of the two IS elements on the cognate short target sequences located at the end of the transposon. Although the transposase gene is a component of naturally occurring transposons, work carried out by the groups of Kleckner on Tn*10* and Berg on Tn*5* has shown that the enzyme still works efficiently when its gene is outside of the mobile unit, though preferably placed *in cis* to the cognate terminal sequences.[8,9] This finding has permitted the construction of recombinant transposons in which only those elements essential for transposition (i.e., IS terminal sequences and transposase gene) have been retained and arranged such that the transposase gene is adjacent to but outside of the mobile DNA segment.[9,10] Because the elimination of nonessential sequences leads to a substantial reduction of transposon size, the resulting recombinant minitransposons are much simpler to handle than natural transposons.

Other advantages of minitransposoons are mentioned later, but two properties are worth emphasizing here. First, once inserted in a target sequence minitransposons are inherited in a stable fashion and, unlike natural transposons, do not provoke DNA rearrangements or other forms of genetic instability because they lack the cognate transposase gene and the greater parts of the IS elements present in the wild-type transposons. Even if host cells later acquire a natural transposon of the same type, the Tn*10* and Tn*5* transposases work poorly *in trans*[7,8] and will generally not stimulate retransposition of the minitransposon. A second advantage of the loss of the transposase gene during transfer to a new replicon is that the host cell does not become immune to further rounds of transposition.

[7] D. E. Berg, *in* "Mobile DNA" (D. E. Berg and M. M. Howe, eds.), p. 185. American Society for Microbiology, Washington, D.C., 1989.

[8] N. Kleckner, *in* "Mobile DNA" (D. E. Berg and M. M. Howe, eds.), p. 227. American Society for Microbiology, Washington, D.C., 1989.

[9] K. W. Dodson and D. E. Berg, *Gene* **76**, 207 (1989).

[10] J. Way, M. A. Davis, D. Morisato, D. E. Roberts, and N. Kleckner, *Gene* **32**, 369 (1984).

This permits the organism to be remutagenized with the same system, provided that subsequent transposons contain distinct selection markers.[3]

General Minitransposon Delivery System

Exploitation of a transposon as a genetic tool requires a system of delivering it into the target strain in such a way that its selectable phenotype is expressed under the experimental conditions employed only through integration of the transposon into a replicon of the new host. The system developed for the mini-Tn*10* elements described in detail elsewhere in this series[11] (and exploited for other transposons and minitransposons[12]) is based on infection of the target strain with a defective lambda (λ) phage carrying the minitransposon, and therefore its utilization is virtually limited to *Escherichia coli* and related bacteria.

For the mini-Tn*10* and mini-Tn*5* transposons described below we have used a more general suicide delivery system based on the narrow-host range plasmid R6K.[13] Plasmids having the R6K origin of replication require the R6K-specified replication protein π and can be maintained only in host strains producing this protein.[13] R6K derivative pGP704,[14] which was developed in the laboratory of Mekalanos, has in addition to an R6K origin of replication RP4*oriT*, the origin of transfer sequence of the auto-conjugative promiscuous plasmid RP4. pGP704 is maintained stably in λ*pir* lysogens and can be mobilized into target bacteria through RP4 transfer functions. Bacteria receiving pGP704 but lacking the π protein are nonpermissive and do not maintain the transferred plasmid. Because mini-Tn*5* and mini-Tn*10* transposons and their cognate transposase genes are constructed as inserts in pGP704 (Figs. 1 and 2), selection of exconjugants stably expressing the marker of the minitransposon selects for clones in which they have been transferred to a replicon of the recipient. With these plasmid series (pLOF for Tn*10* derivatives and pUT for Tn*5* derivatives), transposition is promoted by the cognate transposase encoded on the same plasmid proximal but external to the transposon.

Methods

The basic structures of the mini-Tn*5* and mini-Tn*10* transposons depicted in Figs. 1, 2, 3[15] have been developed to satisfy a number of

[11] N. Kleckner, J. Bender, and S. Gottesman, this series, Vol. 204, p. 139.
[12] S. H. Phadnis and D. E. Berg, *Proc. Natl. Acad. Sci. U.S.A.* **84,** 9118 (1987).
[13] R. Kolter, M. Inuzuka, and D. Helinski, *Cell* (*Cambridge, Mass.*) **15,** 1199 (1988).
[14] V. Miller and J. Mekalanos, *J. Bacteriol.* **170,** 2575 (1988).
[15] J. H. Miller, "A Short Course in Bacterial Genetics." Cold Spring Harbor Laboratory, Cold Spring Harbor, New York, 1992.

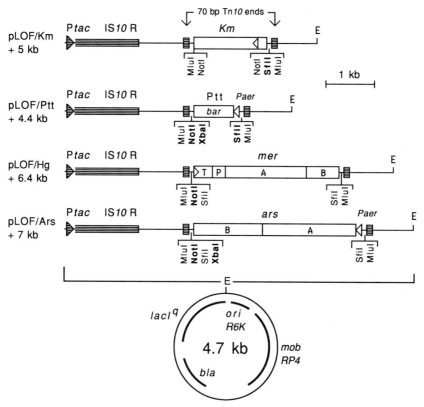

FIG. 1. Mini-Tn*10* delivery plasmids. The common portion of the constructions (a pG P704 derivative plasmid containing a *lacI*^{q} gene; 4.7 kb) is shown at the bottom. The elements of the transposition system (IS*10*R transposase gene expressed from the P*tac* promoter and the Tn*10* inverted repeat ends), the resistance genes of the transposons (Km, kanamycin; *bar*, Ptt/bialaphos; *mer*, HgCl$_2$/phenylmercuric acetate; *ars*, NaAsO$_2$), and relevant restriction sites (E, *Eco*RI) are indicated. Unique sites in the delivery plasmid are indicated by boldface type. The *bar* and *ars* genes are expressed from a heterologous promoter, namely, the promoter P*aer* (see Herrero et al.[3]).

requirements for applications discussed below. The basic experimental design for their use is identical in all cases regardless of the application or the target strain used. These transposons have found broad utilization in the analysis and/or manipulations of a variety of bacteria including *E. coli, Klebsiella, Salmonella, Proteus, Vibrio, Bordetella, Actinobacillus, Rhizobium, Acinetobacter, Rhodobacter, Agrobacterium, Alcaligenes, Aeromonas, Chromohalobacter,* and several pseudomonads.

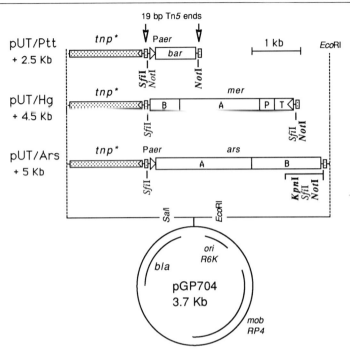

FIG. 2. Delivery plasmids for mini-Tn5 elements with non-antibiotic selection determinants *bar*, *mer*, and *ars*. The common portion of the constructions corresponding to pGP704 (3.7 kb) is shown at the bottom. The transposition system includes the Tn5 terminal ends and a IS50R *tnp* gene devoid of *Not*I sites (*tpn**). Unique sites for the insertion of foreign DNA fragments are indicated in boldface type. For a more detailed map of the components of delivery system, see Fig. 5.

Setting Up Mating Mixture

In a typical mating experiment, donor strains *E. coli* S17-1 λ*pir* or *E. coli* SM10 λ*pir* (Table I) carrying the desired pLOF or pUT derivative are grown overnight with shaking at 37° in 2 ml of LB medium[15] containing 200 µg/ml ampicillin (Ap) and an adequate concentration of the antibiotic appropriate to the transposon marker (see below) to ensure maintenance of the delivery plasmid. The recipient is separately grown under the same conditions (at a different temperature, if necessary) but preferably without selection (see Troubleshooting). Then 10–50 µl of the donor and recipient cultures (~10⁸ cells) are added to 5 ml of 10 m*M* MgSO₄, vortexed for a few seconds, transferred to a 5-ml disposable syringe, and filtered through a Millipore (Bedford, MA) membrane, 13–25 mm diameter type HA (0.45 µm) or equivalent, placed on a reusable filter case. If too many donor

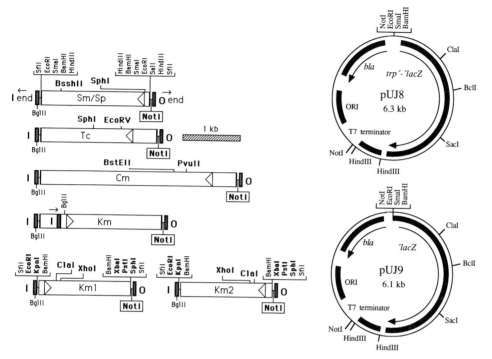

FIG. 3. Mini-Tn5 elements. The mobile units shown are present in the delivery plasmids (pUT series, Table I) as XbaI–EcoRI restriction fragments of different sizes (see Fig. 5). The XbaI site is external to the I end of Tn5, and the EcoRI site external to that of the O end (not shown); neither site is carried by the minitransposon into a different replicon (the BglII site present within the I end is, however, retained after transposition). The first four minitransposons shown carry on both sides of the element, just inside of the inverted repeats (indicated by a thicker line), strong terminators and a collection of restriction sites (displayed only for mini-Tn5 Sm/Sp). Restriction sites unique to the mobile unit (but not necessarily to the delivery plasmid) are indicated in boldface type. The NotI site near the O end is unique to both the minitransposon and the delivery plasmid. At right two auxiliary plasmids for generating in vitro lacZ fusions of the type I (pUJ8) and the type II (pUJ9) are shown. Fusions constructed with these plasmids can be transferred to any of the minitransposons as NotI restriction fragments.[4] Because of the lack of transcriptional terminators upstream of the trp'–'lacZ reporter, a significant level of β-galactosidase activity is present in pUJ8-containing bacteria even in the absence of a promoter cloned in the plasmid. The basal level of the trp'–'lacZ reporter varies with different hosts. In some cases, addition to the medium of the competitive inhibitor of β-galactosidase phenylethyl-β-D-thiogalactoside (TPEG)[15] helps to suppress background LacZ activity. For the pedigree of each construction, see de Lorenzo et al.[4]

TABLE I
Bacteria, Phage, and Plasmids

Species and strain	Relevant genotype/characteristics	Ref.
E. coli CC118 λ*pir*	Δ(*ara-leu*), *araD*, Δ*lacX*74, *galE*, *galK*, *phoA*20, *thi*-1, *rpsE*, *rpoB*, *argE* (Am), *recA*1, λ*pir* phage lysogen	*a*
E. coli SM10 λ*pir*	Km^r, *thi*-1, *thr*, *leu*, *tonA*, *lacY*, *supE*, *recA* :: RP4-2-Tc :: Mu, λ*pir*	*b*
E. coli S17-1 λ*pir*	Tp^r Sm^r *recA*, *thi*, *pro*, *hsdR*$^-$*M*$^+$ RP4 : 2-Tc : Mu : Km Tn7, λ*ph*	
E. coli HB101	Sm^r *recA*, *thi*, *pro*, *leu*, *hsdR*$^-$*M*$^+$	
E. coli LE392	*supF*, *supE*, *hsdR*, *galK*, *trpR*, *metB*, *lacY*, *tonA*	
P. putida KT2442	*hsdR*, Rif^r	*a*

Plasmid	Genotype/phenotype/characteristics	
p18 Sfi	Ap^r; identical to pUC18 but with *Sfi*I/*Eco*RI/*Sal*I/*Hin*dIII/*Sfi*I as MCS (Fig. 4)	*a*
p18 Not	Ap^r; identical to pUC18 but with *Not*I/*Eco*RI/*Sal*I/*Hin*dIII/*Not*I as MCS (Fig. 4)	*a*
pUC18 Sfi	Ap^r; identical to pUC18 but with *Sfi*I/polylinker of pUC18/*Sfi*I as MCS (Fig. 4)	*a*
pUC18 Not	Ap^r; identical to pUC18 but with *Not*I/polylinker of pUC18/*Not*I as MCS (Fig. 4)	*a*
pGP704	Ap^r; *ori* R6K, *mob* RP4, MCS of M13 tg131	*b*
pLOF Km	Ap^r Km^r; delivery plasmid for mini-Tn*10*Km	*a*
pLOF Ptt	Ap^r Ptt^r; delivery plasmid for mini-Tn*10*Ptt	*a*
pLOF Hg	Ap^r Hg^r; delivery plasmid for mini-Tn*10*Hg	*a*
pLOF Ars	Ap^r Ars^r; delivery plasmid for mini-Tn*10*Ars	*a*
pUT Ptt	Ap^r Ptt^r; delivery plasmid for mini-Tn*5* Ptt	*a*
pUT Hg	Ap^r Hg^r; delivery plasmid for mini-Tn*5* Hg	*a*
pUT Ars	Ap^r Ars^r; delivery plasmid for mini-Tn*5* Ars	*a*
pUT mini-Tn*5* Km	Ap^r Km^r; delivery plasmid for mini-Tn*5* Km	*c*
pUT mini-Tn*5* Km1	Ap^r Km^r; delivery plasmid for mini-Tn*5* Km1	*c*
pUT mini-Tn*5* Km2	Ap^r Km^r; delivery plasmid for mini-Tn*5* Km2	*c*
pUT mini-Tn*5* Sm/Sp	Ap^r Sm^r; delivery plasmid for mini-Tn*5* Sm/Sp	*c*
pUT mini-Tn*5* Tc	Ap^r Tc^r; delivery plasmid for mini-Tn*5* Tc	*c*
pUT mini-Tn*5* Cm	Ap^r Cm^r; delivery plasmid for mini-Tn*5* Cm	*c*
pUT mini-Tn*5* *lacZ1*	Ap^r Km^r; delivery plasmid for mini-Tn*5* *lacZ1*	*c*
pUT mini-Tn*5* *lacZ2*	Ap^r Km^r; delivery plasmid for mini-Tn*5* *lacZ2*	*c*
pUT mini-Tn*5* *phoA*	Ap^r Km^r; delivery plasmid for mini-Tn*5* *phoA*	*c*
pUT mini-Tn*5* *xylE*	Ap^r Km^r; delivery plasmid for mini-Tn*5*, *xylE*	*c*
pUT mini-Tn*5* *luxAB*	Ap^r Tc^r; delivery plasmid for mini-Tn*5* *luxAB*	*c*
RP4	Ap^r Km^r Tc^r; Tra$^+$ IncP1 replicon	*d*
RK2013	Km^r, *ori* ColE1, RK2-Mob$^+$ RK2-Tra$^+$	*e*
RK600	Cm^r, *ori* ColE1, RK2-Mob$^+$ RK2-Tra$^+$	*f*
pUJ8	Ap^r, *trp'*–'*lacZ* promoter probe plasmid vector, *lacZ* fusions type I	*c*
pUJ9	Ap^r, *lacZ* promoter probe plasmid vector, *lacZ* fusions type II	*c*
pCNB5	Ap^r Km^r; delivery plasmid for mini-Tn*5*, *lacI*^{q}/P*trc*	*g*

^{a} M. Herrero, V. de Lorenzo, and K. N. Timmis, *J. Bacteriol.* **172**, 6557 (1990).

^{b} V. Miller and J. Mekalanos, *J. Bacteriol.* **170**, 2575 (1988).

and recipient cells are used and the membrane clogs, it is better to refilter a more diluted mix. After filtering, the drained membrane is carefully removed from the case with sterile tweezers (preferably with curved tips), placed on the agar surface of an LB plate (cell side up), and incubated at 30°–37° for 8–18 hr. Air bubbles should be avoided between the filter and agar surface of the plate. In the case of the mini-Tn*10* derivatives, a substantial increase in transposition frequencies can be obtained if, after 4–5 hr of incubation of the filter on an LB plate, the membrane with the mating mix is transferred to the surface of another LB plate containing 50–100 μM isopropyl-β-D-thiogalactoside (IPTG) to increase expression of the transposase[10,11] (Fig. 1). Unfortunately, the same approach seems not to work when applied to Tn*5* derivatives because hyperexpression of the Tn*5*-encoded transposase gene *tnp* results in the simultaneous hyper-production of a transposase inhibitor.[7]

Although this mating protocol should work in most cases, several alternatives are possible, according to individual needs. Donor/recipient ratios can be changed as well as temperatures and the time of mating to suit specific requirements of the recipient. Furthermore, if only a few insertions are needed, 10-μl drops of cultures of donor and recipient strains can simply be mixed and spotted on an LB plate, which is then dried and incubated for several hours before the mating mixture is streaked out on selective medium. Also, in some cases it has been possible to generate mini-Tn*10* insertions by just transforming or electroporating the pLOF derivative into the target bacterium, followed by a short period of expression in LB plus IPTG and subsequent plating on selective medium. So far, however, we have never achieved the same with mini-Tn*5* derivatives, probably because of the low efficiency of the combined transformation/transposition process.

Selection of Exconjugants

The next and critical step is the selection of clones with the minitransposon inserted in the chromosome. For this, the filter with the mating mix is resuspended in 5 ml of 10 mM $MgSO_4$ and 100–500 μl of this suspension (which can be kept at 5° for several weeks) is plated on selective

---

[c] V. de Lorenzo, M. Herrero, U. Jacubzik, and K. N. Timmis, *J. Bacteriol.* **172**, 6568 (1990).

[d] C. M. Thomas, *Plasmid* **5**, 10 (1980).

[e] D. Figurski and D. Helinski, *Proc. Natl. Acad. Sci. U.S.A.* **76**, 1648 (1979).

[f] B. Kessler, V. de Lorenzo, and K. N. Timmis, *Mol. Gen. Genet.* **233**, 293 (1992).

[g] V. de Lorenzo, L. Eltis, B. Kessler, and K. N. Timmis, *Gene* **123**, 17 (1993).

agar. The plates are incubated at the optimal growth temperature of the recipient strain until colonies become visible. It is generally prudent to first plate out a 200-μl aliquot of the mating suspension to estimate the volume that gives 200–400 exconjugants per plate.

The nature of the selection will depend on the characteristics of the recipient strain, but most commonly it will consist of a nutritional (donor strains are auxotrophs) or antibiotic counterselection of the donor. A good selection medium is M9 or M63[15] minimal medium with 0.1% of a single carbon source containing the antibiotic whose resistance is encoded by the minitransposon. If the recipient cannot be selected on minimal medium it can usually be selected on rich medium provided that it has a selectable phenotype that distinguishes it from the donor. If not, it is possible in most cases to isolate spontaneous and stable rifamycin-resistant mutant derivatives of the recipient by plating the cells of an overnight culture (1–2 ml) of the strain of interest on an LB plate with 70 μg/ml rifamycin followed by incubation at 30°–37° for 48 h.

When the marker carried by the transposon is an antibiotic resistance determinant, selection of insertions is straightforward, although optimal concentrations of antibiotics do vary among strains; the minimal inhibitory concentration (MIC) should therefore be predetermined in each case in advance. Orientative concentrations in selection plates as follows: chloramphenicol (Cm) 5–50 μg/ml; kanamycin (Km) 25–75 μg/ml; streptomycin (Sm) 50–100 μg/ml; spectinomycin (Sp) 50–100 μg/ml (*warning*: selection for spectinomycin does not work well in minimal medium); tetracycline (Tc) 2.5–15 μg/ml.

Selection conditions appropriate to nonantibiotic markers such as resistance to mercury, arsenite, or the herbicide bialaphos present in a number of mini-Tn*10* and mini-Tn*5* derivatives (Figs. 1 and 2) deserve some specific comments. Most gram-negative bacteria are sensitive to mercuric salts and organomercurial compounds, which makes the *mer* system engineered in pLOF/Hg and pUT/Hg a marker of choice when available antibiotic resistances are inappropriate or undesirable.[2,3] However, the window of concentrations in which the *mer* system is useful for selection purposes is rather narrow, and depending on the strain the MIC can lie between 1 and 15 μg/ml HgCl$_2$. Because minor differences in concentration can have a substantial impact on the cleanness of selection, a careful determination of the MIC of the target strain for HgCl$_2$ (obtained by increasing concentrations in plates by steps of 0.5 μg/ml) is required prior to the selection of exconjugants from the mating mix. Selection for the resistance to arsenite specified by the minitransposons of pLOF/Ars and pUT/Ars plasmids should be made on a low-phosphate medium such as 121 salts[16] containing

[16] K. Kreuzer, C. Pratt, and A. Torriani, *Genetics* **81**, 4559 (1975).

nor more than 100 μM phosphate and concentrations of NaAsO$_2$ in the range of 5–20 mM, the optimal level of which should be determined, as before, prior to the mating. It is further recommended to precultivate the mating mixture suspension washed from the filter for 4–5 hr in low-phosphate nonselective medium[16] prior to plating out on selective agar. Resistance to the herbicide bialaphos (phosphinotricin triphosphate or Ptt) encoded by the *bar* gene present in pLOF/Ptt and pUT/Ptt should be selected on M9 or M63[15] minimal medium which may contain up to 0.1% casamino acids and concentrations of Ptt in the range of 25–50 μg/ml. In the cases of arsenite and bialaphos resistance, the genes are expressed from a promoter which is repressed in *E. coli* by intracellular iron(II).[3] As it is unclear whether the same regulation operates in other gram-negative bacteria, it is prudent to add to the selection plate 50–200 μM of the iron chelator 2,2'-bipyridyl to ensure full expression of the selective marker.

One aspect of the use of the *bar* gene as selection marker that should be noted is the high frequency of spontaneous mutants that are tolerant of the herbicide.[3] Although it is therefore necessary to confirm that clones on the selection plates are the result of authentic transposon insertions, the Ptt marker is attractive to engineer strains for environmental applications or live vaccine development,[3] that is, situations where the selection marker should not constitute any selective advantage due to the high level of natural resistance. High-quality bialaphos can be requested from Dr. Kozo Nagaoka [Biotechnology Research Lab, Meiji Seika Kaisha Ltd. Pharmaceutical Research Center, Morooka-cho, Kohoku-ku, Yokohama, 222 Japan, phone 045 (541)2521, fax 045 (543)9771].

Confirmation and Further Analysis of Insertions

The first task in the analysis of insertions is to make sure that the acquisition of the selected phenotype by the target bacterium is due to an authentic transposition event and not to the integration of the whole delivery plasmid in a recipient replicon or its illegitimate replication in the new host. In fact, most exconjugants are the result of true transposition events (>95% in the case of *Pseudomonas putida*). Given the fact that the pUT and the pLOF plasmids carry a *bla* gene external to the minitransposon, the absence of the delivery plasmid is readily confirmed by patching exconjugants onto medium containing a β-lactam antibiotic. Authentic transposition results in the loss of the portion of the delivery plasmid containing the *bla*, gene and therefore exconjugants should be sensitive to ampicillin (50–100 μg/ml) and other β-lactams. A convenient but expensive alternative to patching is the use of nitrocefin, which stains *bla*$^+$ colonies a red-brown color. For this, prepare a 0.1 mM solution of nitrocefin (Oxoid) in 50 mM phosphate buffer, pH 7, and spray 100 μl on the surface of the selection plate containing the exconjugants to be examined. Carbenicillin at 50–300 μg/ml or piperacillin at 25–75 μg/ml are active against many strains which do not respond to ampicillin. Some gram-negative bacteria, however, are completely insensitive to a variety of β-lactam antibiotics, and in such cases it may be necessary to confirm the insertion through a Southern blot of the chromosomal DNA of the exconjugants. This type of analysis will also confirm the uniqueness of the transposition events and the degree of randomness of sites of insertion.[3]

Applications of Minitransposons

Mini-Tn5 and mini-Tn10 elements have been designed[3,4] as multipurpose genetic tools. As such, they were constructed in a modular fashion to permit users to adapt conveniently the basic structures to specific purposes. We describe below some of the most common applications made, along with some practical hints for utilization of the minitransposons.

Insertional Mutagenesis

The most common application of minitransposons is insertion mutagenesis, the procedure for which is described above. If different insertions leading to the same phenotype are desired, it is recommended that the filter be cut with a scalpel on the plate into a number of pieces which are then separately processed. To ensure a high proportion of exconjugants resulting from independent transposition events, it is possible to lay the filter with the mating mixture on the surface of an agar plate with a mineral medium (M9 to M63[15]) devoid of carbon source to allow conjugation but not further cell growth during the mating. Sometimes it is possible to use a selective medium that reveals distinct mutations (e.g., nonhemolytic mutants or siderophore minus mutants) or to enrich particular mutants (e.g., auxotrophs), but in most cases the desired phenotypes have to be scored in individual exconjugants which arise from the mating.

It is sometimes desirable to determine insertion frequencies obtained during transposon mutagenesis. With the system described above this is difficult since the number of exconjugants reflects both transposition frequency and the transfer frequency. Only if the RP4-based mobilization system of the delivery plasmids achieved a transfer frequency of 10° would the frequency of exconjugants per generation reflect the actual frequency of transposition. However, because the delivery plasmid does not replicate in the recipient, this frequency cannot be rigorously calculated. The efficiency of the mutagenesis can nevertheless be approximated as an *operational frequency*, which is the ratio of the number of exconjugants to the total number of recipients after 8 hr of mating with a 1 : 1 ratio of donor to recipient. Although the resulting figure does not have an authentic quantitative meaning, it turns out to be very informative in optimizing the mutagenesis procedure.

The selective marker used for isolation of exconjugants influences very significantly the operational frequency. The highest efficiencies are observed with the Km^r and Sm^r markers, whereas the lowest are seen with Tc^r and the nonantibiotic markers (see Troubleshooting), but this does vary among genera. In our hands, mini-Tn5 derivatives with a Km^r marker show operational insertion frequencies of 10^{-5}–10^{-6} in *Pseudomonas*, but these drop by more than one order of magnitude if equivalent transposons with a Tc^r marker are used.

Although the minitransposons described in this chapter should permit the mutagenesis of not only chromosomal genes but also those present on multicopy plasmids, we have found this last possibility to be impractical because of the large number of cointegrates formed between donor and target replicons. In this case, it is better to introduce the target plasmid into an *E. coli* host and, if possible, to use one of the many protocols available for plasmid mutagenesis.[11] However, large low-copy-number plasmids can be subjected to insertional mutagenesis without much problem. If such a plasmid is conjugative, it is even possible to preselect the insertions in it by pooling all exconjugants of the mating and using the mixture as a donor culture for a second mating, from which only transconjugants arise through acquisition of transposon-mutant plasmids.

Minitransposons as Cloning Vectors

Minitransposons are particularly useful for engineering strains which are required to stably maintain a recombinant phenotype in the absence of antibiotic selection, as is required in the construction of live vaccines and genetically modified microorganisms for environmental release. The pLOF and pUT derivatives (Figs. 1–3) are maintained in λpir lysogens such as *E. coli* CC118 λpir, because they cannot replicate otherwise. pLOFKm, pLOFPtt, pLOFHg, pLOFArs, pUTPtt, pUTHg, pUTArs, and pUT/mini-Tn5Sm/Sp, pUT/mini-Tn5Tc, pUT/mini-Tn5Cm, pUT/mini-Tn5Km, pUT/mini-Tn5Km1, and pUT/mini-Tn5Km2 contain unique *Not*I and/or *Sfi*I sites within the minitransposon, into which can be cloned DNA fragments obtained by *Sfi*I or *Not*I digestion. This permits such

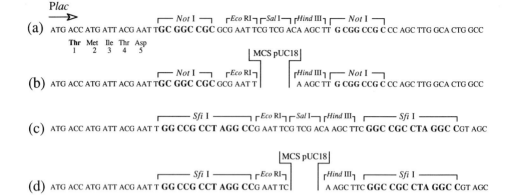

Fɪɢ. 4. Polylinker regions of auxiliary plasmids. p18Not (a), pUC18Not (b), p18Sfi (c), and pUC18Sfi (d) are identical to pUC18 (see Yanisch-Perron *et al.*[17]), excepting the nucleotide sequence of the multiple cloning site (MCS), as shown. The orientation of the *lac* promoter (P*lac*) present in the pUC series is indicated.

DNA fragments to become part of the transposon and therefore to become stably inserted into the chromosome of the target bacterium. A simple means of flanking a fragment of DNA with a *Not*I or *Sfi*I ends is to clone it into the multiple cloning site (MCS) polylinker of the pUC18Not or pUC18Sfi plasmids (Fig. 4)[17] and then to excise the insert by digestion of the hybrid plasmid with *Sfi*I or *Not*I. These pUC-derived plasmids replicate in any *E. coli* strain and exhibit the blue/white screening property in conjunction with 5-bromo-4-chloro-3-indolyl-β-ᴅ-galactopyranoside (X-Gal)-containing plates with the α-*lac* complementation system when used in *E. coli* strains. On such media, colonies turn out a little paler than those carrying standard pUC plasmids.[17] Note that because of the degeneracy of the *Sfi*I cleavage sequence, the *Sfi*I sites present in the pUT/pLOF vectors were designed to be compatible with those of p18Sfi and pUC18Sfi; *Sfi*I fragments from other sources may not therefore be compatible with the *Sfi*I sites in the minitransposons. For the same reason, p18Sfi- and pUC18Sfi-derived fragments can be cloned in only one orientation in pUT/pLOF vectors. Note also that pUT or pLOF derivatives devoid of any DNA fragment between the terminal sequences of the mobile element are not stable.[3]

Ligation mixtures consisting of pLOF or pUT vectors and *Not*I or *Sfi*I inserts should be transformed or electroporated into the *E. coli* CC118 λ*pir* strain (Table I). Attempting to save time by using the delivery strains

[17] C. Yanisch-Perron, J. Vieira, and J. Messing, *Gene* **33,** 103 (1985).

directly as recipients for the ligation mixtures is futile, because they transform with efficiencies orders of magnitude below those of *E. coli* CC118 λ*pir*. In our hands, the latter strain is very stable, and we use it routinely as the recipient for all constructions. Bear in mind that pUT- and pLOF-based plasmids should not replicate in non-λ*pir* lysogens. Once a construction established in *E. coli* CC118 λ*pir* has been analyzed, it can be transformed into delivery strains *E. coli* SM10 λ*pir* or *E. coli* S17-1 λ*pir* for introduction into the target strain by the protocol given above. Although it was the first strain available for this purpose,[3] *E. coli* SM10 λ*pir* forms opaque colonies, sometimes of different sizes, is very slow growing, and can be difficult to transform. *Escherichia coli* S-17-1 λ*pir* seems to be less fastidious, and therefore it is the best choice unless counterselection of the donor is with streptomycin. Long-term storage of constructions in delivery strains is not recommended: for unknown reasons both *E. coli* SM10 λ*pir* and *E. coli* S17-1 λ*pir* seem to lose plasmids very quickly at freezing temperatures.

We have not studied systematically the effect of fragment size on transposition frequencies, but we have inserted up to 12 kilobases (kb) of heterologous DNA in mini-Tn5 derivatives which then transposed at frequencies in the range of those observed with insert-lacking transposons. The possibility of cloning restriction fragments encoding various properties into the *Not*I/*Sfi*I sites of the minitransposons offers many opportunities to construct distinct mobile elements for specific applications. One of these is the study of gene regulation using, for example, *lacZ* fusions present in monocopy by cloning at the *Not*I site of the mobile element of *lacZ* transcriptional fusions generated *in vitro* and their subsequent insertion into bacterial chromosomes. This is facilitated with the use of vectors pUJ8 and pUJ9 (Fig. 3), in which *lacZ* fusions of types I (transcriptional) and II (translational/transcriptional) can be generated and excised as *Not*I restriction fragments for cloning into the delivery vectors. Another useful application of the hybrid minitransposons is the introduction at the cloning sites DNA sequences which can be cleaved by infrequently cutting restriction enzymes such as *Swa*I or *Pac*I. Transposition of the hybrid transposon into the chromosome of a target bacterium generates clones having convenient reference points for the physical mapping of bacterial chromosomes by pulsed-field electrophoresis.[18]

Promoter Probing

A number of mini-Tn5 derivatives have been adapted for promoter probing (Fig. 5). The basis of this application is the absence of transcrip-

[18] K. Kwok-Wong and M. McClelland, *J. Bacteriol.* **174,** 3807 (1992).

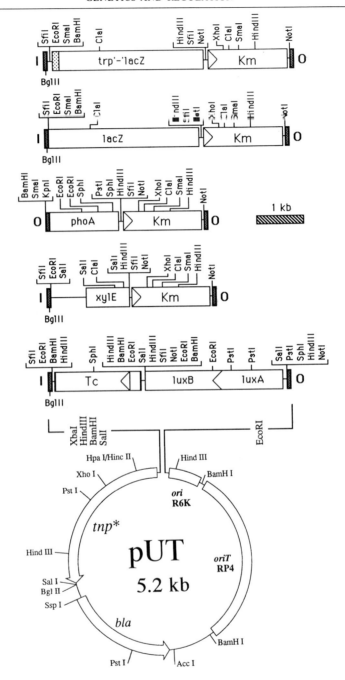

tional terminators within the I and O terminal sequences of Tn5, and therefore insertion of the minitransposons in the appropriate orientation downstream of chromosomal promoters may activate the expression of a promoterless gene placed within the mobile unit. At present transposons with the lacZ, phoA, xylE, and luxAB reporter genes are available (Fig. 5). The modular structure of the mini-Tn5 derivatives also facilitates promoter probing for other purposes. For instance, it is possible to screen for a promoter giving a particular desired level of expression under any given conditions by placing the promoterless gene of interest at one of the extremes of the minitransposon. Mutagenesis of a target strain with such a construction generates exconjugants which express such gene at virtually all possible levels that depend on the nature of the promoter downstream of which the transposon has inserted.[19]

Procedures to measure the activity of the standard reporters in the mini-Tn5 transposons of Fig. 5 have been described in detail elsewhere,[15,20,21] but some comments regarding their use in non-E. coli gram-negative bacteria are worth mentioning here. LacZ is the most convenient reporter available to the geneticist, but it might be troublesome in some cases. We have found that differential permeability to X-Gal among different strains may cause misleading estimations of promoter activity. In most cases, the fluorescent substrate of β-galactosidase, methylumbelliferyl-β-D-galactopyranoside (MUG), is preferable to X-Gal for screening promoter activity (detailed protocols for utilization of MUG are available[15]). Similarly, the mini-Tn5phoA element can be used to detect genes of secreted proteins in many gram-negative bacteria.[22] In our hands, LB medium

---

[19] M. Walker, M. Rohde, J. Wehland, and K. N. Timmis, *Infect. Immun.* **59**, 4238 (1991).
[20] M. Nozaki, this series, Vol. 17A, p. 522.
[21] R. P. Legocki, M. Legocki, T. O. Baldwin, and A. A. Szalay, *Proc. Natl. Acad. Sci. U.S.A.* **83**, 9080 (1986).
[22] M. R. Kaufman and R. K. Taylor, this volume [33].

FIG. 5. Tn5-based promoter–probe minitransposons. The elements shown carry different reporter genes near one of the I/O termini of the minitransposons, and are all oriented (as are the other mini-Tn5 elements; see legend to Fig. 3) as XbaI–EcoRI restriction fragments in the pUT delivery plasmid shown at the bottom. Mini-Tn5 lacZ1 produces type I fusions (transcriptional), whereas mini-Tn5 lacZ2 generates type II fusions (transcriptional/translational). Because of the predigree (see de Lorenzo et al.[4]) mini-Tn5 phoA has two O sites flanking the transposon and additional restriction sites inherited from one of the parental plasmids, pPHO7 [C. Gutierrez and J. C. Devedjian, *Nucleic Acids Res.* **17**, 3999 (1989)]. For the generation of a productive fusion after insertion of the elements in a new replicon, transcription of the reporter gene must be from left to right as shown, except for that of mini-Tn5 luxAB, which must be from right to left.

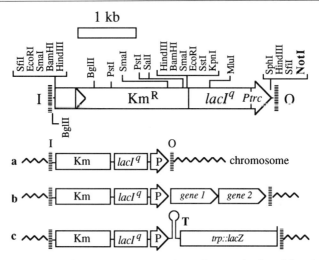

FIG. 6. Mini-Tn5 *lacI*q/P*trc*. The upper part shows the organization of the minitransposon (de Lorenzo *et al.*[5]), which is inserted as an *Xba*I–*Eco*RI restriction fragment in the delivery plasmid pUT (see Fig. 5). The *Not*I site unique in the mobile unit and the delivery plasmid is indicated in boldface type. The lower part represents some applications of the minitransposon: (a) Generation of conditional phenotypes and conditional nonpolar mutations, (b) IPTG-dependent expression of heterologous genes, and (c) *in vivo* analysis of transcription termination caused by a sequence T placed between the promoter and a reporter gene such as *trp'–'lacZ*.

contains enough phosphate in itself to suppress the indigenous PhoA$^+$ phenotype expressed as blue colonies on plates containing 5-bromo-4-chloro-3-indolylphosphate (XP). This makes unnecessary in most cases the generation of a *phoA*$^-$ mutation in the target strain to enable screening for insertions in membrane or exported proteins. Note, however, that the development of the characteristic blue color by colonies grown on XP plates caused by productive PhoA$^+$ insertions is a pH-dependent process and may vary with the genus and strain used, taking several days to appear in some instances. Where possible, increasing the pH of the selection medium to pH 7.5–7.7 may help to accelerate the reaction.

Generation of Conditional Phenotypes and Nonpolar Mutations

A useful member of the pUT family of plasmids is pCNB5[5] (pUT/mini-Tn5 *lacI*q/P*trc*, Fig. 6). This construction carries a minitransposon containing an outward-facing P*trc* promoter (a *trp–lac* hybrid) and a cognate *lacI*q gene, such that promoter activity in the mobile unit can be triggered by addition of IPTG. The promoter is oriented toward a unique

*Not*I site within the minitransposon. In the absence of insertions at this site, random integration of the minitransposon into the chromosome of a target bacterium can generate conditional phenotypes whose expression is dependent on the addition of IPTG to the medium, whereas cloning of heterologous genes into the *Not*I site produces hybrid transposons in which expression of the phenotype encoded by the cloned genes can be triggered by addition of *lac* inducers. This property can be exploited in a number of applications, including generation of nonpolar mutations and analysis of transcriptional terminators (Fig. 6). In these cases, it should be noted, however, that the P*trc* promoter has significant activity even in the absence of induction.[5] Optimal concentrations of IPTG (normally ranging from 0.1 to 5 m*M*) and other conditions such as temperature, time of induction, and medium composition have to be determined in each particular case.

Troubleshooting

Plasmid Derivatives Cannot Be Transformed into Delivery Strains

We have observed that plasmids with transposons carrying certain inserts, which are propagated without problem in *E. coli* CC118 λ*pir*, cannot be maintained in *E. coli* SM10 λ*pir* and/or in *E. coli* S17-1 λ*pir*. In most instances where this problem arises, although the construct may be unstable in one of the two strains it will be stable in the other. In the rare case where the construct cannot be introduced into either of the two strains, it is necessary to mobilize the delivery plasmid directly from *E. coli* CC118 λ*pir* into target cells via a triparental mating with a helper strain such as *E. coli* HB101 (RK2013) or *E. coli* HB101 (RK600) (Table I). To set up the mating, mix equal volumes (10–50 μl) of overnight cultures in LB of each of the three strains in 5 ml of 10 m*M* MgSO$_4$, filter through a Millipore HA membrane, and proceed as described above, counterselecting simultaneously the two *E. coli* strains. Note that if the target strain is an enterobacterium, the helper plasmid may end up stably maintained in the new host.

Low or Zero Yield of Exconjugants

The most trivial explanation of a low or zero yield of exconjugants is that selection conditions were inappropriate. Because sensitivity to antibiotics varies considerably among genera and strains, it is advisable to make an MIC assay with the target cells before the mating/plating is carried out. Some markers like Cm^r, Tc^r, and the nonantibiotic determinants can be particularly troublesome (see Selection of Exconjugants). Unfortunately, most instances of poor yield of exconjugants reflect low frequency of RP4-mediated transfer. Several properties of target cells, such as restriction systems, surface exclusion phenomena, or lack of essential envelope functions, may lead to low frequency or zero transfer. Whether target bacteria are efficient recipients of RP4 can be easily checked by setting up a mating between *E. coli* LE392 (RP4) and the strain of interest, followed by selection of transconjugants resistant to RP4 markers (Table I). If the target strain is unable to act as a recipient for RP4, then the constructions described in this chapter cannot be introduced at high frequency, and an alternative mutagenesis procedure may have to be considered. However, should RP4-containing colonies appear at low frequencies, some changes in the mating protocol may improve the procedure. In addition to increasing the number of matings and selection plates, a change in the antibiotic marker present in the transposon may be tried. Such an exchange is readily accomplished because most resistance genes have been designed as *Sfi*I restriction fragments inserted at the unique *Sfi*I site present in the delivery plasmid (Figs. 1–3 and 5). Pregrowth of the recipient strain on antibiotic medium prior to mating is not recommended since traces of the antibiotic present in the suspension or accumulated by recipient cells can kill donor cells. We have observed, for instance, that *P. putida* KT2442 cells (Table I) pregrown in the presence of rifamycin are totally unable to act as conjugal recipients of pUT-type plasmids.

All (or Most) Exconjugants Are Resistant to Ampicillin/Carbenicillin/Piperacillin

This problem arises frequently when the recipient strain is another *E. coli* since λ*pir* phage released by the donor may lysogenize the recipient, thereby circumventing the suicide mechanism of the delivery system and permitting the propagation of the delivery plasmid in exconjugants. We have observed that such a phenomenon occurs more often when selection of exconjugants is made on plates with tetracycline, streptomycin/spectinomycin, and bialaphos than when clones are selected on media with Km or Cm. *Escherichia coli* CC118 λ*pir* cells harboring plasmids with the Tc^r marker seem to undergo spontaneous induction of phage when selected with the antibiotic. We have also observed that using 1% NaCl instead of 10 mM $MgSO_4$ for making cell suspensions before and after the mating (see above), addition of 1% sodium citrate to the selection plates, growth at 30°, use of lower concentrations of the selective agent in the plates, and shortening mating times help in some, but not all, cases to reduce phage induction.

In non-*E. coli* bacteria, the problem of colonies resistant to β-lactam antibiotics may arise at low frequencies because of illegitimate (i.e., IS-mediated) integration of the whole delivery plasmid into the chromosome of the target cells. In other cases, cryptic replication origins present in the foreign DNA cloned within the transposon vector, which enable the delivery plasmid to replicate in strains devoid of λ*pir*, or homology to chromosomal DNA, can also lead to the formation of ampicillin-resistant exconjugants.

Acknowledgments

This work was supported by Grants BIOT-CT91-0293 (EC BRIDGE Program) to V.D.L. and K.N.T., BIO 89-0497 (Comisión Interministerial de Ciencia y Tecnología) to V.D.L., and BGO/21039433A (Bundesministerium für Forschung und Technologie) to K.N.T.

[32] Use of Transposons to Dissect Pathogenic Strategies of Gram-Positive Bacteria

By ROBERT A. BURNE and ROBERT G. QUIVEY, JR.

Introduction

A wide variety of serious and fatal diseases of humans and animals are attributable to infection by overt and opportunistic gram-positive pathogens. Among humans, for example, the prevalence and severity of diseases elicited by Group A and Group B streptococcal pathogens have increased dramatically over the past few years. The postinfection sequelae from Group A infection, principally rheumatic fever (RF) and acute glomerulonephritis (AGN), are extremely serious diseases which frequently result in death. Despite better detection methods and effective antibiotic therapy to combat Group A streptococci, the incidence of RF and AGN has risen sharply since the 1980s. Nosocomial infection by multidrug-resistant *Staphylococcus aureus,* which cause a variety of infections of skin, surgical sites, and in-dwelling devices, have plagued hospital medicine for decades. In addition, dental caries, caused by the viridans group streptococcus *Streptococcus mutans,* is estimated to cost the public over 25 billion dollars annually, and this disease remains the leading cause of tooth loss in United States. Similarly, intoxication from foodstuffs contaminated with gram-positive organisms such as *Clostridium difficile* and staphylococci is remarkably common, and in some cases can result in irreversible neurologic damage or death. Moreover, although many of these diseases in developed countries have become manageable due to higher quality of and better access to health care, developing nations are suffering from a higher incidence of disease elicited by gram-positive organisms, and the morbidity and mortality caused by these pathogens far exceed that found in the United States.

The ability of the majority of gram-positive bacteria to initiate disease is the result of the possession of a variety of determinants which allow the organisms to colonize, to accumulate in sufficient numbers, to elicit host tissue damage, and to avoid the immune and nonspecific defenses of the host. In Group A streptococci, for example, the binding of epithelial cells, elicitation of various cytotoxins, and production of the antiphago-cytic surface M protein contribute to the pathogenic potential of the organisms. Many Group A and Group B streptococci produce a hyaluronic acid capsule, which is antiphagocytic. The production of hemolysins is also thought to augment the tissue-damaging abilities of these organisms, and synthesis of a C5a peptidase, which inhibits complement function, presumably enhances the survival of the bacteria *in vivo*. The oral streptococci bind tightly to tissues in the mouth through a variety of mechanisms. These bacteria, and in particular *S. mutans,* produce large quantities of acids which can damage oral hard tissues. In addition, the oral streptococci produce a variety of enzymes capable of forming and hydrolyzing polymers of glucose and fructose which enhance the ability of the bacteria to colonize and to produce acids. *Staphylococcus aureus* is capable of attaching to the skin surface and extracellular matrix proteins via specific adhesins and nonspecific interactions, and this organism also produces numerous toxins and proteases which are believed to contribute to the pathology of the infection. Many gram-positive bacteria, and especially the gram-positive cocci, harbor mobile genetic elements which render them resistant to conventional antibiotic therapies.

In addition to the number of virulence attributes these organisms produce, evidence is emerging that the expression of multiple virulence determinants may be differentially regulated in response to environmental stimuli, thus complicating the ability to fully understand pathogenic strategies in these bacteria. Thus, it is clear that dissection of virulence in gram-positive bacteria will require a multidisciplinary approach which will serve to define and characterize not only the virulence determinants, but which will allow for an understanding of how the organisms control expression of these traits to become successful pathogens. Transposon mutagenesis will undoubtedly play a major role in these studies.

Some specific examples of the more recent applications of transposons to the study of gram-positive pathogens are described below. However, to date, transposons have been utilized relatively infrequently for identifying and studying loci involved in virulence in gram-positive bacteria. Instead, it seems that approaches involving recombinant DNA technology to isolate the gene for the virulence determinant of interest have been employed. This approach has facilitated detailed *in vitro* analysis of structure–function relationships and examination of mechanisms of action of virulence components. In many cases, the isolation of the gene has allowed for expression of the gene in a host normally lacking the determinant to evaluate function, or for the inactivation of the determinant *in vitro* or in *Escherichia coli*. For the latter cases, the mutated allele could then be returned to the parent strain, usually by transformation, and integrated into the genome by host recombination machinery. Thus, a strain was created which was identical to the parent in all traits except the expression of the determinant of interest. Mutants derived in this fashion have been tested in appropriate *in vitro* systems or animal models to assess their pathogenic potential with the hope of dissecting the relative contribution of the gene product of interest to virulence.

This approach has proved useful for organisms for which appropriate gene transfer systems have been developed. However, the use of a recombinant DNA approach often has the drawback that genes encoding for nonproteinaceous virulence determinants may not be readily isolated, or may require a tremendous amount of preliminary work to develop the materials for screening for the gene(s) of interest, such as those encoding for capsules. Similarly, use of molecular cloning to isolate genes encoding for proteins which regulate expression of virulence, or to identify and isolate genes which are targets of these regulatory proteins, is largely cumbersome. It is precisely these types of cases where genetic manipulation with transposons has facilitated the study of virulence in gram-positive bacteria. It also seems reasonable to predict that transposons will play a greater role in dissecting virulence in gram-postive organisms given the emphasis which has been placed recently on the regulation of expression of virulence.

A number of investigators have utilized transposons successfully for obtaining mutants with insertions in genes which were thought to contribute to virulence. For the purposes of this discussion, the salient points of these studies include the choice of transposon, the delivery method and vehicles for introduction of the transposon, the randomness with which insertions can be observed, and the stability of the insertions. Specific examples of identification of virulence genes in gram-positive bacteria are presented primarily for the purpose of illustrating the types of methodologies available for applying transposon mutagenesis to the study of pathogenic strategies of these bacteria. The latter portion of the discussion focuses on methods for genetic manipulation and transposon mutagenesis in the oral streptococci because the techniques employed for introducing transposable elements, screening, and subsequently utilizing mutants derived in this fashion are widely applicable to other gram-positive bacteria.

Basic Methods for Introduction of Transposons into Gram-Positive Bacteria

There are three principal methods by which transposons can be introduced into a desired recipient organism; transformation, conjugation, and transduction. Introduction of the transposon into the cells can be accomplished simply by transformation since many gram-positive bacteria, such as pneumococci and many oral streptococci, can be made naturally competent for DNA uptake. In this case, uptake of the transposon on an appropriate delivery vehicle, usually a plasmid, is an active process. In all known cases, the DNA is probably taken up as a single strand, and through interactions with other DNA molecules which have been internalized by the cells, replication, and recombination, the transposon/plasmid molecule recircularizes. Transposition likely occurs following this event. Alternatively, if sequences with homology to the host chromosome are present on the transforming DNA, the transposon may become established in the chromosome by legitimate recombination. Transformation of gram-positive bacteria can be remarkably efficient, as with *Streptococcus sanguis,* and offers an excellent, straightforward approach to establishing a transposon library, that is, a large number of independently isolated clones which have undergone individual, transposition events.

In the absence of a natural transformation system, many gram-positive bacteria can be treated to form protoplasts or spheroplasts. With agents such as polyethylene glycol, the bacteria can then be induced to take up DNA. This method has worked well for bacteria which can regenerate a cell wall under specific conditions, particularly staphylococci and enterococci. Likewise, successful protoplast fusion of a donor harboring the transposon and a suitable recipient have allowed for establishment of transposon-bearing recipients.

Electroporation has also been successfully utilized to introduce DNA into gram-positive organisms, such as some lactococci, Group A streptococci, and certain strains of viridans group streptococci, which have often proved refractile to transformation by natural or artificial means. The variety of bacteria which have been transformed by electroporation and the high efficiency of this process offer great promise for introducing transposons into many different bacterial species (see [30] in this volume).

The most straightforward approach to introducing a transposon by transformation into a strain is to utilize a plasmid carrying the transposon to transform the desired strain. For these experiments, it is desirable either to have a plasmid which cannot replicate in the organism to be studied or to utilize a plasmid which can replicate only under very specific circumstances (e.g., temperatures less than or equal to 30°). Once established in the chromosome, the majority of the transposons used in gram-positive organisms mobilize to a secondary site at a relatively low frequency.[1,2] Therefore, the use of a thermosensitive replicon to deliver the element is more desirable, particularly if the efficiency with which the organism can be transformed is poor. Presumably, by establishing the transposon-bearing plasmid in a given strain and then allowing for expansion of clones carrying the transposon on the plasmid, curing of the plasmid should allow one to obtain more clones derived from independent, initial transposition events. In the case of plasmids which cannot replicate, the number of independent, initial transposition events would be roughly equivalent to the observed number of transformants, and subsequent movement of the transposon would likely be a low-frequency event, judging from existing data for the more commonly used transposons from gram-positive bacteria.

As an alternative to the introduction of DNA by the process of transformation, conjugation has been utilized effectively to establish transconjugants harboring transposon insertions. The conjugative transposon Tn916 has been mobilized from *Enterococcus* to Group A streptococci, viridans streptococci, and even pathogenic *Bacillus* and *Clostridium* species. The majority of the transfers have been accomplished by filter matings. In the usual case, the donor and recipient are incubated at high cell density while immobilized on a filter to allow for cell–cell contact and transfer of the transposon. Transfer of plasmids bearing transposons has also been facilitated by mobilization of the transposon on conjugative plasmids, which frequently have a broad host range. As detailed below, conjugation has proved to be perhaps the most widely used approach for delivering transposons to gram-positive bacteria. In particular, enterococcal strains carrying Tn916 are good donors to streptococci, *Bacillus, Listeria,* and other gram-positive bacteria. The main drawback with this approach is that it is necessary to utilize a recipient strain with some selectable phenotype,

[1] D. B. Clewell, S. E. Flannagan, L. A. Zitzow, Y. A. Su, P. He, E. Senghas, and K. Weaver, *in* "Genetics and Molecular Biology of Streptococci, Lactococci, and Enterococci" (G. M. Dunny, P. P. Cleary, and L. L. McKay, eds.), p. 39. American Society for Microbiology, Washington, D.C., 1991.
[2] J. R. Scott, *in* "Genetics and Molecular Biology of Streptococci, Lactococci, and Enterococci" (G. M. Dunny, P. P. Cleary, and L. L. McKay, eds.), p. 28. American Society for Microbiology, Washington, D.C., 1991.

usually antibiotic resistance, to allow for growth of only the recipient following the mating.

Transducing bacteriophage have been used to mobilize transposons among phage-susceptible hosts. Transduction has been particularly useful for moving a single insertionally inactivated allele to a closely related strain or into the parent strain, since during the initial selection for mutants there often can be multiple transposon insertions detected within a single clone. Nevertheless, transduction has not been widely used for transposon mutagenesis, since the host range of most transducing particles is fairly narrow. All of the general methods, or combinations of these methods discussed above, have been applied to the study of virulence in the gram-positive bacteria. Some of the techniques which provide good examples of the utility of transposons for virulence analysis in gram-positive bacteria are briefly outlined below.

Pathogenic Streptococci

The Group A β-hemolytic streptococci are the etiologic agents of streptococcal pharyngitis as well as numerous other human infections. The postinfection sequelae resulting from primary infection consist of acute rheumatic fever (RF), acute glomerulonephritis (AGN), and a variety of other serious disorders. Although the incidence of RF and AGN declined steadily from 1950–1980 through the prompt treatment of infections with antibiotics, there has been a dramatic increase in the incidence of Group A β-hemolytic streptococci infections and of RF in the United States.[2-4] Likewise, the prevalence of RF and AGN in developing nations is dramatically higher than that found in developed countries.[4] It has been estimated that RF causes 25–40% of all cardiovascular diseases worldwide. More recently, a dramatic increase in the incidence of toxic shock-like syndrome initiated by Group A streptococci in humans, with symptoms such as multiorgan failure and pyrexia, has been observed.[3] The relatively sudden and dramatic increase in both the incidence and severity of disease caused by Group A β-hemolytic streptococci has been postulated to be due to an increase in the expression of a single or multiple virulence attributes of the organisms.[4] In addition to the Group A streptococci, the prevalence and severity of infections by Group B organisms (particularly type III) have risen sharply.[5-7]

[3] L. B. Givner, J. S. Abraham, and B. Wasilauskas, *J. Pediatr.* **118,** 341 (1991).

[4] G. H. Stollerman, *Adv. Intern. Med.* **35,** 1 (1990).

[5] M. H. Rathore, L. L. Barton, and E. L. Kaplan, *Pediatrics* **89,** 743 (1992).

[6] B. Schwartz, A. Schuchat, M. J. Oxtoby, S. L. Cochi, A. Hightower, and C. V. Broome, *J. Am. Med. Assoc.* **266,** 1112 (1991).

[7] J. D. Wenger, A. W. Hightower, R. R. Facklam, S. Gaventa, C. V. Broome, and B. M. S. Group, *J. Infect. Dis.* **162,** 1316 (1990).

A number of components produced by the bacteria have been proposed to function as virulence determinants, including hyaluronic acid capsules and M proteins, which appear to play key roles as antiphagocytic surface structures, as well as streptococcal hemolysins and erythrogenic toxins. Streptokinase, which is believed to function on dissemination of Group A β-hemolytic streptococci in the bloodstream, immunoglobulin G (IgG) Fc-binding protein, and a C5a peptidase which can inhibit the reception of chemotactic signals by polymorphonuclear leukocytes are a few of a large number of attributes produced by pathogenic streptococci which are thought to contribute in major ways to the ability of the organisms to cause human diseases.

Dissection of virulence in Group A β-hemolytic streptococci has proceeded primarily by genetic analysis of cloned virulence loci and testing of mutants in *in vitro* systems or appropriate animal models. However, transposons have been utilized in many instances for the study of specific virulence determinants. (For a detailed description of the methodologies and utility of transpositional mutagenesis in pathogenic streptococci, see (Caparon and Scott[8]). The transposon Tn916, which can function as a conjugative transposon,[9] has been utilized to create insertions in genes thought to contribute directly to virulence. For example, Wessels *et al.*[10] have isolated mutations in the pathway for hyaluronic acid capsule expression following filter mating of Group A β-hemolytic streptococci with the *Enterococcus faecalis* strain CG110, which carries Tn916. The transconjugants proved to be stable. Strains lacking the ability to express functionally the capsular phenotype were unable to grow in fresh human blood, were susceptible to phagocytosis, and were roughly 100-fold less virulent in mice. This technique has also been applied to the study of the capsule of Group B streptococci by Wessels *et al.*[10–12] Perez-Casal *et al.*[13] have utilized a similar strategy, namely, filter mating with enterococci carrying Tn916, to identify the Mry protein, a positive regulator of M protein expression in Group A β-hemolytic streptococci. Although conjugation has proved to be an efficient and productive method to generate Tn916

---

[8] M. G. Caparon and J. R. Scott, this series, Vol. 204, p. 556.

[9] D. B. Clewell and C. Gawron-Burke, *Annu. Rev. Microbiol.* **40,** 635 (1986).

[10] M. R. Wessels, C. E. Rubens, V.-J. Benedi, and D. L. Kasper, *Proc. Natl. Acad. Sci. U.S.A.* **86,** 8983 (1989).

[11] C. E. Rubens, J. M. Kuypers, L. M. Heggen, D. L. Kasper, and M. R. Wessels, *in* "Genetics and Molecular Biology of Streptococci, Lactococci, and Enterococci" (G. M. Dunny, P. P. Cleary, and L. L. McKay, eds.), p. 179. American Society for Microbiology, Washington, D.C., 1991.

[12] M. R. Wessels, R. F. Haft, L. M. Heggen, and C. E. Rubens, *Infect. Immun.* **60,** 392 (1992).

[13] J. Perez-Casal, M. G. Caparon, and J. R. Scott, *J. Bacteriol.* **173,** 2617 (1991).

insertion mutants, Simon and Ferretti [14] and others [15] have described a highly efficient electroporation protocol for some Group A streptococci which may obviate the use of *Enterococcus* to deliver transposons. Moreover, because transformants can be directly selected for by the resistance marker on the transposon, the necessity of preparing host strains which are resistant to some antibiotic to which the donors are sensitive is eliminated.

Manipulation of *Staphylococcus aureus*

Staphylococcus (St.) aureus, coagulase-negative staphylococci, such as *St. epidermidis* and *St. saprophyticus,* and some micrococci cause a wide range of human diseases. *Staphylococcus aureus* produces a variety of factors which are believed to contribute to the pathogenic potential of the organism. Among these are fibronectin-binding proteins which likely mediate initial adherent interactions with the host, exfoliatin, a hyaluronidase, leukocydin, and the potent cytotoxins α, β, γ, and δ. To date, a number of virulence attributes, as well as genetic loci governing virulence have been identified through the use of transposons. With investigations being driven in large part by the prevalence and severity of nosocomial infections by multidrug-resistant staphylococci, a large body of data exists on mobile genetic elements and antibiotic resistance in *St. aureus* and coagulase-negative staphylococci. However, the transposons which have been used for genetic analysis of *St. aureus* usually have been those originally isolated from streptococci. For example, using Tn*918* mobilized from *Enterococcus faecalis* FA378, mutants of *St. aureus* have been isolated which have a decreased ability to bind to human fibronectin.[16] Likewise, *St. aureus* which were defective in capsule production were also isolable using this transposon.[17]

The streptococcal transposon Tn*916* has been transferred from *E. faecalis* by preparing protoplasts of the donor strain and the recipient *St. aureus,* fusing the protoplasts in the presence of polyethylene glycol, regenerating the cell wall, and plating the organisms with appropriate selection for transposon-bearing recipient staphylococci.[18] Recipients of

[14] D. Simon and J. J. Ferretti, *in* "Genetics and Molecular Biology of Streptococci, Lactococci, and Enterococci" (G. M. Dunny, P. P. Cleary and L. L. McKay, eds.), p. 299. American Society for Microbiology, Washington, D.C., 1991.

[15] G. M. Dunny, *in* "Genetics and Molecular Biology of Streptococci, Lactococci, and Enterococci" (G. M. Dunny, P. P. Cleary, and L. L. McKay, eds.), p. 302. American Society for Microbiology, Washington, D.C., 1991.

[16] J. M. Kuypers and R. A. Proctor, *Infect. Immun.* **57**, 2306 (1989).

[17] A. Albus, R. D. Arbeit, and J. C. Lee, *Infect. Immun.* **59**, 1008 (1991).

[18] S. C. Yost, J. M. Jones, and P. A. Pattee, *Plasmid* **19**, 13 (1988).

the transposon displayed evidence of random insertion of Tn916 by Southern hybridization to independently isolated clones. The streptococcal transposon Tn917 has been utilized in St. aureus to identify the sar locus,[19] which appears to control expression of a variety of surface-associated proteins and exoproteins which are related to virulence expression, including proteases and α-hemolysins. In this case Tn917 was delivered by protoplast transformation on a thermosensitive replicon. Insertions of Tn551, which was also introduced into St. aureus on a temperature-sensitive plasmid, have been obtained which resulted in the creation of lipase-deficient mutants.[20] Likewise, Tn551 was also utilized to identify the accessory gene regulator (agr) locus in St. aureus which, like sar, is involved in regulating the expression of a number of virulence determinants.[21]

Listeria monocytogenes

The bacterial pathogen Listeria monocytogenes is among the gram-positive organisms which have proved to be amenable to genetic analysis using transposon mutagenesis. The organisms are found free-living and in association with many animals, including humans. The prevalence of Listeria-initiated disease has been estimated at up to 11.3 cases per million in France, much lower than has been observed for diseases caused by the gram-positive cocci. In most countries, there has been an apparent increase in the prevalence of listeriosis, although this may be due to better reporting, detection, and diagnosis.[22,23] Listerial disease is most commonly observed in pregnant women, neonates, and immunocompromised individuals. Reports of diseases caused by L. monocytogenes include a mild febrile illness, septicemia, meningoencephalitis, and focal infections of the eye, skin, lymph nodes, joints, spine, or brain.

The factors elicited by L. monocytogenes which are thought to contribute to virulence are somewhat poorly defined but include the ability to resist oxidative killing when phagocytosed and the production of antigens which can hypersensitize mice. In addition, production of listerolysin O, a cellular cytotoxin, is believed to contribute to the pathogenic potential of L. monocytogenes in humans. Investigators have succeeded in introducing the streptococcal transposon Tn916 into L. monocytogenes by conjuga-

[19] A. L. Cheung, J. M. Koomey, C. A. Butler, S. J. Projan, and V. A. Fischetti, Proc. Natl. Acad. Sci. U.S.A. 89, 6462 (1992).
[20] M. S. Smeltzer, S. R. Gill, and J. J. Iandolo, J. Bacteriol. 174, 4000 (1992).
[21] P. Recsei, B. Kreiswirth, M. O'Reilly, P. Schlievert, A. Gruss, and R. P. Novick, Mol. Gen. Genet. 202, 58 (1986).
[22] Center for Disease Control, J. Am. Med. Assoc. 267, (1992).
[23] J. McLauchlin, Epidemiol. Infect. 104, 191 (1990).

tion from an enterococcal strain to produce mutations in the major hemolysin, listeriolysin O.[24,25] The mutants have been examined for the ability to survive intracellularly and to initiate disease in an animal model. From these studies, it appears that listeriolysin O augments intracellular survival in macrophages, presumably by inhibiting phagosome/lysosome fusion.[25] Mutants defective in the production of listerolysin O also demonstrated reduced virulence in a mouse model.[24,25] Complementation of the listeriolysin O defect in transposon-derived mutants in *trans* restored virulence in mice.[26] Other examples of transposon mutagensis in *L. monocytogenes* include the use of Tn*1545* to isolate mutants defective in catalase [27] and in phospholipase/metalloproteinase[28] production.

Oral Streptococci

A variety of investigators have utilized transposon mutagenesis in the oral streptococci, members of the viridans group. These organisms include the *S. sanguis* group, *S. mutans,* and *S. mitis,* among others. Principally these bacteria are found as "normal" inhabitants of the human oral cavity. *Streptococcus mutans* is the most common etiologic agent of dental caries. The three most common infectious organisms associated with subacute bacterial endocarditis are the *S. sanguis* group, *S. mitis,* and *S. mutans.* The viridans group streptococci have also been associated with infections of the central nervous system, skin, and intestines. In one study, the bacteria were observed to be responsible for a broad spectrum of infections in 15% of all bone marrow transplantation patients, with *S. mitis* being cultured in 47% of the infections by viridans streptococci.[29]

As with the pathogenic streptococci, recombinant DNA methodology has been the primary tool applied to the study of the genetics of oral streptococci. However, the streptococcal transposons Tn*916* and Tn*917* have also been effectively delivered to a number of oral streptococci and mutants obtained. There is some evidence to indicate that Tn*917* has so-called hot spots for transposition in these organisms. In other words, transposition is not random; rather, when delivered to a bacterium such

[24] J. L. Gaillard, P. Berche, and P. Sansonetti, *Infect. Immun.* **52,** 50 (1986).
[25] D. A. Portnoy, P. S. Jacks, and D. J. Hinrichs, *J. Exp. Med.* **167,** 1459 (1988).
[26] P. Cossart, M. F. Vincente, J. Mengaud, F. Baquero, J. C. Perez-Diaz, and P. Berche, *Infect. Immun.* **57,** 3629 (1989).
[27] M. Le-Blond-Francillard, J.-L. Gaillard, and P. Berche, *Infect. Immun.* **57,** 2569 (1989).
[28] J. Raveneau, C. Geoffrey, J.-L. Beretti, J.-L. Gaillard, J. E. Alouf, and P. Berche, *Infect. Immun.* **60,** 916 (1992).
[29] J. G. Villablanca, M. Steiner, J. Kersey, N. K. C. Ramsey, P. Ferrieri, R. Haake, and D. Weisdorf, *Bone Marrow Transplant.* **6,** 387 (1990).

as *S. mutans*, Tn*917* appears to localize to only a few regions of the chromosome. However, there are a variety of extremely useful delivery vehicles, and Tn*917* derivatives, that contain promoterless reporter genes, which have been constructed primarily in the laboratory of P. Youngman. If the problem of nonrandom insertion in the organ streptococci can be overcome (perhaps by deletion of the preferred target sites), Tn*917* and its derivatives may be particularly useful for genetic analysis in oral streptococci. Unlike Tn*917*, Tn*916* appears to insert quite randomly in the chromosome of a variety of oral streptococci, and the insertions appear to be quite stable once established.[30,31] Thus, it appears that Tn*916* may be of the most use for analyzing virulence in oral streptococci.

Harris *et al.*[32] introduced Tn*916* into *S. mutans* by electroporation on a plasmid vector. In doing so, the investigators were able to generate mutants which were defective in the production of intracellular storage polysaccharide (IPS), a glycogen-like polymer. Such mutants were identified by screening for the loss of a staining reaction with iodine. Following identification of putative IPS-deficient mutants and confirmation of the IPS-defective phenotype, the genetic locus involved in the formation of IPS was isolated by recombinant means. This was accomplished by cloning the Tn*916* [16.4 kilobase pairs (kbp)] and flanking regions as a single *Pst*I fragment from the *S. mutans* chromosome into a cosmid vector in *E. coli*. Because Tn*916* is capable of precise exision in *E. coli*, it was then possible to obtain intact fragments of the genes involved in IPS biosynthesis by passaging the strains in the absence of selective pressure for the transposon-coded antibiotic resistance. This type of so-called reverse cloning experiment was first proposed by Gawron-Burke and Clewell,[33] and as employed by Harris *et al.*[32] it represents an excellent example of utilization of transposon mutagenesis to identify loci that may have phenotypes which cannot be readily selected for when cloned in *E. coli*. M. Chen and D. LeBlanc (personal communication, 1992) have developed a temperature-sensitive replicon, pMC16, which appears to function in a number of streptococci. Using pMC16, these investigators have delivered Tn*916* to *S. sanguis* and *S. mutans,* and, by selection at the restrictive temperature (usually 37°), they have obtained a large number of independent insertions in the chromosomes of these organisms. The ability to deliver Tn*916*, and

[30] J. K. Procino, L. Marri, G. D. Schockman, and L. Daneo-Moore, *Infect. Immun.* **56,** 2866 (1988).
[31] M. C. Chen and D. J. LeBlanc, Univ. of Texas Health Science Center at San Antonio, personal communication (1992).
[32] G. S. Harris, S. M. Michalek, and R. Curtiss III, *Infect. Immun.* **60,** 3175 (1992).
[33] C. Gawron-Burke and D. B. Clewell, *J. Bacteriol.* **159,** 214 (1984).

potentially other transposons, to the streptococci on such a replicon will provide yet another tool for dissecting virulence in these bacteria.

The basic methodologies for transformation or mobilization of Tn916 into oral streptococci provide a broad base of fundamental technologies which can be adapted to other gram-positive bacteria. For these reasons, the remainder of the discussion focuses on specific techniques for utilizing transposons in the bacteria *S. mutans* and *S. sanguis*.

Transformation of Oral Streptococci

Protocols for the transformation of the oral streptococci vary significantly with the bacterium of interest. These can include the treatment of cells to form protoplasts with the intent of reducing the physical barrier (the cell wall) to DNA uptake (*S. faecalis*), as well as the manipulation of strains such that a state of relative competency for DNA uptake is achieved (*S. mutans*). As compared with *E. coli,* the transformation of many gram-positive strains results in low numbers of transformants, as pointed out previously for *St. aureus,*[18] such that additional protocols have been examined, some of which are described below. A procedure that we have used successfully to introduce plasmid and chromosomal DNA into the oral streptococci[34] has been adapted from previously published methods.[35,36]

1. Typically, 5 ml of an overnight culture of the recipient strain (GS-5; UA130; UA 159; LT11) is grown in brain–heart infusion medium (BHI) at 37° in a 5% (v/v) CO_2 incubator. PYG medium, consisting of 1% (w/v) proteose peptone No. 3 (Difco, Detroit, MI), 0.5% (w/v) yeast extract, and 0.1% (w/v) glucose, may also be used with the caveat that the pH of the medium should be set initially to pH 7.4–7.6.[37]

2. Dilute the overnight culture 1 : 20 and 1 : 40 in fresh BHI containing 10% (v/v) horse serum. The horse serum, of necessity, should have been previously heat-treated at 55° for 30 min to inactivate complement proteins. The fresh dilutions are then incubated for an additional 3 hr at 37° and 5% CO_2.

3. One milliliter of each dilution of cells is transferred to sterile polypropylene tubes (Eppendorf) for each transformation experiment. DNA is added to each sample of cells, the suspension is gently mixed (do not vortex at any stage in the protocol: all mixing should be done very gently),

[34] R. G. Quivey and R. C. Faustoferri, *Gene* **116**, 35 (1992).
[35] D. Perry and H. K. Kuramitsu, *Infect. Immun.* **32**, 1295 (1981).
[36] J. W. Gooder and H. Gooder, *J. Bacteriol.* **102**, 820 (1970).
[37] H. K. Kuramitsu, *Mol. Microbiol.* **1**, 229 (1987).

and the mixture is incubated for 2 hr at 37°. The amounts of DNA typically used in our experiments are 10 μg/tube of chromosomal or plasmid DNA.

4. If the selectable marker to be used is erythromycin resistance (Em^r), then 2 hr after the addition of DNA subinhibitory levels of erythromycin (75 ng/ml) are added to induce increased transcription of the *ermB* gene.[38] Outgrowth of the culture is continued for an additional 1 hr at 37°. In the case of Tn*916*, tetracycline is the selectable marker and induction is not required to boost expression of the resistance. Thus, in our experiments, the DNA and cell suspension are incubated for 3 hr if Tc^r is the marker.

5. Following the expression period, cells are collected by centrifugation in a microcentrifuge for 5 min at 10,000 rpm and room temperature. The cell pellet is then resuspended in 100 μl of BHI broth. The addition of antibiotic at this stage is unnecessary.

6. The entire contents of each tube are then plated on a single agar plate containing selective antibiotic (10 μg/ml Em or Tc) and incubated at 37° in a CO_2-enriched atmosphere, usually for 2 days to allow sufficient growth for the unambiguous identification of colonies.

We find that yields with this protocol can range from 200 to 700 colonies per plate.

Use of Temperature-Sensitive Vectors

An alternative to the problems of low transformation efficiencies is to introduce a transposon into a strain via a plasmid that is temperature sensitive for replication. The advantage to this strategy is that only a single transformant, positive for the plasmid markers, need be isolated during a transformation experiment. Subsequently, the transformed strain can be grown in batch culture at the permissive temperature and then shifted to the higher, nonpermissive temperature. The plasmid will cease to replicate and will begin to segregate out of the growing cell population. Transpositional events can then be observed by plating the culture on medium selective for the markers of the transposon with subsequent screening on media selective for the plasmid. Successful transpositions will be revealed by growth on the former medium and the absence of growth on the latter. Techniques similar in concept have been used with the transposon Tn*551*, encoding an Em^r determinant, carried on the temperature-sensitive plasmids pRN3208[39] and pI258*repA36*.[20] These experiments lead to the isolation and characterization of mutations in the *St.*

---

[38] E. Murphy, *in* "Mobile DNA" (D. E. Berg and M. M. Howe, eds.), p. 269. American Society for Microbiology, Washington, D.C., 1989.

[39] T. Oshioda and A. Tomasz, *J. Bacteriol.* **174**, 4952 (1992).

aureus autolysin and extracellular lipase genes, respectively. The frequency of insertion for Tn*551* was reported at 6×10^{-4}, a significantly higher efficiency than previously observed for transformation of *St. aureus* protoplasts (10^{-8}).[18] As described above, a recently constructed vector, pMC16, may be able to act similarly, as a temperature-sensitive vector for Tn*916*-mediated mutagenesis in the oral streptococci.[31]

Electroporation of DNA into Streptococci

The efficiencies of introducing DNA into *E. coli* via electroporation[40] are higher and the technique far less tedious than previous methods of transformation; hence, it is not surprising that considerable effort has gone into finding conditions similarly useful for gram-positive bacteria. Detailed protocols have been reported for the lactococci,[41–43] *B. anthracis*,[42] *St. aureus*,[44] *S. pyogenes*[14,15] and *S. mutans*.[32] The conditions for introducing DNA into these strains are by nature empirically determined. Nevertheless, many of the protocols are similar in that they universally include the use of high electric field strengths which have been achieved with narrow-gap electrode cells (0.2 mm gap width) and high voltages. It has also been shown that the inclusion of glycine in the growth medium of cells being prepared for electroporation is helpful in raising the efficiency of transformation for *S. faecalis* but lethal for the group A streptococci and *B. anthracis*.[15] Other considerations have included the osmotic stabilization of transformants with high concentrations of nonmetabolizable sugars, such as sucrose (*St. aureus*)[44] or raffinose (*S. mutans*)[32] in the electroporation buffers.

Filter Mating with Conjugative Transposons

The acquisition of a transposon by conjugation, namely, the passage of the transposon from a donor strain to a recipient, has been used extensively with the gram-positive bacteria. This has been particularly true with those elements of the Tn*916* family (Tn*916, 918, 919, 925*, and *1545*, which all encode tetracycline resistance of the *tetM* type; see summary descriptions in Murray[38]) and Tn*551* (Em^r) (see, e.g., Stout and Iandolo[45]). Interestingly, transfer of the Tn*916* group from one strain to another is not replication or recombination dependent.[1] The Tn*916* family of transpo-

[40] W. J. Dower, J. F. Miller, and C. W. Ragsdale, *Nucleic Acids Res.* **16**, 6127 (1988).
[41] H. Holo and I. F. Nes, *Appl. Environ. Microbiol.* **55**, 3119 (1989).
[42] G. M. Dunny, L. N. Lee, and D. J. LeBlanc, *Appl. Environ. Microbiol.* **57**, 1194 (1991).
[43] B. M. Chassy and J. L. Flicklinger, *FEMS Microbiol. Lett.* **44**, 173 (1987).
[44] B. Oskouian and G. C. Stewart, *J. Bacteriol.* **172**, 3804 (1990).
[45] V. G. Stout and J. J. Iandolo, *J. Bacteriol.* **172**, 6148 (1990).

sons can mediate the mobilization of other genes, whereas the mobilization of plasmids and chromosomal genes by Tn*551* does require an intact *rec* function in *St. aureus*.[45] The methodology for mating one strain with another has been outlined in detail previously for the streptococci and is applicable to virtually all bacterial strains.[8] The key elements are that two strains, a donor and recipient, are brought into close physical contact, typically by mixing aliquots of two cultures in a plastic syringe and passing the suspension through a 0.2- or 0.45-μm filter membrane. The cells collect on the membrane surface and can then be removed from a Swinnex apparatus (Millipore Corp., Bedford, MA) and incubated on nonselective medium. The mixed cultures are subsequently washed from the membrane and plated on medium selective for the transposon and the new host.

Isolation and Characterization of Mutant Strains Following Transposition

Following the insertion of a transposon into the chromosome of a target organism, a population of strains, arising on medium selective for the presence of the transposon, must be screened for the trait of interest. Screens can include the production of color (or the absence of color production). For example, transposition of Tn*917–lacZ* in *S. faecalis* could be visualized by growth on X-Gal (5-bromo-4-chloro-3-indolyl-β-D-galactoside).[46] Tn*916* mutations in the *S. mutans* chromosome were screened via iodine staining to reveal a mutation in the *glgR* locus.[32] In addition to the use of chromogenic reagents, it is also possible to enrich for mutant phenotypes prior to plating. Fibrinogen binding has been used to enrich for mutants strains following insertion of Tn*917* into the chromosome of *St. aureus*.[19] In that example, *St. aureus* strains which had undergone a temperature shift to facilitate transposition were incubated with fibrinogen for 15 min. Bacteria capable of aggregating in the presence of fibrinogen were collected by centrifugation, and the cells remaining in the supernatant were screened further as possible mutant strains, potentially unable to bind fibronectin.[19] Binding of radiolabeled fibronectin has also been used as a screen for transposon-mediated mutations in *St. aureus*.[16]

Antibiotics may also be used to enrich for mutant strains defective in catabolism of a substrate following transpositional events. In the case where insertion has resulted in the creation of a mutant dependent on additional growth factors, antibiotics can be used to kill growing, wild-type cells in culture. Streptozotocin has been used effectively to isolate carbohydrate uptake mutants of *S. mutans*.[47] It seems likely that this

[46] K. E. Weaver and D. Clewell, *J. Bacteriol.* **170**, 4343 (1988).
[47] G. R. Jacboson, F. Poy, and J. W. Lengeler, *Infect. Immun.* **58**, 543 (1990).

technique could be broadly applicable to the isolation of catabolic defects in populations of cells following transposon insertions into the chromosome of the oral streptococci.

The characterization of strains, identified as mutant strains, must also include an examination of the chromosome via Southern blotting. Numerous reports have been published indicating the presence of multiple insertions of transposons within chromosomes following insertional events (e.g., multiple Tn916 insertions were observed in original isolates of group B streptococci capsule mutants[10]). Protocols for the preparation of streptococcal DNA have been detailed previously[8] as have protocols for the endonuclease digestion of DNAs, blotting, labeling of probes, and treatment of data. The most commonly used probe for Tn916 insertion has been the plasmid pAM120.[48] The fact that the enzyme EcoRI does not cut within Tn916 makes the digestion of chromosomal DNA with EcoRI, followed by probing with labeled pAM120, a useful diagnostic for the number of insertions within a given chromosome. Digestion with BamHI (one site within Tn916) provides an additional tool that can also assist in the evaluation of the number of insertions into a strain.

Preparation of DNA from Small Samples of Oral Streptococci

1. Bacteria (10–20 ml) are grown in BHI containing 20 mM DL-threonine.[49]

2. The cells are collected by centrifugation (10,000 rpm for 5 min at 4°) and resuspended in 1.0–2.5 ml of buffer containing 50 mM Tris-HCl (pH 8.0), 10 mM EDTA, and 10% (w/v) polyethylene glycol 8000.

3. Lysozyme (1 mg/ml) and mutanolysin (50 U) are added, and the mixture is incubated at 37° for 1 hr.

4. Cells are then lysed by adding sodium dodecyl sulfate (SDS) to a final concentration of 1% (w/v) and NaCl to a final concentration of 1 M.

5. DNA is purified from the resulting cell extract with two extractions by phenol–chloroform–isoamyl alcohol (25 : 24 : 1, v/v), followed by one chloroform–isoamyl alcohol (24 : 1, v/v) extraction.

6. RNase is added to a final concentration of 1 μg/ml, and the mixture is incubated at 65° for 15 min.

7. The DNA is precipitated with ethanol, dried, and resuspended to 1 mg/ml in 10 mM Tris-HCl (pH 7.4), 1 mM EDTA.

[48] J. M. Jones, C. Gawron-Burke, S. E. Flannagan, M. Yamamoto, E. Senghas, and D. B. Clewell, in "Streptococcal Genetics" (J. J. Ferretti and R. Curtiss III, eds.), p. 54. American Society for Microbiology, Washington, D.C., 1987.

[49] D. L. Wexler, J. E. C. Penders, W. H. Bowen, and R. A. Burne, Infect. Immun. 60, 3673 (1992).

An additional feature of Tn916 insertional mutagenesis that makes it a very useful element is its ability to excise itself. In the absence of tetracycline selection in clones containing Tn916 insertions in *E. coli,* it was found that Tn916 would excise at high frequency.[33] These observations gave rise to the idea that excision could be precise. The result of precise excision of Tn916 would be the restoration of a disrupted gene to its natural configuration. Thus, the Tn916 family of transposons were thought of as reverse cloning tools, that is, an insertion into the DNA of a target organism could be recovered by digestion of the DNA of the mutant strain with *Eco*RI and cloning of the DNA into *E. coli* via cosmid vectors. Precise excision of the transposon element would thus restore the gene, in effect a reverse clone. Subsequent work by Caparon and Scott has demonstrated that excision of Tn916 is not always precise and hence cannot be relied on to restore a gene completely.[50] Nevertheless, the Tn916 family can be used to reverse-clone gene fragments that can themselves be used as powerful probes for authentic genes isolated in genomic libraries. A recent example of this kind of use for Tn916 in *S. mutans* has been the cloning of the *glgR* locus.[32]

Transposon Analysis of Genes Cloned into *Escherichia coli*

Genes that have been cloned on large pieces of DNA may be analyzed by transposon insertion. The advantage of this technique is that genes with an identifiable product or activity may be mapped to a specific region following the isolation of a panel of insertional mutants, dispersed throughout a cloned DNA fragment. The positive selection of the transposon, by virtue of its antibiotic resistance, coupled with an assay for the gene product thus facilitates DNA analysis prior to a potentially time-consuming sequencing project. An additional advantage is that if the transposon carries a marker that can be expressed in the native strain of the clone, the insertionally inactivated gene can be returned to the original host for analysis of its phenotype after the majority of the manipulations have been done in *E. coli.* One technique that we have used in our laboratories is the insertional analysis of DNA using a derivative of the bacteriophage Mu, which is capable of transposing with high frequency and low discrimination of insertion sites.[51] Derivatives of Mu, carrying a variety of antibiotic resistance genes or genes encoding enzymes for the construction of gene fusions, have been used with considerable success in *E. coli.*[51] These

[50] M. G. Caparon and J. R. Scott, *Cell (Cambridge, Mass.)* **59,** 1027 (1989).
[51] M. L. Pato, *in* "Mobile DNA" (D. E. Berg and M. M. Howe, eds.), p. 23. American Society for Microbiology, Washington, D.C., 1989.

technologies have been extended to the gram-positive bacteria, with engineered transposons being used to construct fusions in *B. subtilis*[52] and *S. faecalis*.[46] A technique involving a modified Mu phage carrying two markers, one selectable in *E. coli,* the other selectable in the oral streptococci, has been used for analyzing the glucosyltransferase[37] and fructanase[49] coding regions of the *S. mutans* chromosome.

Mini-Mu Analysis of Cloned Genes

The occasion often arises that it is necessary or desirable to inactivate a cloned gene, located on a genetically undefined fragment of DNA, and then return it to the original host for analysis of the mutant phenotype. Relatively small, easily manipulatable elements such as mini-Mu can greatly facilitate such an experimental evaluation. The method can be divided into two parts: the introduction of the plasmid-borne clone into a strain containing a chromosomally located Mu phage and the induction of the phage by temperature shock. The resulting lysate will contain phage particles with an insertion in the plasmid that was introduced into the strain. The subsequent procedural step is the infection of a permissive strain with the Mu lysate and the isolation and characterization of the plasmid, now carrying a Mu insertion in a site within the cloned gene. These techniques have been adapted from those previously reported.[37]

Transformation of Escherichia coli with Plasmid Containing Cloned Gene

1. Start a 5.0-ml overnight culture of *E. coli* strain HK730 growing at 30° in Luria–Bertani (LB) medium containing kanamycin at 50 μg/ml (LB/Km$_{50}$).[53] Strain HK730 contains a mini-Mu lysogen carrying an erythromycin resistance gene selectable in the oral streptococci and a kanamycin resistance marker selectable in *E. coli.*[37]

2. In the morning, 0.2 ml of the overnight culture is used to inoculate a prewarmed flask containing 20 ml of LB/Km$_{50}$. The flask is shaken at 30° for 2 hr.

3. Following 2 hr of growth, the cells are collected by centrifugation (5000 rpm, 5 min, 4°), and the culture supernatant is decanted. The cells are washed with 10 ml of ice-cold 10 m*M* NaCl and repelleted as before.

4. The salt solution is carefully poured off, and the cells are resuspended in 10 ml of ice-cold 30 m*M* CaCl$_2$ and maintained on ice for 20 min.

[52] P. J. Youngman, *in* "Plasmids: A Practical Approach" (K. Hardy, ed.), p. 79. IRL Press, Oxord, 1987.
[53] M. Mandel and A. Higa, *J. Mol. Biol.* **53,** 159 (1970).

5. The cells are collected again by centrifugation, the supernatant fluid is poured off, and the cells are resuspended in 1.5 ml of ice-cold 30 mM CaCl$_2$. The cells should now be competent for transformation.

6. Plasmid DNA, containing the target gene, in a volume of 10 ml is added to 0.2 ml of the competent cells and maintained on ice for 1 hr.

7. The cells are shifted to 30° and shaken for 5 min.

8. The DNA and cell suspension is transferred to a 13 × 100 mm sterile, glass tube, and 2 ml of prewarmed LB medium is added to the suspension. The mixture is then shaken for 2 hr at 30°.

9. Following outgrowth of the transformants for 2 hr, 0.1-ml aliquots of the suspension are plated on LB medium containing antibiotic selective for the plasmid used to transform HK730 plus kanamycin to select for the presence of the transposon.

10. Incubate plates overnight at 30°.

11. Pick 50 transformants to two fresh agar plates, one containing erythromycin and the other containing the antibiotic selective for the marker of the plasmid.

12. To verify that the plasmid containing the target DNA is in the cells, it is advisable to pick some colonies, grow them in 2.0 ml cultures overnight, and then prepare plasmid DNA using previously described protocols.[54] The DNA recovered from those procedures should then be examined by restriction endonuclease digestion to assess the conformational stability of the clones.

Preparation of Mini-Mu Lysate

1. Inoculate 2 ml of LB/Km$_{50}$, containing in addition antibiotic selective for the vector, with a single transformant identified above and incubate overnight with shaking at 30°.

2. In the morning, use 0.1 ml of the overnight to inoculate 10 ml of the same medium and shake for 2 hr at 30° to permit the culture to enter the logarithmic growth phase.

3. Mini-Mu constructs contain a temperature-sensitive c gene, the repressor of lytic phage development. To induce the lytic cycle of the phage and hence transposition of the mini-Mu element from the *E. coli* chromosome into the plasmid-borne target gene, the culture is transferred to a shaking water bath at 42°, and growth is permitted to continue for 25 min.

4. Following the 42° heat shock, the cells are shifted to 37°. Our approach to this, taken from the bacteriophage λ literature, is to simply turn

[54] H. C. Birnboim and J. Doly, *Nucleic Acids Res.* **7**, 1513 (1979).

the temperature setting of the bath to 37° and allow the bath and culture to come down to temperature over a 2-hr period.

5. The incubation at elevated temperatures induces Mu phage synthesis and subsequent lysis of the host cell. Following the incubations at 42° and 37°, 0.1 ml of chloroform is added to the culture, and the two phases are mixed vigorously on a vortex mixer. Intact cells are removed from suspension by centrifugation for 5 min at 5000 rpm.

6. The supernatant fluid is aseptically pipetted to a fresh, sterile glass tube, and a second 0.1-ml aliquot of chloroform is added. Particulate material is again collected by centrifugation.

7. Repeat Step 6.

8. Filter the supernatants through a 0.2- or 0.45-μm filter to remove remaining cell debris. Store lysates at 4° in a glass tube.

Preparation of Transduction Recipient Strain

1. Grow a loopful of E. coli M8820 (Mucts) overnight in 2 ml of LB medium at 30°C.

2. Transfer 0.1 ml of the overnight culture to 2 ml of fresh LB. Grow the culture for 3 hr at 30°.

3. Collect the cells by centrifugation at 5000 rpm for 10 min in sterile, glass centrifuge tubes.

4. Gently pour off the supernatant fluids. Resuspend the cell pellet in 1 ml of 10 mM MgCl$_2$ and 5 mM CaCl$_2$. The cells are now ready to act as transduction recipients for the plasmid that has acquired a mini-Mu insertion.

Transductions

1. Add 0.2 ml of the HK730-derived lysate to the E. coli strain M8820 cell suspension. Mix by gently shaking the mixture at 30° for 15 min.

2. Add 4 ml of LB and shake the culture vigorously for 2 hr at 30°.

3. Plate 0.2-ml aliquots on LB/Km$_{50}$ (plus other appropriate antibiotic) plates and incubate 30° overnight.

4. Of the transformants arising on the plates, we usually screen 40 colonies. Grow loopfuls of the samples in 2.0-ml cultures, prepare plasmid DNA, and examine the plasmid for insertions via restriction endonuclease mapping.

5. In addition to more case-specific mapping, BamHI digestion of plasmid digests will also reveal a 600-base pair (bp) fragment arising from the mini-Mu, thus providing a fingerprint pattern that can simplify the identification of an insertion into the plasmid.

6. Following the identification of an insertion in a region of interest, clones may then be examined for the presence of gene products, for example, enzymatic activities or antibody-reactive material. Coupling insertional mutagenesis with assays for activities can thus provide a rapid method for DNA mapping.

The benefits of this technique extend beyond the identification of gene position within a cloned DNA fragment. In the case of the streptococci, mini-Mu carrying the Em^r determinant can be expressed in a streptococcal background. Thus, plasmid carrying the insertions can be transformed into the streptococci. The action of the RecA enzyme promotes the alignment of homologous DNA, and the mini-Mu insertion can be carried into the host chromosome, thereby creating an *in vivo* mutant strain. These strains would then become available as site-directed mutants for studies *in vitro* and with animal models. This approach may also be useful for defining the function of multiple genes in parent organisms.

Considerations and Limitations of Transposon Mutagenesis

Transposon mutagenesis will likely prove to be a powerful tool in dissecting virulence in gram-positive organisms, but the approach is not without a number of limitations. The first consideration regarding evaluating the role of virulence determinants, particularly in experiments involving serial passage or animal models, is the reversion frequency of the mutation as a result of precise excision of the transposon. If there is a strong selection for reversion, as may occur in an animal model, the relative contribution of the virulence determinant will undoubtedly be underestimated. It is particularly important to examine the phenotype and genetic composition of mutants to be introduced into experimental systems and of organisms recovered from these systems.

The second important consideration involves potential polar effects or pleiotropic effects on expression of virulence as a result of transposon insertion. There are numerous examples of cotranscribed genes which encode for proteins important to virulence. Likewise, inactivation of a locus involved in positive regulation of virulence may elicit the phenotype sought for in the experiment, namely, lack of production of the factor of interest, but could also affect gene expression at other virulence loci. It is therefore essential that the organisms be thoroughly characterized before initiating elaborate *in vitro* or *in vivo* studies of virulence. Also, one final consideration would be close examination of mutant strains for secondary transposition events which may have resulted in insertional inactivation of two or more unlinked genetic loci.

The utility of transposons in gram-positive bacteria to study virulence has been clearly demonstrated. This has been particularly true for the gram-positive cocci. However, it seems that the techniques developed for these organisms may be widely applicable to a variety of gram-positive pathogens, as has been observed for *Listeria monocytogenes*. Moreover, transposons from the streptococci have been introduced into bacteria such as *Bacillus anthracis*[1] and *Clostridium tetani*,[55] and they are capable of transposition in *E. coli*,[1] albeit at relatively low frequency, indicating that the utility of the transposons currently being used in gram-positive bacteria is probably not restricted to species which are relatively closely related on an evolutionary scale. Clearly, too, the advances made in the development of technologies for gene transfer to bacteria which have been retractile to more traditional methods promise to provide new and more efficient mechanisms for constructing strains which will have applications in basic research and medicine. Manipulation of transposons such as Tn*916*, Tn*917*, and Tn*1545* to allow for random generation of gene fusion libraries, as has been accomplished in *E. coli* and *B. subtilis*, will provide new opportunities for examining pathogenic strategies and regulatory circuits controlling virulence expression in gram-positive bacteria.

[55] M. Kuhn, M.-C. Prévost, J. Mounier, and P. J. Sansonetti, *Infect. Immun.* **58**, 3477 (1990).

[33] Identification of Bacterial Cell-Surface Virulence Determinants with TnphoA

By Melissa R. Kaufman and Ronald K. Taylor

Introduction

Gene fusion technology has provided bacterial geneticists various innovative and resourceful approaches for analysis of an extensive array of biological mechanisms. By creating fusions with reporter molecules such as β-galactosidase, an abundance of data concerning the regulation, localization, secretory signals, topology, and function of numerous molecules has been elucidated.[1,2] However, the cytoplasmic localization of β-galactosidase limits its efficacy regarding the identification of secreted proteins. To extend the translational fusion application to molecules encountered in the periplasmic space or beyond, hybrid proteins containing

[1] J. M. Slauch and T. J. Silhavy, this series, Vol. 204, p. 213.
[2] K. L. Bieker, G. J. Phillips, and T. J. Silhavy, *J. Bioenerg. Biomembr.* **22**, 291 (1990).

METHODS IN ENZYMOLOGY, VOL. 235

the reporter gene alkaline phosphatase (*phoA*) are preferred.[3-5] Alkaline phosphatase displays enzymatic activity only when exported across the cytoplasmic membrane into the oxidizing environment of the periplasm, where the molecule achieves its operative dimeric state.[6] Thus, fusion molecules containing the carboxyl-terminal portion of alkaline phosphatase serve as reporters for export signals.

Proteins involved in bacterial pathogenesis are particularly amenable to investigation with the *phoA* gene fusion approach considering they are frequently secreted or are cell-surface molecules. Identification of virulence genes *in vivo* has been accomplished by combining *phoA* fusion technology and the versatility of transposons using Tn*phoA*.[4] Insertion mutagenesis with Tn*phoA* confers the ability to generate randomly and to enrich for defined mutations in genes that encode membrane-associated and secreted virulence factors in the absence of any selection or previous knowledge of function. Outlined in this chapter are the rationale and methodologies for utilizing Tn*phoA* and additional *phoA* derivatives for the identification and characterization of various genetic elements that affect pathogenic properties in diverse species of bacteria.

Structure and Function of Tn*phoA*

Tn*phoA* is a derivative of the composite transposon Tn*5* containing a truncated *phoA* gene inserted in the left IS*50* element[4] (Fig. 1). In addition to a deletion of the promoter and translational start sequences, this *phoA* molecule lacks the region encoding the amino-terminal leader peptide required for transmembrane transport. Thus, hybrid molecules are detected only if secretion signals are coded for by the target gene that compensate for the secretion defect in the abbreviated alkaline phosphatase protein. Tn*5* and the derivative Tn*phoA* are premier genetic elements for random mutagenesis owing to their efficient transposition frequency, low target specificity, and broad host range.[7] Additionally, the central element of the transposon encodes the gene for kanamycin resistance (Km^r), thus providing a direct selection for transposition and cloning of the resulting insertion mutation. Tn*phoA* also encodes the gene for bleomycin resistance and a streptomycin resistance gene that is cryptic in *E. coli* and most enteric bacteria but may be of some utility in nonenteric

[3] C. S. Hoffman and A. Wright, *Proc. Natl. Acad. Sci. U.S.A.* **82,** 5107 (1985).

[4] C. Manoil and J. Beckwith, *Proc. Natl. Acad. Sci. U.S.A.* **82,** 8129 (1985).

[5] C. Manoil, J. J. Mekalanos, and J. Beckwith, *J. Bacteriol.* **172,** 515 (1990).

[6] A. Derman and J. Beckwith, *J. Bacteriol.* **173,** 7719 (1991).

[7] D. Berg, *in* "Mobile DNA" (D. E. Berg and M. M. Howe, eds.), p. 185. American Society for Microbiology, Washington, D.C., 1989.

TnphoA:

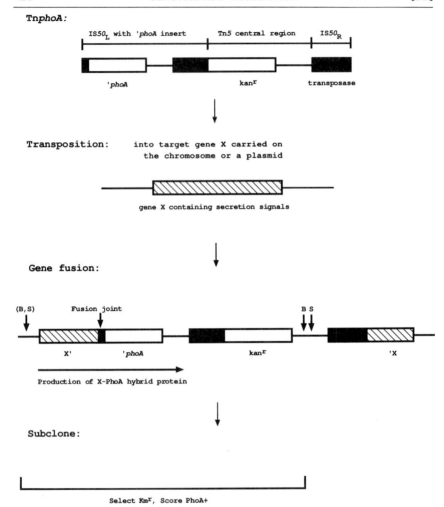

FIG. 1. Events leading to the formation and cloning of an active *phoA* gene fusion by using Tn*phoA* [Redrawn from R. K. Taylor, C. Manoil, and J. J. Mekalanos, *J. Bacteriol.* **171,** 1870 (1989)]. The transposon is a derivative of Tn*5* with a region encoding *E. coli* alkaline phosphatase, minus the secretion and expression signals, inserted into the left IS*50* element.[4] In-frame insertion into gene *X*, encoding a secreted product, interrupts the gene and results in production of an active hybrid protein from the *X–phoA* fusion. One scheme for isolating the gene fusion is to utilize the *Bam*HI (B) or *Sal*I (S) sites that lie distal to the gene encoding Km[r] and a hypothetical site (B or S) upstream of the gene to which *phoA* is fused. Restriction enzymes such as *Xba*I, *Stu*I, *Sac*I, and *Eco*RV, which do not cut within Tn*phoA*, are useful for isolating cloned fragments that carry the insertion plus DNA flanking both ends of the fusion joint.

organisms. Cloned fusion joints may quickly be sequenced to identify the target gene with use of a *phoA* primer.[8]

Detection and Quantitation of Alkaline Phosphatase

Assays for Alkaline Phosphatase Activity

Organisms carrying active PhoA fusions are readily identified by screening for blue colonies on agar supplemented with 40 μg/ml of the chromogenic substrate 5-bromo-4-chloro-3-indolyl phosphate *p*-toluidine salt (XP) solubilized in *N,N*-dimethylformamide. Quantitative assays for alkaline phosphatase activity rely on the rate of hydrolysis of *p*-nitrophenyl phosphate (PNPP) in permeabilized cells.[9]

Spectrophotometric Assay for Alkaline Phosphatase Activity

1. Determine the OD_{600} of an overnight culture.
2. Centrifuge 0.5 ml of the culture to collect the bacterial pellet; aspirate the supernatant and save if an extracellular assay will be performed.
3. Resuspend the pellet in 0.5 ml of 1 *M* Tris, pH 8.0.
4. Add 0.05 ml of pellet resuspension or culture supernatant to 1 *M* Tris, pH 8.0, such that the final volume is 1.0 ml.
5. Add 0.05 ml of 0.1% (w/v) sodium dodecyl sulfate (SDS) and 0.05 ml chloroform, vortex, and incubate at 37° for 5 min.
6. Add 0.1 ml of 0.4% (w/v) *p*-nitrophenyl phosphate (PNPP) solubilized in 1 *M* Tris, pH 8.0, vortex briefly, and incubate at 37° while timing the reaction.
7. On appearance of yellow color, stop the reaction with the addition of 0.1 ml of 1 *M* KH_2PO_4.
8. Determine the time of reaction, OD_{420}, and OD_{550}.
9. A specific activity relative to cell number can be determined with the following equation[10]:

Units of alkaline phosphatase activity =

$$1000 \times \frac{OD_{420} - (1.75 \times OD_{550})}{\text{time (min)} \times \text{volume (ml)} \times OD_{600}}$$

[8] R. K. Taylor, C. Manoil, and J. J. Mekalanos, *J. Bacteriol.* **171,** 1870 (1989).

[9] S. Michaelis, H. Inouye, D. Oliver, and J. Beckwith, *J. Bacteriol.* **154,** 366 (1983).

[10] J. H. Miller, *in* "Experiments in Molecular Genetics." Cold Spring Harbor Laboratory, Cold Spring Harbor, New York, 1972.

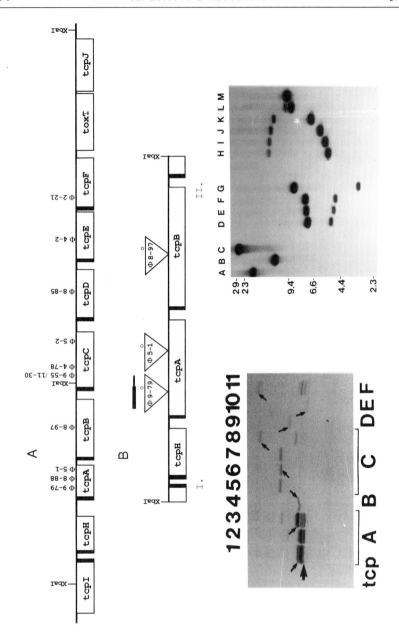

Detection of Hybrid Proteins

Hybrid molecules are also easily visualized by employing antibodies directed against alkaline phosphatase for immunoprecipitation or Western blot analysis of whole cell extracts or subcellular fractions[4,11] (Fig. 2). Additionally, by identifying the nucleotide sequence of target genes, peptide antigenic determinants can be synthesized and used for production of antisera to facilitate study of the newly isolated gene. Visualization of PhoA fusion molecules can be accomplished without specific antisera by performing an activity stain with XP (40 μg/ml) and nitro blue tetrazolium (1 mg/ml) in 1 M Tris, pH 9.0, of proteins separated by Triton X-100 polyacrylamide gel electrophoresis.[12]

Vehicles and Experimental Procedures for Tn*phoA* Delivery

Broad-Host-Range Plasmid Vectors

Broad-host-range plasmids harboring Tn*phoA* serve as efficient delivery systems for the introduction of Tn*phoA* into various species of gram-

[11] R. Taylor, C. Shaw, K. Peterson, P. Spears, and J. Mekalanos, *Vaccine* **6**, 151 (1988).
[12] V. L. Miller, R. K. Taylor, and J. J. Mekalanos, *Cell (Cambridge, Mass.)* **48**, 271 (1986).

FIG. 2. Mapping linked Tn*phoA* fusions by Southern and Western analysis.[11] (A) Map of the TCP gene cluster with the approximate locations of Tn*phoA* gene fusions designated. Relevant restriction sites (*Xba*I) are also indicated. (B) Schematic diagram of rationale used in mapping linked fusions in the TCP gene cluster and the corresponding Southern and Western blots. (The bottom two panels are reproduced from Ref. 11 by permission of Butterworth Heinemann, Ltd.) (I) Western blot probed with anti-alkaline phosphatase of Tn*phoA* fusion strains grown to mid-log phase (OD$_{600}$ ~0.8). Large arrow shows the position of native PhoA, and small arrows show positions of fusion proteins. Other bands are degradation products or cross-reactive bands. The tentative cistrons where each fusion originates are indicated at the bottom. Lane (1) 9-79, (2) 8-88, (3) 5-1, (4) 8-97, (5) 9-55, (6) 11-30, (7) 4-78, (8) 5-2, (9) 8-85, (10) 4-2, (11) 2-21. (II) Southern blot of Tn*phoA* insertion mutants. DNA in lanes A–C was digested with *Xba*I and that in lanes D—M with *Bam*HI and *Xba*I. The blot was probed with a 1.9-kb *Hind*III–*Bam*HI Tn*phoA*-derived fragment that in *Bam*HI–*Xba*I double digests will hybridize to the upstream fragment (relative to transcription of the fusion) more strongly than the downstream fragment (diagrammed above the *tcpA* fusion 9-79 where ° represents the *Bam*HI site internal to Tn*phoA*). Lane (A) 8-96, (B) 5-1, (C) 2-21, (D) 9-79, (E) 8-88, (F) 5-1, (G) 8-97, (H) 9-55, (I) 4-78, (J) 5-2, (K) 8-85, (L) 4-2, (M) 2-21. As outlined for *tcpA* fusions 9-79 and 5-1, multiple insertions within a single cistron will produce hybrid proteins of increasing size as the proximity of the insertion from the 5' end of the gene increases (I, lanes 1 and 3; II, lanes B and D). When the gene fusion maps to an adjacent cistron on the same restriction fragment, as shown for *tcpB* fusion 8-97, the hybrid protein size will decrease, but the band hybridizing strongly to the probe will increase in size (I: lane 4; II: lane G).

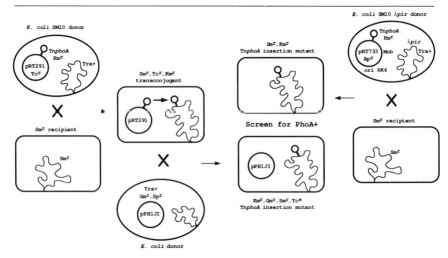

FIG. 3. Two methods for delivery and selection of transposition of Tn*phoA* from *E. coli* to other gram-negative bacteria. At left, pRT291, which carries Tn*phoA*, is mobilized into a streptomycin-resistant (Sm^r) recipient strain. Exconjugants are then utilized as the recipients for a second plasmid, pPH1JI, of the same incompatibility group as pRT291 and encoding gentamycin resistance (Gm^r). By maintaining selection for the presence of Tn*phoA* (Km^r), colonies are obtained in which Tn*phoA* has transposed and the vector plasmid has been lost. Colonies resulting from a fusion that expresses active alkaline phosphatase appear blue on agar supplemented with XP. The second method, shown on the right-hand side, utilizes the suicide vector pRT733, which allows for selection of transpositions in a single step. The vector is mobilized into a Sm^r recipient, and selection for the acquisition of Tn*phoA* (Km^r) directly yields transpositions since pRT733 cannot replicate without the functions provided in *trans* by λpir. Redrawn with permission from Ref. 8.

negative bacteria (Fig. 3). IncP vectors derived from RK2 plasmids carrying Tn*phoA* readily mobilize to numerous strains when provided transfer functions from an appropriate host strain or conjugative helper plasmid.[8,13] To enrich for Tn*phoA* transposition, transconjugants are superinfected with an additional IncP plasmid while selection for Tn*phoA* is maintained. This strategy displaces the initial IncP vector due to incompatibility and substantially increases the probability of recovering transposition events to a chromosomal or resident plasmid locus.

Plasmid pRT291 (IncP, *tra*^-, *mob*^+, Tc^r, Km^r) is composed of Tn*phoA* inserted into IncP, RK2 derivative pRK290.[8] This construct is capable of being mobilized either by triparental mating with conjugative helper plasmid pRK2013[13] or by biparental mating when using a mobilizing donor strain such as *Escherichia coli* SM10, which supplies P-group transfer

[13] G. Ditta, S. Stanfield, D. Corbin, and D. R. Helinski, *Proc. Natl. Acad. Sci. U.S.A.* **77**, 7347 (1980).

functions expressed from RP4-2Tc::Mu stably integrated into its chromosome.[14] When employing SM10 for conjugation of plasmids carrying TnphoA, note that the strain is kanamycin resistant; thus, tetracycline resistance is utilized to select for pRT291.

TnphoA Mutagenesis with pRT291

1. Streak out SM10 (pRT291) on Luria-Bertani (LB) agar (15 g agar, 10 g tryptone, 5 g yeast extract, 5 g NaCl per liter distilled, deionized water)[10] supplemented with Tc, Km, and XP.[14a]
2. Streak out the Sm^r recipient strain (counterselection for mating) on LB agar containing Sm.
3. Following overnight incubation, scrape approximately one loopful of cells of each type and mix together on the surface of an LB agar plate lacking antibiotics.
4. Incubate for 10–24 hr, scrape the growth, and streak or spread to LB agar containing either Sm, Tc or Sm, Km to select for transconjugants. After overnight incubation, colonies that appear should be the recipient carrying pRT291 and therefore display both Tc^r and Km^r.
5. Repeat plate mating as above, now mating pRT291 transconjugants with the E. coli strain MM294 (pro, endA, hsdR, supF)[15] containing IncP plasmid pPH1JI (Sp^r, Gm^r, low-level Sm^r).[16]
6. Incubate for 10–24 hr, make a suspension of the mating mix, and spread on LB agar supplemented with either Km, Gm, Sm, XP or Km, Sp, Sm, XP.
7. Colonies that acquire a blue color within about 1 to 3 days possess potential gene fusions to phoA that encode active, and thus presumably exported, hybrid proteins. Approximately 1% of random transposon insertions result in active fusions.

The pPH1JI plasmid is not maintained efficiently during growth at 42°. In Vibrio cholerae exconjugants after one round of overnight growth in liquid broth at 42°, approximately 20% of the colonies were Gm^s and had lost the plasmid as determined by agarose gel analysis of DNA isolated from several such derivatives.[8]

---

[14] R. Simon, U. Priefer, and A. Puhler, Bio/Technology 1, 784 (1983).

[14a] Abbreviations and concentrations for use of antibiotics and agar supplements are as follows: Ap, 100 μg/ml ampicillin; Gm, 30 μg/ml gentamicin; Km, 45 μg/ml kanamycin; Sm, 100 μg/ml streptomycin; Sp, 75 μg/ml spectinomycin; Tc, 15 μg/ml tetracycline; XP, 40 μg/ml 5-bromo-4-chloro-3-indolyl phosphate p-toluidine salt solubilized in N,N-dimethylformamide.

[15] G. B. Ruvkun and F. M. Ausubel, Nature (London) 289, 85 (1981).

[16] J. E. Beringer, J. L. Beynon, A. V. Buchanan-Wollaston, and A. W. B. Johnston, Nature (London) 276, 633 (1978).

Suicide Vectors for TnphoA Delivery

Derivatives of the suicide vector pJM703.1, originally designed to facilitate marker exchange experiments in *Vibrio cholerae*,[17] are effective devices for delivering Tn*phoA* to a variety of strains, with isolation of fusions occurring in a single step.[8] In these pBR322-based suicide plasmids, the ColE1 origin of replication has been replaced by the *cis* required components of the R6K *ori*, and thus the plasmids can only replicate when the necessary replication functions are provided *in trans*, as in the case of an *E. coli* λ*pir* lysogen.[18] Because these vectors fail to replicate in the recipient organism without the π-replication protein encoded by λ*pir*, selection for the acquisition of Tn*phoA* (Km[r]) directly yields transpositions. Conjugation of the plasmids into various gram-negative species is again achieved by use of the RP4 mobilization site by helper plasmid pRK2013[13] or *E. coli* strain SM10.[14]

TnphoA Mutagenesis with Suicide Vector pRT733

1. Streak SM10 λ*pir* (pRT733) to LB agar containing Ap and XP.
2. Streak out Sm[r] recipient to LB Sm.
3. Perform plate mating as described above for pRT291, finally plating cells on LB agar supplemented with Sm, Km, XP.
4. Blue, Km[r], Ap[s] colonies that appear in 1 to 3 days likely result from Tn*phoA* transposition events. However, some organisms produce a substantial frequency of Km[r], Ap[r] colonies. Such strains are presumed to arise from cointegrate formation between the vector and the chromosome.

Phage λ Vectors for TnphoA Delivery

Delivery of Tn*phoA* by infection with the phage λ derivative λTn*phoA* has been used extensively for mutagenesis of either chromosomal or plasmid-encoded factors in *E. coli* hosts.[19–21] Lambda-resistant species can often be engineered to undergo infection with λTn*phoA* by prior introduction of plasmids such as pAMH62[22] or pTROY9,[23] which express the cell-surface λ receptor LamB. The following protocol for isolating fusions on

[17] V. L. Miller and J. J. Mekalanos, *J. Bacteriol.* **170,** 2575 (1988).
[18] R. Kolter, M. Inuzuka, and D. R. Helinski, *Cell (Cambridge, Mass.)* **15,** 1199 (1978).
[19] C. Gutierrez, J. Bardoness, C. Manoil, and J. Beckwith, *J. Mol. Biol.* **195,** 289 (1987).
[20] D. S. Schifferli, E. H. Beachey, and R. K. Taylor, *Mol. Microbiol.* **5,** 61 (1991).
[21] S. Long, S. McCune, and G. C. Walker, *J. Bacteriol.* **170,** 4257 (1988).
[22] A. Harkki and E. T. Palva, *FEMS Microbiol. Lett.* **27,** 183 (1985).
[23] G. E. deVries, C. K. Raymond, and R. A. Ludwig, *Proc. Natl. Acad. Sci. U.S.A.* **81,** 6080 (1984).

high-copy-number plasmids using λTnphoA[24] is easily adaptable for obtaining chromosomal insertions by excluding antibiotics selective for the target plasmid, using 30–45 μg/ml kanamycin for selection of single-copy chromosomal insertions, and ending with the isolation of blue colonies.

Isolating Active PhoA Fusions to Plasmid Gene Products with λTnphoA

1. Target plasmids must be carried as a monomer in order to observe the phenotype conferred by the insertion. This is routinely accomplished by maintenance in an appropriate recA host such as CC118 (recA, phoA, Km^s, λ^s).[4]
2. Inoculate 2 ml LB broth[10] containing 10 mM MgSO$_4$ and antibiotic selective for the plasmid (e.g., ampicillin) with 0.02 ml of an overnight culture (1 : 100). Grow cells to early stationary phase (OD$_{600}$ ~1.0) and add λTnphoA at a multiplicity of approximately 1. Incubate without agitation at 30° for 15 min, then dilute aliquots 1 : 10 into LB broth to allow outgrowth. Grow a number of separate cultures from each infection to help guarantee isolation of insertions arising from independent transposition events.
3. Grow cultures for 4–15 hr at 30° with aeration.
4. Plate 0.2-ml aliquots on LB agar containing plasmid-selective antibiotic (e.g., ampicillin), 300 μg/ml kanamycin, and 40 μg/ml XP. Incubate for 2–3 days at 30°. For chromosomal insertions, plate cultures on LB agar containing 45 μg/ml kanamycin and 40 μg/ml XP.
5. Scrape colonies (either all colonies or just blue colonies) and prepare plasmid DNA from the pool using an alkaline lysis method.[25] Use this preparation to transform CC118, selecting transformants (after at least 45 min of outgrowth at 37° in LB broth) on LB agar supplemented with 30 μg/ml kanamycin and 40 μg/ml XP.
6. After 1–2 days of incubation at 37°, purify blue colonies on LB agar containing 30 μg/ml kanamycin and 40 μg/ml XP and concurrently score these bacteria on medium selective for the plasmid and supplemented with 40 μg/ml XP. If the active insert is in the antibiotic resistance gene of the plasmid rather than the gene of interest, there will be no growth on the second type of medium. If there is growth on the second medium, but colonies show blue and white sectors, the original colony was probably the result of a double transforma-

[24] C. Manoil, Methods Cell Biol. 34, 61 (1991).
[25] J. Sambrook, E. F. Fritsch, and T. Maniatis, "Molecular Cloning: A Laboratory Manual," 2nd Ed. Cold Spring Harbor Laboratory, Cold Spring Harbor, New York, 1989.

tion event with one plasmid (blue) likely containing a *phoA* fusion to *bla*, for example, and the second plasmid (white) still conferring Ap^r.

7. After purification, prepare plasmid from cells producing blue colonies. Check the purity of the plasmid preparation by transforming cells with selection for plasmid resistance on agar containing XP. Only blue colonies should result if the preparation contains a single species of plasmid. If the preparation is impure, plasmid can be isolated again from cells giving blue colonies in this transformation. If the preparation is pure, it can be analyzed by restriction mapping and nucleotide sequencing to determine the site of Tn*phoA* insertion.

Stocks of λTn*phoA* are prepared and the titer determined with an appropriate amber suppressor strain such as LE392[26] using standard protocols.[27]

Mini-TnphoA Transposon Derivatives

The versatility of Tn*phoA* has been expanded with a minitransposon that stabilizes resultant *phoA* fusions.[28] With deletion of nonessential sequences, mini-Tn5 *phoA* contains all the necessary functions for transposition and fusion generation, including the gene encoding Km^r, yet occupies half the space as the original Tn*phoA*. The minitransposon is located on an R6K-based suicide delivery plasmid, pUT,[29] that provides the IS50R transposase *tnp* gene in *cis* but external to the mobile element and whose conjugal transfer is mediated by the RP4 mobilization origin. Thus, after transposition with loss of the plasmid vector, the insertion is stable owing to the lack of transposase. The presence of a unique *Not*I site external to the Km^r cartridge facilitates the cloning of chromosomal insertions.

Insertional Mutagenesis Using Mini-Tn5 phoA

1. Grow recipient and donor strain *E. coli* SM10 λ*pir*[18] (pUT*phoA*) overnight at 30° to 37° in LB medium[10] in the presence of the required antibiotics.[29]
2. Mix 0.05 to 0.1 ml of each culture in 5 ml of sterile 10 m*M* MgSO$_4$ and filter through a 1.3-mm-diameter Millipore (Bedford, MA) type HA 0.45-μm filter (or equivalent).

[26] F. De Brujin and F. Lupski, *Gene* **27**, 131 (1984).
[27] T.J . Silhavy, M. L. Berman, and L. W. Enquist, "Experiments with Gene Fusions." Cold Spring Harbor Laboratory, Cold Spring Harbor, New York, 1984.
[28] V. de Lorenzo, M. Herrero, U. Jakubzik, and K. N. Timmis, *J. Bacteriol.* **172**, 6568 (1990).
[29] M. Herrero, V. de Lorenzo, and K. N. Timmis, *J. Bacteriol.* **172**, 6557 (1990).

3. Incubate the filter on the surface of LB agar lacking antibiotics for a minimum of 8 hr.
4. Resuspend cells grown on the filter surface in 5 ml sterile 10 m*M* MgSO$_4$.
5. Plate 0.1 to 0.5 ml of the suspension on a medium that counterselects the donor strain and selects recipient cells carrying the transposon marker (Km^r). Include 40 μg/ml XP in the medium to screen for active PhoA fusions.

Analysis of *Pseudomonas putida* exconjugants[28] reveals that approximately 90% of antibiotic-resistant colonies arise from authentic transposition of the mini-Tn5 element rather than from cointegration of the entire delivery plasmid into the recipient chromosome, with a corresponding operational transposition frequency higher than 10^{-6}. Optimal mutagenesis conditions may have to be adjusted when other strains or species are used as recipients as described below in the section on methodological considerations.

Vectors for in Vitro Construction of PhoA Fusions

Detailed analysis of cloned putative virulence determinant genes can be accomplished by precise construction of *in vitro phoA* fusions as an alternative to random mutagenesis. Vectors designed to facilitate creation of translational fusions with *phoA* include pCH2,[3] pPHO7,[30] and pKK-*phoA*.[31] Flanked by polylinkers containing diverse restriction sites, *phoA* is easily excised from pPHO7 to allow generation of a defined in-frame insertion in a cloned target gene. Additionally, a unique *Hin*dIII site allows restriction fragments of interest to be incorporated into pPHO7.

Application of Tn*phoA* in Divergent Species of Bacterial Pathogens

The following sections present examples of bacterial systems in which Tn*phoA* has been successfully utilized to identify virulence factors. This is not designed as a comprehensive review, but rather as a thought-provoking outline of the variety of organisms that are amenable to study with Tn*phoA*.

Colonization Factors of Vibrio cholerae

Adherence to the epithelium of the small intestine is an essential feature of the pathogenic mechanism of *Vibrio cholerae*. Colonization of the brush

[30] C. Gutierrez and J. C. Devedjian, *Nucleic Acids Res.* **17,** 3999 (1989).
[31] J. Kohl, F. Ruker, G. Himmler, D. Mattanovich, and H. Katinger, *Nucleic Acids Res.* **18,** 1069 (1990).

border permits *V. cholerae* to multiply and thus secrete a potent enterotoxin capable of inducing in infected individuals the extraordinary diarrhea
characteristic of cholera.[32] Genes encoding factors involved in the colonization process have been discovered by performing random Tn*phoA* mutagenesis on the *V. cholerae* chromosome with the broad-host-range vector
pRT291 as described previously, using media supplemented with 0.2%
(w/v) glucose to inhibit endogenous phosphatase activity.[33,34] Fusions to
phoA permitted identification of a pilus colonization factor (*tcp*) and accessory colonization factor (*acf*). Mutations in either determinant confer a
dramatic reduction in the ability of the organism to colonize the intestines
of suckling mice. Even more intriguing is the observation that expression
of numerous *V. cholerae* virulence elements, as monitored by PhoA fusion
activity, are coordinately regulated at both the physiological and transcriptional levels by the transmembrane DNA binding protein ToxR.[35] There
now exists a significant quantity of evidence indicating that coordinate
regulation of virulence factors is a common stratagem for many pathogenic
bacteria,[36,37] and Tn*phoA* mutagenesis provides an exceedingly powerful
scheme for detecting coordinately expressed proteins implicated in pathogenesis.

Symbiotic Loci of Rhizobium meliloti

Development of nitrogen-fixing nodules on leguminous plants by *Rhizobium meliloti* requires a complex series of interactions presumably partially mediated by cell-surface components of the bacteria. Symbiotic loci
of rhizobia are thus relevant targets for Tn*phoA* mutagenesis.[21] Introduction of Tn*phoA* to *R. meliloti* is accomplished via conjugation with plasmid
pRK609[21] (a derivative of pRK600) which carries an insertion of Tn*phoA*
created by infection with λTn*phoA*.[24] Originating from pRK2013,[13]
pRK600 is a chloramphenicol and neomycin resistance hybrid plasmid
with the ColE1 replicon allowing replication in *E. coli* but not in *R. meliloti*.
Plasmid pRK600, and its derivative pRK609, can be conjugally transferred
from *E. coli* to *R. meliloti* with the broad-host-range (IncP) transfer functions. Although the plasmid fails to replicate in *R. meliloti,* it persists for
a sufficient period of time to allow transposition,[38] making it an ideal suicide

[32] M. J. Betley, V. L. Miller, and J. J. Mekalanos, *Annu. Rev. Microbiol.* **40,** 577 (1986).
[33] R. K. Taylor, V. L. Miller, D. B. Furlong, and J. J. Mekalanos, *Proc. Natl. Acad. Sci.
U.S.A.* **84,** 2833 (1987).
[34] K. M. Peterson and J. J. Mekalanos, *Infect. Immun.* **56,** 2822 (1988).
[35] V. L. Miller and J. J. Mekalanos, *Proc. Natl. Acad. Sci. U.S.A.* **81,** 3471 (1984).
[36] J. F. Miller, J. J. Mekalanos, and S. Falkow, *Science* **243,** 916 (1989).
[37] V. J. DiRita and J. J. Mekalanos, *Annu. Rev. Genet.* **23,** 455 (1989).
[38] G. F. De Vos, G. C. Walker, and E. R. Singer, *Mol. Gen. Genet.* **204,** 485 (1986).

vector. To eliminate endogenous activity that could affect screening for TnphoA recipients, Sm^r Rm1021 was chemically mutagenized with 2.5% ethyl methanesulfonate, creating the pho host Rm80021. This TnphoA-based strategy catalyzed discovery or more detailed analysis of 25 loci involved in symbiosis, 17 of which were located on the megaplasmid of Rm8002.

Invasion and Intracellular Survival of Salmonella typhimurium

Ability to transcytose epithelial barriers is an important virulence characteristic of invasive Salmonella species and the molecular mechanisms governing the invasion process are now beginning to be understood.[39] To identify surface proteins utilized by the bacterium to gain entry into susceptible hosts, random insertions into the chromosome of a Sm^r S. cholerasuis strain were generated by introduction of TnphoA[40] on the donor plasmid pRT733.[8] Forty-two PhoA$^+$ insertions were isolated that were deficient in the ability to trancytose through polarized epithelial monolayers of Madin-Darby canine kidney (MDCK) cells. These mutant organisms also failed to adhere to MDCK cells, although the insertions were demonstrated by Southern analysis to be unlinked from established adherence factors. Mutations were grouped into six classes, five of which displayed significantly reduced virulence in the mouse model.

Survival and growth of S. typhimurium within the macrophage phagolysosome are essential components of typhoid pathogenesis.[41] Strains carrying mutations in the positive regulatory locus phoP are markedly attenuated in virulence for BALB/c mice.[42] By executing random TnphoA mutagenesis of S. typhimurium, a gene implicated in intracellular growth, pagC, was identified by virtue of its dependence on the PhoP regulator for expression. A S. typhimurium strain with a phoN mutation to suppress background XP hydrolysis (CS019) was constructed by Tn10 insertion mutagenesis and used as the parent strain for TnphoA introduction.

Hemolytic Determinant of Serratia marcescens

Serratia marcescens elaborates a membrane-bound hemolysin that, via inflammatory mediators, may increase vascular permeability, edema formation, and granulocyte accumulation, thus significantly contributing

[39] S. Falkow, R. R. Isberg, and D. A. Portnoy, Annu. Rev. Cell Biol. 8, 333 (1992).
[40] B. B. Finlay, M. N. Starnbach, C. L. Francis, B. A. D. Stocker, S. Chatfield, G. Dougan, and S. Falkow, Mol. Microbiol. 2, 757 (1988).
[41] P. I. Fields, R. V. Swanson, C. G. Haidairis, and F. Heffron, Proc. Natl. Acad. Sci. U.S.A. 83, 5189 (1986).
[42] S. Miller, A. M. Kukral, and J. J. Mekalanos, Proc. Natl. Acad. Sci. U.S.A. 86, 5054 (1989).

to the pathogenicity of the bacterium.[43] To characterize this hemolytic determinant genetically, Tn*phoA* mutagenesis of plasmids containing hemolytic loci was performed successfully to produce nonhemolytic colonies, 6% of which displayed a blue phenotype on agar supplemented with XP.[44] By defining the polarity of the genes with Tn*phoA*, subcloning and a T7 promoter-mediated expression system allowed identification of two protein factors involved in hemolysis. Optimal conditions for Tn*phoA* insertion in *S. marcescens* cloned genes were obtained with the following protocol.

1. *Escherichia coli* CC202 (a derivative of CC118[4] carrying Tn*phoA* on an F plasmid) is transformed with plasmids pES2 (pBR322 derivative) or pUC191 (pUC19 derivative) containing *S. marcescens* cloned genes and selected on TY (0.8% tryptone, 0.5% yeast extract, 0.5% NaCl, pH 7) agar supplemented with 50 μg/ml ampicillin.
2. Plate transformants on TY agar containing ampicillin and 500 μg/ml kanamycin. The high kanamycin concentration is necessary since strain CC202 shows a rather high kanamycin-resistant phenotype without Tn*phoA* transposition into the multicopy plasmids.
3. Isolate plasmids from pooled colonies and transform *E. coli* CC118, selecting on TY agar containing ampicillin, 50 μg/ml kanamycin, and supplemented with XP to screen for translational fusions.

Iron-Regulated Virulence Factor of Vibrio cholerae

Induction under limiting iron conditions is a common theme in the regulation of many virulence-associated genes. Insertion mutagenesis of *V. cholerae* with Tn*phoA*[8] identified a fusion with increased alkaline phosphatase activity when cultured in low-iron conditions.[45] Strains containing fusions to this gene, designated *irgA,* lack a 77-kDa iron-regulated outer membrane protein and show a significant decrease in virulence, as determined by the median lethal dose (LD$_{50}$) and competitive index in suckling mice. Subsequent nucleotide sequence analysis of the *irgA::phoA* fusion revealed that the amino-terminal portion of IrgA is homologous to the ferrienterochelin receptor (FepA) of *E. coli,* indicating that IrgA may be the iron–vibriobactin outer membrane receptor.[46] Additionally, Northern blots demonstrated that iron regulation of *irgA* occurs at the transcriptional

[43] V. Braun, H. Gunther, B. Neuss, and C. Tautz, *Arch. Microbiol.* **141,** 371 (1985).
[44] V. Braun, B. Neuss, Y. Ryan, E. Schiebel, H. Schoffler, and G. Jander, *J. Bacteriol.* **169,** 2113 (1987).
[45] M. B. Goldberg, V. J. DiRita, and S. B. Calderwood, *Infect. Immun.* **58,** 55 (1990).
[46] M. B. Goldberg, S. A. Boyko, and S. B. Calderwood, *J. Bacteriol.* **172,** 6863 (1990).

level and involves an operator binding site analogous to Fur binding sites of *E. coli* and a positive transcriptional activator, IrgB, encoded directly upstream of *irgA*.[47]

Invasion Loci of Enteropathogenic Escherichia coli

Cellular attachment and invasion appear to be crucial components of enteropathogenic *E. coli* (EPEC) infection. Localized invasion of HEp-2 cells *in vitro* is linked to the presence of a high molecular weight plasmid that contains the EAF adherence factor as detected by DNA probes.[48] Identification of the genes required for HEp-2 invasion was accomplished by mutagenizing the EPEC genome with Tn*phoA*.[49] To obtain a marker for counterselection in conjugation experiments, plasmid pUR222[50] (ampicillin resistant) was introduced into EPEC strain E2348p-69 by electroporation. Biparental mating with SM10 pRT291[8] transferred Tn*phoA* to the recipient. pUR222 was then cured from the resulting transconjugant by three passages on agar containing tetracycline and kanamycin, but no ampicillin. Selection for mutants was accomplished by conjugal transfer of incompatible pPH1JI[16] into E2348-69 (pRT291) with simultaneous selection for gentamicin and kanamycin resistance. Gene fusions were then detected on agar supplemented with XP. This strategy elucidated 22 noninvasive derivatives containing insertions both plasmid and chromosomally localized.[48]

Attaching and effacing lesions on the surface of the intestinal epithelium *in vivo* can be produced by EPEC strains devoid of the EAF plasmid.[51] Further understanding of the attaching and effacing invasion process was again generated by Tn*phoA* mutagenesis. Isolation of insertion mutants in EPEC strain JPN15 using the suicide vector pRT733[8] revealed a single genetic locus, designated *eae,* critical for formation of pathogenic lesions.[52]

Pectate Lyase Secretion in Erwinia chrysanthemi

Pathogenicity of the bacterium *Erwinia chrysanthemi,* responsible for soft rot of many plant species, is related to its secretion of plant cell wall-

---

[47] M. B. Goldberg, S. A. Boyko, and S. B. Calderwood, *Proc. Natl. Acad. Sci. U.S.A.* **88,** 1125 (1991).

[48] J. P. Nataro, M. M. Baldini, J. B. Kaper, R. E. Black, N. Bravo, and M. M. Levine, *J. Infect. Dis.* **152,** 560 (1985).

[49] M. S. Donnenberg, S. B. Calderwood, A. Donohue-Rolfe, G. T. Keusch, and J. B. Kaper, *Infect. Immun.* **58,** 1565 (1990).

[50] U. Ruther, M. Koenen, K. Otto, and B. Muller-Hill, *Nucleic Acids Res.* **9,** 4087 (1981).

[51] S. Knutton, D. R. Lloyd, and A. S. McNeish, *Infect. Immun.* **55,** 69 (1987).

[52] A. E. Jerse, J. Yu, B. D. Tall, and J. B. Kaper, *Proc. Natl. Acad. Sci. U.S.A.* **87,** 7839 (1990).

degrading enzymes such as pectate lyases, cellulases, and proteases.[53] Open reading frames within a cloned locus known to play a role in pectate lyase secretion, *outJ*, were detected by mutagenesis with λTn*phoA* in *E. coli*.[54] Exchange recombination of these Tn*phoA* insertions from the pBR325 plasmid vector into the *E. chrysanthemi* chromosome for phenotypic assay was performed by successive culture in phosphate-limited, kanamycin containing medium.[55] The pBR series plasmids are unstable under these conditions and are lost from the cell; thus, the kanamycin selects for cells in which exchange recombination has occurred.

Antigenic Modulation in Bordetella pertussis

Virtually all recognized virulence factors encoded by the causative agent of whooping cough, *Bordetella pertussis,* are coordinately regulated by a phenomenon designated antigenic modulation.[56] An activator encoded by the gene *vir* is essential for expression of these virulence genes.[57] By isolating Tn*phoA* fusions in *B. pertussis* controlled by environmental modulation signals, regulated genes that are both activated and repressed by *vir* have been discovered, including a novel regulatory gene, *mod*.[58] To generate a Tn*phoA* insertion library, either suicide plasmid pRT733[8] or IncP plasmid pRT291[8] is introduced by conjugation into *B. pertussis* as described in previous sections with the following modifications. Matings are performed with 5×10^8 mid-log phase cells of both donor and recipient, which are mixed and plated on Bordet–Gengou (BG) agar plates. Following 6 hr of incubation at 37°, cells are harvested, washed with saline, and spread to BG plates supplemented with the appropriate antibiotics to select transconjugants and counterselect donors. Colonies are lifted from BG plates onto a dry nitrocellulose filter and washed three times by soaking for 10 min per wash on filter paper saturated with 0.14 *M* NaCl. This wash removes the red coloration of the colonies resulting from incubation on BG medium. The washed nitrocellulose filter is then transferred to a stack of filter paper soaked in 1 *M* Tris, pH 8.0, containing 160 μg/ml XP. After an incubation of several minutes to several hours, PhoA-positive (blue) colonies are detected, which can then be purified on BG plates. Employing replica filters may be useful when working with organisms that require

[53] A. Kotoujansky, *Annu. Rev. Phytopathol.* **25,** 405 (1987).
[54] J. Ji, N. Hugouvieux-Cotte-Pattat, and J. Robert-Baudouy, *Mol. Microbiol.* **3,** 285 (1989).
[55] D. L. Roeder and A. Collmer, *J. Bacteriol.* **164,** 51 (1985).
[56] A. A. Weiss and E. L. Hewlett, *Annu. Rev. Microbiol.* **40,** 661 (1986).
[57] A. A. Weiss and S. Falkow, *Infect. Immun.* **43,** 263 (1984).
[58] S. Knapp and J. J. Mekalanos, *J. Bacteriol.* **170,** 5059 (1988).

growth on agar media that does not permit detection of PhoA-positive colonies due to dark background color.

Methodological Considerations

It is imperative to recognize the potential obstacles that may be encountered when using Tn*phoA*. For instance, insertions of Tn*5* in an operon are often polar on downstream gene expression.[7] This polarity could cause assignment of an incorrect phenotype if Tn*phoA* has inserted upstream in an operon from the actual gene of interest. One strategy for circumventing this polarity relies on creation of a nonpolar deletion and is presented in the next section. Although Tn*5* displays little insertion specificity in most organisms,[7] some difficulty with Tn*phoA* insertion "hot spots" or clustering has been observed in *Salmonella* spp.[40]

When screening for gene fusions on agar supplemented with XP, background levels of alkaline phosphatase activity can easily produce false positives, particularly if the target genes are expressed in low amounts. A variety of strategies for obtaining *pho* mutants or repressing endogenous alkaline phosphatase activity in diverse organisms have been presented in this chapter. It would be beneficial to attempt one of these methods prior to Tn*phoA* insertion to eliminate the possibility of potentially misleading PhoA levels. In fact, insertion within a gene encoding a repressor of endogenous alkaline phosphatase is a potential problem when not using a *phoA* background.

It may be difficult to discern in the spectrophotometric alkaline phosphatase assay between fusions to low-copy secreted molecules versus highly expressed cytoplasmic proteins. Activation of cytoplasmic forms of alkaline phosphatase can be eliminated by the addition of 1 mM iodoacetimide to the assay buffers.[24]

Although infrequent occurrences, excision and transposition of Tn*phoA* subsequent to the primary event are possibilities. Maintaining kanamycin selection should prevent recovery of cells that have undergone excision of the transposon. Although secondary transpositions have not been a problem in our analysis of Tn*phoA* generated mutants, the mini-Tn*phoA* transposon on an R6K-based delivery plasmid provides the transposase external to the mobile element such that, after loss of the vector, the insertion is stable owing to the absence of transposase.[28] However, use of mini-Tn*5* elements has been limited in *E. coli* K12 strains. Between 10 to 90% of *E. coli* exconjugants contain lysogenized λ*pir* transferred from the presently available donor strain SM10 λ*pir*, resulting in a significant impairment of the suicide donation system. To lower the frequency of recipient phage infection, perform conjugations on agar plates as opposed to mating in liquid media.

Significant difficulties will conceivably be encountered when attempting Tn*phoA* mutagenesis on targets that are not present in a single copy in the genome, or genes that incur repeated antigenic variation. Fusions to certain genes may generate hybrid proteins toxic to the bacterium, resulting in recovery of uncharacteristically small colonies if insertions can be detected.[59]

Tn*5* may fail to transpose efficiently in some bacteria such as *Haemophilus* species, rendering chromosomal Tn*phoA* mutagenesis impossible.[60] As demonstrated for *Erwinia chrysanthemi*, virulence genes in hosts resistant to Tn*phoA* transposition can be mutagenized after cloning of the locus into *E. coli*.[54] Insertions can then be recombined back into the host bacterium for phenotype analysis. This should be approachable on a random insertion basis, with recombination of any active fusion back into the chromosome of the species under study.

Many PhoA fusion proteins experience significant proteolytic cleavage surrounding the fusion junction, currently limiting the potential purification of hybrid molecules, although Western analysis is nearly always possible. However, fusion proteins created between the *E. coli* heat-stable enterotoxin (STb) and PhoA are exceptionally stable, even retaining STb biological activity in a rat ligated intestinal loop model.[61]

Further Uses for Tn*phoA* in Analysis of Pathogenesis

Construction of Nonpolar Deletions to Replace TnphoA Insertion

Chromosomal Tn*phoA* insertions can easily be converted to deletion mutations to overcome transcriptional polarity or create strains lacking the Tn*phoA*-encoded kanamycin resistance.[62] As outlined in Fig. 4, restriction sites flanking the Tn*phoA* insertion in the cloned gene can be used to excise Tn*phoA*. Religation, with or without appropriate oligonucleotide linkers, then yields an in-frame deletion in the target gene. Recombination of this deletion into a strain carrying the chromosomal Tn*phoA* insertion is screened by searching for white colonies on agar supplemented with XP, and subsequently scoring for kanamycin sensitivity. This strategy should prove particularly useful in the creation of nonpolar deletion muta-

[59] C. Manoil, *J. Bacteriol.* **172**, 1035 (1990).
[60] G. J. Barcak, M. S. Chandler, R. J. Redfield, and J.-F. Tomb, this series, Vol. 204, p. 321.
[61] R. G. Urban, L. A. Dreyfus, and S. C. Whipp, *Infect. Immun.* **58**, 3645 (1990).
[62] M. R. Kaufman and R. K. Taylor, manuscript in preparation.

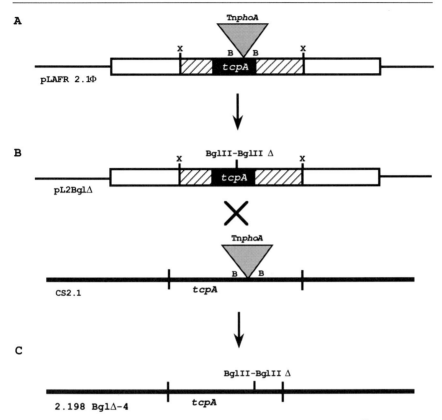

FIG. 4. Construction of a nonpolar deletion to replace a TnphoA insertion.[62] (A) A plasmid clone carrying a TnphoA gene fusion isolated from the chromosome, such as Tc^r, Km^r pLAFR 2.1Φ diagrammed here, is digested at sites that restrict within the gene and flank the TnphoA insertion. (B) Religation, with the addition of oligonucleotide linkers if necessary, creates an in-frame deletion in the target gene that is easily recovered by screening transformants for white colonies on agar containing XP and scoring for kanamycin sensitivity (pL2B-glΔ). The deletion is then recombined into the Sm^r, Km^r strain containing the original chromosomal TnphoA fusion (CS2.1). (C) Exconjugants are used as the recipient for mating with an E. coli strain carrying a plasmid of the same incompatibility group as the plasmid containing the deletion (i.e., Gm^r pPH1JI). Sm^r, Gm^r, Tc^s, Km^s colonies that are white when screened on agar supplemented with XP have replaced the TnphoA insertion with the in-frame deletion (2.198 BglΔ-4). Deletion mutants exhibiting the appropriate phenotypes should be confirmed by Southern blot analysis.

tions in operons encoding proteins involved in multimeric complexes. Additionally, because a nonreverting mutation can be precisely introduced without an accompanying antibiotic selective marker, this tactic provides an excellent mechanism for construction of potential vaccine strains.

Identification of Lipoproteins Involved in Pathogenesis with [³H]Palmitate

Select bacterial outer membrane proteins have been associated with the property of serum resistance by preventing deposition of the complement membrane attack complex into the membrane.[63] Three *V. cholerae* PhoA fusions to virulence-associated proteins, identified by virtue of the dependence on ToxR for regulation,[34] were characterized as lipoproteins with a combination of DNA sequence analysis and [³H]palmitate labeling.[64] Interestingly, PhoA fusion proteins were chosen for [³H]palmitate labeling because the section of the gel encompassing molecules with a mass of 38 kDa or less was heavily labeled by lipopolysaccharide metabolites, making analysis of individual lipoproteins in this region impossible. Hybrid molecules, migrating at 48 kDa or above, thus serve as excellent tools for [³H]palmitate labeling experiments. Mutations in one such lipoprotein, TcpC, rendered *V. cholerae* 10^4–10^6 times more sensitive to the vibriocidal activity of antibody and complement. Because *V. cholerae* is a noninvasive pathogen, this observation raises the possibility that complement-like bactericidal factors exist on the luminal side of the intestine, and evasion of this activity may contribute to the colonization process for numerous bacteria.

Fusion Switching

Conversion of a Tn*phoA* gene fusion to a chloramphenicol acetyltransferase (*cat*) operon fusion is a convenient method for quantitative measurement of transcription.[58] Plasmid pSKCAT is a derivative of pJM103.1[17] containing the promoterless *cat* gene located adjacent to a 5' fragment of Tn*phoA*. Homologous recombination of pSKCAT with chromosomal *phoA* sequences generated with Tn*phoA* insertion concludes with integration of pSKCAT. This merger generates insertional inactivation of *phoA* and simultaneous fusion of upstream sequences to the promoterless *cat* gene (Fig. 5). Chloramphenicol acetyltransferase activity is easily monitored in permeabilized cells containing pSKCAT. This *cat* conversion mechanism provides a straightforward method for creating a transcrip-

[63] K. A. Joiner, *Annu. Rev. Microbiol.* **42,** 201 (1988).
[64] C. Parsot, E. Taxman, and J. J. Mekalanos, *Proc. Natl. Acad. Sci. U.S.A.* **88,** 1641 (1991).

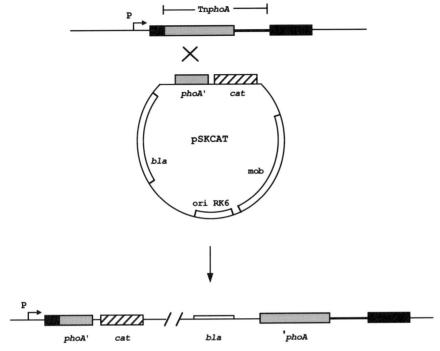

FIG. 5. Conversion of Tn*phoA* fusions to *cat* transcriptional fusions by using pSKCAT. A Tn*phoA* translational fusion is shown on the top line for a gene (solid bar) expressed from a promoter (P→) which is then converted to a *cat* transcriptional fusion by integration of the replication-defective pSKCAT plasmid. Homologous recombination (×) with internal *phoA* sequences disrupts the gene fusion *phoA* sequences (stippled bar) on integration of pSKCAT (below) to give a PhoA⁻ phenotype and simultaneous fusion to a promoterless *cat* gene fragment (hatched bar). Redrawn with permission from Ref. 58.

tional fusion to further determine the level at which production of a given virulence determinant is regulated.

To assist in the description of normal subcellular location or membrane topology of a protein, vectors that carry Tn*lacZ* are applied in interconversion experiments to replace Tn*phoA* fusions.[59] The complementary properties of β-galactosidase and alkaline phosphatase make fusion switching with Tn*lacZ* a model technique for detailed analysis of the functional regions of membrane proteins that contribute to pathogenesis.

*Tn*phoA* as Probe for Genes Linked to Identified Virulence Factors*

Tn*phoA* has been implemented with exceptional success in the analysis of genes required for formation of complex structures on the bacterial cell surface. A cluster of six genes linked to the major subunit of the toxin-coregulated pillin (TCP) of *V. cholerae* was discovered by mapping, with Southern and Western blot analysis, chromosomal Tn*phoA* insertions that alter pilus expression[11] (Fig. 2). This technique allowed identification of a TCP locus encoding proteins implicated in pilus biogenesis and regulation.

Cloning and analysis of novel genes linked to *fasA* responsible for production and function of 987P fimbriae of *E. coli* was assisted by Tn*phoA* tagging.[20] Physical restriction mapping of Tn*phoA* insertions and expression of the corresponding protein products with the coupled T7 polymerase/promoter system distinguished eight genes and a Tn*1681*-like transposon located in the 987P fimbrial gene cluster.

Summary

Insertion mutagenesis using Tn*phoA* has proved to be a potent device for the creation of easily screened knockout mutations in genes encoding virulence determinants in a variety of pathogenic bacteria. Initial identification of genes with Tn*phoA* directly initiates more sophisticated genetic and biochemical studies on these factors essential to our understanding of bacterial pathogenesis.

Acknowledgments

The authors gratefully acknowledge Colin Manoil for critical reading of the manuscript, and Jean-Francoise Tomb for helpful discussions.

[34] Temperature-Sensitive Mutants of Bacterial Pathogens: Isolation and Use to Determine Host Clearance and *in Vivo* Replication Rates

By ANNE MORRIS HOOKE

Introduction

The rate of clearance of a pathogen from the body of an animal cannot be determined quantitatively by simply injecting animals with the pathogen and taking samples over time. As the bacteria are being cleared they are also replicating in the host, so the colony-forming units (cfu) recovered from the animals represent only the net values. Similar difficulties are also encountered when *in vitro* bactericidal assays are performed with

METHODS IN ENZYMOLOGY, VOL. 235

phagocytic cells: as some bacteria are being phagocytosed and killed, others are replicating.[1] On the other hand, quantitative replication rates of bacteria *in vivo* have also been impossible to measure because, again, if animals are injected with the organism and samples are taken over time, as the organism is replicating it is also being cleared. Previous efforts to measure replication *in vivo* were limited to rather specialized systems (*Mycobacterium tuberculosis,* which retain acid-fast staining properties when they are dead,[2] and lysogenized *Salmonella typhimurium* superinfected with mutant bacteriophage[3]) or to experiments with radiolabeled bacteria which probably raise more problems than they solve.[4]

In recent years, however, temperature-sensitive (*ts*) mutants have been used to study host–pathogen interactions, both *in vitro* and *in vivo*. Temperature-sensitive mutants of pathogens with lesions in essential genes (e.g., ribosomal proteins, DNA polymerase, or RNA polymerase) are relatively easy to obtain, with or without the assistance of chemical or other mutagens. Mutants with "tight" phenotypes, that is, those that cease replication immediately after transfer to the nonpermissive temperature (NPT), are readily identified and characterized with respect to revertant frequency, cutoff temperature, and maintenance of surface antigens (it is very important to conserve the native conformation of surface molecules because of interactions between the organism and host clearance mechanisms such as complement).

Mutagenesis

Although it is sometimes possible to isolate spontaneous *ts* mutants of pathogenic bacteria by cycling and recycling cultures through alternating rounds of enrichment, it is usually necessary to enhance the yield of mutants by treating cultures with chemical or physical mutagens. Mutants used for *in vivo* studies should have mutations in essential genes because the effect of mutations in biosynthetic pathways can be abrogated by nutrients available in the host. Mutations in essential genes will most likely result from single base changes, so alkylating agents such as *N*-methyl-*N*'-nitro-*N*-nitrosoguanidine (nitrosoguanidine) or ethyl methanesulfonate should be used. Although frame-shift mutagens and ultraviolet irradiation could generate the desired mutations, the probability of success is not very high because the targets would be very limited. It is also

[1] A. Morris Hooke, M. P. Oeschger, B. J. Zeligs, and J. A. Bellanti, *Infect. Immun.* **20,** 406 (1978).
[2] P. D. Hart and R. J. W. Rees, *Br. J. Exp. Pathol.* **41,** 414 (1980).
[3] G. G. Meynell, *J. Gen. Microbiol.* **21,** 421 (1959).
[4] R. Freter and P. C. M. O'Brien, *Infect. Immun.* **34,** 222 (1981).

important to perform all mutagenesis and enrichment steps in very rich media in order to minimize the isolation of *ts* auxotrophs.

The first step in generating *ts* mutants of a bacterial pathogen involves titration of the mutagenic agent to determine the concentration which causes the most mutations with the least amount of killing. Nitrosoguanidine (Sigma, St. Louis, MO), freshly prepared in acetone (nitrosoguanidine should be kept desiccated at $-20°$ and should not be used after the container has been open for a month), is added to log phase cultures (10 ml in the richest broth appropriate) of the organism at concentrations ranging from 0 through 1, 3, 10, 30, to 100 μg/ml, and the cultures are incubated without aeration for 5, 10, or 15 min at 37°. The mutagen is removed by centrifugation and two or three cycles of washing, and 1 ml of the original culture volume is suspended in 20 volumes of fresh broth and cultured until the cells recover and undergo two or three divisions (for segregation of the induced mutations). Samples can then be plated on agar containing streptomycin (or another appropriate antibiotic) to determine the number of streptomycin-resistant mutants generated by the treatment. Viable cell counts are determined on the cultures both before and after mutagenesis, and at the time they are plated on streptomycin. The enhancement of the spontaneous mutation rate to streptomycin resistance is calculated and plotted against the percentage of cells surviving at each concentration of nitrosoguanidine. Concentrations of nitrosoguanidine that we have found useful for generating *ts* mutants in a number of different bacterial genera are shown in Table I.

Optimum concentrations of the cell wall antibiotics (typically penicillin and D-cycloserine) to be used in the enrichment cycles are determined by incubating broth cultures of the bacteria at the nonpermissive temperature (usually 37°), adding various concentrations of the antibiotic, and monitoring viability over time (usually 3 to 5 hr). To maximize enrichment, 99 to 99.9% of the replicating bacteria should be eliminated by the antibiotic treatment.

Isolation of Temperature-Sensitive Mutants

Once the optimum concentrations for all reagents have been determined, mutagenesis and enrichment for *ts* mutants can begin. A flow sheet for the general procedure is shown in Fig. 1. Nitrosoguanidine is added (at the appropriate concentration) to a broth culture in log phase at 37° (10 ml, $\sim 10^8$ cfu/ml) for the optimal time, without aeration. The mutagen is removed by three cycles of centrifugation, and samples of the cells are suspended in 20 volumes of fresh broth distributed into five or six separate flasks (to minimize the isolation of "siblings") and incubated overnight

TABLE I

TEMPERATURE-SENSITIVE MUTANTS GENERATED BY NITROSOGUANIDINE AND ISOLATED
AFTER TWO CYCLES OF ENRICHMENT

Bacterial strain	Nitrosoguanidine concentration (μg/ml)[a]	Mutants recovered (%)	Ref.
Haemophilus influenzae	2	5–10	b
Salmonella typhi	3	10–15	c
Escherichia coli	10	10–20	d
Pseudomonas aeruginosa	20	10–15	e
Staphylococcus aureus	10–20	3–4	f
Listeria monocytogenes	50	2–5	g

[a] Treatment was in all cases at 37° for 10 min, without aeration.
[b] A. Morris Hooke, J. A. Bellanti, and M. P. Oeschger, Lancet 1, 1472 (1985).
[c] A. Morris Hooke, Z. Wang, M. C. Cerquetti, and J. A. Bellanti, Vaccine 9, 238 (1991).
[d] A. Morris Hooke, M. P. Oeschger, B. J. Zeligs, and J. A. Bellanti, Infect. Immun. 20, 406 (1978).
[e] A. Morris Hooke, P. J. Arroyo, M. P. Oeschger, and J. A. Bellanti, Infect. Immun. 38, 136 (1982).
[f] D. O. Sordelli, M. F. Iglesias, M. Catalano, and A. Morris Hooke, Abstr. Annu. Meeting Am. Soc. Microbiol., E81 (1992).
[g] F. Gervais, A. Morris Hooke, T. A. Tran, and E. Skamene, Infect. Immun. 54, 315 (1986).

with appropriate aeration and at the permissive temperature (PT, limited only by the temperature range of the particular strain, but typically 27° to 30°). When the cells reach a density of approximately 5×10^7 cfu/ml, the cultures are transferred to a shaking water bath at the nonpermissive temperature, and a cell wall antibiotic is added at the optimal concentration. The culture is incubated until 99–99.9% of the cells have been killed; the antibiotic is removed by centrifugation (or treated with, e.g., β-lactamase). The surviving cells are suspended in the same volume of fresh broth and incubated at the permissive temperature until the density again reaches about 5×10^7 cfu/ml. The culture is subjected to another cycle of enrichment, this time with a different cell wall antibiotic (the first cycle also enriches for mutants resistant to the antibiotic used then).

The surviving cells are washed free of antibiotic, diluted appropriately, plated and incubated at the permissive temperature, or suspended again in fresh broth for incubation and another cycle of enrichment (the third enrichment should use the first antibiotic). Colonies which arise after incubation at the permissive temperature are then replica-plated onto chocolate or blood agar (to avoid isolation of ts auxotrophs) and incubated at the nonpermissive and permissive temperatures to identify the ts clones.

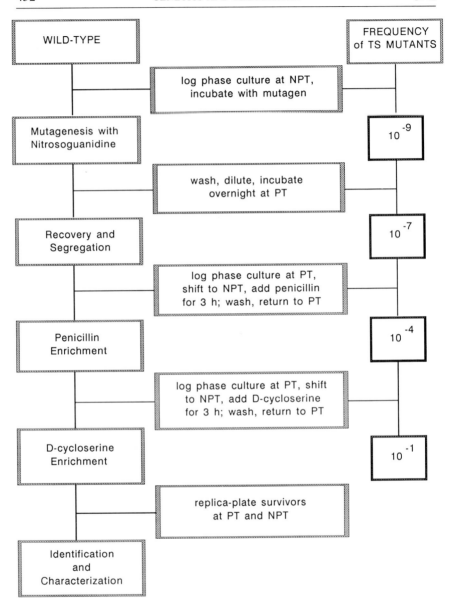

FIG. 1. Flowchart for the generation and isolation of temperature-sensitive mutants of bacterial pathogens. The approximate expected frequency of *ts* mutants in the population is given for each step on the right-hand side. NPT, Nonpermissive temperature; PT, permissive temperature.

Characterization of Temperature-Sensitive Mutants

Mutants identified as *ts* on rich media are characterized with respect to cutoff temperature by streaking on several plates and incubating at different temperatures (28°, 30°, 32°, 34°, 36°, 38°). The next temperature above the highest temperature at which single colonies are visible within 24 hr (depending, of course, on the cultural characteristics of the strain) is considered the cutoff. The reversion rate is estimated by plating large numbers ($>10^7$–10^8 cfu suspended from a single colony) on chocolate agar and incubating at the nonpermissive temperature. Accurate determinations of the reversion rate can only be made with the fluctuation test,[5] but an approximation is usually sufficient for mutants to be used in the assays described here. Mutants with revertant frequencies of less than 10^{-7} are acceptable, as the presence of 1 to 10 revertants in the doses inoculated into animals will, for most short-term (less than 1 day) experiments, have negligible effects on the data. Should it be necessary, however, to reduce the probability of the presence of revertants to even lower numbers, mutations of identical phenotype can be combined in one strain.[6] This may be desirable for experiments requiring long-term exposure to macrophages *in vitro*, as these cells in culture produce mutagenic compounds which could enhance the reversion rate.

The *ts* phenotype is determined by incubating the mutants in broth culture at the permissive temperature until the cell density reaches approximately 10^8 cfu/ml, shifting the culture to the nonpermissive temperature, and monitoring increases in absorbance at 600 nm. Viable counts should also be performed as cell division mutants may cause increases in absorbance without concomitant increases in colony-forming units, in which case the cultures should also be monitored microscopically. Generally only mutants that cease replication immediately after transfer to the nonpermissive temperature are useful, although mutants that continue replication for one or two generations, and then cease, can be held at the nonpermissive temperature for the appropriate time before use. In fact, under certain circumstances these mutants (which we call "coasters") might be preferable. For example, if the expression of surface antigens only seen *in vivo* (e.g., iron-binding proteins or some heat-shock proteins) were important to the experimental situation, then incubation of the mutant in appropriate media (iron-depleted) (at both the PT and NPT) would ensure their synthesis.

Nitrosoguanidine is excellent for inducing *ts* mutations in essential genes, but it does have one disadvantage, namely, the tendency to induce

[5] S. E. Luria and M. Delbrück, *Genetics* **28**, 491 (1943).
[6] A. Morris Hooke, J. A. Bellanti, and M. P. Oeschger, *Lancet* **1**, 1472 (1985).

multiple mutations at the replication forks.[7] It is important, therefore, to check the *ts* mutants for secondary mutations affecting surface molecules which could interfere with both clearance and replication assays. Comparison of outer membrane protein profiles (for gram-negative bacteria) of wild type and mutant may reveal major changes in important surface antigens. Western blots, agglutination,[8] or phagocytosis assays[9] with antibodies to wild type surface antigens could also detect changes which might be important.

Quantitative Clearance Studies

Tight *ts* mutants of pathogenic bacteria can be used to quantitate clearance of the organisms from many areas of a host or to perform quantitative bactericidal assays with macrophages or polymorphonuclear leukocytes.[8,9] Temperature-sensitive mutants of *Pseudomonas cepacia*,[10] *P. aeruginosa*,[11] *Salmonella typhimurium*,[12] and *Escherichia coli*[13] have been used to study clearance of the pathogens from the peritoneal cavities, blood, spleens, and lungs of mice, and the lungs and air sacs of chickens. Typical results from experiments measuring the clearance of *ts E. coli* from the lungs of chickens are shown in Fig. 2. For the assay, White Leghorn chickens are inoculated intratracheally with 10^8 cfu *ts* and 10^7 cfu wild-type *E. coli*. At 0, 2, 3, and 4 hr postinoculation groups of three to four birds are sacrificed, and the lungs are removed aseptically, homogenized in 1 ml sterile water in a Tekmar Stomacher (Cincinnati, OH), diluted in sterile water, plated, and incubated at 30° (for total numbers) and 40° (for wild-type numbers). The numbers of the *ts* mutants are obtained by difference. The difficulty of making any quantitative statement about the clearance of the wild-type *E. coli* is obvious; within the first 2 hr the numbers of wild type recovered from the lungs increased, but for the next 2 hr replication and clearance appeared to cancel one another. In contrast,

[7] E. A. Adelberg, M. Mandel, and G. C. C. Chen, *Biochem. Biophys. Res. Commun.* **18**, 788 (1965).
[8] F. Gervais, A. Morris Hooke, T. A. Tran, and E. Skamene, *Infect. Immun.* **54**, 315 (1986).
[9] A. Morris Hooke, M. P. Oeschger, B. J. Zeligs, and J. A. Bellanti, *Infect. Immun.* **20**, 406 (1978).
[10] A. Morris Hooke, R. C. Palecek, T. M. Palecek, and D. M. Miller, *Curr. Microbiol.* **22**, 129 (1991).
[11] A. Morris Hooke, D. O. Sordelli, M. C. Cerquetti, and J. A. Bellanti, *Infect. Immun.* **55**, 99 (1987).
[12] R. N. Swanson and A. D. O'Brien, *J. Immunol.* **131**, 3014 (1983).
[13] T. S. Agin, H. T. Heeg, R. R. Harper, and A. Morris Hooke, *Abstr. Annu. Meeting Am. Soc. Microbiol.*, E12 (1991).

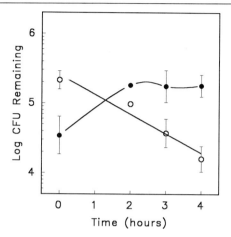

FIG. 2. Clearance of *ts* (○) and wild-type (●) *E. coli* from the lungs of White Leghorn chickens. Experimental details are given in the text. Each point represents the mean colony-forming units (±SEM) recovered from the lungs of three to four birds.

the clearance of the *ts* mutants from the lungs is readily quantitated, and the rate can be calculated from the slope of the regression line on the graph.

In Vivo Replication

We have used mixtures of tight *ts* mutants and parental wild types to measure the *in vivo* replication of *E. coli*,[14] *P. aeruginosa*,[14,15] and *P. cepacia*[10] in mice, and *E. coli* in chickens.[13] We have also used mixtures of tight and coasting mutants of *P. aeruginosa*, each marked with a chromosomal antibiotic resistance (for differentiating the two strains in the recovered samples), to measure the residual *in vivo* replication of the coasting mutant.[11] The model is based on two assumptions: first, that the tight *ts* mutant cannot replicate *in vivo* and, second, that both strains (mutant and wild-type) will be cleared at the same rate. The first assumption is difficult to prove, but in years of experience with the system we have never seen any evidence to the contrary. The second assumption is more problematic, as it has long been suspected that replicating organisms are more susceptible to the bactericidal mechanisms of the host. We have used a variation of the system to test this hypothesis.[13] Those studies, performed with *E. coli* in avian lungs, support the hypothesis, but only

[14] A. Morris Hooke, D. O. Sordelli, M. C. Cerquetti, and A. J. Vogt, *Infect. Immun.* **49**, 424 (1985).
[15] A. Morris Hooke, unpublished data (1985).

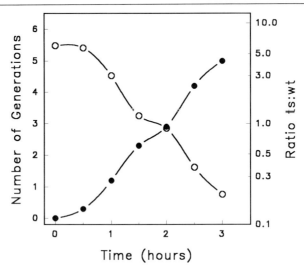

FIG. 3. Replication of *E. coli* in the peritoneal cavities of mice. Experimental details are given in the text. Each point represents the mean of the ratio of *ts* to wild type (○) or the number of generations (●), calculated from the formula given in the text, for three to four mice. The mean generation time (33 min) was determined from the inverse of the slope of a regression line through the points for $t_{0.5}$, t_1, $t_{1.5}$, t_2, $t_{2.5}$, and t_3. (Adapted from A. Morris Hooke *et al.*,[14] with permission.)

in the later stages of infection (7–18 hr postinoculation), so the basic system described here for the quantitative measurement of *in vivo* replication is almost certainly valid for the first few hours after inoculation. A more detailed discussion of the two assumptions and the mathematical derivation of the formula used in the model are given elsewhere.[14]

In the first published description of the model[14] we inoculate groups of ICR mice intraperitoneally with mixtures of *ts* and wild-type *E. coli* at ratios of 5 *ts* to 1 wild type. Immediately after inoculation (t_z) and every 30 min thereafter (t_t) for 3 hr, three to four mice are sacrificed by cervical dislocation, and the peritoneal cavities are lavaged with 5 ml warm sterile saline. The samples recovered are diluted in sterile distilled water (to lyse phagocytes and release viable bacteria), plated, and incubated at 29° (for total numbers) and 37° (for wild-type numbers). The numbers of *ts* mutants are obtained by difference. The ratios of the two organisms are calculated and the number of generations of the wild type determined from the formula[14]

$$n = [\log(\text{ratio at } t_z/\text{ratio at } t_t)]/\log 2$$

The *in vivo* mean generation time (MGT) is calculated from the inverse of the slope of the regression line.

The results of a representative experiment are shown in Fig. 3. As would be expected, a short lag period occurred while the organism adjusted to the *in vivo* environment (time required, perhaps, for the synthesis of siderophores), and this was followed by logarithmic growth typical of *in vitro* cultures. The experiment was performed three separate times with three different batches of mice, and the calculated MGTs for the wild-type *E. coli* in the peritoneal cavity were 32, 33, and 34 min, respectively. Later studies have shown that ratios of up to 20 *ts* to 1 wild type can be used, thus increasing the time available for the experiment before the sensitivity of detection is lost. For reasons that are not entirely clear, however, inoculation ratios of greater than 20 *ts* to 1 wild type (at least with *P. aeruginosa* and some strains of *E. coli*) lead to less than satisfactory results.

It should also be noted that it would be wise to perform experiments with mixed *ts* and wild-type cultures *in vitro* before embarking on the *in vivo* studies in order to eliminate the possibility of untoward interactions between the replicating and nonreplicating bacteria.[14] Although we have never encountered any such problem, it is, for example, theoretically possible that the nonreplicating *ts* mutant (especially at high ratios) might inhibit the replication of the wild type at the nonpermissive temperature, or the replicating wild type might reduce the viability of the mutant at the nonpermissive temperature. Also, as with the mutants used for the clearance experiments, the maintenance of important surface antigens should be confirmed.

Summary

Temperature-sensitive mutants of bacterial pathogens are relatively easy to obtain and characterize. We have used *ts* mutants of a number of bacterial pathogens: *E. coli*[1] and *Listeria monocytogenes*[8] to determine quantitatively the bactericidal activities *in vitro* of macrophages and polymorphonuclear leukocytes; *E. coli*[13] and *P. cepacia*[10] to study clearance of the bacteria *in vivo*; and *Salmonella enteritidis*,[15] *Listeria monocytogenes*,[15] *E. coli*,[13,14] *P. aeruginosa*,[14] and *P. cepacia*[10] to determine quantitatively the replication rates in the spleens, lungs, and peritoneal cavities of mice, and the lungs of chickens.

[35] Use of Conditionally Counterselectable Suicide Vectors for Allelic Exchange

By SCOTT STIBITZ

Introduction

A major goal of research on a pathogenic organism is the complete identification of the virulence determinants of that organism. Virulence determinants may be defined as those genes which enable an organism to colonize the host successfully, and which may then result in host pathology. An operational definition of a virulence determinant might then be as follows: a gene belonging to a pathogen whose inactivation leads to a decrease in virulence of that pathogen. Such a definition may be troublesome in that many genes involved in normal cellular processes may be included. Nevertheless, meeting such a requirement would seem to be a prerequisite for a bona fide virulence determinant. To address these issues experimentally, it is desirable that one be able to inactivate specific chromosomal genes to then be able to assess the contribution of these genes to the overall virulence of an organism.

A second goal of research on pathogenic organisms is an understanding of the regulation of virulence determinants in response to the changing environments often associated with a pathogenic life-style. Powerful tools in studies of these phenomena are gene fusions of regulated genes to reporter genes encoding β-galactosidase, alkaline phosphatase, chloramphenicol acetyltransferase, luciferase, etc., which provide an easily determinable measure of gene expression under different environmental conditions. One would thus like to be able to introduce gene fusions in a directed (i.e., nonrandom) way to particular genes in a regulon.

For the above reasons we have developed methods which allow the easy replacement of bacterial chromosomal alleles with those which have been cloned and modified *in vitro*. These methods were originally developed for use in studies on the human pathogen *Bordetella pertussis*, but it is our hope and expectation that they will be utilizable in other genera and species as well.

Principle of the Method

The replacement of a chromosomal gene with its *in vitro*-altered counterpart by homologous recombination requires two crossovers between

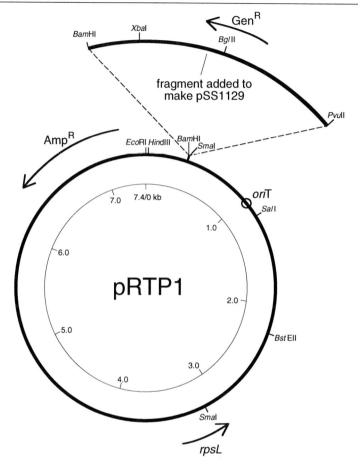

FIG. 1. Restriction map of pRTP1 and pSS1129. The construction of pRTP1 has been previously described [S. Stibitz, W. Black, and S. Falkow, *Gene* **50**, 133 (1986)]. The 2.4-kb *Bam*HI–*Pvu*II fragment added to pRTP1 to make pSS1129 is shown. This fragment was derived from pSK6 [S. Kagan, Ph.D. Thesis, Univ. of Wisconsin, Madison (1981)]. Note that the *Pvu*II site and *Sma*I site involved in the cloning are both destroyed.

the two segments of DNA. We designed a vector, pRTP1 (Fig. 1), which allows one to introduce a cloned gene into a bacterial recipient by conjugation with an *Escherichia coli* donor, and then to select for two successive crossovers between the cloned gene and the recipient chromosome. Transfer to the recipient bacterium can be accomplished because the plasmid contains the origin of transfer from RP4, a promiscuous, IncP plasmid. When Tra functions are provided *in trans,* pRTP1 will be transferred into the recipient bacterium via conjugation. Selection for the first crossover is accomplished simply by imposing selection for the vector or a marker within the cloned sequences. Because the vector, a member of the ColE1 family of plasmids, is incapable of replication in *B. pertussis* and many other bacterial species, this will select for those bacteria in which the plasmid has integrated into the chromosome. If a cloned segment homologous to the chromosome is contained within this plasmid, integration will most likely occur by homologous recombination.

The second crossover is obtained by selecting for the loss of the integrated plasmid. This is accomplished by virtue of the wild-type *E. coli rspL* gene in pRTP1 which confers a streptomycin-sensitive phenotype on strains otherwise resistant due to a mutation in the chromosomal *rpsL* homolog. Thus streptomycin-resistant survivors will be those which will have lost the plasmid copy of *rpsL*. The mechanism by which this loss occurs is homologous recombination between the direct repeats which flank the vector as a result of its integration.

At the end of these manipulations, two crossovers will have occurred between the plasmid-encoded and chromosomal copies of the gene in question. If these crossovers occur on opposite sides of the site of alteration in the cloned gene, the net result is that the alteration will be incorporated. If these occur on the same side, the original allele will remain. In many cases the alteration being introduced is a "knockout" insertion mutation with a selectable gene inserted into the gene in question. In such cases, maintaining selection for this mutation, while selecting for loss of the pRTP1 vector, ensures that the crossovers occur on opposite sides of the insertion mutation. In fact such a replacement can be accomplished as a single step simply by selecting for the mutation and for streptomycin resistance simultaneously after conjugation. If the alteration does not confer a selectable phenotype, then one must screen for its introduction after selecting for two consecutive crossovers. In this case, if there is not an easily scorable phenotype (e.g., loss of hemolysis) associated with the mutation, then one may cross this mutation into a strain in which one has previously crossed in a scorable marker such as kanamycin resistance at the same location. Then after two crossovers one may screen for the introduction of the desired alteration by scoring for loss of the kanamycin resistance marker.

Note that when alleles are introduced which do not contain a selectable marker, one must rely on selectable markers contained within the vector in the conjugation leading to the first crossover. In the original formulation the only such marker on pRTP1 was ampicillin resistance. This was found to be a very poor marker in conjugations, not unexpectedly since the mechanism of resistance in this case involves inactivation of the antibiotic in the medium, allowing for background growth of nonresistant colonies.

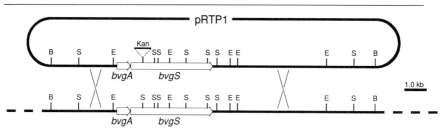

Fig. 2. Diagram of the replacement of the chromosomal *vir* locus with an allele containing an insertion mutation with a selectable phenotype. The *vir* locus encodes two genes: *bvgA* and *bvgS*. B, *Bam*HI; E, *Eco*RI; S, *Sal*I.

A derivative of pRTP1, pSS1129, was therefore constructed through the addition of a DNA fragment containing a gene specifying resistance to gentamicin via a cytoplasmic acetyltransferase enzyme (Fig. 1). Use of the pSS1129 vector has eliminated the problem of background in conjugations.

Procedures

Shown below are examples illustrating different applications of the method which have been performed in the author's laboratory in the course of studies on the *vir* locus of *B. pertussis*. The *vir* locus regulates many virulence determinants in this organism, including the gene for the extracellular adenylate cyclase which gives rise to a scorable phenotype (hemolysis).

Replacement of Chromosomal vir Locus with "Knockout" Insertion of Selectable Kanamycin Resistance Marker

A 14.3-kilobase (kb) *Bam*HI fragment containing the *vir* locus is cloned into pRTP1 (Fig. 2). Insertions of a kanamycin resistance cassette derived from pKanπ[1] are introduced at random *Sal*I sites by partial digestion of this plasmid and ligation of *Sal*I-linearized plasmid with the *Sal*I kanamycin resistance fragment. Particular insertions are transformed into *E. coli* strain SM10,[2] which contains the RP4 Tra genes inserted in the chromosome, and then transferred to *B. pertussis* strain BP536 (Tohama I, *str, nal*).[3] Conjugations are initiated by swabbing together the *E. coli* donor and the *B. pertussis* recipient on Bordet–Gengou agar containing 10 mM MgCl$_2$. After 3 hr of incubation at 37°, the bacteria are recovered with a

[1] W. Black and S. Falkow, *Infect. Immun.* **55**, 2465 (1987).
[2] R. Simon, U. Priefer, and A. Puhler, *Bio/Technology* **1**, 784 (1983).
[3] S. Stibitz and M.-S. Yang, *J. Bacteriol.* **173**, 4288 (1991).

sterile dacron swab and reswabbed onto Bordet–Gengou agar containing 10 μg/ml kanamycin and 50 μg/ml nalidixic acid. Colonies which arise are still Vir⁺ (hemolytic) owing to the presence of an intact copy of the *vir* locus in addition to the mutant copy. The exconjugants are then restreaked onto Bordet–Gengou agar containing 100 μg/ml streptomycin. Colonies which arise are either Hly⁻, Kan^R (mutation incorporated) or Hly⁺, Kan^S (mutation not incorporated). If the exconjugants are instead streaked onto Bordet–Gengou medium containing 10 μg/ml kanamycin and 100 μg/ml streptomycin, all of the colonies which arise are nonhemolytic. Survivors of streptomycin selection in either case are ampicillin sensitive, showing that they have indeed lost the vector. Southern analysis is performed to verify correct incorporation of the mutation.

Introduction of Unmarked In-Frame Linker Insertion Mutations into Chromosomal vir Locus

The 12-base pair (bp) oligonucleotide 5' [PO₄]GCTCGAGCTGCA 3', which contains an internal *XhoI* site and which self anneals to produce *PstI* sticky ends, is used to generate in-frame insertion mutations (Fig. 3). A 2.5-kb *EcoRI* fragment containing part of the *vir* locus is first cloned into pBR322. The oligonucleotide is then introduced at random *PstI* sites by ligation to plasmid DNA which has been linearized by partial digestion with *PstI* and then treated with calf intestinal phosphatase. After transformation and screening for desired insertion sites, possible multiple insertions are resolved by digestion with *XhoI* and religation. The modified *EcoRI* fragments are then cloned into pSS1129, and the resulting plasmids are transformed into SM10. After conjugation with BP536, exconjugants are selected on Bordet–Gengou agar containing 10 μg/ml gentamicin and 50 μg/ml nalidixic acid. Colonies arising are either Hly⁺ or Hly⁻, depending on whether the first crossover occurs upstream or downstream, respectively, of the site of the mutation. On restreaking on medium containing

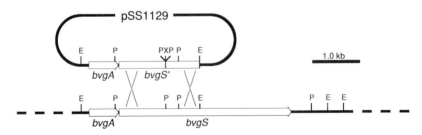

FIG. 3. Diagram of the introduction of an unmarked in-frame linker insertion into the chromosomal *vir* locus. E, *EcoRI*; P, *PstI*; X, *XhoI*.

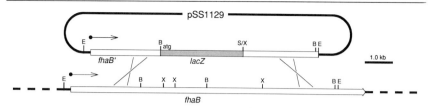

FIG. 4. Diagram of the replacement of the chromosomal *fhaB* gene with an *fhaB* : *lacZ* transcriptional fusion. B, *Bam*HI; E, *Eco*RI; S, *Sal*I; X, *Xho*I.

streptomycin, however, both types of colonies give rise to both hemolytic and nonhemolytic colonies. The hemolytic colonies are those in which the second crossover occurs on the same side as the initial crossover such that the mutation is not incorporated, while the nonhemolytic colonies are those in which the two crossovers occur on opposite sides of the lesion. Correct incorporation of the mutation in the nonhemolytic derivatives is verified by Southern analysis.

Replacement of Chromosomal fhaB Locus with fhaB : lacZ Gene Fusion

A 10-kb *Eco*RI fragment containing the *fhaB* gene (the structural gene for filamentous hemagglutinin) is cloned into pTTQ8.[4] A *Bam*HI–*Sal*I fragment containing the *lacZ* gene of *E. coli* and derived from pRS155[5] is cloned into the *fha* fragment between the *Bam*HI and most downstream *Xho*I site. The *fhaB* : *lacZ Eco*RI fragment is then cloned into pSS1129, and the resulting plasmid (Fig. 4) is transformed into SM10. After conjugation with BP536, selection is imposed for gentamicin and nalidixic acid resistance, and colonies which arise are restreaked onto medium containing streptomycin. Colonies surviving this selection are tested for β-galactosidase activity in a microtiter assay. Approximately 10% of the recombinants arising after streptomycin selection are Lac[+]. Southern analysis is performed to verify correct insertion of the gene fusion. Quantitative β-galactosidase assays show that *lacZ* expression is repressed approximately 50-fold when these *B. pertussis* strains are grown in the presence of 50 m*M* MgSO$_4$ (conditions known to repress synthesis of virulence determinants).

[4] M. J. R. Stark, *Gene* **51,** 255 (1987).
[5] R. W. Simons, F. Houman, and N. Kleckner, *Gene* **53,** 85 (1987).

Discussion

The vectors described here have been invaluable in our studies on the *vir* locus of *B. pertussis*. The potential usefulness of these specific vectors in other bacterial species depends on the proper function of their two primary aspects, namely, that they be "suicide vectors" capable of transfer to, but incapable of replication in, the intended target species and that they contain some marker which confers conditional lethality. Species where either pRTP1 or an analogous plasmid utilizing the same two modules of the ColE1 replication origin and streptomycin counterselection has been used successfully for allelic replacement include *Pseudomonas aeruginosa*[6] and *Legionella pneumophila*.[7]

In cases where one or both of these modules does not operate as needed, a vector fulfilling the same requirements may be constructed using other modules. For example, suicide vectors based on the origin of replication from plasmid R6K in conjunction with the *pir* gene *in trans* have been constructed and work well in species in which ColE1-based plasmids can replicate.[8] Other counterselectable markers are also available. The *sacB* gene confers a lethal phenotype in the presence of sucrose and has been used with success in several systems.[7] On an experimental basis we have had some success in *B. pertussis* using the *ksgA* gene of *E. coli* as a counterselectable marker. This gene confers sensitivity to the aminoglycoside antibiotic kasugamycin in strains which are resistant due to a mutation destroying the function of the chromosomal *ksgA* gene (encoding a ribosomal RNA methylase[9]). The attractiveness of using two different counterselectable markers in the same species becomes apparent when one is attempting repeated substitution within a particular gene of interest. In this case one may construct a strain harboring an insertion of one conditionally lethal marker within that gene through the use of a replacement vector based on the second counterselectable marker. Once such a strain has been constructed, one can achieve replacements of the gene of interest by using the inserted counterselectable marker to select against maintenance of the chromosomal allele after allowing for recombination with the incoming allele to occur. Such an approach has been used

---

[6] M. J. Gambello and B. H. Iglewski, *J. Bacteriol.* **173**, 3000 (1991).

[7] N. P. Cianciotto, R. Long, B. I. Eisenstein, and N. C. Engelberg, *FEMS Microbiol. Let.* **56**, 203 (1988).

[8] V. L. Miller and J. J. Mekalanos, *J. Bacteriol.* **170**, 2575 (1988).

[9] T. L. Helser, J. E. Davies, and J. E. Dahlberg, *Nature (London) New Biol.* **235**, 6 (1972).

successfully for repeated mutagenesis of the *ptx* locus which encodes pertussis toxin.[10]

The question of how much flanking homology is required to cross in a desired mutation successfully is one frequently encountered. While we have not determined the lower limit through a systematic study of this question, it appears that 500 bp on each side is sufficient. This comes from the observation that a 1-kb fragment from the *vir* locus which contains the site of a frameshift mutation responsible for "phase variation" was capable of rescuing the Vir$^+$ phenotype when used in a pRTP1 derivative.[11]

It should also be noted that some background can be demonstrated in each recombinational event. For example, when selecting for the first crossover, if no cloned homologous segment is present in the vector, one can still isolate, at a much reduced frequency, insertions of the vector into the chromosome. We had previously assumed that these insertions occurred via homologous recombination as a result of sequence similarity between the plasmid and chromosomal homologs of the *rpsL* gene. However, since the construction of a physical map of a *B. pertussis* chromosome we have been able to determine the locations of some of these "illegitimate" insertions. It was found that they occurred at random sites in the chromosome, suggesting that our original assumption was incorrect and that these insertions really do constitute examples of "illegitimate recombination." In the case of the second crossover, if one applies selection for and against the vector simultaneously, for example, by imposing gentamicin and streptomycin selection with a pSS1129 derivative, one can recover, at a much reduced frequency, survivors of such a selection. The mechanism by which these colonies arise is unknown. However, if one scores these survivors for a second plasmid marker such as ampicillin resistance, one finds that some contain this marker and some do not, suggesting that they may have arisen through some less defined mechanism involving spontaneous deletion. These exceptions to the "rules" are discussed here to aid in the interpretation of results obtained when attempting allelic exchange under less than optimal conditions. They occur at such a low frequency that they are not normally seen if the flanking homology is sufficient, and if a reasonable screen exists for incorporation of the mutation.

[10] G. R. Zealey, S. M. Loosmore, R. K. Yacoob, S. A. Cockle, L. J. Boux, L. D. Miller, and M. H. Klein, *Bio/Technology* **8**, 1025 (1983).

[11] S. Stibitz, W. Aaronson, D. Monack, and S. Falkow, *Nature (London)* **338**, 226 (1989).

[36] Gene Replacement in *Pseudomonas aeruginosa*

By DEBBIE S. TODER

Introduction

Replacement of the wild-type allele on the bacterial chromosome with a plasmid-borne mutation requires, first, a system for introducing the mutant allele into the bacterium; second, integration of the allele into the chromosome in place of the wild-type, preferably with loss of plasmid sequences; and third, detection and verification of the desired mutant. This powerful approach can be used to create isogenic mutants in which changes in phenotype are attributable to the defined change in genotype. These isogenic mutants can be used to define the contribution of a given gene product to virulence. Additionally, strains with a known and characterized interruption in the structural gene are useful for expression of plasmid-encoded mutants in structure–function studies. In the area of regulation, integration of a single-copy mutant allele or gene fusion in the appropriate chromosomal location may be advantageous compared to study of these constructs on plasmids which are extrachromosomal and often present in multiple copies.

Early Systems

Initial efforts in bacterial gene replacement employed vectors which were capable of transferring to and replicating in the host bacterium. In 1981, Ruvkun and Ausubel reported a method for returning to the chromosome of *Rhizobium meliloti* a segment of DNA which had been mutagenized *in vitro* with Tn5.[1] They cloned this DNA into pRK290, which can be transferred by conjugation to *R. meliloti*. They then introduced an incompatible plasmid by conjugation while maintaining selection for the Tn5-encoded neomycin. The resultant clones had lost the tetracycline resistance (Tc^r) marker encoded by pRK290. Southern analysis confirmed that the mutagenized gene had indeed replaced the wild type on the chromosome.

Lee and Saier used a similar strategy to introduce deletion mutations into the *Escherichia coli* chromosome.[2] A ColE1 plasmid harboring the desired deletion, which would result in a *mtl* mutation, was introduced

[1] G. B. Ruvkun and F. M. Ausubel, *Nature (London)* **289**, 85 (1981).
[2] C. A. Lee and M. H. Saier, Jr., *J. Bacteriol.* **153**, 685 (1983).

into the cell by transformation. After screening for the desired phenotype, the plasmid was removed by introduction of an incompatible plasmid with a different antibiotic marker. This plasmid, pRK2013, is not stably maintained without antibiotic selection, and the strain could thus be cured of plasmid. In *Pseudomonas aeruginosa*, a similar strategy was used by Ostroff and Vasil,[3] who interrupted the *plcS* gene with a tetracycline resistance cartridge and cloned this gene into pR751, which is self-transmissible and capable of replicating in *P. aeruginosa*. They selected for the tetracycline resistance marker used to inactivate the *plcS* gene and screened for the expected nonhemolytic phenotype. A plasmid of the same incompatibility group but with a different antibiotic marker, pME301, was chosen to eliminate the original plasmid. Because pME301 does not replicate above 43°, it was eliminated by raising the temperature.

To circumvent the necessity of curing the recipient strain of the plasmid used to introduce a mutant gene into the bacterium, Gutterson and Koshland used ColE1-based plasmids in a *polA* mutant *E. coli* strain.[4] Because ColE1-based plasmids cannot replicate in such strains, those clones which have acquired the tetracycline resistance marker must have plasmid sequences integrated into the chromosome. To promote loss of vector sequences, cultures were grown without antibiotics and then treated successively with tetracycline and cycloserine to enrich for *tet*^{s} clones. Constructs were verified by phenotypic selection or by Southern analysis.

This system cannot be employed in *P. aeruginosa* because ColE1 plasmids cannot be introduced into *P. aeruginosa* by either transformation or conjugal mating. Theoretically, it should be possible for them to gain entry via electroporation, but our laboratory has not been able to introduce mutagenized DNA to the chromosome in this manner. We have speculated that the mechanism of DNA entry, that is, single stranded versus double stranded, may be responsible for this observation. However, other investigators have created an *anr* mutant by electroporation into *P. aeruginosa* of pACYC184 containing a small internal fragment of this gene. The resulting mutant had the entire plasmid integrated into the chromosome and was not stable in the absence of antibiotic selection.[5]

Suicide Vectors for Use in *Pseudomonas aeruginosa*

To satisfy the need for a vector which can be introduced into *P. aeruginosa* but which cannot replicate independently in the cytoplasm,

[3] R. M. Ostroff and M. L. Vasil, *J. Bacteriol.* **169**, 4597 (1987).

[4] N. I. Gutterson and D. E. Koshland, Jr., *Proc. Natl. Acad. Sci. U.S.A.* **80**, 4894 (1983).

[5] A. Zimmermann, C. Reimmann, M. Galimand, and D. Haas, *Mol. Microbiol.* **5**, 1483 (1991).

suicide plasmids have been constructed. These include pRZ102, pEMR2, pJM703.1, and a pJM703.1 derivative containing a polylinker site, pGP704, each of which has ColE1 or R6K replicon allowing extrachromosomal replication in *E. coli* but not in *P. aeruginosa*. The unique feature of these plasmids is the presence of the RK2 *oriT* site which allows transfer to *P. aeruginosa* by mating when transfer functions are provided *in trans,* either by pRK2013 or integrated into the chromosome of *E. coli* (e.g., SM10). These mobilizable suicide vectors are described in Table I.

Goldberg and Ohman developed the technique of "excision marker rescue" in which the desired mutation is cloned between insertion sequence (IS) elements in a mobilizable suicide vector. The presence of the IS elements is believed to facilitate genetic exchange.[6] This system, using pRZ102, has been employed to mutagenize *algB* in *P. aeruginosa* strains PAO1 and FRD,[6] *toxA* in strain PA103,[7] and *recA*[8] and *orpF*[9] (protein F) in PAO1. Flynn and Ohman have extended this technique, creating the vector pEMR2[10] which contains a *cos* site to allow gene bank construction.

We have used pGP704[11] to construct *lasB* and *lasA* mutants. The R6K origin of replication requires the *pir* gene product, necessitating the use of appropriate λ lysogens of *E. coli* for cloning. Combining the RP4*mob* site with a ColE1 replicon allows use of standard laboratory *E. coli* strains for cloning and yields a plasmid which is mobilizable to *P. aeruginosa* but unable to replicate extrachromosomally. Saiman *et al.* inserted the RP4*mob* site (isolated from pJM703.1) into the *Ssp*I site of pUC19, creating pUCm19. An interrupted *pil* gene was then inserted into pUCm19 and the recombinant used to create a *pil* mutant in *P. aeruginosa* PAK.[12]

It has been our observation and that of others that double crossovers of interrupted genes in these vectors seem to require approximately 1 kilobase (kb) of flanking *P. aeruginosa* DNA on each side of the insertion. Because the vectors have antibiotic resistance markers, loss of vector sequences in these systems leads to acquisition of antibiotic sensitivity. This is useful as a phenotypic screen but does not allow selection against maintenance of vector sequences in the chromosome. It is possible that such selection could promote conversion of a single crossover event involving the entire plasmid to the desired double crossover/replacement with excision of vector. Two approaches designed to select for loss of

[6] J. B. Goldberg and D. E. Ohman, *J. Bacteriol.* **169,** 1593 (1987).
[7] M. J. Wick, J. M. Cook, and B. H. Iglewiski, *Infect. Immun.* **60,** 1128 (1992).
[8] J. M. Horn and D. E. Ohman, *J. Bacteriol.* **170,** 1637 (1988).
[9] W. A. Woodruff and R. E. W. Hancock, *J. Bacteriol.* **170,** 2592 (1988).
[10] J. L. Flynn and D. E. Ohman, *J. Bacteriol.* **170,** 3228 (1988).
[11] V. L. Miller and J. J. Mekalanos, *J. Bacteriol.* **170,** 2575 (1988).
[12] L. Saiman, K. Ishimoto, S. Lory, and A. Prince, *J. Infect. Dis.* **161,** 541 (1990).

TABLE I

MOBILIZABLE SUICIDE VECTORS FOR *P. aeruginosa*

Vector	Special features	Antibiotic markers	Selection against vector	Usable sites	Ref.
pRZ102	Tn5 IS elements	Km^r	None	SalI, BamHI	a
pEMR2	Tn5 IS elements, *cos* site	Km^r, Ap^r	None	BamHI	b
pJM703.1	*oriR6K* requires λ*pir* gene product	Ap^r	None	EcoRI	c
pGP704	Like pJM703.1 but with polylinker	Ap^r	None	BglII, SalI, XbaI, SphI, EcoRV, SacI, EcoRI, KpnI	c
pRTP1	*cos* site	Ap^r, *rpsL*	Via selection against dominant Str^s allele; requires Str^r host	BamHI, EcoRI, SalI, HindIII	d
pNOT19/pMOB3	Gene cloned into pNOT, then 5.8-kb *NotI* fragment of pMOB inserted into same site in pNOT recombinant	Ap^r on pNOT; Km^r, Cm^r on pMOB	Via sucrose; Cm^r for insertion of *mob* and *sacB*	Multiple cloning site of pUC19	e

[a] J. B. Goldberg and D. E. Ohman, *J. Bacteriol.* **169**, 1593 (1987).

[b] J. L. Flynn and D. E. Ohman, *J. Bacteriol.* **170**, 3228 (1988).

[c] V. L. Miller and J. J. Mekalanos, *J. Bacteriol.* **170**, 2575 (1988).

[d] S. Stibitz, W. Black, and S. Falkow, *Gene* **50**, 133 (1986).

[e] H. P. Schweizer, *Mol. Microbiol.* **6**, 1195 (1992).

vector sequences have been developed and used successfully in *P. aeruginosa*.

Stibitz and co-workers[13] designed, for use in *Bordetella pertussis*, pRTP1 which, like the vectors described above, has a ColE1 replicon and RK2 *oriT*. The use of pRTP1 in *B. pertussis* is described in [35] in this volume. The vector carries the *E. coli rpsL* gene which encodes the dominant streptomycin-sensitive allele of ribosomal protein S12. Selection for streptomycin resistance should favor the double crossover event with loss of vector sequences. Because streptomycin-resistant *P. aeruginosa* are easily isolated and pRTP1 is mobilizable into *P. aeruginosa*, this system has been used successfully for gene replacement in this host. Gambello and Iglewski[14] inserted a tetracycline cartridge at the site of an internal deletion of the *lasR* gene and cloned this gene into pRTP1. Tetracycline-resistant products of a mating with this recombinant and a spontaneously occurring streptomycin-resistant *P. aeruginosa* PAO1 were picked onto media containing both tetracycline and streptomycin. Resultant clones were verified by Southern analysis.

More recently Schweizer[15] devised a method for allelic exchange in which the interrupted gene is inserted into pNOT19, which is pUC19 with a unique *Not*I site. A cassette containing the RP4 *oriT*, the *B. subtilius sacB* gene, and a chloramphenicol resistance gene can be isolated as a *Not*I fragment and inserted into this site in the pNOT19 recombinant. The unique feature of this system is the use of the *sacB* gene as a counterselectable marker. In the presence of 5% (w/v) sucrose, only *P. aeruginosa* strains which have lost this vector-encoded gene will survive. If selection is maintained for the antibiotic marker used to interrupt the gene of interest, a high degree of efficiency may be obtained. This system can be used with the Ω cartridge (described below), yielding a mutant strain with only streptomycin resistance, whereas strains created with pRTP1 carry both streptomycin resistance and the resistance used to interrupt the gene interest.

Experience in our laboratory is limited to insertional mutagenesis. The systems described above should, however, be applicable to the replacement of a wild-type gene with an internal deletion, a point mutation, or a gene fusion. Such experiments require careful planning, particularly if the expected phenotype is not known. If phenotypic screening is not possible, the pNOT/pMOB system could be employed with initial selection for chloramphenicol resistance and subsequent counterselection with su-

---

[13] S. Stibitz, W. Black, and S. Falkow, *Gene* **50**, 133 (1986).

[14] M. J. Gambello and B. H. Iglewski, *J. Bacteriol.* **173**, 3000 (1991).

[15] H. P. Schweizer, *Mol. Microbiol.* **6**, 1195 (1992).

crose. If a point mutation is to be introduced, it may be desirable to choose a mutation that will alter a restriction enzyme site. This will make confirmation by Southern analysis possible, whereas sequencing might otherwise be required.

Insertional Mutagenesis

The systems which we have used successfully (pGP704, pRZ102, pRTP1, and pNOT19/pMOB3) all involve cloning of a gene interrupted with a selectable marker into a vector which can be mobilized into *P. aeruginosa* but which cannot replicate in this host. We have used antibiotic markers to interrupt genes and to serve as a selection system for recombinants. The choice of antibiotic should be made carefully, keeping in mind the intended use of the mutant. There is often background streptomycin resistance in *P. aeruginosa;* however, we have noticed strain differences, and this disadvantage can be overcome if phenotypic screening for the desired gene replacement is not cumbersome. Streptomycin resistance does not preclude the use of cosmids and vectors which carry either tetracycline or ampicillin (carbenicillin) resistance. The Ω fragment[16] which encodes streptomycin/spectinomycin resistance flanked by transcriptional and translation stop signals can be isolated from pHP45Ω or pUC18Ω (Ω fragment ligated into the *Sma*I site in pUC18). Compared to streptomycin, tetracycline is preferred for efficacy (i.e., the lack of background resistance), although tetracycline-resistant strains cannot be used with pLAFR vectors. The tetracycline resistance gene from pBR322 can be isolated as a *Eco*RI–*Bal*I fragment or from pUC18T2 with a variety of enzymes.[12] Useful antibiotic cassettes are listed in Table II.

The suicide plasmids described all have RP4*mob* and can be mobilized into *P. aeruginosa* by supplying RP4*tra in trans*. This is easily accomplished by biparental mating with *E. coli* SM10[17] or triparental mating with *E. coli* harboring pRK2013.[18] We check for loss of vector-encoded markers because, although unstable, single crosses into the chromosome can occur. This presumes that the antibiotic marker on the suicide plasmid differs from that used to interrupt the gene. Southern analysis of the recombinant *P. aeruginosa* should then demonstrate the expected change in size or restriction pattern.

Our insertional mutants have been stable through repeated passage in the absence of antibiotics. There is always, however, the theoretical con-

[16] P. Prentki and H. M. Krisch, *Gene* **29,** 303 (1984).
[17] R. Simon, U. Priefer, and A. Puhler, *Bio/Technology* **1,** 784 (1983).
[18] D. H. Figurski and D. R. Helinski, *Proc. Natl. Acad. Sci. U.S.A.* **76,** 1648 (1979).

TABLE II
USEFUL RESISTANCE MARKERS FOR INSERTIONAL MUTAGENESIS

Cartridge	Plasmid	Sites	Selection (drug, mg/ml)	Ref.
Tet[r]	pUC18T2	PstI,BglII	E. coli (tetracycline, 25), P. aeruginosa (tetracycline, 130)	c
Omega (str[r])	pHP45Ω	EcoRI, SmaI, BamHI, HindIII[a]	E. coli (streptomycin, 20), P. aeruginosa (streptomycin, 300–1000)	d
Omega-Tc	pHP45Ω-Tc	EcoRI, SmaI, HindIII[a]	E. coli (tetracycline, 25), P. aeruginosa (tetracycline, 130)	e
Omega-Km	pHP45Ω-Km	EcoRI, BamHI, HindIII[a]	E. coli (kanamycin, 50), P. aeruginosa (Kn, 500–1000, or neomycin, 600)	e
Omega-Cm	pHP45Ω-Cm	EcoRI, BamHI, HindIII[a]	E. coli (chloramphenicol, 20), P. aeruginosa (chloramphenicol, 500–1000)	e
Omega-Hg	pHP45Ω-Hg	SmaI, BamHI, HindIII[a]	E. coli (HgCl₂, 12), P. aeruginosa (HgCl₂, 12)	e
Omega-Ap	pHP45Ω-Ap	EcoRI, SmaI, BamHI[b], HindIII[a]	E. coli (ampicillin, 100), P. aeruginosa (carbenicillin, 200–400)	e
Cm[r]	pUC18CM	BamHI, double digest with HincII and SmaI	E. coli (chloramphenicol, 20), P. aeruginosa (chloramphenicol, 500–1000)	f

[a] Will eliminate the translation stop box, but not the transcriptional terminators flanking the resistance marker.
[b] Will eliminate the transcriptional terminators and translational stop box on one end of the Ap[r] marker.
[c] L. Saiman, K. Ishimoto, S. Lory, and A. Prince, J. Infect. Dis. **161**, 541 (1990).
[d] R. Fellay, J. Frey, and H. Krisch, Gene **52**, 147 (1987).
[e] P. Prentki and H. M. Krisch, Gene **29**, 303 (1984).
[f] H. P. Schweizer, BioTechniques **8**, 614 (1990).

cern that the inserted antibiotic selection cassette could be somehow excised and a wild-type gene reconstituted. An elegant way to avoid even this possibility is to delete some of the gene when the insertion is created. Unfortunately, the combination of this deletion strategy with the available clones and restriction sites may limit the size of the DNA fragments flanking the insertion. Using the pGP704 system, we have obtained crossovers readily when 1 kb or more of chromosomal DNA was present on either side of an antibiotic resistance cassette. In these same constructs, we were unsuccessful when fragments of only 600–800 bp flanked the insertion.

Insertional Mutagenesis with pJM703.1 or pGP704

The mobilizable, suicide vector pJM703.1 was developed by Miller and Mekalanos[11] for use in *Vibrio cholerae* (pGP704 is a pJM703.1 derivative with a polylinker site). The R6K origin of replication requires the *pir* gene product available in λ*pir* lysogens of *E. coli* including SM10 and SY327. Mating is accomplished by virtue of RP4*mob* present on pGP704 and transfer functions from RP4-2-Tc::Mu integrated into the chromosome of SM10.

Procedure

1. Keeping in mind the general considerations outlined above, design a cloning strategy. Much of the sequence of pGP704, including the ampicillin resistance gene from pBR322, is known, but these are areas for which the sequence has not been published. The small size of pGP704 and polylinker make necessary restriction mapping easy. There are at least eight usable polylinker sites including the blunt-end *Eco*RV site.

2. The plasmid-borne DNA fragment of interest is cut with the appropriate restriction endonuclease(s) and purified. Similarly, the fragment to be inserted is prepared and purified. Using standard laboratory procedures, the ligation which will interrupt the gene is carried out, and *E. coli* is transformed and the products plated onto medium with antibiotic selection for the vector and the insertion. Plasmid DNA is prepared from recombinants and the construct verified by restriction analysis.

3. The interrupted gene is cut from the plasmid, and the fragment is purified and ligated into pGP704. Because, in our hands, SM10 is difficult to transform or electroporate efficiently (sometimes even with intact plasmid), we transform *E. coli* SY327 with the ligation mixture. Once we have verified the construct, we transfer the pGP704 recombinant plasmid to *E. coli* SM10[17] because it supplies the transfer functions necessary for mating with *P. aeruginosa*.

4. Overnight cultures of SM10 carrying the pGP704 recombinant (with appropriate antibiotics) and the host *P. aeruginosa* strain are grown with shaking at 37°. If SM10 is not the λ*pir* lysogen used, also grow MM294(pRK2013) in L broth with kanamycin, 50 μg/ml. In the morning, each culture is subcultured 1 : 20 in L broth as above and grown for 3 hr. For biparental mating we use 4 ml of each culture, and for triparental matings 2 ml of each *E. coli* and 4 ml of the *P. aeruginosa* culture are used. The cultures are filtered through a 0.45-μm filter. The filter is removed with a sterile scalpel and forceps and placed overnight on an L plate at 37°. Although colonies come up more slowly on Vogel–Bonner minimal medium, this medium limits *E. coli* growth. Antibiotic levels we typically use for *P. aeruginosa* are as follows: streptomycin, 500–1000 μg/ml; tetracycline, 100 μg/ml; carbenicillin, 400 μg/ml. The filter is next placed, with sterile forceps, into a sterile 15-ml tube containing 5 ml of prewarmed, sterile phosphate-buffered saline (PBS) and the tube vortexed briefly. The sample is then plated onto the selective medium. It is often necessary to dilute the suspension. The suspension can be saved in the refrigerator for a few days in case the optimal dilution is not achieved.

5. When colonies are obtained, they should be screened for lack of carbenicillin resistance, characterized phenotypically, and their chromosomal DNA subjected, alongside the parental *P. aeruginosa* DNA, to Southern analysis to verify gene replacement.

Acknowledgment

Supported by a Cystic Fibrosis Foundation Physician–Scientist Award.

[37] Systems of Experimental Genetics for *Campylobacter* Species

By Patricia Guerry, Ruijin Yao, Richard A. Alm, Donald H. Burr, and Trevor J. Trust

Introduction

The thermophilic campylobacters, *Campylobacter jejuni* and *Campylobacter coli,* are among the most frequently isolated causes of bacterial diarrhea worldwide. These microaerophilic, gram-negative spiral rods are phylogenetically very distant from the Enterobacteriaceae and other bacteria more classically associated with diarrheal disease. Most of the experimental genetic tools used in other enteric pathogens have not been successfully adapted for campylobacters. Although both conjugative and nonconjugative plasmids have been identified in *Campylobacter* spp., these plasmids cannot replicate in *Escherichia coli*. Furthermore, *E. coli* plasmids, including those from the wide-host-range incompatibility groups, cannot replicate in campylobacters. Although *C. jejuni* and *C. coli* have bacteriophages, no system of phage transduction has been developed. No transposons have been identified in campylobacters, and transposons from *E. coli* have not been shown to function in campylobacter hosts.

Despite these difficulties, progress has been made in genetic manipulation of these spiral organisms in several aspects. Labigne-Roussel and co-workers developed the first shuttle vectors,[1] which allow for genes cloned into *E. coli* to be returned to *Campylobacter* spp. on autonomously replicating plasmids, as well as the first suicide vectors,[2] which have been used for site-specific mutagenesis of cloned genes.[2-5] These plasmids can be transferred from *E. coli* into campylobacters by either conjugation[1-4] or electroporation.[5,6] *Campylobacter jejuni* and *C. coli* have recently been shown to be naturally transformable by mechanisms which seem to resemble the natural transformation systems of other gram-negative mucosal pathogens such as *Neisseria* and *Haemophilus*.[7] The natural transformation system is only beginning to be exploited as a genetic tool in studying campylobacters.[8] The purpose of this chapter is to describe basic procedures used in our laboratories for genetic manipulation of *C. jejuni* and *C. coli*.

Suicide and Shuttle Vectors for *Campylobacter jejuni* and *Campylobacter coli*

Labigne-Roussel and co-workers[1] observed that P incompatibility plasmids were capable of being conjugally transferred from *E. coli* into *Campylobacter,* but that they were unable to replicate in the alien host. Thus, the origin of transfer (*oriT*) but not the origin of replication (*oriV*) was functional. A hybrid shuttle vector was constructed which was composed of (1) pBR322, (2) a cryptic plasmid (plPI455) from *C. coli,*[9] (3) the *oriT*

[1] A. Labigne-Roussel, J. Harel, and L. Tompkins, *J. Bacteriol.* **169,** 5320 (1987).
[2] A. Labigne-Roussel, P. Courcoux, and L. Tompkins, *J. Bacteriol.* **170,** 1704 (1988).
[3] P. Guerry, S. M. Logan, S. Thornton, and T. J. Trust, *J. Bacteriol.* **172,** 1853 (1990).
[4] P. Guerry, R. A. Alm, M. E. Power, S. M. Logan, and T. J. Trust, *J. Bacteriol.* **173,** 4757 (1991).
[5] T. Wassenaar, N. M. C. Bleumink-Pluym, and B. A. M. van der Zeijst, *EMBO J.* **10,** 2055 (1991).
[6] J. F. Miller, W. J. Dower, and L. S. Tompkins, *Proc. Natl. Acad. Sci. U.S.A.* **85,** 856 (1988).
[7] Y. Wang and D. E. Taylor, *J. Bacteriol.* **172,** 949 (1990).
[8] R. A. Alm, P. Guerry, M. E. Power, H. Lior, and T. J .Trust, *J. Clin. Microbiol.* **29,** 2438 (1991).
[9] T. Lambert, G. Gerbaud, P. Trieu-Cuot, and P. Courvalin, *Ann. Inst. Pasteur Microbiol.* **136B,** 134 (1985).

from a P incompatibility plasmid, and (4) a kanamycin resistance gene encoding a 3′-aminoglycoside phosphotransferase type III from C. coli plasmid pIP1433.[10] This vector, called pILL550, was capable of being mobilized from E. coli donors containing a P incompatibility group conjugative plasmid into campylobacter recipients. Suicide vectors, lacking the campylobacter replicon, were also constructed[2] and have proved useful for gene replacement mutagenesis of C. jejuni and C.coli.[2,3]

In addition to the original pILL550 vector and derivatives constructed by Labigne-Roussel et al., several others have been developed, notably the pUOA series.[7,11] Table I summarizes the salient features of some currently available suicide and shuttle vectors. A number of general features should be mentioned. No antibiotic resistance genes from E. coli have been shown to function in C. jejuni or C. coli, so all shuttle vectors must include a campylobacter selective marker. All of the campylobacter drug resistance genes function in E. coli. The pILL series of plasmids from Labigne-Roussel utilizes the kanamycin resistance gene from C. coli,[10] a pBR322-based E. coli replicon, and a 3.24-kilobase (kb) campylobacter replicon from pIP1455.[9] The pUOA series uses the same C. coli replicon, but cloned onto pUC13, thus generating some unique restriction sites and blue/white color selection due to alpha complementation of the lacZ gene. This series also has alternate campylobacter markers, namely, tetracycline and chloramphenicol resistance, in addition to kanamycin. Although these vectors are an improvement, there are still some limitations. The plasmids are relatively large due to the size of the campylobacter replicon, and this size may contribute to instability problems. In addition, the actual number of usable restriction sites in the pUC13 polylinker is limited by the presence of restriction sites in the campylobacter plasmid. Newer vectors, such as the pRY series[12] shown in Table I, offer a downsized C. coli replicon and the addition of newer lacZ polylinkers with more unique restriction sites.

A necessary component of the suicide vector system developed by Labigne-Roussel is the campylobacter kanamycin resistance cassette used to insertionally inactivate campylobacter genes cloned in E. coli.[2] This cassette is bracketed by a polylinker region containing unique restriction sites for PstI, BamHI, SmaI, EcoRI, and ClaI. A new cassette based on the campylobacter chloramphenicol acetyltransferase gene[12] has been developed which is bracketed by BamHI, SmaI, KpnI, PstI, EcoRI, and PvuII sites. The availability of an alternate marker for generation of mutations should prove useful in terms of performing complementation analy-

[10] P. Trieu-Cuot, G. Gerbaud, T. Lambert, and P. Courvalin, EMBO J. 4, 3583 (1985).
[11] Y. Wang and D. E. Taylor, Gene, 94 (1990).
[12] R. Yao, R. A. Alm, T. J. Trust, and P. Guerry, Gene 130, 127 (1993).

TABLE I
PLASMID VECTORS FOR *Campylobacter* SPECIES[a]

Plasmid	Phenotypic markers	Size (kb)	Unique cloning sites	*lacZ* screening	Ref.
pILL560	Ap, mob$^+$	4.5	*Eco*RI, *Cla*I, *Hin*dIII, *Xba*I, *Bgl*II, *Pst*I, *Bam*HI	−	*b*
pGK2003	Ap, mob$^+$	3.4	*Kpn*I, *Sac*I, *Sal*I, *Acc*I, *Pst*I, *Hin*dIII, *Sma*I, *Bam*HI, *Xba*I	−	*c*
pILL550	Km, mob$^+$	8.5	*Bam*HI, *Cla*I, *Eco*RI, *Sal*I	−	*d*
pUOA15	Ap, Tc, mob$^+$	11.8	*Bam*HI, *Sac*I, *Xba*I	+	*e*
pUOA13	Ap, Km, mob$^+$	8.25	*Bam*HI, *Sac*I, *Xba*I	+	*e*
pUOA18	Cm, mob$^+$	7.4	*Bam*HI, *Sac*I, *Xba*I	+	*f*
pRY109/110	Km, mob$^+$	6.75	*Kpn*I, *Sac*I, *Sal*I, *Acc*I, *Cla*I, *Pst*I, *Hin*cII, *Apa*I, *Eco*RI, *Bam*HI, *Spe*I, *Xba*I, *Not*I, *Sma*I, *Xho*I, *Sac*II, *Bst*XI	+	*g*
pRY113/114	Cm, mob$^+$	6.45	*Kpn*I, *Sac*I, *Sal*I, *Acc*I, *Apa*I, *Pst*I, *Eco*RV, *Eco*RI, *Sma*I, *Bam*HI, *Spe*I, *Xba*I, *Not*I, *Hin*cII, *Xho*I, *Sac*II, *Dra*II	+	*g*

[a] Plasmids are listed with their phenotypic markers, molecular size in kilobase pairs, the unique restriction sites, and the presence of the *lacZ* indicator gene. Plasmids pILL550 and pGK2003 are suicide vectors which do not replicate in *Campylobacter* spp. Mob, Ability to be conjugally mobilized by P incompatibility plasmids from *E. coli* into *Campylobacter* spp.; Ap, ampicillin resistance (which is expressed only in *E. coli*); Km, kanamycin resistance; Tc, tetracycline resistance; Cm, chloramphenicol resistance.

[b] A. Labigne-Roussel, J. Harel, and L. Tompkins, *J. Bacteriol.* **169**, 5320 (1987).
[c] P. Guerry, R. A. Alm, M. E. Power, S. M. Logan, and T. J. Trust, *J. Bacteriol.* **173**, 4757 (1991).
[d] A. Labigne-Roussel, P. Courcoux, and L. Tompkins, *J. Bacteriol.* **170**, 1704 (1988).
[e] Y. Wang and D. E. Taylor, *J. Bacteriol.* **172**, 949 (1990).
[f] Y. Wang and D. E. Taylor, *Gene,* **94** (1990).
[g] R. Yao, R. A. Alm, T. J. Trust, and P. Guerry, *Gene* **130**, 127 (1993).

ses and cloning into campylobacter backgrounds which have already been mutated. Such antibiotic resistance cassettes are inserted into cloned genes and returned to campylobacters on suicide vectors by the same conjugal transfer protocol used for shuttle vectors.

Conjugative Transfer of Shuttle or Suicide Vectors from *Escherichia coli* to *Campylobacter jejuni* or *Campylobacter coli*

1. *Escherichia coli* donors containing the conjugative plasmid RK212.2[13] and the shuttle or suicide vector are grown in 3 ml of LB broth

[13] D. H. Figurski and D. R. Helinski, *Proc. Natl. Acad. Sci. U.S.A.* **76**, 1648 (1979).

without antibiotics with shaking (200 rpm) for 10–12 hr to a density of approximately 10^9 bacteria/ml (OD_{600} of 1.2).

2. Campylobacter recipients are grown on Mueller–Hinton (MH) agar under microaerobic conditions for 8–14 hr and resuspended in MH broth to an OD_{600} of 1.0 (or ~2–3 × 10^9 cells/ml).

3. The donor and recipient cells are mixed in a ratio of 1 : 3 to 1 : 6, such that approximately 5 × 10^8 donors (~0.5 ml) is mixed with approximately 1.5–3.0 × 10^9 (0.5–1.0 ml) of recipients. The cells are pelleted by centrifugation, resuspended in 0.2 ml of MH broth, and spotted onto the surface of an MH agar plate.

4. The plate containing the mating mixture is incubated, right-side-up, in a microaerobic atmosphere at 37° for 5–6 hr.

5. The bacteria are harvested in 1.0–2.0 ml of MH broth and plated directly and with 10-fold dilution in MH broth onto MH agar containing appropriate antibiotics.

Comments

Transfer frequencies are reduced when the campylobacter recipients are grown longer than 14 hr. *Escherichia coli* donors can be counterselected with complete Blaser–Wang supplement,[1,2] although we have found that addition of 10 μg/ml trimethoprim is effective.[3,4] The levels of other antibiotics are as follows: tetracycline, 8–12 μg/ml; chloramphenicol, 20 μg/ml; and kanamycin, 16–100 μg/ml, depending on the levels of resistance of the recipient strain.

Microaerobic conditions can be achieved either by (1) growth in polybags filled with a gas mixture of 10% CO_2, 85% N_2, and 5% O_2, by volume, (2) growth in a CO_2 incubator (5% CO_2), or (3) growth in an atmosphere containing 5% oxygen, 10% CO_2 produced with a gas-generating kit for campylobacters (BBL, Becton Dickinson Microbiology Systems, Cockeysville, MD). In our experience, however, not all strains grow optimally in 5% CO_2; in these cases, growth in 10% CO_2 environments by either method 1 or 3 is preferable.

Shuttle and Suicide Plasmid Transfer by Electroporation

The first reported use of electroporation in *C. jejuni* was by Miller *et al.*[6] in 1988 (see [30] in this volume). Parameters examined to maximize the transformation frequency included pulse amplitude and duration, the effect of divalent cations, cell growth conditions, and DNA concentration. Although Wassenaar *et al.*[5] have used electroporation to introduce insertionally inactivated flagellin genes into *C. jejuni* and thus construct isogenic flagellin mutants, the technique has not been widely used. This is because of problems with restriction of incoming plasmid DNA. This restriction barrier seems operable regardless of whether the plasmid is purified from *E. coli* or *Campylobacter* spp.[6,14] However, some strains of *C. jejuni* have been identified which seem better able to serve as recipients in electroporation experiments.[6,14] Similar observations have been reported by Labigne *et al.*[15] for *Helicobacter pylori*. The protocol described below is an adaptation of that of Labigne *et al.*[15] for *H. pylori* and is similar to that used by Wassenaar *et al.* for *C. jejuni*.[5] It has been successfully used in our laboratories for electroporation of *C. jejuni* and *C. coli*.

1. Recipient campylobacters are grown for 18 hr at 37° microaerobically on 5 plates of Trypticase soy agar with 5% sheep blood (TSAII, BBL prepared media, Becton Dickinson Microbiology Systems).

2. Cells are harvested with disposable plastic loops, resuspended in 1 ml of ice-cold 15% glycerol–9% sucrose (w/v), and then centrifuged for 4 min at 4° in a microcentrifuge.

3. Cells are washed in 500 μl of the ice-cold glycerol–sucrose solution, centrifuged, and resuspended in 200–250 μl of ice-cold glycerol–sucrose. The final concentration of cells should be 10^9–10^{10}/ml.

4. Plasmid DNA [1 μl of a 500 ng/μl solution in Tris–EDTA buffer (TE) or distilled water] and 50 μl of the cell suspension are added to prechilled polypropylene tubes (17 × 100 mm) and gently mixed by pipetting.

5. Ice the DNA–cell mixture for 1 min and transfer to prechilled 0.2-cm electrode gap cuvettes. Gently tap the sides of the cuvette to ensure that the entire volume of cells is brought to the bottom of the cuvette.

6. Place the cuvette into a Bio-Rad (Richmond, CA) Gene Pulser electroporation apparatus and apply an electrical pulse set at 25 μF, 200 Ω, and 2.5 kV. Time constants should range between 4 and 5 msec.

7. Immediately after the pulse add 100 μl of SOC medium (2% (w/v) tryptone, 0.5% (w/v) yeast extract, 10 mM NaCl, 2.5 mM KCl, 10 mM MgCl$_2$, 10 mM MgSO$_4$, and 20 mM glucose), mix gently with a sterile Pasteur pipette, and spread the cells onto a TSAII blood agar plate. Incubate the plate at 37° under microaerobic conditions for 2 hr to allow expression of plasmid markers.

8. Harvest the cells in 200 μl of Mueller–Hinton broth. Plate 100-μl aliquots of the cell suspension onto Mueller–Hinton agar plates supplemented with the appropriate antibiotic and incubate for 48 hr at 37° under microaerobic conditions.

[14] D.H. Burr, unpublished observation (1992).
[15] R. L. Ferrero, V. Cussac, P. Courcoux, and A. Labigne, *J. Bacteriol.* **174**, 4212 (1992).

Comments

Although Mueller–Hinton plates can be used to grow the campylo-bacter cells, many strains of campylobacters often form dry colonies on this medium, which makes harvesting more difficult than from blood agar. Plasmid DNAs can be purified by cesium chloride–ethidium bromide centrifugation or by alkaline extractions.

We have used a Bio-Rad Gene Pulser exclusively. Although a 0.2-cm electrode gap cuvette (Bio-Rad, No. 165-2086) has been used routinely, transformation frequencies can be increased by using a 0.1-cm gap cuvette (Bio-Rad, No. 165-2089). However, there is an increased likelihood of arcing when the pulse is applied in the smaller cuvettes. When using the 0.1-cm cuvettes, it is recommended that the cell volume be increased to 80 μl, while the DNA concentration is kept constant. It may also be necessary to reduce the voltage to eliminate arcing.

As noted above, uptake of DNA varies among strains, presumably as a result of restriction–modification systems. However, some strains have been identified which can readily accept DNA purified from both *E. coli* and *Campylobacter* spp. hosts.[5,6,14] The incubation time in Step 6 to allow expression of plasmid markers can be varied from 2 to 18 hr for plasmids when selection is for kanamycin resistance.

Natural Transformation of *Campylobacter* Species

It has been demonstrated that most *C. coli* and some *C. jejuni* strains are naturally competent for uptake of DNA during the logarithmic phase of growth.[7] There seems to be a preference for uptake of campylobacter DNA, presumably due to the presence of, as yet unidentified, specific uptake sequences, similar to the mechanism described for *Neisseria* and *Haemophilus*.[7] The natural transformability of *C. jejuni* and *C. coli* was first used to transfer chromosomally encoded antibiotic resistance markers among strains,[7] and more recently to move flagellin genes which have been mutated by insertion of a kanamycin resistance gene from one heat-labile serogroup to strains of *C. jejuni* and *C. coli* belonging to other serogroups.[8] In addition, we have performed gene replacement mutagenesis by transforming a plasmid clone of a *C. coli* gene insertionally inactivated with the kanamycin cassette directly from *E. coli* into *C. jejuni* and *C. coli*.[16] Less work has been done on transformation of shuttle plasmids into campylobacters. Wang and Taylor[7] observed that plasmids transform poorly, most likely because they are linearized during the natural transfor-

[16] P. Guerry and R. A. Alm, unpublished observation (1992).

mation process. However, Wang and Taylor[7] also showed that the system could be adapted to transform plasmids efficiently if alternatively marked plasmids are used. For example, if the recipient cell contained pUOA15 (tetracycline resistant), this plasmid could rescue an incoming homologous plasmid such as pUOA18 (chloramphenicol resistant).

Although two methods for trnsformation were originally described for campylobacters, we have found it easier and more reproducible to use the biphasic method, as described below.

1. Recipient campylobacter cells are grown on MH agar at 37°C under microaerophilic conditions for 18–20 hr.

2. Cells are resuspended in MH broth to an OD_{600} of 0.6, and 250 μl of the cells are added to 1 × 10 cm polypropylene tubes containing 1.5–2.0 ml of Mueller–Hinton agar.

3. The tubes are incubated at 37° for 4 hr in a CO_2 incubator.

4. Then 2–4 μg of DNA is added, and the cell–DNA mixture is incubated for another 4 hr in a CO_2 incubator.

5. After incubation aliquots of 50–100 μl are plated onto selective medium and incubated microaerobically for 48 hr.

Comments

Freshly made (<24 hr) MH agar plugs should be used, since aged plugs will absorb liquid from the cell suspension during incubation. It is important to use cells grown for less than 20 hr, as the transformation frequency is reduced with older cells. Although it has been postulated that early logarithmic cells are more competent,[7] growth for less than 4 hr prior to DNA addition results in fewer transformants in our hands. Finally, the incubations in Steps 2 and 4 can also be performed in a 37° air incubator, although this decreases the number of transformants by about 30%.

[38] *In Vivo* Expression Technology for Selection of Bacterial Genes Specifically Induced in Host Tissues

By James M. Slauch, Michael J. Mahan, and John J. Mekalanos

Introduction

To understand the molecular mechanisms by which a pathogen circumvents the host immune system, replicates, and causes disease, one must identify the gene products that are specifically required for these processes. It has become apparent that the genes for most virulence factors are not constitutively expressed. Indeed, many virulence factors are coordinately regulated by *in vitro* environmental signals that presumably reflect cues encountered in host tissue.[1] The identification of virulence factors is usually dependent on, and limited by, the ability to mimic these host environmental signals *in vitro*. We have developed a novel approach to identify genes important for pathogenesis. Our system, termed IVET (*in vivo* expression technology) is based not on the ability to mimic host factors *in vitro*, but rather on direct selection of those genes that are induced *in vivo*. Presumably, a subset of the genes that are induced in host tissues will include virulence genes that are specifically required for the infection process.

The IVET selection system is based on the fact that a mutation in a biosynthetic gene can dramatically attenuate the growth and persistence of a pathogen in host tissues. Growth of the auxotrophic mutant in the host can then be complemented by operon fusions to the same biosynthetic gene, thus selecting for those promoters that are expressed in host tissues. Several biosynthetic genes can, theoretically, be used in this type of selection system. We have initially concentrated on a *purA* system in *Salmonella typhimurium*. Below, we describe this system in detail in order to discuss the various aspects of IVET. We then describe various modifications that may facilitate the application of the IVET approach in other organisms.

pIVET1: Rationale and Design

The *purA* gene of *Salmonella typhimurium* encodes an enzyme, adenylosuccinate synthase, required for synthesis of adenosine 5'-monophosphate (AMP). *Salmonella typhimurium* strains mutant in the *purA* gene are extremely attenuated in the ability to cause mouse typhoid or to persist in animal tissues.[2] Indeed, in mixed infection experiments with both *purA* and wild-type strains, there is a greater than 10^8-fold selection for the Pur$^+$ strain. Thus, one can select, in the animal, genes that are transcriptionally active *in vivo* by fusing their promoters to the *purA* gene. Only those strains that contain fusions to promoters that are transcriptionally active enough to overcome the parental PurA deficiency in the animal will be able to survive and replicate in the host. The fusion strains that answer this selection are then screened for those that are transcriptionally inactive on normal laboratory medium. This subset of strains contain fusions to

[1] J. J. Mekalanos, *J. Bacteriol.* **174**, 1 (1992).
[2] W. C. McFarland and B. A. Stocker, *J. Microb. Pathol.* **3**, 129 (1987).

genes that are "on" in the mouse and "off" outside the mouse. In other words, these genes are specifically induced in the host.

In constructing our fusion system, we must meet several criteria that have proved to be very important for the success of the selection. First, we wanted to construct the fusions in single copy in the chromosome. This avoids any complications that can arise from the use of plasmids. Second, we wanted to construct the fusions without disruption of any chromosomal genes. If the gene of interest encodes a product required for the infection process, then a fusion that disrupts the gene would not be recovered in the selection. Third, we wanted a convenient way to monitor the transcriptional activity of any given fusion both *in vitro* and *in vivo*.

We met the above criteria by constructing a synthetic operon comprising a promoterless *purA* gene and a promoterless *lac* operon in the suicide plasmid pGP704[3] (pIVET1; Fig. 1). Cloning of chromosomal fragments 5' to the *purA* gene results in transcriptional fusions in which *S. typhimurium* promoters drive the expression of a wild-type copy of *purA* and *lacZY*. Replication of pGP704 derivatives requires the replication protein, Pi, the product of the *pir* gene, supplied *in trans*. The construction of the fusions is done in an *Escherichia coli* strain, SM10 λ*pir*, which contains the replication protein and also the functions required for broad-host-range conjugal transfer of pGP704 derivatives.[4] Therefore, once constructed, the pool of transcriptional fusion plasmids can be mated into the strain of interest.

Construction of *purA–lac* Fusions

Random *Sau*3AI fragments of *S. typhimurium* chromosomal DNA, isolated from a Δ*purA* strain, MT168 (DEL2901[*purA874*::IS*10*]), are cloned into the *Bgl*II site, 5' to the *purA* gene in pIVET1. The plasmids are then electroporated into SM10 λ*pir* to create a pool of *purA–lac* fusions.[5] Conjugal introduction of the pool of fusions into a Δ*purA* strain of *S. typhimurium* that lacks the *pir* gene, and selection for resistance to ampicillin, demands the integration of the plasmids into the chromosome by homologous recombination with the cloned *Salmonella* DNA. This results in single-copy diploid fusions in which the chromosomal promoter drives the expression of the *purA–lac* fusion and the cloned promoter drives the expression of the wild-type gene (Fig. 2).

[3] V. L. Miller and J. J. Mekalanos, *J. Bacteriol.* **170**, 2575 (1988).
[4] R. Simon, U. Priefer, and A. Puhler, *Bio/Technology* **1**, 784 (1983).
[5] M. J. Mahan, J. M. Slauch, and J. J. Mekalanos, *Science* **259**, 686 (1993).

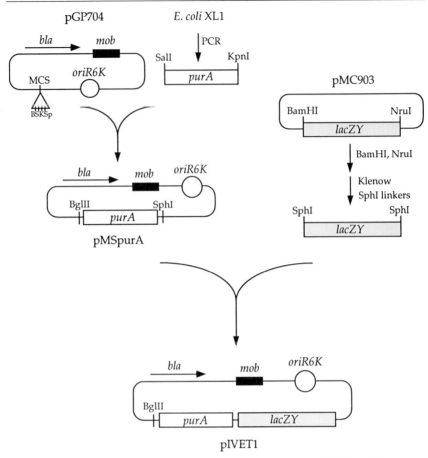

FIG. 1. Construction of pIVET1. The plasmid is a derivative of the broad-host-range suicide vector, pGP704, whose replication is dependent on the Pi protein supplied *in trans* [V. L. Miller and J. J. Mekalanos, *J. Bacteriol.* **170,** 2575 (1988)]. This Ap^r plasmid can be mobilized by *trans*-acting RP4 conjugative functions operating at the *mob* site. Restriction sites in the multiple cloning site (MCS) are defined as follows: B, *Bgl*II; S, *Sal*I; K, *Kpn*I; and Sp, *Sph*I. A promoterless *purA* gene was obtained using the polymerase chain reaction (PCR; Perkin-Elmer Cetus, Norwalk, CT) from the chromosome of *E. coli*, strain XL1, by methods described previously [J. M. Slauch and T. J. Silhavy, *J. Bacteriol.* **173,** 4039 (1991)]. The primers used for the PCR were 5' GAATCCAgTcgacAGCAAACGGTG 3' and 5' CAGGgGTACCAGAATTACGCGTC 3'. These sequences correspond to the sense strand from base pairs 468 to 491 and the antisense strand from base pairs 1814 to 1792, respectively, of the published sequence of the *purA* gene [S. A. Wolfe and J. M. Smith, *J. Biol. Chem.* **263,** 19147 (1988)], with changes (denoted in lowercase letters) to introduce restriction sites near the ends of the amplified fragment. The PCR was carried out in a buffer containing 2.0 mM Mg^{2+} for 25 cycles of 1 min at 94°, 2 min at 55°, and 3 min at 72° with a 5-sec increment added to the 72° elongation step at every cycle. The resulting *purA*-containing fragment was

There are several important points about this integration event. First, the cloned chromosomal sequences provide the only site of homology for integration into the chromosome. The *purA* gene is obtained by the polymerase chain reaction (PCR) from the *Escherichia coli* chromosome, and *E. coli* and *S. typhimurium* chromosomes are sufficiently divergent to prevent recombination.[6] Also, *S. typhimurium* does not contain a *lac* operon. Second, only those clones that contain the 5' end of the operon of interest (i.e., the promoter) will generate both a functional fusion and a duplication that maintains synthesis of the wild-type gene (the event drawn in Fig. 2). Other types of clones will not result in the desired product. For example, constructs that contain an internal fragment of the operon will generate a fusion under the appropriate regulation, but they will disrupt the expression of the wild-type gene. This type of construct can potentially be selected against in the animal if the product of the wild-type gene is required for the infection process. In other cases, there will not be a properly placed promoter to drive the expression of *purA*. Therefore, integration of the clone into the chromosome does not result in a functional fusion. For example, the fusion can be in the wrong orientation with respect to the gene. This fusion will never be expressed and will, therefore, be selected against in the animal.

Red Shift

The transcriptional activity of any given fusion can be determined by assaying the ability of the fusion strain to utilize lactose as a carbon

[6] H. Ochman, and A. C. Wilson, in "*Escherichia coli* and *Salmonella typhimurium*: Cellular and Molecular Biology" (F. C. Neidhardt, ed.), p. 1649. American Society for Microbiology, Washington, D.C., 1987.

---

digested with *Sal*I(5' end of *purA*) and *Kpn*I (3' end of *purA*) (New England Biolabs, Beverly, MA) and ligated into the corresponding sites in pGP704, resulting in plasmid pMS*purA*. A promoterless *lac* operon was obtained on a *Bam*HI, *Nru*I fragment of the plasmid pMC903 [M. J. Casadaban, J. Chou, and S. N. Cohen, *J. Bacteriol.* **143**, 972 (1980)]. This fragment contains the W205 *trp–lac* fusion that effectively removes the transcription start site of the *lac* operon, resulting in a *lacZ*+, *lacY*+ transcriptional fusion [D. H. Mitchell, W. S. Reznikoff, and J. R. Beckwith, *J. Mol. Biol.* **93**, 331 (1975)]. The ends of the restriction fragment were filled in with Klenow (J. Sambrook, E. F. Fritsch, T. Maniatis, "Molecular Cloning: A Laboratory Manual." 2nd Ed. Cold Spring Harbor Laboratory, Cold Spring Harbor, New York, 1989) followed by ligation to *Sph*I linkers. The resulting fragment was digested with *Sph*I and cloned into the *Sph*I site of pMS*purA*, 3' to the *purA* gene, resulting in pIVET1. The *Bgl*II site provides a convenient place to insert random fragments of *Sau*3AI-digested chromosomal DNA.

Pool of random fusions

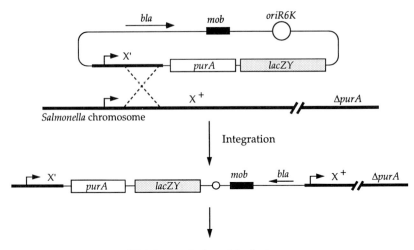

Integration

I.P. inject pool of *purA-lac* fusions
into a BALB/c mouse

Incubate three days;
Selects Pur$^+$ *in vivo*

Recover bacterial cells from spleen

Repeat selection in second mouse

Screen for Lac$^-$ *in vitro*

Fig. 2. Positive selection for genes that are specifically induced in the host. Random, partial *Sau*3AI-restricted *S. typhimurium* chromosomal DNA fragments were cloned into the *Bgl*II site of pIVET1. The ligated pool of plasmid DNA was electroporated into *E. coli* strain SM10 λ*pir*, which contains the Pi protein required for plasmid replication and the RP4 conjugal transfer functions [R. Simon, U. Priefer, and A. Puhler, *BioTechnology* **1**, 784 (1983)]. The pool of *purA–lac* transcriptional fusion plasmids was mobilized into a *purA* deletion strain of *S. typhimurium* that lacks the Pi replication protein {MT168 (DEL2901[*pur-A874::IS10*])} [M. J. Mahan, J. M. Slauch, and J. J. Mekalanos, *Science* **259**, 686 (1993)]. Selection for ampicillin resistance in this strain demands integration of the plasmid into the bacterial chromosome, generating a duplication of *S. typhimurium* material in which the native chromosomal promoter drives the *purA* gene and the cloned promoter drives the expression of the putative virulence gene, X. The chromosomal fusion strains were pooled and injected intraperitoneally into a BALB/c mouse (10^6 cells total). After 3 days, bacterial cells were recovered from the spleen and grown overnight in rich medium (LB), and 10^6 cells were injected intraperitoneally into a second mouse. Bacterial cells, which were Pur$^+$ *in vivo*, were recovered from the spleen and plated on MacConkey lactose indicator medium, and colonies were screened for those that subsequently had reduced expression (Lac$^-$) *in vitro*.

source. We routinely use lactose MacConkey agar. In fusions constructed with pIVET1, the Lac activity measured on a MacConkey plate correlates very well with the PurA activity. In other words, fusion strains that are Lac$^+$ (red) are Pur$^+$, fusions that are Lac$^{+/-}$ (pink) are semiauxotrophic, and fusions that are Lac$^-$ (white) are Pur$^-$.

The initial preselected pool of *purA–lac* fusion strains is plated on lactose MacConkey agar; we have shown that 33% of the colonies are Lac$^+$, 16% are Lac$^{+/-}$, and 50% are Lac$^-$.[5] The vast majority of Lac$^-$ fusions in this pool represent cases in which the fusion is in the wrong orientation with respect to the promoter. This pool of fusion strains is injected (10^6 organisms total) intraperitoneally into a BALB/c mouse. Three days after infection, we remove the spleen, one of the major sites of systemic infection for *S. typhimurium*. The bacteria recovered from the splenic extract are then grown overnight, and the process is repeated. The bacteria recovered from the spleen of the second mouse are plated on lactose MacConkey agar (Fig. 2). In the mouse-selected pool, 86% of the fusion strains are Lac$^+$, 9% are Lac$^{+/-}$, and 5% are Lac$^-$. This dramatic shift toward Lac$^+$ cells (termed the red shift) indicates that there is the expected selection for Pur$^+$ in the mouse. The majority of these strains contain fusions to genes that are constitutively expressed at high levels. However, because we were interested in genes that are specifically induced *in vivo*, we chose the Lac$^-$ strains for further analysis. The fusions in these cells are sufficiently transcriptionally active in the mouse to overcome the parental PurA deficiency. *In vitro*, however, they are transcriptionally inactive. Thus, the promoters driving these fusions are specifically induced in mouse tissues. In *S. typhimurium,* we have termed these operons *ivi*(*in vivo* induced).[5]

Observation of the so-called red shift is the only indication that the selection took place. This is important for any IVET selection and exemplifies the importance of the *lac* operon in the system. The various parameters that affect any given selection, for example, the length of incubation time in the host or the number of times the pool is taken through the selection, should be determined empirically, with the success of the selection monitored by the red shift.

It should be noted that the level of Lac activity in fusions constructed with vectors other than pIVET1 may not coincide with the activity of the gene that is the basis of the selection. In other words, the profile of Lac$^+$ to Lac$^-$ in the preselected pool will vary depending on the IVET vector. In this regard, the red shift observed after selection with other vectors may be more or less than that observed with pIVET1. However, the genes that presumably show the greatest *in vivo* induction are still those that have the least transcriptional activity *in vitro*. In cases where the host-

selected pool still has a significant number of white colonies on Mac-
Conkey agar, it may be necessary to screen for Lac activity using a
different assay medium, such as one that contains the chromogenic substrate X-Gal (5-bromo-4-chloro-3-indolyl-β-D-galactoside). In this case,
one observes a blue shift after passage through the host, where the fusions
that are the lightest blue among the blue-shifted survivors are the fusions
of interest.

In Vivo Assays

We have developed a method to directly measure the transcriptional
activity of the isolated fusions *in vivo*. Comparison of the *in vivo* β-
galactosidase activity with the activity of the same strain grown *in vitro*
proves that the fusions are induced in the animal. An example of this
shown in Fig. 3. Briefly, BALB/c mice are injected intraperitoneally with

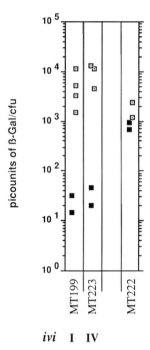

Fig. 3. β-Galactosidase (β-Gal) expression levels from bacterial cells recovered from
mouse spleen versus the same strain grown on rich medium. The vertical axis depicts the
picounits of β-Gal activity per colony-forming unit, where units of β-Gal are expressed as
micromoles *o*-nitrophenol formed per minute. The open boxes denote the β-Gal activity of
cells recovered from the spleen. The filled boxes denote the β-Gal activity from cells grown
overnight in rich medium (LB).

approximately 10^5 cells of an individual *ivi* fusion strain. Six days after infection, the mice are sacrificed and the spleens removed and homogenized. The bacterial cells are largely separated from the splenic material by differential lysis of animal cells in deionized water. The β-galactosidase activity in each sample is determined by a kinetic assay, using the fluorescent substrate fluorescein di-β-D-galactopyranoside (Molecular Probes, Inc., Eugene, OR) and a Model SPF-500c spectrofluorometer (SLM Aminco, Urbana, IL). The activity is reported per colony-forming unit (cfu) in the bacterial suspension. The units of β-galactosidase are obtained by comparing the activity to a standard curve determined with purchased β-galactosidase (Sigma, St. Louis, MO). The units of the purified enzyme are designated by the manufacturer and are defined as micromoles of *o*-nitrophenol formed per minute, using *o*-nitrophenylgalactopyranoside as substrate. Although there is tremendous mouse-to-mouse variability in the assay, it can be seen that fusions to the *ivi* genes *iviI* (MT199) and *iviIV* (MT223)[5] are highly induced in animal tissues relative to growth in laboratory media. As a control for the experiment, we chose a random Lac$^+$ fusion strain, MT222, from the preselected pool. The fusion in this strain is not significantly induced in animal tissues compared to growth in laboratory medium; the fusion is highly expressed in both conditions.

Cloning the Fusions

To identify the *in vivo* induced genes, we chose to clone the fusions and determine the DNA sequence of the region of chromosome adjacent to the fusion joint. In *S. typhimurium,* we devised a method to clone the fusions, in a single step, without the use of restriction enzymes, using the generalized transducing bacteriophage, P22. Briefly, a bacteriophage P22 lysate is made on the fusion strain of interest and used to transduce a recipient strain that contains the replication protein, Pi, which is required for autonomous replication of pIVET1. Presumably, after introduction of the chromosomal fragment containing the integrated fusion construct, the plasmid circularizes by homologous recombination at the region of duplication. This results in a plasmid clone of the fusion. In other organisms where cloning by transduction is not possible, the fusions can be cloned by more standard methods.

Primers homologous to the 5' end of the *purA* sequences are used to sequence 200–400 base pairs (bp) of chromosomal DNA adjacent to the fusion joint (United States Biochemical, Cleveland, OH). These sequences are then used to search the DNA database for homologous sequences. This analysis showed that the *ivi* genes represent both known and unknown genes in *S. typhimurium.*[5]

Null Mutations in *ivi* Operons

An important question is whether these *in vivo* induced genes have any role in pathogenesis. As described, the fusions were constructed in diploid, to maintain the function of the wild-type gene. To test the role of *ivi* genes in pathogenesis, we have isolated insertion mutations that affect the expression of the *ivi* operons. The insertion mutations disrupt the gene at the point of insertion. In addition, they are polar on downstream genes in the operon. The initial fusion strains are phenotypically Lac⁻ but are, nonetheless, blue on the sensitive chromogenic substrate X-Gal. We isolated polar insertion mutations that turn the fusions from blue to light blue.[5] Recombination of these insertion mutations into the chromosome results in an otherwise wild-type strain with an insertion mutation in the *ivi* operon. These insertions, by definition, affect the expression of the operon that was defined by the fusion. These insertion mutants can then be tested for virulence. For example, an insertion mutation in either *iviI* or *iviIV* (shown in Fig. 3) confers a virulence defect.[5]

Regulatory Elements

Once potential virulence genes have been identified, it is important to determine the factors that regulate their expression. The system is designed to facilitate this analysis. As described, the fusions that are of the most interest are phenotypically Lac⁻ on laboratory medium. Therefore, to isolate mutations that result in constitutive expression of these operons, one can simply select Lac⁺.[7] This type of analysis can yield both *cis*- and *trans*-acting regulatory elements. The procedure is slightly complicated by the fact that the fusions are diploid. Therefore, selection for increased expression of the *lac* operon can result in amplification of the fusion construct that starts with recombination between the duplicated chromosomal fragment. However, these amplification mutants can be distinguished by the fact that the Lac⁺ phenotype is genetically unstable.

Variations

The pIVET1 system is based on the *purA* gene. Although *purA* auxotrophy is known to attenuate a number of pathogenic organisms,[2,8–11] this

[7] J. M. Slauch and T. J. Silhavy, this series, Vol. 204, p. 213.
[8] V. S. Baselski, S. Upchurch, and C. D. Parker, *Infect. Immun.* **22**, 181 (1978).
[9] G. Ivanovics, E. Marjai, and A. Dobozy, *J. Bacteriol.* **85**, 147 (1968).
[10] H. B. Levine and R. L. Maurer, *J. Immunol.* **81**, 147 (1958).
[11] S. C. Straley and R. L. Harmon, *Infect. Immun.* **45**, 649 (1984).

may not always be the case. In addition, a *purA* mutation may be difficult to obtain in some organisms.

Any number of biosynthetic genes can, in theory, be used in the IVET selection. For example, we have constructed an analogous plasmid based on the *thyA* gene (pIVET2). This system has several advantages. First, *thyA* mutants can be selected with trimethoprim (Tmp).[12] This is particularly useful in organisms where the isolation of attenuating mutants is problematic. One can positively select a *thyA* mutation (Tmp^r), score for those that are nonreverting, and then determine if the mutation can be complemented with a *thyA* clone. It is useful to clone the *thyA* gene from a heterologous organism, to prevent recombination during integration of the fusion plasmids. Resistance to trimethoprim can also be used to select for those fusions that are transcriptionally inactive *in vitro*, either before or after selection in the host. For example, fusions can be selected that are "on" in the animal. Then the postselected pool can be plated on trimethoprim to select those fusions that are "off" *in vitro*. Repetition of this process should greatly enrich for fusions that are specifically induced in host tissues.

In addition to IVET systems based on complementation of biosynthetic genes, we have also developed plasmids based on drug resistance. For example, we have constructed a synthetic operon in which chromosomal promoters drive the expression of the CAT gene, encoding chloramphenicol acetyltransferase and a promoterless *lac* operon (pIVET8). Mice are treated with chloramphenicol (Cm) and challenged with the pool of fusion strains. Successful selection is monitored, as always, by plating the selected pool of fusion strains on lactose MacConkey agar and noting the red shift. In our experiments with a BALB/c mouse animal model, we have noted that there can be tremendous variability with this type of selection. Therefore, we have treated mice with a wide range of chloramphenicol concentrations and empirically determined the selection efficiency by monitoring the degree of red shift. We anticipate that this choramphenicol selection will be of general use, particularly in systems where it is difficult to obtain auxotrophic mutations or in tissue culture systems where depletion of nutrients is less feasible.

Reporter System

All of the IVET systems to date contain a promoterless *lac* operon to allow convenient monitoring of transcriptional activity *in vitro* and *in*

---

[12] J. H. Miller, "A Short Course in Bacterial Genetics." Cold Spring Harbor Laboratory, Plainview, New York.

vivo. The *lac* system has many advantages, including accommodation of a variety of substrates for use in assays and selections.[7,12] The main advantage is the ability to select Lac⁺ on plates. However, this may not be feasible in all organisms. Indeed, in some organisms it may be necessary to use a reporter system other than *lac.* However, the importance of a convenient method to monitor transcriptional activity cannot be overemphasized.

Summary

We have developed a genetic system, termed IVET (*in vivo* expression technology), designed to identify bacterial genes that are induced when a pathogen infects its host. A subset of these induced genes should include those that encode virulence factors, products specifically required for the infection process. The system is based on complementation of an attenuating auxotrophic mutation by gene fusion, and it is designed to be of use in a wide variety of pathogenic organisms. In *Salmonella typhimurium,* we have successfully used the system to identify a number of genes that are induced in BALB/c mice, and that, when mutated, confer a virulence defect.

The IVET system has several applications in the area of vaccine and antimicrobial drug development. The technique was designed for the identification of virulence factors and thus may lead to the discovery of new antigens useful as vaccine components. The IVET system facilitates the isolation of mutations in genes involved in virulence and, therefore, should aid in the construction of live attenuated vaccines. In addition, the identification of promoters that are optimally expressed in animal tissues provides a means of establishing *in vivo* regulated expression of heterologous antigens in live vaccines, an area that has been previously problematic. Finally, we expect that our methodology will be used to uncover many biosynthetic, catabolic, and regulatory genes that are required for growth of microbes in animal tissues. The elucidation of these gene products should provide new targets for antimicrobial drug development.

Acknowledgments

This work was supported by National Institutes of Health Research Grant AI18045 (J.J.M.), Damon Runyon-Walter Winchell Cancer Research Fund Fellowship DRG-1016 (J.M.S.), and National Research Service Award AI08245 (M.J.M.).

[39] Regulation of Alginate Gene Expression in *Pseudomonas aeruginosa*

By Nicolette A. Zielinski, Siddhartha Roychoudhury, and
A. M. Chakrabarty

Introduction

Alginate is an exopolysaccharide produced by *Pseudomonas aeruginosa* during infection of the lungs of cystic fibrosis (CF) patients. The polysaccharide is rarely produced by *P. aeruginosa* outside the CF lung environment (e.g., during eye, urinary tract, or burn infections), and it constitutes a major virulence factor in CF lung infections. The alginate capsule is believed not only to allow *P. aeruginosa* cells to adhere to the epithelial cells of the CF lung,[1,2] but also to protect the infecting cells from the host immune systems and antibiotic therapy by forming a gel surrounding the cells that are found as microcolonies in the form of a biofilm on the epithelial cells.[3,4] The heavily mucoid form of *P. aeruginosa* isolated from the CF lung is an unstable phenotype, which undergoes transition to nonmucoidy spontaneously at a rapid rate during culturing in the laboratory. This suggests that the mucoid phenotype arising from alginate production is triggered specifically in the CF lung environment. The mechanism of spontaneous switching on of the alginate biosynthetic (*alg*) genes in the CF lung and its switching off outside this environment is, however, little understood.[5]

Organization of *alg* Genes

The pathway of alginate biosynthesis is depicted in Fig. 1A. Alginate is a linear polymer of mannuronic acid and its C-5 epimer guluronic acid, linked via β-1,4-glycosidic bonds. It is not known if the epimerization occurs at the level of the monomer (GDP-mannuronic acid) or polymer (polymannuronic acid), but an epimerase functional at the

---

[1] P. Doig, N. R. Smith, T. Todd, and R. T. Irvin, *Infect. Immun.* **55**, 1517 (1987).

[2] N. R. Baker and C. Svanborg-Eden, *Antibiot. Chemother. (Basil)* **42**, 72 (1989).

[3] J. Lam, R. Chan, K. Tam, and J. W. Costerton, *Infect. Immun.* **28**, 546 (1980).

[4] J. W. Costerton, K. J. Cheng, G. G. Gissey, T. I. Ladd, J. C. Nickel, M. Dasgupta, and T. J. Marrie, *Annu. Rev. Microbiol.* **41**, 435 (1987).

[5] D. E. Ohman, J. B. Goldberg, and J. L. Flynn, *in* "*Pseudomonas*: Biotransformations, Pathogenesis, and Envolving Biotechnology" (S. Silver, A. M. Chakrabarty, B. Iglewski, and S. Kaplan, eds.), p. 28. American Society for Microbiology, Washington, D.C., 1990.

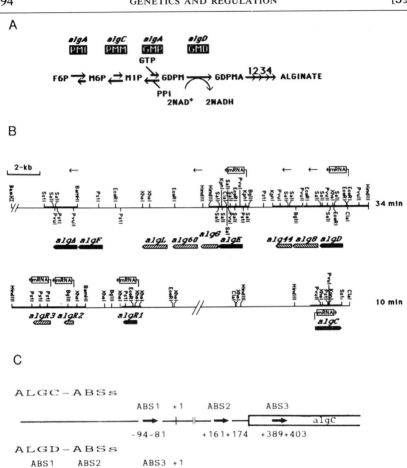

FIG. 1. (A) Alginate biosynthesis pathway. Enzymes catalyzing the known reactions in the pathway are as follows: PMI, phosphomannose isomerase; PMM, phosphomannomutase; GMP, GDPmannose pyrophosphorylase; and GMD, GDPmannose dehydrogenase. The genes encoding these enzymes are indicated above the respective enzyme name. The equilibria for the reactions are shown by the relative sizes of the arrows in both directions of the reactions [algA, D. Shinabarger, A. Berry, T. B. May, R. Rothmel, A. Fialho, and A. M. Chakrabarty, *J. Biol. Chem.* **266**, 2080 (1991); algC, P. J. Padgett and P. V. Phibbs, *Curr. Microbiol.* **14**, 187 (1986); algD, S. Roychoudhury, T. B. May, J. F. Gill, S. K. Singh, D. S. Feingold, and A. M. Chakrabarty, *J. Biol. Chem.* **264**, 9380 (1989)]. The known reaction intermediates in the pathway are as follows: F6P, fructose 6-phosphate; M6P, mannose 6-phosphate; M1P, mannose 1-phosphate; GDPM, GDP-mannose; and GDPMA, GDP-mannuronic acid. The steps between the formation of GDPMA and that of alginate include polymerization/epimerization, acetylation, and export (shown by arrows 1, 2, 3, and

polymer level has been implicated in this process.[6] The mode of alginate biosynthesis has been delineated from fructose 6-phosphate, which is converted to GDP-mannose via mannose 6-phosphate and mannose 1-phosphate catalyzed by the enzymes phosphomannose isomerase (PMI, mannose-6-phosphate isomerase), phosphomannomutase (PMM), and GDPmannose pyrophosphorylase (GMP) (Fig. 1A).[7] GDP-mannose is then converted to GDP-mannuronic acid by a four-electron transfer dehydrogenase (GDPmannose dehydrogenase, GMD). Little is known regarding the enzymes involved in the polymerization and epimerization of the mannuronic acid units or the secretion or acetylation of the polymeric product. A gene encoding an acetylation enzyme has been

---

[6] M. J. Franklin, C. Chitnis, and P. Gacesa, *Abstr. 93rd Gen. Meet. Am. Soc. Microbiol.* **D254**, 140 (1993).

[7] T. B. May, D. Shinabarger, R. Maharaj, J. Kato, L. Chu, J. D. DeVault, S. Roychoudhury, N. A. Zielinski, A. Berry, R. K. Rothmel, T. K. Misra, and A. M. Chakrabarty, *Clin. Microbiol. Rev.* **4**, 191 (1991).

---

4) and are not yet well characterized. (B) Organization of genes involved in alginate synthesis. The locations of the gene clusters on the *P. aeruginosa* chromosome are indicated in minutes. The location of *algE* [L. Chu, T. B. May, A. M. Chakrabarty, and T. K. Misra, *Gene* **107**, 1 (1991)], *algG* [C. Chitnis and D. E. Ohman, *J. Bacteriol.* **172**, 2894 (1990)], and others [T. B. May, D. Shinabarger, R. Maharaj, J. Kato, L. Chu, J. D. DeVault, S. Roychoudhury, N. A. Zielinski, A. Berry, R. K. Rothmel, T. K. Misra, and A. M. Chakrabarty, *Clin. Microbiol. Rev.* **4**, 191 (1991)] has been described. In addition, genes such as *algL* encoding alginate lyase or *algF* encoding an acetylation enzyme have been mapped in the alginate biosynthetic gene cluster at 34 min [N. L. Schiller, S. R. Monday, C. M. Boyd, N. T. Keen, and D. E. Ohman, *J. Bacteriol.* **175**, 4780 (1993); D. Shinabarger, T. B. May, A. Boyd, M. Ghosh, and A. M. Chakrabarty, *Mol. Microbiol.* **9**, 102 (1993)]. The *algC* gene [V. D. Shortridge, M. L. Pato, A. I. Vasil, and M. L. Vasil, *Infect. Immun.* **59**, 3596 (1991)] maps on a 60-kb *Dra*I fragment along with the *algR1, algR2,* and *algR3* cluster, but its relative distance and orientation are not known (as indicated by the broken line). Restriction enzyme sites are indicated by vertical lines. Sequenced regions of DNA are indicated by thick horizontal lines. Thin lines denote unsequenced regions of DNA. Arrows indicate the direction of transcription with the transcriptional start site (mRNA), if known. The stippled areas indicate proteins encoded by the indicated gene and confirmed by N-terminal amino acid sequencing. The hatched areas denote protein-coding regions predicted from DNA sequence analysis of the indicated gene. Open areas indicate proteins encoded by the respective gene for which the exact coding region has not been determined. (C) Location of AlgR1 binding sites (ABS) upstream of the *algD* and *algC* genes as well as one site (ABS-3) within the *algC* gene. The arrows indicate the binding sites. Positions of sites are numbered from the transcriptional start point (+1) of the *algC* and *algD* genes.

cloned[8] and sequenced,[9] which allows extensive acetylation of the mannuronate residues.

The PMI and GMP enzyme activities are associated with a single polypeptide which constitutes a bifunctional enzyme.[10] PMI–GMP has been purified, and some of its properties such as stability, cofactor or metal ion requirements, and K_m and V_{max} values have been determined.[10] The gene *algA*, encoding the bifunctional enzyme PMI–GMP, has been sequenced.[10,11] The gene for the second step of the pathway involving PMM, termed *algC*, has been sequenced, and some of the properties of the enzyme have been described.[12] The gene for the enzyme GMD has been sequenced,[13] and properties of the enzyme such as K_m, V_{max}, substrate and cofactor binding sites, and range of substrates and inhibitors have been described.[14,15] The *algE* gene product is believed to be involved in the polymerization and export of alginate by forming alginate-permeable pores on the cell membrane.[16,17] Another gene *algG*, which is involved in the insertion of guluronate residues on alginate,[18] has also been described. In addition, a number of other genes such as *alg8*, *alg44*, and *alg60* have been cloned and mapped on the *P. aeruginosa* chromosome (Fig. 1B). The functions of the products of these genes are not known. All the genes, except *algC*, map in a cluster at the 34 min region of the *P. aeruginosa* chromosome with *algD* at one terminal (Fig. 1B). The direction of transcription for all the genes is from *algD* to *algA*, which is present at the other end of the cluster.[7,19]

The *algC* gene maps on a 60-kb *Dra*I fragment[20] along with three other regulatory genes *algR1*, *algR2*, and *algR3* at the 10 min region of the

[8] M. J. Franklin and D. E. Ohman, *J. Bacteriol.* **175,** 5057 (1993).

[9] D. Shinabarger, T. B. May, A. Boyd, M. Ghosh, and A. M. Chakrabarty, *Mol. Microbiol.* **9,** 1027 (1993).

[10] D. Shinabarger, A. Berry, T. B. May, R. Rothmel, A. Fialho, and A. M. Chakrabarty, *J. Biol. Chem.* **266,** 2080 (1991).

[11] A. Darzins, B. Frantz, R. I. Vanags, and A. M. Chakrabarty, *Gene* **42,** 293 (1986).

[12] N. A. Zielinski, A. M. Chakrabarty, and A. Berry, *J. Biol. Chem.* **266,** 9754 (1991).

[13] V. Deretic, J. F. Gill, and A. M. Chakrabarty, *Nucleic Acids Res.* **15,** 4567 (1987).

[14] S. Roychoundhury, T. B. May, J. F. Gill, S. K. Singh, D. S. Feingold, and A. M. Chakrabarty, *J. Biol. Chem.* **264,** 9380 (1989).

[15] S. Roychoudhury, K. Chakrabarty, Y.-K. Ho, and A. M. Chakrabarty, *J. Biol. Chem.* **267,** 990 (1992).

[16] E. Grabert, J. Wingender, and U. K. Winkler, *FEMS Microbiol. Lett.* **68,** 83 (1990).

[17] L. Chu, T. B. May, A. M. Chakrabarty, and T. K. Misra, *Gene* **107,** 1 (1991).

[18] C. Chitnis and D. E. Ohman, *J. Bacteriol.* **172,** 2894 (1990).

[19] S. K. Wang, I. Sa-Correia, A. Darzins, and A. M. Chakrabarty, *J. Gen. Microbiol.* **133,** 2303 (1987).

[20] V. D. Shortridge, M. L. Pato, A. I. Vasil, and M. L. Vasil, *Infect. Immun.* **59,** 3596 (1991).

chromosome (Fig. 1B). The genes *algR1*,[21,22] *algR2*,[23] and *algR3*[24] have all been sequenced. Konyecsni and Deretic[25] subsequently sequenced the same regulatory gene cluster and described two genes, *algQ* and *algP*, which appear to be analogous to *algR2* and *algR3*. An intriguing feature of the alginate genes is that the levels of the gene products of *algA*, *algC*, and *algD* (PMI–GMP, PMM, and GMD, respectively) are extremely low.[26] In the mucoid CF isolate *P. aeruginosa* strain 8821, these enzymes are barely detectable. In the nonmucoid spontaneous revertant strain 8822, the levels of PMI–GMP, PMM, and GMD are even lower. Thus two important questions arise concerning alginate biosynthesis: (1) why are the levels of PMI–GMP, PMM, or GMD so low, and how are these levels increased in mucoid cells in comparison to nonmucoid ones, and (2) why are alginate genes specifically activated in the CF lung environment to confer a mucoid phenotype to the cells?

Environmental Activation of Alginate Promoters

The organization of the *alg* genes shows a cluster of biosynthetic genes at 34 min with *algD* as a terminally located gene, whereas another biosynthetic gene, *algC*, is transcribed independently at 10 min (Fig. 1B). The regulatory genes *algR1*, *algR2*, and *algR3*, also located in the 10 min region, are independently regulated. The regulatory gene *algR1* as well as the structural genes *algC*, *algE*, and *algD* have GG-N10-GC type of promoters reminiscent of σ^{54} promoters.[7] Such promoters require activators with an ATPase domain.[27] The role of σ^{54} in the activation of *algD* or *algC* promoters is not clear.[28,29] Even though both the *algC*[28] and the

[21] V. Deretic, R. Dikshit, W. M. Konyecsni, A. M. Chakrabarty, and T. K. Misra, *J. Bacteriol.* **171,** 1278 (1989).

[22] J. D. DeVault, A. Berry, T. K. Misra, A. Darzins, and A. M. Chakrabarty, *Bio/Technology* **7,** 352 (1989).

[23] J. Kato, L. Chu, K. Kitano, J. D. DeVault, K. Kimbara, A. M. Chakrabarty, and T. K. Misra, *Gene* **84,** 31 (1989).

[24] J. Kato, T. K. Misra, and A. M. Chakrabarty, *Proc. Natl. Acad. Sci. U.S.A.* **88,** 1760 (1991).

[25] W. M. Konyecsni and V. Deretic, *J. Bacteriol.* **172,** 2511 (1990).

[26] I. Sa-Correia, A. Darzins, S.-K. Wang, A. Berry, and A. M. Chakrabarty, *J. Bacteriol.* **169,** 3224 (1987).

[27] D. S. Weiss, J. Batut, K. E. Klose, J. Keener, and S. Kustu, *Cell (Cambridge, Mass.)* **67,** 155 (1991).

[28] N. A. Zielinski, R. Maharaj, S. Roychoudhury, C. E. Danganan, W. Hendrickson, and A. M. Chakrabarty, *J. Bacteriol.* **172,** 7680 (1992).

[29] S. Fujiwara, N. A. Zielinski, and A. M. Chakrabarty, *J. Bacteriol.* **175,** 5452 (1993).

TABLE 1
ENVIRONMENTAL FACTORS ALLOWING ALGINATE GENE ACTIVATION[a]

Environmental signal	Promoter–reporter gene fusion	Host bacterium	Level of activation	Ref.
Osmolarity	algC–lacZ	8821	5- to 6-fold	28
		8822	5- to 6-fold	
	algD–xylE	8821	4- to 5-fold	31
		PA01	3-fold	
Ethanol	algD–xylE	8821	3- to 4-fold	32
		PA01	12-fold	
Nitrogen limitation	algD–xylE	8821	2-fold	22
		PA01	3- to 4-fold	
Phosphate limitation	algD–xylE	8821	2-fold	22
		PA01	2-fold	
Adherence to solid	algC–lacZ	8830	3- to 5-fold	33
surface (Biofilm)	algD–lacZ	PA0579	2- to 5-fold	34

[a] Host bacteria are strains of *P. aeruginosa*; strains 8821, 8830, and PA0579 are mucoid, whereas strains 8822 and PA01 are nonmucoid (non-alginate producers).

algD[30] promoters have been shown to require the participation of AlgR1 for activation, AlgR1 does not have an ATPase domain.

The presence of specific environmental factors greatly enhances the level of activation of both the *algC* and the *algD* promoters (Table I[31–34]). For example, high osmolarity due to the presence of high concentrations of NaCl leads to activation of the *algD* promoter,[31] as does starvation of nitrogen or phosphate.[22] The presence of a membrane perturbing, dehydrating agent such as ethanol also leads to the activation of the *algD* promoter.[32] Ethanol allows genotypic switching to mucoidy,[32] suggesting that a loss of membrane integrity is a strong signal for alginate synthesis via activation of the *algD* promoter. Such membrane perturbations may be key characteristics of the highly dehydrated mucous environment with high electrolyte concentrations in the CF lung. Similarly, the *algC* promoter is activated strongly in presence of high concentrations of NaCl,[28] as well as by adherence to solid surfaces to form a biofilm.[33] Formation of biofilms has also been shown to trigger *algD* gene activation.[34] The

[30] J. Kato and A. M. Chakrabarty, *Proc. Natl. Acad. Sci. U.S.A.* **88,** 1760 (1991).

[31] A. Berry, J. D. DeVault, and A. M. Chakrabarty, *J. Bacteriol.* **171,** 2312 (1989).

[32] J. D. DeVault, K. Kimbara, and A. M. Chakrabarty, *Mol. Microbiol.* **4,** 737 (1990).

[33] D. G. Davies, A. M. Chakrabarty, and G. G. Geesey, *Appl. Environ. Microbiol.* **59,** 1181 (1983).

[34] B. D. Hoyle, L. J. Williams, and J. W. Costerton, *Infect. Immun.* **61,** 777 (1993).

methodologies pertaining to the determination of the activation of the *algC* promoter is given below.

Growth Conditions

Pseudomonas aeruginosa and *Escherichia coli* strains harboring an *algC–lacZ* transcriptional fusion plasmid are used for determination of *algC* activation in media of different osmotic strengths. The strains are grown at 37° in YTG medium (5 g of yeast extract, 10 g of tryptone, 2 g glucose per liter) containing either no NaCl or 0.3 *M* NaCl as specified. Cultures of 100 ml are inoculated with 1 ml of overnight seed cultures grown in YTG or YTG containing 0.3 *M* NaCl. Antibiotic concentrations used for the plasmid-containing strains are as follows: carbenicillin (*P. aeruginosa*), 350 μg/ml, and ampicillin (*E. coli.*), 75 μg/ml. Cultures are aerated by shaking at 250 rpm for 18 hr. Cells are then harvested by centrifugation, washed with 0.9% sterile saline, recentrifuged, and stored as frozen pellets at $-70°$.

Extract Preparation

Crude extracts of *P. aeruginosa* or *E. coli* to be used for β-galactosidase assays are prepared by thawing cell pellets in 7 ml of 50 m*M* phosphate buffer (pH 7.0) containing 5 m*M* 2-mercaptoethanol and 1 m*M* MgSO$_4$. Cells are disrupted by sonication three times for 20 sec at 120 W each time on ice. The sonicated suspensions are then centrifuged at 40,000 *g* for 30 min at 4°. The resultant supernatant is used directly for β-galactosidase assays by the procedure of Miller.[35] Protein concentrations are determined by the method of Bradford.[36] Typical assay conditions are described below.

β-Galactosidase Assay

In a test tube, combine lysate (varying dilutions depending on protein concentration) and Z buffer (60 m*M* Na$_2$HPO$_4$, 40 m*M* NaH$_2$PO$_4$, 10 m*M* KCl, 1 m*M* MgSO$_4$, and 50 m*M* 2-mercaptoethanol) to a final volume of 1 ml. For each sample use three different lysate concentrations. Place tubes at 28° for 10 min in a water bath. Add the substrate, 0.2 ml of *o*-nitrophenyl-β-D-galactopyranoside (ONPG; 4 mg/ml), to each tube, vortex briefly, and return to the 28° water bath. When sufficient yellow color has developed ($\sim$0.3 to 1.0 absorbance units at 420 nm), the reaction is stopped

[35] J. H. Miller, "Experiments in Molecular Genetics," p. 352. Cold Spring Harbor Laboratory, Cold Spring Harbor, New York, 1972.
[36] M. M. Bradford, *Anal. Biochem.* **72**, 248 (1976).

with the addition of 0.5 ml of 1 M Na_2CO_3 and the sample vortexed. The formation of o-nitrophenol is spectrophotometrically determined at an absorbance of 420 nm. β-Galactosidase specific activities are defined as nanomoles of o-nitrophenol produced per minute per milligram of crude extract protein at 28°, pH 7.0.

Signal Transduction in Alginate Synthesis

An important question in the activation of the *algD* or *algC* promoters is how the signals mediated by the environmental factors are transduced in *P. aeruginosa* leading to the activation of the promoters. As previously mentioned, AlgR1 is important for the activation of the *algD* and *algC* promoters. It has been further demonstrated that not only AlgR1, but two other regulatory proteins, AlgR2 and AlgR3, are also important in the activation of the *algD* promoter.[23,24] For example, the AlgR2⁻ mutant strain 8882 is phenotypically Alg⁻, suggesting the important role of AlgR2 in alginate synthesis. It is interesting to point out that the mutation in 8882 is not only complemented by *algR2*, but can also be complemented slowly by another gene, *algR3*, when it is present in a copy number of 4 to 6 (but not by a single chromosomal copy) and very efficiently when *algR3* is expressed from a strong promoter such as *tac*.[24] Thus AlgR2 can be replaced in part by an excess of AlgR3.

Both *algR2*[23] and *algR3*[24] have been sequenced. The deduced amino acid sequence of AlgR2 does not show any significant homology with other proteins, whereas the sequence of *algR3* shows that its gene product has significant homology with eukaryotic histone H1 type proteins. Based on the role of other histone type proteins as DNA-bending proteins, it has been postulated that AlgR3 as well as a *P. aeruginosa* homolog of the *E. coli* cAMP receptor protein (CRP), another DNA-bending protein, may be involved in the bending of the *algD* upstream region for its transcriptional activation. Because activation of *algD*,[31,32] or *algC* (N. Zielinski, unpublished observation, 1993), is strongly inhibited in presence of inhibitors of DNA gyrase, it is clear that supercoiling of the promoter regions is also important for their activation. This supercoiling may then be stabilized by a basic protein such as AlgR3. A model of *algD* promoter activation requiring participation of a putative CRP analog, AlgR3, integration host factor (IHF), and phosphorylated AlgR1 has previously been postulated.

The nucleotide sequence of the *algR1* gene has demonstrated the homology of AlgR1 with the response regulators such as OmpR, NtrC, and SpoOA[21,22] of the two-component signal transduction systems. Indeed, AlgR1 has been shown to bind the far upstream and downstream sequences of *algC* and *algD* promoters.[29] There are three copies of the AlgR1 binding site (ABS) upstream of the *algD* gene, whereas there are two copies of the ABS upstream of the *algC* gene and a copy within the *algC* gene (Fig. 1C). Such binding sites behave typically as eukaryotic enhancerlike elements, since they confer activation of the promoters irrespective of their position or orientation.[29]

Concluding Remarks

The activation of the *alg* genes is accomplished via activation of at least two critical *alg* promoters, namely, *algC* and *algD*. Both promoters are modulated by environmental factors such as high NaCl concentrations (high osmolarity) or adherence to solid surfaces (for biofilm formation) which are characteristic of the CF lung environment. Activation of the promoters is accomplished by two-component signal transducing regulatory proteins such as AlgR1 and AlgR2. In addition, auxiliary proteins such as DNA gyrase,[22] AlgR3,[24] *E. coli* CRP-like proteins,[37] and *E. coli* IHF-like proteins[38,39] appear to be involved in the activation process because of their ability to introduce negative supercoiling in the promoter region and allow bending of the far upstream regions harboring the bound phosphorylated AlgR1 to make contact with the RNA polymerase–σ factor complex bound at the promoter region. Other regulatory proteins affecting *algD* promoter activation, such as AlgB, having homology with response regulators such as NtrC, have been described,[40,41] but the mechanism of action of AlgB has not been delineated.

The detailed mechanism of the signaling process leading to phosphorylation of AlgR1 is poorly understood at present. Not only is AlgR1 critical for alginate synthesis, but it may also participate in other aspects of the pathogenic process, since the expression of genes encoding enzymes such as neuraminidase, which might facilitate adherence of *P. aeruginosa* to the CF lung epithelial surface, is significantly enhanced by high osmolarity and is dependent on the presence of AlgR1.[42] In addition, another protein, AlgT,[5] which is negatively regulated by AlgN,[43] controls the level of AlgB

[37] J. D. DeVault, W. Hendrickson, J. Kato, and A. M. Chakrabarty, *Mol. Microbiol.* **5**, 2503 (1991).

[38] D. J. Wozniak, *Abstr. 92nd Annu. Meet. Am. Soc. Microbiol.* **D-232**, 134 (1992).

[39] C. D. Mohr and V. Deretic, *Biochem. Biophys. Res. Commun.* **189**, 837 (1992).

[40] D. J. Wozniak and D. E. Ohman, *J. Bacteriol.* **173**, 1406 (1991).

[41] J. B. Goldberg and T. Dahnke, *Mol. Microbiol.* **6**, 59 (1992).

[42] G. Cacalano, M. Kays, L. Saiman, and A. Prince, *J. Clin. Invest.* **89**, 1866 (1992).

[43] J. B. Goldberg, W. L. Gorman, J. L. Flynn, and D. E. Ohman, *J. Bacteriol.* **175**, 1303 (1993).

and AlgR1, which are involved in *algD* and *algC* gene activation. A gene *algU,* which appears to be the same as *algT,* has homology to that of an alternative sigma factor σ^H of *Bacillus subtilis,* and is postulated to encode a sigma factor for *algD* transcription.[44] The overall infection of the CF lung tissues by mucoid *P. aeruginosa* is, therefore, likely to be a complex process with global regulatory networks like AlgR1 playing a critical role.

Acknowledgments

This work was supported by U.S. Public Health Service Grant AI 16790-13 and in part by AI 31546 from the National Institutes of Health. N.A.Z. and S.R. are supported in part by student traineeship grants from the Cystic Fibrosis Foundation. We are indebted to Dr. T. B. May for help with the figures.

[44] D. W. Martin, M. J. Schurr, M. H. Mudd, and V. Deretic, *Mol. Microbiol.* **9,** 497 (1993).

[40] Regulation of Expression of *Pseudomonas* Exotoxin A by Iron

By Dara W. Frank and Douglas G. Storey

Introduction

Analysis of pathogenic bacteria suggests that expression of certain genes selects for members of a population that can adhere to cells or tissue, evade the host immune response, and reproduce. By definition, this reproduction leads to pathologic changes in the host. The products of genes that allow bacterial reproduction, within the host, are referred to as virulence determinants. What conditions within the host induce bacterial virulence gene expression? This key question forms the basis for examining not only the host environment, but also the mechanism by which bacteria sense the environment and then change gene expression to ensure survival and reproduction. To dissect complex processes into component parts that are amenable to experimental manipulation, a simple model system is required. The model system that is the focus of this chapter involves the study of the induction of a virulence determinant, exotoxin A (ETA) in the opportunistic pathogen *Pseudomonas aeruginosa,* in response to a host environmental condition, iron limitation.

Bacterial Growth and Maximizing Exotoxin A Yields

Strain Selection

In general, analysis of the regulatory pathways controlling a virulence gene product begins with experiments that involve the selection of specific strains for examination. Strain selection for ETA production was initiated because of the production of proteases which destroyed ETA biological activity.[1] Strains that produced less protease or growth conditions that inhibited protease production or activity without affecting ETA production or activity were sought.[2] These studies resulted in the selection of strain PA103 which in subsequent years has been found to be deficient in the regulatory pathway controlling elastase synthesis in *P. aeruginosa*.[3] Strain PA103 is capable of producing approximately 10-fold more ETA than the prototypical strain PA01.[4] As more information became available, the difference in ETA yields between PA103 and PA01 was found to relate less to protease degradation of the product and more to the transcriptional regulation of the *toxA* gene. Comparison of the regulatory genes controlling ETA production between PA01 and PA103 demonstrated that part of the hypertoxigenic phenotype of PA103 was due to the activity of a regulatory gene product, RegB, which enhanced early transcription of a positive activator of ETA transcription, RegA.[5,6] Thus an initial survey of strains may yield useful hosts to examine differences in regulatory or structural genes that may contribute to the final yield, stability, or cellular localization of a virulence factor.

Growth Conditions for Maximal Yields of Exotoxin A

Following strain selection, growth conditions that optimize production were investigated. For ETA production several factors were shown to affect the final yield of toxic activity of strain PA103. These include aeration, growth temperature, carbon and nitrogen source, and elimination of inhibitory substances present in the complex medium.[7] These studies were followed by a set of experiments which demonstrate that, like diphtheria and *Shigella* toxins, ETA production is repressed by the addition of iron

[1] P. V. Liu, *J. Infect. Dis.* **116,** 112 (1966).
[2] P. V. Liu, *J. Infect. Dis.* **116,** 481 (1966).
[3] M. J. Gambello and B. H. Iglewski, *J. Bacteriol.* **173,** 3000 (1991).
[4] D. E. Ohman, R. P. Burns, and B. H. Iglewski, *J. Infect. Dis.* **142,** 547 (1980).
[5] M. J. Wick, D. W. Frank, D. G. Storey, and B. H. Iglewski, *Mol. Microbiol.* **4,** 489 (1990).
[6] D. G. Storey, T. L. Raivio, D. W. Frank, M. J. Wick, S. Kaye, and B. H. Iglewski, *J. Bacteriol.* **173,** 6088 (1991).
[7] P. V. Liu, *J. Infect. Dis.* **128,** 506 (1973).

to the growth medium.[8] A 1 μM Fe^{2+} concentration is optimal for growth and ETA production. Concentrations of Fe^{2+} at 5 μM or above inhibit ETA production 90–95%. Thus, complex medium must be deferrated for optimal expression of ETA. Using a defined medium and a two-stage production process, Blumentals et al. demonstrate that other divalent cations inhibit ETA production. These include Co^{2+}, Cu^{2+}, and Mn^{2+}.[9] The inhibition mediated by these metals, however, requires much higher concentrations (200 μM to 1 mM) than that observed for Fe^{2+}. The inclusion of Ca^{2+} in the defined medium at concentrations of 500 μM increases yields by 3-fold but fails to compensate for the inhibitory effect of iron.[9]

Our studies of ETA production have used the complex medium originally developed by Liu because of the consistency in both ETA production and iron repression of ETA synthesis.[7] This medium is prepared as a stock of trypticase soy broth by adding 150 g to 450 ml of 18 mΩ water. The stock solution is treated with 50 g of Chelex 100, minus 400 mesh (Bio-Rad, Richmond, CA), for 5–6 h at room temperature followed by centrifugation and filtration to remove the Chelex. High molecular weight inhibitory substances are removed by Amicon (Danvers, MA) filtration at 5° using a PM30 membrane. The filtrate is collected, and approximately 25–30 ml of the retentate is discarded. This 10× stock can be kept at −20° for many months without significant changes in ETA yield or iron repression of ETA synthesis.

For ETA production, the 10× stock is diluted to 1× with 18 mΩ water and autoclaved for 15 min. Supplementation of the medium with glycerol to 1% [filter sterilized, 50% (w/v) stock solution] as a carbon source and 50 mM monosodium glutamate (filter sterilized, 2 M stock solution) as a nitrogen source permits high levels of ETA production. Maximum ETA yields are achieved by cultivation in 1× complex medium with supplements, overnight at 32°, with a 10:1 ratio between flask volume and medium volume. The flasks should be shaken to increase aeration. This growth phase is followed by a production culture grown essentially under the same conditions. Inoculation of the production culture so that the starting $OD_{540\ nm}$ is 0.02 will result in maximal toxin yields 16–18 hr later. Maximal repression of ETA production is achieved by adding a 10 mM stock solution of $FeCl_3$ made in 1 N HCl to 5–10 μM in the growth and production media.

[8] M. J. Bjorn, B. H. Iglewski, S. K. Ives, J. C. Sadoff, and M. L. Vasil, Infect. Immun. 19, 785 (1978).
[9] I. I. Blumentals, R. M. Kelly, M. Gorziglia, J. B. Kaufman, and J. Shiloach, Appl. Environ. Microbiol. 53, 2013 (1987).

Growth Curve Analysis

Exotoxin A accumulates in the supernatant of *P. aeruginosa* cultures during the production phase. Thus, the amount of ETA measured in cultures is highly dependent on the time of sampling. Comparison of ETA production in different strains or in strains grown under different conditions is difficult because of variation in growth and accumulation of ETA. To remedy the fluctuation associated with a single point analysis, the pattern of ETA expression is followed by sampling cultures at different intervals during the growth of the cells in production phase.[10,11]

After subculture to the production medium, the OD_{540} of cells is monitored, and samples consisting of 2×10^{10} cells are removed when the OD_{540} or time reaches the following target values: 0.1 (250 ml of cells), 0.3 (85 ml), 0.6 (42 ml), 1.0 (25 ml), 2.0 (12.5 ml), 3.0 (8.33 ml), 4.0 (6.25 ml), 5.0 (12 hr, 5.0 ml), 14 hr (4.0 ml), and 16–17 hr. The ADP-ribosyltransferase activity is assayed from both supernatant samples and cell lysates. Cell lysates are prepared by harvesting 2×10^{10} cells for each time point, washing the cells twice in an ice-cold 50 mM Tris buffer, pH 7.4, and resuspending the pellet in a 2-ml volume of Tris buffer containing 2.5 μg of DNase I and 7000 U of RNase T1. The suspension is passed through an Aminco French pressure cell (SLM Instruments, Urbana, IL) twice at maximal pressure, and the debris and unbroken cells are removed by a 12 min centrifugation at 14,000 g (at 5°).

The standard assay for exotoxin A activity consists of 10 μl of test material preincubated with an equal volume of 8 M urea and 2% (w/v) dithiothreitol (DTT) for 15 min at 25°. Each assay mixture consists of 25 μl of wheat germ EF2, prepared by the methods of Chung and Collier,[12] 25 μl of 125 mM Tris-HCl (pH 7.0), 100 mM DTT, and 5 μl of [^{14}C]NAD (530 mCi/mmol). Duplicate enzyme assays are incubated for 10 min at 25° and stopped with 200 μl of a 10% (w/v) solution of trichloroacetic acid. The precipitated proteins are collected on filters, and radioactivity is measured in a liquid scintillation counter. Data obtained from multiple time points suggest that ETA synthesis occurs in two phases, corresponding to an early cell-associated phase followed by a late extracellular phase of production. Cells grown in high-iron medium fail to accumulate ETA in either cell-associated or extracellular compartments.

[10] D. W. Frank, D. G. Storey, M. S. Hindahl, and B. H. Iglewski, *J. Bacteriol.* **171**, 5304 (1989).
[11] D. W. Frank and B. H. Iglewski, *J. Bacteriol.* **170**, 4477 (1988).
[12] D. W. Chung and R. J. Collier, *Infect. Immun.* **16**, 832 (1977).

Regulatory Level at Which Iron Affects Yield of Exotoxin A

Isolation and mRNA Probing

The detection of an early peak of ETA activity indicated that an early phase of *toxA* transcription must occur. If *toxA* transcription occurs in two phases, does the transcription of the positive regulatory gene for ETA synthesis, *regA,* also reflect this pattern? Total RNA is isolated throughout the production phase from 2×10^{10} cells per sample by the following methods.[11]

1. Quickly harvest (10,000 g, 5 min, 5°) and resuspend cells in 5 ml of ice-cold 10 mM sodium acetate, pH 4.8, 0.15 M sucrose, 0.1 mg/ml heparin, and 10 mM vanadyl ribonucleoside complexes.

2. Remove the wash buffer and resuspend the pellet of cells in 3 ml of buffer without vanadyl ribonucleoside complexes.

3. Add 75 μl of a 20% solution of sodium dodecyl sulfate (SDS) per extraction. Vortex and bring to 65° immediately. Quickly add 3 ml of phenol (equilibrated with acetate buffer, pH 4.8, 65°), vortex, and incubate at 65° for 5 min.[13]

4. Incubate on ice for 5 min, then centrifuge at 12,000 g for 15 min at 5°. Collect the aqueous phase and reextract at 65° with acetate-equilibrated phenol. Remove the aqueous phase and extract with an equal volume of chloroform.

5. Remove the aqueous phase and precipitate nucleic acids with 2.2 volumes of absolute ethanol at $-20°$.

6. Collect the precipitate by centrifugation at 12,000 g for 20 min and wash the pellet with 70% (v/v) ethanol. Dry the pellet briefly.

7. Resuspend the total RNA precipitate in 400 μl of 0.1 M Tris-HCl, pH 7.4, 50 mM NaCl, 10 mM Na$_2$EDTA, and 0.2% (w/v) SDS with 200 μg/ml of proteinase K. Incubate at 37° for 1 hr. Extract the reaction once with an equal volume of phenol and once with chloroform, then precipitate with 2.2 volumes of absolute ethanol.

8. Collect the precipitate by centrifugation at 12,000 g, wash with 70% ethanol, and briefly dry the pellet. Resuspend the pellet in 400 μl of 10 mM Tris-HCl, pH 7.6, 10 mM magnesium acetate, 10 mM DTT, with 0.8–1.0 unit/ml of RNasin (Promega, Madison, WI) and 100 μg/ml of RNase-free DNase (Worthington Biochemicals, Freehold, NJ). Incubate for 30 min at 37°. Extract with phenol, then chloroform, and to the final

---

[13] A. von Gabain, J. G. Belasco, J. L. Schottel, A. C. Y. Chang, and S. N. Cohen, *Proc. Natl. Acad. Sci. U.S.A.* **80,** 653 (1983).

aqueous phase add sodium acetate to 0.1 M and 2.2 volumes of absolute ethanol. The RNA content is quantitated spectrophotometrically by absorbance at A_{260}.

Fragments corresponding to internal sections of the *toxA* and *regA* genes are isolated, labeled with [^{32}P]dCTP in a primer extension reaction, and used as probes to detect mRNA in a dot-blot analysis. Five micrograms of total RNA (per 4-mm-diameter well) is added to 30 μl of 20× standard saline citrate (SSC) and 20 μl of 37% formaldehyde and incubated at 60° for 15 min. After heating, the samples are filtered onto nitrocellulose using a minifold apparatus. Probes are hybridized at high stringency, washed, and the dots corresponding to wells are cut out and counted in scintillation fluid. The mRNA expression patterns for cells grown in low- and high-iron media are examined. The data demonstrate that when cells are grown for maximal ETA production (low iron), the transcript accumulation pattern for both the *regA* and *toxA* genes is biphasic.[11] Each gene shows a transcript accumulation peak early in the growth curve, a decrease, followed by another round of transcript accumulation during the late phase of growth. The response to growth in high-iron medium differentiated the two phases for both genes. The early phase of transcript accumulation is inhibited and the late phase appears not to initiate under high-iron growth conditions.

The size of early and late message for the *toxA* and *regA* genes is determined by Northern blot analysis.[11] Total RNA (10 μg per lane) is denatured with glyoxal and dimethyl sulfoxide (DMSO) and separated on 1.2% agarose gels in a 10 mM phosphate buffer.[14] To detect mRNA, the nucleic acid is transferred to nitrocellulose and hybridized to the identical probes used for dot-blot analysis. *ToxA* mRNA is found not to change size when early and late samples are probed; however, two size classes of *regA* mRNA are detected. The early *regA* mRNA, T1, appears as a smear that migrates in the area of the gel corresponding to 1200–1500 base pairs (bp). RNA samples from late phase growth only demonstrate a smaller 700- to 800-bp *regA* message (T2). A probe was designed to determine whether the large, early, *regA* mRNA, T1, contains sequences 5′ of the proposed transcriptional start site. This upstream probe hybridizes to T1 in Northern blot analysis and reproduces the same inhibition pattern in dot-blot analysis when RNA is extracted from cells grown in high-iron medium. Hybridization with the smaller *regA* message, T2,

[14] T. Maniatis, E. F. Fritsch, and J. Sambrook, *in* "Molecular Cloning: A laboratory Manual," pp. 122 and 200. Cold Spring Harbor Laboratory, Cold Spring Harbor, New York, 1982.

expressed late in the growth cycle, is not detectable with the upstream probe.[11]

Analysis of many samples during the growth of *P. aeruginosa* clarifies the pattern of expression of both the regulator, *regA,* and the virulence gene product, ETA. The large *regA* transcript, T1, is expressed first, early in the growth cycle, followed by a *toxA* transcript peak. ETA made early in the growth cycle is predominantly cell associated, with little activity being located in the supernatant fraction. When cells are grown in high-iron medium the early phase of both *regA* and *toxA* mRNA accumulation is reduced but still detectable. A second phase of transcript accumulation occurs as the cells make the transition between late logarithmic growth and stationary phase. Under low-iron growth conditions the smaller, *regA* T2 and *toxA* transcript accumulation peaks 8–10 h after subculture. Neither the *regA* T2 nor *toxA* mRNA is detectable when cells are grown in high-iron medium.

Copy Number Effects

Do multiple copies of the *regA* or *toxA* genes alter transcript accumulation or iron regulation patterns? Multiple copies of the *toxA* gene have no effect on the iron-mediated inhibition of ETA production. In contrast, multiple copies of the *regA* gene relieve some of the iron inhibition of ETA production. The copy number of *regA* is inversely proportional to the amount of iron inhibition of ETA production.[15] Dot-blot and Northern blot analysis of growth curve samples where the *regA* locus is present in multiple copies demonstrates that T1 accumulation is no longer inhibited by growth in high-iron medium.[10] Like T1, the first phase of *toxA* transcript accumulation is not inhibited when cells are grown in high-iron medium, and the *regA* locus is present in multiple copies. T2 transcript accumulation does not initiate when cells are grown under high-iron conditions regardless of the copy number of the *regA* locus. The lack of initiation of T2 transcription results in a severe reduction in the second phase of *toxA* transcript accumulation. These data support the notion that the iron regulation of *toxA* transcription is the direct result of regulatory events governing *regA* transcription and expression. From data regarding the temporal expression, differences in transcript size, and differential regulatory mechanisms that relate to copy number we conclude that at least two promoters, regulated by different iron-sensitive mechanisms, control *regA* transcription.[10]

[15] M. S. Hindahl, D. W. Frank, and B. H. Iglewski, *Antibiot. Chemother.* (*Basil*) **39,** 279 (1987).

Transcriptional Start Sites

Our model predicts that *regA* should have two start sites for transcription. The transcript corresponding to T1 should have a start site located 5' of the T2 start site, based on the size of the transcripts and on the hybridization analysis with upstream probes. In addition, the start sites for each transcript should correlate to RNA samples from cells harvested at specific times in the growth cycle and to the iron regulatory pattern of those samples. The RNA template for primer extension reactions is isolated from cells and grown under low- and high-iron conditions, which contain multiple copies of the *regA* locus. The following solutions are required for start site analysis:

> $4\times$ RT buffer: 400 mM Tris-HCl, pH 8.3, 520 mM KCl, 100 mM DTT, 40 mM MgCl$_2$ (add actinomycin D to a final concentration of 320 μg/ml just prior to use)
> 0.5 mM Deoxyribonucleoside triphosphates (dNTPs)
> Sequencing stop buffer: 95% formamide, 10 mM EDTA, 10 mM NaOH, 0.3% each of xylene cyanol, bromphenol blue, and orange G

Procedure

1. Total RNA (10 μg) is pelleted and resuspended in 10 μl of water.
2. The annealing reaction consists of 10 μg of the total RNA and labeled oligonucleotide primer [1.5 $\times$ 10^6 counts/min (cpm)] in 20 μl. The oligonucleotide primer is designed to span the AUG codon of *regA* mRNA and is end-labeled using T4 polynucleotide kinase.[14] Sephadex G-25 column chromatography is used to purify the labeled primer.
3. To 4 μl of the annealing reaction, add 2.5 μl of $4\times$ RT buffer containing actinomycin D and 2.5 μl of 0.5 mM dNTP stock solution. Incubate at 42° for 30 min.
4. Add 1 μl of avian myeloblastosis virus (AMV) reverse transcriptase (20 U) and incubate at 42° for 1 hr. Reverse transcriptase that has been mutagenized to eliminate RNase H activity, namely, SuperScript RNase H$^-$ (Bethesda Research Laboratories, Gaithersburg, MD), has been used at 30 U per reaction.
5. Add 30 μl of ethanol and precipitate the nucleic acids on ice for 1 hr. Pellet the sample by centrifugation and resuspend in 4 μl of water and 4 μl of sequencing stop buffer.
6. Heat samples to 95° for 5 min prior to loading on a sequencing gel. Start sites should be sized by running sequencing reactions generated with the same oligonucleotide primer on the same gel.

The major start site for the T1 transcript is located at -164 relative to the AUG codon. This start site is detectable in mRNA from cells grown in high-iron medium as predicted from the data demonstrating iron deregulation of the T1 transcript with the *regA* locus present in multiple copies. Stronger stops for the T1 mRNA are detected in RNA samples extracted at the early time point. The start site for the T2 mRNA is only detectable in RNA extracted from cells grown in low-iron medium regardless of the copy number of the *regA* locus.

Translational and Transcriptional Fusions

To dissect the activity of each *regA* promoter region, a direct assay for *regA* transcription and/or translation is required. To accomplish this goal we have constructed a series of translational fusions using *lacZ* as a reporter gene or transcriptional fusions with a chloramphenicol acetyltransferase reporter (CAT).[6] β-Galactosidase activity is measured as previously described.[16] For CAT assays 1.6×10^9 cells per point are used. The following target optical density values and volumes of cultures are used: OD_{540} of 0.1, 10 ml; 0.3, 6.7 ml; 0.6, 3.3 ml; 1.0, 2.0 ml; 2.0, 1.0 ml; 3.0, 0.67 ml; 4.0, 0.5 ml; and 5.0, 0.4 ml. At each point the cells are harvested and washed with 1 ml of ice-cold 100 mM Tris-HCl, pH 7.8, in a 1.5-ml microcentrifuge tube. Washed cells are resuspended in 0.5 ml Tris-HCl buffer and sonicated with a microtip, power level 3, 50% duty cycle, for 40 sec on ice. Unbroken cells and debris are removed by centrifugation for 10 min at 5°, 12,000 g. The CAT reactions are performed directly in scintillation vials with 1 μl of lysate or dilutions of lysate, 10 μl of [^{14}C]acetyl-CoA (0.1 μCi), 238 μl of 100 mM Tris-HCl, pH 7.8, and 1 μl of a 250 mM stock solution of chloramphenicol in ethanol. The reaction mixture is gently overlaid with a water-immiscible scintillation fluor and incubated at 37° for 50 min.[17] Standard curves with dilutions of purified chloramphenicol acetyltransferase are used to calculate units of activity.

The growth of the cells is somewhat different in fusion experiments. *Pseudomonas* cells with reporter constructs are grown in the presence of iron, washed, and resuspended in either low- or high-iron medium for the production phase to a final OD_{540} of 0.02. This technique, which we refer to as iron synchronization,[18] eliminates ETA or RegA growth phase accumulation from affecting reporter gene synthesis during the production

[16] J. H. Miller, *in* "Experiments in Molecular Genetics," p. 352. Cold Spring Harbor Laboratory, Cold Spring Harbor, New York, 1972.

[17] J. R. Neumann, C. A. Morency, and K. O. Russian, *BioTechniques* **5**, 444 (1987).

[18] D. G. Storey, D. W. Frank, M. A. Farinha, A. M. Kropinski, and B. H. Iglewski, *Mol. Microbiol.* **4**, 499 (1990).

phase (refer to feedback inhibition by ETA and *regB* requirement for T1 initiation in following sections). With the entire *regA* upstream region fused to *lacZ*, a biphasic, iron regulated pattern of β-galactosidase activity can be demonstrated. Deletion of the T1 start site results in expression of β-galactosidase during late logarithmic growth only when cells are cultivated in low-iron medium. Each promoter region is subcloned into a vector in which translational signals are provided by the reporter gene (CAT). This analysis demonstrates that P1, a promoter which initiates the T1 transcript, is expressed early and is not tightly regulated by iron. P2, the promoter that initiates the T2 transcript, functions to produce CAT activity only in a low-iron environment.

Analysis of Strains Differing in Exotoxin A Yield

Cloning and Sequence Analysis

In the hypertoxigenic strain PA103, ETA yields correlate directly with the expression of a positive activator of ETA transcription, *regA*. Down-regulation of *regA* transcription at both the P1 and P2 promoters by growth in high-iron medium accounts for the 90–95% inhibition of ETA yields. Could strains that differ in ETA yields differ at the *regA* locus? The *regA* loci of three strains with different capacities to produce ETA have been cloned and sequenced.[5,6,10,19] This analysis demonstrates that strain PA01, which produces approximately 10-fold less ETA than PA103, has a single base change in the *regA* open reading frame and multiple changes in a downstream region. The most prominent of the changes is the alteration of a translational start codon which begins a small open reading frame (228 bp) located 6 bp downstream of the stop codon for *regA*. The contribution of this downstream region, *regB*, to ETA yield is shown by assembling hybrid operons between the low- and high-producing strains.[5] Because ETA synthesis from PA01 and PA103 is inhibited by iron to the same extent (90–95% decrease), *regB* appears not to be directly involved in the iron repression of *toxA* transcription. These data correlate with a growth curve analysis of *regB* expression using an oligonucleotide probe specific for *regB* sequence. With the *regAB* locus in multiple copies, *regB* mRNA is found to be present in the T1 transcript, expressed early in growth, and not regulated by iron.[6]

Strain PA103-29, which produces little if any ETA, shows two base changes in the *regAB* operon as compared to the parental strain, PA103.

[19] M. S. Hindahl, D. W. Frank, A. Hamood, and B. H. Iglewski, *Nucleic Acids Res.* **16,** 5699 and 8752 (1987).

One of the changes occurs well upstream of the P1 and P2 promoters at position -367 relative to the start codon.[6] It is likely that this change does not influence ETA production. The more critical change occurs at position $+687$. The C to T transition at this position introduces a premature stop codon at amino acid position 231 of RegA. Complementation of PA103-29 with a parental regAB operon restores ETA production that is fully iron regulated.[10] One unexplained observation concerning the phenotype of PA103-29 is that regAB mRNA is not detectable, regardless of growth conditions or time of sampling.[20] Introduction of a premature stop codon in the regA open reading frame may alter the stability of the regAB mRNA.

Expression of regAB Transcriptional and Translational Fusions in Different Hosts

A powerful approach for dissecting the regulation of the two regAB promoters is the use of transcriptional or translational fusions in strain PA103.[18] These experiments have been extended to strains with different genetic backgrounds to test for autologous regulation by either the regAB or toxA loci. From the nucleotide sequence information, PA103 demonstrates a RegA$^+$B$^+$, ETA$^+$ phenotype and produces detectable transcripts containing the early regAB or late phase regA mRNA. In PA103-29 regAB or toxA transcription is not detectable, resulting in a RegA$^-$B$^-$, ETA$^-$ phenotype. Strain PAO1 possesses only the regA open reading frame[5] and produces the T2 transcript encoding RegA,[21] making this strain RegA$^+$B$^-$, ETA$^+$. These strains and a PA103 ETA$^-$ strain[22] constructed by insertion mutagenesis and gene replacement (PA103::toxΩ) have been utilized to analyze the iron regulation of the regAB P1 and P2 promoters supplied in trans.

In all four strains, PA103, PA103-29, PAO1, and PA103::toxΩ, the relative level of activity from the P2 promoter and the timing of activation in low-iron conditions are identical.[6,22] We conclude from these results that neither RegA, RegB, nor ETA are involved in the iron regulation of the P2 promoter.

In contrast to the P2 promoter, the regAB P1 promoter is dramatically influenced by the genetic background of the host. A P1–CAT transcrip-

[20] D. W. Frank and B. H. Iglewski, unpublished observations (1988).
[21] J. Ali and D. G. Storey, in "Abstracts of the American Society for Microbiology 92nd General Meeting," p. 32 B-38.
[22] M. J. Wick, D. W. Frank, D. G. Storey, and B. H. Iglewski, Annu. Rev. Microbiol. 44, 335 (1990).

tional fusion (pP11) is not expressed in either strain PA01 or PA103-29 regardless of the iron content of the production medium or time of RNA isolation. The common phenotype in both strains is the lack of *regB* transcription. Complementation analysis confirms that *regB* is required for early transcriptional initiation from the P1 promoter.[5,6]

As described earlier, P1 activity as measured from a P1–CAT fusion in strain PA103 (pP11) is not influenced by the iron content of the medium.[18] One explanation for the lack of iron regulation in these conditions may be that multiple copies of the P1 promoter region are titrating a limited concentration of a repressor molecule. This hypothesis is supported by transcript accumulation data. When the *regAB* operon is present in a single chromosomal copy, growth in medium containing iron reduces T1 accumulation 3–4-fold.[11] In contrast multiple copies of *regAB* result in only a slight difference in T1 accumulation in high-iron medium.[10] Additional support for a repressor controlling the P1 promoter comes from an analysis of strain Fe18, a high ETA-producing derivative of strain PA01.[23] Fe18 demonstrates enhanced P1 activity but a wild-type *regAB* locus nucleotide sequence, suggesting that the mutation could be in another gene, perhaps a repressor controlling P1 promoter activity.[21] These data argue that the regulation of the *regAB* P1 promoter involves both an unidentified repressor and an activator (RegB).

To test for an ETA feedback inhibitory mechanism, activities of reporter genes of *regAB* transcriptional and translational fusions are compared in an ETA$^-$ host, PA103::toxΩ, and the parental ETA$^+$ strain, PA103. Little difference is seen in the relative level of β-galactosidase activity when early samples are analyzed in either host containing a RegA–LacZ translational fusion. In addition, the iron content of the medium makes no difference during this phase of growth. However, later in the growth curve, much higher levels of LacZ expression are observed in the ETA$^-$ host. This effect occurs only after growth in low-iron medium, which suggests that initiation from the P2 promoter may be altered in the presence or absence of ETA. The P1–CAT and P2–RegA–LacZ fusions have been examined in both host backgrounds. Surprisingly, β-galactosidase activity (under P2 control) is not related to ETA production of the host. In an ETA$^-$ host, however, we observe a large increase in CAT activity (under P1 control) late in the growth curve when cells are grown in low-iron medium. These results suggest that production of ETA in low-iron medium exerts a feedback inhibition of P1 transcriptional initiation at specific points in the growth curve.

[23] P. A. Sokol, C. D. Cox, and B. H. Iglewski, *J. Bacteriol.* **151**, 783 (1982).

Deregulation of Exotoxin A Production

Mutagenesis of toxA Promoter

Studies of the iron regulation of ETA production suggest that one can deregulate ETA synthesis by mutagenesis or substitution of either the *toxA* or *regAB* P1 and P2 promoter regions. To date linker scanning mutagenesis has been used to mutagenize and analyze the expression of only the *toxA* promoter.[24] These experiments were performed using *P. aeruginosa* strain MAM. Mutations in three areas of the *toxA* promoter greatly reduce ETA synthesis. Two of these areas centered at −6 and −41 bp would be consistent with an RNA polymerase binding site.[24] The third area, centered at −63 bp, would likely be a binding site for a positive activator protein.[24] These results suggest that an iron-regulated positive activator (RegA?) may bind to this region.

Promoter Substitution at regAB Locus

Addition of the *lacZ* promoter, upstream of the *regAB* promoters, has been used to override iron regulation of ETA.[25] The *lacZ* promoter–*regAB* construct is compared to multiple copies of the *regAB* locus under the control of the natural pair of promoters in a strain containing at least two chromosomal copies of an ETA–LacZ translational fusion, PA103C. β-Galactosidase activity in this case is a measure of the effect of multiple copies of iron-regulated (P1 + P2 control) *regAB* synthesis or iron-independent (*lacZ* promoter control) *regAB* synthesis. Expression of β-galactosidase activity is inhibited by growth of the transformants in high-iron medium when multiple copies of *regAB*, regulated by P1 and P2 are compared to vector controls late in cell growth. Iron regulation of β-galactosidase activity, however, is overcome by placing *regAB* under the control of the *lacZ* promoter. Similar comparisons using three other hosts (PA103-29, 388, and WR5) demonstrated that iron regulation of ETA–LacZ is lost only when *regAB* is regulated by the *lacZ*, iron-independent promoter. These results support the hypothesis that the iron regulation of *toxA* transcription results directly from the iron regulation of *regAB* transcription.

Influence of Fur and lasR on Pseudomonas aeruginosa toxA Regulation

The regulation of *P. aeruginosa* ETA, elastase, and alkaline protease[26] in response to iron limitation has led to the notion of a coordinated pathway

[24] M. L. Tsaur and R. C. Clowes, *J. Bacteriol.* **171,** 2599 (1989).
[25] M. L. Vasil, C. C. R. Grant, and R. W. Prince, *Mol. Microbiol.* **3,** 371 (1989).
[26] M. J. Bjorn, P. A. Sokol, and B. H. Iglewski, *J. Bacteriol.* **138,** 193 (1979).

controlling these virulence determinants. Two candidates for the regulator of coordinated responses in *P. aeruginosa* might be *lasR* and *fur*. Both these regulators influence the production of exotoxin A.[27-29]

To address the effect of *fur* on global iron regulation in *P. aeruginosa*, expression studies have been done with *E. coli fur*[27] as well as more recent studies with *P. aeruginosa fur*.[28] When the *E. coli fur*, cloned on a multicopy plasmid, is transformed into *P. aeruginosa*, exotoxin A production is strongly inhibited.[27] Transcript accumulation analysis reveals that *toxA* transcript accumulation is much lower in the presence of Fur when compared to the vector control. Multiple copies of *E. coli fur* also have an inhibitory influence on activity from the *regAB* P1 promoter and a slight inhibitory influence on the P2 promoter as measured by Northern blot analysis of T1 and T2 transcripts.[27] These results suggest that *E. coli* Fur influenced *toxA* transcription by inhibiting early transcription of the *regAB* locus.

Pseudomonas aeruginosa Fur appears to have a different influence on the production of exotoxin A. The *P. aeruginosa fur* gene is selected for and cloned by its ability to complement a *fur* mutation in *E. coli* (strain 1618).[28] When the cloned *P. aeruginosa fur* is introduced into strain PA103 on a multicopy plasmid, exotoxin A production is not altered. However, *P. aeruginosa fur* mutants selected by manganese-resistance constitutively produce exotoxin A. Iron regulation of exotoxin A is restored in these mutants in the presence of a multicopy plasmid carrying *P. aeruginosa fur*.[28] Multiple copies of *fur* also seem to have a stimulatory influence on exotoxin A production. The differences in biological effect between *E. coli* and *P. aeruginosa fur* could be partially explained by differences in the protein sequences. The *P. aeruginosa* Fur has 53% identity but differs considerably in the carboxyl terminus when compared to the *E. coli* Fur. Further research is needed to determine whether *P. aeruginosa fur* influences transcription of the *regAB* operon.

Another potential global regulator in *P. aeruginosa* is *lasR*. This gene is required for transcription of *lasB*, the gene encoding elastase and *apr*, the gene encoding alkaline protease.[29] LasR has been implicated as an enhancer of *toxA* transcription.[29] This enhancing effect does not seem to be mediated through the *regAB* operon. Further, it is unclear whether *lasR* itself responds to iron. It is possible that global iron regulation is independent of the regulation mediated by *lasR*.

[27] R. W. Prince, D. G. Storey, A. I. Vasil, and M. L. Vasil, *Mol. Microbiol.* **5**, 2823 (1991).
[28] R. W. Prince, C. D. Cox, and M. L. Vasil, *J. Bacteriol.* **175**, 2589 (1993).
[29] M. J. Gambello, S. Kaye, and B. H. Iglewski, *Infect. Immun.* **61**, 1180 (1993).

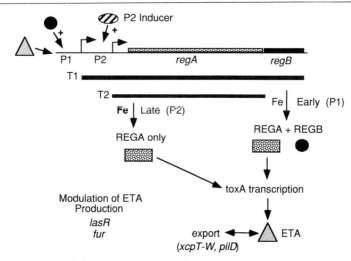

FIG. 1. Summary of the regulatory events that govern regAB and toxA transcription. P1 and P2 are promoter regions regulating regAB transcription. Transcripts synthesized by the regAB locus are shown as thick lines. Different proteins postulated to influence transcription from the regAB locus either in a positive or negative manner (+ or −) are shown as symbols (stippled box, RegA; filled circle, RegB; gray triangle, exotoxin A; striped oval, postulated regulatory molecule required for P2 transcriptional initiation). Transcripts whose synthesis is regulated by iron are labeled with Fe. Numerous products are probably involved in ETA export (xcpT-W, pilD[30]). The lasR and fur genes are modulators of ETA production.

In summary it appears that lasR and fur are modulators of exotoxin A production but further work is needed to determine their exact role in the iron regulation of toxA. In particular it would be interesting to determine if the lasR and fur regulatory pathways intersect or if they represent alternative systems responding to other environmental cues.

Summary and Perspectives

The pattern of regulatory and structural gene expression during different in vitro growth conditions was used as a model system to mimic the molecular events that may occur in vivo when P. aeruginosa is faced with surviving in an iron-limited environment. The studies indicated that the induction or repression of ETA synthesis correlated with transcription of the regAB regulatory locus. This result shifted the focus of iron regulation from the examination of toxA transcription to the study of regAB transcription. Two independently iron-regulated promoters were found that control regAB transcription (Fig. 1). The P1 promoter functions early in growth

[30] D. N. Nunn and S. Lory, J. Bacteriol. 175, 4375 (1993).

to produce an mRNA that encodes both RegA and RegB. ETA is produced but remains cell associated, indicating that RegA is functional. RegB also appears functional as it is required for transcriptional initiation at P1. Thus at early stages of cell growth, the *regAB* locus regulates itself and *toxA* at the level of transcription. This appears to be a limited loop as the *regAB* mRNA is highly unstable and ETA exhibits a feedback inhibitory influence on transcription from the *regAB* P1 promoter. Toxin production is shut off and will not occur again unless the cells are growing in an iron-limited environment in which they have reached late logarithmic phase. We postulate that an additional positive regulatory element (Fig. 1, P2 inducer), unrelated to RegA, RegB, or ETA, is required to promote transcription from the *regA* P2 promoter. Messenger RNA expressed at this point is large enough to encode only RegA, indicating that RegB is not required for late ETA transcription. RegA functions to induce *toxA* transcription by an unknown mechanism, and ETA is synthesized and secreted.

It is clear that only a part of the pathway for iron regulation of ETA production has been studied and that many questions remain unresolved. Does the temporal expression of *regAB* and *toxA* relate to another *in vivo* signal? Do cells have to reach a certain density for ETA to be synthesized in high amounts? What other signals (e.g., low O_2 tension, bacterial factors, an immune response) may be influencing ETA expression? Future projects will require a thorough functional study of the *regAB* gene products. Additionally, the postulated repressor or activator molecule that is actually responsible for late expression and iron regulation of ETA by controlling *regA* P2 transcriptional initiation has yet to be discovered.

[41] Regulation of Cholera Toxin by Temperature, pH, and Osmolarity

By Claudette L. Gardel and John J. Mekalanos

Introduction

The successful bacterial pathogen must interact with the host and survive through constantly changing environments.[1] Pathogens produce factors that facilitate their survival amid the varying external conditions. The external environment that influences the expression of these virulence factors may provide insight into the complex milieu that comprises host tissues.

[1] S. H. Richardson, *J. Bacteriol.* **100,** 27 (1969).

For the gram-negative bacterium *Vibrio cholerae,* the causative agent of Asiatic cholera, the disease process involves ingestion of the bacteria by the human host, passage through the severe low pH of the stomach, penetration by motility through the mucus gel layer of the intestine, adherence to the small intestine, growth of the vibrios, toxin production, and finally severe diarrhea. During the infectious cycle, the vibrios manufacture several factors required for a productive infection including colonization factors, toxins, proteases, and other extracellular enzymes. These products are not made constitutively in the laboratory and are therefore controlled by signals in the host. These environmental stimuli provide the organism with information concerning their location within the host and stage in the infectious cycle. The pathogen presumably senses its surroundings through a highly sophisticated regulatory system that activates the expression of certain gene products and turns off the expression of others.

One approach to understand the signals that regulate gene expression *in vivo* has been to observe the effects of specific external environments *in vitro*. This chapter focuses on the environmental control of toxin expression by the laboratory media parameters pH, temperature, and osmolarity.

Structure of Cholera Toxin

Cholera toxin is an excreted protein composed of six subunits of two different types: one A subunit (molecular weight 27,215) encoded by the *ctxA* gene and five B subunits (each having a molecular weight of 11,677) encoded by the adjacent *ctxB* gene.[2,3] After synthesis, the A subunit is processed by proteolytic cleavage and disulfide bond formation to produce a pair of linked fragments (A1 and A2). The five B subunits form a ring in which the disulfide-linked A1 and A2 peptides are nestled. The holoenzyme has two functions: the pentamer of B subunits specifically binds to the G_{M1} ganglioside receptor fround on the host cell surfaces, and the A1 subunit contains the toxic activity.[4,5] The internalization of the A1 subunit into the host cell causes several reactions initiating with ADP-ribosylation of G_S, a regulatory unit of adenylate cyclase. This modification results in the activation of adenylate cyclase which in turn causes increased cAMP levels, resulting in the massive efflux of chloride and other electrolytes

[2] J. J. Mekalanos, D. J. Swartz, G. D. N. Pearson, N. Harford, F. Groyne, and M. de Wilde, *Nature (London)* **306,** 551 (1983).
[3] J. J. Mekalanos, *in* "Current Topics in Microbiology and Immunology. Genetic Approaches to Bacterial Pathogenicity" (W. Goebel, ed.), p. 97. Springer-Verlag, Berlin and Heidelberg, 1985.
[4] J. Holmgren, *Infect. Immun.* **8,** 851 (1973).
[5] J. J. Mekalanos, R. J. Collier, and W. R. Romig, *J. Biol. Chem.* **254,** 5855 (1979).

into the lumen of the small intestine.[6] The action of cholera toxin results in a profuse osmotic diarrhea. This "rice watery" discharge during the infection may aid in the dissemination of colonized bacteria into the external environment.

Genetic Regulation of Cholera Toxin Genes

The *ctxA* and *ctxB* genes are adjacent genes in one operon that is directly activated by ToxR, a 34-kDa integral membrane DNA-binding protein.[7,8] The cytoplasmic N terminus of the ToxR protein, which shares homology with the transcriptional activation domains of several regulators, directly binds to a repetitive DNA sequence upstream of the *ctxA* locus and stimulates the transcription of the *ctx* locus.[2,9,10] ToxR is a regulator of at least 17 genes important for virulence, including those encoding two outer membrane proteins, the toxin-coregulated pilus (TCP) and less well-defined accessory colonization factors.[10-16] The expression of some of these factors requires a coregulator, ToxT, whose expression depends on ToxR.[17] ToxR or ToxT alone can activate expression of *ctx* genes in *Escherichia coli*.

The activity of ToxR is enhanced by ToxS, another integral membrane protein.[10,18,19] A deletion of *toxS*, the gene encoding ToxS, reduces expression of ToxR-regulated genes. There are differences among classical 01

[6] M. Field, in "Cholera and Related Diarrheas" (O. Ouchterlony and J. Holmgren, eds.), p. 46. Karger, Basel, 1980.

[7] V. L. Miller, R. K. Taylor, and J. J. Mekalanos, *Cell (Cambridge, Mass.)* **48**, 271 (1987).

[8] V. L. Miller and J. J. Mekalanos, *Proc. Natl. Acad. Sci. U.S.A.* **81**, 3471 (1984).

[9] J. J. Mekalanos, V. L. Miller, R. K. Taylor, I. Goldberg, and G. D. N. Pearson, in "Advances in Research on Cholera and Related Diarrheas" (S. Kuwahara and N. F. Pierce, eds.), p. 247. KTK Scientific Publ., Tokyo, 1988.

[10] J. J. Mekalanos and V. J. DiRita, in "ADP-Ribosylating Toxins and G Proteins: Insights into Signal Transduction" (J. Moss and M. Vaughan, eds.), p. 117. American Society for Microbiology, Washington, D.C., 1990.

[11] S. Calderwood, S. Knapp, K. Peterson, R. Taylor, and J. J. Mekalanos, in "Bacterial Protein Toxins" (F. J. Fehrenbach, J. E. Alouf, P. Falmagne, W. Goebel, J. Jeljaszewicz, and D. Jürgens, eds.), p. 169. Gustav Fischer, Stuttgart and New York, 1988.

[12] R. K. Taylor, V. L. Miller, D. B. Furlong, and J. J. Mekalanos, *Proc. Natl. Acad. Sci. U.S.A.* **84**, 2833 (1987).

[13] C. Parsot, E. Taxman, and J. J. Mekalanos, *Proc. Natl. Acad. Sci. U.S.A.* **88**, 1641 (1991).

[14] C. Parsot and J. J. Mekalanos, *J. Bacteriol.* **173**, 2842 (1991).

[15] J. F. Miller, J. J. Miller, and S. Falkow, *Science* **243**, 916 (1989).

[16] K. M. Peterson and J. J. Mekalanos, *Infect. Immun.* **56**, 2822 (1988).

[17] V. J. DiRita, C. Parsot, G. Jander, and J. J. Mekalanos, *Proc. Natl. Acad. Sci. U.S.A.* **88**, 5403 (1991).

[18] V. J. DiRita and J. J. Mekalanos, *Cell (Cambridge, Mass.)* **64**, 29 (1991).

[19] V. L. Miller, V. J. DiRita, and J.J. Mekalanos, *J. Bacteriol.* **171**, 1288 (1989).

strains in the expression and regulation of cholera toxin. Although a deletion of *toxS* constructed in the classical strain 0395 reduces expression of the ToxR regulon, the classical strain 569b carries a deletion of *toxS* and still exhibits high expression of toxin even under certain inhibitory conditions.[1,20] The reason for the constitutive nature of toxin expression in 569b has not been elucidated.

Laboratory Growth Conditions for *Vibrio* Biotypes

The response of the ToxR regulon to environmental stimuli is well documented. Expression of toxin, and other coregulated members of the ToxR regulon, is affected by culture conditions including temperature, pH, aeration, osmolarity, inoculum, and the presence of certain amino acids.[3,7,16,17,21–29] Laboratory studies of *Vibrio* growth have shown that *Vibrio* strains vary considerably in the ability to express cholera toxin and in the response to these signals. Initial studies in maximizing the yield of toxin have summarized the following optimal growth conditions for the classical biotype of *Vibrio cholerae*. Essentially, the critical requirements include a small inoculum and an aerated incubation at 30° in liquid medium with appropriate starting pH (6.5–7.0) and osmolarity ($\sim$66 mM).[1,24,30] Classical strains grown this way produce toxin in late logarithmic and stationary phases.[1,21,23]

The above conditions have not proved successful for eliciting toxin production in El Tor biotype strains. Studies have identified alternate media and growth conditions that stimulate toxin expression in El Tor backgrounds. Iwanaga *et al.*[31–33] found growing El Tor strains in AKI

[20] V. L. Miller and J. J. Mekalanos, *J. Bacteriol.* **163**, 580 (1985).

[21] D. J. Eans and S. H. Richardson, *J. Bacteriol.* **96**, 126 (1968).

[22] R. A. Finkelstein and J. J. LoSpalluto, *J. Infect. Dis.* **121** (Suppl.), S63 (1970).

[23] L. T. Callahan III, R. C. Ryder, and S. H. Richardson, *Infect. Immun.* **4**, 611 (1971).

[24] P. B. Fernandes and J. H. L. Smith, *J. Gen. Microbiol.* **98**, 77 (1977).

[25] V. L. Miller and J. J. Mekalanos, *J. Bacteriol.* **170**, 2575 (1988).

[26] J. J. Mekalanos, *J. Bacteriol.* **174**, (1992).

[27] V. J. DiRita and J. J. Mekalanos, *Annu. Rev. Genet.* **23**, 455 (1989).

[28] V. J. DiRita, K. M. Peterson, and J. J. Mekalanos, *in* "The Bacteria: A Treatise on Structure and Function" (B. H. Iglewski and V. L. Clark, eds.), p. 355. Academic Press, San Diego, 1990.

[29] K. M. Peterson, V. J. DiRita, R. K. Taylor, C. Shaw, and J. J. Mekalanos, *in* "Advances in Research on Cholera and Related Diarrheas" (R. B. Sack and Y. Zinnaka, eds.), p. 247. KTK Scientific Publ., Tokyo, 1990.

[30] L. T. Callahan III and S. H. Richardson, *Infect. Immun.* **7**, 567 (1973).

[31] M. Iwanaga and K. Yamamoto, *J. Clin. Microbiol.* **22**, 405 (1985).

[32] M. Iwanaga, K. Yamamoto, N. Higa, Y. Ichinose, N. Nakasone, and M. Tanabe, *Microbiol. Immunol.* **30**, 1075 (1986).

[33] M. Iwanaga and T. Kuyyakanond, *J. Clin. Microbiol.* **25**, 2314 (1987).

medium with an initial 4-hr nonaerated incubation followed by intense aeration would stimulate cholera toxin production to levels near to those of classic strains as measured by a G_{M1} enzyme-linked immunosorbent assay (ELISA).[34] However, classical strains have been used for most studies on the control of toxin gene expression and are the focus of this chapter.

Environmental Growth Conditions Affecting Toxin Production

Temperature Effects

Cholera toxin production and other genes of the ToxR regulon respond to temperature. More toxin (and TCP pili) is made *in vitro* at 30° than at 37°.[1] Consistent with this observation is the fact that *toxR* gene itself is expressed *in vitro* at higher levels at 30° than at 37°. This phenomenon may be explained by the observation that directly upstream of the *toxR* gene is the heat-shock gene *htpG,* which is transcribed divergently from *toxR.* As the temperature increases from 22° to 37° the expression of a *toxR::lacZ* transcriptional fusion on a low-copy plasmid decreases 5-fold, whereas expression of an *htpG* fusion increases 5-fold. Thus, expression of *toxR* may be the result of steric competition between the heat-shock σ factor and the σ factor that recognizes the *toxR* promoter.[35]

The greater expression of ToxR and toxin at 30° seems paradoxical as one might expect the host temperature to stimulate expression of virulence determinants. However, in the infectious cycle, the *Vibrio* bacteria encounter 37° much sooner than they may want to produce cholera toxin. Premature expression of the toxin might interfere with colonization and result in vibrios being flushed out by the secretory process before multiplication can occur. Thus, *in vivo* at 37°, some additional signal, such as the absence or presence of a nutrient or the interaction of host tissues, may induce cholera toxin expression. Indeed, it has been demonstrated that the addition of 0.1% (w/v) of the bile salt sodium deoxycholate to media stimulated cholera toxin production dramatically in 569b cultures grown at 37°. However, the level of toxin produced did not reach the level obtained by growth at 30° without bile salts.[24]

Osmolarity

Vibrio cholerae is exquisitely sensitive to osmotic pressure and will lyse rapidly in hypotonic media.[36] Growth and survival are enhanced by

[34] R. S. Dubey, M. Lindblad, and J. Holmgren, *J. Gen. Microbiol.* **136,** 1839 (1990).
[35] C. Parsot and J.J. Mekalanos, *Proc. Natl. Acad. Sci. U.S.A.* **87,** 9898 (1990).
[36] A. Lohia, S. Majumdar, A. Chatterjee, and J. Das, *J. Bacteriol.* **163,** 1158 (1985).

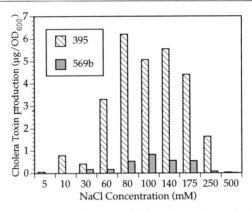

FIG. 1. Toxin expression at varying osmolarities. The effect of the NaCl concentration of the starting medium on the expression of cholera toxin was studied for classical strains 0395 and 569b. Cultures were grown at 30° on Rollodrums as described in the text in tryptone broth (pH 6.6) with varying NaCl concentrations. Supernatants were measured for cholera toxin by ELISA (expressed as micrograms per OD_{600} unit of an overnight culture).

the addition of salt. The expression of cholera toxin is influenced by the concentration of salts (KCl and NaCl) or unmetabolizable sugars (lactose and melibiose) in the medium, suggesting that osmolarity and not ionic strength is the effector.[16,25] The optimal concentration of NaCl for toxin expression has been reported as 66 mM, but the satisfactory range is somewhat broad, such that LB medium, at 86 mM NaCl, works well. High levels of NaCl (>250 mM) are inhibitory for the expression of toxin and other products in the ToxR regulon. Figure 1 shows the effect of the NaCl concentration in tryptone medium on toxin production in the classical strains 0395 and 569b. Expression of toxin is inhibited in both strains at low and high salt concentrations (less than 60 mM or greater than 250 mM).

It is presently unclear as to the cause of osmoregulation of the ToxR regulon. It has been postulated that the periplasmic C terminus of ToxR may be important in sensing or signaling osmolarity as one fusion construction, in which the C terminus of ToxR is replaced by the alkaline phosphatase of *E. coli*, activates *ctx* expression constitutively in the presence of high salt.[7,15,25] However, further fusion constructions now indicate that the periplasmic portion of ToxR may be less important in osmosensing.[37]

pH

Cholera toxin levels in an overnight culture are affected by the pH of the medium. To manufacture toxin optimally *in vitro*, the starting pH of

[37] K. M. Ottemann and J. J. Mekalanos, unpublished results, 1993.

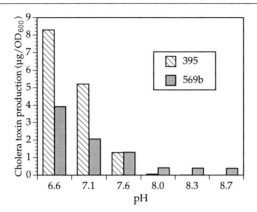

FIG. 2. Effect of medium pH on toxin expression. The effect of the starting pH on toxin expression was studied in classical strains 0395 and 569b. Cultures were grown as described in the text. Tryptone broth (starting at ph 6.6) was supplemented with 1, 2, 3, 4, or 5 μl of 2 N NaOH per milliliter to adjust the pH of the starting medium to 7.1, 7.6, 8.0, 8.3, or 8.7, respectively. Supernatants were analyzed by ELISA for cholera toxin (expressed as micrograms per OD_{600} unit of an overnight culture).

the medium must be approximately pH 6.5. (A starting pH lower than 6.0 does not allow substantial growth.[30,37]) After overnight growth at 30°, the culture pH rises to approximately 8.5. This rise in pH correlates with the rise in toxin production and is absolutely required for optimal toxin production, as little toxin is made in media buffered at pH 6.5 nor in media with a starting pH of 8.5–9.0.[1,16,21,25,30] Unlike temperature, pH does not affect the regulation of the *toxR* gene.[35] Figure 2 shows the production of toxin in two classical *Vibrio* strains when incubated in media with different starting pH values. Note that whereas strain 0395 ceases toxin expression in tryptone broth with a starting pH of 8.0, strain 569b continues toxin production in this medium, and in media with higher starting pH values.

Other Parameters Affecting Toxin Expression

Aeration, size of inoculum, and the presence of specific amino acids significantly influence toxin production. Nonaerated overnight cultures grown at 30° have dramatically decreased toxin levels compared with aerated cultures.[1,24] Heavily inoculated cultures express toxin and other members of the ToxR regulon poorly. The presence of the amino acids asparagine, arginine, aspartic acid, glutamic acid, and serine stimulates toxin production, whereas the addition of the amino acid tryptophan elimi- nates active toxin.[16,23,25] Interestingly, tryptophan induces production by

25- to 30-fold of an extracellular *Vibrio cholerae* protease (Prc) that may contribute to reducing toxin titers.[38]

Assaying Cholera Toxin Production

The following is a simple assay for cholera toxin. Briefly, the procedure requires (1) growing the strains under the appropriate conditions, (2) harvesting the supernatants, and (3) measuring toxin titers by performing ELISAs in G_{M1} ganglioside-coated microtiter dishes.[4,39]

Preparation G_{M1} Ganglioside-Coated Plates

1. Per well of a 96-well flat-bottomed microtiter plate (Linbro Titertek No. 76-331-05, ICN Flow, Horsham, PA), add 200 μl of 10 μg/ml G_{M1} gangliosides (type III, Sigma, St. Louis, MO) in 60 mM sodium carbonate, pH 10.[4,39]
2. Incubate at 37° for 4 hr or overnight.
3. Empty wells and wash three times with phosphate-buffered saline (PBS), pH 7.2 (10× PBS buffer is 1.4 M NaCl, 10 mM NaH$_2$PO$_4 \cdot$H$_2$O, 50 mM Na$_2$HPO$_4 \cdot$7H$_2$O).
4. Add 200 μl of 4 mg/ml bovine serum albumin (BSA, Boehringer Mannheim, Indianapolis, IN) in PBS (PBS–BSA) to each well.
5. Incubate at 37° for at least 4 hr.
6. Empty cells and wash three times with PBS.
7. Store dry coated microtiter plates at 4° for up to 6 months.

Bacterial Growth Conditions

Media and Inocula. For classical strains, overnight cultures for toxin assays are prepared by inoculating 5 ml of liquid medium in 16 × 150 mm glass tubes either directly from a frozen glycerol stock (20% (v/v) glycerol) or from a colony on a freshly incubated rich plate. Care should be taken to not overinoculate since this can prevent toxin production. If one strain is to be examined under varying conditions, a liquid suspension of a colony can be aliquoted in even amounts into all of the tubes. For maximal toxin production, classical strains are grown in LB medium [10 g tryptone, 5 g yeast extract, and 5 g NaCl (86 mM) per liter].

Minimal medium (M9) can be used for investigating the effects of addition of specific amino acids, or tryptone broth can be used for experi-

---

[38] S. R. Bortner, Ph.D. Thesis, Harvard Univ., Cambridge, Massachusetts (1988).
[39] J. J. Mekalanos, this series, Vol. 165, p. 169.

ments focusing on the effect of osmoactive salts.[16,25] Note that *Vibrio* strains should never be refrigerated as viability drops drastically.

Incubation Conditions. The starting cultures are placed on a Rollodrum (New Brunswick Scientific, East Brunswick, NJ, Model TC-7) at full speed (~1 revolution per second) and incubated overnight (at least 16 hr).[1] For all experiments except those involving temperature variation, the cultures are incubated at 30°. Note that some wild-type *Vibrio cholerae* strains grown overnight at 30° may autoagglutinate, as a result of the production of TCP pili. This indicates that cultures were grown properly and will not interfere with the toxin assay.

Experimental Considerations. To vary temperature, the culture tubes are placed on Rollodrums at the temperature to be analyzed. To vary pH, sodium hydroxide is added to the media to raise the pH. Addition of 5 μl of 2 N NaOH per milliliter of LB medium converts the starting permissive pH of 6.5 to a nonpermissive pH of 8.5. To vary osmolarity, NaCl, KCl, or unmetabolizable sugars can be added to any concentration in the liquid medium.

Cholera Toxin Assay

Sample Preparation. The culture samples are prepared by removing the cells by centrifugation and then harvesting the supernatants. Overnight cultures are prepared as described above. Owing to the quantity of toxin in the supernatants and the high sensitivity of the assay, the samples may be diluted 3 to 6-fold before application onto the G_{M1}-coated microtiter plate. Supernatants may be stored overnight at 4° or at −20° for 72 hr.[40] Recentrifuge refrigerated supernatants immediately before assaying.

Enzyme-Linked Immunosorbent Assay

1. Incubate cultures overnight (see above).
2. Spin down 1 ml of culture for 5 min in a microcentrifuge.
3. Remove 300 μl of each supernatant and place in a well of an uncoated microtiter plate. Supernatants may be placed in the desired alignment, and further manipulations may be facilitated with a multichannel pipettor.
4. Dilute samples 1 : 3 by removing 100 μl and mixing with wells having 200 μl of 4 mg/ml bovine serum albumin in phosphate-buffered saline (PBS–BSA, as described above) in the top wells of a G_{M1} ganglioside-coated microtiter dish (Dilution 1).

[40] W. M. Spira and P. J. Fedorka-Cray, *Infect. Immun.* **42**, 501 (1983).

5. Serially dilute 100 μl of Dilution 1 sample down a row of wells, each having 200 μl PBS–BSA (add 100 μl to a 200-μl well, mix by pipetting up and down, and remove 100 μl into next well with 200 μl of PBS–BSA, etc.).

6. Incubate the dish at 37° with gentle shaking for at least 30 min.

7. Empty wells and wash three times with PBS. (Fill wells completely and empty vigorously each time.)

8. Quickly add 200 μl of a 1/1000 dilution of rabbit antitoxin serum in PBS–BSA, to each well of the dish.

9. Incubate as in Step 6.

10. Empty wells and wash as described in Step 7.

11. Quickly, add 200 μl of a 1/5000 dilution of anti-rabbit immunoglobulin G (IgG)–alkaline phosphatase (Boehringer Mannheim, Cat. No. 605 230) in PBS–BSA, to each well of the dish.

12. Incubate as in Step 6.

13. Wash as in Step 7.

14. Per well, add 200 μl of 1.0 M Tris, pH 8.0, containing 2 mg/ml p-nitrophenyl phosphate (Sigma).

15. Incubate at room temperature until the yellow color develops (usually less than 30 min).

16. The assay can be stopped by addition of 50 μl of 0.1 M K_2PO_4 per well.

17. Well color development may be read with an ELISA reader (Series 700 Microplate Reader, Cambridge Technology, Inc., Cambridge, MA), or end points may be marked by eye and the concentration calculated by comparison with toxin control standards. (Pure cholera toxin can be purchased from List Biological Laboratories, Inc., Campbell, CA).

18. Units can be defined as micrograms toxin per OD_{600} unit of initial overnight culture cell density or as micrograms per milliliter of culture supernatant.

[42] Posttranslational Processing of Type IV Prepilin and Homologs by PilD of *Pseudomonas aeruginosa*

By MARK S. STROM, DAVID N. NUNN, and STEPHEN LORY

Introduction

Pseudomonas aeruginosa is an opportunistic pathogen that causes serious and often fatal systemic infections in people with severe burns or immunocompromising conditions, as well as chronic pulmonary infections in those with cystic fibrosis.[1] A number of factors produced by *P. aeruginosa* have been shown to contribute to virulence. These include pili, which mediate attachment to epithelial cells,[2,3] and a number of extracellularly secreted hydrolytic enzymes and toxins, which contribute to tissue destruction.[4]

The monomeric protein subunit of *P. aeruginosa* pili, pilin, is a member of the Type IV *N*-methylphenylalanine class,[5] which includes pilins from *Neisseria gonorrhoeae*,[6] *N. meningiditis*,[7] *Moraxella bovis*,[8] *Dichelobacter* (formerly *Bacteroides*) *nodosus*,[9] and *Vibrio cholerae*.[10] The precursor forms of the pilins undergo two distinct posttranslational modification steps: first the leader sequence is removed, followed by N-methylation of the amino-terminal residue of the mature polypeptide after cleavage of the leader sequence.

The assembly of functional pili requires the activities of three proteins, PilB, PilC, and PilD.[11] PilD is an endopeptidase required for the cleavage of the leader sequence from the *P. aeruginosa* pilin (PilA) monomer as well as a group of related proteins, PddA–D, which are components for

[1] G. P. Bodey, R. Bolivar, V. Fainstein, and L. Jadeja, *Rev. Infect. Dis.* **5**, 279 (1983).
[2] D. E. Woods, D. C. Straus, W. J. Johanson, V. K. Berry, and J. A. Bass, *Infect. Immun.* **29**, 1146 (1980).
[3] E. Chi, T. Mehl, D. Nunn, and S. Lory, *Infect. Immun.* **59**, 822 (1991).
[4] P. V. Liu, *J. Infect. Dis.* **116**, 112 (1966).
[5] K. Johnson, M. L. Parker, and S. Lory, *J. Biol. Chem.* **261**, 15703 (1986).
[6] T. F. Meyer, E. Billyard, R. Haas, S. Storzbach, and M. So, *Proc. Natl. Acad. Sci. U.S.A.* **81**, 6110 (1984).
[7] W. J. Potts and J. R. Saunders, *Mol. Microbiol.* **2**, 647 (1988).
[8] C. F. Marrs, G. Schoolnik, J. M. Koomey, J. Hardy, J. Rothbard, and S. Falkow, *J. Bacteriol.* **163**, 132 (1985).
[9] N. M. McKern, I. J. O'Donnell, A. S. Inglis, D. J. Stewart, and B. L. Clark, *FEBS Lett.* **164**, 149 (1983).
[10] C. E. Shaw and R. K. Taylor, *Infect. Immun.* **58**, 3042 (1990).
[11] D. Nunn, S. Bergman, and S. Lory, *J. Bacteriol.* **172**, 2911 (1990).

METHODS IN ENZYMOLOGY, VOL. 235

an extracellular protein secretion apparatus.[12,13] *Pseudomonas aeruginosa* mutants in *PilD* not only show reduced adherence to epithelial cells, but they are also unable to export exotoxin A, phospholipase C, elastase, and alkaline phosphatase to the extracellular milieu. Instead, these enzymes accumulate in the periplasmic space.[14] PilD was independently isolated as XcpA, in which mutations were also shown to have a pleiotropic export defect.[15] Additional Xcp proteins identical to PddA–D (XcpT–W) have also been reported.[16] The Xcp nomenclature will be used for the remainder of this report.

The prepilin and precursors of Xcp substrates of PilD range in size from 14 to 26 kDa and are characterized by a short, 6 or 8 amino acid, basically charged leader peptide with cleavage occurring between the glycine and phenylalanine residues, in the sequence Gly ↓ Phe-Thr-Leu-(Leu,Ile)-Glu (Fig. 1A). The 16–18 amino acids immediately following the glutamate residue are conserved in hydrophobic character and are thought to play a role in membrane localization and subunit interactions. The sequences of the Type IV pilins and Pdd proteins diverge following this hydrophobic region.

A comparison of the PilD amino acid sequence to protein databases has shown a high degree of homology to PulO of *Klebsiella oxytoca,*[17] ComC of *Bacillus subtilis,*[18] and TcpJ of *V. cholerae*[19] (Fig. 2). Like PilD, TcpJ appears to be a prepilin peptidase, as mutations in the *tcpJ* gene lead to an accumulation of the TCP prepilin. PulO is one of a number of proteins involved in the expression and extracellular localization of pullulanase, a lipoprotein which degrades complex starches.[20] Four other proteins required for pullulanase export, PulG–J, have amino termini that are homologous to the Type IV pilins and Xcp proteins (Fig. 1B).[21] Processing of at least one of the pilinlike proteins, PulG,[17] has been shown to be dependent on the expression of the *pulO* gene product. The export of proteins in at least two other organisms, *Erwinia chrysanthemi*[22] and

[12] D. N. Nunn and S. Lory, *Proc. Natl. Acad. Sci. U.S.A.* **88,** 3281 (1991).
[13] D. N. Nunn and S. Lory, *Proc. Natl. Acad. Sci. U.S.A.* **89,** 47 (1992).
[14] M. S. Strom, D. Nunn, and S. Lory, *J. Bacteriol.* **173,** 1175 (1991).
[15] M. Bally, G. Ball, A. Badere, and A. Lazdunski, *J. Bacteriol.* **173,** 479 (1991).
[16] M. Bally, A. Filloux, M. Akrim, G. Ball, A. Lazdunski, and J. Tommassen, *Mol. Microbiol.* **6,** 1121 (1992).
[17] A. P. Pugsley and B. Dupuy, *Mol. Microbiol.* **6,** 751 (1992).
[18] S. Mohan, J. Aghion, N. Guillen, and D. Dubnau, *J. Bacteriol.* **171,** 6043 (1989).
[19] M. R. Kaufman, J. M. Seyer, and R. K. Taylor, *Genes Dev.* **5,** 1834 (1991).
[20] A. P. Pugsley, C. d'Enfert, I. Reyss, and M. G. Kornacker, *Annu. Rev. Genet.* **24,** 67 (1990).
[21] I. Reyss and A. P. Pugsley, *Mol. Gen. Genet.* **222,** 176 (1990).
[22] S. Y. He, M. Lindeberg, A. K. Chatterjee, and A. Collmer, *Proc. Natl. Acad. Sci. U.S.A.* **88,** 1079 (1991).

A

▼
P. aeruginosa	MetLys	AlaGlnLysGly	PheThrLeuIleGluLeuMetIleValVal-
M. bovis	MetAsn	AlaGlnLysGly	PheThrLeuIleGluLeuMetIleValIle-
N. gonorrhoeae	MetAsnThrLeuGlnLysGly		PheThrLeuIleGluLeuMetIleValIle-
B. nodosus	MetLysSerLeuGlnLysGly		PheThrLeuIleGluLeuMetIleValIle-
V. cholerae	Met-(21aa)--GlnGluGly		MetThrLeuLeuGluValIleIleValLeu-

P. aeruginosa

PddA	LeuGlnArgArgGlnGlnSerGly	PheThrLeuIleGluIleMetValValVal-
PddB	MetArgAlaSerArgGly	PheThrLeuIleGluLeuMetValValMet-
PddC	MetLysArgAlaArgGly	PheThrLeuLeuGluValLeuValAlaLeu-
PddD	MetArgLeuGlnArgGly	PheThrLeuLeuGluLeuLeuLeuIleAlaIle-

B

K. oxytoca

PulG	MetGlnArgGlnArgGly	PheThrLeuLeuGluIleMetValValIle-
PulH	ValArgGlnArgGly	PheThrLeuLeuGluMetMetLeuIleLeu-
PulI	MetLysLysGlnSerGly	MetThrLeuIleGluValMetValAlaLeu-
PulJ	MetIleArgArgSerSerGly	PheThrLeuValGluMetLeuLeuAlaLeu-

B. subtilis
comG ORF3

ORF3	MetAsnGluLysGly	PheThrLeuValGluMetLeuIleValLeu-
ORF4	MetAsnIleLysAsnGluGluLysGly	PheThrLeuLeuGluSerLeuLeuValLeu-
ORF5	MetTrpArgGluAsnLysGly	PheSerThrIleGluThrMetSerAlaLeu-

FIG. 1. Comparison of the amino-terminal sequences of the Type IV pilins and proteins involved in macromolecular transport. (A) Amino-terminal sequences of the Type IV pilins[5,6,8-10] and the Pdd proteins[13] required for extracellular protein secretion by *P. aeruginosa*. (B) Amino-terminal sequences of proteins required for extracellular secretion of pullulanase by *K. oxytoca* (PulG–J) [I. Reyss and A. P. Pugsley, *Mol. Gen. Genet.* **222,** 176 (1990)] and for competence by *B. subtilis* (ComG open reading frames 3–5) [M. Albano, R. Breitling, and D. A. Dubnau, *J. Bacteriol.* **171,** 5386 (1989); R. Breitling and D. Dubnau, *J. Bacteriol.* **172,** 1499 (1990)].

Xanthomonas campestris,[23] has been shown to be dependent on the expression of proteins with pilinlike amino termini, although gene products homologous to PilD have yet to be described. ComC and three proteins having Type IV pilinlike amino termini (ComG ORF3–5) are required for DNA uptake by *B. subtilis*,[24,25] although cleavage of the ComG proteins by ComC has yet to be demonstrated. In light of these observations, PilD can be considered a prototype of a group of endopeptidases responsible not only for processing the Type IV pilin precursors, but also for the processing of proteins required for macromolecular translocation.

[23] F. Dums, J. M. Dow, and M. J. Daniels, *Mol. Gen. Genet.* **229,** 357 (1991).
[24] M. Albano, R. Breitling, and D. A. Dubnau, *J. Bacteriol.* **171,** 5386 (1989).
[25] R. Breitling and D. Dubnau, *J. Bacteriol.* **172,** 1499 (1990).

```
PilD   ....MPLL.DYLASHPLAFVLCAILL...GLLVGSFLNVVVHRLP..KMM
PulO   MVENIALLPEFAAQYP..FLWGSFLF.LSGLAFGSFFNVVIHRLP..LMM
TcpJ   .......M.EYV......YL...ILFSIVSLILGSFSNVVIYRLPRKILL
ComC   .................ML.SILF.IFGLILGSFYYTAGCRIP....L

PilD   ERNWKAEAREALGLEPEPKQATYNLVLPNSACPRCGHEI.RPWENIPLVS
PulO   E...QAE.....GI........NLCFPASFCPQCREPIAW.RDNIPLLG
TcpJ   KNHF......FYDIDS.......N....RSMCPKCGNKISW.YDNVPLLS
ComC   H.........LSI...........IAPRSSCPFCRRTLT.PAELIPILS

PilD   YLALGGKCSSCKAAIGKRYPLVELAT.AL..LSGYV.AWHFGFTWQA.GA
PulO   FLFLKGRSRCCGQPISPRYPLMELATGALFVLAGYLMA..PGVPL.L.GG
TcpJ   YLLLHGKCRHCDEKISLSYFIVEL...SFFIIA.FPI.YWLSTDW.V.DS
ComC   FLFQKGKCKSCGHRISFMYPAAELVTACLFAAAGI....RFGISLELFPA

PilD   MLLLT..WGLLAMSLIDADHQLLPD.VLVLPLLWLGLIA...NHFGLFAS
PulO   LILLS..L.LLILAAIDAQTQLLPD.GLTLPLMWAGLLF...NLSATYVP
TcpJ   FVLLGLYFILFNLFVIDFKSMLLPN.LLTYPIFMLAFIYVQPNQ.AL..T
ComC   VVFISL...LIIVAVTDIHFMLIPNRIL...IFFLPFLAAAR....LISP

PilD   LDD...ALFGA....VFGYLSLWSVFWLFKLVTGKEGMGYGDFKLLAMLG
PulO   LAE...AVVGA....MAGYLSLWSVYWVFRLLSGKEALGYGDFKLLAALG
TcpJ   VES...SIIGGFAAFIITYVSNFIV.RLFKRI...DVMGGGDIKLYTAIG
ComC   LDSWYAGLLGA....AAGFL.FLAV...IAAIT.HGGVGGGDIKLFAVIG

PilD   AWGGWQILPLTILLSSLVGAILGVIMLRLRNAESGT..PIPFGPYLAIAG
PulO   AWLGWQALPQTLLLAS.PAA
TcpJ   TLIGVEFVPYLFLLSSII.AFIHWFFARV.SCRYCL..YIPLGPSI.IIS
ComC   FVLGVKMLAAAFFFSVLIGALYG..AARVLTGRLAKRQPLPFAPAIAAGS

PilD   WIALLWGDQITRTYLQFAGFK
PulO
TcpJ   FVIVFFSIRLM
ComC   ILAYLYGDSIISFYIKMALG
```

FIG. 2. Comparison of amino acid sequences of PilD with the other Type IV prepilin peptidase-like proteins PulO,[17] ComC,[18] and TcpJ,[19] from *K. oxytoca*, *B. subtilis*, and *V. cholerae*, respectively. The conserved cysteine residues are in boldface type and underlined. The alignment was performed using the Wisconsin GCG Pileup program.

Previous reports from our laboratory have described the initial characterization, purification, and kinetics of the PilD endopeptidase.[12,26] Here we describe the approaches used in the isolation and purification of PilD, the development of an optimum *in vitro* assay for PilD activity, and the use of the assay to characterize the kinetics of the enzyme. These methods should be applicable to PilD counterparts in other species.

[26] M. S. Strom and S. Lory, *J. Bacteriol.* **174,** 7345 (1992).

Wt PilD1 PilD2 PilC PilB

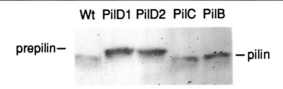

prepilin—   —pilin

FIG. 3. Immunoblot using antisera against pilin of whole-cell extracts of wild-type (Wt) *P. aeruginosa* and against strains carrying transposon Tn*5* insertions in the genes encoding PilD (PilD1, PilD2), PilC, and PilB.[11]

Source of Prepilin Substrates for Leader Peptidase Assays

In *P. aeruginosa* the 4-kilobase (kb) region upstream of the *pilA* contains three genes, *pilB, pilC,* and *pilD,* that encode proteins involved in pilus biogenesis.[11] Membranes are prepared from mutants in each of the three genes, and proteins are separated by sodium dodecyl sulfate–polyacrylamide gel electrophoresis (SDS–PAGE), blotted to nitrocellulose, and probed with antiserum against PilA. As seen in Fig. 3, levels of pilin expression and membrane localization are found to be similar to wild type for each of the mutant classes. Pilin from the *pilD* mutant, however, is of a higher molecular weight form than that of the wild type, suggesting that the *pilD* gene product is required for the cleavage of the leader sequence from the precursor prepilin.[11] Mutants of *P. aeruginosa* in *pilD,* therefore, are a suitable source of unprocessed pilin for biochemical studies. Alternatively, expression of *pilA* in bacterial species lacking Type IV pilin and a homologous export machinery could potentially be used to provide starting material for purification of substrates for leader peptidase cleavage.

The method of choice for preparation of prepilin substrates is by overexpression of *pilA* in *pilD* mutants of *P. aeruginosa* strain PAK. A subclone of the *pilA* gene in the expression vector pMMB66EH[27] yielded pMS*tac*27PD. Induction of the *tac* promoter by addition of isopropylthiogalactoside (IPTG) leads to a dramatic overexpression and accumulation of pilin in the cell membranes. Plasmid pMS*tac*27PD is introduced into both a wild-type and *pilD* mutant strain of PAK, and total membranes are prepared from induced cultures of each. In both cases pilin or putative prepilin constitutes at least 50% of the total membrane protein and is easily purified by gel filtration of membranes solubilized by SDS (see below). To confirm that the pilin isolated from *pilD* mutants indeed contains its leader sequence, the purified proteins are blotted to a polyvinyli-

[27] M. S. Strom and S. Lory, *J. Biol. Chem.* **266,** 1656 (1991).

dene difluoride (PVDF) membrane,[28] and the amino-terminal protein sequence is determined for both. The sequence of the pilin synthesized in the wild-type strain, N-methylPhe-Thr-Leu-Leu-Glu-, is as expected for the mature form. In contrast, the pilin isolated from the *pilD* mutant starts instead with the sequence Met-Lys-Ala-Gln-Lys-, as predicted from the nucleotide sequence for the pilin precursor.[12]

Assay for Prepilin Peptidase Activity

To characterize the prepilin peptidase activity, an *in vitro* prepilin peptidase assay has been developed.[12] Substrate for the assay is prepared by overexpressing prepilin from pMStac27PD, in a *pilD* mutant background. PAK *pilD* (pMStac27PD) is grown on minimal A medium containing 50 mM monosodium glutamate/1% (v/v) glycerol supplemented with 0.5% (w/v) casamino acids and 100 μM IPTG. Cultures are harvested at stationary phase, resuspended in 100 mM triethanolamine hydrochloride, pH 8.0, 500 mM NaCl (1/50 original culture volume). Cells are broken by ultrasonication. After a low-speed centrifugation (3000 g, 10 min, 5°) to remove unbroken cells, the membrane-containing supernatant is subjected to high-speed centrifugation (200,000 g, 1 hr, 5°) to pellet total membranes. The membranes are resuspended in 25 mM triethanolamine hydrochloride (Sigma Chemical Co., St. Louis, MO), pH 7.5, 10% (w/v) glycerol, recentrifuged (200,000 g, 1 hr, 5°), and resuspended in the same buffer to approximately 25 mg/ml total protein (crude or membrane-associated substrate). Purification of the substrate is accomplished by solubilizing the membrane fraction with 2% (w/v) SDS followed by gel filtration on a Superose 12 HR10/30 fast protein liquid chromatography (FPLC) column equilibrated with 25 mM triethanolamine hydrochloride, pH 7.5/1% (w/v) SDS. The prepilin-containing fractions, as determined by SDS–PAGE, are precipitated with acetone and resuspended in water to a protein concentration of 10 mg/ml.

For initial characterization of prepilin peptidase activity, soluble (cytoplasmic and periplasmic) and membrane fractions are prepared by freeze–thaw lysis of lysozyme-treated cells of *P. aeruginosa* PAK-NP. This strain has an insertion in the *pilA* gene and makes no detectable pilin protein as determined by Western blot analysis. Fractions prepared in this manner are mixed with 2 μl of the crude substrate. The mixtures are solubilized with various detergents, and the reaction is carried out using different temperature and pH conditions. The reaction is stopped by the addition of SDS–PAGE sample buffer [0.125 M Tris-HCl, pH 6.8/4% SDS/5% (v/v) 2-mercaptoethanol/10% glycerol/0.02% bromphenol blue], and a portion of the reaction is analyzed by SDS–PAGE.

[28] P. Matsudaira, *J. Biol. Chem.* **262**, 10035 (1987).

Initial experiments show that prepilin cleavage activity is found only in the membrane fraction and is optimal at 37°, pH 7.5, with 0.5% Triton X-100 as the solubilizing detergent. Higher concentrations of Triton X-100 (>2%), however, have an inhibitory effect on cleavage activity. A number of other nonionic detergents, including octanoyl-N-methylglucosamide, n-dodecyl-β-D-maltoside,   3-[(3-cholamidopropyl)dimethylammonio]-1-propane sulfonate (CHAPS), and 3-[(3-cholamido-propyl)dimethylam-monio]-2-hydroxy-1-propane sulfonate (Boehringer Mannheim, Indianapolis, IN) (CHAPSO), are ineffective. Concentrations of SDS as low as 0.01% completely inhibit prepilin cleavage, and the effect is not reversible by dilution into buffer containing 1% Triton X-100. In addition, it is necessary to add acidic phospholipids (cardiolipin or phosphatidylglycerol) to the reaction for optimal cleavage. In all subsequent experiments the reaction conditions are as follows: 1 to 6 μl of pilin substrate mixed with 1 μl of 0.5% (w/v) cardiolipin (Sigma, St. Louis, MO), 1 μl of sample to be tested for activity, and water to a volume of 8 μl. The reactions are started by solubilization of the components with the addition of 2 μl of 5× assay buffer [125 mM triethanolamine hydrochloride, pH 7.5/2.5% (v/v) Triton X-100] and incubated at 37° for 10 min to 1 hr. After the incubation, the reactions are stopped with 30 μl sample buffer. Samples are then boiled 5 min followed by electrophoresis of 5-μl aliquots on SDS/tricine/15% polyacrylamide gels. Maximal separation of precursor and mature pilin is obtained using 0.75 mm × 7 cm × 8.5 cm gels run at a constant 35 mA for 3 hr. Gels are then stained with 50% ethanol/10% (v/v) acetic acid/0.2% (w/v) Coomassie Blue R-250 followed by destaining with 10% (v/v) methanol/10% (v/v) acetic acid.

A typical assay, demonstrating the cleavage of a constant amount of membrane-associated prepilin with PilD (from membranes isolated from *P. aeruginosa* containing the *pilD* gene on the expression vector pMMB66EH as described below) over time, is shown in Fig. 4. As shown, more than 75% of the prepilin is converted to pilin in approximately 10 min. Under these conditions, PilD cleavage of prepilin is linear as a function of enzyme concentration and time when the substrate is at saturating levels.

Purification and Properties of PilD

Using the assay described above, it has been determined that membranes purified from a strain of PAK-NP containing a PilD-overexpressing plasmid (pRBS-L)[12] are 30- to 40-fold more active, per milligram protein, in cleaving prepilin to pilin, as compared to membranes isolated from wild-type *P. aeruginosa*. Moreover, the same plasmid, when introduced into *Escherichia coli* DH5α, confers detectable prepilin peptidase activity

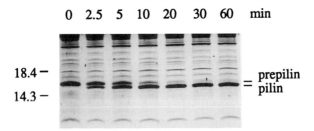

FIG. 4. Coomassie blue-stained Tricine–SDS–polyacrylamide gel of *in vitro* cleavage reactions using membrane-associated PilD and prepilin substrate over a time course. Membranes containing the two reactants were mixed in the presence of 0.5% Triton X-100 as described in the text and incubated at 37° for the times indicated. The molar ratio of substrate to enzyme was 500 : 1. The numbers on the left-hand side indicate the positions of molecular weight markers ($\times 10^{-3}$).

to membranes of the recombinants, whereas membranes prepared from *E. coli* DH5α containing only the corresponding vector are unable to cleave prepilin. The specific activity of leader peptidase recovered from membranes of *E. coli* overexpressing *pilD* is significantly lower than the activity recovered from *P. aeruginosa* carrying the same plasmids, suggesting that expression of *pilD* or the activity of enzyme is suboptimal in a bacterial host other than *P. aeruginosa*.

The purification of PilD is accomplished by detergent solubilization of membranes of *P. aeruginosa* (pRBS-L) and immunoaffinity chromatography. From the deduced PilD amino acid sequence, a hydrophilic region of 20 amino acids has been selected for the design of a synthetic peptide (Fig. 5). The peptide is synthesized to include an additional cysteine

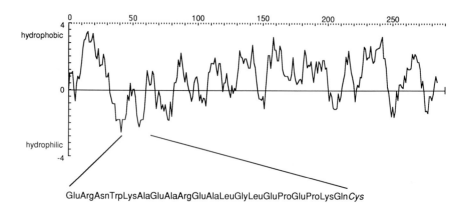

FIG. 5. Hydrophobicity plot of PilD and location and sequence of synthetic peptide used for immunization of rabbits to raise antisera.

residue at the C terminus for coupling through the thiol group. For the production of polyclonal antibodies to the peptide, the peptide is first coupled to keyhole limpet hemocyanin (KLH) using the heterobifunctional reagent *m*-maleimidobenzoyl-*N*-hydroxysuccinimide ester (MBS). Two New Zealand White rabbits are initially injected intramuscularly with 1 mg of the peptide–KLH conjugate and then boosted at 1-month intervals.

Immunoglobulin G (IgG) is purified from antisera by ammonium sulfate precipitation and DEAE chromatography. The PilD-derived peptide is conjugated to a Sulfo-link column (Pierce, Rockford, IL) through the C-terminal cysteine residue. Total IgG is applied to the affinity column and the application repeated three times. After extensive washing with phosphate-buffered saline (PBS: 50 mM sodium phosphate, pH 7.2, 150 mM NaCl), PilD-specific IgG is eluted from the column with 100 mM glycine hydrochloride, pH 2.5, immediately neutralized with 1/10 volume of 1 M Tris-HCl, pH 8.0, and dialyzed against PBS. Two milligrams of the PilD-specific IgG preparation is adsorbed to 1 ml of protein A–Sepharose CL-4B (Sigma) in PBS and cross-linked to the matrix with dimethyl pimelimidate.

Three milliliters of total membranes (25 mg/ml protein) prepared from *P. aeruginosa* PAK-NP (pRBBS-L), is solubilized by the addition of an equal volume of 25 mM triethanolamine hydrochloride, pH 7.5, 10% (v/v) glycerol, and 4% (v/v) Triton X-100. After centrifugation at 150,000 g for 1 hr, at 5°, the clarified supernatant is applied to a DEAE-Sephacel (Sigma) column (3 ml bed volume), washed with 25 mM triethanolamine hydrochloride, pH 7.5, 10% (v/v) glycerol, 1% (v/v) Triton X-100, and eluted with the same buffer containing 200 mM NaCl. Three milliliters of the protein eluted from the DEAE-Sephacel column is incubated with 1 ml of the anti-PilD IgG–protein A–Sepharose conjugate in 25 mM triethanolamine hydrochloride, pH 7.5, 10% (v/v) glycerol, 1% (v/v) Triton X-100, and eluted with the same buffer containing 200 mM NaCl. Three milliliters of the protein eluted from the DEAE-Sephacel column is incubated with 1 ml of the anti-PilD IgG–protein A–Sepharose conjugate in 25 mM triethanolamine hydrochloride, pH 7.5, 10% (v/v) glycerol, 1% (v/v) Triton X-100 overnight at 4° with gentle agitation. After the incubation, the suspension is placed in a 10 × 70 mm plastic column and washed with buffer [25 mM triethanolamine hydrochloride, pH 7.5, 10% (v/v) glycerol, 1% (v/v) Triton X-100, and 200 mM NaCl]. Bound protein is eluted with 100 mM glycine hydrochloride, pH 2.5, 1% Triton X-100 and immediately neutralized with 1/5 volume of 1 M triethanolamine hydrochloride, pH 7.5. The eluted protein is stored at −20° in 50% (w/v) glycerol. A quantitation of prepilin peptidase activity, and the resultant yield and enrichment during the course of purification, is shown in Table I. The

TABLE I
PURIFICATION OF PilD

Fraction	Total protein (mg)	Units[a]	Yield (%)	Specific activity (U/mg)	Enrichment (-fold)
Total membranes	75.0	220,000	100.0	2933	—
Clarified supernatant	33.6	192,000	87.3	5714	2.0
DEAE eluate	24.3	192,000	87.3	7901	2.7
Immunoaffinity eluate	0.08	32,000	14.5	400,000	109.0

[a] One unit of activity was arbitrarily defined as that required to cleave 50% of the substrate in 30 min at 37°.

final product migrates as a single-band polypeptide when analyzed by SDS–PAGE.

Because of the inherent difficulties in the purification of integral membrane proteins by conventional methods, the peptide antibody-based immunoaffinity technique remains the method of choice for obtaining PilD-like enzymes in pure form. Other Type IV prepilin leader peptidases, shown in Fig. 1, contain reasonable stretches of polar or charged amino acids which could be used to design synthetic peptides to generate antibodies for immunoaffinity purification.

A variety of known protease inhibitors and sulfhydryl reagents were tested for the ability to inhibit prepilin cleavage by PilD. The protease inhibitors phenylmethylsulfonyl fluoride (PMSF) and N-tosyl-L-phenylalanine chloromethyl ketone (TPCK; serine protease inhibitors), ethylenediaminetetraacetic acid (EDTA; metalloprotease inhibitor), and pepstatin (acid protease inhibitor) were tested and had little or no effect on prepilin cleavage. The additions of the sulfhydryl reagents N-ethylmaleimide (NEM) or p-chloromercuribenzoate (PCMB) did significantly affect cleavage activity, although the PCMB inhibition could be relieved by the subsequent addition of dithiothreitol (DTT).

Kinetics of PilD Activity in Vitro

The kinetics of prepilin cleavage by PilD have been determined in vitro under a number of conditions.[26] These include combinations of both membrane-bound and purified PilD with both crude and purified prepilin substrate. The amount of PilD in the membrane-bound fractions is determined by comparison of samples to purified PilD of known concentration on immunoblots. Varying concentrations of PilD, substrate concentrations, and time are used for initial characterizations of the cleavage reaction. Enzyme and substrate concentrations that give cleavage reactions which are linear as a function of time and enzyme concentration are used for subsequent kinetic analyses. After SDS–PAGE of an aliquot of the reaction, the resulting Coomassie blue-stained prepilin and pilin bands are scanned with a densitometer. The areas of the peaks resulting from the scan are then used to measure the fraction of precursor cleaved, a value which is used to calculate the velocity of each reaction. All analyses are carried out by measuring the initial reaction velocities with a minimum of six different substrate concentrations and at least three separate reactions per concentration. Values for K_m (μM) and V_{max} ($\mu mol/min/\mu mol$ enzyme) are then determined from substrate concentrations and measured velocities, using standard kinetic calculations.

In this manner, measured apparent values for K_m and V_{max} (or K_{cat}) have been determined for PilD and its cleavage of prepilin. As seen in Table II, the highest turnover rate (K_{cat}) is obtained using both membrane-bound enzyme and substrate (180 min^{-1}), whereas K_m values are approximately equivalent and within experimental error under all conditions. It is not known at this time whether this difference is due to a missing cofactor in the purified preparations, which is needed for optimal enzyme activity, or simply due to a loss of activity or a change in conformation of the enzyme which occurs during the purification procedure and subsequent storage.

The kinetics of cleavage of prepilin from *N. gonorrhoeae* has also been tested using substrate prepared by overexpressing the cloned *N. gonorrhoeae pilE* gene in the *pilD* mutant of *P. aeruginosa*. As shown in

TABLE II
KINETICS OF PREPILIN CLEAVAGE BY PilD

PilD enzyme	Substrate[a]	K_{cat} (min^{-1}) ± SE	K_m(μM) ± SE
Purified	wt, purified	0.9 ± 0.3	184 ± 109
Purified	wt, membrane-associated	9 ± 2	393 ± 177
Membrane-associated	wt, purified	54 ± 4	459 ± 70
Membrane-associated	wt, membrane-associated	180 ± 103	655 ± 512
Membrane-associated	F^{+1}M, membrane-associated	47 ± 8	430 ± 92
Membrane-associated	F^{+1}S, membrane-associated	13 ± 4	166 ± 96
Membrane-associated	F^{+1}N, membrane-associated	6 ± 1	53 ± 14
Membrane-associated	F^{+1}C, membrane-associated	0.7 ± 0.4	176 ± 160
Membrane-associated	wt, *N. gonorrhoeae*	199 ± 36	585 ± 204

[a] wt, Wild-type prepilin from *P. aeruginosa* PAK; F^{+1}M, F^{+1}S, F^{+1}N, and F^{+1}C are prepilin substrates where the Phe at the +1 position is changed to Met, Ser, Asn, and Cys, respectively.

Table II, the measured kinetics of cleavage of this Type IV prepilin by PilD is identical within experimental error to that seen for *P. aeruginosa* prepilin. Therefore, the differences in length and net charge of the *N. gonorrhoeae* prepilin leader sequence (Fig. 1A), do not affect the overall rate of cleavage by PilD.

Cleavage of Substrates Altered at +1 Position

Previous experiments directed at identifying the recognition sequence for PilD have shown that substitutions for the phenylalanine residue at the +1 position relative to the cleavage site in prepilin are well tolerated *in vivo*.[27] A number of the mutated substrates have been subjected to kinetic analysis. In general, substitutions at this site result in a lowered K_{cat} without affecting the affinity of PilD for the substrate (Table II). A methionine residue in place of the phenylalanine results in a K_{cat} closest to that measured for wild-type substrate, which is interesting from the standpoint that this is the only other amino acid found at this position in substrates processed by the PilD class of endopeptidases (Fig. 1). One substitution, an asparagine, results in a greatly decreased K_{cat} with a significantly lower K_m, indicating a higher affinity. Identification of such substrates that have a higher affinity for PilD but that are hydrolyzed at a slower rate could aid in the design of peptide-based inhibitors of the enzyme.

PilD as *N*-Methyltransferase as well as Peptidase

As mentioned previously, the amino-terminal residue of the Type IV pilins is *N*-methylated following cleavage of the leader sequence. We have shown that PilD is able to methylate *in vitro P. aeruginosa* (Fig. 6) and *N. gonorrhoeae* pilins, using radiolabeled *S*-adenosyl-L-methionine (Ado-

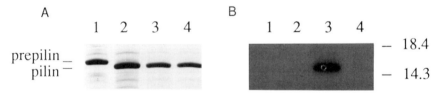

FIG. 6. *In vitro* methylation of pilin by PilD using ³H-labeled *S*-adenosyl-L-methionine as the methyl donor. (A) Coomassie-stained gel: lane 1, prepilin with no PilD; lane 2, prepilin with PilD, after *in vitro* cleavage reaction; lane 3, prepilin, PilD, with adenosyl-L-methionine, *S*-[methyl-³H]; lane 4, reactants in lane 3 in the presence of 10 μM sinefungin. (B) Fluorogram of same gel shown in (A).

Met) as the methyl donor.[29] Only the cleaved form of pilin is methylated, suggesting that the amino-terminal residue of the mature pilin is the site of this modification. Preliminary characterizations demonstrate that methylation requires the presence of reducing agents such as dithiothreitol, indicating that cysteine residues of PilD play a role in the reaction. In fact, two pairs of cysteine residues, at positions 72, 75 and 97, 100 of the 290 amino acid protein, lie within the region of highest homology between PilD and PulO, ComC, TcpJ, suggesting these proteins may also be methyltransferases as well as peptidases (Fig. 2). Further work utilizing site-directed mutagenesis of the codons for these cysteine residues has shown them to be required for full peptidase and methyltransferase activities, and may form part of the methyl donor binding site.[30]

A number of Ado-Met analogs have been tested for the ability to inhibit PilD methyltransferase activity, including sinefungin, S-adenosyl-L-homocysteine (Ado-Cys), and S-adenosyl-L-ethionine (Ado-Eth). Sinefungin appears to be the most potent inhibitor, whereas Ado-Cys and Ado-Eth are slightly less effective competitive inhibitors than unlabeled Ado-Met. Interestingly, sinefungin does not affect the endopeptidase activity of PilD (Fig. 6), suggesting that the sites of Ado-Met and prepilin binding are spatially separated on the protein.

Summary

We have described the characterization of a protein initially identified as having an essential function in biogenesis of polar pili of *P. aeruginosa* by processing precursors of pilin. Other findings have also expanded the range of substrates for PilD to include a set of proteins that are essential components of the extracellular secretion machinery. Direct demonstration of prepilin processing necessitates use of purified substrates and enzymes, and we present general protocols for purification of both enzymes and substrates, as well as an assay for prepilin peptidase activity. For a source of enzyme and substrates, mutants of *P. aeruginosa* defective in pilin processing as well as clones overexpressing the pilin gene and PilD were developed. These methods are applicable to other bacterial systems that express Type IV pili and/or possess the PilD-dependent machinery of extracellular protein secretion.

PilD is a bifunctional enzyme, which carries out not only cleavage but also amino-terminal methylation of the mature pilin. Cleavage and N-methylation of the pilin-like Xcp proteins involved in extracellular pro-

[29] M. S. Strom, D. N. Nunn, and S. Lory, *Proc. Natl. Acad. Sci. U.S.A.* **90**, 2404 (1993).
[30] M. S. Strom, P. Bergman, and S. Lory, *J. Biol. Chem.* **268**, 15788 (1993).

tein secretion have also been shown to be dependent on PilD.[29,31] The leader peptidase activity of PilD is inhibited by sulfhydryl blocking reagents such as NEM and PCMB, whereas the methyltransferase activity of the purified enzyme is dependent on reduction with dithiothreitol. The conserved region containing the cysteine residues lies within the largest hydrophilic domain of the protein as predicted from hydrophobicity analysis, and it is probably exposed to the cytoplasmic side of the cytoplasmic membrane. Identification of the active site residues involved in recognition of the substrates for processing and subsequent methylation is currently underway. Studies on substrate specificities of PilD, with respect to its leader peptidase and methyltransferase activity, may prove to be useful in designing inhibitors which would interfere with maturation of Type IV prepilins and components of the extracellular protein secretion machinery. In light of the fact that an increasing number of both mammalian and plant pathogens are being shown to have extracellular secretion pathways homologous to that seen for *P. aeruginosa,* such inhibitors may be useful tools in the study of the role these peptidases play in bacterial virulence.

Acknowledgments

Work on the isolation and biochemical characterization of PilD from our laboratory was supported by U.S. Public Health Service Grant AI21451 from the National Institutes of Health.

[31] D. N. Nunn and S. Lory, *J. Bacteriol.* **175,** 4375 (1993).

Section VI

Enzyme and Toxin Assays

[43] Bacterial Immunoglobulin A₁ Proteases

By MARTHA H. MULKS and RUSSELL J. SHOBERG

Introduction

The immunoglobulin A_1 (IgA_1) proteases are extracellular bacterial enzymes that specifically cleave human IgA of the IgA_1 subclass to yield intact Fabα and Fcα fragments.[1-4] Bacteria which produce IgA_1 proteases include the causative agent of gonorrhea, *Neisseria gonorrhoeae;* the three most common causative agents of bacterial meningitis, *Neisseria meningitidis, Haemophilus influenzae,* and *Streptococcus pneumoniae;* a variety of oral microorganisms including *Streptococcus sanguis, S. mitis, S. oralis, Prevotella* (*Bacteroides*) sp., and *Capnocytophaga* sp.; as well as *Ureaplasma urealyticum* and *Clostridium ramosum.* Because many of the organisms that produce IgA_1 proteases are mucosal pathogens of humans and IgA_1 proteases specifically degrade the major class of immunoglobulin found in mucosal secretions, the production of IgA_1 proteases has been postulated to be a virulence factor for these bacteria.

Each IgA_1 protease cleaves a single peptide bond within the hinge region of human IgA_1 (Fig. 1). Several of the species which produce IgA_1 proteases, including *N. gonorrhoeae,*[5] *N. meningitidis,*[6] and *H. influenzae,*[7] produce at least two different types of IgA_1 proteases, as defined by the specific peptide bond cleaved, although, with very few exceptions, a given isolate produces only one type of protease.

The genes encoding IgA_1 proteases from *N. gonorrhoeae,*[8] *H. influenzae,*[9] and *S. sanguis*[10] have been cloned and sequenced. Studies on nucleotide sequence homology and on inhibition of protease activity by specific antisera have shown that the IgA_1 proteases from *Neisseria* and *Haemo-*

---

[1] M. Kilian, J. Mestecky, and M. W. Russell, *Microbiol. Rev.* **52,** 296 (1988).
[2] S. J. Kornfeld and A. G. Plaut, *Rev. Infect. Dis.* **3,** 521 (1981).
[3] M. H. Mulks, *in* "Bacterial Enzymes and Virulence" (I. A. Holder, ed.), p. 81. CRC Press, Boca Raton, Florida, 1985.
[4] A. G. Plaut, *Annu. Rev. Microbiol.* **37,** 603 (1983).
[5] M. H. Mulks and J. S. Knapp, *Infect. Immun.* **55,** 931 (1987).
[6] M. H. Mulks, A. G. Plaut, H. A. Feldman, and B. Frangione, *J. Exp. Med.* **152,** 1442 (1980).
[7] M. H. Mulks, S. J. Kornfeld, B. Frangione, and A. G. Plaut, *J. Infect. Dis.* **146,** 266 (1982).
[8] J. Pohlner, R. Halter, K. Beyreuther, and T. F. Meyer, *Nature (London)* **325,** 458 (1987).
[9] K. Poulsen, J. Brandt, J. P. Hjorth, H. C. Thøgersen, and M. Kilian, *Infect. Immun.* **57,** 3097 (1989).
[10] J. V. Gilbert, A. G. Plaut, and A. Wright, *Infect. Immun.* **59,** 7 (1991).

A

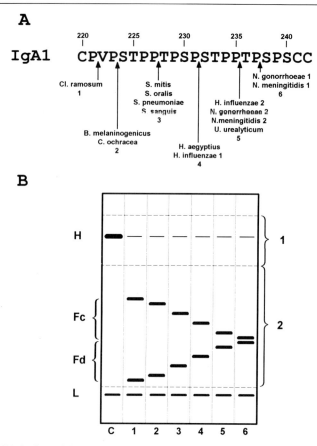

FIG. 1. (A) Amino acid sequence of the human IgA$_1$ hinge region showing peptide bonds cleaved by bacterial IgA$_1$ proteases.[1-4] (B) Diagrammatic representation of IgA$_1$ protease digests of human IgA$_1$ as seen on SDS–PAGE under reducing conditions, showing the relative mobilities of Fcα and Fdα fragments produced by IgA$_1$ proteases of different specificities. Lane C, undigested control; lanes 1–6, cleavage patterns corresponding to digestion by proteases of the six different peptide bond specificities shown in (A). H, IgA$_1$ heavy chain; Fc, Fcα fragment; Fd, heavy chain portion of the Fabα fragment; L, light chain. Horizontal dashed lines indicate where the gel would be cut for quantitative assays; section 1 (brace) contains intact heavy chain, whereas section 2 contains Fcα and Fdα fragments.

philus species are closely related at both the gene and the protein level, whereas the IgA$_1$ protease from *S. sanguis* is distinctly different.[10-14]

[11] R. Halter, J. Pohlner, and T. F. Meyer, *EMBO J.* **8,** 2737 (1989).
[12] J. M. Koomey and S. Falkow, *Infect. Immun.* **43,** 101 (1984).
[13] K. Poulsen, J. P. Hjorth, and M. Kilian, *Infect. Immun.* **56,** 987 (1988).
[14] H. Lomholt, K. Poulsen, D. A. Caugant, and M. Kilian, *Proc. Natl. Acad. Sci. U.S.A.* **89,** 2120 (1992).

Assay Procedures

Many different methods of analyzing a sample for the presence of IgA$_1$ protease have been developed, including immunoelectrophoresis,[15] sodium dodecyl sulfate–polyacrylamide gel electrophoresis (SDS–PAGE),[7] enzyme-linked immunosorbent assay (ELISA),[16,17] and high-performance liquid chromatography (HPLC).[18] All have the following in common: the enzyme sample must be reacted with human IgA$_1$ as substrate and the results of that reaction analyzed for either degradation of intact IgA$_1$ or appearance of IgA$_1$ cleavage products. Synthetic peptide homologs of IgA$_1$ protease cleavage sites have also been tested and can serve as substrates for IgA$_1$ proteases.[19] However, these peptide substrates are not available commercially or readily synthesized. Therefore, currently the only substrate used for the detection and quantitation of IgA$_1$ protease activity is human IgA$_1$.

Preparation of Substrate

Numerous procedures have been described for the purification of IgA from normal human serum, from colostrum, and from plasma, serum, or other fluids of patients with IgA myeloma (see Ref. 20 for a review). The method outlined below is relatively simple and reliable, especially when used to purify myeloma IgA$_1$ from patient sera.

Procedure

1. To 3 ml of IgA$_1$ myeloma serum, slowly add 2 ml of a saturated solution of ammonium sulfate in water, stirring constantly. Continue to stir slowly for 30 min at room temperature. Centrifuge for 20 min at 8000 g and 4°. Discard the supernatant. Wash the precipitate once with 40% saturated ammonium sulfate and resuspend the washed precipitate in 3.0 ml of 0.9% NaCl. Dialyze the sample extensively against 0.9% NaCl to remove all ammonium sulfate.

2. To the dialyzed sample (~3.0 ml), add 6 ml of 60 mM sodium acetate buffer, pH 4.8. Dropwise, add 9.0 ml of caprylic acid, stirring constantly, at room temperature. Continue to stir slowly for 30 min. Centrifuge for 20 min at 8000 g and 4°. Carefully remove the supernatant and discard the

[15] S. K. Mehta, A. G. Plaut, N. J. Calvanico, and T. B. Tomasi, Jr., *J. Immunol.* **111,** 1274 (1973).

[16] J. Reinholdt and M. Kilian, *J. Immunol. Methods* **63,** 367 (1983).

[17] M. S. Blake and C. Eastby, *J. Immunol. Methods* **144,** 215 (1991).

[18] S. B. Mortensen and M. Kilian, *J. Chromatogr.* **296,** 257 (1984).

[19] S. G. Wood and J. Burton, *Infect. Immun.* **59,** 1818 (1991).

[20] J. Mestecky and M. Kilian, this series, Vol. 116, p. 37.

precipitate. Dialyze the supernatant against 5 mM potassium phosphate buffer, pH 8.0.

3. Apply the sample to an anion-exchange column [e.g., Pharmacia (Piscataway, NJ) FPLC (fast protein liquid chromatography) Mono Q or Whatman (Clifton, NJ) DE-52] equilibrated with 15 mM Tris–phosphate buffer, pH 8.0. Elute the column with a linear buffered pH gradient, using 15 mM Tris–phosphate, pH 8.0, as the starting buffer and 0.3 M Tris–phosphate, pH 4.0, as the final buffer.

4. Screen each fraction by Ouchterlony double-diffusion analysis for IgG, IgA, IgM, and albumin, using appropriate antisera and controls. Pool fractions containing IgA, concentrate, and screen again by Ouchterlony to confirm the purity of the IgA. If necessary repeat the chromatography step to remove contaminating material. The major contaminant is usually albumin, which can also be removed by passage over an Affi-Gel blue (Bio-Rad, Richmond, CA) affinity column.

An alternative procedure that is especially useful in the purification of IgA$_1$ from normal human serum is affinity chromatography over a jacalin–Sepharose column. Jacalin, or jack bean lectin, selectively binds human IgA$_1$ but not IgA$_2$.[21]

Quantitative Assay Using Sodium Dodecyl Sulfate–Polyacrylamide Gel Electrophoresis

The following assay technique is based on physical separation of the IgA$_1$ cleavage products (Fabα and Fcα) from the intact IgA$_1$ molecule by electrophoresis.[7,22] Quantitative results are achieved by using trace amounts of radiolabeled IgA$_1$ as part of the reaction mixture.

Reagents

Pure human IgA$_1$, at a concentration of 2.0 mg/ml
^{125}I-labeled human IgA$_1$, approximately 400 μCi/ml
Assay buffer: 50 mM Tris-HCl, pH 7.5, containing 10 mM MgCl$_2$, 10 mM CaCl$_2$, and 0.05% bovine serum albumin (radioimmunoassay grade)
SDS–PAGE sample buffer: 62.5 mM Tris-HCl, pH 6.8, containing 12.5% glycerol, 1.25% SDS, 1.25% 2-mercaptoethanol, and 0.006% bromphenol blue

[21] D. L. Skea, P. Christopoulos, A. G. Plaut, and B. J. Underdown, *Mol. Immunol.* **25,** 1 (1988).
[22] D. A. Simpson, R. P. Hausinger, and M. H. Mulks, *J. Bacteriol.* **170,** 1866 (1988).

Procedure

1. Mix 2.5 μl of the enzyme sample to be assayed, 2.5 μl of human IgA₁ (5 μg), 2.5 μl of [^{125}I]IgA₁ (~1 μCi), and 5 μl of assay buffer, and incubate the mixture in a 37° water bath. After incubation, add 50 μl of SDS–PAGE sample buffer and boil for 5 min to stop the reaction and reduce the IgA₁ to polypeptide monomers.

2. Electrophorese the samples on a 9% polyacrylamide gel containing 0.1% SDS in a Laemmli buffer system[23] to separate the IgA₁ cleavage products. Note that we generally use 50 mM Tris, 384 mM glycine, 0.1% SDS as electrode buffer rather than the standard Laemmli recipe.

3. Stain the gel with 0.05% Coomassie Brilliant Blue in 35% methanol–10% acetic acid and destain in 35% methanol–10% acetic acid. Dry the stained gel onto Whatman 3MM paper and autoradiograph onto Kodak (Rochester, NY) XAR-5 film to localize the labeled IgA₁ bands. Using the autoradiograph as a cutting guide, excise the sections of each lane of the gel containing (1) the intact IgA₁ heavy chain (~56 kDa) and (2) the Fcα and Fdα fragments (~25–35 kDa), as shown in Fig. 1b. The Fdα is defined as the heavy chain portion of the Fabα fragment. Regions of the gel containing light chains (~22 kDa) or secretory component (~80 kDa) are discarded. The amount of radioactivity [counts per minute (cpm)] in each band is measured using a γ counter.

4. The percent cleavage of the IgA₁ in each sample is calculated as follows:

$$\frac{\text{cpm in Fd and Fc fragments}}{\text{cpm in intact heavy chain} + \text{cpm in Fd and Fc fragments}} = \% \text{ fragments}$$

5. The amount of IgA₁ cleaved is then calculated as

$$\frac{\% \text{ fragments in sample} - \% \text{ fragments in negative control}}{\% \text{ fragments in positive control}}$$

$$\times \ \mu\text{g IgA}_1 \text{ per digest} = \mu\text{g IgA}_1 \text{ cleaved}$$

The positive control is a sample in which complete cleavage of the available IgA₁ has occurred, and it provides a value for the maximum percentage of total radioactivity that is found in IgA₁ fragments. This value is usually 75–80% rather than 100%.

6. Units of IgA₁ protease activity are expressed as micrograms IgA₁ cleaved per minute.

[23] U. K. Laemmli, *Nature (London)* **227**, 680 (1970).

Example. An undiluted sample of supernatant from a late log phase broth culture of *N. gonorrhoeae* yields 25% fragments in a 20-min incubation. The percentage of fragments in the negative control is 5% and that in the positive control is 80%. The assay reaction mix included 2.5 μl enzyme sample and 5 μg IgA$_1$. For the calculation,

$$\frac{(25\% \quad 5\%) \times 5 \ \mu g \ IgA_1}{(80\%) \times 20 \ min \times 0.0025 \ ml} = 25 \ \mu g \ IgA_1 \ cleaved/min/ml$$

$$= 25 \ units/ml \ of \ sample$$

Comments. The SDS–PAGE procedure is the only method for the assay of IgA$_1$ protease activity which permits both quantitation of units of IgA$_1$ protease and determination of the type of cleavage, based on the size of the Fdα and Fcα fragments produced. The method can also be used for qualitative assays on culture supernatants or on suspensions of bacterial colonies. For qualitative assays, we recommend using only the radiolabeled substrate to yield increased sensitivity.

The electrophoresis-based assays are the method of choice in use today by most laboratories studying IgA$_1$ proteases. The technique allows for multiple samples to be processed simultaneously, offers consistency, is amenable to qualitative or quantitative studies, and is relatively rapid. The use of radiolabeled IgA$_1$ limits the amount of IgA$_1$ necessary for assays, but it can present a problem for laboratories that do not wish to employ radioisotopes. Detection of IgA$_1$ bands by immunoblotting with antiserum against the α chain is an alternative,[24] although quantitation by densitometric scanning of developed blots is generally less sensitive and less reproducible than assay with radiolabeled substrate.

Qualitative Plate Assay for Immunoglobulin A$_1$ Protease

Solid-phase plate assays can be used for screening large numbers of isolates simultaneously and are most often used to select an IgA$_1$ protease-producing isolate from a population of nonproducers. The first described method utilizes radiolabeled IgA$_1$ immobilized on polyacrylamide beads via an anti-Fcα antibody.[25] The technique involves overlaying bacterial colonies on an agar plate with a layer of immobilized IgA$_1$ in top agar or agarose. Secreted IgA$_1$ protease from producing colonies will cleave the immobilized IgA$_1$, liberating Fabα, which is diffusible. Diffused Fabα fragments are blotted onto a nitrocellulose membrane, which is then autoradiographed. IgA$_1$ protease-producing isolates are identified by a location-corresponding radioactive signal.

[24] T. Ahl and J. Reinholdt, *Infect. Immun.* **59,** 563 (1991).
[25] J. V. Gilbert and A. G. Plaut, *J. Immunol. Methods* **57,** 247 (1983).

Reagents

^{125}I-Labeled human IgA$_1$, approximately 400 μCi/ml
Immunobead (Bio-Rad) reagent specific for human IgA
Phosphate-buffered saline, pH 7.5 (PBS)
2% Agarose in PBS
Nitrocellulose filters

Procedure

1. Reconstitute the lyophilized polyacrylamide beads at a concentration of 1 mg/ml in PBS. Let stand at least 30 min at room temperature.
2. For each 100-mm petri plate to be assayed, mix 2 ml bead suspension plus 4–10 μCi [^{125}I]IgA$_1$ (10–20 μl). Incubate for 90 min at 37° with gentle shaking. Wash the beads three times with PBS to remove unbound IgA. Resuspend the beads to a concentration of 1 mg/ml in PBS and warm to 50° for 5 min.
3. For each plate to be assayed, melt 2 ml of sterile 2% agarose in PBS and cool to 50°.
4. Rapidly mix 2 ml of the ^{125}I-labeled bead suspension plus 2 ml agarose and pour onto the surface of the plate to be assayed. Swirl gently to distribute an even layer of agarose, being careful not to disturb the bacterial colonies. It is helpful to have the culture plates at 37° to prevent the agarose from solidifying too rapidly. Incubate at 37° in a moist chamber, such as a sealed box lined with damp paper towels. The incubation time will vary with the bacteria to be assayed: *N. gonorrhoeae* colonies need approximately 30 min, whereas *Escherichia coli* clones expressing *iga* genes may need 12–16 hr.
5. Lay a dry nitrocellulose filter over the agarose, being careful to avoid trapping any air bubbles, and mark the orientation of the filter. Incubate at 37° for 20 min in a moist chamber. Lift off the filter and wash twice in PBS to remove any adherent agarose or other debris. Air-dry the filter and autoradiograph onto Kodak XAR-5 film.
6. IgA$_1$ protease-producing colonies release radiolabeled IgA$_1$ fragments, which are picked up by the nitrocellulose. Therefore, exposed areas on the autoradiograph correspond to IgA$_1$ protease-producing colonies.

Comments. It is necessary to adjust both incubation times and exposure times to achieve optimal sensitivity with minimal background. Strong IgA$_1$ protease producers such as *N. gonorrhoeae* will need only 15–30 min of incubation for protease activity to be detectable, whereas *E. coli* clones, which are often very weak protease producers, may need 12–16 hr of incubation as well as increased exposure time for the autoradiograph. In any case, incubation with nitrocellulose filters should be for only 20–30 min, or background levels may become too high for interpretation of results.

It is possible to recover viable bacteria from under the overlay agarose if exposure to radiation is kept to a minimum.[26] Alternatively, bacteria or phage to be tested can be spotted in a grid pattern onto two plates, with one plate assayed by this method and the other held in reserve for the isolation of positive colonies.

Two variations have arisen from this method. One involves prior immobilization of the radiolabeled IgA_1 on the nitrocellulose membrane and incubation on a bacteria-covered surface.[27] After allowing for diffusion of the liberated $Fab\alpha$, IgA_1 protease producers can be detected by a decrease in the amount of radioactive signal. This method is not as sensitive as the original assay procedure, since it is much more difficult to detect a negative area than a positive one. A second variation incorporates α-light chain antibodies bound to the nitrocellulose membrane to improve the efficiency of $Fab\alpha$ recovery.[28] Detection of bound $Fc\alpha$ fragments with biotinylated antibodies to $Fc\alpha$ eliminates the need for using radioactive materials.

Immunoassay Procedure for Quantitation of Immunoglobulin A_1 Protease Activity

Several ELISAs for the quantitation of IgA_1 protease activity have been developed.[16,17] These methods are not optimal for screening bacterial samples for IgA_1 protease activity, since they do not permit identification of protease type nor do they distinguish between IgA_1 protease activity and nonspecific degradation of immunoglobulin A_1. However, these assays generally are very sensitive and reproducible, and they have been especially useful in detection of monoclonal antibodies capable of inhibition of IgA_1 protease activity.[17] Reinholdt and Kilian[16] developed an ELISA that utilizes antibody to human light chain to coat the wells and to bind IgA_1 by its Fab end. After digestion with IgA_1 protease, horseradish peroxidase-conjugated antibody against $Fc\alpha$ is used to quantitate the remaining uncleaved IgA_1. In this assay, decreased absorbance correlates with increased protease activity.

Blake and Eastby[17] developed a similar procedure that utilizes the IgA-binding protein from group B streptococci[29] to coat the wells and to bind

[26] R. J. Shoberg, Ph.D. Thesis, Michigan State Univ., East Lansing (1991).
[27] R. Halter, J. Pohlner, and T. F. Meyer, *EMBO J.* **3**, 1595 (1984).
[28] T. A. Brown and I. G. Leak, *J. Immunol. Methods* **123**, 241 (1989).
[29] G. J. Russell-Jones, E. C. Gotschlich, and M. S. Blake, *J. Exp. Med.* **160**, 1467 (1984).

IgA$_1$ substrate by its Fc end. After digestion with IgA$_1$ protease, remaining Faba is detected and quantitated using alkaline phosphatase-conjugated antibodies to human light chain. Decreased absorbance correlates with increased cleavage of IgA$_1$ and increased protease activity.

Both of these ELISA techniques offer the ability to perform multiple assays rapidly and simultaneously, using very small amounts of human IgA$_1$ and no radioactivity, with very high sensitivity. However, the methods may not be useful for assay of IgA$_1$ proteases from oral streptococci, which may cleave only one heavy chain rather than both and thus fail to release cleaved fragments.[16,20]

Enzyme Production and Isolation

Several methods have been described for the production and partial purification of neisserial IgA$_1$ proteases.[17,22,27,30] All rely on a combination of chromatographic procedures, including ion-exchange, chromatofocusing, molecular sieve, and hydrophobic interaction chromatography. The procedure we use to purify the gonococcal type 2 IgA$_1$ protease is a modification of a previously published method[22] that gives a higher yield of active IgA$_1$ protease.

Procedure

1. Inoculate brain–heart infusion (BHI) agar containing 10 ml/liter Kellogg's supplement[31] from a freezer stock of *N. gonorrhoeae* stored at −70° in 2% tryptone–20% glycerol. Grow overnight at 35° under 5% CO_2.

2. Prepare brain–heart infusion broth. Filter through a 10,000 molecular weight cutoff Amicon (Danvers, MA) membrane. Collect and autoclave the filtrate; discard the retentate. Immediately before use, add 10 ml/liter Kellogg's supplement plus 10 ml/liter of 4.2% sodium bicarbonate (BHIK).

3. Inoculate 30 ml of this medium (BHIK) from the overnight plate culture. The starting OD$_{520}$ should be approximately 0.05. Incubate the culture at 35° in a water bath shaker at 150 rpm for 3–4 hr, until the OD$_{520}$ reaches 0.8–1.0. Use the broth to inoculate 500 ml of prewarmed BHIK. Incubate at 35° for 3–4 hr, to an OD$_{520}$ of 0.8–1.2 (late log phase). Centrifuge for 30 min at 6000 g and 4° to pellet the bacteria.

4. Carefully decant the supernatant into a clean, ice-cold flask. Discard the cell pellet. Add 0.5 M EDTA, pH 8.0, to a final concentration of 50 mM.

[30] A. G. Plaut, this series, Vol. 165, p. 117 (1988).
[31] D. S. Kellogg, Jr., W. L. Peacock, W. E. Deacon, L. Brown, and C. I. Pirkle, *J. Bacteriol.* **85**, 1274 (1963).

5. Add approximately 200 ml of DEAE-cellulose (Whatman DE-52) resin, equilibrated in 50 mM Tris-HCl, pH 7.5, to the supernatant. Stir slowly at 4° for at least 1 hr. Centrifuge for 30 min at 6000 rpm and 4° to pellet the resin. Decant and save the supernatant. Wash the pelleted resin with 50 mM Tris-HCl, pH 7.5, plus 5 mM EDTA; decant and save the supernatant. Pool the supernatants and discard the used resin.

6. Concentrate the pooled supernatant by positive pressure dialysis in an Amicon stirred cell using a 50,000 molecular weight cutoff membrane. If the material is to be stored, add 20% sterile, ultrapure glycerol (BRL, Gaithersburg, MD) to the concentrated sample.

7. Using an Amicon or similar membrane filtration device (*not* dialysis tubing), dialyze the sample into at least 2 volumes of distilled water, to dilute the salt concentration. Bring the sample to 47.5 ml with sterile distilled water. Add 2.0 ml of ampholytes (Pharmacia), pH 8.0–10.5 range, plus 0.5 ml of ampholytes, pH 5.0–8.0 range. Separate the sample in a Rotofor (Bio-Rad) isoelectric focusing cell, according to the manufacturer's instructions.

8. Pool the fractions with IgA$_1$ protease activity and exchange into 50 mM Tris-HCl, pH 7.5, plus 5 mM EDTA, 20% glycerol, and 0.2% Nonidet P-40 detergent. Chromatograph over a Pharmacia FPLC Superose 12 HR 10/30 column equilibrated with the same buffer. Screen fractions for IgA$_1$ protease activity by SDS–PAGE assay and for purity by Coomassie blue–silver staining.[32] Pool fractions containing IgA$_1$ protease and exchange into 50 mM Tris-HCl, pH 7.5, plus 5 mM EDTA and 20% glycerol, and store at −70°.

Comments. We have successfully purified type 2 gonococcal IgA$_1$ protease to homogeneity by the above method. Type 1 gonococcal protease can be purified by similar methods, although it is much less stable.[22] Use of 50 mM Tris–acetate, pH 7.5, plus 5 mM EDTA, 1 mM dithiothreitol (DTT), and 20% glycerol as a stabilization buffer is necessary for purification of the type 1 gonococcal protease. The IgA$_1$ proteases from *N. meningitidis*[6] and *H. influenzae*[7] have been partially purified by a combination of anion-exchange and molecular sieve chromatography. The method described here, with modifications such as optimization of appropriate stabilization buffers for each enzyme, may also be applicable to purification of IgA$_1$ proteases from these bacteria.

---

[32] A. Gorg, W. Postel, J. Weser, H. W. Schivaragin, and W. M. Boesken, *Sci. Tools* (*Sweden*) **32**, 5 (1985).

TABLE I
PROPERTIES OF BACTERIAL IMMUNOGLOBULIN A₁ PROTEASES

Species	Enzyme class[a]	M_r ($\times 10^{-3}$)	pI	K_m^b (μmol)	Refs.
Bacteroides melaninogenicus	T	62[c]	5.0	3.4	18,33
Capnocytophaga spp.	M	NR[d]	NR	NR	33
Haemophilus influenzae (type 1)	S	90[c], 108[e]	NR	0.7	9,34,35
Haemophilus influenzae (type 2)	S	100[c]	NR	1.5	34,36
Neisseria gonorrhoeae (type 1)	S	112[c]	NR	0.6	22,34
Neisseria gonorrhoeae (type 2)	S	106–110[c,e]	8.6	1.8	8,22,34
Neisseria meningitidis (type 2)	S	100[c]	NR	NR	36,37
Streptococcus oralis	NR	100[c]	NR	NR	38
Streptococcus pneumoniae	M	NR	NR	NR	37
Streptococcus sanguis	M	100[c], 186[e]	5.45	23	10,34,38,39

[a] M, Metalloprotease; S, serine protease; T, thiol protease.
[b] Determined against human IgA₁.
[c] Determined by chromatography or SDS–PAGE.
[d] NR, Not reported.
[e] Determined by sequence analysis.

Properties of Bacterial Immunoglobulin A₁ Proteases

Table I[33–39] summarizes some of the information known about the chemistry of the bacterial IgA₁ proteases, including the class of protease to which the given IgA₁ protease belongs and the molecular weight, where reported. Data are not available for all of the IgA₁ proteases that have been described. Some of the highlights of the data are the facts that (1) three of the four classes of proteases are represented, with aspartic proteases being the exception, and (2) these enzymes show a wide range of molecular weights and isoelectric points. It appears that this set of proteases with similarly narrow ranges of substrate specificity is actually a grouping of very diverse enzymes which were clustered into a group based solely on that one characteristic.

[33] E. V. G. Frandsen, J. Reinholdt, and M. Kilian, *Infect. Immun.* **55**, 631 (1987).
[34] W. W. Bachovkin, A. G. Plaut, G. R. Flentke, M. Lynch, and C. A. Kettner, *J. Biol. Chem.* **265**, 3738 (1990).
[35] J. Pohlner, C. Maercker, H. Apfel, and T. F. Meyer, in "Frontiers of Mucosal Immunology" (M. Tsuchiya, H. Nagura, T. Hibi, and I. Moro, eds.), Vol. 1, p. 567. Elsevier Science Publ., Amsterdam, 1991.
[36] M. H. Mulks, unpublished data, 1982.
[37] M. Kilian, J. Mestecky, R. Kulhavy, M. Tomana, and W. T. Butler, *J. Immunol.* **124**, 2596 (1980).
[38] J. Reinholdt, M. Tomana, S. B. Mortensen, and M. Kilian, *Infect. Immun.* **58**, 1186 (1990).
[39] A. G. Plaut, J. V. Gilbert, and I. Heller, *Adv. Exp. Med. Biol.* **107**, 489 (1978).

Substrates. By definition, IgA$_1$ proteases proteolyze human IgA$_1$ and not IgA$_2$, IgM, IgG, or other serum proteins such as albumin.[2,3] For the proteases that have been tested against IgA from other mammals, IgA$_1$ from higher primates (i.e., chimpanzee and gorilla) can be hydrolyzed, but not IgA$_1$ from other primates or from a wide range of other mammalian species.[2,3]

The type 2 IgA$_1$ protease from *N. gonorrhoeae* contains three sequences within the 169-kDa proenzyme that are cleaved by autoproteolysis during secretion of the mature protease.[8] Synthetic peptides homologous to two of the sequences are also cleaved by the gonococcal type 2 protease, although similar peptides homologous to the hinge region of human IgA$_1$ are not.[19] In addition, we have demonstrated that several proteins found in the isolated membranes of gram-negative bacteria are degraded by the type 2 IgA$_1$ protease of *N. gonorrhoeae*.[40]

[40] R. J. Shoberg and M. H. Mulks, *Infect. Immun.* **59**, 2535 (1991).

[44] Elastase Assays

By LYNN RUST, CALVIN R. MESSING, and BARBARA H. IGLEWSKI

Introduction

The elastin protein is the primary constituent of the elastic fiber, the predominant connective tissue of the lung, blood vessels, and skin. Characteristic of elastin is the desmoisine cross-linking structure, formed by the covalent binding of four lysyl residues. Elastin includes a high concentration of the small hydrophobic amino acids including glycine, alanine, and valine. It is the most insoluble protein in the human body.[1]

Owing to the unusual structure and insolubility of elastin, few proteases have elastolytic activity, whereas elastases frequently have proteolytic activity against a variety of substrates. The neutral metalloproteinase of *Vibro vulnificus* is active against azocasein and elastin.[2] The metalloenzymes *Bacillus thermoproteolyticus* thermolysin and *Pseudomonas aeruginosa* elastase also have general proteolytic as well as elastolytic activity. In addition to elastin, *P. aeruginosa* elastase is capable of degrading

[1] J. M. Davidson, *in* "Connective Tissue Disease: Molecular Pathology of the Extracellular Matrix" (J. Uitto and A. J. Perejda, eds.), p. 423. Dekker, New York, 1987.
[2] M. H. Kothary and A. S. Kreger, *J. Gen. Microbiol.* **133**, 1783 (1987).

immunoglobulins IgG, IgA, and secretory IgA (S-IgA),[3] several serum complement factors,[4] α_1-proteinase inhibitor,[5] collagen,[6] and fibrin.[7]

Pseudomonas aeruginosa elastase has been implicated in the pathogenesis of P. aeruginosa infections. Increases in antibody titers to elastase have been found in patients recovering from P. aeruginosa infections of the lung.[8] When inoculated intravenously, pure elastase causes pulmonary hemorrhages in mice similar to those seen during P. aeruginosa lung infection.[9] Intracorneal administration of purified elastase into rabbits causes severe corneal damage.[10] Finally, elastase enhances bacterial growth in vivo.[11] Likewise, leukocyte elastase, a serine proteinase present in the azurophil granule of human neutrophils, has been linked to the pathogenesis of pulmonary emphysema.[12]

In this chapter, we describe two methods for purifying elastase from P. aeruginosa supernatants. We also describe three methods to assay P. aeruginosa elastase activity involving elastin–Congo red, fluorogenic substrate, and elastin–nutrient agar plates. Elastin–Congo red and elastin–nutrient agar plates are general elastase assay methods useful for detecting elastin solubilization. Use of [³H]elastin and rhodamine–elastin have been reviewed previously in this series.[13]

Sample Preparation: *Pseudomonas aeruginosa* Supernatants

Purification

Although each of the assays described here does not require purification of elastase from crude supernatants or lysates, there are situations in which purification is desired. For example, purified enzyme is crucial in developing standard curves. The methods below are specific for P. aeruginosa elastase purification and may have to be adapted for other bacterial elastase purifications. Method 1 is adapted from Morihara et al.,[7]

[3] G. Doring, H.-J. Obernesser, and K. Botzenhart, *Zentralbl. Bakteriol. Mikrobiol. Hyg. Abt. Orig. A* **249**, 89 (1981).
[4] D. R. Schultz and K. D. Miller, *Infect. Immun.* **10**, 128 (1974).
[5] K. Morihara, H. Tsuzuki, and K. Oda, *Infect. Immun.* **24**, 188 (1979).
[6] L. W. Heck, K. Morihara, W. B. McRae, and E. J. Miller, *Infect. Immun.* **51**, 115 (1986).
[7] K. Morihara, H. Tsuzuki, T. Oka, H. Inoue, and M. Ebata, *J. Biol. Chem.* **240**, 3295 (1965).
[8] J. D. Klinger, D. C. Strauss, C. B. Hilton, and J. A. Bass, *J. Infect. Dis.* **138**, 49 (1978).
[9] L. D. Gray and A. S. Kreger, *Infect. Immun.* **23**, 159 (1979).
[10] K. Morihara and J. Y. Homma, in "Bacterial Enzymes and Virulence" (I. A. Holder, ed.), p. 41. CRC Press, Boca Raton, Florida, 1985.
[11] J. F. Cicmanec and I. A. Holder, *Infect. Immun.* **25**, 177 (1979).
[12] J. B. Karlinsky and G. L. Snider, *Am. Rev. Respir. Dis.* **117**, 1109 (1978).
[13] M. J. Banda, Z. Werb, and J. H. McKerrow, this series, Vol. 144, p. 288.

whereas Method 2 incorporates high-performance liquid chromatography (HPLC) to purify elastase.

Method 1

1. Fresh supernatant from an overnight Luria-Bertani broth culture is recovered by centrifugation at 15,000 g at 4° for 10 min.
2. Supernatants are filtered through 0.15 μm filters (Gelman Sciences, Ann Arbor, MI).
3. Supernatants are concentrated by centrifugal ultrafiltration at 4° in Centricell 20 units (30,000 molecular weight cutoff, Polysciences, Warrington, PA) according to the manufacturer's recommendations.
4. The concentrated, viscous residues are washed twice with 20 mM sodium phosphate buffer, pH 8.0, by centrifugal ultrafiltration in Centricon 30 devices (Amicon Div., W. R. Grace and Co., Beverly, MA). Concentrates may be stored at −70°.
5. A DEAE-Sepharose column (Sigma, St. Louis, MO, Fast Flow, DFF-100; 1.6 × 23 cm, ~60 ml) is washed successively with 350 ml of 0.1 M NaOH and 350 ml of 1 M NaCl at 4° and then equilibrated with 20 mM sodium phosphate, pH 8.0. The supernatant concentrate in the same buffer is applied to the column and eluted with the same buffer at 1 ml/min with 10-ml fractions collected. The elastase elutes as a well-defined peak in 5–7 fractions starting 70 to 100 ml after elution begins.
6. The fractions are concentrated by ultrafiltration in Centricon 30 units at 4° and assayed by sodium dodecyl sulfate–polyacrylamide gel electrophoresis (SDS-PAGE). Appropriate fractions are combined and aliquoted. Aliquots may be stored at −70°.
7. Elastase concentrations are determined by UV spectroscopy, using an extinction coefficient $E_{280}^{1\%}$ of 14.52.[7]

Method 2. Method 2 is from Ref. 14. Steps 1 through 6 are adapted from Morihara *et al.*[7]

1. *Pseudomonas aeruginosa* PAO1 supernatant is precipitated at 4° overnight with saturated ammonium sulfate added dropwise to 60% (v/v) final concentration.
2. The precipitate is pelleted by centrifugation for 10 min at 8800 g (4°) and dissolved in distilled water.
3. An additional centrifugation (10 min at 8800 g, 4°) yields a clear supernatant.
4. Acetone is slowly added to the supernatant to 30% (v/v) final concentration and stirred at 4° for 2 hr.

[14] C. D. Cox, personal communication (1992).

5. The pellet is centrifuged out and saved. Acetone is again added to the supernatant to a final concentration of 66% (v/v). The supernatant is oily and highly pigmented; removal of the pigment is necessary to avoid interference with monitoring protein elution at 280 nm from the column.

6. The precipitate is pelleted as above and resuspended in 20 mM HEPES buffer, pH 7.0. This fraction contains the majority of elastase activity as well as other proteolytic activities.

7. Molecular sieving is carried out with a Bio-Rad Bio-Sil SEC 250 column (600 × 7.5 mm) with an SEC 250 guard column. A solvent of 20 mM HEPES buffer (pH 7.0) with 10% (v/v) methanol at a flow rate of 0.7 ml/min elutes an elastase peak with a retention time of 41 to 42 min.

8. Fractions are assayed for proteolytic or elastolytic activity.

9. Active fractions are then passed through a reversed-phase or hydrophobic interaction column (Bio-Rad Hi-Pore RP-304, 250 × 4.6 mm; Hi-Pore guard column, 40 × 4.6 mm). A mobile phase of 20 mM HEPES buffer (pH 7.0) with 2% (v/v) dimethyl sulfoxide (DMSO) is enacted at 1.0 ml/min with an increasing gradient of acetonitrile from 0 to 50% beginning at 10 min and spanning 30 min. Elastase elutes toward the end of the gradient, at approximately 40 min.

Elastase Assays

Elastin–Congo Red Assay

Originally developed by Naughton and Sanger,[15] the colorimetric elastin–Congo red assay uses bovine neck ligament (elastin) covalently linked to a red dye (Congo red) as substrate. When digested with elastase, the red dye is released and becomes soluble in aqueous buffer. Elastase activity is quantified by measuring the absorbance of the assay supernatant at 495 nm and interpolating from a standard curve.

Reagents

Assay buffer: 10 mM sodium phosphate buffer, ph 7.0, *or* 30 mM
    Tris buffer, pH 7.2 (see Table I)
Elastin–Congo red (Sigma)
Purified elastase (Nagase Biochemicals Ltd., Kyoto, Japan)

Method

1. Two milliliters of assay buffer is added to 20 mg elastin–Congo red and vortexed immediately to prevent clumping of substrate.

[15] M. A. Naughton and F. Sanger, *Biochem. J.* **78,** 156 (1961).

TABLE I
ELASTASE ACTIVITY IN VARIOUS BUFFERS[a]

Buffer	pH	A_{495}
0.03 M Tris	7.2	1.89 ± 0.03
0.03 M Tris	8.0	1.58 ± 0.04
0.10 M Tris	7.2	1.07 ± 0.06
0.10 M Tris	8.0	0.80 ± 0.05
0.01 M Sodium phosphate	7.0	1.90 ± 0.05
0.01 M Sodium phosphate	8.0	0.11 ± 0.01

[a] Ammonium sulfate-precipitated elastase (Nagase) was serially diluted to 1.6×10^{-3} (~25 µg/ml) in water. Then 100 µl of this dilution was incubated in triplicate with elastin–Congo red as described for 2 hr. Absorbance of the supernatant at A_{495} was determined spectrophotometrically.

2. Culture supernatant or purified elastase is added at 0.05 to 1.25 µg elastase per 2 ml assay buffer and rotated (to keep the substrate well suspended) at 37° for 2 to 6 hr, depending on the amount of elastase added. (The standard curve may not be linear with longer incubations.[16])

3. Insoluble elastin Congo red is pelleted at 1200 g for 10 min at room temperature in a Beckman Model TJ-6 or equivalent centrifuge. The absorbance of the supernatant is measured at 495 nm.

4. The concentration of unknown samples is determined by interpolating from a standard curve of known elastase concentrations (Fig. 1).

Advantages. The assay is a relatively facile colorimetric assay specific for elastolytic activity. The substrate, elastin–Congo red, is easily obtainable, and the assay may be accomplished using standard laboratory equipment. Also, no arduous sample preparation is required. Crude culture supernatants and lysates may be assayed directly for elastase activity. Unknown samples are readily quantified using a standard curve. Finally, this method is not highly labor intensive. Once the assay is set up, the long incubation period allows the researcher freedom to accomplish other goals.

Disadvantages. Sample reproducibility is less than ideal. For this reason, using multiple-sample replicates (3–5) is strongly advised. Also, owing to the insoluble nature of the substrate, the assay is not linear with time. The incubation period should remain constant; unknown elastase samples should lie within the linear range of the assay at the end of the

[16] C. R. Messing, personal observation (1991).

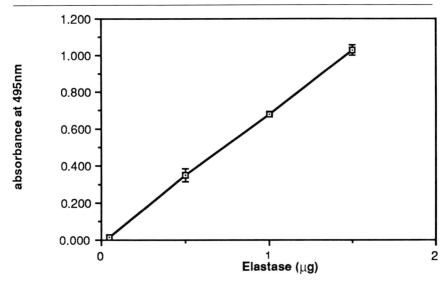

FIG. 1. Standard curve of elastase activity in the elastin–Congo red assay. Ten microliters of ammonium sulfate-precipitated elastase (Nagase) was serially diluted to 10^{-3} in buffer. The absorbance at 280 nm was determined spectrophotometrically. The concentration (C, μg/ml) was determined by the equation $C = A_{280}/E_{280}^{1\%} \times 10^4$, where $E_{280}^{1\%} = 14.52$. From 0 to 1.5 μg of elastase was incubated in duplicate with elastin–Congo red as described for 6 hr, and the absorbance at 495 nm was determined.

incubation period for concentrations to be accurately determined. When assaying *P. aeruginosa* supernatants, LasA elastase as well as LasB elastase will contribute to the total elastolytic activity of the sample.[17,18]

Fluorogenic Substrate Assay

Benzyloxycarbonyl (Z)-Ala-Gly-Leu-Ala-OH has been shown to be among the most susceptible of elastase substrates.[19] In a search for a useful *P. aeruginosa* elastase assay, Nishino and Powers[20] synthesized this peptide with a fluorogenic group (2-aminobenzoyl; Abz) on the amino terminus and a quenching group (4-nitrobenzylamide; Nba) on the carboxy terminus. Cleavage of the peptide by elastase releases the quenching group and allows fluorescence of the substrate, by which elastase activity is monitored with a fluorescence spectrometer. This substrate may also be useful in assaying the activity of thermolysin, *Streptomyces griseus* prote-

[17] J. E. Peters and D. R. Galloway, *J. Bacteriol.* **172,** 2236 (1990).
[18] S. J. Ferrell, D. Toder, and C. R. Messing, personal observations (1992).
[19] K. Morihara and H. Tsuzuki, *Arch. Biochem. Biophys.* **146,** 291 (1971).
[20] N. Nishino and J. C. Powers, *J. Biol. Chem.* **255,** 3482 (1980).

ase, and *Streptomyces fradiae* protease, as Z-Ala-Gly-Leu-Ala-OH was found to be an optimal peptide substrate of these enzymes as well as of *P. aeruginosa* elastase. Z-Phe-Gly-Leu-Ala-OH was an optimal peptide substrate for *Bacillus subtilis* neutral protease(s).[19]

Reagents

Fluorogenic substrate: Abz-Ala-Gly-Leu-Ala-Nba (Enzyme Systems Products, Livermore, CA) in DMSO
Assay buffer[21]:
50 mM Tricine-HCl, pH 7.5, 10 mM CaCl$_2$, 200 mM NaCl
Purified elastase (Nagase)

Method

1. To 2 ml of assay buffer add 10 μl (40 nM) substrate directly to the cuvette. Establish a level baseline before adding enzyme. (Other substrate concentrations may be used.[20])

2. Add elastase at 0.02 to 0.3 nM final concentration. Monitor hydrolysis (excitation 340 nm, emission 415 nm) using a Perkin-Elmer (Norwalk, CT) LS-3B or equivalent fluorescence spectrophotometer for 2 to 3 min. (The sensitivity and linear range of elastase concentrations may vary depending on the fluorescence spectrophotometer used.)

3. To quantify results, calculate the slope of the initial portion of the curve, representing the rate of hydrolysis, and compare to a standard curve of rates of known concentrations.

Advantages. The fluorescence assay is specific for *P. aeruginosa* elastase. It is approximately 40-fold more sensitive than the elastin–Congo red assay, allowing detection of as little as 0.02 nM elastase. Additionally, with higher substrate concentrations, the range of detection is linear to 0.03 mM.[20] From 0.0125 to 0.2 mM elastase, the initial rate is linear up to 15% hydrolysis.[20] Assay flexibility allows the use of crude sample supernatants and lysates. Finally, the fluorogenic substrate is commercially available.

Disadvantages. The assay requires the use of a fluorescence spectrophotometer. Depending on the fluorescence spectrophotometer used, the range and sensitivity of the assay will vary. The linear range of elastase concentrations using the Perkin-Elmer LS-3B model is relatively narrow, at just over one logarithm$_{10}$. Additionally, as each sample and replicate is assayed individually, this method is labor intensive and requires constant attention. To allow for maximum stability of enzyme, each replicate must

---

[21] R. C. Wahl, personal communication (1990).

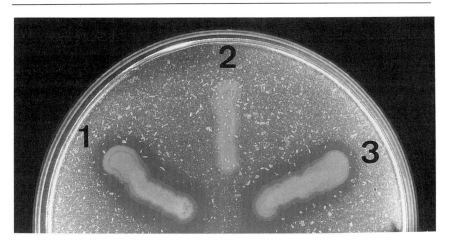

Fig. 2. Elastin–nutrient agar plate assay. Zones of elastolytic clearing are visible around *P. aeruginosa* streaks 1 and 3, whereas elastin particles are visible underneath and surrounding streak 2, indicating that this *P. aeruginosa* culture is negative for elastolytic activity.

also be individually diluted. A stock solution of greater than 100 nM elastase is recommended.

Elastin–Nutrient Agar Assay

The elastin nutrient agar assay is a qualitative assay in which cultures are plated onto elastin-containing agar plates. Zones of elastin clearing following incubation at 37° indicate elastolytic activity. The protocol below is an adaptation of the original method by Rippon and Varadi.[22]

Reagents. Elastin–nutrient agar plates have a nutrient agar base and insoluble elastin-containing agar overlay, and they are made as follows. For the nutrient agar base, to 1 liter water add 8 g nutrient broth (Difco Laboratories, Detroit, MI) and 20 g Agar Noble (Difco Laboratories). Adjust to pH 7.5. Pour plates and allow to harden. Store at 4°. For the elastin overlay, to 300 ml water add 2.4 g nutrient broth (Difco Laboratories), 6 g Agar Noble (Difco Laboratories; Bacto-agar is not suitable for this protocol), and 1 g elastin (Sigma). Adjust to pH 7.5. Autoclave with a stir bar. While stirring, pour 5 ml of overlay per plate and allow to harden. It may take some practice to obtain an even distribution of elastin in the overlay. The overlay contains particulate elastin which is subject to degradation by elastase.

[22] J. W. Rippon and D. P. Varadi, *J. Invest. Dermatol.* **50,** 54 (1968).

Method

1. Streak a fresh culture onto an elastin–nutrient agar plate. Alternatively, serial dilutions of cultures may be plated to assess elastolytic activity of isolated colonies.

2. Incubate for 24 to 48 hr at 37°. Zones of elastolysis will appear as clearings of the elastin surrounding the culture (Fig. 2).

3. Elastolytic activity may be coarsely quantified by measuring the ratio of culture width to the sum of zone of clearing plus culture width.

Advantages. The plate assay is a simple, straightforward method to assess the presence of elastolytic activity of a given culture. Reagents are readily available, and it involves a minimum of labor and time commitment. The sensitivity of the method may be increased by storing plated cultures at 4°; this slows culture growth while allowing elastolysis to proceed. In addition, cell-associated elastolysis may be detected by the clearing of elastin directly beneath the culture. Finally, serial dilutions of cultures allow for screening of a number of clones for a desired elastolytic phenotype.

Disadvantages. The assay is primarily a qualitative assay; differences of less than 5-fold in activity are not easily detectable. LasA elastase will produce a zone of clearing after prolonged (>48 hr) incubation. Also, it often takes an incubation period of 24 to 48 hr to detect elastolytic activity, even for the most active of cultures.

Summary

Two methods of *P. aeruginosa* elastase purification are described: Method 1 involves concentration of sample supernatants, followed by DEAE-Sepharose liquid chromatography, whereas Method 2 involves initial fractionations followed by molecular sieving and hydrophobic interaction high-performance liquid chromatography. The choice of methods depends on the available equipment and supplies.

The methods of assaying elastase activity described are useful for a variety of applications. The elastin–nutrient agar plate method is a qualitative assay to determine the presence of elastase activity produced by a given culture or colony. Use of the quantitative elastin–Congo red assay is appropriate for determining elastase activities of mid-to-high elastase-producing cultures. For more sensitive determinations of *P. aeruginosa* elastase activity, use of the fluorogenic substrate is advisable.

Acknowledgment

Supported by PHS Grant No. 5-T32-AI07362-04.

[45] Zymographic Techniques for Detection and Characterization of Microbial Proteases

By MARILYN S. LANTZ and PAWEL CIBOROWSKI

Introduction

Microbial proteases have been proposed as virulence factors in the pathogenesis of a variety of diseases caused by microorganisms. It has been suggested that microbial proteases may contribute to the pathogenesis of infectious diseases by several mechanisms (for a review, see Ref. 1). Microbial proteases have been shown *in vitro* to degrade (1) host proteins that function in maintaining tissue integrity, (2) proteins that function in host defense, and (3) proteins that function in the regulation of the activity of host proteinases. The mechanism(s) by which they may contribute to the pathogenesis of bacterial infections *in vivo* remains largely unknown. The most compelling evidence to date implicating microbial proteases as virulence factors derives from studies which suggest that protease-deficient mutants are less virulent in animal model infections than are wild-type strains.[2,3] Identification and characterization of microbial proteases are prerequisites for understanding their role in the pathogenesis of infectious diseases. To this end, rapid and sensitive techniques for the detection and characterization of microbial proteases are highly desirable.

Zymographic techniques allow detection of proteases and some other hydrolytic enzymes following electrophoresis in various types of gel matrices. The term zymogram was proposed by Hunter and Markert[4] to refer to starch gel strips in which the location of enzymes is demonstrated by histochemical staining methods following electrophoresis. Sweetman and Ornstein[5] combined the protein-separating capabilities of polyacrylamide gel electrophoresis (PAGE) in nondissociating polyacrylamide gels with these histochemical staining methods to visualize several neutrophil proteases in zymograms. The zymograms of Sweetman and Ornstein[5] were produced following cationic PAGE. Polyacrylamide gels containing separated neutrophil proteases were incubated with specific naphthyl ester

[1] H. Maeda and A. Molla, *Clin. Chem. Acta* **185**, 357 (1989).
[2] J. O. Capobianco, C. G. Lerner, and R. C. Goldman, *Anal. Biochem.* **204**, 96 (1992).
[3] H. N. Shah, S. V. Seddon, and S. E. Garbia, *Oral Microbiol. Immunol.* **4**, 19 (1989).
[4] R. L. Hunter and C. L. Markert, *Science* **125**, 1294 (1957).
[5] F. Sweetman and L. Ornstein, *J. Histochem. Cytochem.* **22**, 327 (1974).

substrates and a diazotized dye. Proteins with esterolytic activity were selectively stained by the insoluble dye formed when the substrate was hydrolyzed. The term zymography was used by Granelli-Piperno and Reich[6] to describe a technique in which proteases separated by sodium dodecyl sulfate (SDS)–PAGE are allowed to diffuse from polyacrylamide gels into an underlying agarose indicator gel containing a protein substrate. In zymograms, zones of lysis, where proteolytically active bands have degraded the substrate in the indicator gel, are visualized as clear zones against an opaque background on dark-field illumination (fibrin–agar gels) or as clear zones against a dark blue background after staining of the gels with Coomassie Brilliant Blue R-250 or amido black.[7] In recent years, the term zymography has also been used to describe related techniques for visualization of proteases directly in substrate-containing polyacrylamide gels.[8,9] In this chapter, we use the term zymography to include all of these techniques for visualization of proteases separated in polyacrylamide gels.

We demonstrate below that a wide variety of commercially available bacterial proteases, as well as some prepared in our laboratory,[9] can be detected and characterized using zymographic techniques. The mammalian pancreatic proteases trypsin and chymotrypsin have been used as controls in some experiments, and they are also readily detected. Our results suggest that zymographic techniques for detection and characterization of bacterial proteases are versatile and have wide applicability. We have been able to detect proteases of several catalytic types usually found in prokaryotic cells, including serine proteases (pronase, subtilisin, and V8 protease), cysteine proteases (clostripain and enzymes of *Porphyromonas gingivalis*), and metalloproteases (serratiopeptidase and achromocollagenase), as well as the mammalian serine proteases trypsin and chymotrypsin. We have used native and denatured proteins as substrates to detect proteases, as well as some low molecular weight synthetic esterase substrates.

With a few exceptions as noted, we did not attempt to optimize conditions for detection of each protease since we wish to make the point that most proteases can be detected using general methods, although not

[6] A. Granelli-Piperno and E. Reich, *J. Exp. Med.* **148,** 223 (1978).
[7] J. L. Westergaard, C. Hackbarth, M. W. Treuhaft, and R. C. Roberts, *J. Immunol. Methods* **34,** 167 (1980).
[8] H. Birkedal-Hansen, R. E. Taylor, J. J. Zambon, P. K. Barua, and M. E. Neiders, *J. Periodontal Res.* **23,** 258 (1988).
[9] M. S. Lantz, R. D. Allen, T. A. Vail, L. M. Switalski, and M. Hook, *J. Bacteriol.* **173,** 495 (1991).

necessarily with maximal sensitivity. Electrophoresis highly concentrates proteases in sample preparations into a narrow zone in a polyacrylamide gel,[6] and, as has already been amply demonstrated, zymographic techniques are very sensitive, allowing detection of nanogram to picogram amounts of proteolytic enzymes under optimal conditions.[10–12] To achieve maximal sensitivity, the conditions must be optimized for each enzyme–substrate pair.

Zymographic techniques offer several advantages over other methods for detecting proteolytic enzymes. When zymographic techniques are used with SDS–polyacrylamide gels, they allow estimation of the relative molecular weight of an enzyme and permit detection and molecular weight estimation of multiple forms of an enzyme or of catalytically active degradation products of an enzyme. In addition, they allow the identification of the catalytically active component in crude mixtures of proteins that have proteolytic activity. Zymographic techniques can be especially useful in detecting proteases that are noncovalently complexed with naturally occurring protease inhibitors in crude extracts, since these complexes frequently dissociate during electrophoresis. These techniques may also be used to characterize enzymes. Although the relative molecular weight of enzymes cannot be estimated when zymographic techniques are used with nondissociating PAGE systems, these systems allow detection and characterization of proteases that are irreversibly denatured by SDS or are active only as multimeric proteins. Rather than presenting every possible modification of these techniques for detection and characterization of proteolytic enzymes, we have chosen to present examples of more widely used techniques and some modifications that we have found useful.

Enzymes and Reagents

Enzymes

With the exception of the proteases of *P. gingivalis,* which are prepared in our laboratory, all the enzymes are obtained from Sigma (St. Louis, MO). The commercially obtained enzymes are dissolved in distilled water and mixed with the appropriate buffers immediately prior to use. Unless otherwise noted by the manufacturer, we assume in our calculations that 1 mg solid is equivalent to 1 mg protein.

[10] T. J. Andary and D. Dabich, *Anal. Biochem.* **57,** 457 (1974).
[11] T. L. Brown, M. G. Yet, and F. Wold, *Anal. Biochem.* **122,** 164 (1982).
[12] P. J. Kelleher and R. L. Juliano, *Anal. Biochem.* **136,** 470 (1984).

Achromocollagenase. Achromocollagenase is collagenase (EC 3.4.24.3) from *Achromobacter iophagus.* One unit liberates 0.01 μmol of Pz-Pro-Leu from Pz-Pro-Leu-Gly-Pro-D-Arg in 5 ml ethyl acetate in 15 min, pH 7.1, 37°. This preparation contains 3900 units/mg solid.

Pronase. Pronase is protease (pronase E) Type XIV, bacterial, from *Streptomyces griseus.* One unit hydrolyzes casein to produce peptide equivalent to 1.0 μmol (181 μg) of tyrosine per minute, pH 11.0, 30°. This preparation contains 5.4 units/mg solid.

V8 Protease. V8 protease is protease glutamyl endopeptidase; EC 3.4.21.19) Type XVII-B from *Staphylococcus aureus* strain V8. One unit is equivalent to approximately 0.004 casein digestion units. One unit hydrolyzes 1 μmol of N-t-BOC-L-glutamic acid α-phenyl ester per minute, pH 7.8, 37°. This preparation contains 770 units/mg solid.

Clostripain. Clostripain (EC 3.4.22.8) is from *Clostridium histolyticum* (clostridiopeptidase B). One unit hydrolyzes 1.0 μmol of BAEE per minute, pH 7.6, 25° in the presence of 2.5 mM dithiothreitol (DTT). This preparation contains 288 units/mg solid, 300 units/mg protein.

Serratiopeptidase. Serratiopeptidase is protease (serratiopeptidase) Type XXVI from *Serratia* species. One unit hydrolyzes casein to produce color equivalent to 1.0 μmol (181 μg) of tyrosine per minute, pH 7.5, 37° (color by Folin–Ciocalteau reagent). This preparation contains 5.2 units/mg solid.

Subtilisin. Subtilisin is protease (subtilisin Carlsberg) Type VIII, bacterial, from *Bacillus licheniformis.* One unit hydrolyzes casein to produce color equivalent to 1.0 μmol (181 μg) of tyrosine per minute, pH 7.5, 37° (color by Folin–Ciocalteau reagent). This preparation contains 7.1 units/mg solid.

Porphyromonas gingivalis proteases. The proteases are extracted from freshly harvested *P. gingivalis* cells using 3-[(3-cholamidopropyl) dimethylammonio]-1-propane sulfonate [CHAPS; final concentration, 0.5% (w/v)] or detergents as described.[9]

Trypsin. Trypsin is from bovine pancreas, type III, dialyzed and lyophilized, essentially salt free. One BAEE unit is defined as change in OD_{253nm} of 0.001/min with BAEE as substrate, pH 7.6, 25°. This preparation contains 10,000–13,000 BAEE units/mg and less than 4 BTEE units/mg protein of chymotrypsin.

Chymotrypsin. Chymotrypsin is from bovine pancreas type I-S and is a 3 times crystallized, lyophilized, and essentially salt-free powder, prepared free of autolysis products and free of low molecular weight contaminants. One unit hydrolyzes 1.0 mmol of BTEE per minute at pH 7.8, 25°. This preparation contains 40–60 units/mg protein.

Reagents

All proteins and reagents are obtained from Sigma unless otherwise noted in the text. *N*-Acetyl-DL-phenylalanyl-β-naphthyl ester is obtained from Bachem, Bioscience Inc., and reagents used for preparation of polyacrylamide gels including acrylamide, bisacrylamide, TEMED, ammonium persulfate, and Coomassie Brilliant Blue R-250 are obtained from Bio-Rad (Richmond, CA).

Methods for Performing Electrophoresis and Developing Zymograms

The method used for polyacrylamide gel electrophoresis in SDS-containing gels is essentially that described by Laemmli[13] except that protein substrates have been copolymerized with the gels[14] as described below. Table I lists the stock solutions and Table II the components of the gels used in our laboratory to perform zymography in SDS-containing polyacrylamide gels. Development of zymograms is performed using a modification of the method described by Heussen and Dowdle.[14] Table III lists the stock solutions used in our laboratory to develop zymograms containing protein substrates. The method used for performing polyacrylamide gel electrophoresis in nondissociating gels is essentially that of Reisfeld *et al.*[15] as modified by Dewald *et al.*[16] and Miyasaki *et al.*[17] Table IV lists the stock solutions and Table V the components of the gels used in our laboratory to perform PAGE in nondissociating gels. Table VI lists the solutions used in our laboratory to detect proteases following PAGE in nondissociating gels.

Zymographic Techniques with Sodium Dodecyl Sulfate-Containing Gels

Specific Details for Preparation of Gels Containing Sodium Dodecyl Sulfate

Casting Gels. Pour the separating gel solution (Table II) into the prepared casts (~4.5 ml each) and overlay with *n*-butanol saturated with water. Polymerization is complete in approximately 30 min, after which the *n*-butanol should be removed with gentle aspiration. The stacking gel

[13] U. K. Laemmli, *Nature (London)* **277**, 680 (1970).
[14] C. Heussen and E. B. Dowdle, *Anal. Biochem.* **102**, 196 (1980).
[15] R. A. Reisfeld, V. J. Lewis, and D. E. Williams, *Nature (London)* **195**, 281 (1962).
[16] B. Dewald, R. Rindler-Ludwig, U. Bretz, and M. Baggiolini, *J. Exp. Med.* **141**, 709 (1975).
[17] K. T. Miyasaki, A. L. Bodeau, and T. F. Flemming, *Infect. Immun.* **59**, 3760 (1991).

TABLE I
STOCK SOLUTIONS FOR ELECTROPHORESIS IN SDS-CONTAINING GELS[a]

Solution	Preparation
Acrylamide–bis acrylamide, 30% : 0.8%	30 g of acrylamide and 0.8 g of bisacrylamide, dilute to 100 ml[b]
Separating gel buffer, 2 M Tris-HCl, pH 8.8	Trizma base 121.1 g; adjust to pH 8.8, final volume 500 ml[b]
Stacking gel buffer, 0.5 M Tris-HCl, pH 6.8	Trizma base 30.28 g; adjust to pH 6.8, final volume 500 ml[b]
Sodium dodecyl sulfate (SDS), 20% (w/v)	SDS, 20 g in 100 ml[c]
TEMED	As provided by Bio-Rad[b]
Ammonium persulfate, 10% (w/v)	Ammonium persulfate, 100 mg in 1 ml; make up fresh
n-Butanol saturated with water	n-Butanol and distilled water, 1 : 1 (v/v), shake vigorously, use upper phase[c]
Running buffer 0.192 M glycine, 25 mM Tris-HCl, 0.1% SDS	14.41 g glycine, 3.02 g Trizma base, 1 g SDS; bring to 1 liter but do not adjust pH[c]
Sample buffer (nonreducing)	Add 10 mg bromphenol blue to 4 ml of 2M Tris-HCl, pH 8.8, add 25 ml of 60% (w/v) sucrose, 17.5 ml of 20% SDS; bring volume to 50 ml[b]
Fibrinogen, 10 mg/ml in distilled water	Dissolve 20 mg fibrinogen in 2 ml of distilled water[b]
Gelatin, 10 mg/ml in distilled water	Dissolve 20 mg gelatin in 2 ml of distilled water[b]
Casein, 10 mg/ml	Dissolve 20 mg casein in 2 ml of distilled water, add 1 drop of separating gel buffer to bring to alkaline pH[b]

[a] Unless otherwise noted, all solutions are prepared in distilled water.
[b] Store at 4°.
[c] Store at 22°.

solution can be added and then the well-forming combs placed. Polymerization of the stacking gel is complete in approximately 30 min. We obtain the best results using freshly prepared gels. Storing gels overnight is not recommended.

Sample Preparation. Enzymes are dissolved in deionized water before use. Prior to electrophoresis, sample buffer is mixed with enzyme solutions (1 : 3, v/v), and allowed to stand at room temperature for 5–10 min before being applied to the gels.

Running Gels. The gels are placed in the gel apparatus, and running buffer is placed in contact with the gels. Electrophoresis is performed with constant voltage (150 V), at 4° unless otherwise noted. Electrophoresis, which takes approximately 45 min, is stopped when the tracking dye is within 2 mm of the bottom of the gel.

TABLE II
PREPARATION OF SDS-CONTAINING GELS

Component	Volume
Separating gel 8%: 10 ml for two minislabs (Mini-PROTEAN II, Bio-Rad)	
Acrylamide–bisacrylamide, 30% : 0.8%	2.67 ml
Tris-HCl buffer, 2 M, pH 8.8	1.88 ml
SDS, 20%	50 μl
Protein substrate stock solution[a]	1 ml
Ammonium persulfate, 10% (w/v)	35 μl
TEMED	5 μl
Distilled water	4.36 ml
Stacking gel 3%: 5 ml for two minislabs (Mini-PROTEAN II, Bio-Rad)	
Acrylamide–bisacrylamide, 30% : 0.8%	0.5 ml
Tris-HCl buffer, 0.5 M, pH 6.8	1.25 ml
SDS, 20%	25 μl
Ammonium persulfate, 10% (w/v)	20 μl
TEMED	7 μl
Distilled water	3.17 ml

[a] Delete substrate and replace volume with distilled water if substrate will be diffused into gel following electrophoresis or if indicator gels will be used to detect proteolytic activity.

Development of Zymograms after Electrophoresis

Copolymerized Substrate. Following electrophoresis, gels are placed in individual polystyrene petri dishes (150 mm diameter, 25 mm deep) and washed successively with 50-ml portions of 2.5% (v/v) Triton X-100 in distilled water (2 times, 10 min each), 2.5% (v/v) Triton X-100 in TB (2 times, 10 min each), and TB (2 times 10 min each) to remove SDS (Table III). After washing is completed, TB (50 ml) is poured into dishes containing the gels, and the dishes are covered and incubated at 37° for 1.5 hr. Incubations can be performed for longer periods (overnight) if required, and protease activators and inhibitors can be added to the TB solution during the incubation period or, in some cases, to washing solutions as required. Increasing the incubation time generally increases sensitivity; however, prolonged incubation can result in diffusion of proteases and substrates with concomitant loss of resolution. The best incubation time for each protease–substrate pair should be determined. Following incubations, gels are fixed for 10 min, stained with Coomassie Brilliant Blue R-250 solution, and destained to reveal zones of substrate lysis.

Substrate Diffused into Gel after Electrophoresis. Protein substrates can be diffused into gels immediately following electrophoresis, during washing, or during the incubation period. TB (50 ml) containing the protein

TABLE III
Stock Solutions for Development of Zymograms Following Electrophoresis[a]

Solution	Preparation
Tris buffer, 50 mM Tris-HCl, pH 7.4 (TB)	Dissolve 6.06 g of Trizma base in distilled water, adjust pH to 7.4 with 50% HCl; bring volume to 1 liter[b]
Triton X-100, 2.5% in distilled water (w/v)	Dissolve 25 g of Triton X-100 in distilled water adjust volume to 1 liter[b]
Triton X-100, 2.5% in TB (w/v)	Dissolve 25 g of Triton X-100 in TB; adjust volume to 1 liter[b]
Fixing/destaining solution, 10% methanol, 10% acetic acid (v/v)	Methanol 400 ml, glacial acetic acid 400 ml; bring to 4 liters with distilled water[b]
Staining solution	Mix 2.3 g Coomassie Brilliant Blue R-250, 58 ml glacial acetic acid, and 300 ml methanol; bring to 1 liter then stir for 1 hr and filter through Whatman (Clifton, NJ) No. 1 filter paper[b]

[a] Unless otherwise noted, all solutions are prepared in distilled water.
[b] Store at 22°.

substrate at a final concentration of 10 mg/ml is added to the dishes containing gels at the appropriate time, otherwise washing and incubation are as described above for gels containing copolymerized substrate. Following incubations, gels are fixed for 10 min, stained with Coomassie Brilliant Blue R-250 solution, and destained to reveal zones of substrate lysis.

Preparation of Indicator Gels for Use with Overlay Techniques

Indicator gels are used to contain protein substrates for certain applications. In these "overlay" techniques, proteases are separated by conventional SDS–PAGE (without protein substrates). The gels are washed following electrophoresis to remove SDS and then are overlaid on indicator gels containing protein substrates to detect proteases. Proteases diffuse from the separating gel into the indicator gel and lyse the substrate. After incubations, indicator gels can be fixed, stained, and destained to reveal zones of lysis in the same manner as described for substrate-containing polyacrylamide gels.

Agarose–Protein Indicator Gels. GEL-BOND film (FMC Bioproducts, Rockland, ME) is cut to the same size as the large glass plate of the Mini-PROTEAN II gel system (Bio-Rad) and fastened to the plate using laboratory tape to secure all sides, according to the manufacturer's directions. The tape is used to limit the spread of the agarose, which is dissolved in TB [1% agarose (w/v), final concentration] by heating to 60°. An aliquot (2.7 ml) of the agarose solution is placed in a polystyrene tube and cooled in a water bath to 40°, after which an aliquot (0.3 ml) of the protein solution is added; the solution is mixed and spread on the prepared GEL-BOND plate using a Pasteur pipette. The gel solidifies at room temperature. When the gel solidifies, the tape is removed, and the GEL-BOND film with the attached indicator gel is placed in a petri dish until used.

Polyacrylamide–Protein Indicator Gels. The indicator gels are identical to the polyacrylamide gels containing copolymerized substrates, except that the acrylamide concentration is 7% and SDS is omitted.

Zymographic Techniques with Nondissociating Gels

Specific Details for Preparation of Nondissociating Gels

Casting Gels. Prepare the separating gel (Table V), pour the solution into the prepared cast (~4.5 ml each), and overlay with *n*-butanol saturated with water. After polymerization is complete (~30 min), the butanol should be removed by gentle aspiration. Separating gels can be stored overnight at 4° overlaid with distilled water. Mix and pour the stacking gel, inserting well-forming combs. Polymerization of the stacking gel is accomplished by exposing the gel to a UV or white light source. After polymerization is complete (~45 min with a strong light source), gels may be stored overnight in a wet chamber at 4°.

Sample Preparation. Samples containing high salt concentrations should be desalted using a PD-10 column (Pharmacia, Piscataway, NJ) equilibrated with sample buffer *without* methyl green and glycerol. Sample buffer can be used for making serial dilutions and for dissolving solid samples.

Running Gels. The gels are placed in the gel apparatus, and running buffer is placed in contact with the gels. Electrophoresis is performed with reversed polarity for 2–8 hr at 4°, with a constant current of 10 mA per gel. The level of running buffer may decrease in the inner chamber during the run. If this occurs, interrupt the run and replenish the buffer.

Development of Zymograms after Electrophoresis in Nondissociating Gels

Copolymerized Protein Substrate. Following electrophoresis, gels are placed in individual polystyrene petri dishes (150 mm diameter, 25 mm deep) and washed twice with 30 ml of PB (Table VI) (5 min each time) at room temperature, to raise the pH of the gel to 7.0. The gels are then

TABLE IV
STOCK SOLUTIONS FOR ELECTROPHORESIS IN NONDISSOCIATING GELS[a]

Solution	Preparation
Acrylamide–bisacrylamide, 30% : 0.8%	30 g of acrylamide and 0.8 g of bisacrylamide dilute to 100 ml[b]
KOH, 1 N	5.6 g KOH in 100 ml[c]
Glacial acetic acid	As provided by manufacturer[c]
TEMED	As provided by Bio-Rad[b]
Triton X-100, 10% (v/v)	10 ml Triton X-100, bring to 100 ml[c]
Digitonin suspension, 1% (w/v)	0.1 g digitonin, 10 ml distilled water; shake vigorously before pipetting (digitonin is not soluble in water)[c]
Ammonium persulfate, 10% (w/v)	Ammonium persulfate, 100 mg in 1 ml distilled water; make up fresh
Riboflavin, 0.1 mg/ml	1 mg of riboflavin, bring to 10 ml[b]
Running buffer	31.2 g β-alanine, 8 ml glacial acetic acid, 5 ml 10% Triton X-100, 10 ml 1% digitonin suspension; bring to 1 liter[c]
Sample buffer	6 ml 1 N KOH, 360 μl glacial acetic acid, 20 ml glycerol, 5 ml 10% Triton X-100, 1 ml 1% digitonin suspension, 2 ml 0.2% methyl green[b,d]
Methyl green, 0.2% (w/v)	4 mg methyl green dissolved in 2 ml distilled water[c]
Protein stock solutions	Same as for PAGE in SDS

[a] Unless otherwise noted, all solutions are prepared in distilled water.
[b] Store at 4°.
[c] Store at 22°.
[d] More concentrated methyl green helps to follow run.

placed in fresh PB containing protease activators as required and incubated at 37° for 1–4 hr. Gels are then fixed, stained with Coomassie Brilliant Blue R-250, and destained to reveal zones of substrate lysis as described for PAGE in substrate-containing SDS gels.

Protein Substrate Diffused into Gel after Electrophoresis. Protein substrates can be diffused into gels immediately following electrophoresis, during washing, or during the incubation period. PB containing protein substrate (10 mg/ml) is added to dishes containing gels at the appropriate time. Otherwise, washing, incubation, and staining are as described above for gels containing copolymerized substrate.

Use of Esterase Substrates to Detect Proteolytic Activity. All solutions should be prepared immediately before use. For one minislab gel, 30 ml of staining solution is sufficient.

TABLE V
PREPARATION OF NONDISSOCIATING GELS

Components of nondissociating gels for cationic PAGE	Volume
Substrate-containing separating gel 15%: 10 ml for two minislabs (Mini-PROTEAN II, Bio-Rad)	
Acrylamide–bisacrylamide, 30% : 0.8%	5.0 ml
KOH, 1 N	0.3 ml
Glacial acetic acid	107 μl
TEMED	25 μl
Triton X-100, 10%	0.5 ml
Digitonin suspension, 1%	100 μl
Ammonium persulfate, 10%	125 μl
Protein substrate stock solution	1.0 ml
Distilled water	2.85 ml
Substrate diffusing into gel	
Separating gel 7.5%: 10 ml for two minislabs (Mini-PROTEAN II, Bio-Rad)	
Acrylamide–bisacrylamide, 30% : 0.8%	2.5 ml
KOH, 1 N	0.3 ml
Glacial acetic acid	107 μl
TEMED	25 μl
Triton X-100, 10%	0.5 ml
Digitonin suspension, 1%	100 μl
Ammonium persulfate, 10%	125 μl
Distilled water	6.34 ml
Stacking gel 3%: 5 ml for two minislabs (Mini-PROTEAN II, Bio-Rad)	
Acrylamide–bisacrylamide, 30% : 0.8%	0.5 ml
KOH, 1 N	225 μl
Glacial acetic acid	22.5 μl
TEMED	3 μl
Triton X-100, 10%	250 μl
Digitonin suspension, 1%	50 μl
Riboflavin, 0.1 mg/ml	0.5 ml
Distilled water	3.45 ml

1. Wash gels in 30 ml of PB (2 times, 5 min each).
2. Dissolve 15 mg of substrates, either naphthol AS-D acetate or N-acetyl-DL-phenylalanyl-β-naphthyl ester, in 1.5 ml of N,N-dimethylformamide. (This is the substrate solution.)
3. Dissolve 20 mg of Fast Blue RR salt in 2 ml of N,N-dimethylformamide for each substrate. (This is the dye solution.)
4. Mix 6 ml of N,N-dimethylformamide with 24 ml of PB. (This is the development buffer.)
5. Add the substrate solution and the dye solution to the development buffer, mixing vigorously.

TABLE VI
SOLUTIONS FOR DEVELOPING ZYMOGRAMS FOLLOWING NONDISSOCIATING ELECTROPHORESIS[a,b]

Solution	Preparation
Sodium phosphate buffer, 0.1 M, pH 7.0 (PB)	Mix 78 ml of 0.5 M NaH$_2$PO$_4$ with 122 ml of 0.5 M Na$_2$HPO$_4$; dilute to 1 liter and adjust to pH 7.0
Naphthol AS-D acetate or N-acetylphenylalanyl-β-naphthyl ester substrate solution[c]	Dissolve 15 mg of substrate in 1.5 ml of N,N-dimethylformamide
Fast Blue RR dye solution[c]	Dissolve 20 mg of Fast Blue RR salt in 2 ml of N,N-dimethylformamide
Acetic acid, 7% (v/v)	Dilute 7 ml of glacial acetic acid to 100 ml

[a] Unless otherwise noted, all solutions are prepared in distilled water.
[b] Protein substrate stock solutions, as well as staining and destaining solutions for gels in which protein substrates are copolymerized or allowed to diffuse into the gels following electrophoresis, are the same as described for PAGE in SDS-containing gels.
[c] Must be prepared immediately before use.

6. Place the washed gel in this solution and incubate at 37° until defined bands appear (5 to 10 min).
7. The reaction may be stopped by placing the gel in 7% (v/v) acetic acid.

Use of Zymographic Techniques to Identify and Characterize Microbial Proteases

Zymography Using Gels Containing Sodium Dodecyl Sulfate

Denaturing Gels Using Copolymerized Protein Substrates. The gel electrophoresis system we have used extensively is essentially the system described by Laemmli[13] with the exception that the protein substrate is added to acrylamide solutions before polymerization and is copolymerized with the gel.[14] For the experiments shown in Fig. 1, protein substrate stock solutions were added to acrylamide mixtures, prior to starting the polymerization reaction, so that the final concentration in the gel was 1 mg/ml. Gels of 0.75 mm thickness were used, and they were cast for use in a Mini-PROTEAN II (Bio-Rad) electrophoresis system (gel dimensions were 83 by 55 mm). The separating gel consisted of an 8% acrylamide gel overlaid with a 3% stacking gel (Table II). Enzyme solutions were diluted in sample buffer (2-fold serial dilutions), and electrophoresis was performed under nonreducing conditions at 4° at constant voltage (150 V). In contrast to the Laemmli system, enzyme samples are not heated prior to electrophoresis. After electrophoresis, gels were washed first with 2.5%

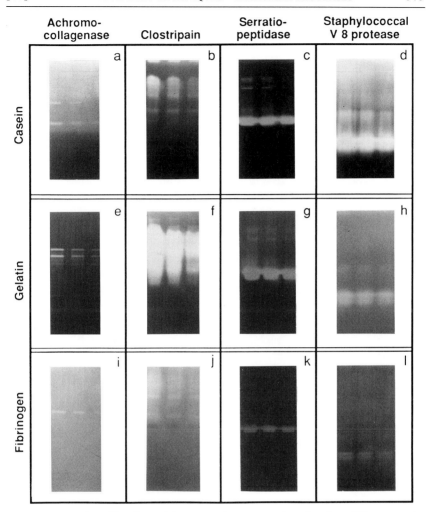

FIG. 1. Detection of four bacterial proteases by zymography using three different copolymerized protein substrates. All polyacrylamide gels are 8%, and all copolymerized proteins are present at 1 mg/ml. Electrophoresis was performed at 4°, and zymograms were developed in TB at 37° for 1.5–4 hr. For detection of clostripain, 10 mM dithiothreitol was added during development. Achromocollagenase (6.0, 3.0, 1.5 μg) in gels with copolymerized (a) casein, (e) gelatin, and (i) fibrinogen; clostripain (0.3, 0.15, 0.075 μg) in gels with copolymerized (b) casein, (f) gelatin, and (j) fibrinogen; serratiopeptidase (4.4, 2.2, 1.1 μg) in gels with copolymerized (c) casein, (g) gelatin, and (k) fibrinogen; staphylococcal V8 protease (2.6, 1.3, 0.65 μg) in gels with copolymerized (d) casein, (h) gelatin, and (l) fibrinogen.

Triton X-100 in distilled water, then with 2.5% Triton X-100 in TB, and finally with TB. The washing procedure was performed over a 1-hr period to remove SDS.

Development of Zymograms. Proteolytic bands were activated in the washed gel by incubation in TB at 37° for 1.5–4 hr. Dithiothreitol (1 mM final concentration) was added during development of gels in which clostripain was used. Otherwise, no other protease activators were added during development of zymograms. Finally, the gels were fixed and stained with Coomassie Brilliant Blue R-250. Lytic bands were visualized against the dark blue background of the gels after destaining.

The data presented in Fig. 1 demonstrate that numerous microbial proteases can be detected after SDS–PAGE in substrate-containing gels, using standard conditions and neutral pH during enzyme activation. We were able to detect all of the bacterial proteases we attempted to use in this study by zymography in denaturing gels containing copolymerized substrates. Even under nonoptimal conditions, we were able to detect clostripain, serratiopeptidase, and V8 protease in the nanogram range with at least one of the substrates. At the limits of visual detection of substrate lysis, obtaining representative photographs is technically impossible, and, for this reason, concentrations of enzymes represented in the figures are substantially higher than those necessary to achieve enzyme detection in the laboratory.

Comments

The data presented in Fig. 1 can be used to illustrate several technical points concerning the use of this technique for detecting protease activity.

Quantitation of Enzyme Activity. The data presented in Fig. 1 demonstrate that, in most cases, the zone of clearing in the zymogram is proportional to the amount of enzyme loaded on the gel. These methods can be used semiquantitatively.

Determination of Relative Molecular Weight(s) of Catalytically Active Component(s). In most of the commercial enzyme preparations proteolytic activity appears to be associated with more than one band. This is most likely due to the presence of catalytically active aggregates of the enzymes, the presence of catalytically active autodigestion products of the enzymes, or to the presence of other contaminating proteases in the commercial preparations.[18,19] The relative molecular weights of each of the active components can be determined by comparing the migration distances with those of standard proteins run on the same gel. Protein standards can

[18] S. A. Lacks and S. S. Springhorn, *J. Biol. Chem.* **255,** 7467 (1980).
[19] V. Keil-Dlouha, *Biochim. Biophys. Acta* **429,** 239 (1976).

easily be visualized against the dark blue background of the zymogram by placing the zymogram on a light box; however, for technical reasons they are not easily photographed. The positions of protein standards on the gels shown in Fig. 1 were not included because of space considerations, but indications of the positions of protein standards on similar zymograms have been included in other figures (see Figs. 3, 4, and 6). We have observed that inclusion of protein substrates in gels does slightly affect the migration of enzymes as well as standard proteins in gels.

We do not recommend adding reducing agents to samples prior to electrophoresis. The reasons are 2-fold. First, cysteine proteases may be activated by reducing agents and extensive autodegradation may occur, even in the presence of SDS. Second, some enzymes require a disulfide-stabilized structure (either intramolecular or between polypeptide chains) in order to be catalytically active.[18] If reducing agents are added routinely to enzyme samples prior to electrophoresis such enzymes may not be detected. This is especially important when new enzymes are being identified and characterized. The effects of adding reducing agents to enzyme samples prior to electrophoresis should be evaluated in a stepwise fashion and may provide important information about both the native structure and catalytic mechanism of an enzyme.

Because samples are not boiled prior to electrophoresis, aggregation of proteases may occur, resulting in detection of some high molecular weight proteolytic species on zymograms. We have found that solubilization of enzymes in SDS without boiling can be enhanced by prolonged incubation of enzymes in sample buffer at room temperature (15–30 min) prior to loading samples on gels.

Heussen and Dowdle[14] have reported that estimates of the relative molecular weights of several plasminogen activators decreased slightly as the percentage of acrylamide decreased from 12 to 8%. This anomalous electrophoretic behavior was not affected by increasing the concentration of plasminogen or SDS in the gels, suggesting that the apparent retardation of the enzymes in the gels with higher total polyacrylamide concentrations was not due to affinity interactions of the enzymes with plasminogen.

In spite of the facts that some enzymes may migrate anomalously, samples are not boiled prior to electrophoresis, reducing agents are not added to samples prior to electrophoresis, and substrate proteins in the gel can affect enzyme migration during electrophoresis, quite accurate estimates of the molecular weights of many proteases have been obtained by zymography. It is usually advisable to verify molecular weight estimates of proteases obtained by zymography using additional methods.

Behavior of Protein Substrates and Substrate Specificity. As can be seen in Fig. 1, all of the enzymes tested degrade all three protein substrates in zymograms, although to varying extents. Casein migrates in the gels during electrophoresis as indicated by the gradient in staining (lighter background at the top of the gel) in some of the casein-containing gels. The migration of casein in the gels is probably due to its solubility at alkaline pH; however, in general, smaller protein substrates tend to migrate during electrophoresis and may diffuse out of gels during development of zymograms. In spite of migration of the substrate during electrophoresis, proteases were still detected in casein-containing gels. Migration and diffusion of substrates can be controlled in several ways including decreasing the pore size of the gel and covalently linking protein substrates to acrylamide.[12] Gelatin and fibrinogen did not migrate during electrophoresis under the experimental conditions used in these studies.

Insights into substrate specificity can be gained using zymographic techniques. For example, gelatin appears to be a better substrate for achromocollagenase and clostripain than is casein (compare Fig. 1a and 1e, and 1b with 1f). Because SDS present during electrophoresis can also denature protein substrates copolymerized in polyacrylamide gels, conclusions about substrate specificity using zymographic techniques should be made with caution. Although substrate specificity sites that are determined only by amino acid sequence are unlikely to be affected, specificity that relies on native substrate conformation may be highly affected. Collagen type I is a good example: SDS irreversibly denatures collagen type I to gelatin, hence gelatinolytic but not collagenolytic activity can be assessed in substrate-containing gels. Many proteins can serve as good protease substrates for zymographic procedures, and it is advisable to test several different substrates when attempting to identify and characterize proteases.

Effects of Sodium Dodecyl Sulfate on Proteases and Necessity for Removal Prior to Development of Zymograms. Most proteolytic enzymes are denatured by treatment with SDS and are inactive when complexed with this detergent. This is convenient during electrophoresis since substrate is not degraded during electrophoretic migration of the enzyme through the gel, and electrophoresis is said to be in the "nonbinding mode."[11] A notable exception in this study was pronase, which actively degrades protein substrates copolymerized in SDS-containing polyacrylamide gels during electrophoresis, regardless of whether electrophoresis is performed at 4° or at 22°. When an enzyme is fully active during electrophoresis, a clear track can be seen from the origin into the gel, the length of which is determined by the amount of enzyme present. This is known as electrophoresis in the "binding mode"[11] and is illustrated for pronase in Fig. 2. In "binding mode" electrophoresis, the enzyme binds to the substrate and hydrolyzes it. Movement of the enzyme through the gel,

1   2   3   4   5

FIG. 2. Detection of pronase by zymography using a polyacrylamide gel containing copolymerized fibrinogen. Electrophoresis of pronase (7.0, 3.5, 1.8, 0.9, 0.45 μg) was performed at 22°, and gels were developed in TB, 37°, for 1.5 hr.

all else being equal, is determined by the rate of hydrolysis of the substrate, and hence by the amount of enzyme loaded onto the gel. The more enzyme, the farther the enzyme can move. Electrophoresis in the "binding mode" has as an advantage the fact that the enzyme remains in the native state during the run, which maximizes the sensitivity of detection. A disadvantage is that multiple forms of the enzyme will not be detected. Multiple forms of pronase can be detected using zymographic techniques when electrophoresis is performed in nondissociating gels at pH 4.3, conditions under which it is minimally active during electrophoresis (see below).

Zymography in SDS-containing gels cannot be used to detect proteases that do not renature following exposure to SDS; however, most proteases do renature, at least to an extent that allows their detection, following removal of SDS from the gels.[6,18] Exceptions to this generalization are multimeric enzymes. For those proteases that do not renature after treatment with SDS, electrophoresis in nondissociating polyacrylamide gels may be an alternative option (see below). Some proteases are partially active in SDS during electrophoresis, and for this reason electrophoresis is generally performed at 4°. Because the extent to which enzyme detection requires removal of SDS from the gels varies for each enzyme,[18] a prolonged, multistep washing procedure is presented here. This procedure can be shortened as appropriate.

Characterization of Proteases

Analysis by SDS–PAGE in substrate-containing gels allows rapid, semiquantitative measurement of proteases in mixtures along with estimation of their molecular weights. In addition, this technique can be helpful in further characterization of an enzyme.

Activators and Inhibitors of Proteolysis. Protease activators and inhibitors can be added during development of zymograms to assess the effect on enzyme activity. The molecules enter the gel by diffusion. Figure 3 illustrates the pronounced activation of achromocollagenase by Ca^{2+}, as is expected for a metalloprotease (compare Fig. 3b and 3c). It is noteworthy that this enzyme was detected even though $CaCl_2$ was not added during development of the zymogram, reinforcing the applicability of these techniques for initial identification of proteases. Figure 3a is of interest in several respects. First, it demonstrates the marked preference of the enzyme for gelatin over casein as a substrate. Second, it suggests that although caseinolytic activity in the preparation is produced by an enzyme that also is gelatinolytic, there is a caseinolytic activity in the preparation that is not gelatinolytic. This contaminating protease has an M_r in the

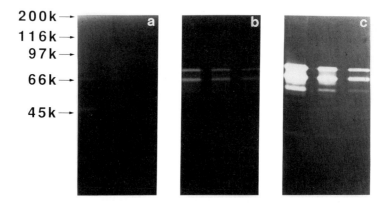

FIG. 3. Effect of substrate and Ca^{2+} on detection of achromocollagenase. (a) Achromocollagenase in casein-containing gel (6, 3, 1.5 μg) developed in TB, 37°, 4 hr, 10 mM $CaCl_2$; (b) achromocollagenase in gelatin-containing gel (6, 3, 1.5 μg) developed in TB 37°, 2 hr; (c) achromocollagenase in gelatin-containing gel (6, 3, 1.5 μg) developed in TB, 37°, 2 hr, 10 mM $CaCl_2$. The numbers and arrows on the left-hand side indicate the molecular sizes (in kilodaltons) of standard proteins.

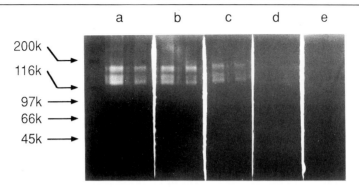

FIG. 4. Inhibition of *P. gingivalis* proteases by TLCK. A *P. gingivalis* extract (0.5% CHAPS) was loaded onto a 6% polyacrylamide gel into which fibrinogen (0.1% final concentration) had been copolymerized. After electrophoresis, the gel was cut into five strips, so that each strip contained two samples, 15 μg protein (left-hand side) and 7.5 μg protein (right-hand side). The gel strips were washed as for zymography except that TLCK was included in the washing solutions at the concentrations indicated below. After washing, the gels were incubated for 90 min at 37° in TB containing 50 mM cysteine and TLCK as indicated below. Gel (a) control, no TLCK; (b) 0.001 mM TLCK; (c) 0.01 mM TLCK; (d) 0.1 mM TLCK; (e) 1 mM TLCK. The numbers and arrows on the left-hand side indicate the molecular sizes (in kilodaltons) of standard proteins.

range of 45,000. Figure 4 illustrates the use of zymography to examine the concentration-dependent inhibition of enzymes from *P. gingivalis* by N^α-*p*-tosyl-L-lysyl chloromethyl ketone (TLCK). We have tentatively classified these enzymes as cysteine proteases because their activity is unaffected by phenylmethylsulfonyl fluoride and diisopropyl fluorophosphate, they require activation by thiol reagents (cysteine and/or dithiothreitol), and they are inhibited by TLCK. It is also possible to discriminate unique proteolytic activities in mixtures containing more than one protease based on differential responses to activators and inhibitors of proteolysis.

Reverse Zymography. The previous section has illustrated how zymography can be used to study protease activation and inhibition by low molecular weight activators and inhibitors of proteolysis that enter the gel by diffusion during development of zymograms. High molecular weight protease inhibitors can also be identified using a modification of the above techniques called reverse zymography.[20,21] In this technique, polyacrylamide gels containing copolymerized substrates are prepared for zymography as described above. However, instead of loading the gel with enzyme

---

[20] J. S. Hanspal, G. R. Bushell, and P. Ghosh, *Anal. Biochem.* **132,** 288 (1983).
[21] G. S. Herron, M. J. Banda, E. J. Clark, J. Gavrilovic, and Z. Werb, *J. Biol. Chem.* **261,** 2814 (1986).

samples, the gel is loaded with a high molecular weight protease inhibitor which migrates into the gel during electrophoresis. Following electrophoresis, the gel is incubated in a solution containing a protease and activators as required, which enter the gel by diffusion, and lyse the substrate, except in the region of the gel to which the inhibitor has migrated.

We attempted to demonstrate this technique using bacterial proteases with soybean trypsin inhibitor (SBTI) and α_1-antitrypsin as inhibitors. We were unable to demonstrate inhibition of any of the bacterial proteases by these inhibitors using reverse zymography. The bacterial enzymes degraded these inhibitors like any protein substrate, and all the gels were clear. Figure 5 (top) illustrates the inhibition of trypsin by SBTI. The stained regions of the gel represent areas where fibrinogen lysis by trypsin

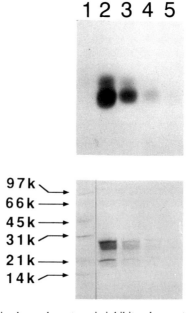

FIG. 5. Trypsin inhibition by soybean trypsin inhibitor demonstrated by reverse zymography. (*Top*) Fibrinogen, 0.1% final concentration (1 mg/ml), was copolymerized in an 11% acrylamide gel. Soybean trypsin inhibitor was dissolved in distilled water, mixed with sample buffer (3 : 1, protein solution/sample buffer), and loaded onto the gel. Following electrophoresis, the gel was washed (as for zymography) and incubated in TB (50 ml) containing trypsin (14 μg/ml, final concentration) for 16 hr at 37°. Following the incubation period, the gel was rinsed with TB and stained with Coomassie Brilliant Blue R-250. (*Bottom*) Soybean trypsin inhibitor after SDS–PAGE in an 11% acrylamide gel (no substrate added). Lane 1, Protein standards (the numbers and arrows on the left-hand side indicate the molecular sizes, in kilodaltons, of standard proteins); lanes 2–5, SBTI (9, 4.5, 2.25, 1.13 μg).

was inhibited by SBTI. The clear areas of the gel are areas where copolymerized fibrinogen was degraded by trypsin. The location and staining intensity of a commercial preparation of SBTI following SDS–PAGE are shown in the bottom gel for comparison. This technique may be particularly useful in identifying inhibitors of microbial proteases in complex mixtures, such as tissue samples or microbial extracts. As is the case with protease detection following SDS–PAGE and zymography, reverse zymography can only detect those protease inhibitors that have the ability to renature after exposure to SDS.

Diffusion of Substrates into Gels Following Electrophoresis. Proteolytic enzymes can be separated by SDS–PAGE in gels devoid of a protein substrate. Following electrophoresis, gels are incubated with a protein substrate. The substrate will diffuse into the gel as a function of time, its size, and the pore size of the gel. In general, substrates should be used at higher concentrations (10-fold) than those used when substrates are copolymerized with the gel.

This method for detection of proteolytic activity is illustrated in Fig. 6. The substrate was diffused into the gel during development of the zymogram. Several points are noteworthy. The background staining is uneven because of uneven diffusion of the substrate, casein, into the gel.

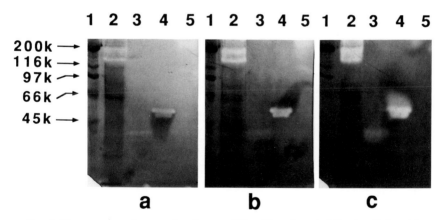

FIG. 6. Zymography using protein substrate diffused into the gels following SDS–PAGE. Proteases were mixed with sample buffer and loaded onto 8% polyacrylamide gels. Following electrophoresis, gels were washed as described in the text to remove SDS, the incubated in TB containing casein (10 mg/ml), cysteine (50 mM), and CaCl$_2$ (10 mM) for (a) 30 min, (b) 1 hr, or (c) 2 hr. Following the incubations, gels were fixed and stained with Coomassie Brilliant Blue R-250. Lanes 1, Protein standards (the numbers and arrows on the left-hand side indicate the molecular sizes, in kilodaltons, of standard proteins); lanes 2, *P. gingivalis* proteases (10 μg protein); lanes 3, V8 protease (2.8 μg); lanes 4, serratiopeptidase (3 μg); lanes 5, pronase (2 μg).

The background is less intensely stained in this gel than in gels containing copolymerized substrate. Nevertheless, the enzymes are detected with high sensitivity. In some cases it may be advantageous to visualize all the proteins in an extract which has proteolytic activity, and this may be difficult to accomplish using copolymerized substrates. Nonproteolytic proteins in a *P. gingivalis* extract can be readily seen against the background of the gels when casein is diffused into the gel during zymogram development (Fig. 6a, lane 2) but cannot be readily distinguished in gels containing copolymerized substrate (Fig. 4). Figure 6 also illustrates that proteases diffuse in the gels as the substrate diffuses in. The proteolytically active bands in all the samples were sharpest after 30 min of incubation and lost resolution as the incubation time was increased (compare Fig. 6a with 6b and 6c). Proteases in gels containing copolymerized substrates diffuse much less rapidly (compare Figs. 6 and 4). More even background staining, comparable to that achieved with copolymerized substrate, can be achieved if substrates are diffused into gels immediately following electrophoresis or during washing procedures.

The "substrate diffusing in" technique can be extremely helpful in identifying proteolytically active bands in protein mixtures. In situations where it is important to distinguish which of several closely migrating bands is proteolytically active, the bands in a substrate diffusion zymogram may be aligned more readily with the protein bands in a sample separated by SDS–PAGE and stained with either Coomassie Brilliant Blue R-250 or silver. When substrate is copolymerized in the gel used for zymography, the migration of sample proteins in the gel can be slightly retarded, making comparisons with stained protein bands in a conventional (non-substrate-containing) gel difficult.

Use of Indicator Gels (Overlay Techniques). In techniques discussed thus far, protein substrates either have been included in the polyacrylamide gels or entered the gels by diffusion after electrophoresis. Zymography can also be performed using a second gel to contain the protein substrate. In this method, proteases are subjected to conventional SDS–PAGE, and the gel is washed to remove SDS and overlaid on a second gel which contains a protein substrate. Protease activators are either added to the final washes of the first gel or incorporated with the protein substrate into the indicator gel. Proteases enter the indicator gel by diffusion and lyse the substrate. Depending on the nature of the substrate, zones of lysis can be detected by visual examination or by staining the indicator gel with a protein stain, i.e., Coomassie Brilliant Blue R-250. Registered on the indicator gel is what has been described as a "contact print"[7] of the proteolytically active proteins in the first gel. Zymograms on agarose–fibrinogen and acrylamide–fibrinogen indicator gels are shown in Fig. 7. The results presented in Fig. 7 suggest that the proteases diffuse more readily into agarose–fibrinogen than into acryl-amide–fibrinogen gels. V8 protease is barely detectable on the fibrino-gen–acrylamide gel (compare Fig. 7a and 7b, lanes 3), and the pronase sample used for the acrylamide–fibrinogen gel appears degraded relative to the sample used in the agarose–fibrinogen gel (compare Fig. 7a and 7b, lanes 1).

Indicator gels have some advantages for specific applications. Some protein substrates, fibrin, for example, form cloudy gels with agar and agarose.[6] When fibrin is lysed, a clear zone forms which can be seen by illuminating the gel overlays against a dark background. In this system, the extent of proteolysis can be followed without stopping the reaction, as is necessary when lysis is assessed using protein staining. If it is import-ant not to denature a particular protein substrate, as might occur during PAGE in SDS, incorporating it into a second gel is a good option. There are two main disadvantages of methods using indicator gels. First, they require that a second gel be prepared. Second, acrylamide gels shrink

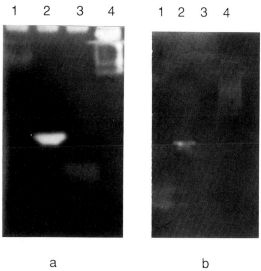

a   b

FIG. 7. Detection of proteolytic activity following SDS–PAGE using indicator gels. Proteases were subjected to electrophoresis in 8% polyacrylamide gels. Following electro-phoresis, gels were washed as described in the text to remove SDS. Cysteine (50 mM) and CaCl$_2$ (10 mM) were added to the final wash. The gel was cut in half, and one half was placed on an agarose–fibrinogen indicator gel (a) and the other placed on an acrylamide–fibrinogen indicator gel (7%) (b). The gel overlays were enclosed in petri dishes as described and incubated at 37° for 1.5 hr. Following the incubation, the indicator gels were stained with Coomassie Brilliant Blue R-250. Lanes 1, Pronase (2 μg); lanes 2, serratiopeptidase (3 μg); lanes 3, V8 protease (2.8 μg); lanes 4, P. gingivalis proteases (10 μg protein).

with fixation and staining, which makes lining up molecular weight markers in the gel after staining with lytic zones in the indicator gel difficult. As a result, molecular weight estimations may be less precise than when markers and enzymes are in the same gel. This problem can be circumvented by using prestained molecular weight standards and marking their location physically in the indicator gel.

Other Adaptations of Zymographic Methods

Substrate Electrophoresis in Polyacrylamide Gradient Gels. Linear polyacrylamide gradient gels can also be used for zymography. They have several advantages over uniform gels for initial characterization of proteases. In general, bands are sharper than on uniform gels, the separation of each band from neighboring bands is better, and molecular weight estimates can be made over a broader range.[22] Protein substrates can be incorporated into gradient gels by adding them to acrylamide solutions at a concentration of 1 mg/ml, prior to casting the gradient gel. We used zymography in linear polyacrylamide gradient gels (3–9%) containing co-polymerized fibrinogen to initially characterize proteases from *P. gingivalis.*[9]

Substrate Electrophoresis in Gels Containing Acrylamide-Conjugated Substrates. Kelleher and Juliano[12] have described a method for covalently linking protein substrates to glutaraldehyde-activated linear polyacrylamide. The use of conjugated substrates has several advantages. It prevents migration of substrates in the gel during electrophoresis as well as diffusion of the substrate out of the gel during zymogram development. This increases the sensitivity of the system because lower substrate concentrations can be used, 200–500 μg protein/ml, as opposed to 1 mg/ml for nonconjugated substrates.

In our experience, conjugation of protein substrates to acrylamide is generally not necessary. It may be particularly useful for small protein substrates, which otherwise may migrate during electrophoresis or diffuse out of the gel during development of zymograms. For most commonly used protein substrates, however, including casein, gelatin, bovine serum albumin, and fibrinogen, copolymerizing the proteins with the gel yields a highly sensitive protease detection system. The small increase in sensitivity that may be achieved using protein–acrylamide conjugates is more than offset by the effort required to prepare substrate-conjugated acrylamide.

Zymogen Activation. Granelli-Piperno and Reich[6] used zymography with agar indicator gels containing fibrin and plasminogen to detect plas-

---

[22] P. J. Blackshear, this series, Vol. 104, p. 242.

minogen activators in cell culture fluids following their separation by SDS–PAGE. Huessen and Dowdle[14] used zymography in SDS–poly-acrylamide gels containing copolymerized plasminogen and gelatin to ana-lyze plasminogen activators. They demonstrated that both the zymogen and the gelatin were retained in the gels during electrophoresis of enzyme samples and that they served as satisfactory sequential substrates for the identification and molecular weight estimation of plasminogen activators. Both studies suggest that zymographic techniques are well suited for both detection and characterization of activators of plasminogen and other zymogens. Zymogen activators can be differentiated from proteases by running, in parallel, zymograms that contain substrate but no zymogen. Zymogen activators will appear as zones of lysis only when a zymogen (i.e., plasminogen) is included in the gel.

Zymography in Nondissociating Gels

The methods described so far have utilized electrophoresis of proteins under denaturing, dissociating conditions, in the presence of SDS. As mentioned above, a few monomeric enzymes do not renature after treat-ment with SDS. Also, enzymes that are oligomeric, and consist of nonco-valently associated subunits, generally do not renature well after PAGE in SDS if the monomers are the same size, and they do not renature at all if the subunits are of different sizes.[18] For many of these enzymes, zymographic techniques may still be used if the proteases are separated in nondissociating gels under conditions where they are not catalytically active, or are minimally catalytically active. Rather than inhibiting cata-lytic activity by causing drastic changes in protein conformation, such as those induced by exposure of proteins to chaotropic agents or SDS, much gentler methods are used to inhibit enzyme activity, such as shifting the pH of the enzyme away from the pH optimum or removing an essential metal ion by treatment with EDTA. In this way, enzyme proteins retain much of the native structure, and these electrophoresis systems are consid-ered to be relatively "nondenaturing."

Separation of proteins in nondissociating gels depends on the size, shape, and charge of the protein: separation by size can be controlled by varying the pore size of the polyacrylamide, and separation by charge is controlled by varying the pH of the system. Suitable combinations of buffers, pH, and gel components can be derived for use with most proteins. Selection of appropriate nondissociating gel systems for use with any given enzyme can be facilitated by knowing in advance the approximate size of the enzyme, the isoelectric point of the enzyme, its activity over a range of pH values, and the extent to which inhibition at extremes of pH is reversible. Some standard procedures for preparing nondissociating acidic, neutral, and basic gels as well as a general review of electrophoretic techniques have been presented by Blackshear.[22]

Considerations for Zymography in Nondissociating Gels

Zymography in nondissociating gels can be performed with protein substrates copolymerized in the gel, by diffusing a protein substrate into the gel following electrophoresis, by using indicator gels, or by using low molecular weight synthetic substrates, the hydrolysis products of which can be coupled to dyes to detect esterase or protease activity following electrophoresis. Electrophoresis can be performed with the protease fully active, partially active, or inactive. With substrate-containing gels, as in the case of SDS–PAGE, it is usually desirable to perform electrophoresis with the enzyme reversibly inactivated to avoid "binding mode" electrophoresis (see Fig. 2). This is most frequently accomplished by performing electrophoresis at a pH where the enzyme is not active. After electrophoresis, the gel is washed in a buffer with an appropriate pH and with appropriate cofactors to activate the enzyme, and substrate lysis can be assessed as in the case of denaturing gels. Alternatively, for some proteases appropriate synthetic substrates are available which permit direct staining of proteolytic bands against the clear background of the gel.

We have found that electrophoresis in acidic gels, cationic PAGE, is useful in the identification and characterization of some microbial proteases. We have essentially used the method of Reisfeld et al.[15] as modified by Dewald et al.[16] and Miyasaki et al.[17] In addition to esterase substrates and dyes, we have used copolymerized protein substrates[11] and soluble protein substrates to detect proteolytic activity.

Copolymerized substrates. The results presented in Fig. 8a suggest that subtilisin (lanes 1 and 2) and chymotrypsin (lanes 5 and 6) are not active at pH 4.3, under conditions of cationic PAGE; however, pronase (lanes 3 and 4) is slightly active, but not nearly as active as in SDS at neutral pH (see Fig. 2). Figure 8b illustrates that subtilisin, pronase, and chymotrypsin are all active after electrophoresis at pH 4.3 in fibrinogen-containing gels, and the enzymes can be detected after increasing the pH of the gels to 7.0 and incubating them at 37°. Multiple forms of pronase (Fig. 8, lane 4) were detected by this method. Figure 9 demonstrates that clostripain can only be detected after cationic PAGE in fibrinogen-containing gels if dithiothreitol is added to incubation mixtures. At least two forms of clostripain can be seen in Fig. 9b (lanes 1–3). Comparison with Fig. 9a (lanes 1–3) suggests that the preparation of clostripain is contaminated with a small amount of a non-DTT-dependent protease (indi-

1   2   3   4   5   6

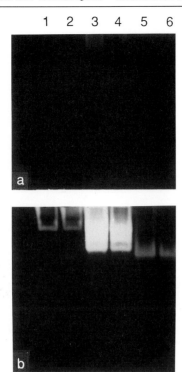

FIG. 8. Determination of enzyme activity during cationic PAGE. Subtilisin, lanes 1 and 2 (3.8, 1.9 μg), pronase, lanes 3 and 4 (3.8, 1.9 μg), and chymotrypsin, lanes 5 and 6 (3.8, 1.9 μg), were subjected to electrophoresis at pH 4.3 in polyacrylamide gels (15%) containing copolymerized fibrinogen at 4°, for 2 hr. (a) The gel was stained with Coomassie Brilliant Blue R-250 immediately after the run with no incubation. (b) The gel was washed in PB for 10 min, then incubated in PB for 2 hr at 37° before staining.

cated by *→). Pronase clearly occurs in three forms, whereas subtilisin appear to occur in one form. Detection of pronase and subtilisin is unaffected by the presence of DTT in the incubation buffer. We have found that band sharpness and resolution following cationic PAGE are mainly functions of the volume of the sample applied to the gel, that is, the smaller the volume, the sharper the bands.

   Soluble Protein Substrates Diffused into Gel Following Cationic Electrophoresis. Gelatin was used as a diffusable substrate for detection of pronase and subtilisin following cationic PAGE (Fig. 10). Electrophoresis was performed in 7.5% acrylamide gels, as opposed to the 15% gels used with copolymerized substrates. The uneven background staining is the result of uneven diffusion of gelatin into the gel. Both enzymes were

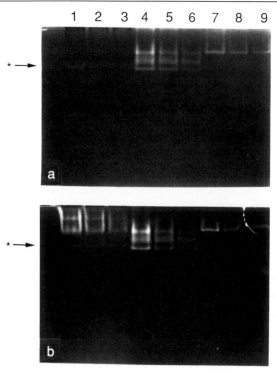

FIG. 9. Effect of addition of dithiothreitol during development of zymograms on detection of clostripain, pronase, and subtilisin, after cationic PAGE in gels containing copolymerized fibrinogen. Clostripain, lanes 1–3 (6.3, 3.3, 1.6 μg), pronase, lanes 4–6 (1.9, 1.0, 0.5 μg), and subtilisin, lanes 7–9 (1.9, 1.0, 0.5 μg), were subjected to electrophoresis at pH 4.3 in polyacrylamide gels (15%) containing copolymerized fibrinogen at 4°, for 2 hr. Following electrophoresis, gels were washed in PB for 10 min, then incubated in PB (a) or PB containing 10 mM dithiothreitol (b) for 1.25 hr at 37°. Gels were stained with Coomassie Brilliant Blue R-250 as described in the text.

readily detected using this technique; however, because of rapid diffusion of the enzymes in the gel during development of the zymogram, the ability to detect multiple forms of pronase was lost.

Use of Synthetic Esterase Substrates to Detect Proteases Following Cationic Electrophoresis. Most proteolytic enzymes exhibit esterase activity, and in some cases proteases hydrolyze amino acid esters more avidly than the corresponding peptides.[5] The ability of some proteases to hydrolyze 1-naphthyl esters of α-halo-substituted and α-amino-substituted carboxylic acids can be used to visualize these enzymes in gels since naphthol released on ester hydrolysis at neutral pH can be simultaneously

1   2   3   4   5   6   7   8

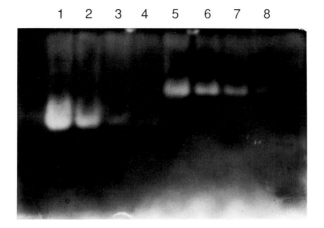

FIG. 10. Zymography following cationic PAGE using protein substrate diffused into the gel following electrophoresis. Pronase, lanes 1–4 (4.0, 2.0, 1.0, 0.5 μg), and subtilisin, lanes 5–8 (4.0, 2.0 1.0, 0.5 μg), were subjected to electrophoresis at pH 4.3 in polyacrylamide gels (7.5%) for 1.5 hr at 4°. Following electrophoresis, the gel was washed in PB for 10 min to raise the pH to 7.0, then incubated in PB containing gelatin (10 mg/ml) for 1.5 hr. The gel was subsequently fixed and stained with Coomassie Brilliant Blue R-250 as described in the text.

coupled to various azo dyes to yield a colored, insoluble product.[5] These reactions were originally developed as cytochemical stains for neutrophil granules[5,23] and were subsequently used to characterize neutrophil "esterases," which were ultimately shown to be proteases.[5,23,24] Peptide substrates (naphthylamide derivatives of amino acids) have also been used with simultaneous azo dye coupling to detect eukaryotic proteases in polyacrylamide gels.[10]

We have used the esterase substrates naphthol AS-D acetate and N-acetyl-DL-phenylalanyl-β-naphthyl ester coupled with Fast Blue RR salt to detect the microbial proteases pronase and subtilisin (Fig. 11) as well as bovine chymotrypsin. The results presented in Fig. 11, suggest that naphthol AS-D acetate is a better substrate for subtilisin than is N-acetyl-DL-phenylalanyl-β-naphthyl ester (Fig. 11, lanes 1 and 2). As when copolymerized protein is used as substrate, one form of subtilisin is detected using the esterase substrates. N-Acetyl-DL-phenylalanyl-β-naphthyl ester appears to be a better substrate for chymotrypsin than naphthol AS-D acetate, which barely detects chymotrypsin (Fig. 11, lanes 5 and 6). Pro-

[23] C. Y. Li, K. W. Lam, and L. T. Yam, J. Histochem. Cytochem. 21, 1 (1973).
[24] R. Rindler-Ludwig, F. Schmalzl, and H. Braunsteiner, Br. J. Haematol. 27, 57 (1974).

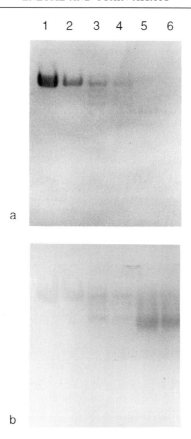

FIG. 11. Zymography following cationic PAGE using proteases detected with synthetic esterase substrates. Subtilisin, lanes 1 and 2 (7.6, 3.8 μg), pronase, lanes 3 and 4 (7.6, 3.8 μg), and chymotrypsin, lanes 5 and 6 (7.6, 3.8 μg), were subjected to electrophoresis at pH 4.3 in polyacrylamide gels (15%) for 1.5 hr at 4°. Following electrophoresis, the gels were washed for 10 min in PB, and the esterolytic activity of the proteases was detected using (a) naphthol AS-D acetate or (b) N-acetyl-DL-phenylalanyl-β-naphthyl ester as substrates. Zymograms were developed as described in the text. The naphthol AS-D acetate substrate yielded bands stained pale blue, and the N-acetyl-DL-phenylalanyl-β-naphthyl ester substrate yielded bands stained pale red.

nase is detected with slightly greater sensitivity by N-acetyl-DL-phenylala-nyl-β-naphthyl ester; however, both substrates sensitively detect four forms of pronase (Fig. 11, lanes 3 and 4). These synthetic esterase substrates, as well as some naphthylamide derivatives, can be extremely useful in identification and characterization of proteases, especially when used in combination with activators and inhibitors of the enzymes. However, not all proteases react readily with available synthetic substrates, necessitating the use of protein substrates for identification and characterization.

Use of Indicator Gels Following Cationic Electrophoresis. Indicator gels can be used following cationic PAGE just as they are used following SDS–PAGE.

Summary

We have presented a variety of zymographic techniques for identification and characterization of microbial proteases, using SDS–PAGE and PAGE in nondissociating gels. Techniques are described using copolymerized protein substrates, diffusable protein substrates, protein substrates incorporated into indicator gels, as well as synthetic esterase substrates. When a newly discovered protease is being characterized, it is advisable to try a variety of techniques, both to determine optimal conditions for enzyme detection and to characterize the protease.

Zymography is a versatile two-stage technique involving protein separation by electrophoresis followed by detection of proteolytic activity. Each particular combination of protease separation and detection techniques had advantages and limitations. Protease separation by SDS–PAGE has as a limitation the fact that some proteases do not renature and hence cannot be detected following treatment with SDS. However, it has as an advantage the fact that it allows estimation of the relative molecular weight of proteases. Protein separation using nondissociating PAGE is performed using much gentler protease inactivation conditions than those produced by treatment with SDS. Like SDS–PAGE, nondissociating PAGE permits detection of multiple forms of enzymes; however, a disadvantage is that it cannot be used to obtain molecular weight estimates of proteases.

The main variable to control during development of zymograms is the length of time of incubations. Increasing incubation (development) time generally increases the sensitivity of protease detection; however, as the length of time of incubation increases so does the extent of diffusion of proteases and substrates. If incubations are prolonged, protease bands will diffuse, decreasing resolution. Additionally, zones of lysis produced by closely migrating proteolytically active species will merge, eliminating the possibility of detecting all proteolytic species in the sample.

Zymographic techniques can be extremely useful in identification and characterization of microbial proteases. If a few properties of a protease are known, such as the pH range over which the enzyme is active, and whether it can renature after exposure to SDS, zymographic techniques can be specifically and readily adapted to optimize conditions for detection and assist in characterization of the enzyme.

Acknowledgments

We gratefully acknowledge the technical assistance of R. D. Allen and discussions with K. Miyasaki. These studies were supported by U.S. Public Health Service Grant DE 07256 and the Alumni Fund of the University of Pittsburgh School of Dental Medicine.

[46] Assays for Bacterial Type I Collagenases

By JOHN D. GRUBB

Introduction

Collagen is the most abundant protein in animal and human tissues and has a central role in defining the structure and function of the extracellular matrix. Therefore, bacteria with the ability to degrade collagen may cause significant tissue destruction and disease progression. In many types of connective tissue, for example, skin, bone, tendon, and cartilage, collagen fibrils form the insoluble framework of the extracellular matrix. The fibrils consist of individual triple helices with a coiled coil conformation held together in a quarter-staggered array by intra- and intermolecular, covalent cross-links located in the nonhelical, telopeptide ends of the collagen molecule. Collagen fibrils *in vivo* are heterotypic structures consisting of either cross-linked Type I, III, and V helices or Type II and XI helices (for review, see van der Rest and Garrone[1]). Our understanding of the relationship between *in vivo* collagen structure and its degradation has been greatly facilitated by the development of methods to solubilize and purify each collagen and to reconstitute fibrils *in vitro* from the individual collagen types.

In the fibrillar or soluble forms, Types I, II, III, and XI collagens are cleaved by mammalian interstitial collagenases at a single site in the triple helical region to form fragments three-quarters and one-quarter of the length of the intact molecule. However, some reaction conditions promote

---

[1] M. van der Rest and R. Garrone, *FASEB J.* **5,** 2814 (1991).

the unfolding of the triple helices of collagen in solution, exposing sites sensitive to nonspecific proteases. Therefore, the most stringent definition of a collagenase is the ability to cleave the native triple helix of collagen with the degradation of Type I considered prototypical of the group, which will be the case in this chapter. Type IV collagen in basement membranes does not form fibrils, and it is cleaved by many proteases but not by the interstitial collagenases. Therefore, although Type IV collagen may be important in colonization by some bacteria, the ability to degrade this collagen does not qualify a protease as a Type I or interstitial collagenase.

Thus far, the best characterized Type I bacterial collagenases are those from *Clostridium histolyticum*[2-4] and *Achromobacter iophagus*.[5] In comparison to their mammalian counterparts, bacterial collagenases possess distinct differences, which have been a source of confusion in the field. Evidence gathered thus far suggests that bacterial collagenases cleave collagen fibrils and constituent triple helices at multiple sites.[6] Furthermore, as has been found for both of the above bacteria, this activity may consist of more than one protease capable of degrading collagen, and the proteases may be difficult to separate.[2,5] Thus, one or more true collagenases may cleave collagen fibrils at enough sites to cause an unfolding of the triple helices under physiological conditions, rendering them susceptible to further degradation by gelatinases. Furthermore, telopeptidase activity may be present which solubilizes collagen fibrils by cleaving in the telopeptide regions and releasing native triple helices. However, under physiological conditions, these triple helices may unfold enough to expose sites sensitive to gelatinases. Therefore, unlike the case for mammalian interstitial collagenases, bacterial degradation of interstitial collagen fibrils may occur via the cumulative activities of several proteases, each with a distinct mode of attack. Descriptions of the assays below are written to take these considerations into account. Assays utilizing peptide analogs of collagen are also described since these are often useful in further characterizing purified collagenases.

Solution Assay of Type I Collagen

The solution assay is the most versatile of the collagenase assays and can be used in different ways to provide much information as long as certain precautions are taken as described below. The collagen for this

[2] M. D. Bond and H. E. Van Wart, *Biochemistry* **23,** 3077 (1984).
[3] M. D. Bond and H. E. Van Wart, *Biochemistry* **23,** 3085 (1984).
[4] M. D. Bond and H. E. Van Wart, *Biochemistry* **23,** 3092 (1984).
[5] N. T. Tong, J. Dumas, and V. Keil-Dlouha, *Biochim. Biophys. Acta* **955,** 43 (1988).
[6] M. F. French, K. A. Mookhtiar, and H. E. Van Wart, *Biochemistry* **26,** 681 (1987).

assay can have intact telopeptide ends, as is true of the acid-soluble preparations available from Sigma (St. Louis, MO) as type VII from rat tail or type VIII from human placenta. Alternatively, Type I collagen can be purified from rat tail tendon in a procedure which has been described in a previous volume of this series.[7] The collagen chains in these preparations are resolved by sodium dodecyl sulfate–polyacrylamide gel electrophoresis (SDS–PAGE) as four bands, with the uppermost bands representing β_{11} (cross-linked α_1 chains) and β_{12} components (cross-linked α_1 and α_2 chains) and the two lower bands α_1 and α_2 chains.[7] Telopeptide-free collagen prepared from acid-soluble collagen by digestion with pepsin is also suitable. Analysis by SDS–PAGE of telopeptide-free collagen resolves two bands representing α_1 and α_2 chains.

Materials

Collagen: dissolved in 13 mM HCl or 5 mM acetic acid

Glycerol: must be added at a final concentration of 10% (v/v) if the collagen concentration exceeds 100 μg/ml to prevent fibril formation

Assay buffer: the pH can be acidic to basic, as the enzyme requires

Incubation temperature: between 15 and 28° depending on the purpose of the experiment and the purity of the enzyme

Procedure. The solution assay is prone to nonspecific proteolysis since local regions of the collagen triple helix unwind as the temperature increases above 28°. Therefore, attention must be paid to the incubation temperature and the inclusion of negative controls using trypsin at 20 μg/ml, which cleaves the triple helix at regions of thermal instability. The reactions can be incubated for up to 20 hr, but shorter times may also be necessary to control for nonspecific proteolysis.

Following incubation for an appropriate amount of time, several methods can be used to analyze the reaction products. Simultaneous use of two or more of the methods may be even more informative. Most simply, the generation of new α-amino groups can be determined colorimetrically using ninhydrin, as described in [47] in this volume. Then SDS–PAGE can be used to separate the reaction products, for which the reactions must be stopped either by a specific inhibitor of the enzyme or by acidification. Collagen cleavage fragments are visualized by Coomassie blue or silver staining if proteins in the enzyme extract are low enough in concentration so as not to interfere. Alternatively, radiolabeled collagen, prepared as described in an earlier volume of this series[7] or by Mookhtiar *et al.*,[8]

[7] H. Birkedal-Hansen, this series, Vol. 144, p. 140.
[8] K. A. Mookhtiar, S. K. Mallya, and H. E. Van Wart, *Anal. Biochem.* **158**, 322 (1986).

can also be used, with detection by autoradiography. According to Mookhtiar *et al.*,[8] collagen molecules with 40 or more lysine plus hydroxylysine residues labeled per molecule cannot undergo fibrillogenesis but do form native triple helices. Collagens prepared in this way remain soluble over a wide range of temperatures and concentrations.

If radiolabeled collagen is used, the addition of dioxane will precipitate native collagen fragments, as described earlier,[7] and the degree of collagen degradation is determined by liquid scintillation spectrometry of the supernatant. Visualization of the initial collagen fragments in the supernatant by SDS–PAGE and autoradiography may also indicate the nature of the cleavages resulting in denaturation of the collagen triple helix. However, the presence of gelatinolytic activity in the enzyme extract may result in rapid degradation of the denatured fragments.

The use of radiolabeled or unlabeled collagen with intact telopeptide ends may permit the detection of telopeptidase activity using SDS–PAGE. As described above, this form of collagen has two components, β_{11} and β_{12}, which result from amino-terminal cross-linkages between the α_1 and α_2 chains. Therefore, telopeptidase activity may be indicated by the disappearance of these components and an increase in α_1 and α_2 chains with short incubation times.

Collagen Fibrils as Substrate

The ability to solubilize collagen fibrils, either reconstituted or from insoluble tendon collagen, is generally regarded as a necessary criterion in the demonstration of collagenolytic activity. The most suitable assay for this purpose is the use of reconstituted collagen fibrils. This assay has gained widespread acceptance owing to the development of techniques to radiolabel acid-soluble collagen to high specific activities without altering its ability to reconstitute fibrils at neutral pH.[7,8] The collagen for this assay can be obtained from Sigma isolated from rat tail tendon (type VII) or human placenta (type VIII) or purified as described in a previous volume in this series.[7] Collagen solubilized by cleavage of the telopeptide ends (e.g., with pepsin), forms fibrils less efficiently and is not as suitable.

Radiolabeling of the collagen to high specific activity and reconstitution of collagen fibrils have been described previously.[7] A second method involves labeling the primary amines in collagen with a fluorescent compound, 2-methoxy-2,4-diphenyl-3(2H)-furanone (MDPF).[9] Fibrils are reconstituted as described for radiolabeled collagen.

[9] R. L. O'Grady, A. Nethery, and N. Hunter, *Anal. Biochem.* **140**, 490 (1984).

Materials

Calf skin collagen (Worthington, Freehold, NJ): 7.5 mg/ml in 75 mM
  sodium citrate, pH 3.7
MDPF: available from Fluka (Ronkonkoma, NY)
Procedure. The collagen solution (5 ml) is dialyzed against 50 mM
borate, 100 mM NaCl (pH 9.0) at 4° overnight. The gel that forms is
broken up by magnetic stirring in the above buffer in a final volume of
30 ml. With continuous stirring in an ice bath, 2 mg MDPF (in 3 ml sodium
sulfate-dried acetone) is added dropwise over a period of 60 min. Stirring
is continued for an additional 60 min. The mixture is centrifuged at 5000
g for 10 min at 4°. The pellet is resuspended in 0.2% (v/v) acetic acid,
adjusted to pH 4 with NaOH, and dialyzed against 5 liters of this solution
overnight at 4°. After dialysis, the solution of labeled substrate is adjusted
to a final volume of 20 ml and stored at $-10°$ in aliquots of 5 ml, where
it is stable for at least 3 months. Alternatively, the collagen may be lyophilized and stored dry. The authors reported a 90% yield with an average
of 6 MDPF molecules bound per collagen molecule. The labeled collagen
is able to form fibrils in which 90% of the substrate is in the gel form.
Trypsin, when incubated with the fibrils at 35°, releases another 10% of
the label.

Comparing the two labeling methods, it appears that collagen can be
radiolabeled to a higher specific activity than fluorescently labeled collagen
and still retain the ability to form fibrils. Mookhtiar *et al.* have shown
that collagen with up to 15 residues labeled is able to form fibrils at 35°.[8]
Therefore, radiolabeled collagen may offer greater sensitivity than the
fluorescently labeled protein.

The main advantages of this assay are the relative ease with which
collagen degradation can be detected and the greater thermostability of
reconstituted fibrils compared to collagen in solution. The assay can be
performed at 37° since collagen fibrils have a higher melting temperature
than collagen in solution.

One variation described previously permits the assay of fibrillar collagen at acidic pH.[7] Collagen fibrils are normally dissolved below pH 5
owing to hydrolysis of the aldolimine cross-links catalyzed by hydrogen
ions. However, reduction of the cross-links with KBH$_4$ to the acid-stable
dihydro form permits fibril formation which is resistant to acid dissolution
but suitable for the detection of collagen breakdown at acidic pH. This
variation may have utility for bacterial proteases active at the lower end
of the pH scale.

A second variation uses collagen fibrils in a zymographic overlay technique to visualize collagenolytic activity by proteins separated by

SDS–PAGE.[7] Ideally, the collagenase must be able to renature following boiling in SDS–PAGE sample buffer. Alternatively, it may be possible to retain activity by incubating the collagenase in sample buffer at lower temperatures.

Insoluble Tendon Collagen as Substrate

The method using insoluble tendon collagen is probably the simplest type of collagenase assay to set up and measures the release of collagen fragments from insoluble collagen. The collagen for this assay can be obtained from Sigma (their type designations of I, II, or V).

Materials

Collagen (dry): 5 mg/tube
Assay buffer: pH greater than 5.0, salt concentration less than 1 M
Procedure. The enzyme preparation is incubated with the collagen at 37° for an appropriate amount of time. The release of collagen fragments into the supernatant is determined using ninhydrin, as described in [47], this volume. Because hydroxyproline is unique to collagen, a specific assay for this amino acid can also be used.[10] Analysis of the supernatant by SDS–PAGE may indicate whether the collagenolytic activity solubilizes large collagen fragments, unless nonspecific proteases are present to degrade the denatured collagen.

The major advantage of this assay is that the collagen is very similar to the native fibrillar form, so that any digestion which occurs is due to a collagenase and/or a telopeptidase. An obvious limitation of the assay is the tedium involved in weighing out the dried collagen into individual assay tubes. However, the major limitation is that the salt concentration must be less than 1 M and the pH greater than 5.0. If either of these conditions are not met, the collagen may be solubilized and at temperatures greater than 30° be susceptible to nonspecific proteolysis. Another limitation is that the enzyme and substrate are in different phases, which may inhibit their interaction. These limitations may be so severe as to restrict the use of the assay to a preliminary identification of collagenolytic activity in an enzyme extract or whole cells.

Synthetic Peptide Analogs of Collagen

The use of synthetic peptides as substrates for collagenases can offer convenience during purification and make possible a kinetic characteriza-

---

[10] K. I. Kivirikko, O. Laitinen, and D. J. Prockop, *Anal. Biochem.* **19**, 249 (1967).

tion of enzyme activity. However, several words of caution are in order. Most importantly, it must be shown that the collagenolytic and peptide hydrolase activities reside in the same protein. Related to that point, the ability to hydrolyze a collagenlike peptide is not conclusive evidence that the enzyme in question is a collagenase. Second, the rate at which a peptide is cleaved will most likely not describe the rate of collagen degradation, in particular if the collagenase cleaves collagen at multiple sites with differing peptide specificities.

2-Furanacryloyl-L-leucylglycyl-L-prolyl-L-alanine

Among collagenase assays using synthetic peptides, the assay involving 2-furanacryloyl-L-leucylglycyl-L-prolyl-L-alanine (FALGPA) has the greatest convenience and sensitivity. This peptide was initially developed as a substrate for *C. histolyticum* collagenase[10] and is also cleaved by *A. iophagus* collagenase. Bond and Van Wart have separated the *C. histolyticum* collagenases into two classes based on the relative activities against collagen and FALGPA.[2]

Materials

FALGPA: 25 mM in dimethyl sulfoxide, final concentration of 0.05 mM

Assay buffer: as appropriate for the enzyme

Procedure. The assay is a continuous one in which the substrate is present below the K_m, and, therefore, initial velocity meaurements are made at short time points (1–4 min) and at room temperature in a spectrophotometer cuvette holder. With time the absorbance at 324 nm decreases as FALGPA is hydrolyzed. At this wavelength and substrate concentration, the background absorbance is 0.695, and the change in absorbance on full hydrolysis is 0.125. Units of enzyme activity are determined as the ratio of A_{324} of partial hydrolysis/A_{324} of full hydrolysis times the substrate concentration. Because it is possible that other bacterial collagenases may have a higher K_m for the peptide, the substrate concentration may need to be increased.

Phenylazobenzyloxycarbonyl-Pro-Leu-Gly-Pro-D-Arg

The peptide phenylazobenzyloxycarbonyl-Pro-Leu-Gly-Pro-D-Arg (Pz-PLGPR) was reported in 1963 as a synthetic substrate for *C. histolyticum* collagenase and, as such, has been used more often as a synthetic substrate for bacterial collagenases.[11,12]

[11] H. E. Van Wart and R. Steinbrink, *Anal. Biochem.* **113**, 356 (1981).
[12] E. Wunsch and H. Heidrich, *Hoppe-Seyler's Z. Physiol. Chem.* **333**, 149 (1963).

Materials

Pz-PLGPR: 50 mM in methanol, final concentration of 1 mM
Citrate: 25 mM
Ethyl acetate
Na$_2$SO$_4$
Glass tubes for assays

Procedure. An appropriate amount of enzyme is added to assay buffer containing Pz-PLGPR in a final volume of 0.25 ml. After incubation, generally at 37° and for up to 1 hr, the reaction is stopped by the addition of 0.5 ml of 25 mM citrate. The cleavage product is extracted by the addition of 2.5 ml ethyl acetate, followed by vigorous vortexing and centrifugation at 500 g to separate the aqueous and organic layers. The upper organic layer is transferred to tubes containing 0.15 g Na$_2$SO$_4$ to dry the ethyl acetate, and the A_{320} is then measured. The amount of Pz-Pro-Leu formed is determined by a standard curve obtained by measuring the A_{320} of various concentrations of Pz-Pro-Leu.

On comparison of the two peptide assays, it is readily apparent that the assay using FALGPA is more convenient. Furthermore, Van Wart and Steinbrink reported that FALGPA was hydrolyzed 3 times as rapidly as Pz-PLGPR, suggesting that replacement of the proline with the furanacryloyl group increases the sensitivity of the peptide to hydrolysis.[10]

Dinitrophenyl Coupled to Collagenlike Peptides

Masui synthesized a series of seven dinitrophenyl (DNP)–peptides having sequences similar to that around the site in collagen cleaved by vertebrate interstitial collagenases.[13] Of these, two were cleaved by tadpole collagenase, namely, DNP-Pro-Leu-Gly-Ile-Ala-Gly-D-Arg-NH$_2$ and DNP-Pro-Gln-Gly-Ile-Ala-Gly-Gln-D-Arg-OH. The latter peptide was specifically cleaved by the collagenase, whereas the former was also cleaved by a human serum peptidase. To date, no bacterial Type I collagenases have been reported to hydrolyze these peptides. The assay is similar to that with Pz-PLGPR in that the peptide product has to be extracted into an organic solvent in order to determine enzyme activity.

Materials

DNP-PLGIAGR: available from Schweitzerhall
DNP-PQGIAGQR: available from Sigma, Serva (Heidelberg, Ger-

---

[13] Y. Masui, T. Takemoto, S. Sakakibara, H. Hori, and Y. Nagai, *Biochem. Med.* **17**, 215 (1977).

many), and Fluka; both peptides can be made as a 5 mM stock solution in dimethyl sulfoxide and are used at final concentration of 0.25 mM.

1 M HCl

Ethyl acetate, to extract DNP-PLGIAGR

Ethyl acetate/n-butanol (1 : 0.15, v/v), to extract DNP-PQGIAGQR

Na_2SO_4

Procedure. The peptide, assay buffer, and enzyme in a final volume of 0.25 ml are incubated at 37° for up to 1 hr. The reaction is stopped by the addition of 1 ml of 1 M HCl, and the peptide product is extracted by 1 ml of the appropriate organic solvent, vigorous vortexing, and centrifugation at 500 g to separate the two layers. The A_{365} of the upper organic layer is measured and the amount of peptide product determined using a standard curve.

[47] Purification and Assays of Bacterial Gelatinases

By JOHN D. GRUBB

Introduction

Bacterial gelatinases may have a significant role in pathogenesis by degrading Type IV collagen in basement membrane and exposing underlying tissue. They may also degrade partially denatured collagen fragments generated either by interstitial collagenase during normal connective tissue metabolism or by bacterial collagenase. The purpose of this chapter is to give a brief description of the chromatographic matrices which may be useful for the selective purification of gelatinases and enzymatic assays using gelatin as substrate.

Purification of Gelatinases by Affinity Chromatography

Affinity chromatography exploits to varying degrees the selective but reversible interaction of a protein with a ligand which has been immobilized on a chromatographic support. The ligands described below can be divided into two groups defined by their interaction with gelatinases. In the first, the binding of gelatinases to gelatin or other extracellular matrix molecules is most likely determined by three-dimensional structural features present in both proteins. In the second, the ligands are small molecules which bind to specific amino acids or metal cofactors. The principles

METHODS IN ENZYMOLOGY, VOL. 235

and the wide variety of practical applications of affinity chromatography have been described in detail previously.[1-3] Although many of the ligands described below are available already coupled to a support, a previous volume in this series has described the methods available to immobilize ligands through different functional groups.[3]

Gelatin–Agarose

Gelatin–agarose is available commercially, or it can be prepared by coupling gelatin to agarose preactivated with cyanogen bromide.[4] The loading capacity of this material will vary depending not only on the amount of immobilized gelatin but also on the number and type of interactions between the extract proteins and gelatin. Bound proteins are eluted from the gelatin using 1 M NaCl, 1 M arginine,[1] or 7.5% (v/v) dimethyl sulfoxide.[5] Prior to assaying the eluted fractions for enzymatic activity, it may be necessary to remove the eluting agent, in particular the latter two, by dialysis.

Protease Inhibitor–Agarose Matrices

Chromatography on inhibitor–agarose matrices relies on the specific interaction of protease inhibitors with individual amino acid residues in the gelatinase active site. Therefore, if the gelatinase is inhibited by a reversible protease inhibitor, binding the gelatinase to the immobilized inhibitor may result in a significant enhancement of purity. Many of these matrices will not be available commercially and, therefore, must be prepared. Preactivated matrices are available or can be prepared as described previously.[3]

Because most of the ligands will be small molecules, the accessibility and orientation may be very important. Attachment of the ligand to the support through a spacer molecule may make the ligand more accessible for binding. It may be necessary to bind the ligand through different functionalities in order to determine the optimum orientation of the ligand. Amino acids or peptides which act as reversible inhibitors might also be useful as ligands. As the interaction between the protein and the ligands will most likely be 1 : 1, the capacity will be primarily dependent on the

[1] S. Osgrove, this series, Vol. 182, p. 357.
[2] S. Osgrove and S. Weiss, this series, Vol. 182, p. 371.
[3] T. M. Phillips, *in* "Chromatography" (E. Heftmann, ed.), p. A309 Elsevier, New York, 1992.
[4] S. C. March, I. Parikh, and P. Cuatrecasas, *Anal. Biochem.* **60,** 149 (1974).
[5] J. G. Lyons, B. Birkedal-Hansen, W. G. I. Moore, R. L. O'Grady, and H. Birkedal-Hansen, *Biochemistry* **30,** 1449 (1991).

concentration of bound ligand and steric interference between bound proteins. Elution may be accomplished in three ways: (1) high salt, (2) a change in pH, and (3) competition by free ligand for the ligand binding site.[3]

Thiol–Disulfide Covalent Chromatography

Thiol–disulfide covalent chromatography relies on the presence of cysteine residues in protein, in particular, one reactive thiol per molecule, as found in cysteine proteases. The technique has been described previously so only a brief description follows.[6,7] Typically, glutathione–agarose is used, in which the glutathione is coupled to the agarose via the amino group so that the thiol group is available for binding. The thiol groups of glutathione–agarose are reacted with 2,2'-dipyridyl disulfide to produce a polymer containing glutathione-2-pyridyl disulfide residues with the release of 2-thiopyridone. Attachment of the thiol-containing protein to the column results in the release of 2-thiopyridone, which is detectable spectrophotometrically at A_{343}. At pH 8, most thiol-containing molecules will react readily. Further selectivity for thiol groups with very low pK_a values, such as found in many cysteine proteases, may be achieved in acidic media (e.g., pH 4). In both cases, thiol-containing proteins are then eluted with a gradient of cysteine.

Immobilized Metal Affinity Chromatography

The immobilized metal affinity chromatography (IMAC) technique is particularly useful in the purification of proteases known to bind a divalent metal ion, for example, Zn^{2+} or Ca^{2+}. A collagenase[8] and a gelatinase,[9] both of which are zinc-dependent proteases, have been purified using the resin. In addition, the resin was used in the purification of calcium-binding proteins.[10] The IMAC technique was developed in the mid-1970s by Porath, and the principles and procedure have been described previously.[2]

[6] K. Brocklehurst, J. Carlsson, M. P. J. Kierstan, and E. M. Crook, this series, Vol. 34, p. 531.

[7] K. Brocklehurst, J. Carlsson, and M. P. J. Kierstan, *Top. Enzyme Ferment. Biotechnol.* **10,** 146 (1985).

[8] T. E. Cawston and J. A. Tyler, *Biochem. J.* **183,** 647 (1979).

[9] L. Smilenov, E. Forsberg, I. Zeligman, M. Sparrman, and S. Johansson, *FEBS Lett.* **302,** 227 (1992).

[10] J. A. Campbell, J. D. Biggart, and R. J. Elliott, *Biochem. Soc. Trans.* **19,** 387S (1991).

Assays of Gelatinase

Gelatin Preparation

Gelatin for the assays described below can be purchased, although commercial preparations can have a large size range of polypeptide chains. A homogeneous gelatin preparation can be obtained by heating acid- or pepsin-solubilized collagen at 56° for 15 min. Type IV collagen may be substituted for gelatin to determine if the gelatinase degrades basement collagen.

Electrophoresis Assay

Gelatin, at 50 μg/ml, is incubated with the enzyme fraction for an appropriate amount of time. The reactions should be stopped either by an inhibitor of the enzyme or by acidification since some gelatinases are capable of renaturing following boiling in sodium dodecyl sulfate–polyacrylamide gel electrophoresis (SDS–PAGE) sample buffer. The degradation of gelatin is visualized by subjecting the reactions to 10% SDS–PAGE followed by silver staining. Detection may be visualized by Coomassie staining if the gelatin concentration is increased. The greatest advantage to this assay is its simplicity, whereas its usefulness is limited by the inability to quantitate degradation.

Ninhydrin Detection of α-Amino Groups

Successive cleavages of the gelatin polypeptide chains result in an increasing number of new α-amino groups, which are detectable colorimetrically using ninhydrin. Among the several variations of this procedure,[11-13] the one described here[13] is the simplest to perform and does not have the unpleasant odor associated with the use of organic solvents in other procedures, while retaining high sensitivity.

Materials

0.5 M Sodium citrate (pH 5.5)
1% Ninhydrin in 0.5 M citrate
Glycerol
Procedure. For each sample, mix together 0.5 ml of 1% ninhydrin, 1.2 ml of glycerol, and 0.2 ml of 0.5 M citrate. Dispense 1.9 ml/tube, add

[11] H. Rosen, *Arch. Biochem. Biophys.* **67,** 10 (1957).
[12] J. V. Singh, S. K. Khanna, and G. B. Singh, *Anal. Biochem.* **85,** 581 (1978).
[13] Y. P. Lee and T. Takahashi, *Anal. Biochem.* **14,** 71 (1966).

sample in 0.1 ml, and heat in a boiling water bath for 12 min. Cool in a water bath at room temperature. Shake each tube and read A_{570} within 1 hr.

Assays Using Labeled Gelatin

Gelatin can be radiolabeled with [^{3}H]acetic anhydride or [^{3}H]formaldehyde, as done for collagen.[14,15] Alternatively, it can be fluorescently labeled with fluorescein isothiocyanate (FITC).[16]

Incubate enzyme with radiolabeled gelatin or FITC–gelatin in a volume of 50 μl for an appropriate amount of time at 37°. The reactions are stopped by adding 100 μl of 10 mg/ml unlabeled gelatin and 50 μl of 50% trichloroacetic acid (TCA). Let the reactions stand at 4° for 30 min and then pellet the protein precipitate by centrifugation in a microcentrifuge. In the case of radiolabeled gelatin, the supernatant is counted to determine the counts per minute (cpm). The FITC-labeled peptides in the supernatant are detected by adding 100 μl to 1.0 ml of 0.5 M Tris (pH 8.75) and measuring the fluorescence using excitation at 490 nm and monitoring emission at 525 nm. Rather than incubating the labeled gelatin in solution, it can also be covalently attached to the preactivated surface of a 96-well plate (Costar, Cambridge, MA), with release of label into the supernatant being measured.

[14] H. Birkedal-Hansen, this series, Vol. 144, p. 140.
[15] K. A. Mookhtiar, S. K. Mallya, and H. E. Van Wart, *Anal. Biochem.* **158**, 322 (1986).
[16] U. Tisljar and H.-W. Denker, *Anal. Biochem.* **152**, 39 (1986).

[48] Assays for Hyaluronidase Activity

By WAYNE L. HYNES and JOSEPH J. FERRETTI

Introduction

Hyaluronidases are a group of enzymes [hyaluronate 4-glycanohydrolase (EC 3.2.1.35, hyaluronoglucosamidase), hyaluronate 3-glycanohydrolase (EC 3.2.1.36, hyaluronoglucuronidase), and hyaluronate lyase (EC 4.2.2.1)] that catalyze the breakdown of hyaluronic acid, a mucopeptide composed of alternating N-acetylglucosamine and glucuronic acid residues. Hyaluronidases are produced by a variety of pathogenic organisms including group A and C streptococci, pneumococci, staphylococci, and clostridia. The enzymes are also found in leeches, snake and insect venoms, and malignant tissues. In the pathogens, hyaluronidases are frequently termed spreading factors because of their ability to break down the hyaluronic acid component found in the cement substance of host tissues. A variety of methods have been employed over the years to assay hyaluronidase activity, but many of the assays appear to be rarely used today and are discussed only briefly. All methods cited should be applicable to hyaluronidases from either microbial or mammalian sources; however, confirmation of such applicability has not always been done and may be dependent on the mechanism of hyaluronic acid breakdown. As there are many different assays for hyaluronidase activity we have separated the methods into groups based on the type of assay performed, namely, (1) spectrophotometric, (2) radiochemical, (3) fluorogenic, (4) enzymoimmunological, (5) plate (solid media), (6) chemical, (7) physicochemical, and (8) zymographic analysis.

Spectrophotometric Assay

A sensitive method for the detection of hyaluronidase activity was developed by Benchetrit et al.[1] based on a shift in maximal absorbance following interaction of anionic mucopolysaccharides with a carbocyanine dye. Hyaluronic acid is freed from protein contamination by digestion with pronase, treatment with chloroform–isoamyl alcohol, and precipitation with NaCl-saturated ethanol. The assay as described by Benchetrit et al. is performed in a final volume of 10 μl. Hyaluronic acid (1 μg) and enzyme are incubated at 37° for 60 min. The reaction is stopped by addition of 9 μl of water and freezing. The dye, Stains-all {1-ethyl-2-[3-(1-ethylnaphtho[1,2-d]thiazolin-2-ylidene)-2-methylpropenyl]naphtho[1,2-d]thiazolium bromide; Eastman Organic Chemicals, Rochester, NY} is prepared at 0.1 mM in water containing 50% dioxane (final concentration of dioxane in the dye solution is 5%), 1 mM acetic acid, and 0.5 mM ascorbic acid. The dye is added to the reaction mixture to give a final volume of 1.0 ml and the absorbancy determined at 640 nm against a blank containing dye (0.9 ml) and water (0.1 ml). One unit of hyaluronidase activity is defined as the amount of enzyme required to decrease the absorbance of the hyaluronic acid–dye complex by 10% after 1 hr incubation at 37°, pH 5.0.

In our hands, this method has been useful for the detection and quantitation of purified hyaluronidase; however, it is not reliable in the detection of activity in crude (culture supernatants) preparations where interfering substances may be present. A modification of the assay carried out in

[1] L. C. Benchetrit, S. L. Pahuja, E. D. Gray, and R. D. Edstrom, Anal. Biochem. **79**, 431 (1977).

100-μl volumes has been used by Hotez *et al.*[2] to detect hyaluronidase activity in hookworm larvae. These authors noted that differences between substrate and products of the reaction were detected only by addition of glacial acetic acid to the dye. The requirement for the acid may be due to the different way the Stains-all dye is prepared.

A further modification to this procedure was recently described for the assay of chondroitin sulfate depolymerase and hyaluronidase activity in viridans streptococci[2b]. In this procedure, the dioxane contains 25 ppm 2,6-di-tert-butyl-4-methylphenol as a stabilizing agent[2a,2b]; prepared in this way the dye solution is stable for at least 2 weeks. Assays are performed in microcentrifuge tubes containing 100 μl of bacterial suspension in 50 mM Tris-Cl buffer pH 7.5, 350 μl 0.2 M Na$_2$HPO$_4$ buffer pH 6.5, and 200 μl of 2.0 mg/ml hyaluronic acid. Following incubation at 37°, samples (20 μl) of the reaction mix are removed, mixed with 180 μl of dye solution, followed by 100 μl of distilled water and the absorbance determined at 620 nm within 30 min of addition of the dye. Removal of the bacterial cells is not necessary as their presence does not appear to interfere with development of the hyaluronic acid–dye complex[2b]. It was reported in these studies that a variety of salts (CaCl$_2$, FeCl$_3$, MgCl$_2$, MnSO$_4$, NaCl, and ZnCl$_2$) at concentrations as low as 1 mM reduce the absorbance of the hyaluronic acid–dye complex, while addition of EDTA enhances the absorbance values.[2a] Difficulties due to the presence of these salts, in particular Fe^{3+}, can be overcome by routinely including chelators such as EDTA in the assay.[2a]

An alternative colorimetric assay based on the interaction of Alcian Blue 8GX with hyaluronic acid has been described[3] in which the absorbance of the digest is inversely proportional to the concentration of hyaluronidase. All solutions are prepared in 50 mM sodium acetate buffer, pH 5.0, containing 50 mM MgCl$_2$. The hyaluronidase-containing solution (0.5 ml) to be assayed is added to 200 μl of substrate containing 60 μg of hyaluronic acid and incubated with shaking at 37° for 25 min. After incubation, 1 ml of Alcian Blue solution (0.02%) is added, and the tubes are shaken and immediately centrifuged. The absorbance of the supernatant is determined at 603 nm.

Recently Pritchard and Lin reported on the use of the thiobarbituric acid assay, based on the detection of sialic acid, for the assay of hyaluroni-

[2] P. J. Hotez, S. Narasimhan, J. Haggerty, L. Milstone, V. Bhopale, G. A. Schad, and F. F. Richards, *Infect. Immun.* **60**, 1018 (1992).

[2a] K. A. Homer, L. Denbow, and D. Beighton, *Anal. Biochem.* **214**, in press.

[2b] K. A. Homer, L. Denbow, R. A. Whiley, and D. Beighton, *J. Clin. Microbiol.* **31**, 1648 (1993).

[3] R. H. Pryce-Jones and N. A. Lannigan, *J. Pharm. Pharmacol.* **31**(Suppl.), 92P (1979).

dase activity.[3a] The assay is carried out in 200 μl volumes consisting of 100 μl of 2 mg/ml substrate, 20 μl of 10X buffer (0.5 M ammonium acetate buffer [pH 6.5], 0.1 M CaCl$_2$), and enzyme diluted in 2 mg/ml BSA. The volume is made up to 200 μl with distilled water. After incubation at 37° the reaction is terminated by placing the tubes briefly in a boiling water bath. The tubes are then assayed as described by Skoza and Mohos.[3b] To the reaction tube 0.1 ml of 6% (w/v) thiobarbituric acid, adjusted to pH 9.0 with NaOH, is added to give a final concentration of at least 1%. Development of the chromophore is by heating the reaction in a boiling water bath for 7.5 min. An equal volume of dimethyl sulphoxide is added to intensify the color and the absorbance measured at 549 nm. Development of the color detected in this assay is due to the action of the hyaluronidase breaking the glycosidic bond by elimination (introduction of a double bond) rather than hydrolysis (addition of water). Acid-hydrolyzed hyaluronic acid does not react in this assay due to the absence of double bonds in the products[3a].

Radiochemical Assay

The principle of the radiochemical method described by Coulson and Girken[4] in studies of tissue hyaluronidases is that cetylpyridinium chloride precipitates hyaluronic acid, but not smaller molecular weight polysaccharides. In this procedure, hyaluronic acid is partially deacylated using hydrazine and then reacylated in the presence of [³H]acetic anhydride. The ³H-labeled hyaluronic acid is purified by precipitation, dialysis, and column chromatography and diluted with cold hyaluronic acid to give a final concentration of 1.2 mg/ml. Hyaluronic acid (60 μg in 50 μl of 0.1 M formate buffer, pH 3.5, containing 0.15 M NaCl) is incubated with 50 μl of enzyme solution at 37° for 20 min. Termination of the reaction is achieved by the addition of 50 μl of 0.25 M disodium hydrogen phosphate. To precipitate undigested substrate, 50 μl of a 1% cetylpyridinium chloride solution is added and the mixture incubated for 30 min at 37°. Following centrifugation, a 50-μl aliquot of supernatant is added to 10 ml of scintillation fluid and the radioactivity of the sample determined. Radioactivity of the blanks, in which the enzyme preparation is added immediately prior to the disodium hydrogen phosphate (reaction terminator), is subtracted from the test readings to determine the radioactivity solubilized by the action of hyaluronidase.

[3a] D. G. Pritchard and B. Lin, *Infect. Immunol.* **61**, 3234 (1993).
[3b] L. Skoza and S. Mohos, *Biochem. J.* **159**, 457 (1976).
[4] C. J. Coulson and R. Girkin, *Anal. Biochem.* **65**, 427 (1975).

An alternative assay which follows the breakdown of ^{3}H-labeled hyaluronic acid is described for the assay of hookworm hyaluronidase.[2] In this procedure radiolabeled hyaluronic acid is purified following synthesis by human keratinocytes. The labeled substrate is incubated with the sample and aliquots removed at various time periods and placed into sodium dodecyl sulfate–polyacrylamide gel electrophoresis (SDS–PAGE) sample buffer. After all samples have been collected, the degraded hyaluronic acid is applied to a polyacrylamide gel. Following electrophoresis, the gel is fixed and prepared for autoradiography. Hyaluronidase activity is observed as a decrease in the size of the labeled hyaluronic acid.

Fluorogenic Assay

Hyaluronic acid labeled with the fluorogenic reagent 2-aminopyridine has been used as the substrate in a rapid, simple, and sensitive assay for the detection of testicular hyaluronidase.[5] Reactions are set up in 50-μl volumes containing 1.5 μg of pyridylaminohyaluronate and enzyme in 50 mM sodium acetate buffer containing 0.15 M NaCl. After incubation at 37° for 60 min, exactly 200 μl of NaCl-saturated ethanol is added and the mixture is left at 0° for 30 min. Following centrifugation, 300 μl of 0.5 M sodium acetate buffer, pH 4.0, is added to 100 μl of supernatant and the fluorescence of the solution measured at an excitation wavelength of 320 nm and an emission wavelength of 400 nm. The increase in pyridylamino products is linearly correlated with enzyme concentration under these conditions.[5] The fluorogenic substrate has also been used for determination of hyaluronidase activity in crude liver extracts, overcoming some of the problems associated with other assays resulting from the presence of interfering compounds.

Indirect Enzymoimmunological Assay

Hyaluronectin is a hyaluronic acid-binding proteoglycan which can be used as a probe in an indirect enzymoimmunological hyaluronidase assay developed for the detection of small amounts of hyaluronidase in preparations from the leech, bovine testes, hepatoma cell lines, bee venom, and streptomyces species.[6] Microtiter plates are coated with hyaluronic acid (100 mg/liter in 0.1 M sodium bicarbonate) and left at 4° overnight prior

[5] T. Nakamura, M. Majima, K. Kubo, K. Takagaki, S. Tamura, and M. Endo, *Anal. Biochem.* **191,** 21 (1990).

[6] B. Delpech, P. Bertrand, and C. Chauzy, *J. Immunol. Methods* **104,** 223 (1987).

to rinsing with distilled water. Diluted samples are added to wells, in duplicate, and incubated at 37° for up to 24 hr. After incubation, the wells are rinsed and incubated with hyaluronectin immune complexes conjugated with alkaline phosphatase, diluted in 0.1 M sodium phosphate, pH 7.0, for 4 hr. Alternatively, the procedure can be carried out as a two-step reaction. Hyaluronectin [150 ng/ml in phosphate-buffered saline (PBS) supplemented with bovine serum albumin (BSA)] is added to the wells for 4 hr. After rinsing, the wells are incubated with diluted conjugated antibodies overnight, whereon the wells are rinsed with PBS and incubated with substrate (p-nitrophenyl phosphate 1 mg/ml in 1 M diethanolamine with 1 mM MgCl$_2$ at pH 9.8) at 37° for 1–2 hr. Hyaluronidase activity is indicated by a decrease in absorbance, measured at 405 nm.

Plate (Solid Media) Assays

A number of approaches have been developed for the assay of hyaluronidase using solid media. One of the simplest plate assays is that described by Smith and Willett[7] to screen bacteria for the ability to produce hyaluronidase. One hundred milliliters of brain–heart infusion broth containing 1% agar is sterilized by autoclaving and allowed to cool to 46°. An aqueous 2 mg/ml solution of sterilized hyaluronic acid is added to a final concentration of 400 μg/ml. A 5% filter-sterilized solution of bovine albumin fraction V is added with constant stirring to give a final concentration of 1% in the medium. The agar mix is poured into petri dishes, allowed to solidify, and stored at 4° until required. Bacteria are inoculated either onto the surface or as stabs into the media prior to incubation at 37°. Following overnight incubation, the plates are flooded with 2 N acetic acid and allowed to stand for 10 min. Hyaluronidase activity is detected as a zone of clearing around the bacterial colonies/stab in a cloudy background, resulting from acetic acid precipitation of an albumin–nondegraded hyaluronic acid complex.

This assay medium can also be used to test liquid samples, such as bacterial culture supernatants and purified or partially purified preparations, for hyaluronidase activity by addition of sodium azide (0.1%,w/v) to the medium prior to pouring into petri dishes. Wells are cut into the solidified assay medium and filled with the sample to be assayed. After incubation, the undigested hyaluronic acid/albumin complex is precipitated with acetic acid. Zones of clearing around wells indicate hyaluronidase activity.[8]

[7] R. F. Smith and N. P. Willett, *Appl. Microbiol.* **16,** 1434 (1968).
[8] W. L. Hynes and J. J. Ferretti, *Infect. Immun.* **57,** 533 (1989).

Another plate assay is based on the precipitation of undigested hyaluronic acid by cetylpyridinium chloride.[9] Assay plates are prepared consisting of 1 mg/ml hyaluronic acid in 1.5% (w/v) agarose buffered with 50 mM sodium citrate (pH 5.3) and containing 0.02% sodium azide (buffer A). Equal volumes of stock solutions of 3% agarose and 2 mg/ml hyaluronic acid in buffer A are mixed at 60° with constant stirring for about 1 min to ensure complete mixing prior to pouring into petri dishes. When the medium solidifies, holes are punched into the medium, filled with the sample to be assayed, and incubated at 37° for 18–20 hr. After incubation, the gel is flooded with 10% (w/v) cetylpyridinium chloride in water and examined against a dark background for zones of clearing around the wells. Zone diameters of known standards can be measured and plotted against units of enzyme using a semilogarithmic scale, thereby allowing an estimation of the amount of hyaluronidase activity in a particular sample.

We have used both of the plate assays in our laboratory for detection of streptococcal hyaluronidase[8,10] and found both to be suitable for assay of liquid preparations. The Smith and Willett assay plates[7] have the added advantage of being able to detect production of hyaluronidase by bacterial colonies. A modification of the cetylpyridinium chloride precipitation method was described by Balke and Weiss,[11] who used it to detect production of hyaluronidase during growth of different anaerobic clostridial strains.

Chemical Assays

Liberation of Reducing Sugars

Assay of hyaluronidase by liberation of reducing sugars is based on the hydrolysis of hyaluronic acid into its reducing sugar components. Results are calculated as a percentage of the total reducing sugar and expressed as equivalents of glucose.[12] The measurement of a decrease in viscosity of hyaluronic acid-containing solution as determined in the viscosity reduction method is a rapid reaction (depolymerization), whereas the hydrolysis of hyaluronic acid with liberation of reducing sugars is a slower reaction.[12] Based on the reduction of ferricyanide to ferrocyanide

[9] P. G. Richman and H. Baer, Anal. Biochem. 109, 376 (1980).
[10] W. L. Hynes and J. J. Ferretti, in "Streptococcal Genetics" (J. J. Ferretti and R. Curtiss III, eds.), p. 150. American Society for Microbiology, Washington, D.C., 1988.
[11] E. Balke and R. Weiss, Zentralbl. Bakteriol. Mikrobiol. Hyg., Ser. A 257, 317 (1984).
[12] K. Meyer, E. Chaffee, G. L. Hobby, and M. H. Dawson, J. Exp. Med. 73, 309 (1941).

a "colorimetric" assay was developed and used to assay bacterial hyaluronidases.[13]

Liberation of Acetylamino Sugars

As discussed by Meyer,[14] early procedures used for the detection of acetylglucosamine[12,15] could not be used to follow the hydrolysis of hyaluronic acid. The problem was associated with the release of acetylglucosamine in excess of the weight of polysaccharide added to the initial reaction mix.

Reissig et al.[16] optimized a colorimetric method for the estimation of acetylamino sugars. Later, Rouleau[17] described the effects of divalent cations, cryoprotective agents, and sulfhydryl compounds on the colorimetric assay for acetylglucosamine. The interfering effects reported in the detection of the acetyl sugar emphasized a need for caution in interpretation of results in such assays, as Ca^{2+} actually had an activating effect on bull sperm hyaluronidase. Methods based on those of Reissig et al.[16] have been used to study hyaluronidase activity from *Propionibacterium acnes*[18] and *Streptococcus dysgalactiae*.[19] Hamai[19] et al. modified the method of Ingham et al.[18] for assay using small volumes. To 50 µl of a 0.2% sodium hyaluronate aqueous solution is added 50 µl of 0.1 M phosphate buffer, pH 6.2, containing 0.02% BSA and 20 µl of buffer containing enzyme. After incubation at 37° for 10 min, the reaction mixture is boiled for 2 min and the increase in reducing sugars assayed using N-acetylglucosamine as a standard.

Absorption of Unsaturated Uronides

Production of unsaturated uronides results in an increase in absorption of UV light.[13,20,21] An increase in the UV absorption (230–235 nm) of a hyaluronic acid solution, following incubation with hyaluronidase, suggests elimination rather than hydrolysis of the hyaluronic acid.[21]

[13] A. Linker, this series, Vol. 8, p. 650.
[14] K. Meyer, *Physiol. Rev.* **27**, 335 (1947).
[15] J. H. Humphrey, *Biochem. J.* **40**, 442 (1946).
[16] J. L. Reissig, J. L. Strominger, and L. F. Leloir, *J. Biol. Chem.* **217**, 959 (1955).
[17] M. Rouleau, *Anal. Biochem.* **103**, 144 (1980).
[18] E. Ingham, K. T. Holland, G. Gowland, and W. J. Cunliffe, *J. Gen. Microbiol.* **115**, 411 (1979).
[19] A. Hamai, K. Morikawa, K. Horie, and K. Tokuyasu, *Agric. Biol. Chem.* **53**, 2163 (1989).
[20] H. Greiling, H. W. Stuhlsatz, and T. Eberhard, *Hoppe-Seyler's Z. Physiol. Chem.* **340**, 243 (1965).
[21] T. Ohya and Y. Kaneko, *Biochim. Biophys. Acta* **198**, 607 (1970).

Physicochemical Assay

Mucin Clot Prevention

The mucin clot prevention (MCP) assay is based on the coprecipitation of native hyaluronic acid with protein to form a mucin clot. When hyaluronic acid is digested with hyaluronidase, the quality and character of the clot is reduced. The MCP test originally described by Robertson *et al.*[22] was later modified by McClean[23] and it is this modification which has been used most widely. Meyer[14] discussed the limitations and some of the problems associated with the MCP test. A modification of the MCP test was described by Unsworth for the detection of hyaluronidase production by *Streptococcus milleri.*[24] Broth culture supernatants (25 μl) are double diluted in 25 μl volumes in a microtiter plate using cold distilled water. After dilution, 25 μl of water is added to each well, for a final volume of 50 μl, followed by 50 μl of hyaluronic acid prepared as follows: 8 ml of cold diluted india ink (20 ml distilled water to 0.01 ml of india ink) is added to Bacto-AHT substrate according to the manufacturer's instructions except for the addition of india ink. The microtiter tray is shaken, incubated at 37° for 20 min, then cooled at 4° for 30 min prior to addition of 25 μl of cold acetic acid to each well. The tray is shaken to ensure mixing of all components. The absence of a black clot indicates hyaluronidase activity, with the titer of hyaluronidase being the highest dilution that fails to show clotting. A modification of the MCP test using india ink to follow clot development has been utilized in the detection of hyaluronidase antibody in patients following streptococcal infection.[25,26]

Viscosity Reduction

The viscosity reduction method measures the rate of decrease in viscosity of a solution of hyaluronic acid. Meyer[14] described a standardized procedure to overcome some of the many variations in this assay (substrate and sodium chloride concentration, pH, and temperature) which occurred between different laboratories. Tirunarayanan and Lundblad[27] found the viscosimetric method to be a reliable assay in investigations into the hyaluronidase produced by *Staphylococcus aureus*.

[22] W. Robertson, M. W. Ropes, and W. Bauer, *J. Biol. Chem.* **133**, 261 (1940).
[23] D. McClean, *Biochem. J.* **37**, 169 (1943).
[24] P. F. Unsworth, *J. Clin. Pathol.* **42**, 506 (1989).
[25] S. A. Halperin, P. Ferrieri, E. D. Gray, E. L. Kaplan, and L. W. Wannamaker, *J. Infect. Dis.* **155**, 253 (1987).
[26] R. A. Murphy, *Appl. Microbiol.* **23**, 1170 (1972).
[27] M. O. Tirunarayanan and G. Lundblad, *Acta Pathol. Microbiol. Scand.* **73**, 211 (1968).

Turbidimetric Assay

The turbidimetric assay is based on the observation that acidified hyaluronic acid in the presence of dilute serum forms a stable colloidal suspension. Once depolymerized, the hyaluronic acid–serum mixture remains clear.[14,28–31] Use of a modified turbimetric assay allowed Ohya and Kaneko to describe a hyaluronidase from *Streptomyces hyalurolyticus* nov. sp.[21] A semiquantitative microassay based on the turbidimetric assay is reported to be sensitive, simple, reproducible, and economical.[32]

Zymographic Analysis

Zymography, a procedure allowing visualization of enzyme activity following electrophoretic fractionation, has been used for the qualitative analysis of a number of hyaluronidases.[2,33–38] Zymographic analysis can be carried out on a variety of solid supports such as agar,[33] acrylamide,[34,35,37,38] or cellulose acetate membranes.[36]

Abramson and Friedman[33] used a modification of the MCP assay to detect hyaluronidase activity in concentrated preparations from *Staphylococcus aureus* and *Streptococcus pyogenes* as well as samples of testicular hyaluronidase. Following electrophoresis in 1% Noble agar with barbital buffer, pH 8.6, on a glass slide, hyaluronic acid–horse serum–agar is layered onto the slide. After incubation, the slides are treated with acetic acid and examined for a clear area against a background of precipitated hyaluronic acid substrate.

An alternative to electrophoresis in agar has been described by Herd *et al.*[36] in which electrophoresis is carried out on cellulose acetate membranes. After electrophoresis, the membrane is overlaid with a second membrane saturated with hyaluronic acid, and the sandwich is incubated at 37° for 30 min. Following incubation, the overlay membrane is stained with alcian blue, washed and allowed to dry. Hyaluronidase activity is seen as white bands in a blue background.

---

[28] M. B. Mathews, this series, Vol. 8, p. 654.
[29] J. Komender, A. Golaszewska, and H. Malczewska, *Histochemie* **35**, 219 (1973).
[30] A. Dorfman and M. L. Ott, *J. Biol. Chem.* **172**, 367 (1948).
[31] S. Tolksdorf, M. H. McCready, D. R. McCullagh, and E. Schwenk, *J. Lab. Clin. Med.* **34**, 74 (1949).
[32] A. N. Ibrahim and M. M. Streitfeld, *Anal. Biochem.* **56**, 428 (1973).
[33] C. Abramson and H. Friedman, *Proc. Soc. Exp. Biol. Med.* **125**, 256 (1967).
[34] B. Fiszer-Szafarz, *Anal. Biochem.* **143**, 76 (1984).
[35] B. Fiszer-Szafarz and E. De Maeyer, *Somatic Cell Mol. Genet.* **15**, 79 (1989).
[36] J. K. Herd, J. Tschida, and L. Motycka, *Anal. Biochem.* **61**, 133 (1974).
[37] Von M. Lieflander and H. Stegemann, *Hoppe-Seyler's Z. Physiol. Chem.* **349**, 157 (1968).
[38] B. Steiner and D. Cruce, *Anal. Biochem.* **200**, 405 (1992).

Fiszer-Szafarz[34,35] incorporated hyaluronic acid into polyacrylamide gels prior to electrophoresis. Following electrophoresis, the gel is washed with water and buffer (50 mM acetic acid, 50 mM KH_2PO_4, 0.15 M NaCl, 2.7 mM sodium EDTA, pH 6.0) before incubation at 37° for 16 h to allow the enzyme to degrade the hyaluronic acid included in the gel. The gel is washed with water and soaked in 50% aqueous formamide before staining with Stains-all. A stock solution of 1 mg/ml Stains all in formamide is diluted 20-fold with 9 volumes of formamide and 10 volumes of water. Staining is carried out in the dark for 2 days before gels are washed in water. Hyaluronidase activity is indicated by pink bands (polyacrylamide staining) in a blue background (undegraded hyaluronic acid staining). Utilizing this procedure, polymorphism among various hyaluronidases has been noted,[34,35] including those from a group A streptococcus and from *Streptomyces hyalurolyticus*. Stains-all sensitivity is considered to be an advantage for the detection of smaller amounts of hyaluronidase in comparison with dyes such as alcian blue or toluidine blue.[34,37] Hotez *et al.*[2] modified this procedure to incorporate SDS into the gel allowing for separation of enzymatic activity as a function of molecular weight. Following electrophoresis at low voltage, the gel is washed with Triton X-100 to displace the SDS, then washed with buffer before staining.

A zymographic analysis of hyaluronidase activity from *Propionibacterium acnes, Streptococcus pyogenes, Staphylococcus aureus*, and *Treponema pallidum* has been reported.[38] Electrophoresis is performed using native polyacrylamide gels followed by overlaying the gels with a 1% agarose zymogen containing (0.8 mg/ml) hyaluronic acid and 1% BSA in 0.3 M sodium phosphate buffer, pH 5.3. The substrate zymograph can be prepared ahead of time and stored at 4° until required. Polyacrylamide gels to be assayed for hyaluronidase activity are washed with buffer (0.3 M sodium phosphate buffer, pH 5.3) and placed directly onto the substrate agarose zymograph. After incubation at 37° for up to 2 days, the polyacrylamide gel is removed and the agarose replica developed by immersion in 2 M acetic acid. Areas of hyaluronic acid breakdown, indicating hyaluronidase activity, appear as clear zones in an opaque background. The authors compared their results to those of Fiszer-Szafarz[34] and reported that agarose replica gels are as sensitive as incorporation of substrate into the acrylamide and had the advantage that the gels were much easier to handle. Additionally, the replicas did not have to be stored in the dark and did not fade with age, apparently being stable indefinitely when stored in acetic acid or water.

[49] ADP-Ribosylating Toxins

By LUCIANO PASSADOR and WALLACE IGLEWSKI

Introduction

The transfer of the ADP-ribosyl moiety of NAD to GTP-binding proteins is a common theme in the mechanism of several bacterial toxins. Toxins capable of carrying out ADP-ribosylation include the cholera toxin of *Vibrio cholerae*, diphtheria toxin (DT) of *Corynebacterium diphtheriae*, the exotoxin A of *Pseudomonas aeruginosa*, pertussis toxin of *Bordetella pertussis*, and the heat-labile enterotoxin (LT) of *Escherichia coli*. In addition, exoenzyme S of *P. aeruginosa*, the C2 toxin of *Clostridium botulinum*, and the iota toxin of *C. perfringens* are also capable of ADP-ribosyltransferase activity although the substrates (vimentin, nonmuscle actin, and actin, respectively), while not GTP-binding proteins, share a common role in formation of filaments in the eukaryotic cell. Thus it appears that the use of ADP-ribosylation by toxins is a common mechanism in bacterial pathogenesis. This chapter contains a short overview on the structure and biology of each toxin and presents a method for the measurement of the ADP-ribosyltransferase activity of the toxin. The assay presented for each toxin is not intended to be the definitive assay. It is presented as just one possibility to assay ADP-ribosyltransferase activity and for contrast of similarities and differences between the toxins. This work is intended to bring together assays for the majority of the well-known ADP-ribosylating toxins and to provide a starting point in studies of toxins which might be related to this group.

Diphtheria Toxin

Diphtheria toxin (DT) is a protein of 535 residues which catalyzes the specific ADP-ribosylation of eukaryotic elongation factor 2 (EF2) in a manner similar to that of exotoxin A of *P. aeruginosa*. The ADP-ribosyl moiety is transferred from NAD to the posttranslationally modified histidine residue diphthamide.[1] The ribosylation of EF2 results in the inhibition of protein synthesis and eventually leads to cell death.

The structural gene for DT is carried by temperate phage which lysogenize *Corynebacterium diphtheriae*. The protein is initially released from the lysogen as a single polypeptide which subsequently becomes proteolyt-

---

[1] B. G. van Ness, J. B. Howard, and J. W. Bodely, *J. Biol. Chem.* **255**, 10710 (1980).

METHODS IN ENZYMOLOGY, VOL. 235

ically nicked and reduced to provide two fragments termed A and B. The A fragment consists of the amino-terminal portion of the molecule and contains the ADP-ribosylating activity of the enzyme. The B fragment is responsible for binding of the toxin to cell receptors and the transmembrane transport of the A fragment.

Studies with cultured cells provided the first evidence that the lethal effects of the toxin were due to an inhibition of protein synthesis. Subsequently it was shown that the toxin was able to inhibit protein synthesis in cell-free systems, suggesting a direct effect on some component of the protein synthesis machinery. This effect was only seen when extracts were not dialyzed. Via reconstitution experiments it was demonstrated that the crucial factor was NAD. In the presence of NAD, elongation of the growing polypeptide chain was inhibited owing to inactivation of EF2.

Assay for Diphtheria Toxin

Activation of the toxin is important for the assay since the intact toxin, although toxic to cells, does not exhibit ADP-ribosyltransferase activity *in vitro*. ADP-ribosyltransferase activity is seen when the toxin is cleaved and reduced; however, the activated molecule is no longer toxic to cells. The assay described[2] makes use of EF2 obtained from wheat germ, although the reagent may be obtained from a number of sources including reticulocytes and yeast. The assay follows the transfer of radioactive label from [*adenine*-^{14}C]NAD to EF2 which is then precipitated with trichloroacetic acid (TCA) and quantitated.

Reagents

Buffer 1: 50 mM Tris-HCl (pH 8.0), 5 mM magnesium acetate, 50 mM KCl, 4 mM CaCl$_2$, 5 mM 2-mercaptoethanol

Buffer 2: 50 mM Tris-HCl (pH 7.5), 1 mM EDTA, 1 mM dithiothreitol (DTT), 5% (v/v) glycerol

Buffer R: 0.5 M Tris-HCl, 10 mM EDTA, 0.4 M DTT, pH 8.2 [*adenine*-^{14}C]NAD: over 400 mCi/mmol, 4 μM solution in water

TCA paper: Whatman (Clifton, NJ) 3MM filter paper is ruled into 1 inch squares, numbered, and impregnated with TCA by immersing the sheets into diethyl ether containing 10% (w/v) TCA

EF2 (see below)

Toxin-containing samples

Preparation of EF2 from Wheat Germ. For this preparation[3] 30 g of raw wheat germ is suspended in 240 ml of cold buffer 1. The mixture is

[2] S. F. Carroll and R. J. Collier, this series, Vol. 165, p. 218.
[3] D. W. Chung and R. J. Collier, *Infect. Immun.* **16,** 832 (1977).

allowed to soak on ice for 2 min. The suspension is then homogenized in a Waring blendor at top speed for 10 sec at 4°. The solution is then rehomogenized 4 more times for 10 sec with the addition of ice to maintain temperature. The homogenized material is filtered through cheese cloth or gauze and the filtrate centrifuged at 21,000 g for 15 min at 4° to remove residual cellular debris. The supernatant is brought to pH 7.6 by the addition of acetic acid and is recentrifuged (21,000 g, 15 min). Solid KCl is added to the supernatant to a final concentration of 100 mM and completely dissolved. A final centrifugation (250,000 g for 1 hr) is done to pellet ribosomes. A crude preparation of EF2 is obtained by collection of material that precipitates between 30 and 50% saturation at 4° with ammonium sulfate. The precipitate is dissolved in 30 ml of Buffer 2 and dialyzed twice against 300 volumes of the same buffer. The dialyzed material is centrifuged to remove any precipitate which might form during dialysis, and the purified EF2 solution is aliquoted (~250 μl/tube) and stored at −70° until required. Chung and Collier[3] report that 10 μl of the solution contains approximately 15 to 30 pmol of the acceptor protein.

ADP-Ribosyltransferase Assay. The typical reaction mixture includes the following:

   10 μl Buffer R
   10 μl EF2 (see above)
   10 μl [*adenine*-14C]NAD
   Sample containing toxin
   Water to adjust volume to 100 μl

In practice the reaction mixture is prepared to a total of 90 μl and the assay initiated via the addition of the 10 μl of labeled NAD. Incubations are routinely carried out at 25° for 15 min. The reactions are terminated by spotting duplicate 40-μl aliquots onto the TCA paper. The paper is washed twice with 5% TCA (10 ml/square for 10 min) and once with methanol for 5 min. The washed paper is then dried under a stream of hot air, and the individual squares are excised and counted in a liquid scintillation counter. The use of a negative control (no toxin) and a positive control (0.5 pmol DT) allow the determination of both background counts and normal levels of incorporation. If radioactivity falls within the linear range of incorporation, the amount of toxin present in samples can then be determined via comparison to the positive control.

Pseudomonas aeruginosa Exotoxin A

Exotoxin A (ETA) has been shown to be the most toxic protein exoproduct of *Pseudomonas aeruginosa*.[4] The toxicity of ETA is due to its

---

[4] B. H. Iglewski and J. C. Sadoff, this series, Vol. 60, p. 780.

ability to inhibit protein synthesis in both cultured cells[5,6] and in mice.[7-9]

The ETA protein is synthesized as an enzymatically inactive proenzyme (613 residues) which must undergo structural modification for ADP-ribosylating activity to be expressed.[10] Like diphtheria toxin, ETA is produced as a single polypeptide which is toxic to cells but which lacks enzymatic activity *in vitro*. The activation of the toxin can occur by one of two methods. ETA may be activated via enzymatic cleavage of the proenzyme to release a 26-kDa fragment which is nontoxic but does express ADP-ribosyltransferase activity. This fragment has been seen in *P. aeruginosa* supernatants and presumably has been generated via the action by proteolytic activities in the culture itself. The fragment may also be obtained by chemical cleavage. Alternatively, ETA may be simultaneously treated with a denaturing agent, such as urea or guanidine hydrochloride, and an agent which reduces disulfide bonds (such as DTT). Treatment with either agent alone is insufficient for activation. The requirement for activation and the fact that forms of ETA which are toxic to cells and animals express no enzymatic activity *in vitro* suggest that the active center must be unavailable and is exposed only after conformational alteration. Although methods of activation for *in vitro* activity are quite well defined, the actual *in vivo* mechanism is not yet understood.

Allured and co-workers[11] have solved the crystalline structure of ETA and defined three domains within it. Domain I is composed primarily of antiparallel β sheets. Domain II is composed of six consecutive α helices, and domain III contains a prominent extended cleft which is believed to be the most likely location of the active center. Domain I consists of two subdomains (residues 1–252 and 365–404) within the amino-terminal half of the molecule and is believed to be responsible for recognition of the receptor and binding to eukaryotic cells.

After binding to cells the ETA enters via receptor-mediated endocytosis, is internalized through clathrin-coated pits, and enters endosomes. To release the active portion of ETA the toxin has to escape the endosomes, and it is presumed that this happens via a translocation event. It is believed that acidification of the endosome is required for toxin to

[5] J. L. Middlebrook and R. B. Dorland, *Can. J. Microbiol.* **23**, 183 (1977).

[6] O. R. Pavlovskis and F. B. Gordon, *J. Infect. Dis.* **125**, 631 (1972).

[7] B. H. Iglewski, P. V. Liu, and D. Kabat, *Infect. Immun.* **15**, 138 (1977).

[8] O. R. Pavlovskis, B. H. Iglewski, and M. Pollack, *Infect. Immun.* **19**, 29 (1978).

[9] O. R. Pavlovskis and A. H. Shackleford, *Infect. Immun.* **9**, 540 (1974).

[10] M. J. Wick, A. N. Hamood, and B. H. Iglewski, *Mol. Microbiol.* **4**, 527 (1990).

[11] V. S. Allured, R. J. Collier, S. F. Carroll, and D. B. McKay, *Proc. Natl. Acad. Sci. U.S.A.* **83**, 1320 (1986).

efficiently bind, insert into, and translocate the endosomal lipid bilayer owing to conformational changes in the ETA mediated by low pH. Studies using deletion analysis and oligonucleotide mutagenesis indicate that domain II (residues 253–364) is involved in the translocation process. Molecules which contain deletions in domain II maintain ADP-ribosyltransferase activity but have reduced cytotoxicity. Furthermore, structural features within domain II support the idea: two of the α helices in this region are long enough to span a membrane, and, although no clear hydrophobic areas are evident, regions which are enriched with hydrophobic residues are present.

Following translocation, the active form of ETA enters the cytoplasm and carries out its enzymatic reaction. Domain III (residues 405–613) is believed to be responsible for this process. Biochemical and genetic techniques have provided evidence for the interaction of domain III with the NAD and EF2 substrates. The mechanism of protein synthesis inhibition is virtually identical to that of diphtheria toxin in that it occurs via the transfer of the ADP-ribosyl moiety of NAD^+ to the diphthamide residue of EF2, resulting in the interruption of polypeptide chain elongation.[7,12] Interestingly, although the synthesis, activation, and mechanism of ETA and DT are similar, there are also significant differences between the two toxins. Unlike DT, ETA holotoxin does not require cleavage to be toxic for eukaryotic cells. Furthermore, in ETA the active site is in the carboxy-terminal portion of protein, whereas in DT the active domain is in the amino portion of the molecule. Last, both toxins exhibit different cell line specificities, suggesting that the two toxins contain different binding domains and most likely utilize different cell receptors.

Assay for Exotoxin A

Reagents

Toxin-containing bacterial supernatants (see below)
Urea–DTT solution: 8 M urea, 0.1 M DTT
Assay buffer: 125 mM Tris, pH 7.0, 100 mM DTT
[adenine-^{14}C]NAD (>400 mCi/mmol)
10% TCA
EF2

Preparation of Bacterial Supernatants Containing Toxin. The bacterial supernatants to be used are obtained from cultures of *P. aeruginosa* grown in TSBD medium[13] at 32° for 18 hr. The cells are pelleted by centrifugation

[12] B. H. Iglewski and D. Kabat, *Proc. Natl. Acad. Sci. U.S.A.* **72**, 2284 (1975).
[13] D. E. Ohman, J. C. Sadoff, and B. H. Iglewski, *Infect. Immun.* **28**, 899 (1980).

in a microcentrifuge for 3–5 min, and the supernatants are transferred to another tube and placed immediately on ice. It is important to maintain all reagents on ice when not incubating.

Preparation of EF2. Given the similarities between DT and ETA, it is not surprising that the assay for each of the toxins is extremely similar. Both toxins make use of EF2 as the acceptor protein, and the procedure outlined for EF2 preparation for the diphtheria toxin assay will provide EF2 for the ETA assay.

Activation of Toxin. To a 10-μl volume of a cleared bacterial supernatant add 10 μl urea–DTT solution and incubate at 25° for 15 min.

ADP-Ribosyltransferase Activity. To 20 μl of activated toxin, add 25 μl of assay buffer and 25 μl of EF2 preparation. Mix well. To start the assay add 5 μl of the [*adenine*-^{14}C]NAD, place the reaction mixture at 25°, and allow incubation to continue for 10 min (incubation times as long as 45 min may be required). After the desired time, add 200 μl of cold 10% TCA to terminate the reaction. The contents are filtered through a 0.45-μm nitrocellulose filter to collect the acid-insoluble precipitate, and the filter and tube are washed 3 times with cold 5% TCA. Filters are then air-dried, and radioactivity is determined by liquid scintillation counting.

The inclusion of a positive control with a known amount of commercially available toxin (List Laboratories, Campbell, CA) will allow determination of the amount of toxin in a given sample. A negative control (no toxin) should also be included to determine background counts.

Pseudomonas aeruginosa Exoenzyme S

In addition to exotoxin A, *P. aeruginosa* produces another protein, exoenzyme S, which has ADP-ribosyltransferase activity. Although exoenzyme S is important in pathogenesis, very little is known about the protein. It does possess ADP-ribosyltransferase activity but is different from ETA in that the acceptor molecule is not specifically EF2 but rather a number of cellular proteins. The intermediate filament protein vimentin and its proteolytic fragments are a major substrate of this reaction.

Assay for Exoenzyme S

The assay makes use of the transfer of the radiolabeled ADP-ribosyl moiety from NAD$^+$ to extracts of eukaryotic cells. The assay described here is essentially that of Coburn *et al.*[14] and makes use of purified exoen-

---

[14] J. Coburn, S. T. Dillon, B. H. Iglewski, and D. M. Gill, *Infect. Immun.* **57,** 996 (1989).

zyme S. The purification of the enzyme is described by Nicas *et al.*[15] In this assay NIH 3T3 cells are used as source of extract, but the assay need not be limited to that cell line.

Reagents

Eukaryotic cell extracts (see below)
Purified exoenzyme S (see below)
Buffer A: 50 mM Tris-HCl, pH 8.0, 150 mM NaCl, 1 mM phenylmethylsulfonyl fluoride (PMSF)
Solution 1: 5 μM [^{32}P]NAD, 0.3 μg/ml exoenzyme S [*adenylate*-^{32}P]NAD (1000 Ci/mmol)
2% Sodium dodecyl sulfate (SDS)
Toxin-containing samples

Preparation of Eukaryotic Cell Extracts. Frozen pellets of NIH 3T3 cells with a volume of 50 μl are lysed by thawing in 100 μl of buffer A. Once thawed, the cells are refrozen and thawed once more. The cell lysates are incubated at 25° for 5 min to allow the consumption of endogenous NAD. After the incubation, poly(ADP-ribose) polymerase is inhibited via the addition of thymidine to a final concentration of 20 mM.

ADP-Ribosyltransferase Assay. A 10-μl aliquot of the cell lysate prepared above is mixed with 12 μl of solution 1 and incubated at 25° for 30 min. Termination of the reaction is achieved by the addition of 20 μl of 2% SDS and heating at 100° for 2 min. Two-dimensional gel analysis is carried out, and the labeled protein products are identified.

Cholera Toxin

Cholera toxin (CT) is composed of two distinct subunits termed A and B. The A subunit itself is initially synthesized as a single polypeptide which is then cleaved to produce a pair of fragments (A1 and A2) linked by disulfide bonds. There are five identical polypeptides present in the B subunit to provide the holotoxin with a subunit configuration of A : B$_5$. The A and B subunits are held together by noncovalent interactions. Neither subunit alone demonstrates toxicity. It is proposed that the B protomer is involved in binding cell surface receptors since competition experiments indicate that it can block the toxicity of the holotoxin. The A subunit, once reduced, results in the release of the A1 fragment which is solely responsible for the ADP-ribosyltransferase activity of the toxin.

---

[15] T. I. Nicas, D. W. Frank, P. Stenzel, J. D. Lile, and B. H. Iglewski, *Eur. J. Clin. Microbiol.* **4,** 175 (1985).

The toxin binds to receptors on the intestinal mucosal cell and activates adenylate cyclase. The cell surface receptor is believed to be the ganglio-side G_{M1} for which the B subunit has a high binding affinity. The α subunit of the guanine–nucleotide-binding protein G_s (the membrane bound posi-tive regulator of adenylate cyclase) is the target of the toxin. ADP-ribosyla-tion of G_s is enhanced by the presence of endogenous cellular factors that have been identified as ADP-ribosylation factors (ARFs) which must interact with the toxin. The ARFs are capable of binding GTP and appear in membrane forms (mARFs sometimes called S) and soluble forms (sARFs sometimes called CF). The activation of the adenylate cyclase leads to an elevated level of cyclic AMP (cAMP) which interferes with ion transport in the intestinal mucosa. The disruption of ion transport leads to a loss of fluid from the tissue and the production of diarrhea in infected individuals.

Assay for Cholera Toxin

The assay described is essentially that of Gill and Coburn.[16] It makes use of a sample with a known amount of toxin. However, supernatants from cultures can be tested by substitution of the known sample with the supernatant in question. In this assay it is necessary to prepare a stock solution of sARF. The procedure for this is outlined in the original paper[16] and is not repeated here. It is also necessary to prepare stocks of pigeon erythrocyte ghosts as the source of mARF and G_s. The assay can be carried out using either insoluble mARF, as outlined below, or the mARF can be solubilized (see original publication[16]).

Reagents

CF (see above)
Medium A: 10 mM HEPES, pH 7.3, 130 mM NaCl, 0.01% (w/v) sodium azide, 0.01 trypsin inhibitory units/ml aprotonin
2 mM CaCl$_2$
Micrococcal nuclease (300 U/ml)
Buffer 1: 100 mM Tris-HCl, pH 8.8, 5 mM DTT, 0.5% (w/v) SDS, 1 mM EDTA
500 mM Iodoacetamide
Buffer 2: 0.1% (w/v) SDS, 130 mM NaCl, 10 mM HEPES, pH 7.3
1 M Guanosine 5'-O-thiotriphosphate (GTPγS)
Carboxymethylated toxin (see original publication[16]) or toxin-contain-ing samples
[*adenylate*-^{32}P]NAD (1000 Ci/mmol)

[16] D. M. Gill and J. Coburn, *Biochemistry* **26**, 6364 (1987).

1 M Thymidine

Reaction mix: 50 μg/ml carboxymethylated toxin, 25 μM [^{32}P]NAD, 50 mM thymidine

Saline

Preparation of Erythrocyte Ghosts. Erythrocytes are washed in 1 volume of medium A and lysed by freeze–thaw. The ghosts obtained are washed once and repelleted. To remove DNA, the ghosts are incubated for 15 min at 37° in 2 mM CaCl$_2$ and 3 units/ml micrococcal nuclease. Treated ghosts can be stored at $-70°$ at this stage until required. Just prior to use, an aliquot is thawed and washed twice in 5 volumes of medium A.

Activation of Toxin. A solution of 1 mg/ml toxin in buffer 1 is incubated at 37° for 10 min, following which iodoacetamide is added to 10 mM and the mixture incubated for 20 min more. After incubation the mixture is dialyzed against two changes of buffer 2. The dialyzed material is then brought up to 5 ml total volume and stored as 0.5-ml aliquots at $-70°$.

Activation of ARF. Ten microliters of CF solution is added to a 5-μl pellet of ghosts. To this mixture is added GTPγS to a final concentration of 100 μM. The pellet is completely resuspended by vortexing, and the mixture is incubated at 37° for 15 min.

ADP-Ribosylation. To the 15 μl of solution from the above step, add 4 μl of reaction mix and incubate at 25° for the desired length of time. After incubation dilute the reaction in 2 ml of saline and centrifuge to recover the membranes. The membranes are dissolved in SDS sample buffer and analyzed on 7.5–10% polyacrylamide gels containing 0.1% SDS. After autoradiography, the bands of interest in the M_r 42,000 range or the whole track can be excised and quantitated. The amount of ADP-ribose incorporated can be calculated from the counts obtained.

Escherichia coli Heat-Labile Enterotoxin

The heat-labile enterotoxin (LT) of several strains of *E. coli* is an enzyme which is both structurally and functionally similar to cholera toxin.[17] Like cholera toxin, LT consists of two subunits termed A and B, and the holotoxin has a structure of A : B$_{4-6}$. As in the case of CT, the B subunits are responsible for binding to cell surface receptors and transmission of the A subunit into the cell. The cell surface receptor has been identified as the ganglioside G$_{M1}$ just as for CT. Also similar to CT is the fact that the A subunit is initially synthesized as a single polypeptide but is proteolytically nicked to result in two polypeptides linked via disulfide

---

[17] J. Moss and M. Vaughn, *Adv. Enzymol.* **61,** 303 (1988).

bonds. The nicked A subunit must then be reduced in order to demonstrate activity. The free A1 fragment then ADP-ribosylates G_s, a regulatory component of adenylate cyclase, resulting in activation of adenylate cyclase activity in the same way as CT. The assay for *E. coli* LT is identical to that of CT and can be performed in the same way.

Clostridial Toxins

There are various clostridial toxins such as *Clostridium botulinum* C2 toxin and *C. perfringens* iota toxin which possess ADP-ribosyltransferase activity.[18] The C2 toxin is produced by many strains of *C. botulinum* types C and D and consists of two components labeled I and II.[19] Unlike cholera and pertussis toxins, the two subunits of C2 toxin are not linked. As a result, the activity of the toxin is dependent on the interaction of two independent proteins. Studies suggest that component II is responsible for the binding of the toxin to the cell surface and that component I contains the ADP-ribosyltransferase activity. Component II appears to require cleavage by trypsin in order to efficiently bind component I. It is presumed that the trypsin cleavage exposes the component I binding site. The target of the C2 toxin ADP-ribosylation is nonmuscle actin. However, skeletal muscle actin can be ADP-ribosylated to a certain extent.

Although it has a similar structure, iota toxin differs from C2 toxin primarily in substrate specificity. Iota toxin is capable of ADP-ribosylating both muscle and nonmuscle actin.[20] Both toxins ribosylate actin at the same arginine residue, resulting in a reduction in the ability of actin to polymerize.[21-24] Assays for both toxins are given below and are essentially those of Geipel *et al.*[25] for iota toxin and Aktories *et al.*[21] for C2 toxin.

[18] K. Aktories, M. Barmann, S. G. Chhatwal, and P. Presek, *Trends Pharmacol. Sci.* **8,** 158 (1986).

[19] K. Aktories and I. Just, *in* "ADP-Ribosylating Toxins and G Proteins" (J. Moss and M. Vaughan, eds.), p. 79. American Society for Microbiology, Washington, D.C., 1990.

[20] B. Schering, M. Barmann, G. S. Chhatwal, U. Geipel, and K. Aktories, *Eur. J. Biochem.* **171,** 225 (1988).

[21] K. Aktories, M. Barmann, I. Ohishi, S. Tsuyama, K. H. Jacobs, and E. Habermann, *Nature (London)* **322,** 390 (1986).

[22] J. Vandekerchove, B. Schering, M. Barmann, and K. Aktories, *FEBS Lett.* **225,** 48 (1987).

[23] B. Schering, M. Barmann, G. S. Chhatwal, U. Geipel, and K. Aktories, *Eur. J. Biochem.* **171,** 225 (1988).

[24] J. Vandekerchove, B. Schering, M. Barmann, and K. Aktories, *J. Biol. Chem.* **263,** 696 (1988).

[25] U. Geipel, I. Just, B. Schering, D. Haas, and K. Aktories, *Eur. J. Biochem.* **179,** 229 (1989).

Purification procedures for C2[26] and iota toxin[27] have been previously described.

Assay of C2 Toxin

The assay of C2 toxin makes use of cytosol from human platelets obtained via a freeze–thaw procedure, although purified actin may be used.[28] Platelets are isolated as described.[29] Radiolabeled NAD is used as a source of the ADP-ribosyl moiety.

Reagents

Human platelet cytosol (see below)
Hypotonic buffer: 10 mM triethanolamine hydrochloride, pH 7.4, 5 mM EDTA
[adenylate-^{32}P]NAD (1000 Ci/mmol)
Reaction mixture: 10 mM thymidine, 5 mM MgCl$_2$, 1 mM EDTA, sample containing toxin, 0.5 μM [^{32}P]NAD, 0.5 mM ATP, and 50 mM triethanolamine hydrochloride, pH 7.4
Samples containing toxin
20% TCA
SDS sample buffer: 10% (w/v) glycerol, 5% (w/v) 2-mercaptoethanol, 2.5% (w/v) SDS, 50 mM Tris-HCl, pH 6.8, 0.05% bromphenol blue
30% (v/v) Hydrogen peroxide

ADP-Ribosylation Assay. Platelet cytosol is obtained via freezing and thawing in hypotonic buffer. Lysates are centrifuged at 30,000 g for 15 min and the supernatant collected.

The ADP-ribosylation is carried out in a reaction mixture which is incubated at 37° for 1 hr and terminated by the addition of 1 ml of 20% TCA. The acid-insoluble material is pelleted and resuspended in 50 μl of SDS sample buffer, boiled, and electrophoresed on SDS–polyacrylamide gels as described for other toxins in this chapter (see above). Gel bands which contain radioactive material are then excised, solubilized in the presence of 30% hydrogen peroxide, and counted in a liquid scintillation counter.

[26] I. Ohishi, W. Iwasaki, and T. Sakaguchi, *Infect. Immun.* **30,** 668 (1980).
[27] B. G. Stiles and T. D. Wilkins, *Infect. Immun.* **54,** 683 (1986).
[28] K. H. Jakobs, W. Sauer, and G. Schultz, *Naunyn-Schmiedeberg's Arch. Pharmacol.* **302,** 285 (1981).
[29] K. Aktories, T. Ankenbauer, B. Schering, and K. H. Jakobs, *Eur. J. Biochem.* **161,** 155 (1986).

Assay for Iota Toxin

In the iota toxin assay actin is obtained from rabbit skeletal muscle as previously described.[30]

Reagents

[*adenylate*-^{32}P]NAD (1000 Ci/mmol)
Samples containing toxin
Skeletal muscle actin (see above)
Reaction mixture: 1000 μM MgCl$_2$, 1 mM DTT, 50 μM [^{32}P]NAD, sample containing iota toxin, 133 μM CaCl$_2$, 133 μM ATP, 3 μM NaN$_3$, 1.3 mM Tris-HCl, pH 7.4, 1.5 mg skeletal muscle actin, and 10 mM triethanolamine hydrochloride, pH 7.4, in a total volume of 1.5 ml
Termination mix: 2% SDS, 1 mg/ml bovine serum albumin (BSA)
30% TCA

Procedure. The reaction mixture is allowed to incubate at 37° for 1 hr, and the reaction is terminated by the addition of 0.4 ml of termination mix and 0.5 ml of 30% TCA and placement on ice for 30 min. The precipitated material is collected via filtration through 0.45-μm nitrocellulose filters, and the filters are washed with 16 ml of 6% TCA prior to air drying and counting in a liquid scintillation counter.

Pertussis Toxin

Pertussis toxin is produced by strains of *Bordetella pertussis* and in its native state exists as a hexamer of five different subunits which make up the enzymatically active A subunit (S1) and a B subunit (S2, S3, two S4, S5) which is reported to be involved in binding of the toxin.[31] If the complex is treated with mild denaturing agents, the S1 subunit is dissociated from the rest of the complex. When incubated with disulfide reducing agents, S1 exhibits ADP-ribosyltransferase activity. The dissociated, reduced form of the toxin is used in toxin assays *in vitro*. Because the B component is required for binding to cell surfaces, studies on intact cells must use the holotoxin. Pertussis toxin catalyzes the ADP-ribosylation of the α subunit of guanine nucleotide-binding proteins (G proteins) which are, for the most part, associated with the plasma membrane. More specifically pertussis toxin is involved in the ADP-ribosylation of G$_i$, transducin, and G$_o$. The net effect is an uncoupling of the G proteins from their

[30] J. A. Spudich and S. Watt, *J. Biol. Chem.* **246**, 4866 (1971).
[31] M. Ui, *in* "ADP-Ribosylating Toxins and G Proteins" (J. Moss and M. Vaughan, eds.), p. 45. American Society for Microbiology, Washington, D.C., 1990.

receptors, resulting in an interruption of intracellular signal transduction. Unlike cholera toxin, which requires ARFs, there have not been any requirements for cofactors other than NAD demonstrated for pertussis toxin.

Assay for Pertussis Toxin

The assay is similar to others previously mentioned in that radiolabeled NAD is used as a source of the ADP-ribosyl moiety.[32] The assay described makes use of the ability of pertussis toxin to ADP-ribosylate G proteins. However, other proteins such as bovine serum albumin (BSA) have been used as the acceptor protein.[33] In general, following incubation the reaction mixture is electrophoresed on polyacrylamide gels and appropriate regions excised and quantitated. If a positive control containing a known amount of toxin is used, then the amount of toxin in an unknown sample may be deduced. A negative control should also be included to determine the background radioactivity. Kopf and Woolkalis[32] suggest the inclusion of three control incubations to aid in defining the specificity of the toxin-catalyzed transfer of ADP-ribose to a G protein: (1) a sample incubated without toxin but with the toxin activation vehicle to allow the determination of nonspecific radiolabeling of protein due to other enzymes which may be present in the extracts; (2) a sample incubated in the presence of excess nonradioactive NAD^+ to eliminate the detection of proteins which may be susceptible to the enzymatic transfer of ADP-ribose from [32P]NAD; and (3) a sample incubated in the presence of guanosine 5'-O-3-thiotriphosphate, which is expected to result in the diminution of the [32P]ADP-ribosylated G protein α subunit since GTP or its nonhydrolyzable analogs cause the dissociation of the α subunit from other subunits. In this manner the α subunit cannot be ADP-ribosylated by pertussis toxin.

Reagents

Tissue extracts (see below)
Samples containing toxin
[*adenylate*-32P]NAD (1000 Ci/mmol)
Toxin activation reaction mixture: 100 μg/ml pertussis toxin (List Biological Laboratories) or samples containing toxin, 50 mM HEPES, pH 8.0, 1 mg/ml BSA, 20 mM DTT, 0.125% SDS
ADP-ribosylation reaction mixture: Tissue extract (not more than 100 μg protein), 20 μg/ml activated toxin (see below), 5 mM DTT, 10

[32] G. S. Kopf and M. Woolkalis, this series, Vol. 195, p. 257.
[33] H. R. Kaslow, L.-K. Lim, J. Moss, and D. D. Lesikar, *Biochemistry* **26,** 123 (1987).

mM HEPES, pH 8.0, 0.2 mg/ml BSA, 1 mM EDTA, 10 mM thymidine, 5 μM [^{32}P]NAD

Preparation of Tissue Extracts. Any standard procedure[34] for obtaining tissue extracts appears to be suitable. However, it is important to minimize proteolysis. The addition of protease inhibitors such as phenylmethylsulfonyl fluoride to extraction buffers does not appear to inhibit ADP-ribosylation catalyzed by pertussis toxin. Furthermore, it is wise to choose protocols that will enrich for membrane fractions since the target G proteins are most often associated with cell membranes.

Activation of Toxin. The activation of the toxin is not strictly required when using broken cell extracts but is suggested to increase the efficiency of the ADP-ribosylation reaction.[35] The activation of pertussis toxin is achieved via the use of DTT. The following protocol is that described by Enomoto and Gill.[36] In the original protocol a commercial source of pertussis toxin is used. However a sample to be tested may be substituted for the pure toxin. The activation reaction mixture is incubated at 30° for 30 min. Activated toxin is diluted 1 : 5 in the ADP-ribosylation assay to maintain SDS concentrations below 0.025%; greater concentrations of SDS may prove to be inhibitory.

ADP-Ribosylation Reaction. The ribosylation reaction mixture is incubated at 30° for up to 1 hr and is terminated by placing into standard SDS sample buffer and boiling for 3 min. Samples are then electrophoresed on 10 to 12% polyacrylamide gels. The desired bands are excised and quantitated.

Concluding Remarks

The intact toxic forms of most of the ADP-ribosylating toxins normally lack enzyme activity in the *in vitro* assay systems. The ADP-ribosyltransferases must be activated *in vitro* by cleavage of the toxin, reduction of disulfide bonds, and/or altering the configuration of the protein with denaturing agents such as urea, depending on the particular toxin being studied.

All of the assays use radioactively labeled NAD as the donor of the ADP-ribosyl moiety. NAD is either in the form of [U-*adenine*-^{14}C]NAD or [*adenylate*-^{32}P]NAD. The [^{32}P]NAD has the advantage of being available at higher specific activities and lower cost than the [^{14}C]NAD but has the

[34] J. D. Dignam, P. L. Martin, B. S. Shastry, and R. G. Roeder, this series, Vol. 101, p. 582.
[35] J. Moss, S. J. Stanley, D. L. Burns, J. A. Hsia, D. A. Yost, G. A. Meyers, and E. L. Hewlett, *J. Biol. Chem.* **258**, 11879 (1983).
[36] K. Enomoto and D. M. Gill, *J. Biol. Chem.* **255**, 1252 (1980).

disadvantage of a short radioactive half-life. Transfer of the radioactively labeled ADP-ribosyl moiety to the target protein is most easily monitored by acid precipitation of the labeled acceptor protein. If multiple proteins serve as acceptors, SDS–polyacrylamide gel electrophoresis may be used for analysis. The primary target is most readily labeled and thus is the first protein saturated as the toxin or NAD concentration is increased. A more rapid but less sensitive assay involves the acid precipitation of all the ADP-ribosylated proteins that are formed in the assay. However, under these assay conditions the primary target may be greatly outnumbered by secondary products. Often an assay system can be established with a model acceptor, such as the cholera toxin ADP-ribosylation of poly(L-arginine)[37] which is acid precipitated for quantitation of radioactivity transferred from the NAD donor.

Partially purified preparations may contain extraneous enzymes and acceptor proteins which can also utilize the radioactively labeled NAD. Frequent contaminants are endogenous glycohydrolase and poly(ADP)-ribosyltransferase. Thymidine is often used to inhibit poly(ADP)-ribosyltransferase. An alternative inhibitor is 10 mM 3-aminobenzamide. More detailed descriptions of potential artifactual results in the assay systems are provided by Gill and Woolkalis.[38]

[37] J. J. Mekalanos, R. J. Collier, and W. R. Romig, *J. Biol. Chem.* **254,** 5849 (1979).
[38] D. M. Gill and M. Woolkalis, this series, Vol. 165, p. 235.

[50] Photoaffinity Labeling of Active Site Residues in ADP-Ribosylating Toxins

By Stephen F. Carroll and R. John Collier

Introduction

The ADP-ribosylating toxins are a class of enzymes which catalytically transfer the ADP-ribosyl moiety of NAD into covalent linkage with selected acceptor amino acids on specific target proteins (TP). For the bacte-

$$NAD + TP \rightleftharpoons ADP\text{-ribosyl-}TP + nicotinamide + H^+$$

rial proteins diphtheria toxin and *Pseudomonas aeruginosa* exotoxin A, the target amino acid is diphthamide, a posttranslationally modified histidine residue on elongation factor 2 (EF-2). Functionally, the ADP-ribosylation of EF-2 results in the inhibition of protein synthesis and, ultimately,

TABLE I
BACTERIAL TOXINS AND RELATED PROTEINS THAT ACT BY ADP-RIBOSYLATION MECHANISMS

Toxin	Target protein	Modified residue	Effect of modification
Diphtheria toxin	Elongation factor 2	Diphthamide (modified histidine)	Lethal; inhibits protein synthesis
Pseudomonas exotoxin A	Elongation factor 2	Diphthamide (modified histidine)	Lethal; inhibits protein synthesis
Cholera toxin	Adenylate cyclase, $G_{s\alpha}$ subunit	Arginine-201	Stimulates adenylate cyclase activity
Escherichia coli LT1 and LT2 (heat-labile toxins)	Adenylate cyclase, $G_{s\alpha}$ subunit	Arginine-201	Stimulates adenylate cyclase activity
Pertussis toxin	Adenylate cyclase, $G_{i\alpha}$ subunit	Cysteine-352	Stimulates adenylate cyclase activity
Botulinum C2 toxin	Nonmuscle actin	Arginine-177	Inhibition of actin polymerization
Botulinum C3 exoenzyme	Rho	Asparagine-41	Unable to enter cells
Pseudomonas exoenzyme S	Vimentin, Ras, Rab3, Rab4, and certain other small GTP-binding proteins	Arginine	Unable to enter cells

cell death. For other members of this class of enzymes (Table I), the acceptor amino acid, the target protein, and the consequence of ADP-ribosylation differ greatly.

Despite these differences in action, each of the ADP-ribosylating toxins possesses a specific binding site for NAD, the required cofactor for the ADP-ribosylation reaction (see Table I). In our studies to identify active site residues of several ADP-ribosylating toxins,[1,2] we have utilized direct photoaffinity labeling with native NAD as a means to localize the NAD binding site within the primary (and tertiary) structure of the toxins. This labeling approach differs from the more commonly employed use of photo-activatable ligand analogs,[3] in that the chemical structure of NAD has not been altered by the introduction of a photolabile group (such as a diazo or arylazide derivative). Instead, the intrinsic photolability of toxin-bound NAD to UV light provides the energy necessary for covalently cross-linking a portion of the NAD molecule into the cofactor binding site. Because no chemical modifications of NAD are required prior to photola-

[1] S. F. Carroll and R. J. Collier, *Proc. Natl. Acad. Sci. U.S.A.* **81**, 3307 (1984).
[2] S. F. Carroll and R. J .Collier, *J. Biol. Chem.* **262**, 8707–8711 (1987).
[3] H. Bayley and J. R. Knowles, this series, Vol. 46, p. 69.

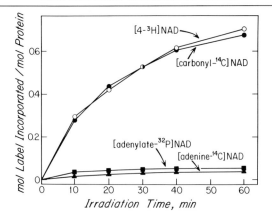

FIG. 1. UV-induced incorporation of radiolabel from NAD to an enzymatically active fragment of *P. aeruginosa* exotoxin A (ETA$_{frag}$). Reaction mixtures containing ETA$_{frag}$ (20 μM) and NAD (40 μM) radiolabeled in the nicotinamide, adenylate phosphate, or adenine moiety were prepared in 50 mM Tris-HCl (pH 7.2) at 0° and irradiated on ice with a G15-T8 germicidal lamp (predominantly 253.7 nm). The intensity of UV light received by the samples was 3 mW/cm^2. At intervals up to 60 min, aliquots were removed and mixed with carrier ovalbumin, and TCA-precipitable radioactivity was determined. (Reprinted from Carroll and Collier[2] with permission.)

beling, high-affinity ligand binding is assured and reagents are commercially available.

The conditions outlined here have proved useful for the photoaffinity labeling and identification of active site residues of diphtheria toxin,[1] *Pseudomonas aeruginosa* exotoxin A,[2] and pertussis toxin.[4,5] The general approach involves irradiating mixtures of toxin and NAD radiolabeled in either the nicotinamide, phosphate, or adenine moiety with UV light (254 nm), followed by a determination of acid-precipitable radioactivity. For each of the proteins, (1) label is most efficiently transferred from the nicotinamide moiety of NAD to the toxin (Fig. 1), and (2) the transferred radiolabel is associated with a specific glutamic acid residue (Glu-148, Glu-553, Glu-129, respectively). An action spectrum of the photolabeling showed a wavelength maximum of approximately 260 nm (Fig. 2). In the case of diphtheria toxin, the structure of the photoproduct at position 148 was determined primarily by nuclear magnetic resonance (NMR) and fast atom bombardment mass spectrometry,[6] and is shown in Fig. 3. Based

[4] J. T. Barbieri, L. M. Mende-Mueller, R. Rappuoli, and R. J. Collier, *Infect. Immun.* **57**, 3549 (1989).
[5] S. A. Cockle, *FEBS Lett.* **249**, 329 (1989).
[6] S. F. Carroll, J. A. McCloskey, P. F. Crain, N. J. Oppenheimer, T. M. Marschner, and R. J. Collier, *Proc. Natl. Acad. Sci. U.S.A.* **82**, 7237 (1985).

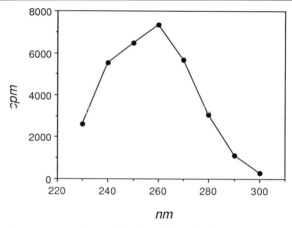

FIG. 2. Action spectrum of photolabeling of diphtheria toxin fragment A (DTA) with [*carbonyl*-^{14}C]NAD. Mixtures (125 μl) of DTA (5 μM) and [*carbonyl*-^{14}C]NAD (3 μM) in 50 mM Tris-HCl (pH 7.2) were irradiated at 4° for selected periods in 3 × 3 mm quartz cuvettes in an SLM-Aminco spectrofluorimeter (Model SPF-500C) with slits giving a bandpass of 5 nm. Irradiation times (ranging from 58 min at 230 nm to 9 min at 300 nm) were chosen on the basis of actinometric measurements (S. L. Murov, "Handbook of Photochemistry," p. 119. Dekker, New York, 1973) to give constant excitation energy at the chosen wavelengths. After irradiation, the contents of the cuvette were removed, and two 50-μl aliquots were prepared and spotted on TCA–paper and processed.

on the efficiency of nicotinamide transfer (0.6 to 0.8 mol/mol), it is likely that these glutamic acid residues reside within the nicotinamide subsite of the enzymes, and that the photoproducts at Gly-553 of exotoxin A and Glu-129 of pertussis toxin are similar to that determined for diphtheria toxin Glu-148. Although similar techniques have been used to photolabel

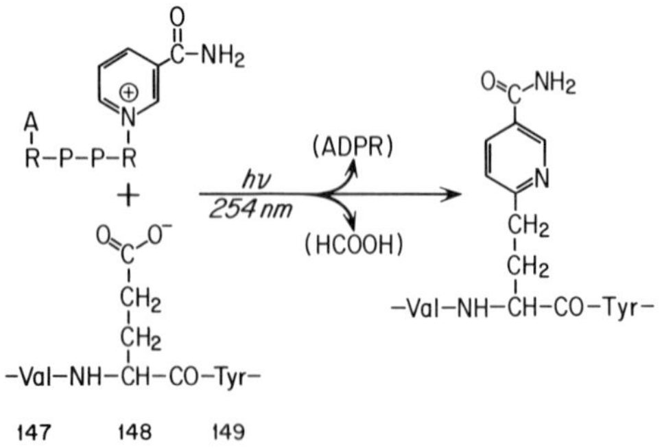

FIG. 3. Structure of the photoproduct at position 148 of diphtheria toxin fragment A, as determined by NMR and fast atom bombardment mass spectrometry. (Reprinted from Carroll et al.,[6] with permission.)

cholera toxin,[7] neither the nature of the reaction nor the amino acid residue(s) labeled have been characterized.

That the glutamic acid residues photolabeled by NAD are indeed located within the catalytic centers of the toxins has been demonstrated by studies involving site-directed mutagenesis of the cloned toxin genes. For example, in diphtheria toxin substitution of Glu-148 with Asp substantially reduces ADP-ribosyltransferase activity without affecting NAD binding,[8] and similar results have been obtained with mutants of exotoxin A[9] and pertussis toxin.[10]

Outlined below are two photolabeling protocols we have found useful. In the first, reaction mixtures are applied as droplets (10–200 μl) to the surface of a plastic plate, allowing numerous samples to be irradiated at one time. Additionally, this system can easily be scaled up such that the entire surface of the plate is flooded with a single sample, providing sufficient material for subsequent protein fragmentation and sequencing. The second protocol is an adaptation of the first technique, which allows samples as small as 2 μl to be irradiated and analyzed. This method is particularly useful for evaluating a variety of proteins, ligands, or reaction conditions, as well as for ligand concentrations (>100 μM) that show appreciable absorption of the incident light energy and might therefore interfere with the photochemical reaction. Because of the short path length, this system should be amenable to the study of ADP-ribosyltransferases with relatively low binding affinities (millimolar K_d values) for NAD.

By way of illustration, we describe analytical and preparative methods for the photoaffinity labeling of diphtheria toxin fragment A (DTA). The efficient labeling of other ADP-ribosyltransferases would likely require purified proteins (or protein fragments) capable of binding NAD. Typically this is accomplished by the isolation of catalytically active toxin fragments by purification,[1,5] proteolysis,[2] or recombinant[4] techniques. In addition, knowledge of the relative binding affinity of NAD for the protein is helpful in designing initial reaction conditions.

Photolysis

NAD (as well as other nucleotides and nucleosides) exhibits a strong absorbance in the UV range, with a maximum near 260 nm. For this reason, irradiation of toxin–NAD complexes with UV light was originally

[7] B. A. Wilson, K. A. Reich, B. R. Weinstein, and R. J. Collier, *Biochemistry* **29**, 8643 (1990).

[8] C. M. Douglas and R. J. Collier, *Biochemistry* **29**, 5043 (1990).

[9] M. Pizza, A. Bartoloni, A. Prugnola, S. Silvestri, and R. Rappuoli, *Proc. Natl. Acad. Sci. U.S.A.* **85**, 7521 (1988).

[10] T. S. Galloway, R. M. Tait, and S. van Heynigen, *Biochem. J.* **242**, 927 (1987).

selected to induce photochemical cross-linking. This is easily accomplished by irradiating samples with a low-pressure mercury lamp (15 W G15-T8, General Electric), which emits almost exclusively at 253.7 nm and fits in a standard fluorescent lamp reflector housing. Irradiation with a Spectroline ENF-24 lamp has also been used.[4] Although other systems (such as a fluorimeter) can be employed, a low-pressure mercury lamp provides sufficient energy for these reactions and allows many samples to be analyzed simultaneously. Appropriate protection for eyes, face, and hands should be worn when manipulating or viewing samples under the UV light source.

To ensure the reproducibility of sample irradiation from experiment to experiment, the output of the lamp should be regularly monitored. In practice, it is advisable to measure the intensity of UV light received by the samples prior to each experiment. This can be varied by adjusting the distance separating the lamp and samples. UV radiometry at 254 nm (Model UVX, UV Products, San Gabriel, CA) has proved simple and convenient for this purpose, although photochemical methods (such as potassium ferrioxylate actinomery[11]) have also been used. In our experiments, irradiation at 3–4 mW/cm^2 appears optimal (Fig. 4).

Reagents

Carbonyl-labeled NAD [*carbonyl*-^{14}C]NAD (30–50 mCi/mmol, CFA.372) or [4-^{3}H]NAD (0.5–2.5 Ci/mmol, TRA.298) is available from Amersham (Arlington Heights, IL). The concentration of labeled or unlabeled NAD can be confirmed by using an E_{1cm}^{1mM} of 17.8 at 259 nm.

DTA: Methods for the purification of diphtheria toxin and its enzymatically active A fragment (DTA) have been described elsewhere in this series.[12] Purified DT holotoxin is also available from List Laboratories (Campbell, CA).

10% TCA in ether: Fifty grams of trichloroacetic acid (TCA) is dissolved in 500 ml of diethyl ether and stored in a glass bottle. The solution is stable at room temperature for several months and can be reused many times.

TCA–paper: A sheet of Whatman (Clifton, NJ) 3MM filter paper (ruled in pencil into 1-inch squares) is trimmed to a size reflecting the number of samples to be analyzed. For example, for 20 samples to be analyzed in duplicate, a 5 × 8 inch ruled sheet (40 squares) would be prepared. Each square is then numbered (in pencil), and

[11] A. Markovitz, *Biochim. Biophys. Acta* **281**, 523 (1972).
[12] S. F. Carroll, J. T. Barbieri, and R. J. Collier, this series, Vol. 165, p. 68.

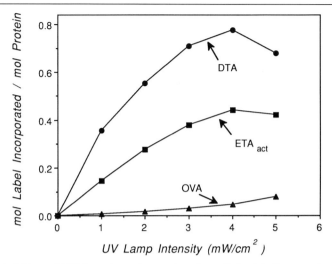

FIG. 4. Effect of UV lamp intensity on the photoaffinity labeling reaction. Reaction mixtures containing 20 μM protein and 50 μM [*carbonyl*-^{14}C]NAD in 50 mM Tris-HCl (pH 7.2) were irradiated at the indicated UV intensities as described in the legend for Fig. 1. After 30 min, aliquots were removed, and TCA-precipitable radioactivity was determined. The proteins examined were diphtheria toxin fragment A (DTA) *P. aeruginosa* exotoxin A previously activated[2] by exposure to urea and dithiothreitol (ETA$_{act}$), and ovalbumin (OVA).

the sheet is placed into a glass baking dish and briefly immersed in 10% TCA in ether. The ether solution is then decanted and the TCA-impregnated paper is allowed to air. The entire operation should be performed under a fume hood.

5% TCA: Fifty grams of TCA is dissolved in 1 liter of water and stored at room temperature.

Method I

Analytical. Solutions containing DTA (20 μM) and carbonyl-labeled NAD (40 μM) are prepared at 0° and applied as 100-μl droplets to the surface of a prechilled, inverted microtiter plate lid (Falcon 3071 or Corning 25803) floating in a shallow ice bath. For multiple samples, the droplets are applied to the plate so as to create a series of parallel rows along the length of the tray. In this manner, more than 20 different replicates, proteins, or reagent concentrations can be evaluated. A control protein (such as ovalbumin) should be included as a negative control. The tray is then covered with a UV-opaque shield and carefully placed under the UV lamp. For this and all subsequent operations near the UV source, appropriate protective clothing and goggles should be worn. The shield is then removed from the samples and irradiation ($\sim$3 mW/cm^2, or roughly 5 cm from the lamp surface) is allowed to proceed for up to 60 min.

For an evaluation of reaction kinetics and extent, the tray is covered at various times and removed from the UV source, and 10-μl aliquots are removed from the droplets and added to 40 μl of carrier protein (0.2 mg/ml ovalbumin in 50 mM Tris-HCl, pH 7.2). Aliquots are also taken at t = 0 (prior to irradiation) to provide background values, and all samples are maintained on ice until the end of the experiment; samples (40 μl) are then spotted onto TCA–paper. The addition of denaturants (such as guanidine hydrochloride) to the samples prior to the determination of TCA-precipitable radioactivity has not been necessary for the proteins we have studied, suggesting that noncovalently bound but not cross-linked NAD does not remain associated with the toxins during the precipitation process. The TCA–paper sheet is immediately washed twice in 5% TCA (10 ml/square) for 15 min, followed by a brief (5 min) wash in methanol and then air drying. The numbered squares are then cut up and counted in a beta counter with appropriate scintillation fluid. An aliquot taken from each reaction mixture, diluted with carrier protein and spotted onto an untreated filter square, is counted directly (no washes) and serves as a measure of input radioactivity. Duplicate blank, untreated paper squares are counted to determine input background.

Preparative. Mixtures of DTA (100 μM) and carbonyl-labeled NAD (200 μM) in 50 mM Tris-HCl (pH 7.2) are prepared at 0° and transferred to a prechilled, inverted plate (Falcon 3071) floating in a shallow ice bath. For this purpose, we have used both microtiter plate lids (or petri dishes), which hold approximately 20 ml (or 11 ml) for a film 2 mm thick. The thickness of the film must be kept to a minimum, since the absorbance of NAD at 200 μM (A_{254} of 3.5) is quite high, and optical shielding could occur. If it becomes necessary to reduce the specific activity of the photolabeled protein, appropriate amounts of unlabeled NAD can be included in the photolysis mixture. The plate is then irradiated as described above for 45–60 min, and aliquots are removed at intervals to monitor the progress of the reaction. The mixture is then recovered from the tray, and the reaction by-products, as well as any unreacted NAD, are removed by size-exclusion chromatography at room temperature on a 0.7 $\times$ 20 cm column of Sephadex G-50 fine, equilibrated in 5% (v/v) acetic acid. Following lyophilization, approximately 60 nmol of photolabeled protein suitable for fractionation and sequencing is obtained.

Method II

To reduce sample size (and possible quenching of UV light energy), samples are irradiated in small fused quartz capillaries (0.5 mm i.d. $\times$ 0.7

mm o.d. × 10 cm long). Prior to use, the capillaries are scored with a diamond pencil and broken into lengths of approximately 2.5 cm (volume 5 μl). For larger (or smaller) sample volumes, the length of the capillaries can be adjusted accordingly. The capillaries are then cleaned by brief immersion in a small volume of HPLC-grade acetone and air dried on a piece of Whatman 3MM paper. All manipulations of the capillaries should be performed with forceps to avoid contamination of the quartz surface.

Reaction mixtures, prepared as described in the analytical example above, are tranferred to the quartz tubes by capillary action, and each capillary is weighed before and after being loaded with sample. These and other transfers of the capillary contents can be facilitated by the use of a standard microcapillary aspirator or dropper. Care should be taken to wipe any excess sample from the exterior surface of the capillary with Kimwipe tissues, since the calculated weight is used as a measure of actual sample volume. The filled capillaries are then placed on a precooled aluminum block resting in a shallow ice bath, and irradiated. On completion, the contents of each capillary are ejected (with rinsing) into a 1.5-ml microcentrifuge tube containing 30 μl of 50 mM Tris-HCl, pH 7.2, 0.2% (w/v) ovalbumin. Twenty-five microliters of this diluted mixture is then spotted onto TCA–paper and processed for acid-precipitable radioactivity as before.

Calculations

The calculation of moles per mole labeling can be determined by conversion of all concentrations, volumes, and the amount of radioactivity transferred to molar values. It is more convenient, however, to utilize the concentration ratio of the two components (NAD and toxin), together with the background ($t = 0$) and input radioactivities. To determine specific radioactivity at each time point, $t = 0$ values are subtracted from the measured counts per minute (cpm). Similarly, corrected input radioactivity is determined by subtracting the counts obtained with an untreated paper square from the measured input values. From these data the amount of radioactivity corresponding to 1 mol NAD per mol toxin is calculated as follows:

$$1 \text{ mol/mol} = \frac{[\text{toxin}]}{[\text{NAD}]} \text{(corrected input cpm)}$$

where [toxin] and [NAD] represent the molar concentrations of toxin and NAD in the reaction mixture, respectively. The fractional mole/mole labeling is then calculated by dividing the specific counts for each time point by the value for 1 mol/mol.

[51] Activation of Cholera Toxin by ADP-Ribosylation Factors

By JOEL MOSS, S.-C. TSAI, and MARTHA VAUGHAN

Introduction

Cholera toxin (CT) and the closely related *Escherichia coli* heat-labile enterotoxin (LT) are responsible in part for the pathogenesis of cholera and "traveler's diarrhea," respectively.[1] The toxins are oligomeric proteins, consisting of one A (CTA) and five B subunits.[1] The B subunits are responsible for toxin binding to the cell surface via ganglioside G_{M1} [galactosyl-*N*-acetylgalactosaminyl-(*N*-acetylneuraminyl)-galactosylglucosyl-ceramide].[1] The A subunit (CTA) catalyzes the ADP-ribosylation of $G_{s\alpha}$, the α subunit of a guanine nucleotide-binding (G) protein responsible for stimulation of adenylate cyclase and regulation of ion flux.[1,2]

The catalytic activities of CTA and LTA are latent; activation requires proteolytic nicking at a site near the carboxyl terminus and reduction of a single disulfide bond that links the two resulting polypeptides, CTA1 and CTA2.[3,4] CTA1 possesses ADP-ribosyltransferase activity[5]; it transfers the ADP-ribose moiety from NAD to, in addition to $G_{s\alpha}$, other acceptors, which include simple guanidino compounds (e.g., arginine, agmatine), proteins unrelated to $G_{s\alpha}$, CTA1 itself (auto-ADP-ribosylation), or water (NAD$^+$ glycohydrolase reaction).[6-11]

[1] J. Moss and M. Vaughan, *Adv. Enzymol.* **61**, 303 (1988).
[2] L. Birnbaumer, R. Mattera, A. Yatani, J. Codina, A. M. J. Van Dongen, and A. M. Brown, *in* "ADP-Ribosylating Toxins and G Proteins: Insights into Signal Transduction" (J. Moss and M. Vaughan, eds.), p. 225. American Society for Microbiology, Washington, D.C., 1990.
[3] J. J. Mekalanos, R. J. Collier, and W. R. Romig, *J. Biol. Chem.* **254**, 5855 (1979).
[4] J. Moss, J. C. Osborne, Jr., P. H. Fishman, S. Nakaya and D. C. Robertson, *J. Biol. Chem.* **256**, 12861 (1981).
[5] J. Moss, S. J. Stanley, and M. C. Lin, *J. Biol. Chem.* **254**, 11993 (1979).
[6] J. Moss, V. C. Manganiello, and M. Vaughan, *Proc. Natl. Acad. Sci. U.S.A.* **73**, 4424 (1976).
[7] J. Moss and M. Vaughan, *J. Biol. Chem.* **252**, 2455 (1977).
[8] J. B. Trepel, D. M. Chuang, and N. H. Neff, *Proc. Natl. Acad. Sci. U.S.A.* **74**, 5440 (1977).
[9] J. Moss and M. Vaughan, *Proc. Natl. Acad. Sci. U.S.A.* **75**, 3621 (1978).
[10] J. Moss, S. J. Stanley, P. A. Watkins, and M. Vaughan, *J. Biol. Chem.* **255**, 7835 (1980).
[11] J. K. Northup, P. C. Sternweis, M. D. Smigel, L. S. Schleifer, E. M. Ross, and A. G. Gilman, *Proc. Natl. Acad. Sci. U.S.A.* **77**, 6516 (1980).

Activation of Cholera Toxin by ADP-Ribosylation Factors

A family of 20-kDa guanine nucleotide-binding proteins, termed ADP-ribosylation factors or ARFs, stimulate all cholera toxin-catalyzed reactions and appear to be allosteric activators of CTA1.[12-15] In the presence of GTP or nonhydrolyzable GTP analogs, but not GDP or adenine nucleotides, ARFs enhance the ADP-ribosyltransferase activities of cholera toxin.[12,16,17]

The ARFs are ubiquitous, highly conserved proteins distributed widely in eukaryotes from *Giardia* to mammals.[18-23] At least six mammalian ARF genes have been identified.[19,20,22,24,25] Based on size, deduced amino acid sequences, gene structure, and phylogenetic analysis, they fall into three classes.[22,26-28] Class I consists of ARFs 1, 2, and 3; class II, ARFs 4 and 5; and class III, ARF 6.[22] The ARFs from all three classes expressed as

[12] R. A. Kahn and A. G. Gilman, *J. Biol. Chem.* **259,** 6228 (1984).

[13] D. A. Bobak, S.-C. Tsai, J. Moss, and M. Vaughan, *in* "ADP-Ribosylating Toxins and G Proteins: Insights into Signal Transduction" (J. Moss and M. Vaughan, eds.), p. 439. American Society for Microbiology, Washington, D.C., 1990.

[14] M. Noda, S.-C. Tsai, R. Adamik, D. A. Bobak, J. Moss and M. Vaughan, *Biochemistry* **28,** 7936 (1989).

[15] M. Noda, S.-C. Tsai, R. Adamik, J. Moss, and M. Vaughan, *Biochim. Biophys. Acta* **1034,** 195 (1990).

[16] S.-C. Tsai, M. Noda, R. Adamik, J. Moss, and M. Vaughan, *Proc. Natl. Acad. Sci. U.S.A.* **84,** 5139 (1987).

[17] S.-C. Tsai, M. Noda, R. Adamik, P. Chang, H.-C. Chen, J. Moss, and M. Vaughan, *J. Biol. Chem.* **263,** 1768 (1988).

[18] R. A. Kahn, C. Goddard, and M. Newkirk, *J. Biol. Chem.* **263,** 8282 (1988).

[19] S. R. Price, M. S. Nightingale, S.-C. Tsai, K. C. Williamson, R. Adamik, H.-C. Chen, J. Moss, and M. Vaughan, *Proc. Natl. Acad. Sci. U.S.A.* **85,** 5488 (1988).

[20] J. L. Sewell and R. A. Kahn, *Proc. Natl. Acad. Sci. U.S.A.* **85,** 4620 (1988).

[21] S.-C. Tsai, R. Adamik, M. Tsuchiya, P. O. Chang, J. Moss, and M. Vaughan, *J. Biol. Chem.* **266,** 8213 (1991).

[22] M. Tsuchiya, S. R. Price, S.-C. Tsai, J. Moss, and M. Vaughan, *J. Biol. Chem.* **266,** 2772 (1991).

[23] J. J. Murtagh, Jr., M. R. Mowatt, C.-M. Lee, F.-J. S. Lee, K. Mishima, T. E. Nash, J. Moss, and M. Vaughan, *J. Biol. Chem.* **267,** 9654 (1992).

[24] D. A. Bobak, M. S. Nightingale, J. J. Murtagh, S. R. Price, J. Moss, and M. Vaughan, *Proc. Natl. Acad. Sci. U.S.A.* **86,** 6101 (1989).

[25] L. Monaco, J. J. Murtagh, K. B. Newman, S.-C. Tsai, J. Moss, and M. Vaughan, *Proc. Natl. Acad. Sci. U.S.A.* **87,** 2206 (1990).

[26] S.-C. Tsai, R. S. Haun, M. Tsuchiya, J. Moss, and M. Vaughan, *J. Biol. Chem.* **266,** 23053 (1991).

[27] R. S. Haun, I. M. Serventi, S. C. Tsai, C.-M. Lee, E. Cavanaugh, L. Stevens, J. Moss, and M. Vaughan, *Clin. Res.* **40,** 148A (1992).

[28] C.-M. Lee, R. S. Haun, S.-C. Tsai, J. Moss, and M. Vaughan, *J. Biol. Chem.* **267,** 9028 (1992).

recombinant proteins in *E. coli* exhibit the characteristic GTP-dependent stimulation of CTA-catalyzed reactions.[29,30]

ADP-ribosylation factors have been identified in both membrane and soluble fractions.[12,16,17] Two ARFs, sARF I and II, purified from bovine brain cytosol, were identified by sequencing as the ARF 1 and 3 gene products, respectively.[31] In a tissue homogenate, when ARF is in the GDP-bound form, most is in the soluble fraction and only a small percentage is membrane associated.[16–18] The ARFs appear to be localized in part to the Golgi where they participate in protein and vesicular trafficking.[32–34] Membrane and soluble ARFs, as well as a recombinant ARF expressed in Sf9 insect cells, are myristoylated.[18,35] Recombinant ARFs synthesized in *E. coli* can be myristoylated when coexpressed with *N*-myristoyltransferase.[36]

Assays described here assess CT-catalyzed ADP-ribosylation of $G_{s\alpha}$, other cellular proteins, and simple guanidino compounds, as well as the auto-ADP-ribosylation of CTA1. In addition, conditions are outlined to evaluate the effect of ARF, in the presence of GTP or analogs and phospholipids and detergents, on CTA ADP-ribosyltransferase activity.

Effects of Other Assay Components

Phospholipids (e.g., dimyristoylphosphatidylcholine, phosphatidylserine, phosphatidylinositol, cardiolipin) and detergents [e.g., cholate, sodium dodecyl sulfate (SDS)] enhance activation of CT by some native and recombinant ARFs.[16,17,30,37] Cytosolic ARFs in the presence of GTPγS [guanosine 5′-*O*-(γ-thio)triphosphate] bind to phosphatidylser-

[29] R. A. Kahn, F. G. Kern, J. Clark, E. P. Gelmann, and C. Rulka, *J. Biol. Chem.* **266**, 2606 (1991).

[30] S. R. Price, C. F. Welsh, R. S. Haun, S. J. Stanley, J. Moss, and M. Vaughan, *J. Biol. Chem.* **267**, 17766 (1992).

[31] S.-C. Tsai, R. Adamik, R. S. Haun, J. Moss, and M. Vaughan, *Proc. Natl. Acad. Sci. U.S.A.* **89**, 9272 (1992).

[32] T. Stearns, M. C. Willingham, D. Botstein, and R. A. Kahn, *Proc. Natl. Acad. Sci. U.S.A.* **87**, 1238 (1990).

[33] T. Serafini, L. Orci, M. Amherat, M. Brunner, R. A. Kahn, and J. E. Rothman, *Cell (Cambridge, Mass.)* **67**, 239 (1991).

[34] J. G. Donaldson, R. A. Kahn, J. Lippincott-Schwartz, and R. D. Klausner, *Science* **254**, 1197.

[35] B. C. Kunz, K. A. Muczynski, C. F. Welsh, S. J. Stanley, S.-C. Tsai, R. Adamik, P. P. Chang, J. Moss, and M. Vaughan, *Biochemistry* **32**, 6643 (1993).

[36] R. S. Haun, S.-C. Tsai, R. Adamik, J. Moss, and M. Vaughan, *J. Biol. Chem.* **268**, 7064 (1993).

[37] D. A. Bobak, M. M. Bliziotes, M. Noda, S.-C. Tsai, R. Adamik, and J. Moss, *Biochemistry* **29**, 855 (1990).

ine.[38] NAD$^+$ glycohydrolase (NADase) activities that hydrolyze NAD may interfere in assays that contain crude tissue fractions. Cibachrome blue (60 μM) inhibited greater than 80% of tissue NAD$^+$ glycohydrolase activities and, to a lesser extent, CTA activity, but, under similar conditions, did not inhibit ARF activation of CTA.[31] The dye is therefore included in assays using crude cellular fractions. Cellular poly(ADP-ribose) polymerase and mono-ADP-ribosyltransferase may increase background ADP-ribosylation.[39,40] These activities are blocked in part by thymidine.[39] At low protein concentrations, depending on detergent and phospholipid content of the assay, CT and/or ARF may be unstable. Addition of ovalbumin or phospholipids/detergents may stabilize both.

ADP-ribosylation in membranes may not always be quantitative and should be evaluated in that light (e.g., with immunological quantification of $G_{s\alpha}$). Many alternative assays can be employed, and, depending on the objective (to measure CTA, ARF, or $G_{s\alpha}$), different assays may be preferable. Assays using simple guanidino compounds (e.g., agmatine) permit the determination of toxin activity in the absence of tissue components (e.g., $G_{s\alpha}$) and have been used for the quantitative estimation of ARF activation of toxin. Auto-ADP-ribosylation, determined in the presence of SDS, is a sensitive measure of ARF activity in crude cell lysates and is independent of $G_{s\alpha}$.[16,17] Readers are referred to prior discussions of ADP-ribosylation in this series.[41]

Assay of ADP-Ribosylation

Stock Solutions

Stock solutions include the following: 1 M potassium phosphate, pH 7.5, refrigerated; 0.1 M magnesium chloride, refrigerated; 1 M dithiothreitol (DTT), frozen in small portions; 0.2 M thymidine, refrigerated; 20 mM NAD, frozen in small portions; CTA, 1 μg/μl (add 250 μl water to vial containing 0.25 mg CTA), refrigerated; ovalbumin, 10 mg/ml, frozen;

[38] M. W. Walker, D. A. Bobak, S.-C. Tsai, J. Moss, and M. Vaughan, *J. Biol. Chem.* **267,** 3230 (1992).

[39] K. Ueda, *in* "ADP-Ribosylating Toxins and G Proteins: Insights into Signal Transduction" (J. Moss and M. Vaughan, Eds.), p. 525. American Society for Microbiology, Washington, D.C., 1990.

[40] K. C. Williamson and J. Moss, *in* "ADP-Ribosylating Toxins and G. Proteins: Insights into Signal Transduction" (J. Moss and M. Vaughan, eds.), p. 493. American Society for Microbiology, Washington, D.C., (1990).

[41] J. Moss, S.-C. Tsai, S. R. Price, D. A. Bobak, and M. Vaughan, this series, Vol. 195, p. 243.

0.5 M agmatine, frozen; 20 mM dimyristoylphosphatidylcholine (DMPC), refrigerated; 5% sodium cholate, refrigerated; 10 mM guanine nucleotide and 50 mM ATP, frozen in small portions; and ARF protein, frozen in small portions.

Materials

Cholera toxin A subunit is purchased from List Biologicals (Campbell, CA); dithiothreitol from Schwarz/Mann Biotech (Cleveland, OH); [32P]NAD (30 Ci/mol), [carbonyl-14C]NAD (53 mCi/mmol), and [adenine-U-14C]NAD (303 Ci/mol) from New England Nuclear (Boston, MA); and dimyristoylphosphatidylcholine, GTP, GDP, sodium cholate, agmatine, and thymidine from Sigma (St. Louis, MO). Other chemicals are purchased from several sources.

Sources of ADP-Ribosylation Factors

The ARFs are abundant in the brain of rat, bovine, chicken, or frog.[21] The major portion is cytosolic.[17] As the purification procedure has been published,[17] only a brief description is presented here. The cytosolic fraction from brain homogenate is adjusted to pH 5.3 and centrifuged. The supernatant is applied to a column of CM-Sepharose equilibrated with phosphate buffer, pH 5.3. The column is eluted with a linear gradient of 25 to 250 mM NaCl. Fractions that activate CTA-catalyzed ADP-ribosylag-matine formation are found in two peaks, the first termed sARF I and the second sARF II. Each pool of peak fractions, after exchanging the buffer to Tris, pH 8.0, is applied to a column of hydroxylapatite that is eluted with a phosphate gradient, 0 to 50 mM. Fractions containing ARF activity are pooled, concentrated, and applied to a column of Ultrogel AcA 54 to yield sARF I (ARF 1 gene product) or sARF II (ARF 3 gene product).[17,31]

Recombinant ARFs (ARF 2, ARF 5, and ARF 6) from mammalian ARF cDNAs have been expressed in E. coli.[30] The method of preparation has been published. Myristoylated ARFs are synthesized in E. coli cotransfected with cDNA for N-myristoyltransferase.[36] In Sf9 insect cells, a mixture of myristoylated and nonmyristoylated ARF 2 is synthesized from bovine retinal ARF 2 cDNA using a baculovirus vector.[35]

Procedures for Assay of Cholera Toxin-Catalyzed ADP-Ribosylation Stimulated by ADP-Ribosylation Factors

The reaction mixture (100 or 150 μl) contains the components shown in Table I and additional water as needed for the given volume. Glass tubes that can be centrifuged to collect TCA-precipitated products before SDS–PAGE should be used to assay [32P]ADP-ribosylprotein formation.

TABLE I
ASSAYS FOR ARF-STIMULATED CHOLERA TOXIN-CATALYZED ADP-RIBOSYLATION

Components	ADP-ribosyltransferase[a] (μl)	NAD: agmatine ADP-ribosyltransferase (μl)	NAD$^+$ glycohydrolase (μl)
Potassium phosphate buffer	5	7.5	7.5
MgCl$_2$	5	7.5	7.5
DTT	0.3	3	3
CTA	1	1	1
NAD	1 (2 mM)	1.5 (20 mM)	3 (2 mM)
Labeled NAD	2 μCi [^{32}P]NAD	0.05 μCi [adenine-U-^{14}C]NAD or [carbonyl-^{14}C]NAD	0.05 μCi [carbonyl-^{14}Cl]NAD
Thymidine[b]	10	—	—
DMPC[b]	5	—	—
Cholate	2	—	—
Ovalbumin	—	1.5	1.5
Agmatine[b]	—	3.0	—
ATP	1.0	1.5	1.5
GTP	2	3.0	3.0
ARF protein	1.0 μg	1.0 μg	1.0 μg
Product:	[^{32}P]ADP-ribosylprotein and ADP-ribosyl-CTA1[c]	[adenine-U-^{14}C]ADP-ribosylagmatine or [carbonyl-^{14}C]-nicotinamide	[carbonyl-^{14}C]Nicotinamide
Product isolation:	SDS–PAGE, autoradiography	AG1-X2 column chromatography	AG1-X2 column chromatography

[a] Protein substrate for this assay not included. The total volume of this assay is 100 μl, and that of the other two is 150 μl; the volume is adjusted with water as needed.
[b] Precipitates resulting from refrigeration dissolve with heating to approximately 50°.
[c] If ADP-ribosyl-CTA1 is the product of interest, DMPC and cholate are omitted; ADP-ribosylation of G$_{s\alpha}$ requires the presence of DMPC and cholate. All the incubations are carried out at 30° for 30 to 60 min.

Plastic or glass tubes can be used in other assays. The reaction mixture is added to the assay tube and incubated at 30° for 30 to 60 min. Each assay includes a series of "blank" tubes containing reaction mixtures with or without the protein fraction and/or CTA.

We premix stock reagents in dilute buffer at 0° for easy dispensing by means of an Eppendorf dispenser. An example of preparation of solution for assay of 10 samples follows.

ADP-Ribosylation of $G_{s\alpha}$ or Other Protein. Prepare the following solutions: (a) 20 μl of GTP solution, from stock containing 20 μl of 10 mM GTP plus 180 μl of water; (b) 20 μl of DMPC/cholate solution, from stock containing 50 μl of 20 mM DMPC, 20 μl of 5% cholate, and 130 μl of water; (c) 10 μl of preactivated cholera toxin, from stock containing 10 μg CTA (or 50 μg of cholera toxin) in 10 μl, 3 μl of 1 M dithiothreitol, 30 μl of 200 mM glycine buffer or 50 mM phosphate buffer (pH 7.5), and 50 μl of water incubated at 30° for 10 min; and (d) 30 μl of [^{32}P]NAD solution, from stock containing 50 μl of 1 M phosphate buffer (pH 7.5), 50 μl of 0.1 M MgCl$_2$, 10 μl of 50 mM ATP, 100 μl of 0.2 M thymidine,

10 μl of 2 mM NAD, 5 μl of [^{32}P]NAD (20–30 μCi), and 75 μl of water. To a glass tube containing protein fraction (e.g., ARF, G$_{s\alpha}$ in membrane, or other ADP-ribose acceptor protein), add in order (a), (b), and (c) followed by extra water, and finally (d) to start the reaction at 30° for 30–60 min. Results with 1 and 3 mM DMPC are similar.

ADP-Ribosylagmatine Formation. For assay of 10 samples, prepare the following: (a) 20 μl of ovalbumin, Cibachrome blue solution, from stock containing 15 μl of 10 mg/ml ovalbumin, 9 μl of 10 mM Cibachrome blue, and 176 μl of 10 mM phosphate buffer, pH 7.5; (b) 20 μl of phospholipid such as cardiolipin, from stock containing 480 μg of cardiolipin in 300 μl of chloroform, which is evaporated under N$_2$, and to the lipid residue is added 200 μl of 10 mM phosphate buffer, pH 7.5, followed by sonification for 1 hr in a water bath sonifier; (c) 20 μl of GTP solution, from stock containing 30 μl of 10 mM GTP and 170 μl of water; (d) 20 μl of cholera toxin solution, from stock containing 10 μg of CTA in 10 μl plus 190 μl of 10 mM phosphate buffer, pH 7.5 (it is not necessary to be preactivate CTA for this or the next assay; dithiothreitol in the reaction mix is sufficient); and (e) 30 μl of [^{14}C]NAD solution, from stock containing 75 μl of 1 M phosphate buffer (pH 7.5), 75 μl of 0.1 M MgCl$_2$, 30 μl of 0.5 M agmatine, 15 μl of 50 mM ATP, 15 μl of 20 mM NAD, 30 μl of 1 M dithiothreitol, 20 μl of [^{14}C]NAD, and water to make 300 μl total volume. To a tube kept at 0° add in order (a) ARF preparations, (b), (c), and (d) followed by extra water and finally (e) to start the reaction, which is allowed to proceed at 30° for 60 min. The rate of ADP-ribosylagmatine formation is constant for several hours at 30° with 10 mM agmatine as substrate.

NAD$^+$ Glycohydrolase. Reagents are premixed as for the ADP-ribosylagmatine assay except that agmatine is omitted and 40 μM [*carbonyl-*^{14}C]NAD is used. In assays that include crude tissue preparations, 0.5 mM ATP and 0.2 mM GTP may be added to minimize the effects of GTP hydrolysis and to serve as alternative substrates for pyrophosphatases. In particular, using a crude tissue fraction as the source of ARF, several concentrations of protein should be tested in the presence of Cibachrome blue, ATP, and GTP to establish a linear relationship between protein added and product (e.g., ADP-ribosylagmatine).

Analysis of [^{32}P]ADP-Ribosyl G$_{s\alpha}$ or Other Protein or Auto-ADP-Ribosylated CTA1

To terminate the reaction, add 1 ml of cold 7.5% trichloroacetic acid and 5 μg of bovine serum albumin (omitted if sufficient protein is present in the sample), mix, keep in ice for 30 to 60 min, and centrifuge for 30 min at 2300 g or for 20 min in a microcentrifuge at 14,000 g. Decant the supernatant and wipe off excess from the sides of the tubes. Usually 100 μg of pelleted protein is easily dissolved in 75 μl of SDS–mercaptoethanol solution A.[42] For preparation of 1 ml of this solution mix 200 μl of 0.6 M Tris base, 200 μl of 50% glycerol, 100 μl of 10% SDS, 50 μl of mercaptoethanol, 50 μl of 0.1% bromphenol blue in water, and 400 μl of water. If after addition of the SDS solution to a sample it turns yellow, add extra 0.6 M Tris base dropwise until the solution becomes blue. The sample is heated ($\sim$65°, 10 min) and subjected to electrophoresis in 12 to 14% polyacrylamide gels (14 $\times$ 18 cm glass plates, Hoeffer Scientific Instruments, San Francisco, CA). After electrophoresis, the gel is stained with Coomassie blue, destained, and partially dehydrated in 50% methanol to prevent cracking during drying in a gel dryer or placing in a sealed bag. Gels are exposed to Kodak (Rochester, NY) X-Omat film overnight at $-70°$ with intensifying screen. The autoradiograms can be quantified by densitometry.

Analysis of Products by AG1-X2 Column Chromatography

After incubation as described, the sample is transferred to an ice bath. Duplicate aliquots (each one-third of the assay) are applied to columns (0.8 ml, 0.5 $\times$ 4 cm, prepared in a Pasteur pipette) of AG1-X2. The [^{14}C]ADP-ribosylagmatine synthesized or [^{14}C]nicotinamide released during the incubation is eluted with 5 ml of water into scintillation counting vials for radioassay. The result is expressed as nanomoles of ADP-ribosylagmatine or nicotinamide formed per unit time.

[42] U. K. Laemmli, *Nature (London)* **227,** 680 (1970).

[52] Toxins That Inhibit Host Protein Synthesis

By TOM G. OBRIG

Introduction

Several bacterial toxins are known to be potent inhibitors of eukaryotic protein synthesis.[1–3] Diphtheria toxin and *Pseudomonas aeruginosa* exo-

[1] B. H. Iglewski and V. L. Clark, "Molecular Basis of Bacterial Pathogenesis." Academic Press, New York, 1990.
[2] C. B. Saelinger, "Trafficking of Bacterial Toxins." CRC Press, Boca Raton, Florida, 1990.

toxin A inactivate eukaryotic elongation factor 2 (eEF-2) through an ADP-ribosylation mechanism. These agents are described in [49] of this volume and, thus, will not be emphasized here. The toxins of *Shigella* (Shiga toxin) and *Escherichia coli* (Shiga-like toxin or verotoxin) catalyze the inactivation of eukaryotic cytoplasmic (80 S) ribosomes[4] in the absence of ADP-ribosylation[5] and are related to plant-derived toxins that do the same.[6,7] All of the toxins listed above inhibit the process of peptide elongation. Yet to be discovered are protein toxins from pathogenic bacteria that affect other phases of protein synthesis, such as formation of amino-acyl-tRNAs (aa-tRNAs), initiation and termination phases of eukaryotic cytoplasmic protein synthesis. This chapter addresses the methodologies that have been employed in the study of ribosome-inhibiting toxins; however, the techniques are equally useful to study agents that act at sites other than the ribosome. The methods described would allow one to discern if newly discovered toxin could cause a functional lesion in one or more of the processes of peptide initiation, elongation, or termination.

The Shiga toxins are a family of multisubunit proteinaceous toxins related in primary structure. Shiga holotoxin (M_r 69,000) includes five copies of a B subunit (M_r 7,000 each) and a single copy of an A subunit (M_r 31,000).[8,9] The B subunit facilitates binding of holotoxin to target cells via the neutral sphingoglycolipid receptor, globotriaosylceramide (Gb₃) expressed on the cell surface. Internalization of holotoxin involves formation of coated pits and movement of toxin through the Golgi to the cytoplasm.[10,11] During this process, the A and B subunits become separated, and the A subunit is cleaved in a trypsinlike manner to yield a maximally active A_1 subunit (M_r 27,000) that inactivates ribosomes.[12]

[3] J. Stephen and R. A. Pietrowski, "Bacterial Toxins." American Society for Microbiology, Washington, D.C. 1986.
[4] C. A. Mims, "The Pathogenesis of Infectious Disease." Academic Press, Orlando, Florida, 1987.
[5] M. R. Thompson, M. S. Steinberg, P. Gemski, S. B. Formal, and B. P. Doctor, *Biochem. Biophys. Res. Commun.* **71**, 783 (1976).
[6] T. G. Obrig, J. D. Irvin, and B. Hardesty, *Arch. Biochem. Biophys.* **155**, 278 (1973).
[7] L. Carrasco, C. Fernandez-Puentes, and D. Vasquez, *Eur. J. Biochem.* **54**, 499 (1975).
[8] S. Olsnes, R. Reisbig, and K. Eiklid, *J. Biol. Chem.* **256**, 8732 (1981).
[9] A. Donohue-Rolfe, G. Keusch, C. Edson, D. Thorley-Lawson, and M. Jacewicz, *J. Exp. Med.* **160**, 1767 (1984).
[10] S. Olsnes and K. Sandvig, in "Immunotoxins" (A. Frankel, ed.), p. 39. Kluwer Academic Publ., Boston, 1988.
[11] K. Sandvig, S. Olsnes, J. E. Brown, O. Petersen, and B. van Deurs, *J. Cell Biol.* **108**, 1331 (1989).
[12] R. Reisbig, S. Olsnes, and K. Eklid, *J. Biol. Chem.* **256**, 8739 (1981).

The Shiga toxins and the plant-derived toxins such as ricin catalyze the inactivation of ribosomes through the same mechanism, namely, by depurination of a single nucleotide in 28 S rRNA of 60 S ribosomes.[13,14] The product of the N-glycohydrolase activity is a full-length 28 S rRNA with an intact ribose–phosphodiester backbone, but one which confers an inability of that part of the ribosome containing the depurination site to interact effectively with eukaryotic elongation factor eEF-1 and to a lesser extent eEF-2, whose binding sites overlap on the ribosome. In a functional sense, the toxin-treated ribosome no longer binds efficiently aa-tRNA into the ribosomal A ("acceptor") site during peptide elongation.[15]

Cytotoxicity

The Shiga toxins were demonstrated to be cytotoxic to several mammalian cell types, including monkey kidney Vero cells, which resulted in the term VT for verotoxin.[16–18] To date, the most sensitive cell types have been obtained from mammalian sources, such as Vero cells,[18] HeLa cells[12] and human kidney endothelial cells.[19] The latter cell type may be the primary target of the Shiga toxins in human disease.[20] This cytotoxic response has been demonstrated to be temporally related to toxin inhibition of protein synthesis.[21]

Several assays are available for the measure of cytotoxic activity of bacterial toxins, most using the number of remaining viable cells as a parameter. The assay may be accomplished by the removal of cells from multiwell cell culture plates by treatment with a solution of EGTA and trypsin containing 0.04% (w/v) trypan blue followed by direct enumeration of the cells in a hemocytometer. Trypan blue enters all cells but is secreted from viable cells only via an energy-requiring process.

The dye neutral red is used to measure cytotoxicity where a larger

[13] Y. Endo and K. Tsurugi, J. Biol. Chem. 262, 8128 (1987).
[14] Y. Endo, K. Tsurugi, T. Yutsudo, Y. Takeda, K. Ogasawara, and K. Igarashi, Eur. J. Biochem. 171, 45 (1988).
[15] T. G. Obrig, T. P. Moran, and J. E. Brown, Biochem. J. 244, 287 (1987).
[16] G. Vicari, A. L. Olitzki, and Z. Olitzki, Br. J. Expt. Pathol. 41, 179 (1960).
[17] G. T. Keusch, M. Jacewicz, and S. Z. Hirschman, J. Infect. Dis. 125, 539 (1972).
[18] J. Konowalchuk, J. I. Speirs, and S. Stavric, Infect. Immun. 18, 775 (1977).
[19] T. G. Obrig, C. B. Louise, C. A. Lingwood, B. Boyd, L. Barley-Maloney, and T. O. Daniel, J. Biol. Chem. 268, 15484 (1993).
[20] T. G. Obrig, in "Hemolytic Uremic Syndrome and Thrombotic Thrombocytopenic Purpura" (B. S. Kaplan, R. S. Trompeter, and J. L. Moake, eds.), p. 405. Dekker, New York, 1992.
[21] C. B. Louise and T. G. Obrig, Infect. Immun. 59, 4173 (1991).

number of samples are to be processed.[22] Neutral red stains only viable cells. In this case, neutral red (50 μg/ml) is incubated for 1 hr with cells in multiwell culture plates. Cells are rinsed with a phosphate-buffered saline (PBS) solution, pH 7.0, to remove free extracellular dye, and the cells are lysed with an acidic ethanol solution (0.5N HCl/35% ethanol) to release the dye for direct measurement at 570 nm in an enzyme-linked immunosorbent assay (ELISA) reader.[21] The neutral red assay yields results that are linear over the 3000 to 16,000 cell/well range. A stock solution of neutral red is prepared every 2 weeks, filtered through a 0.22-μm filter, and stored at 4°.

Whole Cell Protein Synthesis

Inhibition of Protein Synthesis

The measure of protein synthesis in whole cells utilizes methodology common to virtually all cell types. The assays are conducted in multiwell culture plates with as few as 4000 cells/well in a 96-well plate. To maximize incorporation of a single radiolabeled amino acid, the cells are preincubated for 2 to 24 hr with a medium deficient in that amino acid. This serves to deplete partially the pool of free amino acid and aa-tRNA. [35S]Methionine is usually utilized as the isotope has a short half-life of 87 days and is monitored with a relatively high counting efficiency compared to tritium. Alternatively, ^{14}C- or ^{3}H-labeled leucine or phenylalanine are employed as these amino acids are incorporated directly into proteins and do not give rise to radioactive secondary metabolites.

To measure whole cell protein synthesis, substratum-attached cells (4000–16,000/well) in a 96-well plate are labeled with 3 μCi of [35S]methionine (specific activity 1000 Ci/mmol) in 0.1 ml of methionine-deficient medium.[21] Following incubation for 3 hr, the radioactive medium is removed and the cells are lysed by incubation with 60 μl of 0.2 M NaOH for 30 min at 20°. Aliquots (5 μl) of the lysate are spotted on 1.5 × 1.5 cm squares marked on a single piece of filter paper (Whatman Clifton, NJ, No. 42). The paper is then immersed sequentially in solutions of 10% (v/v) trichloroacetic acid (4°) for 5 min, 70% (v/v) ethanol (4°) for 10 min, and 100% (v/v) ethanol (20°) for 10 min. The filter is dried under a heat lamp, cut into 1.5 cm squares, and the pieces are placed in scintillation vials with 5 ml of nonaqueous scintillation solution. This result typically yields samples that contain approximately 20,000 counts/min (cpm) per 5 μl of cell lysate. The procedure is similar for nonadherent cells except

[22] R. Riddell, R. Clothier, and M. Balls, *Food Chem. Toxicol.* **24,** 469 (1986).

that the plates are centrifuged at 1000 g for 10 min before changing medium and removing radioactive medium.

Cell-Free Protein Synthesis

Cell-free protein synthesis assays are of value if a toxin has been shown to inhibit protein synthesis in whole cells, but it is not clear if the result is due to a direct effect on protein synthesis or an effect on amino acid transport at the plasma membrane level or, alternatively, a primary effect on RNA or DNA synthesis with a secondary effect on protein synthesis. Should it be established that the toxin directly inhibits protein synthesis in a cell lysate preparation, subsequent *in vitro* assays can be developed which allow one to define the mechanism of the toxin in regard to the overall process of protein synthesis, namely, initiation, elongation, or termination.

Unfractionated Cell Lysates

Total Protein Synthesis. Cytoplasmic protein synthesis is carried out in crude extracts of eukaryotic whole cells from which the nuclei and mitochondria have been removed. Protein synthesis in the S30 (30,000 g supernatant) fraction is due to elongation of peptides on existing polysomes as well as initiation on free mRNAs. Whole cells are suspended at 10^6 to 10^7 cells/ml in hypotonic lysing medium, HKMB 10/10/2/5 (10 mM HEPES–KOH, pH 7.6, 10 mM KCl, 2 mM MgCl$_2$, 5 mM 2-mercaptoethanol), and disrupted with 8–20 strokes of a tight-fitting pestle in a Dounce homogenizer. Lysis is monitored by detection of remaining whole cells in the lysate with a light microscope. Additional KCl to 70 mM and glycerol to 30% are added to the homogenate. The sample is centrifuged sequentially at 1000 g for 10 min at 4° and the resultant supernatant at 30,000 g for 20 min at 4° to remove membranes, nuclei, and mitochondria. This S30 preparation contains the substrates necessary for protein synthesis and may be stored at −80° for several months prior to use. However, once thawed the sample should not be refrozen. The assay for incorporation of amino acids into total protein is carried out in a reaction volume of 0.10 ml containing the following at their final concentration: 20 mM HEPES–KOH, pH 7.5, 75 mM KCl, 2 mM magnesium acetate, 50 μM amino acids (each, except methionine), 1 mM ATP, 0.2 mM GTP, 7 mM creatine phosphate, 0.2 mg/ml creatine phosphokinase, 5 mM 2-mercaptoethanol, 50 μl of lysate, and 5 μCi/ml [^{35}S]methionine (2000 Ci/mol). The mixture is incubated at 37° for 30 min. To stop the reactions, 0.5 ml of 1 M NaOH is added to hydrolyze aa-tRNA and solubilize all proteins. Two milliliters of ice-cold 10% trichloroacetic acid is then added, and the samples are applied to a 25 mm diameter glass fiber filter (type GF/C, Whatman) on a vacuum filter block. The filter is washed three times with 5 ml each time of cold 5% trichloroacetic acid, dried in under a heat lamp, and placed in a scintillation vial with 5 ml of nonaqueous scintillant for the measure of radioactivity.

An alternative to preparation of the above lysate system is the commercially available lysate made from rabbit reticulocytes or wheat germ (Amersham, Arlington Heights, IL; New England Nuclear, Boston, MA; Promega, Madison, WI; or U.S. Biochemical, Cleveland OH). These assay kits may also be obtained with the lysate pretreated with calcium-dependent micrococcal nuclease to selectively eliminate endogenous mRNA for dependency on added exogenous mRNA. In the latter assay system all protein synthesis starts at the initiation of peptide synthesis.

Analysis of Polysomes. In either lysate system, analysis of the polysomes will help reveal if the inhibition occurs at initiation versus elongation or termination phases of protein synthesis. For this purpose, lysates are incubated for 30 min at 37° in the presence and absence of the inhibitor, and 0.2 ml of the lysates is layered onto 4.7-ml 15–45% sucrose gradients prepared in solution TKM 10/50/3 (10 mM Tris-HCl, pH 7.4, 50 mM KCl, and 3 mM MgCl$_2$). The gradients are centrifuged at 280,000 g for 35 to 45 min at 4° and the fluid removed by gravity flow from the tubes while monitoring the contents at 260 nm in a flow cell. A well-resolved gradient will reveal distinct peaks of 40 S, 60 S, and 80 S ribosomes as well as the disome through heptasome complexes. If the inhibitor acts at elongation or termination steps of protein synthesis, the polysomes will accumulate toward the larger sizes. Preferential inhibition of initiation will result in a disappearance of the polysomes and accumulation of 40 S, 60 S, and 80 S ribosomes.

If the polysome profiles indicate that the inhibitor may be acting at the level of peptide elongation, unfractionated lysates are used to further study which of the steps of elongation are being affected. One method to achieve this goal involves the use of cetyltriethylammonium bromide (CTAB) and puromycin following exposure of the lysate system to the inhibitor being studied. This approach was employed to show that Shiga toxin was likely to prevent the binding of aa-tRNA into the A site of ribosomes rather than to inhibit peptidyltransferase or "translocation" of peptidyl-tRNA from the A to the P site of ribosomes.[23] While being convenient, this method provides only indirect evidence for the action of

[23] J. E. Brown, T. G. Obrig, M. A. Ussery, and T. P. Moran, *Microb. Pathog.* **1**, 325 (1986).

an inhibitor on the peptide elongation process and will not be discussed further in this chapter.

Peptide Initiation. The process of eukaryotic peptide initiation has been reviewed.[24] Formation of the final initiation complex, [^{35}S]Met–tRNA$_f$:40 S or 80 S ribosome:mRNA, is monitored in unfractionated S30 lysates with assay components described above. This is accomplished in one of two ways. In the case of the complex with 80 S ribosomes, the initiation [^{35}S]Met–tRNA$_f$ bound into the ribosomal P site reacts with puromycin to yield [^{35}S]Met–puromycin. The [^{35}S]Met–tRNA$_f$–puromycin product is extracted with ethyl acetate in the presence of ammonium bicarbonate, pH 9.0, and the radioactivity in the organic phase is monitored by liquid scintillation.

If the initiation complex with 40 S ribosomes is to be examined, the lysate is layered onto a 4.5-ml gradient of 20–45% sucrose, in solution TKMB 20/100/3/1 (20 mM Tris-HCl, pH 7.4, 100 mM KCl, 3 mM magnesium acetate, 1 mM 2-mercaptoethanol) in a 5-ml tube and centrifuged at 200,000 g for 2 hr at 4°. The gradient is pumped out of the tube, monitored at 260 nm, and fractions collected for measurement of radioactivity by liquid scintillation. Inhibitors of initiation will reduce the amount of radioactivity associated with the 40 S peak. A comparison is conducted by incubation of lysate with an inhibitor of peptide elongation such as cycloheximide[25] which results in the accumulation of radioactivity in the 40 S peak.

It should be noted that the acylated initiator form of tRNA^{Met}, Met-tRNA$_f$ is recognized by initiation factors and not by eEF-1, thus is incorporated only into initiation complexes. [^{35}S]Met–tRNA$_f$ for use in the assay is preformed by incubation of rat liver tRNA (deacylated at pH 9 for 2 h at 4°) with [^{35}S]Met and *E. coli* synthetase enzymes.[23] *Escherichia coli* synthases specifically aminoacylate eukaryoytic tRNA^{fMet} vs. tRNA$_m$.

Fractionated Cell Lysates

Once it is established that a protein toxin is an inhibitor of initiation, elongation, or termination, more refined assays are performed with fractionated lysates of cells to determine if the inhibitor prevents formation of specific complexes between different soluble factors (enzymes, mRNA, aa-tRNA, GTP, etc.) and ribosomes. The primary disadvantage of these assays is their requirement for purified factors and ribosomes.

---

[24] W. C. Merrick, *Enzyme* **44**, 7 (1990).
[25] T. G. Obrig, W. J. Culp, W. L. McKeehan, and B. Hardesty, *J. Biol. Chem.* **246**, 174 (1971).

Source of Components. Ribosomes for the assays are prepared from rabbit reticulocytes as described previously.[15] In most cases, ribosomes are treated with 0.5 M KCl to remove weakly bound elongation factors. Purification of eEF-1 and eEF-2 from reticulocytes is carried out with ammonium sulfate precipitation and column chromatography.[15] Residual eEF-2 activity in eEF-1 preparations is inactivated by incubation of the sample with 20 mM N-ethylmaleimide at 4° for 10 min followed by addition of dithiothreitol to 40 mM final conc.

Peptide Elongation. All three steps of peptide elongation can be tested individually in fractionated cell-free assays. The assays described below are (1) codon-dependent binding of aa-tRNA to the ribosomal A (acceptor) site catalyzed by eEF-1, (2) eEF-2-dependent translocation of aa-tRNA from the A (acceptor) to the P (peptidyl) ribosomal sites, and (3) peptide bond formation catalyzed by peptidyltransferase.

Binding of aa-tRNA to ribosomes is performed with the binding of Phe-tRNA to salt-washed ribosomes directed by the artificial mRNA, polyuridylic acid [poly(U)] and purified eEF-1 protein. Preparation of salt-washed ribosomes from rabbit reticulocytes and of radioactive aa-tRNA is described elsewhere.[15] The binding reaction is carried out in a total volume of 0.1 ml with the following reagents listed at their final concentrations and in order of addition: 25 mM Tris-HCl, pH 7.4, 62 mM KCl, 5 mM magnesium acetate, 2.4 mM dithiothreitol, 40 μM GTP, 20 μg poly(U), toxin, or diluent, 1 A_{260} unit (21 pmol) of salt-washed ribosomes, 3 μg eEF-1 protein, and 5–10 pmol of [³H]Phe–tRNA (2000 Ci/mol). The reaction is incubated at 37° for 5 min, and an ice-cold solution of TKM 50/50/8 (50 mM Tris-HCl, pH 7.4, 50 mM KCl, 8 mM MgCl$_2$) is added to stop the reaction. The contents are applied to a nitrocellulose filter (BA85, 0.45 μm pore size, Schleicher and Schuell, Keene, NH) on a vacuum filter apparatus. The filters are then washed with three times 10 ml each time of solution A and placed in a scintillation vial with 5 ml of aqueous scintillant for the measurement of radioactivity.

Translocation of aa-tRNA from the A to the P site is catalyzed by eEF-2. This reaction is carried out by first binding aa-tRNA into the A site nonenzymatically (i.e., at high magnesium concentrations) and moving the aa-tRNA over to the P site in the presence of eEF-2. By definition, aa-tRNA located in the P site is reactive with puromycin located in the A site, yielding aminoacylpuromycin (aa-puromycin). Puromycin binds selectively into the A site and can do so only if the A site is vacant. For the translocation reaction, [³H]Phe-tRNA is nonenzymatically bound to salt-washed ribosomes in a reaction volume of 52 ml containing the following reagents listed at their final concentrations: 50 mM Tris-HCl, pH 7.4, 120 mM KCl, 16 mM MgCl$_2$, 5 mM dithiothreitol, 10 mg poly(U), 620 A_{260}

units of salt-washed ribosomes, and 3 nmol of [³H]Phe–tRNA (1500 Ci/ mol). The reaction mixture is incubated at 37° for 20 min and chilled on ice. To separate the ribosomes from other reaction components, 26 ml of the mixture is layered over 7 ml of 15% sucrose prepared in a solution of TKMB 50/120/8/5 (50 mM Tris-HCl, pH 7.4, 120 mM KCl, 8 mM MgCl$_2$, 5 mM 2-mercaptoethanol) in a 35-ml centrifuge tube. The contents are centrifuged in at 140,000 g for 12 hr at 4°. Supernatants are decanted and the ribosomal pellets resuspended at 125 A_{260} units/ml in solution TKMB 20/100/5/1 (20 mM Tris-HCl, pH 7.4, 100 mM KCl, 5 mM MgCl$_2$, 1 mM 2-mercaptoethanol) containing 10% (v/v) glycerol.

The translocation step is performed in a final volume of 0.5 ml containing the following, in order of addition: 50 mM Tris-HCl, pH 7.4, 70 mM KCl, 5 mM MgCl$_2$, 5 mM dithiothreitol, 3 A_{260} units of salt-washed ribosomes with approximately 5000 cpm of [³H]Phe–tRNA nonenzymatically bound, toxin or diluent, 0.5 μg eEF-2 protein, and 0.2 mM GTP. The reaction mixture is incubated at 37° for 6 min and cooled to 4°. Puromycin hydrochloride is then added to 1 mm, and the contents are incubated at 4° for 20 min. [³H]Phe–puromycin formed is extracted from the reaction mixture by addition of 0.5 ml of 2 mM ammonium bicarbonate, pH 9.0, and 1 ml of ethyl acetate. The contents are mixed vigorously for 15 sec, centrifuged at 500 g for 5 min, and a portion of the organic (top) phase is monitored for radioactivity in 10 ml of scintillant.

Peptide bond formation is catalyzed by peptidyltransferase. Peptidyltransferase is not a soluble cytoplasmic protein as are eEF-1 and eEF-2, but is an inherent property of the 60 S ribosome. The puromycin reaction which is dependent on peptidyltransferase activity represents the formation of a peptide bond between puromycin, which mimics the 3′ terminus of an aa-tRNA, and the amino acid of aa-tRNA located in the P site. Thus, the puromycin reaction described above is a measure of peptidyltransferase activity. To determine the effect a protein toxin has on peptidyl transferase, one performs a two-step assay, exactly as described for translocation, except the toxin is now added just prior to the second incubation step (puromycin reaction) rather than prior to the first step (aa-tRNA binding).

Peptide Initiation. Most regulation of eukaryotic protein synthesis takes place at the level of peptide initiation. Because of this and the relatively large number of protein factors involved in the process,[24] it is expected that bacterial pathogens would produce some initiation-inhibiting protein exotoxins. To date, however, such inhibitors are yet to be described. Nonetheless, for newly identified inhibitors of peptide elongation it is important to rule out their partial or complete effect on initiation. The following is an abbreviated approach for accomplishing this task.

Observation of polysome profiles from toxin-treated and control lysates is performed with sucrose gradients as described in an earlier section. A toxin which inhibits only peptide elongation will yield polysomes of larger size. As mentioned above, conceptually it is possible that a toxin may be a weak inhibitor of initiation and a strong inhibitor of the elongation process. In this case, one would expect to see a reduction in the total amount of polysome material, with the remaining polysomes skewed toward the larger sizes.

A more definitive procedure for assessing the activity of an initiation inhibitor is to quantitate the final complexes formed in the initiation pathway. A ternary complex that comprises [^{35}S]Met–tRNA$_f$:GPT:eIF-2 (eukaryotic initiation factor 2) is formed as a soluble intermediate prior to its attachment to the 40 S ribosomal subunit. The ternary complex is measured in a cell-free assay by its binding to a nitrocellulose membrane. A less than complete ternary complex is not retained by a nitrocellulose filter. The assay is carried out in a total volume of 50 μl containing 20 mM Tris-HCl, pH 7.6, 0.1 M KCl, 1.0 mM magnesium acetate, 1.0 mM dithiothreitol, 0.5 mM GTP, 3.0 mM phosphoenolpyruvate, 0.1 U pyruvate kinase, 0.1 μCi [^{35}S]Met–tRNA$_f$ (specific activity 1500 Ci/mmol), and approximately 5 μg protein from a 0.5 M KCl ribosomal salt wash fraction. The mixture is incubated for 20 min at 20° and the reaction stopped by addition of 2.0 ml TKM 20/100/5 buffer solution (20 mM Tris-HCl, pH 7.6, 100 mM KCl, 5 mM MgCl$_2$). Samples are applied to nitrocellulose membrane filters on a vacuum filter apparatus and washed three times with 5 ml of buffer solution each time. The filters are dried and measured for radioactivity by liquid scintillation.

Summary

Methods have been described that are sufficient to determine if a bacterial protein toxin is a selective inhibitor of eukaryotic protein synthesis, and, if so, which part of the overall process is affected. More defined assays are presented for studying the steps of peptide elongation as this is where such toxins have been shown to act.

[53] Assays of Hemolytic Toxins

By GAIL E. ROWE and RODNEY A. WELCH

Introduction

Hemolysins are cytolytic toxins found in a broad diversity of organisms. They are named for their capacity to lyse erythrocytes, but many are toxic to other cell types as well. Such cytolysins produced by invertebrates are active against bacterial cells. They are functionally analogous to the lytic terminal components of complement and can protect invertebrates against bacterial infections. Consequently, invertebrate hemolysins have been touted as humoral immune effectors.[1] Microbial hemolysins are generally considered to be virulence factors, although the relative contribution of hemolysins to disease is variable among microbes and different host species. Many microbial hemolysins can lyse leukocytes. This activity may enhance survival of the microbe when confronted with a host immune response.[2]

The importance, *in vivo,* of red blood cell (RBC) lysis by hemolysins is unclear. However, erythrolysis has been proposed as a mechanism for iron acquisition in iron-deficient microenvironments.[3,4] Alternatively, RBC lysis may be a spurious correlate, whereas lysis of other target cells provides a selective advantage to hemolysin-producing organisms.[5,6] Regardless of the biological significance of erythrolysis, per se, hemolysis is a convenient phenotype which affords the investigator a simple, quantitative means of screening for the presence of cytolytic toxins. The following is a review of commonly used methods for identification and quantification of hemolysins and their cytolytic activity. Included also are considerations for adapting general assay procedures to a specific application.

Mechanisms of Cytolysis

Cytolytic toxins can be separated into three categories based on the mechanism of action against target cell membranes: enzymatic, pore-

---

[1] C. Canicatti, *Experientia* **46**, 239 (1990).
[2] W. Goebel, T. Chakraborty, and J. Kreft, *Antonie van Leeuwenhoek* **54**, 453 (1988).
[3] L. Chu, T. E. Bramanti, J. L. Ebersole, and S. C. Holt, *Infect. Immun.* **59**, 1932 (1991).
[4] C. Waalwijk, D. MacLaren, and J. de Graaff, *Infect. Immun.* **42**, 245 (1983).
[5] S. Bhakdi, S. Greulich, M. Muhly, B. Eberspacher, H. Becker, A. Thiele, and F. Hugo, *J. Exp. Med.* **169**, 737 (1989).
[6] S. J. Cavalieri and I. S. Snyder, *Infect. Immun.* **37**, 966 (1982).

forming, or surfactant. The kinetics of cytolysis differ depending on the mechanism involved. Assay conditions, including amount of toxin and target cells, incubation time, and temperature, should be established which are consistent with the lytic nature of the toxin of interest.

Enzymatic Mechanisms

Hemolysins which cause cytolysis by enzymatic disruption of target cell membranes include phospholipases such as the α toxin from *Clostridium perfringens*[7] and the sphingomyelinase (β toxin) of *Staphylococcus aureus*.[8] In keeping with their enzymatic nature, cytolysins in this category show a high degree of substrate specificity and are recycled to react with multiple target cells. Sphingomyelin degradation by *S. aureus* β toxin is an example of such substrate specificity. This and similar toxins require a period of chilling (0–4°) after incubation at 37° for *in vitro* manifestation of the hemolytic phenotype. One explanation for the "hot/cold" hemolysis is the creation of a lipid monolayer where the sphingomyelin has been hydrolyzed from the outer leaflet of the target cell membrane. This monolayer may be able to persist at 37°, but it becomes unstable and collapses at lower temperatures.[9] Because enzymatic hemolysins are recycled, each toxin molecule should be able to act on several target cells. Thus, the percent hemolysis elaborated by a constant amount of toxin should remain unchanged with increasing target cell concentration.[10]

Pore Formation

Pore-forming cytolysins such as the bacterial RTX[11,12] or thiol-activated toxins[13] insert into the target cell membranes to form transmembrane pores. The three general steps in this type of cytolysis are (1) binding of the toxin to the target cell membrane, (2) penetration and disruption of the membrane (pore formation), and (3) alteration of membrane permeability leading to cytolysis. The first step (binding) is temperature independent for many toxins[14,15] and can occur at 0°–4°. Toxins that require oligomeriza-

[7] M. G. Macfarlane and G. C. J. G. Knight, *Biochem. J.* **35**, 884 (1941).
[8] H. M. Doery, B. J. Magnusson, I. M. Cheyne, and J. Gulasekharam, *Nature (London)* **198**, 1091 (1963).
[9] A. W. Bernheimer, *Biochim. Biophys. Acta* **344**, 27 (1974).
[10] A. W. Bernheimer, in "Microbial Toxins" (T. Montie, S. Kadis, and S. J. Ajl, eds.), p. 183. Academic Press, New York and London, 1970).
[11] R. A. Welch, *Mol. Microbiol.* **5**, 521 (1991).
[12] J. G. Coote, *FEMS Microbiol. Rev.* **88**, 137 (1992).
[13] V. Braun and T. Focareta, *Crit. Rev. Microbiol.* **18**, 115 (1991).
[14] M. Pinkney, E. Beachey, and M. Kehoe, *Infect. Immun.* **57**, 2553 (1989).
[15] B. Eberspacher, F. Hugo, and S. Bhakdi, *Infect. Immun.* **57**, 983 (1989).

tion for pore formation, such as streptolysin O, may require higher temperatures (37°) for this step.[14] Putative monomeric pore formers, such as *Escherichia coli* hemolysin,[16] may also require a higher temperature for lysis.[15] The reason for this is not known but may be due to inhibition of transmembrane penetration by the toxin with decreased membrane fluidity at lower temperatures. Because insertional cytolysins are not recycled in the manner of enzymatic toxins, the percent hemolysis by a given amount of toxin is expected to decrease with increasing target cell concentration when the toxin is not in excess. However, other biological factors can influence the kinetics, such as the number of membrane lesions required for lysis of an individual cell and the oligomeric or monomeric nature of the toxin.

Surfactants

The delta (δ) toxin of *Staphylococcus aureus*[17] and the heat-stable hemolysin from *Pseudomonas aeruginosa*[18] are surfactants. These toxins are highly hydrophobic. In the case of *S. aureus* δ toxin, the hydrophobic amino acids are localized in a "core" region which renders the polypeptide amphipathic and surface active.[19] The detergentlike action of surfactant hemolysins causes cytolysis by solubilization of the target cell membrane. These toxins are thermostable (at 100°), can be inhibited by phospholipids or fatty acids, have a relatively low level of hemolytic activity (1–4 orders of magnitude fewer hemolytic units/mg than other hemolysins), have comparable activity against target cells regardless of the host species, and show decreases in percent hemolysis by a given amount of toxin with increased target cell concentration.[9,10,19]

Screening for Hemolytic Activity

The simplest method for determining whether an organism produces a hemolysin is to screen samples for RBC lysis on a blood agar plate (BAP). In some circumstances, however, a liquid hemolysis assay may be more appropriate. These two assays are detailed below, along with their strengths and limitations.

[16] S. Bhakdi, N. Mackman, J. M. Nicaud, and I. B. Holland, *Infect. Immun.* **52**, 63 (1986).
[17] G. M. Wiseman, *in* "Microbial Toxins" (T. Montie, S. Kadis, and S. J. Ajl, eds.), p. 237. Academic Press, New York and London, 1970).
[18] P. V. Liu, *J. Infect. Dis.* **130**(Suppl.), S94 (1974).
[19] J. E. Fitton, A. Dell, and W. V. Shaw, *FEBS Lett.* **115**, 209 (1980).

Blood Agar Plates

Organisms are grown on a complex nutrient agar plate with 5% (v/v) defibrinated blood. It may be necessary to screen for lysis of RBC from several sources since some hemolysins are active against cells from a narrow range of species. The BAP screening method is only effective with organisms which grow well on an agar medium. If several days of incubation are required for growth, the discoloration of the BAP will preclude detection of hemolysis. Instead of direct growth on BAP, cellular fractions may be applied to BAP in discrete spots which permit diffusion of the putative lysin through the agar matrix. Alternatively, the liquid assay (described below) may be used to screen for hemolysis. The incubation temperature should be chosen to best approximate the *in vivo* source of the toxin. The time required for visible hemolysis may vary from a few hours for cellular fractions to a day or more for cultured organisms.

The BAP screening assay is very sensitive where potential artifacts need to be considered. For example, when genomic libraries of *Escherichia coli* strain K12 (a nonhemolytic strain) are expressed in *E. coli* K12 on high-copy cloning vectors and screened on blood agar, rare colonies with β-hemolytic zones can be observed.[20] Thus, identification of a hemolytic toxin should also include genetic or biochemical evidence of the toxin macromolecule. Furthermore, when examining BAP for zones of hemolysis around samples/organisms, care should be taken to distinguish true (β) hemolysis from the nonlytic discoloration (greening) of α "hemolysis."

Liquid Hemolysis Assay

Organisms that are better suited to growth in a liquid culture than on an agar plate may be screened for production of a hemolysin by the liquid hemolysis assay. A sample of actively growing organisms in culture broth is the source of the putative hemolysin. Intracellular cytolysins may require mechanical disruption of the cells to detect the toxin. Secreted hemolysins can be identified by screening cell-free culture supernatants. The sample to be screened will be referred to simply as "toxin" in the remainder of this chapter.

The liquid assay system uses spectrophotometric detection of hemoglobin released from erythrocytes as the indicator of cytolysis. The hemolytic activity of a single concentration of toxin on a given number of erythrocytes is measured and compared to total osmotic lysis of the same number of cells. Variations of the assay are numerous. What follows is a general

---

[20] R. A. Welch, personal observation (1982).

description with sufficient citations to direct the reader to several specific protocols.[21-27]

Defibrinated RBCs are sedimented (12,000 g in a microcentrifuge for 1 min), washed by centrifugation in isotonic buffer until the supernatant is clear, then diluted in buffer to the desired concentration. The target cell concentration (or target cell to toxin ratio) should be chosen to afford the best demonstration of differences in toxic activity. The cell concentration should be low enough to permit detection of weak hemolysis as a significant percentage of total hemolysis. However, the cell concentration should be high enough such that even strongly hemolytic samples do not lyse all the cells in the system. A range of 0.5–2.0% RBCs is recommended, where 100% is defined as the concentration of cells in the washed pellet. Alternatively, RBCs may be enumerated using a hemocytometer, permitting standardization of RBC concentration in cells per milliliter.[28] Adjustments in the amount of toxin used per assay may also be useful.

Erythrocytes and toxin are added to buffer and incubated at an appropriate *in vivo* temperature (usually 37°). The incubation time is variable but generally ranges from 20 min to 2 hr. Unlysed cells and cell membranes are removed by centrifugation. Hemolysis is detectable as the red color of released hemoglobin in supernatants and by the reduced size of cell pellets compared to the same cell concentration in buffer without toxin.

Quantifying Hemolytic Activity

Once the presence of a hemolysin has been determined, further studies may require comparison of the amount of hemolytic activity in different toxin samples, against RBCs from different species, or with different assay conditions. Such comparisons require quantification of hemolytic activity. Methods for measuring the amount of hemolytic activity are described below for several different hemolysis assays.

[21] R A. Welch and S. Pellett, *J. Bacteriol.* **170**, 1622 (1988).
[22] S. Nomura, M. Fujino, M. Yamakawa, and E. Kawahara, *J. Bacteriol.* **170**, 3694 (1988).
[23] J. N. Krieger, P. Wolner-Hanssen, C. Stevens, and K. K. Holmes, *J. Infect. Dis.* **161**, 307 (1990).
[24] J. Vadivelu, S. D. Puthucheary, and P. Navaratnam, *J. Med. Microbiol.* **35**, 363 (1991).
[25] A. Monkiedje, J. H. Wall, A. J. Englande, and A. C. Anderson, *J. Environ. Sci. Health* **B25**, 777 (1990).
[26] M. K. Majumdar, D. P. Sikdar, A. B. Sarma, and S. K. Majumdar, *J. Appl. Bacteriol.* **69**, 241 (1990).
[27] D. J. Beecher and J. D. Macmillan, *Infect. Immun.* **58**, 2220 (1990).
[28] C. Forestier and R. A. Welch, *Infect. Immun.* **59**, 4212 (1991).

Blood Agar

Quantitation of hemolytic activity by comparison of the hemolytic zone diameter on a BAP is a convenient screen, at best. Variation may be caused by the thickness of the agar in the plates or by different growth rate of organisms on the plate. Nevertheless, the convenience of the method affords it some degree of usefulness. The radial-diffusion assay is a variation of the BAP method.[27] Agar with 5% defibrinated blood is poured onto glass slides. Samples are placed in wells punched in the agar. Slides are then incubated for 1–3 hr, with humidity. Comparisons are based on the radii of the hemolytic zones minus the radius of the wells.

Liquid Hemolysis Assay

A less subjective comparison of relative amounts of hemolytic activity can be obtained from the liquid hemolysis assay, as described above. The mixture of toxin and RBCs are centrifuged after incubation to remove unlysed erythrocytes and cell debris. Supernatants containing released hemoglobin are transferred to fresh tubes and diluted in buffer, as necessary. The absorbance of the supernatants should be read within the linear range of a standard curve (A_{540} less than 1.0). Results are reported as "% complete lysis" by dividing the absorbance from test samples by the absorbance of hemoglobin released from the same number of RBCs when lysed by distilled water or detergent.

A variation of the liquid hemolysis assay, which we identify here as the "dilution assay," tests serial (2-fold) dilutions of toxin for lysis of an established percentage (usually 50%) of a set number of RBCs.[15,29,30] Erythrocytes are prepared as for the liquid assay and added to an equal volume of toxin that has been serially diluted (1 : 2) in buffer. The incubation time and temperature are variable, as described for the liquid assay. Unlysed cells and cell debris are removed from the assay mixture, and the released hemoglobin is measured, as above. Results are given in hemolytic units (HU). One unit is the amount of test sample which causes lysis of 50% of the cells in the assay (i.e., the sample has one-half the absorbance reading as the same number of cells lysed by water or detergent). Therefore, the number of hemolytic units in a sample is equal to the reciprocal of the dilution which evokes 50% of maximal erythrolysis.

[29] K. Venkateswaran, C. Kiiyukia, M. Takaki, H. Nakano, H. Matsuda, H. Kawakami, and H. Hashimoto, *Appl. Environ. Microbiol.* **55**, 2613 (1989).

[30] L. A. M. G. Van Leengoed and H. W. Dickerson, *Infect. Immun.* **60**, 353 (1992).

The "microtiter assay" is a simpler variation of the dilution assay described above.[31,32] Erythrocytes and toxin dilutions, prepared as for the dilution assay, are added to a 96-well microtiter plate. After incubation, wells are visually inspected for the degree of hemolysis. The hemolytic titer may be determined from (1) the maximum dilution with hemolysis[31] or (2) the dilution which produces 50% lysis, as determined by comparing RBC pellet size and supernatant color to positive (RBCs in distilled water) and negative (RBCs in buffer) controls.[32]

Kinetics of Hemolysis

The kinetics of hemolysis may be of interest in some studies, such as the characterization of a newly identified hemolytic toxin. The rate of hemolysis over time can be determined by spectrophotometric analysis of aliquots of cell-free supernatants taken from a liquid hemolysis assay at frequent intervals during the course of incubation.[33] Alternatively, cytolysis may be measured spectrophotometrically as the decrease in turbidity over time (corresponding to decreased numbers of unlysed cells).[34] Both methods are amenable to studying the kinetics of a hemolysis reaction with respect to toxin and/or target cell concentration.

Quantifying Cytolytic Activity

Hemolysins are frequently lytic to cells other than erythrocytes. In fact, the greater biological relevance of many hemolytic cytolysins may reside in their toxicity to nucleated cells. For this reason, we have included methods for assessing cytolysis of nucleated cells by hemolysins. These methods are useful with freshly isolated target cells or cultured cell lines. However, presentation of methods for the isolation and maintenance of target cells is beyond the scope of this chapter.

Phase-Contrast Microscopy

Phase-contrast microscopy is more time consuming and subjective than other methods but has the advantages of being a more direct and technically simple procedure. Target cells are diluted to the desired concentration in tissue culture medium and placed in a 96-well microtiter plate. Serially diluted (1 : 2) toxin is added, and the wells and plates are

[31] H. Nakano, T. Kameyama, K. Vankateswaran, H. Kawakami, and H. Hashimoto, *Microbiol. Immunol.* **34**, 447 (1990).

[32] R. A. Welch and S. Falkow, *Infect. Immun.* **43**, 156 (1984).

[33] E. Schiebel and V. Braun, *Mol. Microbiol.* **3**, 445 (1989).

[34] R. P. Rennie, J. H. Freer, and J. P. Arbuthnott, *J. Med. Microbiol.* **7**, 189 (1974).

incubated briefly. After the toxin is removed and wells are washed, the plates are incubated with fresh tissue culture medium. The target cells are then scanned under a phase-contrast microscope and scored for cytotoxic morphology. Results are reported either as "negative" or degrees of "positive" (1+ to 4+) based on the percentage of target cells affected.[29] Alternatively, comparisons may be made based on the reciprocal of the highest dilution which shows cytotoxicity.[29] One unit of toxin activity has been defined as the amount of toxin required to lyse 95% of the cells in the assay.[35]

Chromium Release Assay

The chromium release assay is an indirect but sensitive assay for cytolysis[36] by which the prepared target cells are incubated with sodium [^{51}Cr]chromate before mixing with an aliquot of toxin in the microtiter plate. After 1 hr of incubation, cells are collected by centrifugation, the ^{51}Cr-containing supernatant is transferred to a scintillation vial, and the radioactivity is counted in a γ counter. Results are reported as "% maximal lysis" by dividing the mean test counts per minute (cpm) by the mean maximum counts emitted by total lysis of labeled target cells with 1 N HCl. Spontaneous release from target cells incubated with tissue culture medium alone should be examined since very high background counts can interfere with interpretation of test results. Our recommendation is that the background counts should not exceed 10% of those from total (HCl) lysis.

Neutral Red Assay

The neutral red assay is based on the uptake of vital dye[29,35,37] by viable target cells. The toxin sample is incubated with prepared target cells to permit cytolysis. Unlysed cells are harvested by centrifugation and reincubated in buffer with neutral red dye. Mixtures are then washed and lysed with detergent or ethanol plus sodium phosphate to release vital dye from the cytoplasm of viable target cells. The lysed cells are centrifuged to pellet nuclei, and the supernatant containing released neutral red is analyzed spectrophotometrically. Lysis is measured as the loss of viability by comparing the percent dye uptake (A_{545}) of the test sample to the A_{545} corresponding to 100% uptake by target cells without toxin. Like the hemoglobin release assay for assessing hemolytic activity, this cytolytic

[35] W. T. Cruz, R. Young, Y. F. Chang, and D. K. Struck, *Mol. Microbiol.* **4**, 1933 (1990).
[36] P. E. Shewen and B. N. Wilkie, *Infect. Immun.* **35**, 91 (1982).
[37] C. N. Greer and P. E. Shewen, *Vet. Microbiol.* **12**, 33 (1986).

assay may also utilize 2-fold serial dilutions of toxin with a given number of target cells to determine the cytotoxic units (CU) in a given test sample. As for hemolytic units, one cyotoxic unit is the amount of toxin required to kill 50% of target cells.

Procedural Considerations

Optimal assay conditions for the study of a given hemolytic toxin depends on many variables peculiar to the individual system. The following list is intended as a guide to some parameters that should be considered.

Growth Phase of Toxin Producer

Hemolysin should be harvested from toxigenic organisms at a time when toxin production is high but denaturation or inactivation is minimal. In bacteria that constitutively synthesize the toxin, optimal harvest occurs during mid to late log phase. This may not be the case for organisms where toxin production is regulated temporally or via environmental stimuli, such as hemolysins from *Bordetella pertussis* or *Staphylococcus aureus*.[38–40] Toxin production from parasites may be limited to certain stages of the life cycle.[41]

Toxin Preparation

Some considerations for preparation of experimental samples include the source of the sample (e.g., body fluids versus body extracts of invertebrates[1]) and the degree of purity of the preparation. In some cases, conflicting experimental results may be obtained from crude toxin preparations compared to purified samples (Clinkenbeard *et al.* versus Cruz *et al.* referenced in Ref. 12). While crude extracts may be poorly defined and could contain hemolysis inhibitors or enhancers, purification may exclude important cofactors.[27] Furthermore, hemolytic activity may be lost during purification or storage of the toxin.[42,43] Stability of toxin activity

[38] S. Knapp and J. J. Mekalanos, *J. Bacteriol.* **170,** 5059 (1988).
[39] P. Rescei, B. Kreiswirth, M. O'Reilly, P. Schleivert, A. Gruss, and R. P. Novick, *Mol. Gen. Genet.* **202,** 58 (1986).
[40] S. Stibitz, W. Aaronson, D. Monack, and S. Falkow, *Nature (London)* **338,** 266 (1989).
[41] N. W. Andrews, *Exp. Parasitol.* **71,** 241 (1990).
[42] K. R. Hardie, J. P. Issartel, E. Koronakis, C. Hughes, and V. Koronakis, *Mol. Microbiol.* **5,** 1669 (1991).
[43] H. Ostolaza, B. Bartoleme, J. Serra, F. de la Cruz, and F. Goni, *FEBS Lett.* **280,** 195 (1991).

can, in some cases, be enhanced by including bovine serum albumin (BSA) or chaotropic agents, such as guanidine hydrochloride[35] or urea.[43,44]

Target Cells

By definition, all hemolysins attack erythrocytes. However, the efficiency of lysis may vary depending on the RBC donor species.[15] Many hemolysins are also lytic to other cell types, but the range of target cell specificity may vary greatly, even among closely related toxins.[28,46,47] The target cell concentration (target cell to toxin ratio) should be chosen to afford the best demonstration of differences in toxic activity (see Screening for Hemolytic Activity, Liquid Hemolysis Assay for further discussion).

Buffers

Various researchers use Tris- or phosphate-buffered saline (PBS) solutions for diluting toxin and/or target cells (see references describing specific assays). The basis for this choice has been described elsewhere.[48]

Ion Requirements

Calcium, magnesium, zinc, or other cations are required for activity of many hemolysins.[1,30,34,49,50] Cations may act as stabilizing agents to prevent denaturation or agglutination of toxin molecules.[1] Alternatively, they may be directly involved in hemolysin–target cell interaction, perhaps by binding to the hemolysin and facilitating the necessary folded conformation.[49] By analogy to the complement system, cations may enhance polymerization of toxin molecules on the target cell membrane during pore formation.[1]

Thiol Activation

Some hemolysins require thiol treatment for maximal activity. For more information, see a review of gram-positive bacterial thiol-activated hemolysins.[13]

[44] M. I. Gonzalez-Carrero, J. Zabala, F. d. l. Cruz, and J. M. Ortiz, *Mol. Gen. Genet.* **109,** 106 (1985).
[45] A. W. Bernheimer and L. L. Schwartz, *J. Gen. Microbiol.* **30,** 455 (1963).
[46] C. Forestier and R. A. Welch, *Infect. Immun.* **58,** 828 (1990).
[47] D. R. McWhinney, Y. F. Chang, R. Young, and D. K. Struck, *J. Bacteriol.* **174,** 291 (1992).
[48] A.W. Bernheimer, this series, Vol. 165, p. 213.
[49] D. F. Boehm, R. A. Welch, and I. S. Snyder, *Infect. Immun.* **58,** 1951 (1990).
[50] C. Canicatti and M. Grasso, *Mar. Biol.* **99,** 393 (1988).

Summary

The ability to produce a cytolytic toxin contributes to the success of many organisms in a particular niche by such diverse means as lysis of a phagolysosomal membrane of the macrophage by hemolysin from the intracellular parasite *Trypanosoma cruzi*,[51] disruption of leukocyte activity by the *Escherichia coli* hemolysin,[52] and destruction of invading bacteria by hemolysin from the annelid *Glycera dibranchiata*.[53] The relative contribution of erythrocyte lysis to survival of the cytolysin producer is still under investigation. Nevertheless, the hemolytic phenotype is both a powerful tool for identifying novel cytolysins and a convenient marker for studying cytolytic activity in established toxins.

[51] V. Ley, E. S. Robbins, V. Nussenzweig, and N. W. Andrews, *J. Exp. Med.* **171,** 401 (1990).
[52] S. J. Cavalieri and I. S. Snyder, *Infect. Immun.* **36,** 455 (1982).
[53] R. S. Anderson and B. M. Chain, *J. Invertebr. Pathol.* **40,** 320 (1982).

[54] Identification and Assay of RTX Family of Cytolysins

By ANTHONY L. LOBO and RODNEY A. WELCH

Properties of RTX Cytolysins

Introduction

The RTX (for repeats in toxin, to be explained below) pore-forming cytolysins are products of a gene cluster which has disseminated throughout a number of gram-negative bacterial genera. This is inferred from the conserved operon organization and strong sequence similarity among RTX genes from phylogenetically diverse bacteria.[1,2] It is clear that, following this diaspora, the RTX determinants evolved further to suit the organisms in which they resided. Many of the cytolysins exhibit target cell specificities that reflect the kinds of environments which the RTX-producing organisms inhabit. The cattle pathogen *Pasteurella haemolytica*, for example, produces an RTX toxin which only lyses ruminant leukocytes.[3,4] Some RTX proteins are not cytolysins but proteases, like the metalloproteases

[1] R. A. Welch, *Mol. Microbiol.* **5,** 521 (1991).
[2] R. A. Welch and R. Y. C. Lo, submitted for publication.
[3] P. E. Shewen and B. N. Wilkie, *Infect. Immun.* **35,** 91 (1982).
[4] R. Y. C. Lo, C. Strathdee, and P. Shewen, *Infect. Immun.* **55,** 1987 (1987).

of *Erwinia chrysanthemi*[5] and *Serratia marcescens*.[6] Other confirmed members include the prototype RTX toxin *Escherichia coli* hemolysin,[7] hemolysins from *Proteus vulgaris*,[8] *Morganella morganii*,[8] and *Actinobacillus pleuropneumoniae*,[9] leukotoxin from *A. actinomycetemcomitans*,[10,11] and the adenylate cyclase/hemolysin from *Bordetella pertussis*.[12]

Much of the interest in studying RTX cytolysins lies in understanding their role in the pathogenesis of organisms which produce them. For example, studies have shown that hemolysin-producing strains of *E. coli* are up to 1000-fold more virulent than isogenic hemolysin-negative mutants when tested in a rat model of peritonitis.[13,14] Still unclear, however, are the ways in which RTX toxins participate in the disease process. There is a growing understanding of the effects which sublytic doses of toxin have on immune cells, such as lowering of intracellular ATP levels[15] and interference with antigen processing,[16] and these effects may be more important in disease than overt lysis of target cells.

Another area of interest in the study of RTX cytolysins is the unique mechanism by which they are secreted. All of the known RTX proteins are either secreted extracellularly or are localized to the cell surface. In the case of the *E. coli* hemolysin, the secretion process is different from the classic *sec* pathway, and no cleavable amino-terminal signal sequence is involved.[17,18] Also, it appears that the protein is transported across both the cytoplasmic and outer membranes simultaneously with no periplasmic

---

[5] P. Delepelaire and C. Wandersman, *J. Biol. Chem.* **264**, 9083 (1989).

[6] K. Nakahama, K. Yoshimura, R. Marumoto, M. Kikuchi, I. S. Lee, T. Hase, and H. Matsubara, *Nucleic Acids Res.* **14**, 5843 (1986).

[7] T. Felmlee, S. Pellett, and R. A. Welch, *J. Bacteriol.* **163**, 94 (1985).

[8] V. Koronakis, M. Cross, B. Senior, E. Koronakis, and C. Hughes, *J. Bacteriol.* **169**, 1509 (1987).

[9] Y.-F. Chang, R. Young, and D. K. Struck, *DNA* **8**, 635 (1989).

[10] D. Kolodrubetz, T. Dailey, J. Ebersole, and E. Kraig, *Infect. Immun.* **57**, 1465 (1989).

[11] E. T. Lally, E. E. Golub, I. R. Kieba, N. S. Taichman, J. Rosenbloom, J. C. Rosenbloom, C. W. Gibson, and D. R. Demuth, *J. Biol. Chem.* **264**, 15451 (1989).

[12] P. Glaser, D. Ladant, O. Sezer, F. Pichot, A. Ullman, and A. Danchin, *Mol. Microbiol.* **2**, 19 (1988).

[13] R. A. Welch, E. P. Dellinger, B. Minshew, and S. Falkow, *Nature (London)* **294**, 665 (1981).

[14] R.A. Welch and S. Falkow, *Infect. Immun.* **43**, 156 (1984).

[15] S. Bhakdi, M. Muhly, S. Korom, and G. Schmidt, *J. Clin. Invest.* **85**, 1746 (1990).

[16] H. K. Ziegler, *in* "Microbial Determinants of Virulence and Host Response" (E. M. Ayoub, G. H. Cassell, W. C. Branche, and T.J. Henry, eds.), p. 283. American Society for Microbiology, Washington, D.C., 1990.

[17] T. Felmlee, S. Pellett, E. Y. Lee, and R. A. Welch, *J. Bacteriol.* **163**, 88 (1985).

[18] N. Mackman, K. Baker, L. Gray, R. Haigh, J. M. Nicaud, and I. B. Holland, *EMBO J.* **6**, 2835 (1987).

intermediate.[19,20] Exactly how the components of the apparatus interact with one another and with the secreted protein is still incompletely understood.

The known extent to which the RTX gene family has dispersed and evolved raises questions about how widespread the family is, and whether any new functions can be ascribed to RTX proteins. The purpose of this chapter is to outline methods used to detect RTX cytolysins and to measure their various activities. In this way, previously undiscovered RTX members may be found, thereby adding to the understanding of how these proteins work and contribute to disease.

Operon Organization and Structural Features

As mentioned above, the arrangement of genes within the RTX operons has been conserved. There are five genes involved: the toxin structural gene, designated *A*; a gene whose product is required for modifying the toxin to an active form, called *C*; and three genes involved in transport, namely, *B, D,* and *E*. In the case of the *E. coli* hemolysin, the prototype RTX cytolysin and the most extensively studied member, the transcriptional order of the genes is *CABD*.[21] In *E. coli,* the *E* gene is known as *tolC* and is unlinked to the *CABD* locus[22]; in other cases, such as for the *E. chrysanthemi* metalloprotease[23] and the *B. pertussis* adenylate cyclase/hemolysin,[24] the *E* gene is adjacent to the 3' end of the *CABD* cluster. It has been shown that the *CA* gene locus of the *A. pleuropneumoniae* serotype 1 hemolysin is separated from the *BD* locus by at least 8 kilobases (kb) of intervening DNA[25]; for all other RTX operons studied at the sequence level, however, the RTX genes comprise a transcriptional unit.

Accompanying this conservation of operon organization among RTX members is strong similarity at the level of DNA and amino acid sequences. For example, between the *E. coli* hemolysin and *P. haemolytica* leukotoxin operons, the predicted amino acid sequences of the *C* gene products are 66% similar, the *A* gene products 62%, the *B* gene products 90.5%, and the *D* gene products 75.6% similar.[26] Of interest besides the high degree of homology within the family is the realization that the

[19] T. Felmlee and R. A. Welch, *Proc. Natl. Acad. Sci. U.S.A.* **85,** 5269 (1988).
[20] W. D. Thomas, Jr., S. Wagner, and R. A. Welch, *J. Bacteriol.* **174,** 6771 (1992).
[21] R. A. Welch and S. Pellett, *J. Bacteriol.* **170,** 1622 (1988).
[22] C. Wandersman and P. Delepelaire, *Proc. Natl. Acad. Sci. U.S.A.* **87,** 4776 (1990).
[23] S. Letoffe, P. Delepelaire, and C. Wandersman, *EMBO J.* **9,** 1375 (1990).
[24] P. Glaser, H. Sakamoto, J. Bellalou, A. Ullmann, and A. Danchin, *EMBO J.* **7,** 3997 (1988).
[25] Y.-F. Chang, R. Young, and D. K. Struck, *J. Bacteriol.* **173,** 5151 (1991).
[26] C. Forestier and R. A. Welch, *Infect. Immun.* **58,** 828 (1990).

RTX *B* genes are part of a larger superfamily of membrane-bound, ATP-dependent transporters, with homologs found in cell types from bacteria to mammalian cells.[27,28]

The proteins of the RTX family are characterized by a number of structural features, including the eponymous repeat domain of the toxin molecule. This domain is a 9-amino acid, glycine- and aspartate-rich motif, tandemly repeated (hence the name) from 4 to 14 times. *Bordetella pertussis* adenylate cyclase/hemolysin contains 47 copies of the Gly-Asp sequence, but these are not all in tandem. Evidence suggests that this region of the toxin molecule is essential for binding calcium, which in turn is needed for cytolytic activity.[29] At the level of amino acid sequence similarity, the repeat region is the most highly conserved within the toxin molecule, and it serves as a signature motif for the RTX family. By analysis of amino acid sequences, predictions can be made about other structural features of RTX proteins, such as those responsible for inserting into target cell membranes.

Identification of RTX Cytolysins

Methods for identifying RTX cytolysins fall into two categories: those based on detecting RTX-specific DNA sequences and those aimed at finding RTX proteins. Obviously, the success of either approach in RTX identification depends to a large extent on the suitability of the probes used. This section discusses the use of DNA and antibody probes in studying the spread of the RTX determinant.

DNA Probes

The utility of probes derived from coding sequences within RTX operons in identifying heterologous RTX genes is evident from the success of this approach in cloning some of these toxin determinants. Kolodrubetz *et al.*[10] used a DNA fragment from the *P. haemolytica* leukotoxin operon as a probe in Southern blot analyses to identify and clone the gene encoding the leukotoxin from the human periodontal pathogen *Actinobacillus actinomycetemcomitans*. In their work, they used a probe from within the *lktA* gene of *P. haemolytica* to hybridize under conditions of low stringency [25°, 50% (v/v) formamide, 0.1% (w/v) sodium dodecyl sulfate

[27] C. F. Higgins, I. D. Hiles, G. P. Salmond, D. R. Gill, J. A. Downie, I. J. Evans, I. B. Holland, L. Gray, S. Buckel, A. W. Bell, and M. A. Hermodson, *Nature (London)* **323**, 448 (1986).

[28] M. H. Blight and I. B. Holland, *Mol. Microbiol.* **4**, 873 (1990).

[29] D. F. Boehm, R. A. Welch, and I. S. Snyder, *Infect. Immun.* **58**, 1959 (1990).

(SDS)] with restriction enzyme-digested *A. actinomycetemcomitans* genomic DNA. After isolating the homologous *A. actinomycetemcomitans* fragments and ligating them into vector plasmids, they screened for clones containing the fragment of interest with the same *P. haemolytica* probe in colony blot hybridizations.

Computer-derived alignments of DNA sequences are useful in identifying regions of homology between RTX genes for which sequence data are known. Our laboratory has made use of alignments of DNA or amino acid sequences carried out by Wisconsin Genetics Computer Group software (Madison, WI); other such programs may be useful as well. As is evident from the work of Kolodrubetz *et al.*, the RTX A genes contain regions suitable for use in DNA hybridization analyses. Prominent among these regions are the sequences encoding the repeats-containing portion of the toxins. Both at the DNA sequence level and at the amino acid sequence level, these regions contain the longest stretches within the A genes of sustained similarity needed for hybridization analysis. Even better suited are the sequences encoding the RTX B genes. As mentioned above, the B genes are very highly conserved among RTX members and therefore should make the best probes. A drawback to be considered, however, is that the B gene probes may detect homologous sequences which are not of RTX origin, but from other members of the ATP-dependent transporter superfamily to which the RTX B genes belong. Another consideration when using a probe derived from a B gene is that, as in the case of the genes encoding the serotype 1 *A. pleuropneumoniae* hemolysin, the B and D genes may not be linked to the toxin structural gene.

Antibody Probes

Various investigators have raised useful polyclonal antisera and monoclonal antibodies (MAbs) to RTX proteins. Lo *et al.*[30] used a specific polyclonal antiserum to identify recombinants in cloning the determinant encoding the *P. haemolytica* leukotoxin. Chang *et al.*[9] screened bacteriophage λ libraries of *A. pleuropneumoniae* genomic DNA with affinity-purified swine antiserum against the *A. pleuropneumoniae* serotype 5 hemolysin. These efforts represent one way that antibody probes can be useful in RTX studies.

Another way is to use antibodies raised against one member of the family as a probe of culture material from a variety of other bacteria in order to identify previously undiscovered RTX members. In this type of analysis, the culture material (bacterial cells or 10% (w/v) trichloroacetic

---

[30] R. Y. C. Lo, P. Shewen, C. Strathdee, and C. N. Greer, *Infect. Immun.* **50**, 667 (1985).

acid-precipitable material from culture supernatant) is analyzed by standard immunodetection procedures (Western blot)[31] using an RTX-specific antibody preparation as a probe.

The ideal antibody probe for identifying putative RTX proteins is one which is known to cross-react with every confirmed RTX protein. This antibody presumably would recognize an epitope which is common to all RTX members. Of course, the signature structural motif of the RTX family, the repeats domain, fits the requirement of a common RTX epitope. Our laboratory has mapped the epitopes recognized by a panel of monoclonal antibodies raised against the *E. coli* hemolysin through the use of deletion mutants.[32] One such MAb, designated A10, was found to recognize an epitope within the repeats region of the hemolysin. In immunoblot analysis, this MAb reacted with eight of eight confirmed RTX proteins.[33] Another MAb, G3, whose epitope did not map to within the repeats domain but immediately proximal to it, also cross-reacted with all confirmed RTX members tested. This indicates that there may exist other domains besides the repeats domain which are common to all RTX members. Our experience has been that polyclonal antisera do not exhibit the same broad range of cross-reactivity as these MAbs in detecting heterologous RTX proteins.

An important consideration when using this type of procedure for identifying putative RTX proteins concerns the conditions under which the antibody preparation is allowed to react with the electroblotted proteins. We have found that blocking the nitrocellulose sheet containing the blotted material with 0.5% Tween 20 in phosphate-buffered saline (PBS) and incubating the sheet in the same solution plus antiserum works well for most, but not all, of the antisera we have tested. Some antibody preparations show cross-reactivity only when a blocking agent such as 5% skim milk in Tris-buffered saline is used in place of 0.5% Tween 20.

Assay of RTX Cytolysins

Considering the conservation of sequences and operon organization among the RTX family, the diversity of activities carried out by the various RTX proteins is remarkable. Most of the well-studied RTX proteins are cytolysins, and much of the remainder of this chapter is devoted to discussing methods for assaying cytolytic effects of these proteins, as well as effects caused by doses of the cytolysins too low to lyse target cells. It

---

[31] H. Towbin, T. Staehlin, and J. Gordon, *Proc. Natl. Acad. Sci. U.S.A.* **76**, 4350 (1979).
[32] S. Pellett, D. F. Boehm, I. S. Snyder, G. Rowe, and R. A. Welch, *Infect. Immun.* **58**, 822 (1990).
[33] A. Lobo and R. A. Welch, unpublished data (1991).

is important to keep in mind, however, that not all RTX proteins are cytolysins. Other properties, such as calcium-binding activity, may be looked for in the characterization of a putative RTX protein.

Hemolytic Activity

Many RTX cytolysins have the ability to lyse erythrocytes from a variety of avian and mammalian species. This type of activity is most easily visualized on a solid growth medium containing erythrocytes as a zone of clearing around a bacterial colony producing a hemolytic protein. Typically, the solid medium (blood agar) contains 5% defibrinated blood in a standard blood agar base. Also, because RTX hemolysins require calcium for activity, the blood agar should be supplemented with $CaCl_2 \cdot 2H_2O$ (typically around 10 mM final concentration).

Measurement of hemolytic activity in a liquid assay involves mixing a sample containing hemolysin with a suspension of erythrocytes in physiological saline (0.85% (w/v) NaCl) plus 10 mM Ca^{2+}. The erythrocyte suspension is prepared by collecting erythrocytes by centrifugation from defibrinated blood. Blood may be obtained from a variety of species, depending on availability. The erythrocytes are washed three times in the saline/Ca^{2+} solution, and resuspended to a final concentration of 10% (assuming a packed cell volume of 100%) in saline/Ca^{2+}. Because the RTX cytolysins are secreted extracellularly or are localized to the cell surface, the nature of the hemolysin-containing samples is either cell-free supernatant or a suspension of hemolysin-producing cells. The supernatant is prepared by growing a mid- to late-log phase culture of the bacterium and then removing the cells by centrifugation. The resulting supernatant is passed through a 0.2 μm filter to remove any remaining cells. Care must be taken to use a filter with low protein-binding capability, such as the Acrodisc (Gelman Sciences, Ann Arbor, MI).

For the cell-associated assay, bacterial cells collected by centrifugation are resuspended in the saline/Ca^{2+} solution such that they are 10-fold concentrated from the original culture. In a typical assay, 200 μl of 10% erythrocytes is mixed with 200 μl of supernatant or 25 μl of resuspended bacterial cells in a final volume of 1 ml (made up with saline/Ca^{2+}). The hemolysin-containing samples should be prepared immediately before use. The assay mixtures are incubated at 37° (or at growth temperature of hemolysin-producing organism) for a period of time ranging from 20–30 min to 2 hr or longer. Hemolytic activity by the *B. pertussis* adenylate cyclase/hemolysin, for example, is evident only after 5 hr of incubation.[34]

---

[34] I. Ehrmann, M. Gray, V. Gordon, L. Gray, and E. L. Hewlett, *FEBS Lett.* **278**, 79 (1991).

Following incubation, the blood cells, bacteria, and cell debris are removed by centrifugation, and the supernatant is measured spectrophotometrically for absorbance at 540 nm as a measure of released hemoglobin (for more information, see Rowe and Welch[35]).

The requirement for calcium may be demonstrated by including in the assay a chelator of calcium ions, such as ethylene glycol bis(β-aminoethyl ether) N,N,N',N'-tetraacetic acid (EGTA). Boehm et al.[36] found that, in order to show calcium dependency for E. coli hemolysin activity most clearly, the bacteria had to be grown in broth which had been treated before autoclaving with 7 μM EGTA. Supernatants from cultures grown in this broth had 200-fold greater hemolytic activity in an assay as described above than with 10 mM EGTA in place of 10 mM Ca^{2+}.

Nonerythrocyte Cytolytic Activity

Those RTX cytolysins capable of lysing erythrocytes are able to lyse a variety of nucleated cells as well. For RTX proteins capable of attacking only certain types of leukocytes, however, the converse is not true. Therefore, the relatively easy task of measuring lysis by release of hemoglobin by erythrocytes must be replaced by the more arduous task of assessing lysis of cultured cells or freshly isolated primary cells. The most important consideration when devising experiments to measure nonerythrocyte cytolytic activity is the choice of target cells. For the RTX hemolysins this need not be a difficult choice. The E. coli hemolysin, for example, lyses fibroblasts, leukocytes, and various epithelial and endothelial cells.[37] The targets of RTX leukotoxins must be much more carefully chosen. The A. actinomycetemcomitans leukotoxin acts only against leukemic and lymphoid cell lines of humans and certain non-human primates, but not against other human cell types or similar cell types of other species.[38,39] The P. haemolytica leukotoxin shows a similar pattern of specificity toward ruminant cells.[3] An understanding of the types of environments colonized by the bacterium may aid in selecting the type of cell as the target of its cytolysin.

Most commonly used methods for measuring lysis detect either release of a radioactive compound from lysed cells or the inhibition of uptake of a dye normally taken up by undamaged cells. In the first method, sodium

[35] G. E. Rowe and R. A. Welch, this volume [53].

[36] D. F. Boehm, R. A. Welch, and I. S. Snyder, Infect. Immun. **58**, 1951 (1990).

[37] S. Cavalieri, G. Bohach, and I. S. Snyder, Microbiol. Rev. **48**, 326 (1984).

[38] N. S. Taichman, D. L. Simpson, S. Sakurada, M. Cranfield, J. DiRienzo, and J. Slots, Oral Microbiol. Immunol. **2**, 97 (1987).

[39] D. L. Simpson, P. Berthold, and N. S. Taichman, Infect. Immun. **56**, 1162 (1988).

[^{51}Cr]chromate is the radioactive label of choice for most investigators. Cultured target cells are harvested and washed in growth medium several times, resuspended to a concentration of 10^6 cells/ml, and then incubated with 50 μCi ^{51}Cr per milliliter of cell suspension for 1 hr at 37°. Following incubation, the cells are washed to remove unincorporated label and again are resuspended to 10^6 cells/ml. For these steps, growth medium lacking antibiotics and fetal calf serum should be used, since those components have adverse effects on toxin activity. Typically, 10^5 labeled cells are mixed with the cytolysin-containing sample, and the assay mixture is incubated for 30 min to 1 hr. Longer incubation times may result in increased spontaneous lysis. At the end of the incubation of toxin with target, the cells and debris are removed by centrifugation, and the resulting supernatant is counted in a γ counter. Toxin activity is determined as a percentage of complete lysis achieved by exposing the labeled cells to 1.16 N HCl.

A nonradioactive method of choice for determining cytotoxicity is to measure the uptake by target cells of the vital dye neutral red following exposure of the cells to the toxin preparation. Developed by Kull and Cuatrecasas,[40] this procedure has been used by Greer and Shewen[41] and by Cruz et al.[42] to assess toxicity by P. haemolytica leukotoxin. In this assay, the washed cells (unlabeled) and toxin preparation are mixed and incubated together as for the chromium release assay. Following this, the cells are collected by centrifugation, washed in buffered saline, resuspended in buffered saline containing 0.02% (w/v) neutral red, and incubated for another hour at 37°. After incubation with the dye, the cells are washed and lysed with detergent (0.5% (v/v) Nonidet P-40 or 0.5% (w/v) SDS in 50 mM acetic acid). The lysate is cleared by centrifugation, and the absorbance at 545 nm of the resulting supernatant is measured to quantify the released neutral red.

Sublytic Effects of RTX Cytolysins

Doses of RTX cytolysins too low to directly cause cell lysis have been shown to exert profound effects on target cells. Although these activities cannot yet be considered RTX-specific, their study represents the most recent foray in assessing the role of RTX cytolysins in pathogenesis.

Most work in this area has focused on effects of sublytic doses of E. coli hemolysin and P. haemolytica leukotoxin on various types of immune cells. Some of the cellular consequences of exposure to low concentrations

[40] F. C. Kull and P. Cuatrecasas, J. Immunol. **126**, 1279 (1981).
[41] C. N. Greer and P. E. Shewen, Vet. Microbiol. **12**, 33 (1986).
[42] W. T. Cruz, R. Young, Y. F. Chang, and D. K. Struck, Mol. Microbiol. **4**, 1933 (1990).

of these toxins include increased chemiluminescence by human neutrophils,[43] decreased chemiluminescence by bovine polymorphonuclear leukocytes,[3,44] release of granule constituents,[45] intracellular ATP depletion,[15,45] release of histamine and leukotrienes,[46,47] increased release of interleukin1β by human monocytes,[15] generation of superoxide anions and hydrogen peroxide,[48] and inhibition of macrophage antigen presentation.[16]

Calcium-Binding Activity

The calcium-binding ability of an RTX protein can be demonstrated by a method in which the protein is immobilized by electroblotting on to a nitrocellulose sheet and then incubated with radioactive $^{45}Ca^{2+}$. The sheet is then washed, and the polypeptides to which the isotope has bound are detected by autoradiography. This method, described by Maruyama et al.,[49] has been used by Boehm et al.[29] to show that the wild-type E. coli hemolysin binds ^{45}Ca, but a mutant form of the toxin, in which the repeats region of the protein had been deleted, does not.

In this procedure, samples of toxin are prepared by precipitating protein from culture supernatants with polyethylene glycol (30% (w/v) PEG 3350 added to supernatant and stirred at 4° for 1 hr); alternatively, whole cells may be used as the source of toxin. The samples are mixed with Laemmli electrophoresis sample buffer[50] and subjected to SDS–polyacrylamide gel electrophoresis. The separated polypeptides are transferred to nitrocellulose electrophoretically by the method of Towbin et al.[31] After transfer, the nitrocellulose sheet is soaked in a buffer containing 60 mM KCl, 5 mM MgCl$_2$, and 10 mM imidazole hydrochloride (pH 6.8) for 1 hr, changing the buffer every 15 min. Next the sheet is incubated in the same buffer containing 1.0 μCi/ml $^{45}CaCl_2$ for 10 min. The sheet is then rinsed with distilled water and allowed to dry. When dry, the nitrocellulose sheet is used to expose X-ray film for autoradiography.

Maruyama et al. reported that this method is capable of detecting as little as 2 μg of calcium-binding protein, but Boehm et al. demonstrated

---

[43] S.J. Cavalieri and I. S. Snyder, Infect. Immun. **37**, 966 (1982).
[44] P. A. J. Henricks, G. J. Binkhorst, A. A. Drijver, H. Van Der Vliet, and F. P. Nijkamp, Vet. Microbiol. **22**, 259 (1990).
[45] S. Bhakdi, S. Greulich, M. Muhly, B. Eberspacher, H. Becker, A. Thiele, and F. Hugo, J. Exp. Med. **169**, 737 (1989).
[46] J. Scheffer, W. Konig, J. Hacker, and W. Goebel, Infect. Immun. **50**, 271 (1985).
[47] F. Grimminger, U. Sibelius, S. Bhakdi, N. Suttorp, and W. Seeger, J. Clin. Invest. **88**, 1531 (1991).
[48] S. Bhakdi and E. Martin, Infect. Immun. **59**, 2955 (1991).
[49] K. Maruyama, T. Mikawa, and S. Ebashi, J. Biochem. (Tokyo) **95**, 511 (1984).
[50] U. K. Laemmli, Nature (London) **227**, 680 (1970).

that this method could easily detect 2 μg of *E. coli* hemolysin. Calmodulin (0.5 μg) is a good positive control for the procedure.

Pore Formation

The RTX cytolysins act by forming pores in the membranes of target cells, thereby causing osmotic lysis of the cell due to the influx of water. The influx of water can be prevented by certain compounds, mainly sugars and polysaccharides.[51] The protection against swelling by these carbohydrates is proportional to the molecular size (more accurately, the molecular diameter) and the concentration in solution. For protection to occur, the molecular size of the protecting agent must be larger than the size of the pore formed by the toxin, and the concentration of the agent must be close to isotonic with the intracellular concentration of solutes. By using a series of protecting agents varying in molecular size, one may estimate the size of the pore formed by the toxin in the target cell membrane. This type of experiment has been used to estimate the pore sizes created by RTX hemolysins from *E. coli*[51] and *A. pleuropneumoniae*[52] and leukotoxins from *P. haemolytica*[53] and *A. actinomycetemcomitans*.[54]

The carbohydrate protectants used in these determinations ranged in size from glycerol (molecular weight 92.1) and mannitol (182.1) to dextran 4 (~4000) and inulin (~5000). Bhakdi *et al.*[51] found that 30 mM dextran 4, with a molecular diameter of about 3 nm, prevented lysis of erythrocytes by *E. coli* hemolysin, whereas 30 mM sucrose (0.9 nm) or 30 mM raffinose (1.3 nm) did not prevent lysis. For the leukotoxins, however, compounds as small as sucrose and maltose (0.96 nm) prevented leakage of cytoplasmic enzymes or ^{51}Cr from labeled cells.[53,54] In both cases, the assays used were similar to those described above with the addition of the carbohydrates. To aid in selection of compounds to use as protectants, refer to the publication by Scherrer and Gerhardt[55] describing molecular sizes of carbohydrates used as osmotic protectants.

Another approach to studying pores formed by RTX cytolysins has been the biophysical characterization of channels formed by the toxins in artificial planar lipid bilayers (see [56] in this volume). Solutions of lipids are applied to a hole in a partition separating two aqueous chambers.

[51] S. Bhakdi, N. Mackman, J. M. Nicaud, and I. B. Holland, *Infect. Immun.* **52**, 63 (1986).
[52] G. Lalonde, T. V. McDonald, P. Gardner, and P. D. O'Hanley, *J. Biol. Chem.* **264**, 13559 (1989).
[53] K. D. Clinkenbeard, D. A. Mosier, and A. W. Confer, *Infect. Immun.* **57**, 420 (1989).
[54] M. Iwase, E. T. Lally, P. Berthold, H. M. Korchak, and N. S. Taichman, *Infect. Immun.* **58**, 1782 (1990).
[55] R. Scherrer and P. Gerhardt, *J. Bacteriol.* **107**, 718 (1971).

Toxin is added to one side, and the current flowing through the membrane is monitored. The conductance on formation of channels within the bilayer can be observed under varying conditions of ionic composition of the aqueous phases, lipid composition, and toxin concentration. In this way, information can be gained about pore size and half-life, ion selectivity of the channels, etc.

Studies regarding the behavior of *E. coli* hemolysin in planar lipid bilayers have been done by Menestrina *et al.*[56] and Benz *et al.*,[57] with conflicting results. Menestrina *et al.* reported channel formation in pure phosphatidylcholine and phosphatidylethanolamine preparations, and a 5:1 molar mixture of the two, whereas Benz *et al.* showed almost no activity in pure lipids and defined mixtures but high activity using asolectin, a crude mixture of soybean lipids. The relationship between conductance and concentration of toxin observed by Benz *et al.* suggested that several hemolysin molecules may be needed to form the conductive unit; Menestrina *et al.* found a linear relationship between hemolysin concentration and conductance, indicating channel formation by toxin monomers.

The use of this approach in the study of RTX cytolysins is still in the early stages, but applications of the technique may lie in answering questions about regions of the cytolysin molecules responsible for pore formation, the stoichiometry of pore formation, and the membrane receptors or constituents needed for toxin binding to membranes.

Conclusion

The methods and techniques discussed in this chapter have allowed investigators to answer some of the questions regarding the structure and function of RTX cytolysins and their role in pathogenesis. Still to be elucidated are the bases of target cell specificity found among the various toxins, the full details of the processes by which they are activated and secreted, and the importance of toxin-induced immune cell dysfunction in diseases caused by RTX-producing organisms. Discovering new members of the family, perhaps possessing unique properties, as well as further study of known members, will add to our understanding of RTX cytolysins.

[56] G. Menestrina, N. Mackman, I. B. Holland, and S. Bhakdi, *Biochim. Biophys. Acta* **905**, 109 (1987).
[57] R. Benz, A. Schmid, W. Wagner, and W. Goebel, *Infect. Immun.* **57**, 887 (1989)

[55] Assay of Cytopathogenic Toxins in Cultured Cells

By Monica Thelestam and Inger Florin

Introduction

Different criteria can be used for the classification of bacterial toxins. If classified according to action on mammalian cells, three major types of toxins can be discerned.[1] Some examples representative of these types of toxins are shown in Table I. Type I toxins act analogously to growth factors by binding to a cell surface receptor which mediates a transmembrane signal to the cell interior. Type II toxins act directly on the plasma membrane either by pore formation or by disruption of the lipid bilayer. Toxins of type III are internalized and act by modification of cytosolic targets. Most type II and III toxins are cytopathogenic. The few toxins so far recognized as type I alter the cell metabolism in the absence of cytopathogenic alterations. This chapter is restricted to cytopathogenic toxins (CPTs).

In the literature bacterial toxins are denoted by a variety of names which may sometimes be synonymous or at least in some sense overlapping. Here we use the following definitions (Fig. 1): Cytopathogenic (or cytopathic) toxins cause light-microscopically visible morphological cell alterations, which ultimately may or may not lead to cell death. Cytotoxins (or cytolethal or cytocidal toxins) are toxins that cause the death of cultured cells, that is, irreversibly inhibit the ability of the cells to proliferate. The cytopathogenic effects (CPEs) caused by noncytotoxic toxins are reversible. Therefore these toxins may be denoted as cytostatics (cf. bacteriostatic agents), although this term has traditionally been restricted to certain plant toxins which are being used as antineoplastic drugs. In conclusion, although almost every cytotoxin initially causes a CPE, not all CPTs are cytotoxic.

Based on the subcellular target the CPTs (Fig. 1) can be divided into those that have the plasma membrane as target (type II toxins) and those that have an intracellular target (type III toxins). The type II cytotoxic CPTs (Fig. 1, left-hand side) are denoted as cytolysins, as they cause gross structural damage to the plasma membrane and thereby lyse cells directly. Some type III cytotoxic CPTs cause a slow cell degeneration due to inhibition of macromolecular synthesis or other metabolic disturbances,

---

[1] P. Boquet and D. M. Gill, in "Sourcebook of Bacterial Protein Toxins" (J. E. Alouf and J. H. Freer, eds.), p. 23. Academic Press, London and New York, 1991.

TABLE I
EXAMPLES OF DIFFERENT TYPES OF TOXINS

Type I: Extracellular action	Type II: Membrane damage	Type III: Intracellular action
E. coli heat stable enterotoxin (ST)	E. coli hemolysin	E. coli heat-labile enterotoxin (LT)
S. aureus toxic shock syndrome toxin 1, TSST-1	S. aureus α toxin	Cholera toxin
	C. perfringens perfringo- lysin O	Diphtheria toxin
	C. perfringens enterotoxin	P. aeruginosa exotoxin A
	A. hydrophila aerolysin	C. difficile toxins
	S. pyogenes streptolysins O and S	Clostridial neurotoxins
		Pertussis toxin
		Shiga toxin

such as energy poisoning or defects in the signal transduction system. Such effects, when irreversible, lead to necrotic cell death, characterized by changes of the mitochondria, endoplasmatic reticulum, etc., and finally cell degeneration. Other type III toxins may induce a degradation of the

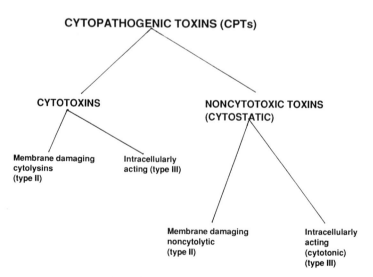

FIG. 1. Classification of cytopathogenic toxins.

DNA structure, leading to so-called programmed cell death (apoptosis).[2] A number of intracellularly acting CPTs disorganize the cytoskeletal microfilaments irreversibly, leading initially to a characteristic CPE and resulting in an inability to proliferate due to lack of a contractile ring. Noncytotoxic (cytostatic) bacterial CPTs (Fig. 1, right-hand side) are the membrane-damaging toxins, which induce limited-size pores that can be repaired, as well as the so-called cytotonic toxins. These are certain enterotoxins which have reversible effects on second messenger systems by ADP-ribosylation of G proteins that regulate adenylate cyclase.[3]

Sensitive antibody-based methods have been developed for the rapid detection and assay of many bacterial CPTs. However, when the biological activity is to be determined, assay of the toxins in cultured cells is still indispensable. Such assays are needed for determination of the neutralizing capacity of antitoxins and the activity of modified toxins, for example, toxin-based vaccines (toxoids) and otherwise modified toxins, as toxin fragments used in studies on structure–activity relationships. In addition, the identification of potential CPTs from "new" pathogenic bacteria obviously requires an assay of biological activity. Here we first briefly review some major types of CPEs, then present a standardized cell culture method for detection of CPTs, and finally give directions for toxin titration in order to quantify the CPE.

Types of Toxin-Induced Cytopathogenic Effects Seen by Light Microscopy

Cellular morphological changes collectively defined as "cytopathogenic effects" can be broadly divided as follows[4]: (1) cell lysis and detachment, (2) change of cell shape and of cell surface (e.g., blebs), and (3) intracellularly visible changes. These effects can be seen by inverted light microscopy or phase-contrast microscopy. Specific staining procedures may enhance details, but it is useful to observe cells directly in unfixed/unstained cultures to be able to follow the progression of the CPE. In the initial screening of the morphological effects it can be helpful to include parallel treatment with reference substances for which the mode of action is known. Then it is sometimes possible to draw conclusions as to the likely major target structure in the cell from mere observation in the light microscope of a characteristic CPE. This may provide a good basis for

[2] I. D. Bowen and S. M. Bowen, "Programmed Cell Death in Tumors and Tissues." Chapman & Hall, New York, 1990.
[3] K. L. Richards and S. D. Douglas, Microbiol. Rev. 42, 592 (1978).
[4] E. Walum, K. Stenberg, and D. Jenssen, "Understanding Cell Toxicology." Ellis Horwood, New York, 1990.

TABLE II
CYTOPATHOGENIC EFFECTS INDUCED BY BACTERIAL TOXINS AND REFERENCE SUBSTANCES

Type of effect	Toxin example	Reference substance example
Lysis, condensed nuclei	*P. aeruginosa* heat-stable hemolysin	Triton X-100
Lysis, enlarged flat nuclei, prominent nucleoli	*C. perfringens* perfringolysin O	Digitonin
Swelling (colloid osmotic)	*C. perfringens* enterotoxin	Hypotonic NaCl
Cytoplasmic retraction, cell rounding	*C. difficile* toxins A and B	Cytochalasins
Cytotonic (round Y1 cells, elongated CHO cells)	Cholera toxin, *E. coli* LT	Dibutyryl-cAMP
Clustered growth of CHO cells	Pertussis toxin	?
Cytosolic granulation and deflation, degeneration	Diphtheria toxin	Cycloheximide
Cytosolic vacuolation	*H. pylori* cytotoxin	Chloroquine
Multinucleation	*E. coli* CNF, *C. difficile* toxin B	Cytochalasins
Chromatin condensation and nuclear fragmentation	*C. difficile* toxin A	Apoptosis-inducing stimuli

subsequent approaches aiming at clarification of the molecular mechanism. Examples of different types of CPEs are detailed below and summarized in Table II.

Cell Lysis and Detachment

Cell lysis caused by cytolysins via gross damage to the plasma membrane structure is one of the effects most easily scored in cultured cells. Plasma membranes are more or less solubilized because of large lesions induced in the membrane, the cytosol is granulated, and the nuclei become clearly visible and are seen as somewhat condensed. A typical feature in confluent layers of cells is that the cell layer appears to remain essentially attached to the substrate because of the intactness of the cytoskeleton. In less dense cultures the cells detach. This type of CPE is induced by detergentlike toxins[5] and certain phospholipases. A mild detergent such as Triton X-100 (0.1% v/v) serves as a suitable reference substance.

A rather similar CPE, but with nuclei appearing as enlarged and flat with prominent contours and nucleoli, occurs on exposure to the so-

[5] K. Fujita, T. Akino, and H. Yoshioka, *Infect. Immun.* **56**, 1385 (1988).

called thiol-activated toxins, for example, perfringolysin O of *Clostridium perfringens* or streptolysin O of *Streptococcus pyogenes*. These toxins bind to membrane cholesterol and induce membrane lesions of variable but more limited size than detergents.[6] Cells treated with high concentrations of such cytolysins detach (unless cultures are very dense), but low doses allow the leaky cells to remain spread on the substrate. As a reference substance digitonin (10–100 μM), which also interacts with cholesterol, may be used.[7]

Cell detachment is caused by many other CPTs as well. This is easy to detect by light microscopy without staining, and it can be quantitated by staining the remaining cells.[8]

Change of Cell Shape and of Cell Surface

Colloid osmotic swelling of cells is caused by membrane-damaging toxins which induce pores of very small size, leaving the membrane unlysed but affecting its permeability barrier function. Influx of water takes place in order to equilibrate the osmotic strength of the cytosol with that of the extracellular environment, and the cells swell. This is caused, for instance, by the *Clostridium perfringens* enterotoxin.[9] Similarly round and swollen cells are seen after a short exposure to hypotonic NaCl (10 mM).[10]

Cell rounding without swelling can occur when the cytoskeleton is disorganized in the absence of plasma membrane damage, as, for example, after exposure to the high molecular weight clostridial cytotoxins such as the α-toxin of *Clostridium novyi*.[11] This cytoskeletal disorganization may be a secondary phenomenon, taking place after some kind of intracellular toxic events. However, bacterial toxins which act directly on the cytoskeleton, for example, the C2 toxin of *Clostridium botulinum* which affects actin,[12] cause a characteristic cytoplasmic retraction and actinomorphic effect in fibroblasts, which is later followed by complete rounding. A good reference substance for this type of effect is cytochalasin, which binds directly to actin.[13]

[6] M. Thelestam and R. Möllby, *Biochim. Biophys. Acta* **557**, 156 (1979).

[7] G. Ahnert-Hilger, W. Mach, K. J. Föhr, and M. Gratzl, *Methods Cell Biol.* **31**, 63 (1989).

[8] J. E. Brown, S. W. Rothman, and B. P. Doctor, *Infect. Immun.* **29**, 98 (1980).

[9] B. A. McClane and J. L. McDonel, *Biochim. Biophys. Acta* **641**, 401 (1981).

[10] M. Thelestam and R. Möllby, *Toxicon* **21**, 805 (1983).

[11] P. Bette, A. Oksche, F. Mauler, C. von Eichel-Streiber, M. R. Popoff, and E. Habermann, *Toxicon* **29**, 877 (1991).

[12] K. Aktories, M. Wille, and I. Just, *Curr. Top. Microbiol. Immunol.* **175**, 97 (1992).

[13] M. Thelestam and R. Gross, *in* "Handbook of Toxinology" (W. T. Shier and D. Mebs, eds.), p. 423. Dekker, New York and Basel, 1990.

Development of plasma membrane blebs is common when the cytoskeleton is affected. Such blebs can be vaguely seen by light microscopy in some cells, but closer inspection and characterization require electron microscopy.[14]

The cytotonic heat-labile enterotoxins (LT) of *Escherichia coli* and *Vibrio cholerae* cause a rounding of certain cells (e.g., mouse adrenocortex Y1 cells) and a characteristic elongation of other cells [Chinese hamster ovary (CHO) cells] owing to the increased level of cytosolic cyclic AMP.[3] This well-known example underlines the finding that different types of cells may respond to the same toxin with widely differing shape alterations. Dibutyryl-cAMP (1 m*M*) can be used as a reference substance.[15]

A characteristic, slowly developing, clustered growth pattern of Chinese hamster ovary cells is induced by pertussis toxin.[16] To our knowledge this type of effect has not been reported for other toxins. It is not due to a simple agglutination, and the molecular details behind it are unknown. Rapid diagnosis of pertussis based on this assay has been reported.[17] We are not aware of any suitable reference substance for this type of effect, except pertussis toxin itself (which is commercially available).

Intracellularly Visible Changes

Cytosolic granulation is an early symptom in many kinds of CPE, for instance, in membrane damage, but also occurs in cytopathogenicity induced by toxins which inhibit protein synthesis. The first cytopathogenic signs in, for example, diphtheria toxin-treated cells appear only after hours or 1 day, and they usually consist of granulation and deflation of the cytoplasm. Later on the cells deteriorate into debris. Substances which inhibit protein synthesis, for example, cycloheximide (50–100 μg/ml), give similar morphological changes.

A prominent cytosolic vacuolation has been described in cells treated with aerolysin from *Aeromonas hydrophila*[18] and with a cytotoxin produced by *Helicobacter pylori*.[19] Vacuolation has not been reported for other type II toxins, and the molecular basis for this effect is not clear. Identical looking vacuolation of cultured cells has long been known to be induced by weak bases which accumulate in acidic cellular compartments

[14] W. Malorni, C. Fiorentini, S. Paradisi, M. Giuliano, P. Mastrantonio, and G. Donelli, *Exp. Mol. Pathol.* **52,** 340 (1990).
[15] A. W. Hsie and T. T. Puck, *Proc. Natl. Acad. Sci. U.S.A.* **68,** 358 (1971).
[16] E. L. Hewlett, K. T. Sauer, G. A. Myers, J. L. Colwell, and R. L. Guerrant, *Infect. Immun.* **40,** 1198 (1983).
[17] S. A. Halperin, R. Bortolussi, A. Kasina, and A. J. Wort, *J. Clin. Microbiol.* **28,** 32 (1990).
[18] M. Thelestam and Å. Ljungh, *Infect. Immun.* **34,** 949 (1981).
[19] T. L. Cover and M. J. Blaser, *J. Biol. Chem.* **267,** 10570 (1992).

(endosomes and lysosomes).[20] These so-called lysosomotropic agents, for instance, chloroquine (100 μM) or ammonium chloride[21] (30 mM), can be used as reference substances.

Nuclei are not normally visible in growing cells but become apparent (as described for cytolytic toxins) when the plasma membranes are severely damaged. Multinucleation can be caused in transformed cells by toxins which interfere with the microfilament part of the cytoskeleton, for example, *Clostridium difficile* toxin B[22] and the cytotoxic necrotizing factor (CNF) produced by *E. coli*.[23] This effect is usually detected only after a few days of toxin exposure. The cells become much larger than control cells, and nuclei can be counted after staining. Cytochalasins give the same effect.[13,22] Nuclear polarization and fragmentation and in certain cells even extrusion of the nucleus from the cytoplasm is seen after exposure of certain cells to *C. difficile* toxin A.[24] The latter events can be seen by light microscopy provided the cells are fixed and stained with DNA-specific dyes.[25] Such alterations are characteristic of apoptosis and can thus be mimicked by apoptosis-inducing stimuli.[26]

Assay of Cytopathogenic Toxins in Cultured Cells

Choice of Target Cells

Cells growing attached and spread onto a solid substrate are more useful than cells in suspension, since cytopathogenic effects often show up as some form of cell rounding. As already mentioned different cell types may show variable responses to the same toxin. Certain bacterial toxins act rather selectively on just a few types of cells, in some cases because of the presence of specific receptors on the cell surface and in other cases because of differences in intracellular targets.

When dealing with well-known toxins it is obvious to choose a target cell known to be sensitive. When a potential new toxin is being searched for it is important to screen the responses of a few different cell lines

[20] C. de Duve, T. de Barsy, B. Poole, A. Trouet, P. Tulkens, and F. Van Hoof, *Biochem. Pharmacol.* **23**, 2495 (1974).

[21] C. C. Cain and R. F. Murphy, *J. Cell Biol.* **106**, 269 (1988).

[22] M. C. Shoshan, P. Åman, S. Skog, I. Florin, and M. Thelestam, *Eur. J. Cell Biol.* **53**, 357 (1990).

[23] C. Fiorentini, G. Arancia, A. Caprioli, V. Falbo, F. M. Ruggieri, and G. Donelli, *Toxicon* **26**, 1047 (1988).

[24] C. Fiorentini, W. Malorni, S. Paradisi, M. Giuliano, P. Mastrantonio, and G. Donelli, *Infect. Immun.* **58**, 2329 (1990).

[25] C. Fiorentini, G. Donelli, P. Nicotera, and M. Thelestam, unpublished (1992).

[26] B. F. Trump and I. K. Berezesky, *Curr. Opin. Cell Biol.* **4**, 227 (1992).

initially. The choice of suitable target cells depends on several factors: Should the cells be preferably of human or animal origin? Which organ/tissue is colonized/infected by the bacterium *in vivo*? Are the natural target cells thought to be of epithelioid, fibroblastic, neuronal, lymphoid, or muscle tissue origin? Important in choosing a cell line are also practical aspects such as how well the cells grow and their appearance *in vitro*, which is crucial for ease of scoring morphological effects by light microscopy. The importance of the latter aspect depends on the type of CPT in question. For example, if the toxin causes cell rounding this effect is easier to see in fibroblasts, which grow elongated, than in epithelioid cells, which already generally have a rather condensed morphology in the normal condition. For detection of cytolytic toxins the shape of the target cell is less important.

There are many human and animal cell lines available from the American Type Culture Collection (ATCC, Rockville, MD). Some of these are established cell lines, that is, they are immortalized by chemical or viral transformation. They usually grow rapidly *in vitro* and can be maintained in culture for long periods of time. A disadvantage may be the low degree of differentiation. However, "normal" untransformed cell lines are available which have a limited lifetime *in vitro* and may exhibit a higher degree of differentiation than the transformed counterparts. The aging of such cell lines on *in vitro* cultivation implies an extra variable to be considered when standardizing an assay system. Older cells may be more sensitive than younger ones to any toxin, and they also grow more slowly than young cells.

Note: A new toxin may be membrane damaging. Because most membrane-damaging toxins are hemolytic, it may be advisable to perform a simple hemolytic titration on red blood cells before initiating cell cultivation (see [53] in this volume). However, there are examples of toxins which can cause membrane damage to cultured cells although they are not hemolytic. Thus, when a hemolysis test is negative the new toxin should also be assayed on cultured cells.

Cultivation of Cells

Detailed instructions for cell cultivation are found in handbooks.[27] In short, for maintenance the cells are cultivated in flasks in the medium recommended by the provider of the cell line. For assay of CPTs cells are most suitably cultivated in 96-well microtitre plates. The seeding density should be adjusted so that the cells at the time of toxin assay are neither

---

[27] R. I. Freshney, "Culture of Animal Cells," Alan R. Liss, New York, 1988.

too sparse nor too confluent. When cells are very sparse early after seeding they have not yet completely developed the characteristic morphology, and it may be difficult to judge whether there is a CPE. The denser the cells, usually the less pronounced is the cytopathogenicity. This is due to at least two factors: (1) the number of toxin molecules per cell will be lower with a higher number of cells, and (2) densely growing cells support one another by contacts between the cells. Thus, cells should be well spread on the plastic substrate, and in the logarithmic growth phase, but still separated from one another. The density at scoring should be approximately 200 cells per field of view (at a 100× magnification).

The actual number of cells to be inoculated depends on the growth properties of the cell type and the length of the time period between seeding and testing. As a rule of thumb we seed approximately 3000 untransformed fibroblasts per well on day 0 and perform the assay on day 2 or 3. Untransformed epithelial cells grow more slowly than fibroblasts. Transformed cells usually grow faster than the normal counterparts and may thus require a lower cell number in the inoculum.

Detection of Cytopathogenic Toxins in Cultured Cells

To detect whether a bacterium produces any CPT at all, supernatants from bacterial cultures are added to cells incubated at 37°, and the response of the cells is followed under an inverted microscope at regular time intervals. The types of morphological changes described above are checked in order to get a preliminary idea of what kind of toxin is present. The cells are scored every 15 min for the first hour, then every half hour for a few hours, and then at longer intervals. The initial frequent scoring may not permit exact counting of the proportion of affected cells. However, relatively good approximations can be made by quick inspection of the cultures.

This type of initial screening gives a rough estimate of both dose–response and time–course relationships, on which further, more exact determinations can be based. The response of the cells is followed for a few days. The time span for appearance of different types of CPEs is large. For instance, in the case of cytolytic detergentlike toxins an effect can usually be seen within a few minutes after toxin addition; with endocytosed toxins there is at least a 25-min latency, as in the case of, for instance, the *C. difficile* toxin B, but it can be much longer as with toxins which inhibit protein synthesis, for example, diphtheria toxin, which may give no visible alterations during the first 24 hr.

The highest sensitivity in detection of cytopathogenicity is achieved by mixing toxin with cells before seeding, that is, directly after trypsinization.[28] However, the procedure suggested above is more practical since the cells otherwise have to be maintained in several flasks and trypsinized before each test. With the procedure above, the same plate can be used for 2–3 days (depending on how fast the cells reach confluency). It has the additional advantage that any toxin receptor of (glyco)protein nature, which might be removed or altered during trypsinization, should have reappeared after a couple of days in culture.

Several points should be kept in mind. (1) The bacterial supernatant should be filter-sterilized before application to cells since any remaining bacteria will grow rapidly in the rich cell culture medium and interfere with the test. (2) Cultured mammalian cells respond adversely to many bacterial culture media. It is thus advisable to dilute the bacterial supernatant at least 1/5 in cell culture medium. Similarly treated media in which no bacteria have grown should be included as negative controls. Another possibility is to cultivate the bacteria directly in cell culture medium so that supernatants can be assayed undiluted or even concentrated. (3) Some toxins are inactivated by components of cell culture media, for example, by serum constituents. The initial screening tests for detection of new toxins should thus be performed not only in complete medium but also in medium without serum and in buffers, such as Tris- or phosphate-buffered saline, pH 7.4 (TBS or PBS). However, one cannot expect cultured cells to thrive in simple buffers for more than a few hours. The time cells in culture can endure TBS or PBS is dependent to a certain degree on the cell density. Cells in confluent monolayers sustain buffers somewhat better than sparsely growing cells. (4) As "positive" controls one may use supernatants from related bacteria known to produce a toxin(s) which might be similar to the one under study. Moreover, neutralization of the CPE with antibodies against toxins from related bacteria may give additional information as to the nature of the new toxin. Various types of reference substances (as suggested above) may serve as further comparative controls.

Titration of Cytopathogenic Toxins in Cultured Cells

We have elaborated the protocol below for the routine assay of C. difficile toxins in different toxin preparations,[29] but it is applicable to any CPT. The cells chosen for the test are seeded at a suitable density (~3000 cells/wells) in 96-well plates using 200 μl/well of a medium recommended for the cell type. The cultures are incubated at 37° and should have a cell

---

[28] I. Florin and F. Antillon, *J. Med. Microbiol.* **37**, 22 (1992).
[29] I. Florin and M. Thelestam, *Infect. Immun.* **33**, 67 (1981).

density suitable for CPT assay after 2–4 days, depending on the growth rate of the cell type.

For the CPT assay, 10-fold dilutions (in medium or buffer) of culture supernatant or isolated toxin are made in a series of at least 12 wells. Duplicate toxin samples should be used, and wells to which the toxin diluent is added serve as negative controls. The plate is incubated at 37°, and the CPE is scored after various time periods, adjusted according to the properties of the specific toxin. In titrations of the *C. difficile* toxins we take the final score after 24 hr but the most suitable time period may vary with the toxin. At the chosen time point the dilution of toxin which has caused a CPE in 50% of the cells is estimated. This dilution corresponds to the 50% tissue culture dose (TCD_{50}). The inverted value of this dilution gives the titer, namely, the number of TCD_{50} units in the stock solution of the toxin. TCD_{100} or TCD_{10} values may also be employed. A more accurate final determination of the titer is made by titrating the toxin in 2-fold dilutions around the region of the approximately determined TCD value and counting affected cells in the wells showing around 10, 50, or 100% CPE for exact determination of TCD_{10}, TCD_{50}, or TCD_{100} values, respectively. We count three randomly selected fields of view per well, scoring approximately 200 cells per field. Means are calculated of the three fields from duplicate samples, that is, six fields corresponding to a total of about 1200 scored cells. Scoring also after 48 hr or more may give additional information, for instance, about the reversibility of the CPE.

When reporting the specific activity of a CPT, namely, the number of TCD units per amount of protein, it is important to detail all the variables of the assay, such as the number of cells seeded per well, the composition and volume of medium, the number of days after seeding, the time of toxin exposure, the type of CPE, and the type of TCD unit determined (TCD_{100}, TCD_{50}, or TCD_{10}). Much confusion has arisen about toxin titers and specific activities because of inadequate reports of these variables. Owing to interlaboratory discrepancies between titrations some authors prefer to report toxin quantities as amount of protein. However, the toxin must then be absolutely free of contaminating proteins, which is not always the case. In addition, a highly purified toxin preparation may lose activity on storage. Therefore, it is advisable to titrate the toxin at regular time intervals and to adjust the concentration according to biological activity rather than absolute amount of protein.

To check that a toxoid (denatured or otherwise inactivated toxin) is properly detoxified, titrations of the toxoid are made in the same way as for toxin, and the titer compared to that of the native toxin. To check the neturalizing capacity of an antitoxin, 10-fold dilutions of the antitoxin are made and mixed with a series of 10-fold dilutions of the toxin. Toxin dilutions without antitoxin are used as controls for comparison.

Additional Approaches

In the protocol described above the toxin is present on the cells continuously. An alternative or complementary type of assay involves exposure of the cells to the toxin in a short binding step (5–30 min at 4°, 22°, or 37°), whereon it is removed and the cells carefully rinsed. Fresh toxin-free medium is added and the development of the CPE followed. This gives additional information regarding the possible development of cytopathogenicity after a binding step. Furthermore, the efficiency of toxin binding to the cells at various temperatures and times is clarified. For example, in the case of *C. difficile* toxin B, contact with cells for a few seconds followed by thorough rinsing will suffice to give irreversible intoxication, indicating efficient binding to the cell surface, although the titer obtained in this modified assay is lower than on continuous exposure.[29] By contrast, certain membrane-damaging toxins are not strongly bound and will be diluted away from the cells on toxin removal and rinsing of the cells.[11]

A principally different and much more laborious approach will be to mix different dilutions of the toxin with cells in suspension, and then seed the mixtures to plates and score the behavior of the cells on incubation at 37°. Will the cells attach to and spread on the substrate? Provided the toxin receptors have not been removed by the trypsinization, this procedure will usually give a higher titer of the toxin, since cells in suspension are generally more susceptible to CPTs than already growing cells. With the exception of strongly cytolytic toxins this method gives no indication as to CPE until after at least 6 hr which is the approximate time it takes for most cells to attach and spread. However, information concerning a possible antiproliferative action (cytotoxicity) will result with this approach, although an estimate of antiproliferative action can be obtained only after 2–3 days. By that time cells in control cultures have doubled a few times, whereas cells treated with an antiproliferative toxin (at efficient doses) will not be able to multiply and thus show a much lower count.

For further characterization of a new CPT biochemical approaches are useful. Toxin-induced effects can be measured, such as inhibited macromolecular synthesis, activation/inhibition of specific enzymes, energy poisoning, changes of levels of second messenger molecules, leakage of cytosolic markers across a permeabilized plasma membrane, etc.[4]

[56] Use of Lipid Bilayer Membranes to Detect Pore Formation by Toxins

By BRUCE L. KAGAN and YURI SOKOLOV

Introduction

The development of planar lipid bilayer membranes by Mueller et al.[1] made it possible to study the properties of ion channels in an *in vitro* system. The bimolecular "artificial" membranes replicate the physiological environment of an ion channel (much as a buffer replicates the physiological environment of an enzyme) without many of the complicating cell-specific factors that affect channel function. Because of the physiological, electrical, and geometrical simplicity of the system, planar lipid bilayers provide an extremely sensitive assay for ion channels. Indeed, the opening and closing of a single ion channel can be readily detected.

The ability of lipids to form highly oriented structures at water–air interfaces has been used by Montal and Mueller[2] to improve this system significantly with the development of "solvent-free" bilayers. These membranes are thinner, have more stable capacitances, and are better targets for the reconstitution of channel-forming toxic proteins (White[3]; B. L. Kagan, unpublished observations). These membranes are also free from the "anesthetic" effects caused by the typical organic solvents used in the thicker Muller–Rudin membranes. Our laboratory and others have exploited these model membranes as an assay for channel formation by a number of soluble proteins, many of them toxins.[4-7] Planar lipid bilayers provide convenience, sensitivity, and straightforward interpretation of results. These characteristics have made them ideally suited to the detection of ion channels formed by toxins.

Membrane Formation

The formation of the bilayers is carried out in a Teflon chamber with two compartments separated by a thin (30 μm) Teflon or polypropylene,

---

[1] P. Mueller, D. O. Rudin, H. T. Tien, and W. C. Wescott, *Nature* (*London*) **194,** 979 (1962).

[2] M. Montal and P. Mueller, *Proc. Natl. Acad. Sci. U.S.A.* **69,** 3561 (1972).

[3] S. H. White, *in* "Ion Channel Reconstitution" (C. Miller, ed.), p. 3. Plenum, New York, 1986.

[4] J. J. Donovan, M. I. Simon, and M. Montal, *Nature* (*London*) **298,** 669 (1982).

[5] D. H. Hoch, M. Romero-Mira, B. E. Erlich, A. Finkelstein, B. R. DasGupta, and L. L. Simpson, *Proc. Natl. Acad. Sci. U.S.A.* **82,** 1692 (1985).

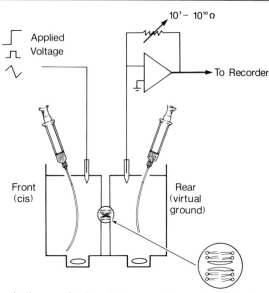

FIG. 1. Schematic diagram of a planar bilayer membrane setup. A basic setup for preparing lipid bilayers is shown. A Teflon chamber is divided by a thin partition into two aqueous compartments. The bilayer is formed in a small hole in the partition, and channel-forming proteins can then be inserted into the bilayer (see large circle). Magnetic fleas stir the aqueous solutions. Electrodes connect the solutions to the stimulating and recording electronics. Syringes are used to raise and lower the solutions.

polyethylene, Saran wrap, or any other similar material film with an aperture diameter of 50–200 μm. Each of the compartments is connected by plastic tubing to syringes filled with aqueous salt solutions (Fig. 1). Millimolar concentrations of divalent cations (e.g., Ca^{2+} or Mg^{2+}) improve membrane formation and stability. Initially, the level of solutions in both compartments is raised to just below the aperture. Then, a 1% solution of lipid (or mixture of lipids, if required) in a highly volatile organic solvent such as pentane, heptane, or hexane is carefully spread at the surfaces of the aqueous phases of both compartments. A small amount of squalene (usually 20 μl of a 1% solution in pentane) is spread at the partition between compartments. The squalene does not make up the bilayer proper, but rather conditions the partition surface to facilitate lipid contact.[3] After solvent evaporation (15–20 min), the bilayer is formed by a gentle sequential raising of the solution surfaces in both compartments to a level above the aperture. After raising the rear solution, bilayer formation usually

[6] B. L. Kagan, A. Finkelstein, and M. Colombini, *Proc. Natl. Acad. Sci. U.S.A.* **78,** 4950 (1981).
[7] B. L. Kagan, *Nature (London)* **302,** 709 (1983).

requires several raisings and lowerings of the front solution. Formation of the bilayer is verified by monitoring the electrical characteristics (capacitance and conductance).

To monitor membrane formation, a triangular wave or square pulse of 10–20 mV and 100 Hz frequency is used. At the time when the levels of solutions are below the aperture level, the capacitance response is very small. In raising the solution levels above the aperture, the capacitance increases when monolayers are opposed. The union of two monolayers into a bilayer is indicated by a sharp increase in capacitance. To measure conductance, a dc voltage of 100 mV is applied to the bilayer. Suitable membranes under these conditions should show a stable current of less than 1 pA (conductance of less than 10 pS).

Electrical Measurements in Bilayers

Electrodes

The measurement of the electrical parameters of lipid bilayers requires electrical connection of the membrane to the recording equipment. Silver–silver chloride electrodes can be purchased or fashioned by chloride plating a silver wire. Calomel electrodes can also be used. Electrodes must be "matched" to reduce electrode asymmetry and should exhibit good voltage stability. Salt bridges are used for measurements in conditions of asymmetric solutions (see below).

Recording Equipment

All the various electrical circuits for the measuring of electrical characteristics of lipid bilayers which have been described in the literature[8–10] have been based on the principles involved in the measurement of small currents and high resistances. For a minimal bilayer system, not all of these sophisticated circuits are necessary. Usually, an operational amplifier with suitable input parameters (Keithley 427, Cleveland, OH; Burr Brown 3523, Tucson, AZ; and similar) is used in the inverting mode, with the negative terminal connected to the "virtual ground" side of the chamber. High-quality resistors are used in the feedback loop (Fig. 1). A square pulse or triangle wave generator and oscilloscope may be employed to monitor membrane capacitance, usually only at the stage of bilayer formation. Typical capacitances are of the order of 0.8 μF/cm^2. The source of dc

[8] T. E. Andreoli, J. A. Bangham, and D. C. Tosteson, *J. Gen. Physiol.* **50,** 1729 (1967).

[9] V. K. Miyamoto and T. E. Thompson, *J. Colloid Interface Sci.* **25,** 16 (1965).

[10] O. Alvarez, *in* "Ion Channel Reconstitution" (C. Miller, ed.), p. 115. Plenum, New York, 1986.

voltage can be a battery with a voltage divider or a standard signal generator. A variety of chart and $X-Y$ recorders can fit the requirements for recording the current responses. For more extensive and sophisticated data acquisition, a digital tape recorder and video cassette recorder can allow recording of large amounts of data. An oscilloscope is useful for monitoring input voltage and output current and capacitance.

Physical Setup

Physically, the basic setup for electrical measurements in lipid bilayers consists of electrical equipment, electrodes, a low-power microscope with light source for monitoring bilayer formation, and a chamber with partitions and aperture. The whole setup must fit some general demands such as being free from mechanical vibration to increase bilayer stability and reduce electrical noise. Because the most vibration-sensitive element is the chamber, it is useful to minimize mechanical contact between the chamber and other equipment. The best results are produced using an air table for the chamber while other equipment is placed on another support. It is also possible to use amortizators and hard support (e.g., heavy steel plate on tennis balls). The electrical circuits and the chamber should be shielded (using a Faraday cage) to eliminate ac 60 Hz interference and the influence of static charges. The measurement of bilayer conductance requires elimination of conductive pathways between the two aqueous compartments except via the bilayer. Practically, it can be achieved by tight contact between compartments and by coating the contacting walls with petroleum jelly or silicone grease.

Quality Control

Most artifacts and difficulties in electrical measurements in lipid bilayers are of a chemical nature. Impure or aged lipids which tend to oxidize can create both difficulties in the formation of bilayers and the appearance of nonspecific membrane conductances. To avoid these problems, opened vials containing lipids should be stored under nitrogen at $-20°$. Prolonged storage of lipids should be avoided if possible, but can be done at $-70°$ if necessary. Poor water quality can create similar problems. Before beginning an experiment, the solutions should be filtered through detergent-free antibacterial filters (e.g., Millipore, Bedford, MA, 0.22 μm).

The same high demands apply to the cleanliness of the chamber and partition. One standard procedure for washing chamber and partition is a three-stage washing in distilled water (to remove salts and proteins), in petroleum ether (to remove petroleum jelly), and chloroform–methanol (2 : 1, v/v) to remove lipids; however, other chemicals like ethanol, sulfuric acid, and chromic acid can be used as well. Some laboratories advocate the use of ultrasonic cleaners to assist this process. We believe that detergents should not be used in the washing procedure, although some laboratories advocate the use of Joy detergent (Procter & Gamble, Cincinnati, OH).

Insertion of Toxins into Bilayers

At least two different methods have been used successfully to reconstitute channel-forming proteins into lipid bilayers. The first method uses the addition of protein to the electrolyte solution bathing the bilayer. This method has been used in many studies of bacterial toxins.[11-14] Despite the apparent simplicity of the method, investigators can encounter difficulties in the proper choice of conditions for reconstitution. In some cases, successful reconstitution depends on the presence of certain lipids, cholesterol, or specific toxin receptors.[15,16] On the other hand, some channel-forming toxins require special conditions in the aqueous solution. There might be a requirement for calcium ions, a certain value of pH, a pH gradient across the bilayer,[5,6] or addition of small concentrations of some detergents (such as octylglucoside.[11,17]

The second method of reconstitution is a fusion of channel-containing vesicles with lipid bilayers.[18] The general prerequisites of this method are the presence of negatively charged lipids in bilayer composition, millimolar amounts of Ca^{2+} in the aqueous solution,[19,20] and an osmotic gradient across the planar membrane.[21] In these experiments, the fusion rate can be controlled by the amount of negatively charged lipids and the Ca^{2+} concentration.

Single-Channel Records

Usually, the addition of small amounts of channel-forming toxins to the solution on one side of a bilayer results in a stepwise increase of the

[11] M. Deleers, N. Beugnier, P. Falmagne, V. Cabiaux, and J. M. Ruysschaert, *FEBS Lett.* **160,** 82 (1983).
[12] V. Davidson, K. R. Brunden, W. A. Cramer, and F. S. Cohen, *J. Membr. Biol.* **79,** 105 (1984).
[13] R. O. Blaustein, T. M. Koehler, R. J. Collier, and A. Finkelstein, *Proc. Natl. Acad. Sci. U.S.A.* **86,** 2209 (1989).
[14] B. L. Kagan, *Nature (London)* **302,** 709 (1983).
[15] H. H. Lah and P. Y. Law, *Annu. Rev. Pharmacol. Toxicol.* **20,** 201 (1980).
[16] M. T. Tosteson and D. C. Tosteson, *Nature (London)* **275,** 142 (1978).
[17] J. O. Bullock and F. S. Cohen, *Biochem. Biophys. Acta* **856,** 101 (1986).
[18] C. Miller and R. Racker, *J. Membr. Biol.* **30,** 283 (1976).
[19] C. Miller, *J. Membr. Biol.* **40,** 1 (1978).
[20] P. Labraca, R. Coronado, and C. Miller, *J. Gen. Physiol.* **76,** 397 (1980).
[21] J. Zimmerberg, F. S. Cohen, and A. Finkelstein, *Science* **210,** 906 (1980).

membrane current, and therefore conductance. The size of these steps depends on the nature of the toxin and the conditions of the experiment (salt concentration, voltage, pH, etc.) and ranges from less than 1 to over 100 pS.[7,22] In some cases, the channels remain open and incorporation of each new channel produces the next step in the record. In other cases, the channels demonstrate a tendency to open and close. These records provide much information about channel properties (see below) and are direct evidence of the channel-forming activity of the toxin under study.

Macroscopic Currents

Addition of high concentrations of the channel-forming toxin results in a large increase of the bilayer conductance. In this case the conductance steps corresponding to insertions of single channels usually cannot be distinguished. The current in these experiments can reach the values of nanoamperes and more. In some cases, the conductance reaches steady-state values within a relatively short time (20–30 min). In other cases, this time is much longer, and for stabilization of bilayer conductance it is necessary to perfuse the toxin-containing compartment with toxin-free solution to obtain a steady-state conductance.

Ion Selectivity

One measure of selectivity of ion channels is based on the measurements of ionic currents in the conditions of a salt gradient across the bilayer (see Eisenman and Horn,[23] for a fuller discussion of ion selectivity). The bilayer is formed in symmetric solutions, and the necessary amount of the toxin is added to one of the compartments. A large conductance (several orders of magnitude greater than the bare bilayer conductance) is desirable to obtain maximal accuracy. After the conductance of the bilayer reaches a steady-state level, the potential applied to the bilayer is set to zero. Correspondingly, the current across the bilayer becomes zero too. (A small offset current may exist at $V = 0$ owing to amplifier offset, electrode asymmetry, etc.). The voltage needed to reverse this current to zero should be measured and subtracted from the reversal potential measured in the next step. The necessary amount of concentrated salt solution is then added to one of the compartments to reach the required ion gradient across the bilayer.

The creation of a salt gradient usually results in a current $I = g(v - E)$ at $V = 0$, where E is the "reversal potential" for the channel in that

[22] S. J. Schein, B. L. Kagan, and A. Finkelstein, *Nature* (*London*) **276,** 159 (1978).
[23] G. Eisenman and R. Horn, *J. Membr. Biol.* **76,** 197 (1983).

gradient. The direction of the current determines the type of the channel selectivity (cation or anion). Then, the opposite polarity voltage is applied to compensate this current to zero. The value of this voltage (reversal potential) is used for quantitative calculation from the Goldman–Hodgkin–Katz equation[24] of cation–anion permeability of the channel. As it follows from the Nernst equation for ideal selectivity, for a 10-fold gradient of ionic activity across the bilayer at room temperature for monovalent ions, the value of the reversal potential is about 59 mV.

Channel Size

The size of the channel is a parameter of crucial interest for understanding the mechanism of entry of active toxin subunits into the cytosol. By using ions of different sizes and shapes as penetrating ions and measuring the relative single-channel conductance and reversal potentials, it may be possible to estimate the effective radius of the pore.[5,6]

Molecularity

Study of the dependence of bilayer conductance on the concentration of toxin in bath solutions can provide information about the molecular organization of the channel. When conductance is plotted as a function of concentration on log–log axes, the slope of the curve suggests the number of protein monomers participating in channel formation.

Interpretation of Results

Two major questions arise in the study of putative channel formation by bacterial toxins. First, is the observed channel real? Second, is the observed channel relevant to *in vivo* toxin action and, if so, how? In the sections which follow we demonstrate some approaches to answering these questions by using the examples of the colicins and diphtheria toxin, two well-established classes of channel-forming bacterial toxins. Although the approaches for the two toxins are instructive, it should be borne in mind that new toxins may subserve as yet unknown functions and that channel formation may play other roles in toxin action besides those described below.

Changes in Conductance

After adding minute or not-so-minute quantities of the toxin to a lipid bilayer, and observing changes in the conductance properties of the mem-

---

[24] B. Hille, "Ionic Channels of Excitable Membranes." Sinauer, Sunderland, Massachusetts, 1984.

brane, one must ask whether the observed conductance changes are real; that is, are they in fact due to channel formation by the toxin? An important prerequisite for answering this question is to have stable, quiet, reliable, high-resistance, and high-capacitance membranes. Careful monitoring of membrane formation by visual, conductance, and capacitance measures, as described above, is vital. Additionally, we recommend subjecting the membrane to test voltages of +100 or −100 mV as well as vigorous stirring for at least 10 min prior to adding toxin. This allows a stable baseline current to be measured and a chance for any impurities or instabilities to manifest themselves.

If changes in conductance are seen only after addition of toxin, this suggests that they are in fact due to the toxin itself. However, it is common for soluble proteins and peptides to interact with membranes. This interaction may be transient or long-lasting, and may depend on a variety of factors such as solution composition, lipid composition, and protein concentration. Such interactions may be electrically silent or may induce changes in bilayer conductance without forming true ion channels. Although no foolproof formula can be given for distinguishing nonspecific protein–lipid interactions from channel formation, the following guidelines should prove helpful.

Stability. Ion channels are generally thought to be permanent protein pathways through the lipid membrane. Channel-forming toxins generally form channels which are stable in the bilayer and last for minutes to hours. Although the open time of a channel may be much shorter than this, the channel itself is usually long-lasting. Transient phenomena must be regarded more suspiciously.

Homogeneity. Channels are usually identified by a fingerprint, namely, the single-channel conductance. Under comparable conditions of salt, lipid, pH, voltage, etc., this quantity is usually invariant for a given toxin. Single-channel conductance measurements are usually distributed in a Gaussian manner around a unitary mean conductance value. For example, the various members of the colicin family (A, E-1, I-A, I-B) exhibit unique single-channel conductances.[6] Occasionally a toxin may exhibit more than one single-channel conductance (e.g., yeast killer toxin 180, 220, pS). This may represent different states of a single species or multiple species. More rarely, multiples of a single value are seen, for example, 4, 8, 12, and 16 pS; this may represent oligomerization of a monomeric toxin such as has been described for diphtheria toxin (Romero-Mira[25]; W. Yuan and B. L. Kagan, unpublished results).

[25] M. Romero-Mira, Ph.D. Thesis, Albert Einstein College of Medicine, Bronx, New York (1989).

Reproducibility. As with all biological measurements, variability exists, but the behavior of a true ion channel should be reproducible with respect to single-channel conductance, ionic selectivity, voltage dependence, etc. Common artifacts, such as oxidized lipid molecules, denatured protein molecules, or contamination, should prove correctable with appropriate attention to technique and cleanliness as mentioned above.

Size. Although a wide range of channel conductances have been reported for channel-forming toxins, the usual range is 1–1000 pS. Conductance changes outside this range may be due to artifactual causes.

Selectivity. Leakage artifacts are typically nonselective although they can be selective, depending on conditions. Nonspecific leaks in negatively charged membranes, for example, will tend to exhibit cation selectivity although ideal selectivity is unusual. Selectivity measurements should be stable and reproducible, which is not likely to be the case with a lipid-induced leak.

Voltage Dependence. Although many ionic channels are not voltage-dependent, the presence of this property can suggest a functional relevance that may be significant. For example, the membrane-active colicins (A, E-1, I-A, I-B) exhibit strongly voltage-dependent behavior, slowly opening at cis-positive voltages and rapidly closing at zero, or negative, voltages.[22] This accords with the experimental observation that colicins only kill their target cells when the target maintains a membrane potential (inside negative). Treatment of target cells with uncouplers [e.g., dinitrophenol (DNP), carbonyl cyanide 3-chlorophenylhydrazone CCCP] or other agents which prevent the development of a membrane potential across the bacterial inner membrane protects the cells from the actions of these colicins.

Controls. Because of the ever-present possibility of artifacts, it is important to perform relevant control experiments. As mentioned above, membranes must be high capacitance, high resistance, and stable to voltages of at least 100 mV before addition of a toxin. Choosing appropriate controls depends on the individual toxin. An inactive or denatured form of the toxin is frequently a good choice. For example, diphtheria toxoid, a denatured form of diphtheria toxin, is inactive in the lipid bilayer system, even at concentrations 1000-fold higher than those required for diphtheria toxin to form channels. Nicking of diphtheria toxin is a first step in separating the A and B subunits and is thought to be an early step in the activation of diphtheria toxin. Unnicked diphtheria toxin is much less active (and, possibly, inactive) in forming channels in the lipid bilayer system. An unprocessed form of toxin or protoxin may, thus, be activated by endogenous factors *in vivo* but may be without activity in the bilayer system owing to the lack of these factors (for example, proteases).

Mutant forms of toxin constitute another kind of control. Nontoxic mutants might be expected to be inactive in the bilayer assay, but the case of diphtheria toxin (DT) shows that this is not always so. Many toxins have more than one functional domain (e.g., diphtheria toxin has an enzymatic domain, a receptor-binding domain, and a translocation domain). Mutations may affect one or several of these domains. Thus, the DT mutant CRM-45, which lacks the C terminal receptor-binding domain of DT, is active in the bilayer assay.[6] The mutant CRM-197, which has a mutation in the active site of the enzymatic domain of DT, is also active in the bilayer assay, although both of the mutants are nontoxic to cells. However, mutants in the channel-forming domain of DT appear to be inactive in the bilayer assay.[26]

Lipid Requirements. Although many toxins can form channels in a wide variety of lipids, there are reports of requirements for specific lipid components such as sterols (amphotericin B), phosphatidylinositol (DT, Donovan *et al.*[4]), and gangliosides (cholera toxin, Tosteson and Tosteson[16]).These lipids may enhance activity or be absolutely required for activity, and they may even be required in a specific leaflet of the bilayer.[4]

Sources of Artifacts. This section could fill volumes, but we mention only the most common. Oxidized lipids are a frequent source of leakiness and membrane instability, and they can even result in lipid channels. Care must be taken in the preparation and storage of lipids to exclude oxygen and water, to prevent oxidation. Use of antioxidants as stabilizers may also help. Lipid channels may also be seen when lipids undergo phase transitions or when lipids prone to forming hexagonal phases are used. Care must be taken as to the particular mixture used and the working temperature to avoid these problems. Proteins can also interact with bilayers nonspecifically, and cause extraneous conductances or channels. The purity of the toxin preparation is critical to valid assessment of the channel-forming properties of the toxin. Phospholipases, a common activity in toxin preparations, must be vigorously excluded as these will cause bilayer leakage and degeneration. Other surface-active proteins such as lipoproteins are also common sources of artifacts. While sterile or ultra-clean conditions are not required for bilayer preparation, it is certainly true that common dirt can contribute to lipid bilayer (and, later, investigator) instability.

Relevance of Channel to Toxin

Single-Channel Conductance. The electrical conductance of a single toxin channel indicates its permeability to ions in the aqueous phase.

[26] V. Cabiaux, J. Mindel, and R. J. Collier, *Biophys. J.* **61**, A211 (1992).

Although this measurement does not denote the physical dimensions of the pore (e.g., the DT channel has a diameter of at least 18 Å but a single-channel conductance of only 10 pS in 0.1 M KCl, whereas the calcium-activated potassium channel has a diameter of about 4 Å and a single-channel conductance of about 300 pS), it does give a measure of ion permeability that may be relevant for channel function *in vivo*. For example, colicin A channels with a single-channel conductance of about 5 pS create a significant ionic leak in bacterial cells (diameter about 1 μm). A yeast cell, diameter about 10 μm, would be unaffected by such a small leak of potassium, but killer toxins of yeast have a single-channel conductance of about 200 pS, which makes a significant leak in this larger cell.[7] Thus, the single-channel conductance can be compared to the physiological disruption seen *in vivo* to judge whether channel formation can plausibly account for the disturbance.

The physical size of the channel can also be measured by using progressively larger ions in ionic selectivity experiments,[5] by using nonelectrolytes and measuring the effect of osmotic gradients on vesicles doped with the toxin channel to be tested,[6] or by using radioactive marker fluxes in vesicles.[27] The physical size can give clues to the role of the channel. In DT, the exceptionally large diameter of the pore lends credence to the hypothesis that the pore plays a role in protein translocation across the membrane[6] as the widest dimension of the extended polypeptide chain of fragment A (the enzymatic toxic chain of DT) is about 16 Å. Wide channels may also allow a range of cellular constituents to leak out, contributing to cytotoxicity.

Ion Selectivity. For the most part, the toxin channels characterized to date have been relatively nonselective when compared to the exquisite selectivity exhibited by the sodium or potassium channels of nerve and muscle. It is likely that this lack of selectivity is functionally relevant to the role of channel formation in cytotoxicity. Although a high degree of ionic specificity is necessary for the channels of excitable membranes to perform their functional roles, a cell-killing channel has a much lesser need for high degree of selectivity. Indeed, a nonselective channel may be more capable of inflicting damage to a target cell since ionic fluxes may include a wide range of critical intracellular ions (e.g., K^+, Mg^{2+}, Cl^-). Nonselectivity also allows counterions to flow, thus facilitating leakage.

It is important to consider ionic selectivity in light of the cellular pathophysiology observed. If a toxin, such as colicin A, causes efflux of intracellular potassium, it is reasonable to expect a significant permeability

[27] L. S. Zalman and B. J. Wisnieski, *Proc. Natl. Acad. Sci. U.S.A.* **81,** 3341 (1984).

of the channel to potassium. Although this is by no means the only explanation for the *in vivo* observed potassium efflux, it is by far the simplest.

Whereas a channel may be permeable to a wide variety of ions, fluxes of these ions may not necessarily be seen *in vivo*. Observation of fluxes will depend on the transmembrane gradient of the particular ion, the membrane voltage, the ability of counterions to cross the membrane, the baseline permeability of the target membrane to the particular ion, and the concentration of the ion. For example, although it is virtually certain that protons are permeable to the colicin A channel, proton efflux is not normally observed in bacterial efflux experiments because of the low concentrations of protons in the medium and in the cell itself (10^{-7} M).

Pathophysiology. A clear knowledge of the effects of the toxin on cell physiology is vital to understanding the role of the channel in toxin action. As an example, the role of colicin A channels in disrupting bacterial energy metabolism could not be fathomed until the Mitchell hypothesis made it clear that bacterial energy metabolism required a transmembrane proton gradient ($\Delta\mu_{H^+}$). Once this was understood, it became clear that channels could disrupt the transmembrane ion gradient and lead to cell death.

The diphtheria toxin (DT) cellular pathophysiology led to a very different understanding of the role of the channel in cytotoxicity. The requirement of low pH (<5.0) for channel activity suggested that DT entered the cell through a low-pH endosomal compartment; this led to the idea that the channel did not play a role in leaking vital ions out of the cell (since channel formation only took place in an internal membrane) but, rather, that the channel had a hand in the entry of the toxic enzymatic A fragment of DT into the cell. Sizing the physical dimensions of the channel with nonelectrolytes lent additional strength to this idea.

Kinetics. The time course of cellular intoxication can be difficult to interpret. While channel-forming activity in a bilayer typically requires only seconds to minutes (usually limited by diffusion across the aqueous unstirred layer bounding the membrane), toxin channel formation *in vivo* may require much longer periods (e.g., hours to days in the case of DT where the toxin must be bound and then endocytosed into an acidic compartment). In the case of the colicins, the *in vivo* time course is changed by the need for colicins to bind to receptors in the outer membrane and then traverse this membrane to gain access to the inner membrane; thus, agents which slow this process, such as low temperature or metabolic inhibitors, will also delay the appearance of colicin channels in the target inner membrane. In spite of these difficulties, the single-hit kinetics of colicin action suggested that a single molecule might be capable of killing a cell and, thus, led to important insights about colicin action and the molecularity of the colicin channel.

Voltage Dependence. Voltage dependence, the ability of channels to open and close in response to transmembrane voltages, can be a defining characteristic of an ion channel. Colicin channels, which open only at cis-positive transmembrane voltages (i.e., when the side of the membrane containing colicin is made positive relative to the other side), exhibit a voltage dependence which appears to be highly relevant to colicin physiology. It is well-known that uncouplers, metabolic inhibitors, or other treatments which eliminated the bacterial transmembrane proton gradient and, thus, the transmembrane voltage, could protect target cells from colicin action. This protection is highly consistent with the *in vitro* voltage-dependent opening and closing of colicin channels.

For DT, the much more complex voltage-dependent behavior of the channel bears no clear relationship to the functional role of the channel. This may be due to our lack of knowledge of the voltage gradient existing across the endosomal membrane where the DT channel inserts.

Lipid Dependence. Although lipid bilayers can be formed from a wide variety of natural and synthetic lipids, we usually choose lipids for their stability and fluidity, which favors protein insertion. Attempts to mimic the target cell membrane lipid composition, while logical, may not always provide the best conditions for assaying toxin channel activity. Colicin A, for example, is nearly an order of magnitude more active in membranes formed from asolectin, a highly fluid soybean lipid, than in membranes formed from *Escherichia coli* phospholipids. This paradox may reflect the fact that the *in vitro* bilayer has no assistance from receptors or cellular machinery to help the toxin channel insert. Thus, the insertion process *in vivo* may differ from that observed *in vitro,* even though the resultant channel is essentially the same.

The presence or absence of cholesterol, sphingomyelin, phosphatidylinositol, or other special lipids should be considered when choosing a lipid mixture. Because protein toxins are inherently destabilizing to these bilayers, a balance must be struck between fluidity and stability. The most stable membranes, for example, those formed from diphytanoylphosphatidylcholine (Avanti Polar Lipids, Birmingham, AL), may be impenetrable to the typical toxins such as colicin or DT (B. L. Kagan, unpublished observations).

Direct and Indirect Cytotoxicity. At least two general modes of channel-forming toxic action are now known, the direct and the indirect. The colicins, yeast killer toxin, *Staphylococcal aureus* toxin, etc., are directly toxic channels. The channel causes cytotoxicity directly, either by leakage of vital intracellular ions, influx of toxic ions, or disruption of ion gradients required for transport or energy production. With indirect toxins, such as diphtheria, tetanus, botulinum, *Pseudomonas,* or anthrax toxins,[5,6,13] the channel plays a role in facilitating entry of another toxic part of the molecule, perhaps by forming a protein tunnel through which the toxic molecule can enter the cell.

Despite these functional differences, both types of channel-forming toxins can be detected using lipid bilayer methods. Intriguingly, the properties of the two classes of channels seem to differ. The direct toxins tend to have a larger single-channel conductance and to show less voltage dependence; the indirect toxins also, typically, have a significant pH dependence, perhaps reflecting their entry into cells via an acidic endosomal compartment. Both classes of channels tend to be rather nonselective among ions and to be fairly large in physical dimension. These features are consistent with the role of these channels in cell killing or in toxin translocation as opposed to the narrow diameter, highly selective channels of excitable nerve and muscle membranes.

Indirectly toxic channels serve to facilitate the translocations and cell entry of a toxic part of the toxin molecule such as an ADP-ribosylase. Directly toxic channels kill cells by altering the permeability of cell membranes to ions and other molecules. It has not been clearly established what specific alterations in membrane permeability are cytotoxic. Rescue experiments with colicins showed that plating colicin-treated cells in media fortified with high potassium and magnesium concentrations could greatly reduce toxicity. This suggested that leakage of potassium and magnesium was important in colicin channel killing, although the rescue was incomplete, hinting that other mechanisms were also at work. Influx of toxic ions such as calcium could be important, and depolarization of the cell membrane which, for bacteria, is a vital energy transducing organelle could be alternate mechanisms of channel-induced pathogenicity. Further *in vivo* experiments are needed to delineate the specific mechanisms more precisely.

Concentrations. Planar lipid bilayer membranes are extremely inefficient assays for ionic channels. Typically, fewer than 1 molecule in 10^5 enters the bilayer to form a channel. This low efficiency is a result of the relatively small area of the bilayer, the relatively large volume of solution, the relatively large volume of monolayer, the geometry of the chamber, and the efficiency of insertion into the bilayer. The power of the technique lies with its sensitivity. A single toxin molecule channel can be detected using simple electronic measuring devices. These functions also tend to make the bilayer highly variable as a quantitative assay. The system is excellent for assessing the properties of a channel such as voltage dependence, selectivity, or single-channel conductance, but it is rather imprecise for determining the channel-forming activity of a sample. In addition to the factors cited above, other sources of variability include bilayer area, microlenses of solvent, bilayer position within the partition, and lipid purity and age. Although some channels are active below 1 ng/ml others require concentrations greater than 50 μg/ml. These activity differences may reflect difficulty of insertion and diffusion through unstirred layers as well as some of the factors mentioned above.

Receptors. In most of the channel-forming toxin studies performed to date, there have not been specific receptors present in the bilayer. Although reports of specific receptor requirements exist (e.g., ganglioside G_{M1} for cholera toxin), the vast majority of studies have shown that channel-forming toxins are at home in a wide variety of lipid environments. Occasionally a specific lipid has been shown to markedly enhance channel-forming activity such as in the case of phosphatidylinositol and diphtheria toxin, but it may not be an absolute requirement for activity. Although proteinaceous receptors can be incorporated into bilayers, they decrease bilayer stability and may cause artifactual conductances which can interfere with detection of the toxin-induced conductance. Still, with the appropriate controls, receptors may enhance the likelihood of detecting channel formation.

Acknowledgments

This work was supported by grants from the U.S. Department of Veterans Affairs and the National Institute of Mental Health (MH43433).

[57] Uptake and Processing of Toxins by Mammalian Cells

By CATHARINE B. SAELINGER and RANDAL E. MORRIS

Introduction

A variety of bacterial and plant toxins, including *Pseudomonas* exotoxin A (PE), diphtheria toxin, *Shigella* toxin, cholera toxin, abrin, and ricin, kill mammalian cells by delivering an enzymatically active toxin fragment to the cell cytosol which then inactivates the appropriate substrate.[1] Many of the toxins initially bind to receptors on the cell surface, are internalized by endocytosis, enter the cell cytosol, and there block an essential cell function. These toxins have separate domains or fragments involved in cell binding, translocation, and enzymatic activity. To under-

---

[1] C. B. Saelinger, *in* "Trafficking of Bacterial Toxins" (C. B. Saelinger, ed.), p. 1. CRC Press, Boca Raton, Florida, 1990.

stand the action of these toxins at the cellular level, it is necessary to have methods to follow their internalization and subsequent intracellular movement.

Several approaches can be taken to monitor toxin internalization. One can use electron microscopy to detect toxin, at the ultrastructural level, within a limited population of cells. Alternatively, one can follow populations of toxin molecules; this can be done using labeled toxins and method ologies routinely used to trace the intracellular movement of biologically relevant ligands, or by altering conditions of the experiment and determining the effect of these changes on expression of toxicity. The work to be discussed deals primarily with the internalization of *Pseudomonas* exotoxin A by sensitive cells in culture, but it is applicable to the study of most toxic molecules.

Ultrastructural Studies

The intracellular trafficking of proteins in mammalian cells after receptor-mediated endocytosis has been carefully scrutinized in many systems. To visualize this movement at the ultrastructural level, one needs a method to localize the protein of interest. We describe here a biotinyl ligand : streptavidin–gold technique which measures binding of toxin to the cell surface and subsequent internalization of toxin into intracellular organelles, via conventional electron microscopy.

Preparation of Streptavidin–Gold Colloids

Gold sols of varying sizes can be prepared by the reduction of gold chloride.[2,3] We routinely use avidin–gold colloids between 5 and 7 nm in the biotinyl ligand : avidin–gold method, as larger size colloids do not bind biotinyl ligands.

The method used to produce 5 nm gold sols by the reduction of gold chloride with white phosphorus has been described in detail.[3] In brief, a gold chloride solution (240 ml of doubly distilled water, 5.4 ml of 0.1 M K_2CO_3, and 6 ml of 0.5% gold chloride solution) is mixed with 2 ml of a white phosphorus–ether solution (1.6 ml diethyl ether and 0.4 ml white phosphorus-saturated diethyl ether) for 30 min at 23°. During this time the solution turns a rust-brown color. The reaction is driven to completion by refluxing for 5–20 min, or until the solution becomes a deep red color. The gold sols are then cooled to 23°, the pH is adjusted to 7.5, and the sols are stabilized with avidin, that is, are converted to gold colloids, by

[2] R. E. Morris and C. B. Saelinger, *J. Microsc.* **143**, 171 (1986).
[3] R. E. Morris and C. B. Saelinger, this series, Vol. 184, p. 379.

the adsorption of streptavidin onto the sols. Succinylated egg white avidin can be used in place of streptavidin.[3]

Prior to absorption, streptavidin (Sigma Chemical Co., St. Louis, MO) is resuspended to a concentration of 1 mg/ml in distilled water (10 ml total volume) and is dialyzed against four 1-liter changes of 5 mM phosphate buffer, pH 7.5 (Na$_2$HPO$_4$ · 7H$_2$O, 1.14 g/liter; NaH$_2$PO$_4$, 0.08 g/liter) at 4° for 72 hr. After dialysis the protein solution is centrifuged at 1500 g for 10 min and then filter-sterilized through an 0.22-μm filter.

The amount of protein required to stabilize a given volume of sols is determined by titration.[3,4] After stabilization, the streptavidin–gold is allowed to sit in the cold for 72 hr. The colloids are collected by centrifugation at 18,000 g for 3 hr at 4°. The resultant soft pellet is resuspended in 5 mM phosphate buffer (to one-half the original volume), then diluted to its original volume with 0.1 M Tris-HCl buffer, pH 7.5, containing 0.3 M NaCl and polyethylene glycol (molecular weight 20,000; 0.4 mg/ml). Prior to storage all colloidal suspensions are filter-sterilized (0.22-μm filter).

Biotinylation of Proteins

Biotin, a 244 Da vitamin, is covalently bonded to proteins via the ε-amino group of lysine residues by the following procedure.[3] Prior to biotinylation, the protein is dialyzed againt 50 mM borate buffer, pH 9.2 (sodium tetraborate, 19.02 g/liter). Sufficient sulfosuccinimidyl-6-(biotinamido)hexanoate (Pierce Chemical Co., Rockford, IL) is added to effect a 5 : 1 molar ratio (5 mol biotinylating reagent to 1 mol protein). The biotinylating reagent is water soluble; thus, it may be added directly to the protein solution. The reaction mixture is incubated at 23° for 4 hr, after which the biotinylated protein is dialyzed against multiple changes of phosphate-buffered saline (PBS), pH 7.4 (NaCl, 8.01 g/liter; Na$_2$HPO$_4$ · 7H$_2$O, 2.14 g/liter; KH$_2$PO$_4$, 0.27 g/liter; KCl, 0.22 g/liter), at 4°. This procedure typically results in the incorporation of 2 mol of biotin per mole of protein. In our experience this has proved to be a very gentle procedure which does not alter the biological activity of the protein.

Endocytosis of Biotinyl Ligands

Internalization of biotinyl toxin–avidin–gold by cell monolayers is as follows.[3,5]

1. Cell monolayers are grown to near confluency in 35-mm plastic petri dishes.

[4] W. D. Geoghegan and G. A. Ackerman, *J. Histochem. Cytochem.* **11**, 1187 (1977).
[5] R. E. Morris and C. B. Saelinger, *J. Histochem. Cytochem.* **32**, 124 (1984).

2. To remove exogenous biotin, monolayers are washed several times with cold Hanks' balanced salt solution containing 1% bovine serum albumin (HBSS/BSA), and then incubated in HBSS/BSA for 2 hr at 37°.

3. Monolayers are cooled to 4° for 20 min.

4. Two milliliters of precooled biotinyl ligand, adjusted to the appropriate concentration, is added to each plate. Routinely a saturating concentration of toxin is used (e.g., biotinyl PE, 100 ng/ml). Control samples include plates with no ligand and plates with biotinyl ligand in the presence of 200 molar excess of native ligand (to show competition binding). Ligand is allowed to bind to cells for 1 hr at 4°

5. Wash monolayers three times with cold HBSS/BSA.

6. Two milliliters of HBSS/BSA containing streptavidin–gold (diluted 1 : 10) is added. The colloid is allowed to bind for 30 min at 4°.

7. All samples are washed three times with cold HBSS/BSA.

8. The control samples and the zero-time sample are washed an additional three times with cold HBSS containing 5% sucrose (HBSS/sucrose), and then fixed for 30 min at 4° with HBSS/sucrose containing 1.5% (v/v) glutaraldehyde.

9. Two milliliters of warm HBSS/BSA is added to all other samples to initiate internalization.

10. After appropriate warming at 37° (5 sec to 120 min), endocytosis is stopped by washing the cells three times with cold HBSS/sucrose, and the samples are fixed as described in Step 8.

11. Wash all samples three times in cold 0.1 M sodium cacodylate buffer, pH 7.4, containing 7.5% (w/v) sucrose (SCB).

12. Fix for 30 min at 4° with SCB containing 4.0% (v/v) paraformaldehyde and 2.5% (v/v) glutaraldehyde.

13. Postfix samples for 1 hr at 4° in SCB containing 1% (v/v) osmium tetroxide which has been reduced with 1.5% (w/v) potassium ferrocyanide.

14. Repeat step 11.

15. Wash all samples three times with distilled, deionized water at 23°.

16. Scrape samples from the petri dishes.

17. Embed in 1% (w/v) ultralow temperature gelling agarose (Sigma) in plastic microcentrifuge tubes.

18. Allow cells to settle to bottom of the microcentrifuge tube by gravity. If they do not settle within 1 hr at 23°, pellet by low-speed centrifugation. Avoid packing the cells too tightly.

19. Allow the agarose to solidify overnight at 4°.

20. Remove plugs and dice into 1-mm³ cubes.

21. Begin dehydration by three washes with 70% (v/v) ethanol.

22. Strain *en bloc* for 10 min at 23° with 0.5% uranyl acetate in 70% (v/v) ethanol.

23. Continue dehydration with 3 washes with 70% ethanol (v/v) followed by three 5-min incubations in 100% ethanol.
24. Infiltrate and embed by standard procedures using epoxy resin.[3]

Owing to significant concentrations of biotin in sera, the above procedure can be used only under *in vitro* conditions. In addition, medium containing biotin or serum cannot be used during experimentation.

One of the advantages of the biotinyl ligand : avidin–gold methodology is that specific binding can be demonstrated by a competition assay.[5] This is done by including a sample with a 100- to 200-fold excess of nonbiotinylated, native ligand during the primary incubation step (Step 4) with the biotinyl toxin.

If needed, endosomes or lysosomes can be identified by preincubation of cell monolayers with horseradish peroxidase (HRP; 1 mg/ml in HBSS/ BSA) prior to initiation of the experiment. To identify endosomes, incubation is overnight at 15°. To identify lysosomes, incubation is at 37° for 60 min followed by an additional 60-min incubation at 37° in medium without HRP. To localize the horseradish peroxidase, prior to Step 11 above, samples are washed three times at 23° with 50 mM Tris, pH 7.4, containing 7.5% sucrose (Tris/sucrose). Two milliliters of Tris/sucrose containing diaminobenzidine (0.5 mg/ml) and H_2O_2 (0.01%) is added for 20 min at 23°. This causes the development of an electron-dense precipitate.[6]

Comments. The biotinyl ligand : avidin–gold method has been used to follow the intracellular routing of several different proteins.[5,7–9] We have shown that the intracellular movement of biotinyl ligands is not altered by the attachment of streptavidin–gold colloids, if the electron-dense marker is added *after* biotinyl toxin has interacted with the cell surface receptor.[10]

We have followed the binding, internalization, and routing of three different biotinyl ligands in mouse LM fibroblasts.[10] The ligands used were biotinyl choleragenoid (B fragment of cholera toxin), biotinyl wheat germ agglutinin, and biotinyl-PE. The ligands were chosen because each binds to a different cell surface moiety. The plant lectin wheat germ agglutinin binds to D-*N*-acetylglucosamine and sialic acid residues,[11] choleragenoid

[6] R. C. Graham and M. J. Karnovsky, *J. Histochem. Cytochem.* **14,** 291 (1966).
[7] B. van Deurs, K. Sandvig, O. W. Petersen, S. Olsnes, and K. Simons, *J. Biol. Chem.* **106,** 253 (1988).
[8] R. E. Morris, A. S. Gerstein, P. F. Bonventre, and C. B. Saelinger, *Infect. Immun.* **50,** 721 (1985).
[9] K. Sandvig, K. Prydz, M. Ryd, and B. van Deurs, *J. Biol. Chem.* **113,** 553 (1991).
[10] R. E. Morris, G. M. Ciraolo, and C. B. Saelinger, *J. Histochem. Cytochem.* **40,** 711 (1992).
[11] V. P. Bharanandan and A. W. Katlic, *J. Biol. Chem.* **254,** 4000 (1979).

binds to the glycolipid G_{M1},[12] and PE binds to a glycoprotein.[13,14] All three ligands show distinct patterns of endocytosis and intracellular movement. Thus, we conclude that the biotinyl toxin : avidin–gold technique represents a valid method to follow binding, entry, and subsequent intracellular routing of ligands.[10]

Biochemical Methods

Several biochemical approaches have been used to examine steps involved in expression of toxicity. These approaches include the use of agents which alter toxicity, comparing the intracellular routing of toxin in sensitive and resistant cells, and determining the role of toxin processing in expression of toxicity.

Modification of Toxicity

The advantage of working with toxins over other protein ligands is that many toxins have easily measured biological activities, such as inhibition of protein synthesis. It is possible to manipulate cells in a variety of ways, using techniques which are known to interfere with normal intracellular movement of ligands, and then determine the effect of these manipulations on expression of biological activity. This approach has been applied to a variety of toxins, including ricin, PE, diphtheria, shiga, and anthrax toxins.[15–18]

While numerous cell lines are susceptible to the action of PE, mouse LM fibroblasts (ATCC, Rockville, MD, CCL 1.2) appear to be the most sensitive.[19] Biological activity is determined by measuring the ability of PE to inhibit the incorporation of radiolabeled amino acids into acid-insoluble protein. That concentration of toxin which inhibits protein synthesis by 50%, relative to untreated controls, is termed the TCD_{50}.[20] A typical TCD_{50} assay for PE follows.

[12] P. H. Fishman, *in* "ADP-Ribosylating Toxins and G Proteins: Insights into Signal Transduction" (J. Moss and M. Vaughan, eds.), p. 127. American Society for Microbiology, Washington, D.C., 1990.

[13] J. J. Forristal, M. R. Thompson, R. E. Morris, and C. B. Saelinger, *Infect. Immun.* **59**, 2880 (1991).

[14] M. R. Thompson, J. J. Forristal, P. Kauffmann, T. Madden, K. Kozak, R. E. Morris, and C. B. Saelinger, *J. Biol. Chem.* **266**, 2390 (1991).

[15] M. H. Marnell, S.-P. Shia, M. Stookey, and R. K. Draper, *Infect. Immun.* **44**, 145 (1984).

[16] K. Sandvig and S. Olsnes, *J. Biol. Chem.* **257**, 7504 (1982).

[17] V. M. Gordon, S. H. Leppla, and E. L. Hewlett, *Infect. Immun.* **56**, 1066 (1988).

[18] R. E. Morris and C. B. Saelinger, *Infect. Immun.* **52**, 445 (1986).

[19] J. L. Middlebrook and R. B. Dorland, *Can. J. Microbiol.* **23**, 183 (1977).

[20] C. B. Saelinger, this series, Vol. 165, p. 226.

1. Seed cells in 24-well tissue culture dishes at a concentration of approximately 5×10^5 cells per well, and incubate for 18 hr at 37° until cells reach near confluency.

2. Prior to use, wash monolayers one time with complete tissue culture medium and cool to 4° for 15 min.

3. Add 1 ml of cold growth medium containing between 10^{-8} and 10^{-10} M toxin. Control wells receive medium with no toxin. All dilutions are assayed in triplicate. Incubate toxin with cells for 60 min at 4°.

4. Remove the toxin-containing medium, wash one time with HBSS, and reincubate the cells in fresh medium for 5 hr at 37°.

5. Remove the medium and replace with medium containing 1/20 the normal amount of L-leucine and 2 μCi/ml L-[^{3}H]leucine. Incubate at 37° for 60 min.

6. Wash monolayers three times in phosphate-buffered saline, pH 7.2, and dissolve cells in 1.0 ml of 0.1 N NaOH. After 5 min at room temperature, collect the NaOH in centrifuge tubes, precipitate the protein with 1.5 ml of cold 12% (w/v) trichloroacetic acid (TCA), and allow to sit on ice for at least 60 min.

7. For processing, wash the acid-insoluble precipitate two times in 6% TCA (3 min, 1500 g) and digest in 1.5 ml of 0.1 N NaOH for 30 min at 56°. Assay one aliquot for protein and another by liquid scintillation.

8. Data are expressed as incorporation of [^{3}H]leucine in counts per minute (cpm) per microgram of protein, or as percentage of inhibition of protein synthesis in toxin-treated cells as compared to no-toxin control cells.

9. For modifications of the assay, if cell lines of reduced toxin sensitivity are used, it is necessary to extend the length of time of incubation at 37° and/or to incubate cells continuously in the presence of toxin.

Several toxins enter cells by receptor-mediated endocytosis and subsequently move to endosomes. Steps in this pathway can be altered by several procedures.[18,21] Cells can be treated with a protease (e.g., trypsin) to remove surface proteins or with antitoxin to neutralize surface-associated toxin. Because endosomes are mildly acidic in nature, incubation with acidotropic agents (e.g., 20 mM methylamine) will alter their function. Incubation of cells at reduced temperatures (below 19°) will slow endosome fusion with lysosomes or with Golgi-associated vesicles.[22] The effect of these treatments on TCD$_{50}$ values is determined.

[21] M. H. Marnell, S.-P. Shia, M. Stookey, and R. K. Draper, *Infect. Immun.* **44,** 145 (1984).
[22] W. A. Dunn, A. L. Hubbard, and N. N. Aronson, Jr., *J. Biol. Chem.* **255,** 5971 (1980).

In addition, these modifications can be used to follow the movement of toxin in cells. In these experiments, toxin is bound to cells at 4°, cells are warmed to 37° to initiate entry, and then at various times antitoxin (to neutralize surface-associated toxin) or methylamine (to alter endosomal function) is added, or cells are shifted to 19° (to block fusion with lysosomes). Incubation is continued for up to 5 hr, and then protein synthesis is measured. The earliest time point at which a procedure does not fully protect cells against the action of toxin suggests that, by this time, the toxin has moved to a step in its intracellular journey beyond which the agent acts.[18,21]

Comments. Cytotoxicity assays measure an end point in toxin action, for example, inhibition of protein synthesis. They depend on successful completion of a series of steps (binding, entry, inactivation of substrate) and measure only a population of toxin molecules. When a sensitive cell line is used these assays are consistent, easy to run, and allow an easy monitoring of toxin action. When using different protocols to alter the action of toxin, it must be kept in mind that the agents/procedures may act at more than one site.

Internalization

To follow internalization, a method must be available to detect small quantities of protein. If the toxin can be radiolabeled without altering biological activity, then methods used to determine binding and internalization of physiologically relevant ligands can be applied to toxins. The standard method of labeling proteins is by iodination. In these experiments, iodinated toxin is incubated with cell monolayers for 1 hr in the presence or absence of 100-fold excess native toxin. Monolayers are washed extensively to remove unbound toxin, and samples are taken at various time points (e.g., 0–60 min) after warming cells to 37°. Cells are washed, solubilized with 0.1 N NaOH, and collected, and aliquots are counted in a γ counter.[23] This method has been used to determine the rate of internalization of several toxins including diphtheria toxin, ricin, and shiga toxin. This methodology is rapid, allows differentiation of specific and nonspecifically bound toxin, and gives an estimate of the amount of toxin located intracellularly and of the kinetics of toxin degradation.

In some instances iodination alters the biological and/or binding properties of the protein. One way to circumvent this problem is to use native

---

[23] R. B. Dorland, *Can. J. Microbiol.* **28**, 611 (1982).

toxin and identify the toxin by means of an enzyme-linked immunosorbent assay (ELISA). An ELISA to quantitate PE follows.[24]

Reagents

Carbonate buffer: 0.16 g Na_2CO_3, 0.37 g $NaHCO_3$, 0.02 g NaN_3, 100 ml distilled water, pH 9.6; make fresh daily

Peroxidase substrate: Sigma Fast OPD (P9187) soluble peroxidase substrate, made up in distilled H_2O according to manufacturer's directions.

Stop buffer: 3.15 g Na_2SO_3, 472.2 ml distilled water, 27.8 ml concentrated H_2SO_4 (prepare in a hood); store in the cold

Wash buffer: 1.53 g KH_2PO_4, 5.0 g Na_2HPO_4, 42.5 g NaCl, 0.5 g thimerosal, 2.5 ml Triton X-100, 2.5 ml Tween 20, make up to 5 liters with distilled H_2O, pH 7.2; store at room temperature

Goat antiexotoxin A (List Biological Laboratories, Campbell, CA): Stock antitoxin (100 μl) is diluted into 100 μl glycerol and stored in a freezer; for the ELISA, dilute 20 μl of this stock into 10 ml carbonate buffer

Biotinylated goat antiexotoxin A (antitoxin from List Biochemicals, Inc; biotinylated by standard procedures[3]): Stock antitoxin is 10 mg/ml, diluted 1 : 1 in glycerol, and stored in freezer; dilute stock antitoxin 1 : 1500 (in 3% (w/v) BSA–PBS) (biotinylated sheep antitoxin can be substituted for biotinylated goat antitoxin)

Streptavidin–HRP (Sigma, Cat. No. S5512): Original vial is diluted 1 : 1 in glycerol and stored in freezer; for use, dilute 1 : 2000 in 3% BSA–PBS.

Procedure

1. Wash flat-bottomed microELISA plates with water.
2. Sensitize plates with 100 μl goat antiexotoxin A (1 : 500) in carbonate buffer. Incubate plates for 1 hr at 37°.
3. Wash plates 3 times with buffer A.
4. Block plates with 300 μl of 7% dry milk in PBS, pH 7.2, for 60 min at room temperature.
5. Add known concentrations (100 μl) of PE or detergent-treated cell fractions and incubate on a shaker overnight at 4°.
6. Wash plates 5 times with buffer A.
7. Incubate with 100 μl biotinyl goat antiexotoxin A (1 : 1500) (List Biochemicals) in 3% BSA–PBS for 1 hr at 37°.

---

[24] M. Z. Kounnas, R. E. Morris, M. R. Thompson, D. J. FitzGerald, D. K. Strickland, and C. B. Saelinger, *J. Biol. Chem.* **267**, 12420 (1992).

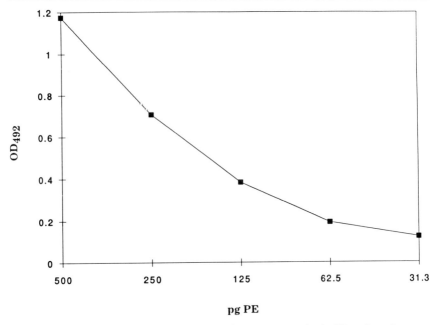

FIG. 1. Use of ELISA to quantitate *Pseudomonas* exotoxin A. Microtiter plates are sensitized with goat antiexotoxin A. Standard concentrations of toxin are added, and plates are processed as described in the text.

8. Repeat Step 6.

9. Incubate plates with streptavidin–HRP (diluted 1:2000 in 3% BSA–PBS) for 1 hr at 23°.

10. Add substrate (OPD and hydrogen peroxide) and allow color development to proceed for 30 min at room temperature.

11. Stop the reaction by addition of stop buffer and measure the absorbance at 492 nm.

Note: to avoid high background readings, it is essential that the BSA is peroxidase-free.

Comments. Carbonate buffer is used in place of goat antiexotoxin A in Step 2 to obtain background readings. Typically determinations are made in triplicate. The assay is linear between 50 and 500 pg PE (see Fig. 1).

Intracellular Localization of Toxin

Similarly, iodinated toxin or native toxin identified by ELISA can be used in conjunction with sucrose density gradient centrifugation to localize toxin to specific organelles following internalization. Although specifics will differ depending on the toxin and cells used, the following method has been employed successfully for tracing the intracellular routing of PE in toxin-sensitive cells. Cell monolayers are maintained in culture for 2 days prior to use.

1. Incubate cells at 4° for 15 min prior to addition of PE (2 μg/ml) in cold HBSS or tissue culture fluid containing 10 mM HEPES, pH 7.2.

2. Incubate cells at 4° overnight with toxin to allow saturation of receptors.

3. Rinse monolayers with cold HBSS–HEPES, and reincubate at 37° for desired time periods.

4. To differentiate surface-associated toxin from intracellular toxin, cool monolayers to 4°, then incubate with 1.25% trypsin–10 mM EDTA (in Ca^{2+}/Mg^{2+}-free HBSS, pH 4.5) for 60 min; this treatment removes greater than 98% of the toxin associated with the cell at 4°.

5. Collect cells by centrifugation for 20 min at 1500 g in a refrigerated centrifuge.

6. Homogenize cells in 0.25 M sucrose containing 5 mM EDTA, pH 7.2, by 100 strokes with a Dounce homogenizer (Wheaton, Millville, NJ) with a tight-fitting pestle (on ice).

7. Spin the homogenate at 1500 g for 20 min, at 4°.

8. Layer the resulting supernatant (1 ml, prepared in 0.25 M sucrose–5 mM EDTA) on a 10–65% sucrose gradient, containing 1 mM EDTA.

9. Centrifuge for approximately 18 hr at 29,000 g in a Beckman SW 41 rotor, at 4°.

10. Collect fractions of 0.5 ml and assay each fraction for toxin by the ELISA just described and for marker enzymes.

Prior to use, all gradient fractions are incubated (1 : 1) with 0.2% Triton X-100 in 3% bovine serum albumin–phosphate-buffered saline, pH 7.2 (BSA–PBS) to solubilize membranes. The following marker enzyme activities are routinely used to identify organelles: Golgi, galactosyltransferase[25]; lysosomes, β-galactosidase and β-glucuronidase.[26] Endosomes are identified by allowing cells to internalize horseradish peroxidase (150 μg/ml) for 18 hr at 15°, followed by washing and reincubation at 15° for 1 hr.[5] Plasma membranes can be identified by the Na^+, K^+-ATPase assay,[27] by the alkaline phosphatase assay,[28] by binding of horseradish peroxidase-labeled concanavalin A to cells at 4° or following surface biotinylation.[29]

[25] K. Brew, J. Sharper, K. Olsens, I. Trayer, and R. Hill, *J. Biol. Chem.* **250,** 1434 (1975).
[26] J. Glaser and W. Sly, *J. Lab. Clin. Med.* **82,** 969 (1973).
[27] M. Esmann, this series, Vol. 156, p. 105.
[28] A. Pitt and A. L. Schwartz, *Exp. Cell Res.* **194,** 128 (1991).
[29] C. M. Goodloe-Holland and E. J. Luna, *Methods Cell Biol.* **28,** 103 (1987).

Comments. It must be remembered that it is difficult to separate organelles easily and cleanly; therefore, subcellular fractionation experiments only provide a trend as to toxin localization. In addition there is some evidence that the type of gradient used can alter the results, but this has not been extensively examined.[30]

We have used subcellular fractionation and electron microscopy to compare the time course of intracellular movement of PE in toxin sensitive mouse fibroblasts and have obtained similar results with both methodologies. The PE is initially seen on the cell surface, is located in endosomes after 5 min of incubation at 37°, and is found in the Golgi apparatus as early as 10–15 min after internalization is initiated.[31] A similar approach has been taken with shiga toxin,[9] ricin immunotoxins,[32] diphtheria toxin,[33] and cholera toxin.[30]

Intracellular Processing of Toxins

Many bacterial toxins are synthesized as enzymatically inactive intact polypeptides and must be converted to an active form to express biological activity. Processing may involve cleavage of the intact toxin molecule to form active fragments, or it may involve a conformational change in the toxin molecule to expose the enzyme active site. Studies on the mechanism of cellular processing are in their infancy, and generalization of methods cannot be made at this point in time.

Several approaches can be taken to investigate the cellular processing of toxins. One can look at the ability of different subcellular fractions to generate an active form of toxin. This allows a determination of the cellular components which have the capacity in a test tube to process toxin. However, this is an artificial situation and may not reflect what actually occurs in an intact cell. Alternatively, one can look for generation of active forms in intact cells. This approach has been used successfully by FitzGerald and colleagues for PE[34] and by Olsnes and colleagues[35] for diphtheria toxin. In addition one can use genetically altered forms of toxin to determine the contribution of different subunits to toxicity.

[30] M. Janicot and B. Desbuquois, *Eur. J. Biochem.* **163,** 433 (1987).
[31] D. FitzGerald, R. E. Morris, and C. B. Saelinger, *Cell (Cambridge, Mass.)* **21,** 867 (1980).
[32] J. M. Manske and D. A. Vallera, *in* "Trafficking of Bacterial Toxins" (C. B. Saelinger, ed.), p. 129. CRC Press, Boca Raton, Florida, 1990.
[33] K. N. Fedde and W. S. Sly, *J. Cell. Biochem.* **37,** 233 (1988).
[34] C. Fryling, M. Ogata, and D. FitzGerald, *Infect. Immun.* **60,** 497 (1992).
[35] J. O. Moskaug, K. Sandvig, and S. Olsnes, *J. Biol. Chem.* **263,** 2518 (1988).

Conclusions

Ultrastructural studies, although highly reproducible, examine only a very small population of cells (0.0001% of the entire population). On the other hand, studies which take a biochemical approach, including following toxin by subcellular fractionation or by measuring toxicity, assess the entire cell population and indicate trends. When carrying out experiments using inhibitors, it must be remembered that most of the agents used act at more than one site in the cell, and therefore caution must be used in interpreting the results. Comparison of events in sensitive and resistant cells may allow identification of factors which are required for the routing and/or processing of toxin (conversion to an enzyme active form) and which are lacking in resistant cells. Use of genetically altered toxins helps to differentiate sites of binding, translocation, and enzyme activity. A combination of methods therefore allows one to more clearly determine the events which occur in an intact cell.

Author Index

Numbers in parenthesis are footnote reference numbers and indicate that an author's work is referred to although the name is not cited in the text.

A

Aaronson, W., 465, 665
Abdillahi, T., 164, 166(17)
Abdullah, T., 357
Abou-setta, M. M., 31
Abraham, J. S., 410
Abramson, C., 615
Acharya, A. S., 289–290
Achen, M. G., 381, 382(17)
Achenbach, L., 221
Ackerman, G. A., 707
Adam, A. R., 254
Adamik, R., 641–642, 642(16, 17), 643(16, 17, 31), 644(17, 21, 31, 35, 36)
Adams, J., 358, 359(17), 360(17), 362(17), 363(17)
Adamus, G., 47
Adelberg, E. A., 454
Adler, J., 230
Agabian, N., 139
Aghion, J., 528, 530(18)
Agin, T. S., 454, 455(13), 457(13)
Ahl, T., 548
Ahnert-Hilger, G., 683
Aida, Y., 277
Airo-Brown, L. P., 84, 93(9)
Akino, T., 682
Akrim, M., 528
Aktories, K., 626–627, 683
Alatt, D. D., 366(cc), 367
Albano, M., 529
Alberro, M. R., 180
Albus, A., 412
Alderete, J. F., 139
Alexander, H. E., 162
Alexander, J. E., 381
Ali, J., 512, 513(21)
Ali, R. W., 110
Allen, A. L., 139
Allen, L. N., 369

Allen, R. D., 564, 566(9), 586(9)
Allen, R. G., 50
Allen, S. P., 381
Allen, S. S., 107
Allmaier, G., 274, 280(74)
Allured, V. S., 620
Alm, R. A., 474–476, 476(4), 477, 478(4), 480, 480(8)
Alouf, J. E., 414
Altman, P. L., 125
Alvarez, O., 693
Åman, P., 685
Amann, R., 221
Ames, B. N., 328, 357
Ames, G.F.-L., 229, 234–236, 238, 240(13), 241, 241(13)
Ames, P., 295, 297(8), 298(8)
Amherat, M., 642
Andary, T. J., 565, 591(10)
Anderle, S. K., 264, 285(39)
Anderson, A. C., 661
Anderson, A. J., 297
Anderson, B. E., 26
Anderson, G. J., 346
Anderson, J. A., 114, 115(40)
Anderson, J. E., 360, 361(20)
Anderson, K. L., 180
Anderson, P., 99–100, 305, 306(7)
Anderson, R. J., 353, 371
Anderson, R. S., 667
Andreoli, T. E., 693
Andrew, P. W., 381
Andrews, N. W., 665, 667
Andrews, S. B., 325
Anilonis, A., 185, 193, 195(5)
Ankenbauer, R., 328
Ankenbauer, T., 627
Anraku, G. K., 236
Antillon, F., 688
Antoine, A. D., 335, 336(13)
Apfel, H., 553

Apicella, M. A., 242, 242(12), 243, 250
Arancia, G., 685
Arbeit, R. D., 412
Arbuthnott, J. P., 121, 122(11), 125, 127(11, 25), 128(25), 132(11), 133, 133(11), 134(11), 137(25), 663, 666(34)
Arceneaux, J.E.L., 363, 370(2), 372(2)
Arciniega, J. L., 37
Arima, K., 258
Arko, R. J., 120–122, 122(13), 124, 124(1, 18), 125, 125(1, 12, 18), 126, 126(18), 128, 128(17), 129, 129(36, 39), 132(12, 41), 137, 137(13, 21)
Armstrong, L. R., 142
Armstrong, S. K., 232, 366(29, f), 367, 371
Arnesen, O., 164
Arnheim, N., 208
Arnold, R. R., 121, 124, 124(2), 127(2), 128(2), 136, 136(2), 137(21)
Arnow, L. E., 332, 353, 357, 371
Aronson, N. N., Jr., 711
Arroyo, P. J., 451
Artenstein, M., 163
Aschaffenburg, P. H., 116
Ashburn, L. L., 48
Ashley, F. P., 108
Ashwell, G., 304
Ashworth, L.A.E., 48
Atkin, C. L., 331
Attstrom, R., 111
Auckenthaler, R., 121, 127, 132(6, 34)
Auerbach, H., 295, 297(8), 302(8)
Austrian, R., 170
Ausubel, F. M., 365, 369, 369(14), 433, 466
Averill, D. R., 95–96, 96(7), 97(11), 98(11), 99(7, 11), 100
Avery, B. E., 107
Azuma, I., 258
Azuma, Y., 118

B

Bachovkin, W. W., 553
Badere, A., 528
Baer, H., 612
Bagg, A., 330, 364, 366(mm), 367
Baggiolini, M., 567, 588(16)
Bakacs, T., 42, 43(9), 45(9), 46

Baker, C. N., 164
Baker, K., 668
Baker, N. R., 493
Bakketeig, L. S., 164
Balch, W. E., 207
Baldini, M. M., 441
Balkc, E., 612
Balkowski, B., 254
Ball, G., 528
Balls, M., 650
Bally, M., 528
Balows, A., 122, 124(18), 125(18), 126(18)
Baltodano, A., 94
Band, J. D., 94
Banda, M. J., 555, 581
Bangham, J. A., 693
Banks, S. D., 169
Bannochie, C. J., 321
Baquero, F., 414
Barbieri, J. T., 633, 635(4), 636, 636(4)
Barcak, G. J., 444
Bardoness, J., 434
Barenkamp, S. J., 59, 61(5)
Barghouthi, S. M., 363, 370(2), 372(2)
Barker, R., 308
Barley-Maloney, L., 649
Barmann, M., 626
Barnes, R. C., 83, 84(1)
Barns, S. M., 209, 212
Baron, L., 40, 41(7)
Barrett, J. F., 265, 269
Barron, A. L., 78, 84–85, 91(6), 92
Bartlett, J. G., 134
Bartoleme, B., 665, 666(43)
Bartoloni, A., 635
Barton, L. L., 410
Barua, P. K., 564
Baselski, V. S., 490
Bass, J. A., 134, 527, 555
Bastouil, 40
Bates, J. H., 196–197
Batson, H. C., 29, 31(2), 35(2)
Batteiger, B. E., 83
Battershil, J. L., 133
Batut, J., 497
Bauer, D., 140, 143(2), 150(2)
Bauer, H. M., 218
Bauer, W., 614
Bayley, H., 632
Beachey, E., 658, 659(14)

Beachey, E. H., 288–289, 290(27), 291(27), 294, 434, 448(20)
Beagley, K. W., 143
Bearden, J., 193
Beattie, D. T., 58
Beck, A., 196
Becker, C. G., 290
Becker, D. M., 375
Becker, H., 657, 676
Becker, J., 221, 222(38)
Beckwith, J., 427, 429, 431(4), 434, 435(4), 440(4)
Beckwith, J. R., 485
Bedi, G., 109, 118(20a), 119(20a)
Beecher, D. J., 661, 662(27), 665(27)
Beighton, D., 608
Beinert, H., 319
Belasco, J. G., 506
Belcher, R., 300
Bell, A., 121, 122(9), 127(9), 134(9)
Bell, A. W., 235, 670
Bellalou, J., 669
Belland, R. J., 70
Bellanti, J. A., 449, 451, 453–454, 455(11), 457(1)
Bellofatto, V., 375
Beltran, P., 183
Beltz, G. A., 365
Benchetrit, L. C., 607
Bender, J., 388, 393(11), 397(11)
Benedi, V.-J., 411, 420(10)
Benkley, F., 243
Bentley, M. D., 335
Benz, R., 225, 678
Berche, P., 414
Berczi, A., 320
Berendt, R. F., 50
Beretti, J.-L., 414
Berezesky, I. K., 685
Berg, C. M., 386, 387(6)
Berg, D. E., 386–387, 387(6), 388, 393(7), 427, 443(7)
Berg, P., 194
Berger-Rabinowitz, S., 288
Bergman, P., 539
Bergman, S., 527, 531(11)
Beringer, J. E., 433, 441(16)
Berman, M. L., 436
Berner, I., 321, 341
Bernheimer, A. W., 658, 659(9, 10), 666

Berry, A., 299–300, 494–496, 496(7), 497, 497(7), 498, 498(22), 500(22, 31), 501(22)
Berry, V. K., 527
Berry, W. C., Jr., 108
Berthold, P., 674, 677
Bertrand, P., 610
Besdine, R. W., 288
Bessen, D., 290
Bessen, D. E., 294
Beste, D. J., 59, 60(6), 62(6), 63(6)
Betley, M. J., 438
Bette, P., 683, 690(11)
Betzner, A. S., 271
Beugnier, N., 695
Beutler, B., 103, 105, 105(36)
Bevandick, I. M., 328
Beynon, J. L., 433, 441(16)
Beyreuther, K., 543, 554(8)
Bhakdi, S., 657–659, 659(15), 662(15), 668, 676, 676(15), 677–678
Bharanandan, V. P., 709
Bhopale, V., 608, 615(2), 616(2)
Bibb, W. F., 164
Bieker, K. L., 426
Biemann, K., 271, 274, 280(70)
Bigazzi, P. E., 84
Biggart, J. D., 604
Bijlmer, H. A., 160, 163(1)
Bik, E. M., 198, 204(13)
Billyard, E., 527, 529(6)
Bindereif, A., 364, 366(30, q), 367, 368(8), 372
Binkhorst, G. J., 676
Biondi, A., 277
Birkedal-Hansen, B., 603
Birkedal-Hansen, H., 564, 596, 597(7), 598(7), 599(7), 603, 606
Birnbaumer, L., 640
Birnboim, H. C., 423
Bisercic, M., 183
Bisno, A. L., 289
Biswas, G. D., 358, 359(17), 360, 360(17), 361(18), 362, 362(17), 363(17)
Bittner, W., 351
Bittner, W. E., 250
Bjorck, L., 286
Bjorn, H., 116, 117(54)
Bjorn, M. J., 504, 514
Bjune, G., 164
Black, R. E., 441

Black, W., 459, 461, 469–470
Blackshear, P. J., 586, 588(22)
Blake, M., 232
Blake, M. S., 232, 545, 550, 550(17), 551(17)
Blakemore, R., 207
Blakemore, R. P., 212
Blanton, K. J., 358, 359(17), 360(17), 362(17), 363(17)
Blaschek, H. P., 375, 380(8, 9), 381, 381(11)
Blaser, M. J., 684
Blaustein, R. O., 695, 703(13)
Bleumink-Pluym, N.M.C., 475, 476(5), 478(5), 479(5), 480(5)
Blieden, T., 111
Blight, M. H., 670
Blitchington, R. B., 209, 212
Bliziotes, M. M., 642
Bloch, W., 221, 222(38)
Bluestone, C. D., 59, 60(6), 62(6), 63(6)
Blumenthals, I. I., 504
Boackle, R. J., 254
Bobak, D. A., 641–643
Bobo, L., 74
Bodeau, A. L., 567, 588(17)
Bodely, J. W., 617
Bodey, G. P., 527
Boehm, D. F., 666, 670, 672, 674, 676(29)
Boelaert, J. R., 354
Boesken, W. M., 552
Bohach, G., 674
Bohannan, H. M., 109
Boivin, A., 243
Bolduc, G., 61, 62(8)
Bolivar, R., 527
Bond, M. D., 595, 600(2)
Bonen, L., 207
Bonventre, P. F., 709
Boquet, P., 679
Borgna-Pignatti, C., 325
Borschberg, U., 103, 104(37)
Bortner, S. R., 524
Bortolussi, R., 97, 684
Botstein, D., 642
Bott, K., 186–187, 190(19)
Bott, K. F., 185
Botzenhart, K., 555
Bouwsma, O., 111
Boux, L. J., 465
Bowen, I. D., 681
Bowen, S. M., 681
Bowen, W. H., 420, 422(49)

Bowler, L. D., 183
Boyd, A., 495–496
Boyd, B., 649
Boyd, C. M., 495
Boyko, S. A., 440–441
Bozzola, J. J., 325
Bracha, R., 265
Bradford, M. M., 499
Bradford, W. L., 162
Brady, T. J., 325
Bramanti, T. E., 347, 657
Brandt, J., 543
Brandtz, P., 105
Braude, A. I., 128
Braun, V., 366(s, w, x, y, aa, bb, dd–hh, jj, kk, nn), 367–368, 372, 440, 658, 663, 666(13)
Braunsteiner, H., 591
Bravo, N., 441
Bray, M. A., 104
Brazilian Purpuric Fever Study Group, 243
Breeding, S.A.L., 328
Breitling, R., 529
Brennan, M. J., 57, 58(20)
Brenner, R. M., 83, 84(1)
Brent, R., 365, 369(14)
Bretz, U., 567, 588(16)
Brew, K., 715
Brickman, T. J., 351, 366(c, e, k), 367
Briles, D. E., 170
Brinton, C. C., Jr., 59, 60(6), 61, 62(6), 63(6)
Brisson-Noël, A., 197, 198(10)
Britigan, B. E., 324
Brock, J. H., 316, 323
Brocklehurst, K., 604
Brokopp, C. D., 167, 168(27)
Bronze, M. S., 289
Broome, C. V., 94, 121, 125(12), 132(12), 164, 410
Brown, A. H., 298
Brown, A. M., 640
Brown, D. C., 257
Brown, G. D., 140, 143
Brown, J. E., 142, 145(7), 151, 648–649, 652, 653(23), 654(15), 683
Brown, L., 551
Brown, L. R., 107
Brown, M.R.W., 121, 137(5)
Brown, R. R., 264, 285(39)
Brown, S. E., 369
Brown, T. A., 550

Brown, T. L., 565, 578(11), 588(11)
Brown, T. M., 246, 247(27), 251(27)
Brown, W. J., 129
Bruch, J. M., 116
Brunden, K. R., 695
Brunham, R. C., 70
Brunner, M., 642
Brunsvold, M., 113
Buchanan-Wollaston, A. V., 433, 441(16)
Buchhotz, S. R., 185, 190(12), 191(12), 192, 194(12)
Buckel, S., 670
Buckel, S. D., 235
Buckmire, F.L.A., 162
Buikema, W. J., 369
Bulen, J. J., 344
Bullard, J. C., 121, 128–129, 129(36), 132(41)
Bullen, J. J., 327
Bullock, J. O., 695
Bundle, D. R., 305, 306(9), 308(9)
Bunschoten, A. E., 197
Bupta-Preeti, 320
Burdett, V., 294
Burgdorf, K., 274, 275(73), 280(73)
Burke, B., 97
Burkhardt, R., 366(bb), 367
Burlingame, A. L., 242
Burne, R. A., 405, 420, 422(49)
Burnet, B. R., 49
Burns, D. L., 57, 630
Burns, R. P., 503
Burr, D. H., 474, 479, 480(14)
Burroughs, M., 103, 104(37), 105
Burstyn, D. G., 49, 51(9), 52(9), 53(9), 54(9), 55(9)
Burton, J., 545, 554(19)
Buscher, H. P., 300
Bushell, G. R., 581
Butler, C. A., 413, 419(19)
Butler, W. T., 553
Butterton, J. R., 352
Buysse, J., 44, 46(10)
Byers, B. R., 363, 370(2), 372(2)

C

Cabellos, C., 105
Cabiaux, V., 695, 700
Cacalano, G., 501

Cady, A. B., 274
Caffesse, R. G., 112
Cain, C. C., 685
Calderwood, S., 519
Calderwood, S. B., 352, 440–441
Caldwell, H. D., 70–71, 79
Callahan, L. T. III, 520, 523(23, 30)
Callard, R. E., 126
Calvanico, N. J., 545
Calvin, N. M., 378
Cameron, D. R., 366, 366(u), 367
Campagnari, A. A., 242, 250
Campbell, D., 142
Campbell, G. L., 288
Campbell, J. A., 604
Campbell, S., 70
Canicatti, C., 657, 665(1), 666, 666(1)
Cantekin, E. Y., 59
Caparon, M. G., 411, 419(8), 420(8), 421
Caparros, M., 274, 280(74)
Capobianco, J. O., 563
Caprioli, A., 685
Caravano, R., 285
Carbonetti, N. H., 366(p), 371
Carlson, A. M., 297, 303(13)
Carlsson, J., 121, 127(3), 128(3), 135(3), 604
Carmel, G., 366(31, u), 367, 372
Caroline, L., 316, 328(6)
Carr, J. H., 26
Carrasco, L., 648
Carroll, S. F., 618, 620, 631–633, 633(1, 2), 634(6), 635(1, 2), 636, 637(2)
Carson, J., 226, 227(6), 229(6), 230(6), 231(6)
Carter, M. J., 59, 60(6), 61, 62(6), 63(6)
Carter, P., 319
Carter, P. E., 185, 195(4), 196(4)
Cartwright, J., 325
Casals-Stenzel, J., 300
Cash, M. A., 134
Castriota-Scanderberg, A., 325
Catalano, M., 451
Cato, E. P., 109
Caton, J., 108, 110–111, 111(25)
Caton, J. G., 106, 112, 113(36)
Catty, D., 197–198
Caugant, D. A., 160, 164
Cavalieri, S., 674
Cavalieri, S. J., 657, 667, 676
Cavanaugh, E., 641
Cave, M. D., 197
Cawston, T. E., 604

Centers for Disease Control, 413
Cerami, A., 105
Cerquetti, M. C., 451, 454–455, 455(11), 456(14), 457(14)
Chaffee, E., 612, 613(12)
Chain, B. M., 667
Chakrabarty, A. M., 295–296, 299, 299(9), 300, 303, 493, 496, 496(7), 497, 497(7), 498, 498(22, 28), 500(21–24, 31, 32), 501, 501(22, 24, 29)
Chakrabarty, K., 496
Chakraborty, T., 657
Chambers, J. C., 218
Chan, R., 493
Chandhoke, V., 354
Chandler, F. W., 121, 125(12), 132, 132(12)
Chandler, M. S., 444
Chandra, S., 325
Chang, A.C.Y., 506
Chang, K. M., 118
Chang, P., 641, 642(17), 643(17), 644(17)
Chang, P. O., 641, 644(21)
Chang, P. P., 642, 644(35)
Chang, Y.-F., 668–669, 671(9)
Chang, Y. F., 664, 666, 675
Chapman, S. S., 165
Chargaff, E., 189
Chassy, B. M., 375, 377, 380, 381(11), 384, 418
Chater, K., 386
Chatfield, S., 439, 443(40)
Chatterjee, A., 521
Chatterjee, A. K., 528
Chatterjee, A. N., 232, 265, 266(44)
Chaudhuri, K., 232
Chauzy, C., 610
Chedid, L., 254
Chen, C.-Y., 128, 129(39)
Chen, C. Y., 122, 128
Chen, G.C.C., 454
Chen, H.-C., 641, 642(17), 643(17), 644(17)
Chen, K., 209, 212
Chen, K. N., 207
Chen, M. C., 415, 418(31)
Chen, P. S., 300
Chen, V. J., 324
Chenault, S. S., 366(o), 367
Cheng, K. J., 493
Cheng, S., 186, 190(18)
Cherbas, P. T., 365

Chetty, C., 264, 285(39)
Cheung, A. L., 413, 419(19)
Chevrestt, A., 185, 190(12), 191(12), 192, 194(12)
Cheyne, I. M., 658
Chhatwall, S. G., 626
Chi, E., 527
Chiang, E. Y., 288, 290(27), 291(27)
Chiang, T. M., 288, 290(27), 291(27)
Childers, C. C., 31
Chimera, J., 218
Chipman, D. M., 269
Chitnis, C., 495–496
Cho, M. I., 108
Choay, J., 254
Chopra, I., 232
Chou, J., 485
Chrane, D. F., 94
Christopoulos, P., 546
Chu, L., 495–496, 496(7), 497, 497(7), 500(23), 657
Chuang, D. M., 640
Chuard, C., 121, 132(6)
Chung, D. W., 505, 618, 619(3)
Cianciotto, N. P., 464
Ciborowski, P., 563
Cicmanec, J. F., 555
Cioffe, C., 105
Ciorbaru, R., 254
Ciraolo, G. M., 709, 710(10)
Cisneros, R. L., 121, 122(14), 125(14), 134–135, 135(10)
Clark, B. L., 527, 529(9)
Clark, E. J., 581
Clark, J., 642
Clark, V. L., 647
Clark-Curtiss, J. E., 197
Clarke, A., 310
Clarridge, J. E. III, 26
Cleary, P. P., 286, 289, 294
Clewell, D., 419, 422(46)
Clewell, D. B., 409, 411, 415, 418(1), 420, 421(33), 426(1)
Cline, G. R., 327
Clinkenbeard, K. D., 677
Closs, O., 164
Clothier, R., 650
Clowes, R. C., 514
Coburn, J., 622, 624
Cochi, S. L., 410

Cockayne, A., 133
Cockle, S. A., 465, 633, 635(5)
Codina, J., 640
Coffman, R. L., 143
Coffman, T. J., 324
Cohen, C., 289
Cohen, F. S., 695
Cohen, J. O., 288
Cohen, M. E., 115
Cohen, M. S., 324
Cohen, S. C., 506
Cohen, S. N., 485
Cole, J. A., 129
Coleman, G., 174
Coleman, K. D., 121, 122(7), 127(7), 138(7)
Collier, R. J., 505, 518, 618, 619(3), 620, 631–633, 633(1, 2), 634(6), 635, 635(1, 2, 4), 636, 636(4), 637(2), 640, 695, 700, 703(13)
Collins, D. M., 197
Collmer, A., 442, 528
Colombini, M., 691(6), 692, 695(6), 697(6), 698(6), 700(6), 701(6), 703(6)
Colwell, J. L., 684
Confer, A. W., 677
Connelly, J., 325
Conrad, M. E., 320
Cook, J. M., 468
Cookson, B. T., 270, 280(62)
Coote, J. G., 658, 665(12)
Corbeil, L. B., 128
Corbeil, R. R., 128
Corbin, D., 432, 434(13), 438(13)
Cornelissen, C. N., 356, 360, 361(18, 20), 362
Coronado, R., 695
Cossart, P., 414
Costerton, J. W., 493, 498
Costich, E. G., 108
Coulson, C. J., 609
Coulton, J. W., 366(31, t, u, cc), 367, 372
Courcoux, P., 475, 476(2), 477, 478(2), 479
Courvalin, P., 475–476, 476(9)
Cover, T. L., 684
Cowan, G. M., 183
Cowell, J. L., 47, 49, 51(9), 52(9, 10), 53, 53(9, 10), 54(9, 10), 55, 55(9, 10, 13, 14), 57(18), 59
Cox, C. D., 315, 324, 328, 513, 515, 556
Coy, M., 349

Crabb, J. H., 121, 122(14), 125(14), 135(10)
Crain, M. J., 170
Crain, P. F., 633, 634(6)
Cramer, W. A., 695
Crane, F. L., 320
Cranfield, M., 674
Crawford, F. G., 139
Crawford, J., 196
Crawford, J. T., 197–198, 204(15)
Creeger, E. S., 229
Cromartie, W. J., 264, 285(39)
Crook, E. M., 604
Crosa, J. H., 316, 352, 364
Cross, G.A.M., 375
Cross, M., 668
Cruce, D., 615, 616(38)
Cruz, W. T., 664, 675
Cryz, S. J., Jr., 48, 138
Csáky, T., 331, 357
Cuagant, D. A., 544
Cuatrecasas, P., 603, 675
Culp, W. J., 653
Cummings, N. P., 254
Cunliffe, W. J., 613
Cunningham, M. W., 288
Curtiss, R. III, 415, 418(32), 419(32), 421(32)
Cussac, V., 479
Cutler, C. W., 121, 124, 124(2), 127(2), 128(2), 136, 136(2), 137(21)
Cventanovic, B., 305
Cwirla, S. E., 376, 384(10)
Cybulski, G. R., 325

D

Dabich, D., 565, 591(10)
Dabrowski, J., 308, 311(14)
Dacey, R. G., 95, 100(8), 102(8)
Dahlbeck, D., 369
Dahlberg, J. E., 186–187, 464
Dahlen, G., 125, 136(24)
Dahnke, T., 501
Dai, W., 254
Dailey, H. A., 320
Dailey, T., 668, 670(10), 671(10)
Dale, J. B., 289
Dale, J. W., 197–198, 204(13, 15)
Dalrymple, J. M., 150, 151(21), 152(21)
Dammin, G., 45

Danchin, A., 668–669
Dancis, A., 346
Daneo-Moore, L., 415
Danganan, C. E., 497, 498(28)
Daniel, A. S., 183
Daniel, T. O., 649
Daniels, M. J., 529
Darougar, S., 69, 79(6), 80, 84
Darveau, R. P., 243, 245(24)
Darzins, A., 296, 299(9), 496–497, 498(22), 500(22), 501(22)
Das, A. K., 111, 115(33), 116(33)
Das, B. C., 270
Das, J., 232, 521
DasGupta, B. R., 691, 695(5), 697(5), 701(5), 703(5)
Dasgupta, M., 493
Daum, R. S., 100
Davenne, D., 271
David, J. R., 282
Davidson, B. E., 381, 382(17)
Davidson, J. M., 554
Davidson, V., 695
Davies, A. M., 288
Davies, D. B., 128, 129(38)
Davies, D. G., 498
Davies, J. E., 464
Davis, M. A., 387, 393(10)
Davis, S. M., 358, 359(17), 360(17), 362(17), 363(17)
Dawes, E. A., 297
Dawson, C., 196
Dawson, C. R., 69, 70(4), 71, 71(4), 80
Dawson, M. H., 612, 613(12)
Day, S.E.J., 121, 122(11), 127(11), 132(11), 133(11), 134(11)
Dayal, P. A., 196
Dayton, S., 300
Deacon, W. E., 551
Deb, S., 229
de Barsy, T., 685
De Brujin, F., 436
Decazes, J. M., 106
de Duve, C., 685
de Graaff, J., 657
de Haas, P., 204, 205(17)
de Haas, P.E.W., 196–198, 199(14), 204(13), 205(14)
de la Cruz, F., 665–666, 666(43)
Delbrück, M., 453

Deleers, M., 695
Delepelaire, P., 668–669
de Lisle, G. W., 197
Dell, A., 659
Dellinger, E. P., 668
DeLong, E. F., 208
de Lorenzo, V., 350, 366(q, r), 367, 386, 388(3), 389(3), 391(4), 392, 392(a, e, f, g), 393, 394(2, 3), 395(3), 396(3, 4), 398(3), 399(3), 401(4), 402(5), 403(5), 436, 437(28), 443(28)
Delpech, B., 610
Delvenne, P., 222
De Maeyer, E., 615, 616(35)
Demuth, D. R., 668
Denbow, L., 608
Den Daas-Slagt, H. J., 328
d'Enfert, C., 528
Denker, H.-W., 606
Denyer, S. P., 133
DePamphilis, M. L., 230
De Pedro, M. A., 266
dePeer, Y. V., 214
Derber, D. B., 61
Deretic, V., 496–497, 500(21), 501–502
DeRijk, P., 214
Derman, A., 427
Derome, A. E., 271
Derrien, M., 254
DerVartanian, M., 328–329
Desbuquois, B., 716
deUzeda, M., 323
DeVault, J. D., 495, 496(7), 497, 497(7), 498, 498(22), 500(22, 23, 31, 32), 501, 501(22)
Devedjian, J. C., 401, 436
De Vos, G. F., 438
deVries, G. E., 434
DeWachter, R., 214
Dewald, B., 567, 588(16)
de Weger, L. A., 351
Dewey, H., 32
de Wilde, M., 518, 519(2)
Dickerson, H. W., 662, 666(30)
Dieckmann, M., 194
Dignam, J. D., 630
Dikshit, R., 497, 500(21)
Dillon, S. T., 622
Dimond, R. L., 249
Dimtriev, B. A., 252
Diorio, C., 316

DiRienzo, J., 674
DiRita, V. J., 438, 440, 519–520, 520(17)
Ditta, G., 432, 434(13), 438(13)
Dixon, W. J., 33
Dmitriev, B. A., 167
Dobozy, A., 490
Doctor, B. P., 151, 648, 683
Dodge, P. R., 94
Dodson, K. W., 387
Doeg, K. A., 320
Doery, H. M., 658
Doig, P., 493
Doly, J., 423
Donaldson, J. G., 642
Donegan, E., 92
Donelli, G., 684–685
Donnenberg, M. S., 441
Donohue-Rolfe, A., 151, 441, 648
Donovan, J. J., 691, 700(4)
Doolittle, R. F., 180
Doopman, W. J., 143
Dorfman, A., 288, 615
Doring, G., 555
Dorland, R. B., 620, 710, 712
Dougan, G., 125, 127(25), 128(25), 137(25), 439, 443(40)
Douglas, C. M., 635
Douglas, S. D., 681, 684(3)
Dow, J. M., 529
Dowdle, E. B., 567, 574(14), 577(14), 587(14)
Dower, W. J., 375, 376(3), 378(3, 9), 380(8, 9), 381(2, 9, 11), 382(2, 3), 383(2, 3), 384(2, 3, 9, 10), 385(2, 9), 418, 475, 478(6), 479(6), 480(6)
Downie, J. A., 235, 670
Doyle, R. J., 257
Doyle, W. J., 59, 60(6), 62(6), 63(6), 68
Draper, R. K., 710–711, 712(21)
Drechsel, H., 354
Drew, M. D., 328
Drews, G., 233
Dreyfus, L. A., 444
Drijver, A. A., 676
Drueckhammer, D., 366(i), 367
D'Souza, T. M., 185, 196(14)
Dubey, R. S., 521
Dubnau, D., 528–529, 530(18)
Dubnau, D. A., 529
DuBow, M. S., 316, 366(t), 367

Dumas, J., 595
Dums, F., 529
Duncan, J. L., 121, 138(4), 139(8)
Duncan, K., 366, 366(b), 367
Duncan, W. P., 128–129, 129(36)
Dunn, P. E., 254
Dunn, W. A., 711
Dunny, G. M., 381, 382(15), 412, 418, 418(15)
Dupuy, B., 528, 530(17)
Dutton, G.G.S., 305
Dworkin, M., 230
Dwyer, B., 197
Dyer, D. W., 357–358, 358(11), 359(17), 360(17), 362, 362(17), 363(17)
Dyer, T. A., 207
Dykhuizen, D. E., 183
Dziarski, A., 282
Dziarski, R., 253, 257, 258(19), 263, 266, 282, 284, 284(47)

E

Eads, M. E., 260
Eagon, R. G., 174
Ealding, W., 143
Eans, D. J., 520, 523(21)
Earhart, C. F., 231, 328
Earhart, C. R., 366(d, g, m, o), 367
Eastby, C., 545, 550(17), 551(17)
Ebashi, S., 676
Ebata, M., 555, 556(7)
Eberhard, T., 613
Ebersole, J., 113, 668, 670(10), 671(10)
Ebersole, J. L., 657
Eberspacher, B., 657–658, 659(15), 662(15), 676
Edeker, B. L., 324
Eden, P. A., 212
Edson, C., 648
Edstrom, R. D., 607
Edwards, M. S., 295, 297(8), 298(8), 302(8)
Egami, F., 315
Egan, W., 305
Egan, W. M., 161
Egelberg, J., 111
Egnor, R. W., 118
Ehrmann, I., 673
Eichenberger, E., 243

Eickbush, T. H., 365
Eick-Helmerich, K., 366(jj, kk), 367
Eiklid, K., 648
Eisenach, K. D., 197
Eisenman, 696
Eisenstein, B. I., 464
Fisner, R. L., 185, 186(9), 194(9)
Eklid, K., 648
Elder, M. G., 84
Eldering, G., 162
Eldridge, J. H., 143
ElKholy, A. M., 288
Elkins, M. F., 366(g, m), 367
Elledge, S. J., 369
Ellegaard, B., 108
Elliott, H. L., 140
Elliott, R. J., 604
Ellison, R. T., 328
Ellouz, F., 254
Elsinghorst, E. A., 41, 46(8)
Elson, C., 143
Eltis, L., 386, 392(g), 393, 402(5), 403(5)
Elwood, H. J., 209
Emmering, T. E., 115
Endo, M., 610
Endo, Y., 649
Eng, J., 164
Engel, H., 271
Engelberg, N. C., 464
Englande, A. J., 661
Ennis, P. D., 220
Enomoto, K., 630
Enquist, L. W., 436
Ericsson, I., 115
Erlich, B. E., 691, 695(5), 697(5), 701(5), 703(5)
Erlich, H. A., 177, 208, 220
Ernst, J. D., 105–106
Erwin, A., 243
Erwin, A. L., 283
Esmann, M., 715
Espeland, M., 111
Espersen, F., 297, 301(10), 302(10)
Espevik, T., 105
Etienne, C., 305
Evans, I. J., 235, 670
Evans, L. R., 295, 297(5), 301(5), 304(5)
Evans, R. T., 109, 118(20a), 119, 119(20a)
Everson, M. P., 143
Ewanowich, C., 48, 138
Expert, D., 347, 351

F

Facklam, R. R., 94, 288, 410
Faingezicht, I., 94
Fainstein, V., 527
Fairwell, T., 289–290, 294
Falbo, V., 685
Falder, P., 77, 84, 93(8)
Falkow, S., 48, 53, 55(10), 210, 212, 219(13), 220(14), 438–439, 442, 443(40), 459, 461, 465, 469–470, 519, 522(15), 527, 529(8), 544, 663, 665, 668
Falmagne, P., 695
Faloona, F., 208
Farazdaghi, M., 75, 76(20)
Farinha, M. A., 510, 512(18), 513(18)
Farmer, J. J. III, 167, 168(27)
Farmer, S. G., 139
Farr, A. L., 247
Farrar, W. E., Jr., 125
Fasching, C. E., 122, 127(19), 129(19), 135(19)
Faul, W. P., 320
Faustoferri, R. C., 416
Fecker, L., 366(s), 367
Fedde, K. N., 716
Federal Register, 1, 4(6), 10(6), 16(6), 25(6)
Fedorka-Cray, P. J., 525
Feigin, R. D., 94
Feingold, D. S., 303, 494, 496
Fekete, F. A., 354
Feldman, H. A., 543, 552(6)
Fellay, R., 472
Felmlee, T., 668–669
Felsenstein, J., 217
Felten, J., 113
Feng, D. F., 180
Ferenci, T., 241
Fernandes, P. B., 520, 521(24), 523(24)
Fernandez-Puentes, C., 648
Ferrell, S. J., 559
Ferrero, R. L., 479
Ferretti, J. J., 412, 418(14), 606, 611–612, 612(8)
Ferrier, J. M., 108, 111(18), 115(18)
Ferrieri, P., 97, 414, 614
Fesik, S. W., 308
Feutrier, J. Y., 183
Fialho, A., 494, 496
Fiedler, F., 258
Field, M., 519

Fields, P. I., 439
Figurski, D., 392(e), 393
Figurski, D. H., 471, 477
Filip, C., 231
Filloux, A., 528
Finberg, R., 121, 135(10)
Finberg, R. W., 121, 122(14), 125(14), 135
Finch, R., 133
Fine, A. S., 118
Fink, F. N., 102
Finkelstein, A., 691, 691(6), 692, 695, 695(5, 6), 696, 697(5, 6), 698(6), 699(22), 700(6), 701(5, 6), 703(5, 6, 13)
Finkelstein, R. A., 328, 357, 520
Finlay, B. B., 439, 443(40)
Finn, T. M., 58, 125, 127(25), 128(25), 137(25)
Finney, D. J., 35
Fiorentini, C., 684–685
Fischer, H., 265, 266(45)
Fischetti, V. A., 171, 286–287, 287(3), 288–289, 289(2), 290, 291(35), 292, 292(35), 294, 413, 419(19)
Fish, W. W., 319
Fisher, R. A., 33(21), 35
Fisher, S. J., 242
Fishman, P. H., 640, 710
Fiss, E. H., 350–351
Fiszer-Szafarz, B., 615, 616(34, 35)
Fitton, J. E., 659
Fitzgeorge, R. B., 48
FitzGerald, D., 716
FitzGerald, D. J., 713
Flaherty, J. T., 324
Flannagan, S. E., 409, 418(1), 420, 426(1)
Flebbe, L., 283
Fleming, T. J., 254
Fleming, T. P., 366, 366(a), 367
Flemming, T. F., 567, 588(17)
Flentke, G. R., 553
Fletcher, D. D., 102
Fletcher, G., 231
Fletcher, P. L., 269
Flicker, P. F., 289
Flickinger, J. L., 375, 384, 418
Florin, I., 679, 685, 688, 690(29)
Flowers, H. M., 269
Flynn, J. L., 468–469, 493, 501, 501(5)
Focareta, T., 658, 666(13)
Föhr, K. J., 683
Folkening, W. J., 263, 272(37)

Forde, A., 222
Forestier, C., 661, 666, 666(28), 669
Formal, S., 40, 41(7), 45
Formal, S. B., 140, 143(4), 648
Forrest, M., 105
Forrester, L. J., 366(f), 367
Forristal, J. J., 710
Forsberg, E., 604
Fournier, J. M., 165
Fowler, R. T., 320
Fox, A., 264, 285(39)
Fox, E. N., 287–288
Fox, G. E., 207, 217, 220
Fox, J. E., 129
Francis, C. L., 366(29), 371, 439, 443(40)
Frandsen, E.V.G., 553
Frangione, B., 543, 545(7), 546(7), 552(6, 7)
Frank, D. W., 317, 502–503, 505, 506(11), 507(11), 508, 508(10, 11), 510, 510(6), 511, 511(5, 6, 10), 512, 512(5, 6, 10, 18), 513(5, 6, 10, 11, 18), 623
Franklin, M. J., 495–496
Franklin, P. S., 103, 105(36)
Franklin, R., 48, 138
Frantz, B., 111, 350, 496
Franza, T., 351
Frasch, C. E., 159, 163–166, 252
Frasch, C. S., 249, 250(30), 251(30)
Fraser, C.E.O., 69
Fraser, D. W., 94
Frederick, C. B., 335
Fredriksen, J. H., 164
Freer, J. H., 663, 666(34)
Freese, S. J., 304
Freij, B. J., 94
French, M. F., 595
Freshney, R. I., 686
Freter, R., 449
Freudenberg, M. A., 282
Freund, S., 354
Freundt, E. A., 84
Frey, J., 472
Friedman, A. M., 369
Friedman, H., 615
Frisken, K. W., 109
Frithz, E., 286
Fritsch, E. F., 189, 194(21), 355, 365, 369(13), 435, 485, 507, 509(14)
Froholm, L. O., 164, 166
Frolik, C. A., 324

Frome, W. J., 107
Fromm, M. L., 375
Frost, G. E., 329, 364
Fryling, C., 716
Fujii, H., 258
Fujino, M., 661
Fujita, K., 682
Fujiwara, S., 497, 501(29)
Fulbright, R. S., 260
Fumarola, D., 277
Fung, K. P., 31
Funk, A., 270, 271(60)
Furlong, C. E., 235, 240, 241(17)
Furlong, D. B., 438, 519
Furman, B. L., 48

G

Gabay, J. E., 232
Gacesa, P., 295, 297, 297(7), 302, 302(7), 303, 303(7, 23), 495
Gad, S. C., 32, 36(15), 37(15)
Gaillard, J.-L., 414
Gaillard, J. L., 414
Galanos, C., 243, 282
Gale, J., 84, 93(8)
Galimand, M., 467
Gallery, F., 221, 222(38)
Galloway, D. R., 559
Galloway, T. S., 635
Gamazo, C., 233
Gambello, M. J., 464, 470, 503, 515
Gander, J. E., 226, 227(6), 229(6), 230(6), 231(6)
Gandrup, J. S., 116
Ganss, M., 48, 138
Garant, P. R., 108, 119
Garbia, S. E., 563
Garcia-Bustos, J. F., 285
Gardel, C. L., 517
Gardner, P., 677
Garibaldi, J. A., 330
Garrone, R., 594
Gaventa, S., 94, 410
Gavrilovic, J., 581
Gawron-Burke, C., 411, 415, 420, 421(33)
Geesey, G. G., 498
Geipel, U., 626
Gelfand, D., 220

Gelfand, D. H., 219
Gelfand, G. H., 177
Gellin, B. G., 164
Gelmann, E. P., 642
Gemski, P., 40, 41(7), 648
Genco, C. A., 120–122, 124, 124(2), 127(2), 128, 128(2), 129(39), 136(2), 137(21)
Genco, R. J., 106, 109, 118, 118(20a), 119(20a)
Geng, C., 254
Gensberg, K., 352
Gentry, M. K., 150, 151(21), 152(21)
Geoffrey, C., 414
Geoghegan, W. D., 707
George, P. M., 320
Gerbaud, G., 475–476, 476(9)
Gerding, D. N., 122, 127(19), 129(19), 135(19)
Gerhardt, P., 677
Gerstein, A. S., 709
Gervais, F., 451, 454, 457(8)
Gfell, M. A., 260, 280(34)
Ghosh, M., 495–496
Ghosh, P., 581
Ghuysen, J. M., 279
Gianella, R., 45
Gibbons, R. J., 110, 118
Gibbs, P., 32
Gibson, B. W., 242
Gibson, C. W., 668
Gibson, F., 329, 341
Gibson, J., 207
Gicquel, B., 197–198, 198(10), 204(15)
Giebink, G. S., 59
Giehl, T. J., 328
Giesbrecht, P., 280
Gilbert, J. V., 543, 544(10), 548, 553
Gill, D. M., 622, 624, 630–631, 679
Gill, D. R., 235, 670
Gill, J. F., 303, 494, 496
Gill, P. R., 354
Gill, S. R., 413, 417(20)
Gillam, A. H., 353
Gilliam, A. H., 371
Gillis, Z. A., 124
Gilman, A. G., 640–641, 642(12)
Gilmore, M. N., 160
Gilsdorf, J., 59
Giovannini, F., 350
Girkin, R., 609

Gissey, G. G., 493
Givner, L. B., 410
Glaser, J., 715
Glaser, P., 668–669
Glauner, B., 229, 262, 274, 278(35)
Gleeson, L. J., 84
Goddard, C., 641, 642(18)
Goebel, W., 657, 676, 678
Goebel, W. F., 243
Golaszewska, A., 615
Gold, M. R., 254
Goldberg, I., 519
Goldberg, J. B., 468–469, 493, 501, 501(5)
Goldberg, M. B., 440–441
Goldblum, R. M., 143
Goldfarb, J., 295, 297(8), 302(8)
Goldhaber, P., 115–116, 116(44)
Goldman, R. C., 242, 247(3), 563
Goldman, W. E., 270, 280(62)
Goldstein, I. M., 105
Golub, E. E., 668
Golub, L. M., 109, 118, 118(20a), 119(20a)
Gomi, H., 286
Goni, F., 665, 666(43)
Gonzalez, C. F., 328
Gonzalez, M. I., 666
Good, R. F., 366(ii), 367
Gooder, H., 264, 285(39), 416
Gooder, J. W., 416
Goodgal, S., 385
Goodloe-Holland, C. M., 715
Goodwin, J. F., 320
Gorbach, S. L., 135
Gordon, F. B., 620
Gordon, H. A., 108
Gordon, J., 672, 676(31)
Gordon, V., 673
Gordon, V. M., 710
Gorg, A., 552
Gorin, P.A.J., 295
Gorman, W. L., 501
Gorziglia, M., 504
Gotschlich, E. C., 162, 232–233, 234(29),
    288, 305, 550
Gottesman, S., 388, 393(11), 397(11)
Gottlieb, P., 185, 186(7), 188, 189(7, 8),
    191(7), 194(7, 8), 195(8)
Gottlow, J., 115
Gould-Kostka, J. L., 57
Govan, J.R.W., 297, 352

Gowland, G., 613
Grabert, E., 496
Graham, R. C., 709
Granelli-Piperno, A., 564, 579(6), 585(6),
    586(6)
Grant, C.C.R., 514
Grasdalen, H., 295, 302, 303(24)
Grasso, M., 666
Gratzl, M., 683
Gray, B. M., 170
Gray, E. D., 607, 614
Gray, H. B., 317
Gray, L., 235, 294, 668, 670, 673
Gray, L. D., 555
Gray, M., 673
Grayston, J. T., 69
Greaves, M., 362, 363(22)
Green, B. A., 59
Green, C. J., 185
Green, L., 183
Greenblatt, J., 254
Greene, J. C., 109
Greene, R. C., 209, 212
Greer, C. E., 218
Greer, C. N., 664, 671, 675
Greiling, H., 613
Greiner, M., 321, 341
Greulich, S., 657, 676
Griffin, D. E., 151
Griffiss, J. M., 242, 242(12), 243, 249(7)
Griffiths, E., 344, 346
Griffiths, G., 340
Grimm, L., 102
Grimminger, F., 676
Grimont, F., 184, 195(3), 196(3)
Grimont, P.A.D., 184, 195(3), 196(3)
Groenen, P., 204, 205(17)
Groenen, P.M.A., 197
Groisman, E. A., 386, 387(6)
Gronnesby, J. K., 164
Gross, H., 288
Gross, R., 366(y), 367, 683, 685(13)
Group, B.M.S., 94, 410
Groyne, F., 518, 519(2)
Grubb, J. D., 594, 602
Gruss, A., 413, 665
Gryfinski, M. E., 325
Guarente, L., 375
Guenounou, M., 283
Guerrant, R. L., 684

Guerry, P., 474–476, 476(3, 4), 477, 478(3, 4), 480, 480(8)
Guesdon, J., 197, 198(10)
Guesdon, J. L., 197
Guiati, S., 61, 62(8)
Guibourdenche, M., 165
Guiliano, M., 684–685
Guillen, N., 528, 530(18)
Guinand, M., 274
Gulasekharam, J., 658
Gunther, H., 440
Gupta, C. K., 31
Gupta, R., 207
Gutell, R. R., 211, 215
Gutierrez, C., 401, 434, 437
Gutteridge, J. M., 316
Gutterson, N. I., 467
Guzzo, A., 316

H

Haake, R., 414
Haanes, E. J., 294
Haanes-Fritz, E., 289
Haas, D., 328, 467, 626
Haas, R., 527, 529(6)
Haase, A. T., 222
Habermann, E., 626, 683, 690(11)
Hackbarth, C., 564, 584(7)
Hackbarth, C. J., 102, 104, 104(34), 105
Hacker, J., 676
Hacking, A. J., 297
Hadlow, W. J., 79
Hadzija, O., 265
Hafez, K., 288
Haft, R. F., 411
Haggerty, J., 608, 615(2), 616(2)
Haidairis, C. G., 439
Haigh, R., 668
Ha-Kyung, C. K., 59
Halbert, S. A., 84
Hale, T., 45
Hale, T. L., 242, 249(7)
Hall, B. G., 183
Hall, C. M., 115
Halling, A., 116, 117(54)
Hallowell, D., 94
Halperin, S. A., 49, 614, 684
Halstensen, A., 105, 164

Halter, R., 543–544, 550, 551(27), 554(8)
Hamai, A., 613
Hamilton, S. R., 143
Hammack, C., 242, 249(7)
Hammond, G. W., 137
Hamood, A., 511
Hamood, A. N., 620
Hamp, S. F., 115, 116(45)
Hanawalt, P. C., 378
Hancock, I. C., 310
Hancock, R., 258
Hancock, R. E., 121, 122(9), 127(9), 134(9)
Hancock, R.E.W., 133, 228, 232, 243, 245(24), 468
Handasyde, K. A., 84
Hanesson, H. H., 305
Hanks, J. H., 335, 336(13), 357
Hansen, E. J., 103, 105, 105(36)
Hansler, S., 107
Hanson, R. S., 369
Hanspal, J. S., 581
Hantke, K., 364, 366(10, 12, 30, v, w, x, dd, ll, nn, oo), 367, 372
Hardesty, B., 648, 653
Hardie, K. R., 665
Hardy, J., 527, 529(8)
Hardy, J. K., 321
Hardy, M. R., 310
Hareide, B., 164
Harel, J., 475, 477, 478(1)
Harford, N., 518, 519(2)
Harkki, A., 434
Harlacher, R., 254
Harlander, S. K., 381
Harmon, R. L., 490
Harper, R. R., 454, 455(13), 457(13)
Harrell, W. K., 288
Harris, G. S., 415, 418(32), 419(32), 421(32)
Harrison, J. S., 303
Harshman, L., 185
Hart, P. D., 449
Hartiala, K., 105
Harvey, S., 185, 186(15)
Harz, H., 274, 275(73), 280(73)
Hase, T., 668
Hasegawa, M., 207
Hash, J. H., 269
Hashimoto, H., 662–663, 664(29)
Haskins, W. T., 243
Hassan, A., 320

Hassell, T. M., 110, 111(29)
Hassid, W. Z., 295
Hattori, S., 286
Haug, A., 303
Haun, R. S., 641–642, 643(31), 644(30, 31, 36)
Hausinger, R. P., 546, 551(22), 552(22)
Havlicek, J., 289
He, P., 409, 418(1), 426(1)
He, S. Y., 528
Heath, D. G., 286
Hebeler, B. H., 260
Heck, L. W., 555
Heden, L.-O., 286
Heeg, H. T., 454, 455(13), 457(13)
Heffron, F., 439
Heggen, L. M., 411
Heidrich, H., 601
Heifetz, S. A., 49
Heijl, L., 107, 108(4), 118
Heilmann, H. D., 262
Heinrikson, R. L., 278
Helenius, A., 231
Helinski, D., 388, 392, 392(e), 393
Helinski, D. R., 432, 434, 434(13), 436(18), 438(13), 471, 477
Heller, I., 553
Hellstern, D., 366(31), 372
Helser, T. L., 464
Hemer, B., 254
Hendricks, P.A.J., 676
Hendrickson, W., 497, 498(28), 501
Hendy, M. D., 216
Hengstler, B., 101–105
Henning, D., 366(31), 372
Henrichsen, J., 170
Heppel, L. A., 236
Herbert, D., 284
Herd, J. K., 615
Hermans, P., 204, 205(17)
Hermans, P.W.M., 196–198, 199(14), 204, 204(13, 15), 205(14)
Hermodson, M. A., 235, 670
Herrero, M., 350, 366(r), 367, 386, 388(3), 389(3), 391(4), 392, 392(a, c), 393, 394(3), 395(3), 396(3, 4), 398(3), 399(3), 401(4), 436, 437(28), 443(28)
Herrmann, B. F., 121, 127(3), 128(3), 135(3)
Herrmann, M., 121, 127, 132(6, 34)
Herron, G. S., 581

Hespell, R. B., 207
Hetherington, C. M., 84
Heussen, C., 567, 574(14), 577(14), 587(14)
Hewlett, E. L., 48, 442, 630, 673, 684, 710
Hewlins, J. E., 303
Hey, E.G.A., 115
Heymann, H., 258
Heymer, B., 254
Higa, A., 422
Higa, N., 520
Higgins, C. F., 235, 670
Hightower, A., 410
Hightower, A. W., 94
Higuchi, R., 210, 219(17)
Higuchi, R. G., 178
Hiles, I. D., 235, 670
Hill, C. W., 185, 186(15)
Hill, G. D., 137
Hill, P., 358
Hill, R., 715
Hille, B., 697
Hillier, A. J., 381, 382(17)
Hilton, C. B., 555
Himmler, G., 437
Himpkamp, P., 271
Hindahl, M. S., 505, 508, 508(10), 511, 511(10), 512(10), 513(10)
Hinnebusch, A. G., 346
Hinrichs, D. J., 414
Hirschman, S. Z., 649
Hirsh, S. K., 94
Hirst, P. N., 352
Hitchcock, P. J., 246, 247(27), 251(27)
Hjorth, J. P., 543–544
Ho, Y.-K., 496
Hobby, G. L., 612, 613(12)
Hobson, A., 235
Hoch, D. H., 691, 695(5), 697(5), 701(5), 703(5)
Hoff, G. E., 305
Hoffman, C. S., 427, 437(3)
Hoffman, S., 189
Hofling, J. F., 121, 127(3), 128(3), 135(3)
Hofmann, P., 311
Hofschneider, P. H., 375
Hohnadel, D., 328
Hoiby, E. A., 164, 166
Høy, N., 297, 301(10), 302(10)
Holbrook, S., 235
Holbrook, S. R., 235

Holdeman, L. V., 109
Holden, I., 354
Holder, I. A., 555
Holland, I. B., 235, 659, 668, 670, 677–678
Holland, K. T., 613
Holland, S. M., 74
Hollingshead, S. K., 286–287, 289, 289(2)
Hollis, M. A., 185
Holmes, K. K., 84, 233, 234(29), 661
Holmes, S. J., 94
Holmgren, J., 518, 521, 524(4)
Holmquist, B., 327
Holo, H., 382, 418
Holt, S. C., 113, 347, 657
Holten, E., 164
Holtje, J. V., 262, 270, 271(58), 274, 275(73), 278(35), 280(73)
Homer, K. A., 608
Homma, J. Y., 555
Honjo, T., 143
Hoogkamp-Korstainje, J.A.A., 328
Hook, M., 564, 566(9), 586(9)
Hooykaas, P.J.J., 384
Hopwood, D., 386
Hori, H., 601
Horie, K., 613
Horn, 696
Horn, G. T., 208
Horn, J. M., 468
Hornibrook, J. W., 48
Horowitz, M. A., 232
Horstmann, R. D., 286
Horvath, I., 121
Hotez, P. J., 608, 615(2), 616(2)
Houck, H., 219
Hough, L., 303
Houghton, P., 114, 115(40)
Houman, F., 461
Howard, J. B., 617
Howell, M. K., 115
Hoyle, B. D., 498
Hoyt, M. J., 94
Hozumi, T., 286
Hsia, J. A., 630
Hsie, A. W., 684
Hubbard, A. L., 711
Hudson, A. P., 74
Hughes, C., 665, 668
Hughes, K., 352
Hugo, F., 657–658, 659(15), 662(15), 676

Hugouvieux-Cotte-Pattat, N., 442, 444(54)
Hultgren, S. J., 121, 138(4)
Humphrey, J. H., 613
Huneke, R. B., 75, 77(19)
Hunter, N., 597
Hunter, R. L., 563
Hunter, S. B., 164
Hurwitch, S., 295, 297(8), 302(8)
Huschka, H. G., 321
Hussein, S., 366(dd), 367
Huynh, B. H., 324
Hynes, W. L., 606, 611–612, 612(8)

I

Iandolo, J. J., 413, 417(20), 418, 419(45)
Ibrahim, A. N., 615
Ichinose, Y., 520
Igarashi, K., 649
Igietseme, J. U., 83
Iglesias, M. F., 451
Iglewski, B. H., 317, 381, 464, 468, 470, 503–505, 506(11), 507(11), 508, 508(10, 11), 510, 510(6), 511, 511(5, 6, 10), 512, 512(5, 6, 10, 18), 513, 513(5, 6, 10, 11, 18), 514–515, 554, 619–621, 621(7), 622–623, 647
Iglewski, W., 617
Ingham, E., 613
Inglewski, B. H., 554
Inglis, A. S., 527, 529(9)
Innis, M. A., 177, 219
Inoue, H., 555, 556(7)
Inouye, H., 429
Introna, M., 277
Inuzuka, M., 388, 434, 436(18)
Inzana, T., 246
Irons, L. I., 48
Irvin, J. D., 648
Irvin, R. T., 493
Irwin, D. M., 220
Ishidate, K., 229
Ishimaru, C. A., 370, 372(23)
Ishimoto, K., 468, 471(12), 472
Issartel, J. P., 665
Issekutz, A. C., 277, 283(77)
Ito, J. A., Jr., 128
Ito, J. I., Jr., 84, 93(9)
Ito, N., 118

Ivanovics, G., 490
Ives, S. K., 504
Iwabe, N., 207
Iwanaga, M., 520
Iwasaki, W., 627
Izumiya, K., 49, 52(10), 53(10), 55(10)

J

Jacewicz, M., 151, 648–649
Jacks, P. S., 414
Jackson, K., 197
Jacobs, K. A., 365
Jacobs, K. H., 626
Jacobs, M., 384
Jacobs, W. R., Jr., 385
Jacobson, G. R., 236, 419
Jacubzik, U., 386, 391(4), 392(c), 393, 396(4), 401(4)
Jadeja, L., 527
Jain, D. K., 31
Jakobs, K. H., 627
Jakubzik, U., 436, 437(28), 443(28)
Jamieson, D. K., 235
Janczura, E., 257, 258(19)
Jander, G., 440, 519, 520(17)
Janicot, M., 716
Jann, B., 304–305, 305(3), 311
Jann, K., 243, 304–305, 305(3), 311
Jansen, J., 115, 117(49)
Jarvis, E. D., 184–186, 186(9–11), 187, 190(12, 18), 191(12), 192, 194(9–12)
Jean, R., 366(u), 367
Jeanes, A., 298
Jeanloz, R. W., 269
Jeffcoat, M. J., 115, 116(44)
Jeffcoat, M. K., 115–116
Jeffcoat, R. L., 116
Jellison, J., 354
Jennings, H. J., 169, 174(31), 242, 305, 306(9), 308, 308(9), 311(13)
Jensen, A., 319
Jensen, S. B., 108
Jenssen, D., 681, 690(4)
Jerse, A. E., 441
Jessee, J., 378
Ji, J., 442
Johannsen, L., 274, 280
Johanson, W. J., 527

Johansson, S., 604
Johns, C. M., 242
Johnson, A. P., 84
Johnson, D. R., 173
Johnson, H. G., 115–116, 116(44)
Johnson, K., 527, 529(5)
Johnson, K. G., 242–243, 244(23)
Johnson, M. S., 115
Johnson, N. W., 108
Johnson, R. H., 289
Johnson, S. L., 69, 70(4), 71, 71(4)
Johnson, W. G., Jr., 134
Johnston, A.W.B., 433, 441(16)
Johnston, K. H., 233, 234(29)
Johnston, R. B., Jr., 254
Joiner, K. A., 446
Jones, B. R., 71
Jones, D., 381
Jones, D. C., 26
Jones, J. M., 412, 416(18), 418(18), 420
Jones, J.K.N., 303
Jones, K. F., 289–290, 292
Jones, R. S., 295, 297(2), 298(2), 301(2), 302(2), 303(2)
Jones, S. J., 294
Jordan, J., 378
Jorgensen, S. E., 319
Joshi, A., 241
Jubelirer, D. P., 94
Juhn, S. K., 59
Juliano, R. L., 565, 578(12), 586(12)
Jung, G., 321, 341, 354
Jurtshuk, P., Jr., 220
Just, I., 626, 683
Juy, D., 254

K

Kabat, D., 620–621, 621(7)
Kadis, S., 232
Kadner, R. J., 350, 366(l), 367
Kadurugamuwa, J. L., 101
Kafatos, F. C., 365
Kagan, B. L., 691, 691(6, 7), 692, 695, 695(6), 696, 697(6), 698, 698(6), 699(22), 700(6), 701(6, 7), 703(6, 13)
Kagan, S., 459
Kageyama, M., 233
Kahn, R. A., 641–642, 642(12, 18)

Kalkwarf, K. L., 108
Kalmar, J. R., 136
Kalve, V. I., 357
Kameyama, T., 663
Kamio, Y., 226
Kamisanago, K., 258
Kandler, O., 207, 253, 256(2)
Kane, J. I., 84
Kaneko, Y., 613, 615(21)
Kang, A. H., 288–289, 290(27), 291(27)
Kanost, M. R., 254
Kant, R., 325
Kantoci, D., 271
Kapczynski, D., 121, 124(2), 127(2), 128(2), 136(2)
Kapczynski, D. R., 124, 128, 129(39), 137(21)
Kaper, J. B., 441
Kapka, R., 99, 100(21)
Kaplan, E. L., 173, 410, 614
Kaplan, M. L., 116
Kaplan, S. L., 94
Kapsimalis, B., 118
Kar, S., 61
Karalus, R., 242
Karasic, R., 59, 60(6), 62(6), 63(6)
Karlinsky, J. B., 555
Karnovsky, M. J., 709
Karnovsky, M. L., 254
Karring, T., 115
Kasina, A., 49, 684
Kaslow, H. R., 629
Kasmala, L., 139
Kasper, D. L., 121, 122(14), 125(14), 134–135, 135(10), 411, 420(10)
Katinger, H., 437
Katlic, A. W., 709
Kato, J., 495, 496(7), 497, 497(7), 498, 500(23, 24), 501, 501(24)
Katz, D. D., 125
Kauc, L., 385
Kauffmann, P., 710
Kaufman, J. B., 504
Kaufman, M. R., 426, 444, 445(62), 528, 530(19)
Kawahara, E., 661
Kawakami, H., 662–663, 664(29)
Kawanishi, H., 142
Kaye, S., 503, 510(6), 511(6), 512(6), 513(6), 515

Kays, M., 501
Kazemier, B., 271
Keck, W., 270–271, 271(60)
Keen, N., 369
Keen, N. T., 495
Keener, J., 497
Keenleyside, W., 310
Keevil, C. W., 128, 129(38)
Keglevic, D., 265, 269, 271
Kehl, K.S.C., 139
Kehoe, M., 658, 659(14)
Kehoe, M. A., 294
Keiderling, W., 243
Keil-Dlouha, V., 576, 595
Keiser, T., 385
Kelleher, J. E., 183
Kelleher, P. J., 565, 578(12), 586(12)
Kellogg, D. S., Jr., 551
Kelly, N. M., 121, 122(9), 127(9), 133, 134(9)
Kelly, R. M., 504
Kelly, W. J., 102
Kendrick, P. L., 162
Kennard, B. D., 94
Kenne, L., 242
Kenny, C. P., 305, 306(9), 308(9)
Kenny, E. B., 108
Kent, T. A., 324
Kent, T. H., 125
Keren, D. F., 140–143, 143(2–4), 145(7), 149(5), 150(2, 3)
Kern, F. G., 642
Kern, S. E., 140, 143(2), 150(2)
Kersey, J., 414
Kessel, M., 57
Kessler, B., 386, 392(f, g), 402(5), 403(5)
Kettner, C. A., 553
Keusch, G., 648
Keusch, G. T., 151, 441, 649
Khandke, K. M., 289–290, 294
Khanna, S. K., 605
Khaw, B. A., 325
Khayam-Bashi, H., 104
Kieba, I. R., 668
Kierstan, M.P.J., 604
Kierule, P., 105
Kiiyukia, C., 662, 664(29)
Kikuchi, M., 668
Kilian, M., 163, 543–544, 544(1), 545, 550(16), 551(16, 20), 553
Kimbara, K., 497–498, 500(23, 32)

Kimura, A., 53, 55(14)
Kimura, M., 175
Kingston, R. E., 365, 369(14)
Kishimoto, F., 286
Kitano, K., 497, 500(23)
Kiviat, N. B., 218
Kivirikko, K. I., 599, 600(10)
Kiyono, H., 142–143
Kjimji, P. L., 336
Klapper, D. G., 264
Klaus, H., 110, 111(29)
Klausen, B., 108–109, 109(8), 117(8), 118(8, 20a), 119
Klausner, R. D., 346, 642
Klebba, P. E., 357
Kleckner, N., 387–388, 393(10, 11), 397(11), 463
Klein, J. O., 94
Klein, M. H., 465
Klempner, M. S., 135
Klinger, J. D., 555
Klose, K. E., 497
Klotz, U., 241
Knapp, J. S., 543
Knapp, S., 442, 446(58), 447(58), 519, 665
Knecht, D. A., 249
Knight, G.C.J.G., 658
Knivel, Y. A., 167
Knobloch, J., 286
Knowles, J. R., 632
Knox, K. K., 139
Knutsen, G. L., 50
Knutson, C. A., 298
Knutton, S., 441
Koch, A. L., 227
Koch, G. G., 358, 359(17), 360(17), 362(17), 363(17)
Kocharova, N. A., 167
Kochetkov, N. K., 167
Koehler, T. M., 695, 703(13)
Koenen, M., 441
Kohl, J., 437
Kolodrubetz, D., 668, 670(10), 671(10)
Kolter, R., 388, 434, 436(18)
Komender, J., 615
Komorowski, R. A., 139
Konetschny-Rapp, S., 321, 354
Konog, W., 676
Konowalchuk, J., 649
Konyecsni, W. M., 497, 500(21)

Koomey, J. M., 413, 419(19), 527, 529(8), 544
Kopecko, D., 39, 44, 46(10)
Kopf, G. S., 629
Korchak, H. M., 677
Kornacker, M. G., 528
Kornfeld, S. J., 543, 544(2), 545(7), 546(7), 552(7), 554(2)
Kornman, K. S., 113
Korom, S., 668, 676(15)
Koronakis, E., 665, 668
Koronakis, V., 665, 668
Koshland, D. E., Jr., 467
Koster, W., 366(y, aa), 367
Kotani, S., 283
Kothary, M. H., 554
Kotoujansky, A., 442
Kounnas, M. Z., 713
Kowalski, C. J., 112
Kowata, S., 283
Kozak, K., 710
Kraig, E., 668, 670(10), 671(10)
Kramarik, J. A., 61
Krampitz, L. O., 288
Kraus, S. J., 132
Kraus, W., 289
Krause, R. M., 258
Kreft, J., 657
Kreger, A. S., 554–555
Kreiswirth, B., 413, 665
Krejci, R. F., 108
Kreuzer, K., 394, 395(16)
Kriauciunas, A., 324
Krieger, J. N., 661
Krisch, H. M., 471–472
Krish, H., 472
Kroll, J. S., 97, 99(14), 100(14)
Kropinski, A. M., 510, 512(18), 513(18)
Krueger, J. M., 254, 271, 274
Kryshtalskyj, E., 108, 111(18), 115(18)
Kuan, W., 211, 214
Kubillus, S. L., 271
Kubo, K., 610
Kuhn, M., 426
Kuhnemund, O., 289
Kukral, A. M., 439
Kulhavy, R., 553
Kull, F. C., 675
Kulshin, V. A., 252
Kuma, K., 207

Kunka, B. S., 328
Kunz, B., 644(35)
Kunz, B. C., 642
Kuo, C., 69
Kuo, C.-C., 83, 84(1)
Kuo, C. C., 84
Kuo, S., 133
Kuramitsu, H. K., 416, 422(37)
Kurata, T., 40
Kuratana, M., 99
Kusche, M., 305
Kusser, W., 271
Kuster, H., 103, 104(37)
Kustu, S., 238, 497
Kusumoto, S., 283
Kuypers, J. M., 411–412, 419(16)
Kuyyakanond, T., 520
Kwaik, Y. A., 242
Kwarecki, K., 257, 258(19)
Kwok, S., 210, 219(17)
Kwok-Wong, K., 399

L

Labigne, A., 479
Labigne-Roussel, A., 475, 476(2), 477, 478(1, 2)
Labischinski, H., 280
Labraca, P., 695
LaBrec, E., 40, 41(7), 45
Lacks, S. A., 576, 577(18), 579(18), 587(18)
Ladant, D., 668
Ladd, T. I., 493
Ladesic, B., 265, 269, 275
Ladwig, R., 233
Laemmli, U. K., 247, 303, 547, 567, 574(13), 647, 676
LaFauci, G., 185, 186(9–11), 187, 194(9–11)
Lah, H. H., 695
Laitinen, O., 599, 600(10)
Lally, E. T., 668, 677
Lalonde, G., 381, 382(18), 385(18), 677
Lam, J., 493
Lam, K. W., 591
Lambert, P. H., 106
Lambert, T., 475–476, 476(9)
Lamm, M. E., 143
Lancefield, R., 287

Lancefield, R. C., 171, 172(36), 173(36), 174, 286–287
Landers, T., 378
Landolfo, S., 277
Landy, M., 243
Lane, D. J., 208–209, 212
Langhans, W., 254
Langman, L., 329
Lankford, C. E., 328–329
Lannigan, N. A., 608
Lantz, M. S., 563–564, 566(9), 586(9)
Lanyi, B., 167
Larrick, J. W., 105
Larsen, B., 295, 302–303, 303(24)
Larsen, N., 211, 214
Larson, E. W., 51, 58(12)
Lascelles, J., 320
Law, P. Y., 695
Laws, A. J., 109
Lazdunski, A., 528
Leak, I. G., 550
Leapman, R. D., 325
Lebel, M. H., 94
LeBlanc, D. J., 381, 382(15), 415, 418, 418(31)
Le-Blond-Francillard, M., 414
Lebman, D. A., 143
Lecar, H., 235
Leclerc, C., 254
Leclerc, C. D., 254
Lederer, E., 254–255
Lederman, L., 254
Lee, A. K., 84
Lee, A. S., 325
Lee, B. C., 358, 362(15)
Lee, C. A., 466
Lee, C.-J., 169
Lee, C.-M., 641
Lee, E. Y., 668
Lee, F.-J.S., 641
Lee, I. S., 668
Lee, J. C., 412
Lee, J. J., 186, 191(17)
Lee, J.-Y., 109, 118(20a), 119(20a)
Lee, L. N., 381, 382(15), 418
Lee, Y. P., 605
Leef, M. F., 55
Lefkowitz, J., 143
Lefrancier, P., 254

Lehner, A. F., 185, 186(15)
Lehrer, S., 49
Leinonen, M., 162
Leive, L., 230, 242–243, 247(3)
Leloir, L. F., 613
Lengeler, J. W., 419
Leonardo, E. D., 384
Lepp, P. W., 212
Leppla, S. H., 710
Lerner, C. G., 563
Lesikar, D. D., 629
Lesse, A. J., 242, 250
Letoffe, S., 669
Lever, J. E., 236, 238, 239(16), 240(16), 241(14)
Levine, H. B., 490
Levine, M. M., 441
Levy, B. M., 107
Lew, D. P., 121, 127, 132(34)
Lewin, J. E., 327
Lewis, A. G., 353, 371
Lewis, B. J., 207
Lewis, V. J., 567, 588(15)
Ley, V., 667
Li, C. Y., 591
Li, J. P., 169
Lian, C. J., 137
Lie, T., 110
Lieberman, H. R., 31
Lieflander, V. M., 615, 616(37)
Lile, J. D., 623
Lim, L.-K., 629
Lin, B., 608(3a), 609
Lin, H., 189
Lin, M. C., 640
Lin, T.-Y., 295
Lin, W., 305
Linbak, A.-K., 164
Lindahl, G., 286, 305
Lindberg, B., 310
Lindblad, M., 521
Lindeberg, M., 528
Lindhe, J., 114–115, 116(43), 118
Lindner, B., 252
Lingwood, C. A., 649
Linker, A., 295, 297(2, 5), 298(2), 301(2, 5), 302(2), 303(2), 304(5), 613
Lior, H., 475, 480(8)
Lippincott-Schwartz, J., 642

Listgarten, M. A., 108
Litchfield, J. T., Jr., 30, 31(3)
Liu, H., 103
Liu, J., 366, 366(b), 367
Liu, P. V., 503, 503(7), 527, 620, 621(7), 659
Liu, T.-Y., 163, 305
Liu, W., 229
Liu, Y., 61, 62(8)
Ljungh, Å., 684
Lloyd, D. R., 441
Lo, R.Y.C., 667, 671
Lobo, A., 672
Lobo, A. L., 667
Loe, H., 108, 110–111, 115, 115(34), 116(43)
Loftin, K. C., 107
Logan, L. C., 129, 132(41)
Logan, S. M., 475, 476(3, 4), 477, 478(3, 4)
Lohia, A., 521
Lomholt, H., 544
Long, R., 464
Long, S., 434, 438(21)
Long, S. R., 369
Long, W. J., 105
Lönngren, J., 310
Loock, C. A., 102
Loosmore, S. M., 465
Lopatin, D. E., 112
Loper, J. E., 370, 372(23)
Lopez, M., 82, 90
Lory, S., 468, 471(12), 472, 527–528, 529(5, 13), 530, 530(12), 531, 531(11), 532(12), 533(12), 536(26), 538(27), 539–540, 540(29)
LoSpalluto, J. J., 520
Loughney, K., 186–187
Louise, C. B., 649, 650(21)
Loutit, J. S., 210, 212, 219(13)
Lowry, O. H., 247
Lowy, I., 254
Lucet, J.-C., 121, 127, 132(6, 34)
Luckey, M., 328, 357
Luderitz, O., 242–243
Ludwig, R. A., 434
Luehrsen, K. R., 207
Luna, E. J., 715
Lund, E., 170, 186–187
Lundblad, G., 614
Lundrigan, M. D., 350, 366(l), 367
Lung, E., 169

Lupski, F., 436
Luria, S. E., 453
Lynch, M., 553
Lyng, K., 70
Lyon, C., 80
Lyons, J. G., 603
Lyons, J. M., 84, 93(9)
Lystad, A., 164

M

MacAlister, T. J., 229
MacConnell, P., 221–222, 222(38)
MacDermott, R. P., 210, 212, 220(14)
MacDonald, A. B., 69
MacDonald, T. T., 142
Macfarlane, M. G., 658
Mach, W.. 683
Macinkenas, M. A., 211, 214
MacIntyre, D., 105
Maciver, I., 121, 137(5)
Macke, T. J., 211, 214, 217
Mackman, N., 659, 668, 677–678
MacLachlan, P. R., 310
MacLaren, D., 657
MacLaren, D. M., 356–357
Maclean, I. W., 70
Macmillan, J. D., 661, 662(27), 665(27)
Madden, T., 710
Madden, T. E., 106
Madsen, S. L., 321
Maeda, H., 563
Maercker, C., 553
MaGee, B. B., 185, 196(14)
MaGee, P. T., 185, 196(14)
Magnusson, B. J., 658
Magrath, D. I., 341
Magrum, L. J., 207
Mahan, M. J., 481, 483, 486, 487(5), 489(5), 490(5)
Maharaj, R., 495, 496(7), 497, 497(7), 498(28)
Mahenthiralingham, E., 128, 129(38)
Mailloux, J. L., 141, 149(5)
Maizels, N., 194
Majima, M., 610
Majumdar, M. K., 661
Majumdar, S., 521
Majumdar, S. K., 661

Makalanos, J. J., 58
Makela, P. H., 242, 247(4)
Makino, S., 40
Malaty, R., 80
Malczewska, H., 615
Mallya, S. K., 596, 597(8), 598(8), 606
Maloney, H., 121, 124(2), 127(2), 128(2), 136(2)
Malorni, W., 684–685
Maloy, S. R., 180
Malta, J. C., 115
Manclark, C. R., 49, 51(9), 52(9, 10), 53(9, 10), 54(9, 10), 55(9, 10), 57
Mandel, M., 422, 454
Mandelco, L., 221
Mandrell, R. E., 166, 242, 242(12), 243
Manganiello, V. C., 640
Maniatis, T., 189, 194(21), 355, 365, 369(13), 435, 485, 507, 509(14)
Maniloff, J., 207
Manjula, B. N., 287, 289–290, 291(35), 292, 292(35), 294
Manniello, J. A., 258
Mannos, M. M., 218
Manoil, C., 427–429, 431(4), 432(8), 433(8), 434, 434(8), 435, 435(4), 438(24), 439(8), 440(4, 8), 441(8), 442(8), 443(24), 444, 447(59)
Manos, M. M., 210
Manoussakis, 40
Manske, J. M., 716
Mantovani, A., 277
Marcelis, J. H., 328
March, S. C., 603
Mardh, P. A., 84
Margulies, L., 194
Marino, J., 96, 97(11), 98(11), 99(11)
Marjai, E., 490
Markert, C. L., 563
Markham, R. B., 135
Markovitz, A., 636
Marmur, J., 189, 191(23)
Marnell, M. H., 710–711, 712(21)
Marri, L., 415
Marrie, T. J., 493
Marrs, C. F., 527, 529(8)
Marschner, T. M., 633, 634(6)
Marsh, T. L., 211, 214
Marshak, G., 59
Marshall, R. B., 132, 133(44)

Martell, A. E., 321
Martin, D. R., 173
Martin, D. W., 502
Martin, E., 676
Martin, P. L., 630
Martin, R. W., 84
Martin, S. A., 271, 274–275, 280(70)
Marugg, J. D., 351
Marumoto, R., 668
Maruyama, K., 676
Maskell, J. P., 328
Mason, P., 366(t, u, cc), 367
Mastrantonio, P., 684–685
Masui, Y., 601
Mathews, M. B., 615
Mathias, C. J., 321
Matsubara, H., 668
Matsuda, H., 662, 664(29)
Matsudaira, P., 532
Matsumoto, T., 283
Mattanovich, D., 437
Mattera, R., 640
Matthews, L. W., 297, 303(13)
Mauceri, A. A., 288
Mauler, F., 683, 690(11)
Maurer, R. L., 490
May, T. B., 295, 303, 494–496, 496(7), 497(7)
Mayrand, D., 110
Mazie, J. C., 165
McAdam, R., 198, 204(15)
McAdam, R. A., 197–198
McBride, B. C., 110
McCarty, M., 258
McCaughey, M. J., 211, 214
McClane, B. A., 683
McClean, D., 614
McClelland, M., 399
McCloskey, J. A., 633, 634(6)
McColl, K. A., 84
McComb, D. E., 69
McCoy, R. D., 305
McCracken, G. H., Jr., 102
McCracken, G.H.J., 94, 102–103, 105, 105(36), 106
McCready, M. H., 615
McCullagh, D. R., 615
McCullough, B., 134
McCune, S., 434, 438(21)
McCutchan, J. A., 128

McDade, J. E., 1
McDade, R. I., Jr., 233
McDaniel, L. S., 170
McDonald, R. A., 140, 142, 143(3, 4), 145(7), 150(3)
McDonald, T. V., 677
McDonel, J. L., 683
McEntee, C. M., 74
McFarland, W. C., 482, 490(2)
McGee, Z. A., 254, 285(6)
McGhee, J., 142–143
McIntosh, M. A., 328, 351, 366(29, a, c, e, f, h, j, k, n), 367, 371
McIntyre, D. A., 381
McKay, D. B., 620
McKeehan, W. L., 653
McKenna, W. R., 357, 358(11)
McKern, N. M., 527, 529(9)
McKerrow, J. H., 555
McLauchlin, J., 413
McLaugh, W., 242
McNamara, T. F., 118
McNeish, A. S., 441
McRae, W. B., 555
McWhinney, D. R., 666
McWilliams, M., 143
Meade, B. D., 57
Medlin, L., 209
Mehl, T., 527
Mehrotra, S. C., 320
Mehta, S. C., 31
Mehta, S. K., 545
Meiwes, J., 354
Mekalanos, J., 388, 392, 392(b), 393, 431, 448(11)
Mekalanos, J. J., 58, 427–429, 431, 432(8), 433(8), 434, 434(8), 438–439, 439(8), 440(8), 441(8), 442, 442(8), 446, 446(17, 34, 58), 447(58), 464, 468–469, 473(11), 481–484, 486, 487(5), 489(5), 490(5), 517–519, 519(2), 520, 520(3, 7, 16, 17), 521–522, 522(7, 16, 25), 523(16, 25, 35, 37), 524, 525(16, 25), 631, 640, 665
Melly, M. A., 254, 285(6)
Melton, R. E., 385
Mende-Mueller, L. M., 633, 635(4), 636(4)
Menestrina, G., 678
Mengaud, J., 414
Mercenier, A., 384
Meridith, S. C., 278

Merrick, W. C., 653, 655(24)
Mertsola, J., 105
Mesrobeanu, L., 243
Messing, C. R., 554, 558–559
Messing, J., 395, 398, 398(17)
Mestecky, J., 543, 544(1), 545, 551(20), 553
Mett, H., 270, 271(60)
Metzger, J., 321, 341, 354
Meyer, J.-M., 328
Meyer, K., 612–613, 613(12), 614(14), 615(14)
Meyer, T. F., 527, 529(6), 543–544, 550, 551(27), 553, 554(8)
Meyers, G. A., 630
Meynell, G. G., 449
Michaelis, S., 429
Michalek, S., 142
Michalek, S. M., 415, 418(32), 419(32), 421(32)
Michel, R. G., 323
Mickelsen, P. A., 356–357, 357(1), 358(11)
Middlebrook, J. L., 620, 710
Mifuchi, I., 258
Mikawa, T., 676
Mikx, F.H.M., 115
Miles, A. A., 336
Miller, C., 695
Miller, C. L., 254
Miller, D. M., 454, 455(10), 457(10)
Miller, D. R., 263, 272(37)
Miller, D.M.I., 241
Miller, E. J., 555
Miller, J., 308
Miller, J. F., 375, 376(3), 378(3), 381, 381(2), 382(2, 3, 18), 383(2, 3), 384(2, 3), 385(2, 18), 418, 438, 475, 478(6), 479(6), 480(6), 519, 522(15)
Miller, J. H., 368, 370(17), 388, 390(15), 391(15), 394(15), 395(15), 396(15), 401(15), 429, 433(10), 435(10), 491, 492(12), 499, 510
Miller, J. J., 519, 522(15)
Miller, K. D., 555
Miller, L., 294, 311(17)
Miller, L. C., 30, 31(3), 35(3)
Miller, L. D., 465
Miller, S., 439
Miller, V., 388, 392(b), 393, 431
Miller, V. L., 434, 438, 446(17), 464, 468–469, 473(11), 483–484, 519–520, 520(7), 522(7, 25), 523(25), 525(25)

Mills, E. L., 59
Mills, R., 59
Milner, K. C., 243
Milstone, L., 608, 615(2), 616(2)
Mims, C. A., 648
Mimura, C., 235
Mimura, C. S., 235
Mindel, J., 700
Ming, L. J., 324
Minshew, B., 668
Mirelman, D., 265, 269–270, 271(58)
Mische, S. M., 289
Mischell, R. I., 254
Mishima, K., 641
Misra, T. K., 495–496, 496(7), 497, 497(7), 498(22), 500(21–24), 501(22, 24)
Mitchell, D. H., 485
Mitchell, M. A., 385
Miyamoto, V. K., 693
Miyasaki, K. T., 567, 588(17)
Miyata, T., 207
Mizuno, T., 233
Mocca, L. F., 164, 166
Mochizuki, H., 283
Model, P., 292
Modun, B., 133
Mohan, S., 528, 530(18)
Mohos, S., 609
Mohr, C. D., 501
Molla, A., 563
Möllby, R., 683
Moller, B. R., 84
Monack, D., 665
Monaco, L., 641
Monday, S. R., 495
Monkiedje, A., 661
Monnickendam, M. A., 69, 79(6), 80
Montal, M., 691, 700(4)
Moody, J. A., 122, 127(19), 129(19), 135(19)
Moody, M. D., 171, 172(39)
Mookhtiar, K. A., 595–596, 597(8), 598(8), 606
Moore, D. D., 365, 369(14)
Moore, W.E.C., 109
Moore, W.G.I., 603
Moos, A. B., 79
Moran, T. P., 649, 652, 653(23), 654(15)
Moravek, L., 289
Morency, C. A., 510
Morgan, C. P., 48
Mori, M., 118

Morihara, K., 555, 556(7), 559, 560(19)
Morikawa, K., 613
Morisato, D., 387, 393(10)
Moriyon, I., 233
Morris, L. J., 360
Morris, R. E., 705–707, 707(3), 709, 709(3, 5), 710, 710(10), 711(18), 712(18), 713, 713(3), 715(5), 716
Morris Hooke, A., 448–449, 451, 453–455, 455(10, 11, 13), 456(14), 457(1, 8, 10, 13–15)
Morrison, D. C., 283
Morrison, G. H., 325
Morrison, N. E., 335, 336(13)
Morrison, R. P., 70
Morse, S. A., 1, 83, 84(1), 121–122, 122(13), 128, 129(39), 137, 137(13)
Mortensen, S. B., 545, 553
Morton, D. J., 356
Moses, E. B., 84, 91(6)
Mosier, D. A., 677
Moskaug, J. O., 716
Moss, J., 625, 629–630, 640–642, 642(16, 17), 643, 643(16, 17, 31), 644(17, 21, 30, 31, 35, 36)
Mosteller-Barnum, L., 142
Motycka, L., 615
Moulder, J. W., 439
Mounier, J., 426
Mount, D. T., 84
Mountzouros, K. T., 53, 55, 55(14), 57(18)
Mouw, A. R., 294
Mowatt, M. R., 641
Moxon, E. R., 95–96, 96(7), 97, 97(11, 12), 98, 98(11, 13), 99, 99(7, 11, 12, 14, 17), 100(13, 14, 20, 21)
Mozo, T., 384
Muczynski, K. A., 642, 644(35)
Mudd, M. H., 502
Mueller, L., 308
Mueller, P., 691
Muench, H., 34
Muhly, M., 657, 668, 676, 676(15)
Mukherjee, S., 111, 115(33), 116(33)
Mulczyk, M., 47
Mulks, M. H., 543, 544(3), 545(7), 546, 546(7), 551(22), 552(6, 7, 22), 553–554, 554(3)
Muller-Hill, B., 441
Mullis, K. B., 208
Munford, R. S., 243, 283

Muntwyler, E., 300
Murayama, S., 40
Murov, S. L., 634
Murphy, E., 417, 418(38)
Murphy, R. A., 614
Murphy, R. F., 685
Murray, E. S., 69
Murray, N. E., 183
Murtagh, J. J., Jr., 641
Musser, J. M., 160, 175
Mustafa, M. M., 103, 105, 105(36)
Mustafa, S., 385
Myers, G. A., 48, 684
Myler, P. J., 139

N

Nagai, Y., 601
Nahlik, M. S., 328, 366(a, c, j), 367
Nairn, C. A., 129
Nakae, T., 225
Nakahama, K., 668
Nakamura, T., 610
Nakano, H., 662–663, 664(29)
Nakano, M., 283
Nakasone, N., 520
Nakaya, S., 640
Napoli, C., 369
Narasimhan, S., 608, 615(2), 616(2)
Narbad, M.J.E., 303
Nash, T. E., 641
Nasjleti, C. E., 112
Nassar, M., 94
Nataro, J. P., 441
Nato, F., 165
Nauciel, C., 283
Naughton, M. A., 557
Naumski, R., 265, 269
Navaratnam, P., 661
Navia, J. M., 82, 90, 108, 117(8), 118(7)
Naylor, M. N., 108
NCCLS Document M2-A4, 3
NCCLS Document M7-A2, 3
NCCLS Document M11-T2, 3
Neefs, J.-M., 214
Neff, N. H., 640
Nei, M., 182
Neiders, M. E., 564
Neilands, J. B., 328–331, 331(5), 333, 338(4), 340, 340(4), 341(5), 342(4),

343(5), 344, 344(4, 5), 346, 349–354, 354(1), 357, 364, 366(q, r, z, mm), 367, 368(8), 371
Neimark, H., 209, 212
Nelson, J., 97
Nelson, J. D., 102
Nelson, K., 174–175, 181(2)
Noe, I. F., 382, 418
Nethery, A., 597
Neumann, E., 375
Neumann, J. R., 510
Neuss, B., 440
Newhall, W. J., 91
Newkirk, M., 641, 642(18)
Newman, K. B., 641
Newman, M. G., 110
Newman, R., 362, 363(22)
Ngassapa, D.N.B., 115
Nguyen, S., 197, 198(10)
Nicas, T. I., 623
Nicaud, J. M., 659, 668, 677
Nickel, J. C., 493
Nicotera, P., 685
Nief, F., 254
Niemöller, U. M., 103, 104(37)
Nigg, C., 90
Nightingale, M. S., 641
Nijkamp, F. P., 676
Nikaido, H., 225–226, 228–229, 230(1), 236, 240(13), 241(13)
Niles, W. D., 232
Nishino, N., 559, 560(20)
Noble, R. C., 160
Noda, M., 641–642, 642(16, 17), 643(16, 17), 644(17)
Noel, K. D., 241
Nokleby, H., 164
Noller, H. F., 211, 215
Nomura, S., 661
Northup, J. K., 640
Nossal, N. G., 236
Novic, B., 300
Novick, R. P., 413, 665
Nunez, H. A., 308
Nunn, D., 527–528, 531(11)
Nunn, D. N., 527–528, 529(13), 530(12), 532(12), 533(12), 539–540, 540(29)
Nuovo, G. J., 221–222, 222(38)
Nussenzweig, V., 667
Nutten, A. J., 300

Nydegger, U. E., 106, 121, 132(16)
Nyman, S., 115

O

Oaks, E., 45
Obal, F., Jr., 274
Obernesser, H.-J., 555
Oberti, J., 285
O'Brien, A., 29
O'Brien, A. D., 454
O'Brien, P.C.M., 449
Obrig, T. G., 647–649, 650(21), 652–653, 653(23), 654(15)
Ochman, H., 178, 485
Oda, K., 555
Oda, M., 49, 51(9), 52(9), 53(9), 54(9), 55, 55(9)
Odio, C. M., 94
O'Donnell, I. J., 527, 529(9)
Oeschger, M. P., 449, 451, 453–454, 457(1)
Ofek, I., 288–289
O'Gara, F., 352
Ogasawara, K., 649
Ogata, M., 716
O'Grady, R. L., 597, 603
O'Halloran, T. V., 350
O'Hanley, P., 381, 382(18), 385(18)
O'Hanley, P. D., 677
Ohishi, I., 626–627
Ohman, D. E., 468–469, 493, 495–496, 501, 501(5), 503, 621
Ohya, T., 613, 615(21)
Oka, T., 555, 556(7)
Oksche, A., 683, 690(11)
Okutani, K., 305
Olitzki, A. L., 649
Olitzki, Z., 649
Oliver, D., 429
Oliver, J. D., 328
Olsen, G. J., 208, 211, 214, 216
Olsen, K. D., 94, 103, 105(36)
Olsens, K., 715
Olsnes, S., 648, 709–710, 716
Olson, J. S., 241
Olson, K. D., 94
O'Malley, J. A., 320
Omland, T., 162

Onderdonk, A. B., 121, 122(14), 125(14), 134–135, 135(10)
Oppenheimer, N. J., 633, 634(6)
Orci, L., 642
O'Reilly, M., 413, 665
O'Reilly, T., 121, 137(5)
Orndorff, P. E., 230
Ornstein, L., 563, 590(5), 591(5)
Orr, M. B., 109
Orskov, F., 160, 305, 310
Orskov, I., 160, 305
Ortiz, J. M., 666
Osawa, S., 207
Osawa, T., 269
Osborn, M. F., 84
Osborn, M. J., 226, 227(6), 229(6), 230(6), 231(6), 300
Osborne, J. C., Jr., 640
Osgrove, S., 603, 610(2)
Oshioda, T., 417
Oskouian, B., 418
Ostapchuk, P., 193
Ostolaza, H., 665, 666(43)
Ostroff, R. M., 467
Ostrow, P. T., 99, 100(20, 21)
O'Sullivan, D. J., 352
Ott, M. L., 615
Ottemann, K. M., 522, 523(35)
Otto, B. R., 356–357
Otto, K., 441
Overbeck, R., 211, 214
Owase, M., 677
Owen, B. D., 328
Owen, R. J., 196
Oxtoby, M. J., 410
Ozenberger, B. A., 328, 366(c, e, j), 367

P

Paabo, S., 218, 220
Pabst, M. J., 254, 277
Pace, B., 208, 219
Pace, N. R., 208, 212, 219
Padgett, P. J., 494
Page, R. C., 108
Pahuja, S. L., 607
Pal, S., 70, 75, 77(19)
Palecek, R. C., 454, 455(10), 457(10)
Palecek, T. M., 454, 455(10), 457(10)

Palmer, G., 324
Palva, L., 242, 247(4)
Pappenheimer, J. R., 254
Paradisi, S., 684–685
Paramov, N. A., 167
Parham, P., 220
Parikh, I., 603
Paris, A. L., 121, 125(12), 132(12)
Paris, M., 94
Parisi, E., 226, 227(6), 229(6), 230(6), 231(6)
Park, J. T., 258
Park, S. F., 381, 382(16), 384(16)
Parker, A., 382, 383(24)
Parker, C. D., 232, 490
Parker, M. L., 527, 529(5)
Parr, T. R., Jr., 232
Parry, D.A.D., 287
Parsonnet, J., 124
Parsons, N. J., 127, 129, 129(33)
Parsot, C., 446, 519, 520(17), 521, 523(35)
Paruchuri, D. K., 360, 361(18)
Passador, L., 617
Patel, M. K., 111, 115(33), 116(33)
Patel, P. V., 129
Pato, M. L., 421, 495–496
Pattee, P. A., 412, 416(18), 418(18)
Patton, D. L., 83–84, 84(1), 85
Paul, S., 232
Paulsen, H., 271
Pavlik, M., 289
Pavlovskis, O. R., 620
Paw, B. H., 366(q), 367
Payne, E. E., 59
Payne, S. M., 328–329, 336, 340, 341(14), 345, 352, 363, 369(2), 372(2)
Peacock, W. L., 129, 551
Pearson, G.D.N., 518–519, 519(2)
Pedersen, S. S., 297, 301(10), 302(10)
Pe'er, J., 76
Pelletier, D. A., 209, 212
Pellett, S., 661, 668–669, 672
Pelton, S. I., 61, 62(8)
Pelva, E. T., 434
Pelzer, H., 253
Penders, J.E.C., 420, 422(49)
Penn, C. W., 127, 129(32, 33)
Pennington, T. H., 185, 195(4), 196(4)
Penny, D., 216
Peppler, M., 48, 138
Peppler, M. S., 242(13), 243

Perez-Casal, J., 411
Perez-Diaz, J. C., 414
Perez-Pinero, B., 68
Perkins, H. R., 257–258, 258(19), 266
Perkins, R. L., 124
Perlman, E., 243
Perlmann, 287
Perry, D., 416
Perry, M. B., 243, 244(23)
Persing, D. H., 220
Peters, J. E., 559
Peters, T., 271
Petersen, B. H., 283
Petersen, O., 648
Petersen, O. W., 709
Peterson, K., 431, 448(11), 519
Peterson, K. M., 438, 446(34), 519–520, 520(16), 522(16), 523(16), 525(16)
Peterson, L. R., 122, 127(19), 129(19), 135(19)
Peterson, S. L., 218
Petit, J. F., 254
Pettis, G. S., 351, 366(f, h, k), 367
Pflugrath, J. W., 241
Phadnis, S. H., 388
Phaff, H. J., 331
Phibbs, P. V., 494
Philips, N. J., 242
Philips, T. M., 603, 604(3)
Phillips, G. J., 426
Phillips, G. N., 289
Phillips-Jones, M. K., 381
Phillips-Quagliata, J. M., 143
Phipps, P. J., 284
Pichot, F., 668
Pier, G. B., 124, 167, 295, 297(8), 302(8)
Pierce, J. R., 338
Pierce, N. F., 143
Pietrowski, R. A., 647(3), 648
Pike, R. M., 1
Pike, W. J., 133
Pilot, T., 115, 117(49)
Pine, L., 288
Pinkney, M., 658, 659(14)
Pinner, R. W., 164
Pirkle, C. I., 551
Pitt, A., 715
Pitt, T. L., 167
Pittenauer, E., 274, 280(74)
Pittman, M., 48, 161–162

Pitts, A., 142
Pizza, M., 635
Plaut, A. G., 543, 544(2, 4), 544(10), 545, 545(7), 546, 546(7), 548, 551, 552(6, 7), 553, 554(2)
Pohlner, J., 543–544, 550, 551(27), 553, 554(8)
Pokorny, M., 265
Polentarutti, N., 277
Pollack, J. R., 328, 357
Pollack, M., 620
Pollock, J. J., 269
Polson, A., 111
Poole, B., 685
Poolman, J. T., 164, 166, 166(17)
Popoff, M. R., 683, 690(11)
Porter, P., 140, 143(2), 150(2)
Porter, T. N., 121, 138(4)
Portnoy, D. A., 414
Postel, W., 552
Postic, D., 165
Postle, K., 364, 366(ii), 367
Potts, W. J., 527
Pou, S., 324
Poulsen, K., 543–544
Powell, I. B., 381, 382(17)
Powell, P. E., 327
Power, M. E., 475, 476(4), 477, 478(4), 480(8)
Powers, J. C., 559, 560(20)
Poxton, I. R., 310
Poy, F., 419
Prasad, F., 105
Pratt, C., 394, 395(16)
Preiss, J., 304
Prendergast, R. A., 69–70, 70(4), 71, 71(4), 75, 76(20)
Prentki, P., 471–472
Presek, P., 626
Pressler, U., 366(ff), 367
Pretzer, E., 269
Prévost, M.-C., 426
Price, S. R., 641–643, 644(30)
Priefer, U., 433, 434(14), 461, 471, 473(17), 483, 486
Prince, A., 468, 471(12), 472, 501
Prince, R. W., 317, 514–515
Prior, R. B., 124
Pritchard, D. G., 608(3a), 609
Procino, J. K., 415, 419(16)

Prockop, D. J., 599, 600(10)
Proctor, D. P., 303, 412
Prody, C., 238
Prody, C. A., 366(z), 367
Projan, S. J., 413, 419(19)
Prugnola, A., 635
Pryce-Jones, R. H., 608
Prydz, K., 709, 716(9)
Pu, Z., 75, 77(19)
Puck, T. T., 684
Pugsley, A. P., 528–529, 530(17)
Puhler, A., 433, 434(14), 461, 471, 473(17), 483, 486
Puthucheary, S. D., 661

Q

Quagliarello, V. J., 105
Que, L., 324
Quie, P. G., 59
Quinn, T. C., 74
Quiocho, F. A., 241
Quivey, R. J., Jr., 405, 416

R

Racker, R., 695
Ragsdale, C. W., 375, 376(3), 378(3), 382(3), 383(3), 384(2, 3), 418
Rainard, P., 328
Raios, K., 197
Raivio, T. L., 503, 510(6), 511(6), 512(6), 513(6)
Ralls, S. A., 115
Ralston, D. M., 350
Ramamurthy, N. S., 109, 118, 118(20a), 119(20a)
Ramijford, S. P., 111, 115(34)
Ramilo, O., 103, 105, 105(36)
Ramos, C. L., 324
Ramphal, R., 49
Ramsey, K. H., 91
Ramsey, N.K.C., 414
Rand, K. H., 219
Randall, L. L., 229
Randall, R. J., 247
Rank, R. G., 69, 78, 83–85, 87(16), 91, 91(6), 92

Rapoza, P. A., 76
Rappuoli, R., 633, 635, 635(4), 636(4)
Rasheed, J. K., 121, 125(12), 132(12)
Rateitschak, E. M., 110, 111(29)
Rathore, M. H., 410
Raveneau, J., 414
Raymond, C. K., 434
Reacher, M., 74, 76
Recsei, P., 413
Reddy, M. S., 115–116
Redfield, R. J., 186, 191(17), 444
Reed, L. J., 34
Rees, R.J.W., 449
Reeves, P. R., 183
Regnery, R. L., 26
Rehfeld, C. E., 115
Reich, C. V., 357
Reich, E., 564, 579(6), 585(6), 586(6)
Reich, K. A., 635
Reichert, J., 366(i), 367
Reid, C.P.P., 327
Reijntjens, F.M.J., 115
Reilly, J. S., 59
Reimmann, C., 467
Reingold, A., 218
Reinholdt, J., 545, 548, 550(16), 551(16), 553
Reinicke, B., 280
Reisbig, R., 648
Reisfeld, R. A., 567, 588(15)
Reissig, J. L., 613
Relf, W. A., 294
Relman, D. A., 53, 55(14), 205, 210, 212, 219(13), 220(14)
Remeza, V., 194
Remold, H. G., 282
Rennie, R. P., 663, 666(34)
Rescei, P., 665
Rest, R., 269
Retzel, E. F., 222
Reutter, W., 282
Rey, M., 305
Reynolds, B. D., 283
Reyss, I., 528–529
Reznikoff, W. S., 485
Rhodes, C., 194
Ribi, E., 243
Rice, P., 242(12), 243
Rice, P. A., 61, 62(8)
Riceaga, J., 316
Rich, R., 104

Richards, F. F., 608, 615(2), 616(2)
Richards, K. L., 681, 684(3)
Richards, M. P., 321
Richardson, S. H., 517, 520, 520(1), 521(1), 523(1, 21, 23, 30), 525(1)
Richman, P. G., 612
Richmond, S. J., 70
Richter, A. G., 124
Richter, I. R., 185, 186(11), 187, 194(11)
Riddell, R., 650
Rietschel, E. T., 252
Rifkin, B. R., 107, 108(4)
Rigby, P.W.J., 194
Riley, M., 185, 193, 195(5)
Rindler-Ludwig, R., 567, 588(16), 591
Riou, J. Y., 165
Rippon, J. W., 561
Risby, T. H., 322
Risser, R. C., 105
Ro, J. Y., 325
Robbins, E. S., 667
Robbins, J., 305
Robbins, J. C., 294
Robert-Baudouy, J., 442, 444(54)
Roberts, D. E., 387, 393(10)
Roberts, I. S., 381
Roberts, R. C., 564, 584(7)
Roberts, R. S., 243
Robertson, D. C., 640
Robertson, W., 614
Robinson, A., 48
Rocha, E. R., 323
Rode, H. N., 366(u), 367
Rodgers, J., 94
Rodriguez-Barradas, M. C., 26
Roeder, D. L., 442
Roeder, R. G., 630
Rogers, H. J., 257, 258(19), 327–328, 336
Rogers, M. B., 382, 383(24)
Rohn, D. D., 59
Rohner, P., 121, 127, 132(6, 34)
Roman, D. G., 346
Romanowska, E., 47
Romero-Mira, M., 691, 695(5), 697(5), 698, 701(5), 703(5)
Romig, W. R., 518, 631, 640
Ronald, A. R., 137
Ropes, M. W., 614
Rosen, G. M., 324

Rosen, H., 605
Rosenberg, E. Y., 229
Rosenberg, H., 329, 364
Rosenberg, H. M., 115
Rosenbloom, J., 668
Rosenbloom, J. C., 668
Rosenbrough, N. J., 247
Rosenbusch, J. P., 236
Rosenfeld, R. M., 68
Rosenqvist, E., 164, 166
Rosenthal, R. S., 253–254, 260, 263, 270–271, 271(61), 272(37), 274, 279, 279(61), 280(34, 70), 283, 285(6)
Rosevear, P., 308
Rosner, A. M., 140, 143(3), 150(3)
Ross, B. C., 197
Ross, E. M., 640
Rothbard, J., 527, 529(8)
Rothfield, L. I., 229
Rothlauf, M. V., 269
Rothman, J. E., 642
Rothman, S. W., 151, 683
Rothmel, R., 494, 496
Rothmel, R. K., 495, 496(7), 497(7)
Rotilie, A., 124
Rotta, J., 173
Rouleau, M., 613
Rout, W., 45
Rouviere, P., 221
Rovin, S., 108
Row, G. E., 657
Rowe, G., 672
Rowe, G. E., 674
Roychoudhury, S., 303, 493–496, 496(7), 497, 497(7), 498(28)
Rubens, C. E., 411, 420(10)
Rubin, L. G., 96–97, 97(12), 98–99, 99(12, 14, 17), 100(14)
Rudin, D. O., 691
Rudner, R., 184–186, 186(7, 9–11), 187–189, 189(7), 190(12, 18), 191(7, 12), 192, 194, 194(7, 9–12)
Ruggieri, F. M., 685
Ruker, F., 437
Rulka, C., 642
Rumore, P., 209, 212
Rusnak, F., 366(i), 367
Russell, H., 288
Russell, L. D., 325

Russell, M. W., 543, 544(1)
Russell, N. J., 295, 297, 297(7), 302, 302(7), 303, 303(7, 23)
Russell, R. J., 121, 122(11), 127(11), 132(11), 133(11), 134(11)
Russell-Jones, G. J., 550
Russian, K. O., 510
Rust, L., 554
Ruther, U., 441
Rutz, J. M., 357
Ruvkun, G. B., 433, 466
Ruysschaert, J. M., 695
Ryan, Y., 440
Ryd, M., 709, 716(9)
Ryder, R. C., 520, 523(23)

S

Sa-Correia, I., 496–497
Sadler, K. N., 212
Sadoff, J. C., 504, 619, 621
Saelinger, C. B., 647, 705–707, 707(3), 709, 709(3, 5), 710, 710(10), 711, 711(18), 712(18), 713, 713(3), 715(5), 716
Saez-Llorens, X., 94
Sagot, N., 165
Saier, M. H., 466
Saiki, R. K., 208
Saiman, L., 468, 471(12), 472, 501
Saito-Taki, T., 283
Saitou, N., 182
Sakaguchi, T., 627
Sakai, T., 40
Sakaitani, M., 366(i), 367
Sakakibara, S., 601
Sakamoto, H., 669
Sakurada, S., 674
Salmond, G. P., 670
Salmond, G.P.C., 235
Salter, R. D., 220
Saltman, P., 315
Sambrook, C. M., 300
Sambrook, J., 189, 194(21), 355, 365, 369(13), 435, 485, 507, 509(14)
Sanborn, W. R., 305
Sande, E. R., 102, 104(34)
Sande, M. A., 95, 100(8), 102, 102(8), 104, 104(34), 105–106

Sande, S., 105
Sanders, M. M., 83, 85, 87(16)
Sandler, M. B., 116
Sandvig, K., 648, 709–710, 716, 716(9)
Sanger, F., 557
Sansonetti, P., 44–45, 414
Sansonetti, P. J., 426
Sarabia, V., 310
Sarasola, P., 297
Sarhed, G., 115
Sarkar, G., 219
Sarma, A. B., 661
Sarvani, G. A., 173
Saskawa, C., 40
Sato, H., 49, 52(10), 53(10), 54, 54(10), 55(10, 15)
Sato, Y., 49, 52(10), 53(10), 54, 54(10), 55(10, 15)
Satta, G., 256
Sauer, K. T., 684
Sauer, M., 366(w, x), 367
Sauer, W., 627
Saukkonen, K., 104–105
Saunders, J. A., 377
Saunders, J. R., 527
Saunders, M., 295, 297(8), 302(8)
Sawyer, W. D., 260
Saxe, S. R., 109
Schaad, U. B., 102
Schachter, J., 69, 70(4), 71, 71(4), 80–81, 85, 86(18), 92–93
Schad, G. A., 608, 615(2), 616(2)
Schade, A. L., 316, 328(6)
Schaefer-Ridder, M., 375
Schaeffer, A. J., 121, 138(4)
Schaffer, S., 366(nn), 367
Schagger, H., 250, 251(33)
Schalla, W. O., 125, 129
Scharf, S., 208
Scharrer, E., 254
Schauer, R., 300
Scheffer, J., 676
Scheifele, D. W., 100
Schein, S. J., 696, 699(22)
Scheld, W. M., 95, 102, 105–106
Schell, S. C., 160
Schenkein, H. A., 136
Schering, B., 626–627
Scherrer, R., 677

Schiebel, E., 440, 663
Schifferli, D. S., 434, 448(20)
Schiller, N. L., 304, 495
Schivaragin, H. W., 552
Schlech, W. F., 94
Schleifer, K. H., 253, 256(2), 257, 258(22)
Schleifer, L. S., 640
Schleivert, P., 665
Schleivert, P. M., 175
Schlievert, P., 413
Schmalzl, F., 591
Schmid, A., 678
Schmidt, A., 304, 305(3)
Schmidt, G., 668, 676(15)
Schmidt, T. M., 205, 208, 210, 212, 219, 219(13), 220(14)
Schnaitman, C. A., 226, 230, 231(18)
Schneerson, R., 305
Schneewind, O., 292
Schneider, C., 362, 363(22)
Schneider, H., 45, 242, 242(12), 243, 249(7)
Schockman, G. D., 415
Schoffler, H., 440
Schoofs, G. M., 31
Schoolnik, G., 527, 529(8)
Schottel, J. L., 506
Schramm, V. L., 269
Schrettenbrunner, A., 258
Schroeder, H. E., 108, 114
Schryvers, A., 358, 360, 362(15)
Schuchat, A., 410
Schuitema, A. R., 197
Schultz, D. R., 555
Schultz, G., 627
Schulze, M. L., 1
Schurr, M. H., 502
Schwab, J. H., 254, 264, 285(39)
Schwartz, A. L., 715
Schwartz, B., 410
Schwartz, L. L., 666
Schwarz, U., 262, 266, 270–271, 271(58, 60), 278(35)
Schweizer, H. P., 469–470, 472
Schwenk, E., 615
Schwyn, B., 333, 353–354, 354(1), 357, 371
Scott, J. R., 286–287, 289, 289(2), 292, 409, 410(2), 411, 419(8), 420(8), 421
Scott, P. J., 140, 143(2, 3), 150(2, 3)
Seddon, S. V., 563
Sedivy, J. M., 384

Seeger, W., 676
Seeliger, H.P.R., 258
Seger, J., 258
Seid, R. C., 242, 249(7)
Seidman, J. G., 365, 369(14)
Seifter, S., 300
Selander, R. K., 160, 164, 174–175, 181(2), 183
Sen, D., 127, 129(33)
Sen, P. K., 358, 359(17), 360(17), 362(17), 363(17)
Senghas, E., 409, 418(1), 420, 426(1)
Senior, B., 668
Serafini, T., 642
Sereny, B., 39–40, 40(1), 41(1), 42(1), 43(1, 4, 5), 44(1), 45(1, 4), 46
Serianni, A. S., 308
Seroky, J., 68
Serra, J., 665, 666(43)
Serventi, I. M., 641
Sesartic, L., 275
Setauchi, Y., 185, 186(11), 187, 194(11)
Sewell, J. L., 641
Seyer, J. M., 288–289, 290(27), 291(27), 528, 530(19)
Sezer, O., 668
Sfintescu, C., 109, 118(20a), 119, 119(20a)
Shackleford, A. H., 620
Shah, H. N., 563
Shahin, R. D., 47, 55, 57–58, 58(20)
Shalaby, R., 105
Shales, S. W., 232
Shaltiel, S., 241
Shand, G. H., 297, 301(10), 302(10)
Shapiro, M. E., 135
Sharon, N., 265, 269–270, 271(58)
Sharp, P. M., 183
Sharper, J., 715
Shastry, B. S., 630
Shaw, C., 431, 448(11), 520
Shaw, C. E., 527, 529(9)
Shaw, W. V., 659
Shea, C., 107
Shea, C. M., 366(n), 367
Shear, M. J., 242
Sheffield, F., 29
Shelton, A. P., 133
Shen, T., 325
Sherbock-Cox, V., 302, 303(23)
Sherry, B., 105

Shewen, P., 667, 671
Shewen, P. E., 664, 667, 674(3), 675, 676(3)
Shia, S.-P., 710–711, 712(21)
Shiba, T., 283
Shibl, A. M., 105
Shiley, J., 320
Shiloah, J., 504
Shimizu, H., 258
Shin, M., 350
Shinabarger, D., 494–496, 496(7), 497(7)
Shinnick, T., 198, 204(15)
Shive, W., 335
Shoberg, R. J., 543, 550, 554
Shockman, G. D., 265, 269
Shofstahl, J. H., 321
Shoham, S., 271
Shoolery, J. N., 354
Shortridge, V. D., 495–496
Shoshan, M. C., 685
Shyamala, V., 235
Sibelius, U., 676
Sievertsen, H. J., 286
Sigel, S., 340
Sikdar, D. P., 661
Silhavy, T. J., 369, 426, 436, 484, 490, 492(7)
Silness, P., 110
Silver, S., 317, 364
Silverman, S. H., 121, 137(5)
Silverstein, A. M., 69, 70(4), 71(4)
Silvestri, S., 635
Simon, D., 412, 418(14), 473(17)
Simon, E. H., 341, 350
Simon, G. L., 135
Simon, M. I., 691, 700(4)
Simon, R., 433, 434(14), 461, 471, 483, 486
Simons, K., 231, 709
Simons, R. W., 463
Simpson, D. A., 546, 551(22), 552(22), 695(5)
Simpson, D. L., 674
Simpson, D. M., 107
Simpson, L. L., 691, 697(5), 701(5), 703(5)
Simpson, L. M., 328
Simpson, W. A., 289
Simpson, W. J., 294
Sims, T. N., 110
Singer, E. R., 438
Singh, G. B., 605
Singh, J. V., 605
Singh, S. K., 303, 494, 496

Singh, S. P., 357
Sinha, R. K., 270, 271(61), 279, 279(61)
Sinn, L. M., 122, 127(19), 129(19), 135(19)
Sivonen, S., 162
Sizensky, J. A., 320
Sjobring, U., 286
Skamene, E., 451, 457(8)
Skaug, N., 110
Skea, D. L., 546
Skjak-Braek, G., 295
Skog, S., 685
Skowronek, K., 385
Skoza, L., 609
Slack, R.C.B., 133
Slatzman, L. E., 142
Slauch, J. M., 369, 481, 483–484, 486, 487(5), 489(5), 490, 490(5), 492(7)
Slavin, W., 322
Slizewicz, B., 165
Slots, J., 106, 125, 136(24), 674
Sly, W., 715
Sly, W. S., 716
Small, P., 198, 204, 204(15)
Small, P. A., 49
Small, P. M., 102, 104, 104(34)
Smeltzer, M. S., 413, 417(20)
Smidsrød, O., 302, 303(24)
Smigel, M. D., 640
Smilenov, L., 604
Smit, J., 226
Smith, A. J., 352
Smith, A. L., 95–96, 96(7), 97(11), 98(11), 99(7, 11), 100
Smith, A. W., 381
Smith, B. A., 112
Smith, C. J., 382, 383(24)
Smith, D. H., 95–96, 96(7), 97(11), 98(11), 99(7, 11), 100, 305, 306(7)
Smith, H., 127, 129, 129(32, 33), 132, 133(44)
Smith, H. O., 186, 191(17)
Smith, I.C.P., 308, 311(13)
Smith, J., 139
Smith, J. A., 365, 369(14)
Smith, J. M., 183, 484
Smith, J. S., 112
Smith, J.H.L., 520, 521(24), 523(24)
Smith, M., 378
Smith, M. J., 354
Smith, N. H., 183
Smith, N. R., 493

Smith, R. F., 611, 612(7)
Snapper, S. B., 385
Snider, G. L., 555
Sninsky, J. J., 177, 219–220
Snyder, I. S., 657, 666–667, 670, 672, 674, 676, 676(29)
So, M., 527, 529(6)
Socransky, S. S., 110, 118
Sodek, J., 108, 111(18), 115(18)
Soderberg, L.S.F., 83, 92
Sogin, M. L., 208–209
Sokol, P. A., 513–514
Sokolov, Y., 691
Solberg, L. K., 164
Soll, D. R., 198, 199(14), 204(13), 205(14)
Soloff, B. L., 84–85, 91(6)
Sommer, S. S., 219
Sordelli, D. O., 451, 454–455, 455(11), 456(14), 457(14)
Sorenson, W. P., 111, 115(34)
Sorrell, R. W., 31
Southern, E., 184, 191(1)
Sowers, A. E., 377
Spanier, J. G., 294
Sparling, P. F., 356–357, 357(1), 358, 358(11), 359(17), 360, 360(17), 361(18, 20), 362, 362(17), 363, 363(17), 365(1), 369(1), 371(1), 372(1)
Sparrman, M., 604
Spears, P., 431, 448(11)
Speert, D. P., 295, 297(8), 302(8)
Speirs, J. I., 649
Spence, M., 139
Spencer, J., 142
Spencer, J.F.T., 295
Spillane, B. J., 128, 129(38)
Spinola, S. M., 242
Spira, W. M., 525
Spiro, T. G., 315
Spratt, B. G., 183
Springhorn, S. S., 576, 577(18), 579(18), 587(18)
Spudich, E. N., 229
Spudich, J. A., 628
Sripakash, K. S., 294
Sriyosachati, S., 328
Staab, J. F., 366(d, g), 367
Stackebrandt, E., 207, 217
Staehlin, T., 672, 676(31)

Stahl, D. A., 207–208, 221
Stahl, U., 384
Stamm, W. E., 84
Stanfield, S., 432, 434(13), 438(13)
Stanislavsky, E. S., 167
Stanley, S. J., 630, 640, 642, 644(30, 35)
Stanley-Samuelson, P., 350–351
Stark, M. J. R., 463
Starnbach, M. N., 439, 443(40)
Staskawicz, B., 369
Staskus, K. A., 222
Staub, A. M., 242
Staudenmaier, H., 366(ff–hh), 367–368
Stavric, S., 649
Stearns, T., 642
Stechberg, B., 94
Steel, M. A., 216
Stegemann, H., 615, 616(37)
Steinberg, M. S., 648
Steinbrink, R., 600, 601(11)
Steiner, B., 615, 616(38)
Steiner, M., 414
Steinman, C. R., 209, 212
Stenberg, K., 681, 690(4)
Stenzel, P., 623
Stephan, C. E., 31
Stephen, J., 647(3), 648
Stephens, D. M., 197
Sternfeld, M. D., 83, 84(1)
Sternweis, P. C., 640
Steurer, F. J., 125–126, 129
Stevens, C., 661
Stevens, L., 641
Stewart, D. J., 527, 529(9)
Stewart, G., 186–187, 190(19)
Stewart, G. C., 185, 418
Stewart, G.S.A.B., 381, 382(16), 384(16)
Stewart, S. M., 94
Stibitz, S., 458–459, 461, 465, 469–470, 665
Stickel, S., 209
Stiles, B. G., 627
Stocker, B. A., 482, 490(2)
Stocker, B.A.D., 439, 443(40)
Stoebner, J. A., 352
Stollerman, G., 288–289, 290(27), 291(27)
Stollerman, G. H., 410
Stone, J. L., 325
Stookey, L. L., 319
Stookey, M., 710–711, 712(21)

753

Storey, D. G., 502–503, 505, 508(10), 510,
510(6), 511(5, 6, 10), 512, 512(5, 6, 10,
18), 513(5, 6, 10, 18, 21), 515
Storzbach, S., 527, 529(6)
Stout, V. G., 418, 419(45)
Straley, S. C., 490
Strance, R. E., 284
Strathdee, C., 667, 671
Straus, D. C., 527
Strausbaugh, L. J., 102
Strauss, D. C., 555
Streips, U. N., 257
Streitfeld, M. M., 615
Strickland, D. K., 713
Strober, W., 142
Strom, M. S., 527–528, 530–531, 536(26),
538(27), 539, 540(29)
Strominger, J. L., 279, 613
Strubel, E., 140, 143(3), 150(3)
Struck, D. F., 664, 675
Struck, D. K., 666, 668–669, 671(9)
Struhl, K., 365, 369(14)
Stuart, K., 139
Studamire, B., 184–185, 190(12), 191(12),
192, 194(12)
Stuhlsatz, H. W., 613
Su, Y. A., 409, 418(1), 426(1)
Subcommittee on arbovirus laboratory
safety for arboviruses and certain other
viruses of vertebrates, 2
Sugiura, Y., 316
Sulkin, S. E., 1
Sundqvist, G. K., 121, 127(3), 128(3), 135(3)
Supance, J. S., 59
Suprun-Brown, L., 139
Sutherland, R., 362, 363(22)
Sutton, A., 305
Suttorp, N., 676
Svanborg-Eden, C., 493
Swanson, J., 260
Swanson, R. N., 454
Swanson, R. V., 439
Swarts, J. D., 68
Swartz, D. J., 518, 519(2)
Sweetman, F., 563, 590(5), 591(5)
Swenson, C. E., 92–93
Swift, H. F., 171, 172(36), 173(36)
Swim, S. C., 260, 263, 272(37), 280(34)
Switalski, L. M., 564, 566(9), 586(9)

Swofford, D. L., 216
Syriopoulou, V. P., 100
Syrogiannopoulos, G. A., 94
Szanezlo, P. J., 327
Szaniszlo, P. J., 335

T

Tagawa, C., 286
Tagg, J. R., 109
Taichman, N. S., 668, 674, 677
Tainter, M. L., 30, 31(3), 35(3)
Tait, R. M., 635
Takacs, B. J., 236
Takada, H., 283
Takagaki, K., 610
Takahashi, T., 605
Takaki, M., 662, 664(29)
Takeda, Y., 649
Takemoto, T., 601
Takita, T., 316
Talkington, D. H., 170
Tall, B. D., 441
Tam, K., 493
Tamura, G., 258
Tamura, S., 610
Tanabe, M., 520
Tanabe, M. J., 283
Tanner, R. S., 207
Tanpowpong, K., 59, 60(6), 62(6), 63(6)
Tashiro, C. J., 218
Tatnell, P. J., 297
Täuber, M. G., 93, 103–104, 104(37), 105–
106
Tautz, C., 440
Taxman, E., 446, 519
Taylor, A., 270
Taylor, D. E., 475–477, 480(7), 481(7)
Taylor, D. J., 297
Taylor, H. R., 69–70, 70(4), 71, 71(4, 12),
74–76, 76(13, 14, 20), 77(13, 14, 19)
Taylor, P., 375
Taylor, R., 431, 448(11), 519
Taylor, R. E., 564
Taylor, R. K., 426, 428–429, 431, 432(8),
433(8), 434, 434(8), 438, 439(8), 440(8),
441(8), 442(8), 444, 445(62), 448(20),

519–520, 520(7), 522(7), 527–528, 529(9), 530(19)
Taylor-Robinson, D., 77, 84, 93(7, 8)
Tessman, I., 341, 350
Theilade, E., 108
Thelestam, M., 679, 683–685, 685(13), 688, 690(29)
Thiele, A., 657, 676
Thierry, D., 197, 198(10)
Thom, J. R., 229
Thomas, B. J., 84
Thomas, C. M., 392(d), 393
Thomas, D. M., 167
Thomas, M. L., 102
Thomas, W. D., Jr., 669
Thompson, M. R., 648, 710, 713
Thompson, S. A., 360, 361(18)
Thompson, S. E., 125–126
Thompson, T. E., 693
Thompson, W. R., 32, 36(12, 13)
Thomson-Carter, F. M., 185, 195(4), 196(4)
Thøgersen, H. C., 543
Thorley-Lawson, D., 648
Thornton, S., 475, 476(3), 478(3)
Thwaits, R. N., 232
Thyberg, H., 116, 117(54)
Thylefors, B., 71
Tien, H. T., 691
Tight, R., 124
Tilbury, A. M., 69, 79(6), 80
Timmins, C., 49
Timmis, K. N., 386, 388(3), 389(3), 391(4), 392, 392(a, c, f, g), 393, 394(2, 3), 395(3), 396(3, 4), 398(3), 399(3), 401(4), 402(5), 403(5), 433(28), 436, 437(28)
Ting, Y., 218
Tipper, D. J., 275, 279
Tirunarayanan, M. O., 614
Tisljar, U., 606
To, A.C.C., 59, 60(6), 61, 62(6), 63(6)
To, S.C.-M., 59, 60(6), 62(6), 63(6)
Todd, T., 493
Toder, D., 559
Toder, D. S., 466
Tokuyasu, K., 613
Tolksdorf, S., 615
Tomana, M., 553
Tomasi, T. B., 545
Tomasic, J., 265, 269, 275
Tomasz, A., 102–105, 265, 266(45), 285, 417

Tomb, J.-F., 444
Tommassen, J., 351, 528
Tomozawa, T., 129
Tompkins, L., 475, 476(2), 477, 478(1, 2)
Tompkins, L. S., 210, 212, 219(13), 375, 381, 381(2), 382(2, 18), 383(2), 384(2, 3), 385(2, 18), 475, 478(6), 479(6), 480(6)
Tong, N. T., 595
Top, F. H., 172
Toribara, T. Y., 300
Torriani, A., 394, 395(16)
Tosteson, D. C., 693, 695, 700(16)
Tosteson, M. T., 695, 700(16)
Toussaint, A., 347
Towbin, H., 672, 676(31)
Tracey, M. V., 298
Tran, T. A., 451, 457(8)
Traub, W. H., 328
Trayer, I., 715
Traynor, H., 320
Trees, D. L., 121, 122(13), 137, 137(13)
Treharne, J. D., 69, 79(6), 80
Trepel, J. B., 640
Treuhaft, M. W., 564, 584(7)
Trevelyan, W. E., 303
Trevors, J. T., 375, 380(8), 381(11)
Triau, R., 305
Trieu-Cuot, P., 475–476, 476(9)
Trouet, A., 685
Troy, F. A., 304–305, 305(2)
Trump, B. F., 685
Trus, B. L., 287, 289
Trust, T. J., 474–476, 476(3, 4), 477, 478(3, 4), 480(8)
Tsai, C.-M., 166, 249, 250(30), 251(30), 252
Tsai, J., 358–359, 359(17), 360, 360(17), 361(18), 362, 362(17), 363(17)
Tsai, S.-C., 641–642, 642(16, 17), 643, 643(16, 17, 31), 644(17, 21, 31, 35, 36)
Tsai, S. C., 641
Tsaur, M. L., 514
Tschida, J., 615
Tseng, J., 143
Tsuchiya, M., 641, 644(21)
Tsui, F.-P., 161
Tsujimoto, M., 283
Tsurugi, K., 649
Tsuyama, S., 626
Tsuzuki, H., 555, 556(7), 559, 560(19)
Tuazon, C. U., 265, 266(44)

Tuffrey, M., 77, 84, 93(7, 8)
Tulknes, P., 685
Tuomanen, E., 102–104, 104(37), 105
Turck, D. C., 96
Tureen, J. H., 104–105
Turner, J. S., 170
Turner, M. W., 126
Tyler, A. N., 270, 280(62)
Tyler, J. A., 604
Tynecka, Z., 265, 266(43)

U

Ueda, K., 643
Ui, M., 628
Ullman, A., 668
Ullmann, A., 669
Ulrich, J. R., 282
Umezawa, H., 316
Underdown, B. J., 546
Unsworth, P. F., 614
Upchurch, S., 490
Upholt, W. B., 195
Urban, R. G., 444
U.S. Department of Health and Human Services, 1, 2(5)
Ussery, M. A., 652, 653(23)

V

Vaara, M., 225, 230(1)
Vaatainen, A., 121, 127(3), 128(3), 135(3)
Vacheron, F., 283
Vadivelu, J., 661
Vail, T. A., 564, 566(9), 586(9)
Valinger, Z., 265, 269, 275
Valisena, S., 256
Vallera, D. A., 716
Vanags, R. I., 496
van Alpen, L., 160, 163(1)
Vandekerchove, J., 626
Vandenberg, P. A., 328
van der Helm, D., 344
van der Rest, M., 594
Van Der Vliet, H., 676
Van Der Weele, L. T., 115, 117(49)
van der Zeijst, B.A.M., 475, 476(5), 478(5), 479(5), 480(5)

van Deurs, B., 648, 709, 716(9)
Van Dijk, L. J., 115, 117(49)
van Dijk, L. J., 108
Van Dongen, A.M.J., 640
van Doorn, J., 356
van Embden, J.D.A., 196–198, 199(14), 204, 204(13, 15), 205(14, 17)
van Heijenoort, J., 270
van Heynigen, S., 635
Van Hoof, F., 685
Van Hove, B., 366(gg, hh), 367–368
Vankateswaran, K., 663
Van Leengoed, L.A.M.G., 662, 666(30)
Vann, W. F., 304–305, 305(3)
van Ness, B. G., 617
van Soolingen, D., 196–198, 199(14), 204, 204(13), 205(14, 17)
Van Wart, H. E., 595–596, 597(8), 598(8), 600, 600(2), 601(11), 606
Varadi, D. P., 561
Varaldo, P. E., 256
Vasil, A. I., 495–496, 515
Vasil, M. L., 317, 467, 495–496, 504, 514–515
Vasli, K. K., 121, 122(11), 127(11), 132(11), 133(11), 134(11)
Vasquez, D., 648
Vaudaux, P., 106, 121, 132(16)
Vaughan, M., 640–642, 642(16, 17), 643, 643(16, 17, 31), 644(17, 21, 30, 31, 35, 36)
Vaughn, M., 625
Veale, D. R., 127, 129(32, 33), 132, 133(44)
Vedros, N. A., 164
Venkatesan, M., 44, 46(10)
Venkateswaran, K., 662, 664(29)
Vermeulen, M. W., 282
Vermillion, J. R., 109
Vermund, S. H., 139
Vernon, N., 99, 100(21)
Versieck, J., 323
Verweij-van Vught, A.M.J.J., 356–357
Vicari, G., 649
Vickrey, P. E., 335
Vieira, J., 395, 398, 398(17)
Villablanca, J. G., 414
Vimr, E. R., 305
Vincente, M. F., 414
Vincent-Lévy-Frébault, V., 197, 198(10)
Viney, J., 142

Vinogradov, E. V., 167
Viscidi, R. P., 74
Vodinelich, L., 362, 363(22)
Vogt, A. J., 455, 456(14), 457(14)
Voino-Yasenetsky, M., 42, 43(9), 45(9), 46
Vold, B. S., 185
Vollger, H. F., 305
von Eichel Streiber, C., 683, 690(11)
von Gabain, A., 506
Von Jager, G., 250, 251(33)
Vose, B. M., 142
Vosti, K. L., 288
Vukajlovich, S. W., 283

W

Waage, A., 105
Waalwijk, C., 657
Wadman, W. H., 303
Wagegg, W., 366(ee), 367
Wagner, S., 669
Wagner, W., 678
Wahl, R. C., 560
Walbot, V., 375
Walderhaug, M., 317, 364
Waldman, R. H., 288
Waldvogel, F. A., 106, 121, 127, 132(6, 16, 34)
Walker, G. C., 369, 434, 438, 438(21)
Walker, J. S., 51, 58(12)
Walker, M. W., 643
Wall, J. H., 661
Wallsmith, D. E., 254
Walmsley, T. A., 320
Walsh, C. T., 366, 366(b, i), 367
Walter, J., 271
Waltman, W. D., 170
Walum, E., 681, 690(4)
Wandersman, C., 668–669
Wang, S.-K., 497
Wang, S. K., 496
Wang, S.-P., 69, 83, 84(1)
Wang, S. P., 84
Wang, Y., 375, 475–477, 480(7), 481(7)
Wang, Z., 451
Wannamaker, L. W., 97, 172, 614
Ward, J. B., 265, 266(43)
Ward, J. I., 94
Wardlaw, A. C., 48

Waring, W. S., 316, 326(5)
Warner, H., 300
Warner, P. J., 364, 368(8), 371
Wasilauskas, B., 410
Wassef, J. S., 141–142, 145(7), 149(5)
Wassenaar, T., 475, 476(5), 478(5), 479(5), 480(5)
Waterman, S. H., 164
Watkins, N. G., 79
Watkins, P. A., 640
Watt, S., 628
Watts, J. C., 132
Way, J., 387, 393(10)
Wayne, R., 328, 357
Weatherwax, R., 235
Weaver, K., 409, 418(1), 422(46), 426(1)
Weaver, K. E., 419
Weaver, R., 164
Webb, J. W., 242
Webster, C. A., 133
Wechter, W. J., 115, 116(44)
Weckesser, J., 233
Wedege, E., 166
Weichter, W. J., 116
Weidel, W., 253
Weil, C. S., 32, 36(13–15), 37(14, 15)
Weinberg, E. D., 316, 344
Weinstein, B. R., 635
Weisbeek, P. J., 351
Weisburg, W. G., 209, 212
Weisdorf, D., 414
Weiser, B., 211, 215
Weiss, A. A., 48, 442
Weiss, D. S., 497
Weiss, R., 612
Weiss, S., 603, 610(2)
Weissleder, R., 325
Welch, M. J., 321
Welch, R. A., 657–658, 660–661, 663, 666, 666(28), 667–670, 672, 674, 676(29)
Welkos, S., 29
Weller, P. F., 100
Welsh, C. F., 642, 644(30, 35)
Wenger, J. D., 94, 410
Wennstrom, J., 118
Werb, Z., 555, 581
Werkman, C. H., 316, 326(5)
Wescott, W. C., 691
Weser, J., 552
Wessels, M. R., 411, 420(10)

West, S. E., 357
West, S. K., 71
West, S.E.H., 363, 365(1), 369(1), 371(1), 372(1)
Westergaard, J. L., 564, 584(7)
Westphal, O., 242–243
Wetterlow, L. H., 121, 122(7), 127(7), 138(7)
Wexler, D. L., 420, 422(49)
Whalley, K., 235
Wheatcroft, M. G., 107
Wheelis, M. L., 207
Whiley, R. A., 608
Whipp, S. C., 444
White, A.-M., 185, 190(12), 191(12), 192, 194(12)
White, H. J., 78, 84–85, 91(6)
White, S. H., 691, 692(3)
White, T. J., 177, 219
Whitfield, C., 310
Whittam, T. S., 160, 175, 181(2)
Whittum-Hudson, J. A., 69–70, 74–76, 76(14, 20), 77(14, 19)
Wick, M. J., 468, 503, 510(6), 511(5, 6), 512, 512(5, 6), 513(5, 6), 620
Widom, R. L., 185, 186(9–11), 187, 194(9–11)
Wilcoxon, F., 30
Wilde, C. E. III, 260, 280(34)
Wilkie, B. N., 664, 667, 674(3), 676(3)
Wilkins, T. D., 627
Wilkinson, H. W., 174
Wilkinson, N. C., 305
Wille, M., 683
Willett, N. P., 611, 612(7)
Willhite, C. C., 31
Williams, D. E., 567, 588(15)
Williams, L. J., 498
Williams, P., 133, 356, 366(p), 367
Williams, P. H., 316, 347, 364, 368(8), 371
Williams, R. C., 115–116, 116(44)
Williams, W. K., 288
Williamson, K. C., 641, 643
Williamson, S., 142
Willims, H. L., 320
Willingham, M. C., 642
Willis, R. C., 240, 241(17)
Wilson, A. C., 220
Wilson, A. T., 171, 172(36), 173(36)
Wilson, B. A., 635

Wilson, E., 171, 172(39)
Wilson, F., 186–187, 190(19)
Wilson, K., 175
Wilson, K. H., 209, 212
Wilt, J. C., 137
Wingender, J., 496
Winkelmann, G., 321, 341, 344, 353–354
Winkelstein, J. A., 96, 98(13), 100(13)
Winkler, U. K., 496
Wiseman, G. M., 659
Wisnieski, B. J., 701
Wisotzkey, J. D., 220
Wispelwey, B., 105
Witkowska, D., 47
Witt, K. A., 132, 133(44)
Wittner, M. K., 288
Wnendt, S., 384
Woese, C. L., 184
Woese, C. R., 206–207, 211, 214–215, 217, 221
Wold, F., 565, 578(11), 588(11)
Wolf, H. F., 110, 111(29)
Wolfe, R. S., 207
Wolfe, S. A., 484
Wolner-Hanssen, P., 84, 661
Wolnik, K. A., 322
Wolpe, S., 105
Wong, I., 80
Wong, K. H., 125–126, 129, 132(41)
Wong, W., 265, 266(44)
Wood, S. G., 545, 554(19)
Wood, S. J., 59, 60(6), 61, 62(6)
Woodland, R. M., 84
Woodruff, W. A., 468
Woods, D. E., 48, 134, 138, 527
Wooldridge, K. G., 316
Woolkalis, M., 629, 631
Woolwine, J. D., 102, 104(34)
Wort, A. J., 684
Wozniak, D. J., 501
Wright, A., 427, 437(3), 543, 544(10)
Wright, A. M., 328
Wright, D. K., 210
Wright, J. G., 350
Wright, R. M., 279
Wright, W. H., 108
Wulff, J. L., 231
Wunderlich, A. C., 128
Wunsch, E., 601

X

Xing, Y., 211, 214

Y

Yacoob, R. K., 165
Yam, L. T., 591
Yamakawa, M., 661
Yamamoto, K., 520
Yamamoto, M., 420
Yamasaki, R., 242
Yanagihara, Y., 258
Yancey, R. J., 328, 357
Yang, M.-S., 461
Yanisch-Perron, C., 398
Yao, R., 474, 476–477
Yaraghi, Z., 366(gg), 367
Yardley, J. H., 140, 143
Yatani, A., 640
Yates, F., 33(21), 35
Yates, P. S., 70
Yeh, L. J., 69
Yet, M. G., 565, 578(11), 588(11)
Yetter, R. A., 49
Yokogawa, K., 283
Yoshikawa, M., 40
Yoshimura, K., 668
Yoshioka, H., 682
Yost, D. A., 630
Yost, S. C., 412, 416(18), 418(18)
Yother, J., 170
Young, E., 70, 76(13), 77(13)
Young, F. E., 260, 265, 266(44)
Young, H. W., 50–51, 58(12)
Young, I. G., 329, 338
Young, R., 664, 666, 668–669, 671(9), 675
Youngman, P. J., 422
Yu, J., 441
Yuan, , 698
Yutsudo, T., 649

Z

Zabala, J., 666
Zablen, L. B., 207
Zabriskie, J. B., 288
Zähner, H., 354
Zahringer, U., 252
Zainuddin, Z. F., 198
Zajic, J. E., 315
Zak, O., 101–105
Zaleznik, D. F., 135
Zalman, L. S., 701
Zambon, J. J., 107, 109, 118(20a), 119(20a), 564
Zander, H. A., 107, 108(4), 112, 113(36)
Zealey, G. R., 465
Zeiger, A. R., 265, 266(44)
Zeleznick, L. D., 258
Zeligman, I., 604
Zeligs, B. J., 449, 451, 454, 457(1)
Zemmour, J., 220
Zhang, Q.-Y., 183
Zhou, J., 183
Ziegler, D. M., 320
Ziegler, H. K., 668
Zielinski, N., 500
Zielinski, N. A., 299–300, 493, 495–496, 496(7), 497, 497(7), 498(28), 501(29)
Zimmerberg, J., 695
Zimmerli, W., 121, 132(16)
Zimmerman, L., 364, 366(12, y, ff), 367
Zimmerman, R. A., 171, 172(39)
Zimmermann, A., 467
Zitzow, L. A., 409, 418(1), 426(1)
Zoeller, C., 40
Zollinger, W., 242(12), 243
Zollinger, W. D., 164, 166, 242, 249(7)
Zon, G., 161
Zrike, J., 229
Zuider, E. R., 308
Zwahlen, A., 93, 96–98, 98(13), 99, 99(14, 17), 100(13, 14), 106

Subject Index

A

Achromobacter iophagus, Type I collagenase, 595
  assay, 600
Achromocollagenase, 566
  zymographic characterization, 580–581
Actinobacillus actinomycetemcomitans
  leukotoxin, 106, 668
    DNA probe for, 670–671
    nonerythrocyte cytolytic activity, assay, 674
    pore formation assay, 677
  periodontitis, 107
  rat model, 117
Actinobacillus pleuropneumoniae
  electrotransformation, cell harvest for, 381
  hemolysin, 668
    antibody probes for, 671–672
    operon organization, 669
    pore formation assay, 677
Actinomyces naeslundii, periodontal pathogenicity, rat model, 117
Actinomyces viscosus, periodontal pathogenicity, rat model, 117
Adenylate cyclase, activation by bacterial toxins, 624, 626
ADP-ribosylagmatine assay, 643, 646
ADP-ribosylation
  by bacterial toxin, 617–632
  cholera toxin-catalyzed, assay, 642–647
ADP-ribosylation factors, 624
  cholera toxin activation by, 641–642
  classes, 641–642
  membrane forms, 624, 642
  myristoylated, 642
  recombinant, 642
  soluble forms, 624, 642
  sources, 644
  species distribution, 641
Aerobactin, 316, 347–348, 357, 364
  inhibition, 329

isolation, 338, 341–342
  purification, 341–342
Aeromonas hydrophila
  aerolysin, cytopathogenic effects, 684
  amoA gene, analog of *Escherichia coli entC* gene, 363
Affinity chromatography
  immobilized metal, microbial gelatinase, 604
  Pseudomonas aeruginosa PilD, 534–536
Agglutination
  for *Haemophilus influenzae* capsular serotyping, 162
  for *Neisseria meningitidis* serogrouping, 164–165
Agrobacterium, electrotransformation, DNA delivery, 384
Alginase
  degradation of alginate, 304
  production by mucoid *Pseudomonas aeruginosa*, 304
  sources, 304
Alginate
  Azotobacter vinelandii, 295
  marine algal, 295
  properties, 295
  Pseudomonas aeruginosa, 295–304
    assay, 298–301
    gene expression, regulation, 493–502
Alkaline phosphatase
  detection, 429–431
  hybrid proteins, detection, 430–431
  quantitation, 429–431
  reporter gene, in bacterial cell-surface virulence determinants analysis, 426
  spectrophotometric assay, 429
  unit of activity, 429
Allelic exchange, suicide vectors for, 458–465
Anguibactin, 352
Animals, *see also specific animals*
  biosafety recommendations, 2, 11–16

chamber implant models for host–parasite interaction studies, 120–140
median infectious dose in, determination, 29–39
median lethal dose in, determination, 29–39
Antibiotic resistance, carried by transposon, selection for, 394
Apoptosis, 681, 685
Arnow assay, siderophores, 332, 353, 371
Arsenite resistance, carried by transposon, selection for, 394–395
Arthrobacter flavescens, see Aureobacterium flavescens
Arthropods, biosafety recommendations, 2
Atomic absorption spectroscopy, iron, 322
Atomic fluorescence spectroscopy, iron, 323
Aureobacterium flavescens, assay for siderophores, 335–336
Azotobacter vinelandii, alginate, 295

B

Baboon, periodontitis model, 112
Bacillus
chromosomal DNA, Southern hybridization, 185–188
peptidoglycans, trichloroacetic acid extraction, 257
restriction relationships, dendrogram analysis, 186, 188
RFLP analysis, 186–196
Bacillus anthracis, laboratory hazards and biosafety recommendations, 17
Bacillus licheniformis, soluble polymeric peptidoglycans secreted by, isolation, 266
Bacillus subtilis
ComC, amino acid sequence, 528–530
neutral protease, substrate, 560
Bacillus thermoproteolyticus, thermolysin
activities, 554
assay, 559
Bacteria, *see also specific bacteria*
acid-fast, peptidoglycans, isolation, 255
capsular polysaccharides
charge, 305
functions, 304
NMR studies, 306
cell envelope, crude fractionation, 226–227
genomic DNA, preparation, 175–177, 199–201
gram-negative
genetic manipulations with minitransposons, 386–405
genomic DNA, preparation, 175–176
with iron-regulated membrane proteins, 347
peptidoglycans
isolation, 255, 260–263
low-molecular-weight oligomers and monomers, isolation, 269–273
soluble polymeric, isolation, 264–265
gram-positive
disease associations, 405–406
electrotransformation, cell harvest for, 381
genomic DNA, preparation, 176–177
pathogenic strategies, transposon dissection of, 405–426
peptidoglycans
isolation, 255–260
low-molecular-weight oligomers and monomers, isolation, 268–269
secreted by cells, isolation, 265–268
soluble polymeric, isolation, 263–264
virulence analysis, transposon mutagenesis studies, 406–426
isolation and characterization of mutant strains, 419–421
virulence determinants, 406
pathogenic, biosafety recommendations, 1–26
replication rates, measurement *in vivo*, 448–449
virulence factors, chamber implant studies, 127
Bacteriophage
λ, vectors for Tn*phoA* delivery, 434–435
and active *phoA* fusions to plasmid gene products, 435–436
mini-Mu, in analysis of cloned genes, 422–425
transposon delivery by, 410

Bacteroides, periodontitis, 107
Bacteroides fragilis
  antimicrobial efficacy against, animal
    chamber studies, 135
  host–parasite interactions, animal cham-
    ber models, 121, 134–135
Bacteroides melaninogenicus
  host–parasite interactions, animal cham-
    ber models, 134–135
  IgA$_1$ protease, properties, 553
Bathophenanthroline disulfonate, in iron
    assays, 320
Beagle, periodontitis model, 107, 114–117
Bialaphos
  resistance, carried by transposon, selec-
    tion for, 394–395
  source, 395
Biosafety, with pathogenic bacteria, 1–26
  animal biosafety levels, 2, 11–16
  laboratory biosafety levels, 1, 5–10
Biotinylation, proteins, 707
Bleomycin, in iron assay, 316
Bordetella pertussis
  antigenic modulation, analysis by
    Tn*phoA* mutagenesis, 442–443
  chromosomal *fhaB* locus, replacement
    with *fhaB:lacZ* gene fusion, 463
  colonization, 58
    in mouse, 53, 55–57
  filamentous hemagglutinin, 53
  gene products, in pathogenesis, 58
  gene replacement in, 470
  hemolysin, 665, 668
    hemolytic activity, assay, 673
    operon organization, 669
    structure, 670
  host–parasite interactions, animal cham-
    ber models, 121, 138
  immunity in mouse, 57–58
  infection, *see* Pertussis
  intranasal administration to mouse, 48–
    49
  laboratory hazards and biosafety recom-
    mendations, 17, 49–50
  lipooligosaccharide, 242
  porin, detergent solubilization, 232
  strain 18323, 54–55
  Tohama I, 54–55
  transglycosylase, 270
  transmission, 48, 53

vir locus, 48, 58
  allelic exchange with plasmid vectors,
    461–465
  replacement with knock-out insertion
    of selectable kanamycin resis-
    tance marker, 461–462
  unmarked in-frame linker insertion
    mutations, 462–463
  virulence
    age effects, 53–54
    determinants, 58
Borellia burgdorferi, laboratory hazards
    and biosafety recommendations, 17
Branhamella catarrhalis, lipooligosac-
    charide, 242
Brevibacterium divaricatum, peptidogly-
    cans, low-molecular-weight oligomers
    and monomers, isolation, 269
Brucella abortus, laboratory hazards and
    biosafety recommendations, 17
Brucella canis, laboratory hazards and
    biosafety recommendations, 17
Brucella melitensis
  laboratory hazards and biosafety recom-
    mendations, 17
  outer membranes, vesicles, 233
Brucella suis, laboratory hazards and
    biosafety recommendations, 17

C

Campylobacter
  experimental genetics, 474–481
    resistance markers for, 476–477
    shuttle vectors for, 475–480
    suicide vectors for, 475–480
    vector transfer
      conjugative, from *Escherichia coli*,
        477–478
      by electroporation, 477–478
    natural transformation, 475, 480–481
    outer membranes, vesicles, 233
Campylobacter coli
  disease associations, 474
  laboratory hazards and biosafety recom-
    mendations, 18
Campylobacter fetus, subsp. *fetus*, labora-
    tory hazards and biosafety recommen-
    dations, 18

Campylobacter jejuni
  disease associations, 474
  electrotransformation, 385
  laboratory hazards and biosafety recom-
    mendations, 18
Capnocytophaga
  IgA₁ protease, 543
  properties, 553
  periodontitis, 107
  rat model, 117
  virulence factors, 106
Cat, *Chlamydia psittaci* urogenital infection, 84
Centrifugation, sucrose density gradient,
  outer membranes, 228–230
Chalaropsis, muramidase, peptidoglycan
  fragments released by, 269–270
Chamber implant models, animal, for
  host–parasite interaction studies, 120–
  140
  animals, 124–126
  applications, 139–140
  chamber characteristics, 123–124
  chamber fluid characteristics, 123–124
  host response studies, 127–128
  implantation technique, 123
  implant types, 121–122
  leukocyte response, effect of implant
    type, 122–123
  parameters examined, 127–128
  pathogens, 128–139
  sampling technique, 123
  virulence factor studies, 127
Chelex, in deferration of laboratory media,
  327
Chicken, bacterial replication in, measurement *in vivo*, 455–457
Chinchilla
  cost, 60
  experimental otitis media, 59–68
  immunization
    active, 60–61
    passive, 61–62
  middle ear inoculation, 63
  nasal challenge, 63–64
  otomicroscopy, 63–64
  otoscopy, 60
  sources, 60
  tympanometry, 60, 63–64
Chlamydia, virulence, evaluation, 70

Chlamydia pneumoniae, laboratory hazards and biosafety recommendations,
  18
Chlamydia psittaci
  guinea pig agent of inclusion conjunctivi-
    tis, *see* Guinea pig agent of inclu-
    sion conjunctivitis
  laboratory hazards and biosafety recom-
    mendations, 18
  urogenital infection, animal models, 84
Chlamydia trachomatis
  agent of mouse pneumonitis, *see* Mouse
    pneumonitis agent
  human oculogenital strains, genital
    infection in mouse, 93
  laboratory hazards and biosafety recom-
    mendations, 18
  ocular infections
    animal models, 69–83
      human biovars for, 71
      clinical disease score, 71–73
      culture assay, 71, 73
      direct fluorescent antibody assay, 71,
        73–74
      guinea pig model, 69, 78–83
      polymerase chain reaction analysis,
        74
      primate models, 69
        animals, 70
        antibody response to, assessment,
          77–78
        cell-mediated immune response,
          assessment, 76–77
        clinical disease assessment, 71–73
        histological evaluation, 74–76
        induction, 70–71
        microbiological assessment, 73–74
        total clinical disease score, 71–73
    serovars
      for animal models of ocular infection,
        71
      as human oculogenital strains, 93
    urogenital infection, 83
      animal models, 83–84
Chloramphenicol, resistance, carried by
  transposon, selection for, 394
Cholera, pathogenesis, 518
Cholera toxin
  activation by ADP-ribosylation factors,
    641–642

assay, 644–646
effects of assay components, 642–643
ADP-ribosyltransferase activity, 617,
632, 640
assay, 625, 642–647
assay, 524–526, 624–625
auto-ADP-ribosylation, assay, 646–647
cytopathogenic effects, 682
ELISA, 525–526
expression
environmental growth conditions
affecting, 521–524
factors affecting, 523–524
osmolarity effects, 521–522
pH effects, 522–523
strain differences, 520–521
temperature effects, 521
genes, genetic regulation, 519–520
mechanism of action, 680
receptor binding, 623–624
regulation by environmental factors,
517–526
structure, 518–519, 640
subunits, 623
synthesis, 623
Chromatography, *see also* Affinity chromatography
AG1-X-2 column, in ADP-ribosylation
assay, 647
thiol–disulfide covalent, microbial gelatinase, 604
Chrome azurol S assay, siderophores,
333–334, 353–356
Chymotrypsin, 566
zymography in nondissociating gels with
copolymerized substrates, 588–589
Citrate, as iron-chelating bacteriostatic
agent, 346
Citrobacter, genomic DNA, preparation,
175–176
Clostridium, neurotoxins, mechanism of
action, 680
Clostridium botulinum
C3 exoenzyme, ADP-ribosyltransferase
activity, 632
C2 toxin
ADP-ribosyltransferase activity, 617,
626, 632
assay, 626–627
cytopathogenic effects, 683

laboratory hazards and biosafety recommendations, 18
Clostridium difficile
disease associations, 405
toxin A, cytopathogenic effects, 682, 685
toxin B, cytopathogenic effects, 682, 685
toxins, mechanism of action, 680
Clostridium histolyticum, Type I collagenase, 595
assay
with 2-furanacryloyl-L-leucylglycyl-L-
prolyl-L-alanine, 600
with phenylazobenzyloxycarbonyl-
Pro-Leu-Gly-Pro-D-Arg, 600–601
Clostridium novyi, α-toxin, cytopathogenic
effects, 683
Clostridium perfringens
α toxin, cytolytic action, enzymatic
mechanisms, 658
electrotransformation, cell harvest for,
381
enterotoxin
cytopathogenic effects, 682–683
mechanism of action, 680
host–parasite interactions, animal chamber models, 134–135
iota toxin
ADP-ribosyltransferase activity, 617,
626–628
assay, 626
perfringolysin O
cytopathogenic effects, 682–683
mechanism of action, 680
Clostridium ramosum, IgA₁ protease, 543
Clostridium tetani, laboratory hazards and
biosafety recommendations, 19
Clostripain, 566
zymography in nondissociating gels with
copolymerized substrates, 588–590
Collagen
enzymatic cleavage, 594–595
fibrils, as collagenase substrate, 597–
599
insoluble tendon, as collagenase substrate, 599
radiolabeling, 597–598
structure, 594
synthetic peptide analogs, as collagenase
substrates, 599–601
Type I, solution assay, 595–597

Collagenase
  activity, 594–595
  bacterial, 595
  Type I, assays, 594–602
    collagen fibril substrate, 597–599
    with collagen synthetic peptide
      analogs, 599–601
    with dinitrophenyl coupled to colla-
      genlike peptides, 601–602
    insoluble tendon collagen as sub-
      strate, 599
    solution, 595–597
  interstitial, 595
Computer programs
  for moving average interpolation, 32
  for probit analysis, 31
Conalbumin, as iron-chelating bacteriostatic agent, 328, 346
Conjugation, transposon delivery by, 409–
  410
Conjunctivitis, see also Keratoconjunctivitis
  experimental, applications, 46
  in guinea pig, 78–83
Containment equipment, 5, 8–9, 11, 14
Corynebacterium, peptidoglycans, hot
  formamide extraction, 258
Corynebacterium diphtheriae
  diphtheria toxin, see Diphtheria toxin
  keratoconjunctivitis, 46
  laboratory hazards and biosafety recom-
    mendations, 19
Coxiella burnetii, laboratory hazards and
  biosafety recommendations, 24
Csáky assay, siderophores, 331–332, 353,
  371
Cynomolgus monkey, see Macaca fascicularis
Cytokines, in rabbit bacterial meningitis
  model, 105
Cytolysin, 679
  cell lysis, 682–683
  hemolytic phenotype, 657–659, 667
  RTX family, 658
    antibody probes for, 671–672
    assay, 672–678
    calcium binding activity assay, 676–
      677
    DNA probes for, 670–671
    hemolytic activity, assay, 673–674
    homology, 669–670

identification, 670–672
nonerythrocyte cytolytic activity,
  assay, 674–675
operon organization, 669–670
pore formation assay, 677–678
properties, 667–670
role in pathogenesis, 668
secretion, 668–669
structure, 670
sublytic effects, 675–676
target cell specificity, 667
Cytotoxins, see also specific cytotoxins
  definition, 679
  Listeria, 413
  staphylococcal, 412

D

Deferration, laboratory media, 326–327,
  346
Deferriferrichrome A, as iron-chelating
  bacteriostatic agent, 346
Dental plaque
  bacteria, virulence factors, 106–107
  clinical indices, 110–111
  toxicity, tests, 110
Dichelobacter nodosus, pilin, 527
Diethylenetriaminepentaacetic acid, in
  deferration of laboratory media, 327
Diphtheria toxin
  active site residues, photoaffinity label-
    ing, 633–639
  ADP-ribosyltransferase activity, 617–
    619, 631–632
  assay, 619
  assay, 618–619
  cytopathogenic effects, 682
  inhibition of host protein synthesis, 647–
    648
  mechanism of action, 680
2,2'-Dipyridyl, as iron-chelating bacteriostatic agent, 328, 346
Dithizone, in deferration of laboratory
  media, 327
DNA, bacterial
  chromosomal, Southern blotting, 185–
    188
  for electrotransformation, 383–384
  genomic
    isolation, 189–190

preparation, 175–177, 199–201
restriction patterns, 186–187, 190–193
hybridization, detection of *Escherichia coli* iron acquisition gene analogs by, 363–365
mycobacterial, Southern blotting, 202
ribosomal
as chronometer, 184–185
in RFLP analysis, 184–196
single-stranded
production, 178–179
sequencing, 179–180
DNA Data Bank of Japan, electronic mail address, 213
DNA fingerprinting, *Mycobacterium tuberculosis*, 196–205
Dog
oral hygiene, 108
periodontal health, effect of diet, 108–109
periodontitis model, 107, 111, 114–117
clinical indices, 111
data analysis, 116–117
disease assessment, 116–117
and human periodontitis, similarities, 114–115
induction, 115
Dose
median effective, 29, 31
determination, staircase method, 33
median infectious, determination, 29–39
median lethal, 29
confidence intervals, 30–32
95%, 37
determination in animals, 29–39
moving average interpolation, 32–34, 36–37, 39
probit analysis, 31–36, 39
Reed–Muench method, 30, 33–35
staircase method, 33, 37–39
inhalation, determination, 37–39
oral, determination, 33–36
median response, 29
Dose–response, animal experiments, 29
dose range for, 29
50% end-point determination, 29–30
genetic stability of host and microbe for, 29
Dose–response curve
characterization, 29, 32

50% end-point determination, 29–30
slope, calculation, 37

E

ED_{50}, *see* Dose, median effective
EDDA, *see* Ethylenediamine-di(*o*-hydroxyphenylacetic acid)
Ehrlichia chaffeensis, laboratory hazards and biosafety recommendations, 25
Eikenella corrodens, periodontitis, 107
rat model, 117
Elastase, assays, 554–562
Elastin, 554
Electron microscopy, in iron assays, 325
Electron paramagnetic resonance, iron, 324
Electrophoresis, *see* Gel electrophoresis
Electroporation, for bacterial transformation, 375–385, 418
cell growth for, 381
cell preparation for, 381–382
devices, 376–378
DNA for, 383–384
electrical parameters, 382–383
genetic factors, 384–385
gram-positive cell walls and, 382
precautions, 380
protocols
establishment, 380–385
optimization, 380–385
pulse generator, 376–378
exponential decay waveform, 376–378
sample parameters, effects during pulse delivery, 383
transposon delivery, 408
vector transfer, in *Campylobacter*, 478–479
Elongation factor 2, ADP-ribosylation, 617–619, 631–632, 648
Emission spectroscopy, inductively coupled plasma-optical, iron, 322–323
Endotoxin, role in peptidoglycan functions, 282–284
Enterobacteriaceae, outer membranes, isolation, 226–230
Enterobactin, 316, 364
isolation, 339–340
purification, 338–341
Enterochelin, 357
inhibition, 328

Enterococcus faecalis, electrotransformation, 382
Enzyme electromorph typing, 160
Erwinia chrysanthemi
  iron-regulated siderophores and outer membrane proteins, 351
  macromolecular transport, proteins required for, amino acid sequences, 528–529
  metalloprotease, 667–668
  operon organization, 669
  pectate lyase secretion, analysis by Tn*phoA* mutagenesis, 441–442
Escherichia coli
  aerobactin, 364
  bacteriostatic agents, iron-chelating, 328–329
  capsular polysaccharides, ^{13}C-labeled, preparation, 308–310
  citrates, 364
  cloned genes in, mini-Mu analysis, 422–425
  cytotoxic necrotizing factor, cytopathogenic effects, 682, 685
  electrotransformation, 378–380
    cell harvest for, 381
    DNA delivery, 384
    genetic factors, 385
    precautions, 380
  enterobactin, 364
    isolation and purification, 338–341
  enteroinvasive
    environmental sources, 46
    virulence, Sereny assay, 45–46
  enteropathogenic, invasion loci, analysis by Tn*phoA* mutagenesis, 441
  extracts, preparation, 499
  ferric enterobactin receptor, 348, 350
    purification, 350–351
  fractionation, 292–293
  gene replacement in, 466–467
  genes
    fur, 349–350
      in iron reporter system, 316
    luxAB, in iron reporter system, 316
  genes cloned into, transposon analysis, 421–422
  genetic manipulations with minitransposons, mating mixture set-up, 390–393
  genomic DNA, preparation, 175–176
  growth, 499
  heat-labile enterotoxin
    ADP-ribosyltransferase activity, 617, 625–626, 632, 640
    assay, 626
    cytopathogenic effects, 682, 684
    mechanism of action, 680
    structure, 625
  heat-stable enterotoxin, mechanism of action, 680
  hemolysin, 659, 667–668
    calcium binding activity, assay, 676–677
    hemolytic activity, assay, 674
    mechanism of action, 680
    nonerythrocyte cytolytic activity, assay, 674
    operon organization, 669
    pore formation assay, 677
    secretion, 668
    sublytic effects, 675–676
  host–parasite interactions, animal chamber models, 121, 125, 137–138
  insertional mutagenesis, resistance markers for, 472
  iron acquisition, 363–372
    genes, 365–366
      analogs, 363–365
      mutants, 365–369
        complementation, 365–370
        complementing clone, identification, 370–372
  iron chelator, 364
  iron-regulated outer membrane proteins, 347–351
    Fur protein and, 349
    genes, iron box sequence, 349–350
  K antigens
    analysis, 310–311
    charge, 305
    detergent precipitation, 305
    purification, 304–311
  K1 polysaccharide, 304–305
  K5 polysaccharide, 305
    purification, 306–308
  K14 polysaccharide, ^{13}C-labeled, ^{13}C–^{13}C COSY experiments, 308–309, 311
  laboratory hazards and biosafety recommendations, 19
  lipopolysaccharides, 159, 161

meningitis
  animal models, 102
  lytic antibiotic therapy, 104
  nucleotide sequence variation, PCR
    analysis, 175, 180–182
  outer membranes
    ferric siderophore receptors, 347–348
    TonB box, 348
    isolation, 226–230
    vesicles, 233
  periplasm, recombinant M protein ex-
    pression, 292–294
  periplasmic binding protein, isolation,
    236–238
  phase variation of type 1 pili, identifica-
    tion as virulence factor, 138
  rRNA, secondary structure, 211
  Shiga-like toxin, inhibition of host pro-
    tein synthesis, 648
  siderophores, 364
    receptors, as colicin and bacterio-
      phage receptors, 371–372
  temperature-sensitive mutants
    applications, 457
    generation by nitrosoguanidine, 451
    quantitative clearance studies with,
      454
    replication, measurement *in vivo*, 455–
      456
  transformation with plasmid containing
    cloned gene, 422–423
  transglycosylase, 270–271
  types, 161
Ethylenediamine-di(*o*-hydroxyphenylacetic
  acid), as iron-chelating bacteriostatic
  agent, 328, 346
European Molecular Biology Laboratory,
  Data Library, electronic mail address,
  213
Excision marker rescue, 468
Exfoliatin, 412
Exotoxin A, *see Pseudomonas aeruginosa*,
  exotoxin A
Eye, infections, *see also* Conjunctivitis;
  Keratoconjunctivitis
  animal models, 69–83
  Chlamydia trachomatis, 69–83

F

Ferrichrome, isolation, 342–343
Ferric perchlorate assay, siderophores, 331

Ferritin, electron transitions, 317
Ferrozine, in iron assays, 319–320
Flame atomic emission spectrometry, iron,
  322
Francisella tularensis, laboratory hazards
  and biosafety recommendations, 19
Fusobacterium nucleatum, periodontitis,
  107
  rat model, 117

G

β-Galactosidase, assay, 499–500
Ganglioside G_{m1}
  as cholera toxin cell surface receptor,
    624
  as heat-labile enterotoxin cell surface
    receptor, 625
  plates coated with
    in cholera toxin assay, 524–526
    preparation, 524
Gelatin
  radiolabeled, gelatinase assay with, 606
  radiolabeling, 606
Gelatin-agarose, 603
Gelatinase, bacterial, 602–606
  activities, 602
  affinity chromatography, 602–604
  assays, 605–606
    labeled gelatin method, 606
  purification, 602–604
Gel electrophoresis
  binding mode, 578
  nonbinding mode, 578
  polyacrylamide, in nondissociating gels,
    567, 572
    gel preparation, 567, 573
  sodium dodecyl sulfate, 567–568
    gel preparation, 567–569
GenBank, electronic mail address, 213
Genes
  bacterial, induced in host tissues, *in vivo*
    expression technology for, 481–492
  bar
    regulation, 395
    as selection marker for minitransposon
      genetic manipulation, 395
  bvg, 48, 58, 461
  fusion
    phoA, *in vitro* construction, vectors
      for, 437
    Tn*phoA*, switching, 446–447

gapA, sequencing by PCR analysis, 180–182

iron-responsive, in iron assay, 316–317

putP, sequencing by PCR analysis, 180–182

replacement in *Pseudomonas aeruginosa*, 466–474

rRNA-encoding, PCR amplification, 207

vir, 48, 58
   allelic exchange with plasmid vectors, 461–465

Glomerulonephritis, acute, 405, 410

Glycera dibranchiata, hemolysin, 667

Gold, sol preparation, 706–707

G proteins
   α subunit, ADP-ribosylation, 628
   G_i, ADP-ribosylation, 628
   G_o, ADP-ribosylation, 628
   G_s
      ADP-ribosylation, 624, 626
      α subunit, ADP-ribosylation, 640
         cholera toxin-catalyzed, 642–647

Granulation, cytosolic, toxin-induced, 684

Guinea pig
   chamber implant model, for host–parasite interaction studies, 125–126
      Neisseria gonorrhoeae, 129–132
      Porphyromonas gingivalis, 135–136
   conjunctiva, inoculation, 41–42
   keratoconjunctivitis model
      bacterial dose, 42
      baseline evaluation, 40–41
      course, 42–43
      immunity, development, 43–44
      inoculum preparation, 41
      Sereny assay, 40–41
      severity, 43
   ocular infection model, *Chlamydia trachomatis*, 69–83
   urogenital infection model, *Chlamydia psittaci*, 84

Guinea pig agent of inclusion conjunctivitis, 69
   culture, 78, 85
      from conjunctiva of infected animal, 81–82
   infection
      genital
         advantages, 85
         animals, 85

         course, 86
         immune response to, evaluation, 90
         inclusion scores, 86–87
         induction, 85–86
         inoculation procedures, 85–86
         in lower tract, clinical disease assessment, 85–86
         pathology, 84–85
         in upper tract, clinical disease assessment, 87–90
      ocular
         animals, 78
         clinical disease assessment, 79–80
         course, 79–81
         immune response to, evaluation, 82–83
         inclusion score, 81
         induction, 78–83
         microbiological assessment, 80–82
      serological testing for, 78
   inoculum preparation, 85

H

Haemophilus ducreyi
   host–parasite interactions, animal chamber models, 121, 137
   implant type, 122
   laboratory hazards and biosafety recommendations, 19
   lipooligosaccharide, 242

Haemophilus influenzae
   capsular polysaccharides, 95–96, 159, 161
      isolation, 162–163, 305
   capsular serotyping, 162–163
   capsular types, 159–160
   disease associations, 160–161
   electrotransformation, genetic factors, 385
   host–parasite interactions, animal chamber models, 137
   IgA_1 protease, 543
      properties, 553
      purification, 552
   laboratory hazards and biosafety recommendations, 19
   lipooligosaccharide, 242
   meningitis, 94, 161
      animal models, 95–96, 102

infant rat model, 95–100
  advantages and disadvantages, 96
  blood culture, 98
  cerebrospinal fluid culture, 98–99
  hematogenous spread of organism, 100
  and human disease, similarities, 96
  ID_{50}, 97
  induction, 97–98
  inoculation, 95
  inoculum preparation, 96–97
  lessons learned from, 99–100
  strains for, 96
  lytic antibiotic therapy, 104–105
nontypable, otitis
  animal models, 59
  chinchilla experimental model
    analysis of ears of challenged animals, 64–68
    applications, 59
    challenge modes for, 59–60
    data analysis, 66–68
    immunization, 60–62
    initial infectious dose, 63
    inoculation, 63–64
    inoculum preparation, 62–63
    middle ear inoculation, 63
    nasal challenge, 63–64
    vaccine effectiveness in, monitoring, 66–68
  outer membrane vesicles, isolation, 233
  temperature-sensitive mutants, generation by nitrosoguanidine, 451
  type b, 96, 160–161
    enzyme electromorph typing, 164
    intraperitoneal challenge, in infant rats, 97–98
    intravenous challenge, in infant rats, 98–99
    nasopharyngeal colonization, infant rats, 97, 99
    strains, characterization for epidemiologic studies, 163
    virulence, 99–100
  types, 159, 161
Helicobacter pylori, cytotoxin, cytopathogenic effects, 682, 684

Heme, microbial receptors, 345
Hemolysin
  assays, 657–667
    buffers, 666
    procedural considerations, 665–666
  blood agar plate assay
    hemolytic, 662
    screening, 659–660
  cytolysis, 657
    chromium release assay, 664
    enzymatic mechanisms, 658
    measurement, 663–665
    mechanisms, 657–659
    neutral red assay, 664–665
    phase-contrast microscopy, 663–664
    by pore formation, 658–659
  hemolytic activity
    kinetics, 663
    measurement, 661–663
  ion requirements, 666
  liquid hemolysis assay, 660–663
  microbial, as virulence factors, 657
  preparation, 665–666
  production, relationship to growth phase of toxin producer, 665
  screening for, 659–661
  surfactant activity, 659
  target cells, 666
  thiol activation, 666
Histidine-binding protein HisJ
  activity assays, 238–240
  isolation, 236–238
  purification, 240–241
Host clearance, quantitative studies with temperature-sensitive mutants, 454–455
Host–parasite interaction
  animal chamber models, 120, 128
  advantages, 120–121
  animal models, validity, 120
Host response, to bacterial infection, chamber implant studies, 127–128
Hot spots
  for *Mycobacterium tuberculosis* IS*6110* integration, 204
  for transposition of streptococcal transposons, 414–415
Hyaluronate 3-glycanhydrolase, 606
Hyaluronate 4-glycanhydrolase, 606
Hyaluronate lyase, 606

Hyaluronidase
  activities, 606
  assays, 606–616
    by absorption of unsaturated uronides,
      613
    by acetylamino sugar liberation, 613
    chemical methods, 612–613
    fluorogenic, 610
    indirect enzymoimmunological
      method, 610–611
    mucin clot prevention method, 614
    physicochemical methods, 613–615
    plate methods, 611–612
    radiochemical, 609–610
    by reducing sugar liberation, 612–
      613
    on solid media, 611–612
    spectrophotometric, 607–609
    turbidimetric, 615
    viscosity reduction method, 614
    zymographic techniques, 615–616
  in pathogens, 606–607
  species distribution, 606
  as spreading factor, 607
  staphylococcal, 412
Hyaluronoglucosamidase, 606
Hyaluronoglucuronidase, 606
8-Hydroxyquinoline, in deferration of
  laboratory media, 326–327

I

ID$_{50}$, see Dose, median infectious
Immunoglobulin, mouse, 126
Immunoglobulin A
  secretion
    animal models, 140–155
    in response to Shigella
      mouse lavage model, 144–147
      rabbit Thiry–Vella loop model, 140–
        143, 155
    secretory, protective effects
      animal model, 150
      against cytotoxicity by Shiga toxin,
        150–154
Immunoglobulin A$_1$ protease, bacterial,
  543–554
  activity, 543

assay, 545–551
  qualitative, 548–550
  quantitative
    immunoassay procedure, 550–551
    SDS-PAGE procedure, 546–548
  substrate preparation, 545–546
  isolation, 551–552
  peptide bonds cleaved by, 543–544
  production, 551–552
  properties, 553–554
  sequence homology, 543–544
  species distribution, 543
  substrates, 554
Indicator gels
  agarose–protein, 570–571
  for overlay techniques, preparation,
    570–571
  polyacrylamide–protein, 571
  in zymography of microbial proteases,
    584–586
Interleukin-1, production, endotoxin and
  peptidoglycan in, 282–283
Interleukin-1β, in rabbit bacterial
  meningitis model, 105
International Antigenic Typing System, for
  Pseudomonas aeruginosa, 167
In vivo expression technology
  definition, 482
  for selection of bacterial genes induced
    in host tissues, 481–492
    applications, 492
    biosynthetic gene complementation,
      491
    drug resistance plasmids in, 491
    purA gene system, 481–490
    reporter system, 491–492
    variations, 490–491
Iron
  assays
    biological, 316–317
    colorimetric, 315
    electron microscopy, 325
  chelates, HPLC, 321
  deficiency, bacterial response, 329–330,
    345
  electronic states, 317–318
  ferric
    colorimetric reagents for, 319–321
    ligands, 317–318

ferrous
  colorimetric reagents for, 319–321
  ligands, 317–318
  intrinsic, removal, 327–329
  isotopes, in iron assays, 323
  magnetic resonance imaging, 325
  metabolism, aerobic, 316
  microbial acquisition, 315–316, 345–346
    genes for, cloning in *Escherichia coli*,
      363–372
    with host Fe-binding proteins, 345,
      356
    mutants, 346, 352–356
  naturally occurring complexes, spectros-
    copy, 317–319
  neutron activation analysis, 323–324
  in pathogenesis, 315
  regulation of *Pseudomonas* exotoxin A
    expression, 502–517
  removal
    from labware, 325–326
    from media, 326–327
    from water, 326
  spectroscopy
    atomic absorption, 322
    atomic fluorescence, 323
    electron paramagnetic resonance, 324
    flame atomic emission, 322
    inductively coupled plasma-optical
      emission, 322–323
    Möbauer, 324

K

Kanamycin resistance, carried by transposon, selection for, 394
Keratoconjunctivitis, experimental
  Sereny assay, 39–47
  shigellosa, 39–47
Klebsiella
  capsular polysaccharides, isolation, 305
  genomic DNA, preparation, 175–176
Klebsiella oxytoca, PulO, amino acid
  sequence, 528–530
Klebsiella pneumoniae
  lipopolysaccharide-based typing, 161
  meningitis, animal models, 102
  types, 161

Koala, *Chlamydia psittaci* urogenital
  infection, 84

L

Laboratory, biosafety levels, 1, 5–10
Lactococcus lactis, electrotransformation,
  cell harvest for, 381
Lactoferrin, 316
  as iron-chelating bacteriostatic agent,
    328
LD_{50}, *see* Dose, median lethal
Legionella-like agents, laboratory hazards
  and biosafety recommendations, 20
Legionella pneumophila
  laboratory hazards and biosafety recom-
    mendations, 20
  porin, detergent solubilization, 232
Leptospira interrogans, laboratory hazards
  and biosafety recommendations, 20
Leukocydin, 412
Limulus amebocyte lysate assay, peptidoglycan preparations, for endotoxin
  contamination, 281–282
Lipid A, 242
  isolation, 252
Lipid bilayer membranes
  electrical measurements, 693–695
  formation, 691–693
  properties, 691
  toxin insertion, 695–697
  toxin pore formation, 691–705
    artifacts, 700
    channel size, 697
    concentrations, 704–705
    conductance properties and, 697–700
    cytotoxic effects, 703–704
    ion selectivity, 696–697, 701–702
    kinetics, 702
    lipid dependence, 703
    macroscopic currents, 696
    molecularity, 697
    pathophysiology, 702
    receptor requirements, 705
    relevance to toxicity, 700–705
    single-channel conductance, 700–701
    single-channel records, 695–696
    voltage dependence, 703

Lipooligosaccharides, 242
in acrylamide gels, silver staining, 251–
252
gel electrophoresis
on acrylamide, 247–249
on Tricine–SDS–polyacrylamide, 250–
251
isolation
phenol–water extraction technique,
243–245
small-scale preparations, 246–247
low-molecular-weight, poorly separated,
resolution with long gels, 249–250
preparations, purity, 247–252
Lipopolysaccharides
in acrylamide gels, silver staining, 251–
252
components, 242
gel electrophoresis
on acrylamide, 247–249
on Tricine–SDS–polyacrylamide, 250–
251
isolation
phenol–water extraction technique,
243–245
rapid micromethod, 246–247
SDS solubilization method, 243–246
small-scale preparations, 246–247
microextraction using proteinase K
digestion, 247
physical properties, 230
preparations, purity, 247–252
in serotyping, 159, 161
structure, 230
Listeria
disease associations, 413
laboratory hazards and biosafety recommendations, 20
peptidoglycans, trichloroacetic acid
extraction, 257
Listeria monocytogenes
electrotransformation, 382
DNA delivery, 384
meningitis, animal models, 102
temperature-sensitive mutants
applications, 457
generation by nitrosoguanidine, 451
virulence analysis, transposon mutagenesis methods, 413–414
virulence factors, 413

Listerolysin O, 413–414
Lysosomotropic agents, 685

M

Macaca fascicularis
ocular infection, Chlamydia trachomatis,
69
advantages, 71
antibody response, assessment, 77–78
cell-mediated immune response, assessment, 76–77
clinical disease assessment, 71–73
histological evaluation, 74–76
induction, 70–71
inoculum, 71
microbiological assessment, 73–74
total clinical disease score, 71–73
periodontitis model, 117
sources, 70
tuberculin testing, 70
Macaca mulatta, see also Monkey, rhesus
periodontitis model, 117
Macaque, pig-tailed, Chlamydia trachomatis urogenital infection, 84
Magnesium carbonate, in deferration of
laboratory media, 327
Magnetic resonance imaging, iron, 325
Marmoset
Chlamydia trachomatis urogenital infection, 84
periodontitis model, 112
Mass spectrometry, siderophores, 344
Mean generation time, in vivo, 457
Media
deferration, 326–327, 346
laboratory, iron content, 346
Membranes, see also Lipid bilayer membranes; Outer membranes
bacterial cytoplasmic, buoyant density,
225, 230
Meningitis, bacterial
animal models, 93–106
epidemiology, 93–94
Escherichia coli
animal models, 102
lytic antibiotic therapy, 104
Haemophilus influenzae, 94, 161
animal models, 95–96, 102
infant rat model, 95–100

immune response to, 105–106
infant rat model, 95–100
Klebsiella pneumoniae, animal models, 102
Listeria monocytogenes, animal models, 102
Neisseria meningitidis, 94, 163–164
neonatal, *Streptococcus agalactiae*, 173
pathogenesis, 94–95
pathogens, 94
pneumococcal, 94–95, 102, 104, 168
Proteus mirabilis, animal models, 102
Pseudomonas aeruginosa, animal models, 102
rabbit model, 95, 100–106
streptococcal, group B, 94, 96, 173
Streptococcus pneumoniae, 94, 168
animal models, 95, 102
lytic antibiotic therapy, 104
Mercury resistance, carried by transposon, selection for, 394
Metallothionein, HPLC, 321
Micrococcus, peptidoglycans, trichloroacetic acid extraction, 257
Micrococcus lysodeikticus, low-molecular-weight peptidoglycan oligomers and monomers, isolation, 269
Minitransposons
as cloning vectors, 397–399
delivery, plasmids for, 388–391, 397–399, 403
delivery system, 388–390
genetic manipulations with, in gram-negative bacteria, 386–405
applications, 389, 396–403
conditional phenotype generation, 402–403
exconjugants
low or zero yield, 403–404
multiple drug resistance in, 404–405
selection, 393–395
insertional mutagenesis with, 396–397
insertions, confirmation and analysis, 395–396
mating mixture set-up, 390–393
nonpolar mutation generation, 402–403
troubleshooting, 403–405
markers carried by, 393–395
promoter-probe, 399–402
properties, 387–388

structure, 388–391
Tn5 *phoA*, 436
insertional mutagenesis with, 436–437
Monkey
cynomolgus, *see Macaca fascicularis*
grivet, *Chlamydia trachomatis* urogenital infection, 84
rhesus
Chlamydia trachomatis urogenital infection, 83–84
periodontitis model, 117
squirrel, periodontitis model, 117
Moraxella bovis, pilin, 527
Moraxella catarrhalis, otitis, 59
Morganella morganii, hemolysin, 668
Mössbauer spectroscopy, iron, 324
Mouse
bacterial replication in, measurement *in vivo*, 455–457
chamber implant model, for host–parasite interaction studies, 125–126
Bordetella pertussis, 138
Escherichia coli, 137–138
Haemophilus ducreyi, 137
leukocyte response, effect implant type, 122–123
Neisseria gonorrhoeae, 129–131
Porphyromonas gingivalis, 136–137
Streptococcus pyogenes, 138–139
endotoxin-resistant, 284
genital infection, *Chlamydia trachomatis*, 90–93
human oculogenital strains for, 93
immunoglobulins, 126
lavage model, for mucosal immunity, 143–144
to *Shigella* antigens, 144–147, 154–155
pertussis respiratory infection model, 47–58
animal strains for, 55
applications, 57–58
bacterial strains for, 54–55
disease course, and animal age, 53–55
initial infective dose, 51–52
variation, 55–57
urogenital infection, *Chlamydia trachomatis*, 84
Mouse pneumonitis agent, genital infection in mouse, 90–93
advantages, 91

clinical disease assessment
   in lower tract, 91–92
   in upper tract, 92–93
   course, 91–93
   induction
      in lower tract, 91
      in upper tract, 92
Moving average interpolation, for LD$_{50}$
   determination, 32–34, 36–37, 39
M protein, streptococcal, 171, 286–294,
   411
   acid extraction, 287–288
   antibodies, 286–287
   antiphagocytic property, 286
   pepsin extraction, 288–292
   preparation, 287–288
   purification, 291–292, 294
   recombinant, 292–294
   structure, 286–287, 289
Mucosal immunity
   mouse lavage model, secretory IgA
      response to *Shigella* in, 143–147
   rabbit Thiry–Vella loop model, secretory
      IgA response to *Shigella* in, 140–
      143
Multinucleation, toxin-induced, 685
Muramidase, peptidoglycan fragments
   released by, 269–270
Muramyl dipeptide, 253–254
   analogs, 254
Mutagenesis
   insertional
      with mini-Tn5 *phoA*, 436–437
      with minitransposons, in gram-nega-
         tive bacteria, 396–397
      Pseudomonas aeruginosa, 471–473
      with pJM703.1 or pGP704, 473–
         474
      resistance markers for, 471–473
   for temperature-sensitive mutants, 449–
      450
   transposon, virulence analysis in gram-
      positive bacteria, 406–426
      applications, 426
      considerations, 425–426
      limitations, 425–426
Mycobacterium, laboratory hazards and
   biosafety recommendations, 20
Mycobacterium bovis
   laboratory hazards and biosafety recom-
      mendations, 20–21

strain *Mycobacterium bovis* BCG, iden-
   tification, 197–198
Mycobacterium leprae, laboratory hazards
   and biosafety recommendations, 20
Mycobacterium smegmatis, electrotransformation, 385
Mycobacterium tuberculosis
   DNA fingerprinting, 196–205
      analysis, 204
      applications, 196, 205
      interpretation, 204
      principles, 198–200
   genomic DNA, isolation, 199–201
   IS6110, 197–199
      DNA fragments containing, hybridiza-
         tion and detection, 203–204
      DNA probe, preparation, 202–203
      integration, hot spots for, 204
   laboratory hazards and biosafety recom-
      mendations, 20–21
   RFLP analysis, 204–205
   strain typing, 196–198
      principles, 198–200
Mycobacterium tuberculosis complex
   insertion sequences, 197–198
   species
      DNA polymorphism, 196–197
      taxonomic relatedness, 196–197
   transposable elements, 197
Myxococcus xanthus, outer membrane,
   buoyant density, 230

N

NAD
   ^{14}C-labeled, in ADP-ribosyltransferasse
      assays, 618–619, 621–622, 630, 643,
      646
   as cofactor for ADP-ribosylation, 632
   in photoaffinity labeling of active-site
      residues in ADP-ribosylating toxins,
      632–639
   ^{32}P-labeled, in ADP-ribosyltransferase
      assays, 623–625, 627–630
NAD$^+$ glycohydrolase, 646
Neisseria, outer membrane vesicles, 233
   isolation, 233
Neisseria gonorrhoeae
   genes
      analog of *Escherichia coli fhuB* gene,
         363

tbpA, 360–361
tbpB, 360–361
host–parasite interactions, animal chamber models, 121, 125, 128–132
implant type, 122
host response, chamber implant studies, 128
IgA$_1$ protease, 543
    assay
        qualitative, 548–550
        quantitative, 546–548
    production, 551–552
    properties, 553
    purification, 551–552
    substrates, 554
iron acquisition
    from *Escherichia coli*-derived siderophores, 357
    from host Fe-binding proteins, 356–357
iron-uptake mutants, 129
laboratory hazards and biosafety recommendations, 21
lipopolysaccharide, 159
peptidoglycans
    isolation, 260–263
    low-molecular-weight oligomers and monomers, isolation, 269
    soluble polymeric, isolation, 264–265
pilin, 527
porin, detergent solubilization, 232
receptor-mediated transferrin-iron uptake mechanism, 356–363
strain FA19, transferrin receptor, 360–363
transferrin-Fe uptake mechanism, 358–362
transferrin-nonutilizing mutants, 359–360
    lrf, 360
    tlu, 360
    trf, 360–361
transglycosylase, 270–271
urogenital infection, 83
virulence, and iron, 128–129
Neisseria meningitidis
    capsular polysaccharides, 159, 161, 163–164
        isolation, 305
    disease associations, 163
    enzyme electromorph typing, 164

IgA$_1$ protease, 543
    properties, 553
    purification, 552
laboratory hazards and biosafety recommendations, 22
lipooligosaccharide, 242
lipopolysaccharides, 159
meningitis, 94, 163–164
pilin, 527
serogrouping, 164–165
serogroups, 160, 163–164
    disease associations, 160
serotyping, 165–166
subtyping, 165–166
types, 159, 161
Neutron activation analysis, iron, 323–324
Nitrilotriacetic acid, as iron-chelating bacteriostatic agent, 329
Nitrosoguanidine, as mutagen, 449–451
Nuclear magnetic resonance
    ^{13}C, bacterial capsular polysaccharides, 306, 310–311
    ^{1}H, bacterial capsular polysaccharides, 306
    siderophores, 344

O

O-antigen, 242
Otitis media
    causative organisms, 59
    chinchilla experimental model, 59–68
        analysis of ears of challenged animals, 64–68
        animals, 60
            active immunization, 60–61
            passive immunization, 61–62
        applications, 59
        challenge procedures, 63–64
        data analysis, 66–68
        Haemophilus influenzae, challenge modes for, 59–60
        inoculation, 63–64
        inoculum preparation, 62–63
        vaccine effectiveness in, monitoring, 66–68
    pneumococcal, 168
Outer membranes
    buoyant density, 225–226, 230
    fragments, 225
    instability, 225

iron-regulated proteins, 344–352
isolation, 225–234
    by differential extraction with deter-
        gents, 230–232
    equilibrium density gradient centrifu-
        gation technique, 230
    principles, 225
    by selective release from intact cells,
        232–234
    separation from cytoplasmic membrane,
        225–226
    siderophore receptors, 345
    structure, 225
    vesicles, 225
        isolation, 232–234
Ovotransferrin, as iron-chelating bacterio-
    static agent, 328

                        P

Papio anubis, periodontitis model, 112
Pasteurella haemolytica, RTX toxin
    antibody probes for, 671–672
    DNA, as probe for *Actinobacillus ac-
        tinomycetemcomitans* genomic
        DNA, 670–671
    nonerythrocyte cytolytic activity, assay,
        674
    pore formation assay, 677
    sublytic effects, 675–676
    target cell specificity, 667
Pasteurella multicoda, laboratory hazards
    and biosafety recommendations, 22
Pathogens, *see also specific pathogens*
    antigenic diversity, 159
    antimicrobial susceptibility, in determi-
        nation of biosafety procedures, 3
    bacterial, uncultured, phylogenetic
        identification with rRNAs, 205–222
    biosafety recommendations, 1–26
    phenotypic heterogeneity, 160–161
    production quantities, in determination
        of biosafety procedures, 3
    serological classification, 159–174
    strains, classification
        methods, 160
        principles and rationale for, 159–160
PAUP computer program, 217
Pectate lyase, secretion in *Erwinia chry-
    santhemi*, analysis by Tn*phoA* muta-
    genesis, 441–442

Peptidoglycan
    activities, 253–254
    amino acid composition, analysis, 278–
        279
    amino sugars, analysis, 278–279
    anhydro monomers, preparation, 272–
        273
    Chalaropsis monomers, reversed-phase
        HPLC, 274–277
    characterization, 278–279
    chemical characterization, 278–279
    fragments
        free peptides, 275
        muramidase-derived, 269–270
        peptide-free glycan chains, 275
        transglycosylase-derived, 270–271
    functions, role of endotoxin, 282–284
    glycan chain length, determination, 279–
        280
    gonococcal, isolation, 260–263
    gram-negative
        isolation, 255, 260–263
        soluble polymeric, isolation, 264–265
    gram-positive
        isolation, 255–260
        soluble polymeric
            isolation, 263–264
            secreted by cells, isolation, 265–268
    hot formamide extraction, 258
    hydrolysis, 279
    insoluble, isolation, 255–263
    low-molecular-weight, fractionation by
        size, 273–274
    low-molecular-weight oligomers and
        monomers, isolation, 268–277
    mass spectrometry, 280
    mycobacterial, isolation, 255
    percent O-acetylation, determination,
        280
    percent peptide cross-linking, determina-
        tion, 279
    physical properties, 253
    preparations
        carbohydrates in, 284–285
        detergents in, 285
        endotoxin contamination, elimination/
            reduction, 277–278, 281–284
        purity, 284
        quality assurance, 281–285
        teichoic acid contamination, 284–285
    soluble fragments, activities, 253–254

soluble polymeric
  isolation, 263–268
  secreted by cells
    isolation, 265–268
    properties, 265
  sonicated, 263
  structural analysis, 278–280
  structure, 253
  trichloroacetic acid extraction, 257–260
Peptostreptococcus anaerobius, host–
    parasite interactions, animal chamber
    models, 134–135
Perchloric acid assay, siderophores, 331
Periodontal disease, animal models, 106–
  119
  pathogens, 106–107
  selection, 107–108
Periodontitis, animal models
  biochemical markers, 111
  clinical indices, 110–111
  culturing methods for, 109–110
  diet effects, 108–109
  disease activity, 111
  Eastman interdental bleeding index, 111
  germ state for, 109–110
  gingival index, 110–111
  housing effects, 109
  hygiene index, 110–111
  oral hygiene for, 108
  papillary bleeding index, 110–111
  plaque index, 110–111
  plaque toxicity tests, 110
Periplasmic binding proteins
  activity assays, 238–240
  deliganding procedures, 241
  isolation, 236–238
  mechanism of action, 235
  purification, 240–241
Permease, periplasmic
  components, 234–235
  mechanism of action, 235
  substrate-binding protein, 234–241
Pertussis
  mouse respiratory infection model, 47–
    58
    advantages and disadvantages, 49, 58
    aerosol infection, 49, 52
      apparatus, 49–52
    animals, 53–55, 58
    applications, 57–58
    bacterial strains for, 54

course, 53–55
    initial infective dose, 48–49, 51–52
      variation, 55–57
    inoculum preparation, 52
    intranasal inoculation, 48–49
    lung colonization, 53, 55–58
    parameters of infection and disease,
      53–55
    procedure, 52–53
    tracheal colonization, 53, 58
  pathogenesis, 48
  pathology, 48
  vaccine, mouse experiments, 57–58
Pertussis toxin
  active site residues, photoaffinity label-
    ing, 633–635
  ADP-ribosyltransferase activity, 617,
    628, 632
    assay, 630
  assay, 629–630, 684
  cytopathogenic effects, 682, 684
  mechanism of action, 680
  structure, 628
Phasmids, for complementation of *Escheri-
    chia coli* mutants, 369–370
Phosvitin, electron transitions, 317
PHYLIP computer program, 217
Phylogenetic analysis
  bootstrap analysis, 216–217
  computer programs, 217
  distance matrix approach, 216
  macromolecular sequences used for,
    206–207
  maximum likelihood technique, 216–217
  parsimony analyses, 216–217
  with PCR-based sequencing, 180–183
  with rRNAs, 205–222
  treeing algorithms, selection, 215–217
  treeing artifacts, 220–221
Phytosiderophores, 354
Pilin, 527
  precursors, posttranslational processing,
    527
  type IV, amino acid sequences, 528–529
Plasmids
  for minitransposon delivery, 388–391,
    397–399, 403
  pACYC184, 467
  pBNR322, 475
  pEMR2, 468–469
  pGK2003, 477

pGP704, 468–469, 471, 473–474, 483–484
pILL, 476–477
pILL550, 476–477
pILL560, 477
pIP1455, 476
pIVET1, 482–484
pIVET2, 491
pIVET8, 491
pJM703.1, 434, 468–469, 473–474
  Tn*phoA* mutagenesis with, 433
pME301, 467
pNOT19/pMOB3, 469–471
pPH1JI, 432–433
p1PI433, 476
p1PI455, 475
pR751, 467
pRK290, 466
pRK2013, 467
pRT291, 432–433
  Tn*phoA* mutagenesis with, 433
pRTP1, 459–460, 469–471
pRY109/110, 477
pRY113/114, 477
pRZ102, 468–469, 471
pSKCAT, 446–447
pSS1129, 459, 461
pUC13, 476
pUOA, 476–477
pUOA13, 477
pUOA15, 477, 481
pUOA18, 477
for transposon delivery, 408–409
  broad-host-range, 431–433
Platelet-activating factor, in rabbit bacterial
  meningitis model, 105
Pneumonia, pneumococcal, 168
Polymerase chain reaction
  amplification
    mycobacterial IS*6110* DNA probe,
      202–203
    rDNA, 207, 209–213
    16 S rRNA, technical issues and
      problems, 218–221
    target sequences, 177–178
  assay, *Chlamydia trachomatis* ocular
    infections, 74
  conditions, 178
  nucleotide sequencing based on, for
    analysis of genetic variation, 174–
    183

primer design, 178
product DNA
  λ-exonuclease treatment, 179
  single-stranded, generation, 178–179
  in situ, in uncharacterized pathogens,
    221–222
Porphyromonas gingivalis
  host–parasite interactions, animal cham-
    ber models, 121, 125, 135–137
  periodontitis, 107
    primate model, 113–114
    rat model, 117
  proteases, 566
    inhibition, zymographic studies, 581
  virulence factors, 106
Prevotella, IgA$_1$ protease, 543
Prevotella intermedia
  periodontitis, 107
  virulence factors, 106
Primates, *see also specific primates*
  oral hygiene, 108
  periodontitis model, 107, 111–114
    advantages and disadvantages, 114
    animals, 112
    applications, 112–114
    clinical disease evaluation, 112–113
    clinical indices, 111
    histologic studies, 112–113
    induction, 112
Probit analysis
  for LD$_{50}$ determination, 31–36, 39
  Statistical Analysis System for, 31, 39
Promoters, probing, minitransposons for,
  399–402
Pronase, 566
  detection, zymographic technique, 578–
    579
  zymography in nondissociating gels with
    copolymerized substrates, 588–590
Propionibacterium, peptidoglycans, trichloroacetic acid extraction, 258
Propionibacterium acnes, hyaluronidase
  assay, 613
  zymographic analysis, 616
Protease inhibitor-agarose matrix, 603–604
Proteases, microbial
  activators, zymographic studies, 580–
    581
  activities, 563
  assay, 560

catalytically active components, relative molecular weights, determination, 576–577
characterization, zymographic techniques, 563–594
detection, zymographic techniques, 563–594
IgA$_1$, see Immunoglobulin A$_1$ protease, bacterial
inhibition, reverse zymography, 582–583
inhibitors, zymographic studies, 580–581
protein substrate behavior, zymographic characterization, 577–578
sodium dodecyl sulfate effects, 578–580
substrate specificity, determination, zymographic techniques, 577–578
V8, 566
as virulence factors, 563
zymographic studies
  adaptations, 586–587
  applications, 574–593
  with detergent-containing gels, 567–571
  enzyme characterization by, 580–586
  in nondissociating gels, 571–574, 587–593
    with copolymerized substrates, 588–589
    indicator gel method, 593
    with substrate diffusion after cationic electrophoresis, 589–591
    with synthetic esterase substrates, 590–593
  overlay techniques, 584–586
    indicator gels for, preparation, 570–571
  quantitation of enzyme activity, 576
  substrate diffusing in method, 583–584
Proteins
  biotinylation, 707
  intracellular routing, monitoring, biotinyl ligand:avidin–gold method, 707–710
Proteus mirabilis, meningitis, animal models, 102
Proteus vulgaris, hemolysin, 668
Pseudomonas, iron-regulated siderophores and outer membrane proteins, 351–352
Pseudomonas aeruginosa
  acetylation enzyme, in alginate synthesis, 494–496

alg genes
  organization, 493–497
  promoters, environmental activation, 497–501
alginate, 295–304
  alginase treatment, 304
  assay, 298–301
    colorimetric, 300–301
    uronic acid method, 298–299
  biosynthesis pathway, 493–494
  chromatography, 303
  functions, 493
  gene expression, regulation, 493–502
  isolation, 296–298
    bacterial strains, 296
    from culture supernatant, 297–298
    removal of cell material, 297
  positive identification, 302–304
  production, 493
  properties, 295
  purification, 295, 301–302
  spectroscopy, 302–303
  structure, 295–296
  synthesis, signal transduction in, 494–495, 500–502
bacteriostatic agents, iron-chelating, 328
disease associations, 167, 527
elastase
  activities, 554–555
  assays, 557–562
    elastin–Congo Red method, 557–559
    elastin–nutrient agar method, 561–562
    fluorogenic substrate method, 559–561
  in pathogenesis, 555
  purification, 555–557
  strains producing, pulmonary persistence, 134
electrotransformation, cell lysis during, prevention, 381
epimerase, in alginate synthesis, 493–495
exoenzyme S
  ADP-ribosyltransferase activity, 617, 622, 632
    assay, 623
  assay, 622–623
exotoxin A
  active site residues, photoaffinity labeling, 633–635

ADP-ribosyltransferase activity, 617,
    631–632
  assay, 622
  assay, 505, 621–622
  cell binding, 621
  cells sensitive to, 710–712
  ELISA, 713–714
  expression, regulation by iron, 502–
    517
    copy number effects, 508
    deregulation, 514–516
    growth curve analysis and, 505
    patterns, 506–511
    regulation, 516–517
    strain selection, 503
    transcriptional start site analysis,
      509–510
    transcriptional and transcriptional
      fusions, 510–511
  inhibition of host protein synthesis,
    647–648
  internalization into cells, 621, 706–715
    monitoring, 712–716
    ultrastructural studies, 706–710
  intracellular localization, 714–716
  intracellular routing, monitoring,
    biotinyl ligand:avidin–gold
    method, 706–710
  mechanism of action, 680
  strains producing, pulmonary persist-
    ence, 134
  structure, 620–621
  synthesis, 620
  toxicity, 619–620
    modification, 710–712
    monitoring, biochemical methods,
      710–712
    TCD$_{50}$ assay, 710–712
  translocation in cell, 621
  yield
    maximal, growth conditions for,
      503–504
    strains differing in, analysis, 511–
      513
exotoxin S, strains producing, pulmonary persistence, 134
extracts, preparation, 499
GDPmannose dehydrogenase, in alginate
  synthesis, 494–495, 497

GDPmannose pyrophosphorylase, in
  alginate synthesis, 494–497
gene replacement in, 466–474
  early systems, 466–467
genes
  fur, in iron reporter system, 316
  regA, in iron reporter system, 316
  regAB, transcription regulation, 502–
    517
  toxA, transcription regulation, 502–
    517
growth, 296, 499
heat-stable hemolysin, cytolysis, mechanism, 659
hemolysin, cytopathogenic effects, 682
host–parasite interactions, animal chamber models, 121, 133–134
insertional mutagenesis, 471–473
  resistance markers for, 471–473
International Antigenic Typing System,
  167
iron-regulated siderophores and outer
  membrane proteins, 351–352
lipopolysaccharides, 159, 161
macromolecular transport, proteins
  required for, amino acid sequences,
  528–529
meningitis, animal models, 102
mucoid, alginate production, quantitation, 299–300
O-type polysaccharides, 167
outer membrane vesicles, isolation, 233
PddA-D proteins, 527–528
phosphomannomutase, in alginate synthesis, 494–495, 497
phosphomannose isomerase, in alginate
  synthesis, 494–497
PilA, 527
PilB, 527
PilC, 527
PilD
  amino acid sequence, 528–530
  antisera production, 534–535
    peptide used for, 534
  bifunctional activity, 539–540
  endopeptidase activity, 527
  hydrophobicity plot, 534
  immunoaffinity chromatography, 534–
    536

N-methyltransferase activity, 538–539
  inhibitor, 539
prepilin peptidase activity, 527
  assay, 531–533
  inhibitors, 536
  kinetics *in vivo*, 536–538
  on *Neisseria. gonorrhoeae* prepilin,
    537–538
  on substrates altered at +1 position,
    538
  purification, 533–536
  substrates, 528, 539
  source, 531–532
pili, as virulence factors, 527
pilin, 527
serotyping, 167–168
suicide vectors for, 467–471
supernatants, preparation, 555–557
temperature-sensitive mutants
  applications, 457
  generation by nitrosoguanidine, 451
  quantitative clearance studies with,
    454
  replication, measurement *in vivo*, 455,
    457
type IV prepilin, posttranslational processing, 527–540
types, 159, 161, 167
virulence factors, 527
Xcp proteins, 528
Pseudomonas cepacia, temperature-sensitive mutants
  applications, 457
  quantitative clearance studies with, 454
  replication, measurement *in vivo*, 455
Pseudomonas pseudomallei, laboratory
  hazards and biosafety recommendations, 22

Q

Quellung reaction, 170

R

Rabbit
  chamber implant model, for host–para-
    site interaction studies, 124–125
  Escherichia coli, 137–138
  Porphyromonas gingivalis, 135–136

experimental keratoconjunctivitis in,
  Sereny assay, 40–41
gut-associated lymphoid tissue, immune
  response in, 142–143
ileal loop
  acutely isolated, *Shigella* uptake by M
    cells and villi, 147–148
  chronically isolated, Thiry–Vella
    model
    preparation, 140–141
    secretory IgA response to *Shigella*
      in, 140–142, 155
  initial processing of *Shigella* prepara-
    tions by, 148–150
intestinal loop secretions, anti-Shiga
  toxin activity, 152–154
meningitis model, 95, 100–106
  advantages and disadvantages, 100
  animal preparation and setups, 101
  applications, 100
  bacteria, 102
  cerebrospinal fluid sampling, 103–104
  cytokines in, 105
  experimental parameters, 103–104
  immune response, 105–106
  initial infective dose, 102
  inoculation, 95
  inoculum preparation, 102–103
  lessons learned from, 104–106
  model for protection by secretory IgA,
    preparation, 150
  mucosal immune response to Shiga
    toxin, 150–154
  toxic shock syndrome model, 132–133
Rat
  infant
    Haemophilus influenzae meningitis
      model, 94–100
    sources, 96
  periodontal health, housing effects, 109
  periodontitis model, 107, 111, 117–119
    advantages, 117–118
    disease assessment, 118–119
    germ state and culturing methods,
      109–110
    and human periodontitis, similarities,
      117
    inoculation, 118
    pathogens, 117

Rattus norvegicus, *see* Rat
Recombination
  intragenic, 181–183
  rate, and functional type of gene product, 181–183
Reed–Muench method, for LD$_{50}$ determination, 30, 33–35
Respiratory infection, models, 47–48
Restriction fragment length polymorphism, analysis in bacteria, 184–196
  analysis of differentiation and similarity coefficients, 195
  applications, 185–186, 195–196
  bacterial strains, 189
  densitometry, 194–195
  genomic DNA
    isolation, 189–190
    restriction patterns, 186–187, 190–193
  hybridization probes, 194
  Mycobacterium tuberculosis, 204–205
  principles, 184–186
  probe selection, 189
  restriction endonuclease selection, 186–189
  Southern blotting, 186–187, 190–193
Rheumatic fever, 405, 410
Rhizobactin, 354
Rhizobium meliloti
  gene replacement in, 466
  symbiotic loci, analysis by Tn*phoA* mutagenesis, 438–439
Rhizoferrin, 354
Rhodobacter capsulatus, outer membrane vesicles, isolation, 233
Ribosomal RNA
  genes
    as chromometers, 184
    RFLP analysis
      applications, 184
      in bacteria, 184–196
  reverse transcriptase sequencing, 208
  sequences
    alignment, 213–215
    comparison, 213–215
    databases, 213–214
  small subunit 16 S
    genes, cloning, 208–209
    PCR amplification, 209–213
      and contamination of tissues or reagents, 219–220

primers, 218–219
tissue quality and type for, 218
in phylogenetic analysis, 207–222
  technical issues and problems, 217–221
secondary structure, 209–211
sequence microheterogeneity, 220
from uncultured pathogens
  phylogenetic identification using, 205–222
    confirmation, *in situ* techniques, 221–222
    procedures, 207–218
    rationale, 206–207
    technical issues and problems, 218–221
  preparation, 207–213
Rickettsia akari, laboratory hazards and biosafety recommendations, 25
Rickettsia canada, laboratory hazards and biosafety recommendations, 25
Rickettsial agents, laboratory hazards and biosafety recommendations, 24–25
Rickettsia prowazekii, laboratory hazards and biosafety recommendations, 25
Rickettsia rickettsii, laboratory hazards and biosafety recommendations, 25
Rickettsia tsutsugamushi, laboratory hazards and biosafety recommendations, 25
Rickettsia typhi, laboratory hazards and biosafety recommendations, 25
RNA, *see* Ribosomal RNA; Transfer RNA
Rochalimaea quintana, laboratory hazards and biosafety recommendations, 25
Rochalimaea vinsonii, laboratory hazards and biosafety recommendations, 25
RTX proteins
  cytolysins, *see* Cytolysin, RTX family
  protease activity, 667–668
  structure, 670

S

Saccharomyces cerevisiae, iron acquisition by, 345–346
Saimiri sciureus, periodontitis model, 117
Salmonella
  bacteriostatic agents, iron-chelating, 328

genomic DNA, preparation, 175–176
laboratory hazards and biosafety recommendations, 22
lipopolysaccharide, 159
nucleotide sequence variation, PCR analysis, 175, 180–182
Salmonella enteritidis
keratoconjunctivitis, Sereny assay, 46
temperature-sensitive mutants, applications, 457
Salmonella typhi
laboratory hazards and biosafety recommendations, 22
temperature-sensitive mutants, generation by nitrosoguanidine, 451
Salmonella typhimurium
invasion and cellular survival, analysis by Tn*phoA* mutagenesis, 439
keratoconjunctivitis, Sereny assay, 46
outer membranes, isolation, 226–230
periplasmic binding protein, isolation, 236–238
purA gene, *in vivo* expression technology for, 482
assays, 488–489
fusion cloning, 489
fusion strains, red shift, 485–488
in vivo assays, 488–489
null mutations in *ivi* operons, 490
plasmid for, 482–484
purA–lac fusion construction, 483–486
regulatory element analysis, 490
variations, 490–491
temperature-sensitive mutants, quantitative clearance studies with, 454
Salpingitis, mouse, from mouse pneumonitis agent, 92–93
Sarkosyl, solubilization of outer membrane, 230–232
Scanning electron microscopy, and energy-loss spectroscopy, in iron assays, 325
Scanning transmission electron microscopy, in iron assays, 325
Septicemia, neonatal, *Streptococcus agalactiae*, 173
Sereny assay, 39–47
advantages and disadvantages, 47
animals, 40–41
applications, 40, 45–47

assessment, 44–45
bacterial dose and course of *Shigella* keratoconjunctivitis, 42–43
inoculation of conjunctiva, 41–42
inoculum, preparation, 41
Serotyping
bacterial pathogens, 159–160
biochemical basis, 159, 161
group A streptococci, 171–173
Haemophilus influenzae, 162–163
meningococcal, 165–166
Streptococcus pneumoniae, 170
Serratia marcescens
bacteriostatic agents, iron-chelating, 328
hemolytic determinant, analysis by Tn*phoA* mutagenesis, 439–440
metalloproteases, 667–668
Serratiopeptidase, 566
Shiga toxin
cytotoxicity, 648
assays, 150, 648–650
protective effects of secretory IgA, 150–154
HeLa assay, 150, 152–153
inhibition of host protein synthesis, 648
internalization, 648
mechanism of action, 680
mucosal immune response to, animal model, 144–147, 150–154
structure, 648
Shigella
antigens, *see also* Shiga toxin
initial processing by rabbit intestinal preparations, 148–150
mucosal immunity to, mouse lavage model, 144–147
secretory IgA response to, animal models, 140–155
culture, for Sereny assay, 41
environmental sources, 46
experimental keratoconjunctivitis
bacterial dose, 42
course, 42–43
histopathology, 42–43, 45
immunity, development, 43–44
macroscopic characteristics, 42–43
Sereny assay, 39–47
severity, 43
ID$_{50}$, for experimental keratoconjunctivitis, 42

immunity, development in guinea pig, 43–44, 47
infection, in humans
    pathogenesis, 44
    pathophysiology, 44
laboratory hazards and biosafety recommendations, 23, 41–42
lipopolysaccharide, 159
toxin, *see* Shiga toxin
uptake by M cells and villi in acutely isolated rabbit ileal loop, 147–148
virulence, *in vivo* assays, 44
Shigella flexneri, aerobactin, isolation, 341–342
Shuttle vectors, for *Campylobacter*, 475–480
    conjugative transfer of vectors from *Escherichia coli*, 477–478
    transfer by electroporation, 478–479
Siderophore receptors, 345, 347
Siderophores, 316, 329–344, 356; *see also specific siderophores*
    assays
        Arnow, 332, 353, 371
        chemical, 331–332, 357, 371
        Chrome azurol S, 333–334, 353–356
        colorimetric, 331–332, 371
        cross-feeding, 371
        Csáky, 331–332, 353, 371
        ferric perchlorate, 331
        functional, 332–338
        perchloric acid, 331
    bioassays, 335–338, 357
        Aureobacterium flavescens method, 335–336
    catechol, 329, 343, 353, 357
        purification, 338–340
    characterization, 343–344
    charged, isolation, 338
    classification, 329
    detection, 330–331
    extraction with organic solvents, 338
    functions, 329, 345
    hydroxamate, 329, 343, 353, 357
        isolation, 338
    isolation, 321, 338–343
    ligands, 347
    mass spectrometry, 344
    nuclear magnetic resonance, 344

purification, 338–343
spectral analysis, 343
structural analysis, 343–344
synthesis, 329–330, 345
    bacterial culture conditions for, 330
X-ray diffraction analysis, 344
Simultaneous multielement atomic absorption continuous source spectrometer, 321
Sinefungin, inhibition of *Pseudomonas aeruginosa* PilD *N*-methyltransferase activity, 539
Southern blotting
    Bacillus chromosomal DNA, 185–188
    mycobacterial DNA, 202
    in RFLP analysis in bacteria, 186–187, 190–193
Spectinomycin resistance, carried by transposon, selection for, 394
Spectrometer, simultaneous multielement atomic absorption continuous source, 321
Spectroscopy, iron, 322–324
Sphingomyelinase, *Staphylococcus aureus*, cytolytic mechanisms, 658
Staircase method
    for ED_{50} determination, 33
    for LD_{50} determination, 33, 37–39
Staphylococci
    disease associations, 405
    peptidoglycans, trichloroacetic acid extraction, 257–260
Staphylococcus aureus
    α toxin, mechanism of action, 680
    bacteriostatic agents, iron-chelating, 328
    β toxin, cytolytic action, enzymatic mechanisms, 658
    δ toxin, cytolytic action, mechanism, 659
    disease associations, 405, 412
    hemolysin, 665
    host–parasite interactions, animal chamber models, 121, 132–133
    hyaluronidase, assay, 614–616
    laboratory hazards and biosafety recommendations, 23
    pathogenic strategies, 406
    peptidoglycan preparations, teichoic acid contamination, 285

peptidoglycans, soluble polymeric
   isolation, 263–264
   secreted by cells, isolation, 266–268
polysaccharide-based typing, 161
temperature-sensitive mutants, generation by nitrosoguanidine, 451
toxic shock syndrome toxin 1, mechanism of action, 680
TSS-producing strains, host–parasite
   interactions, animal chamber
   models, 125
types, 161
virulence analysis, transposon mutagenesis methods, 412–413
virulence factors, 412
Staphylococcus epidermidis, host–parasite
   interactions, animal chamber models,
   132–133
Staphyloferrin A, 354
Statistical Analysis System, 31, 39
Streptavidin–gold colloids, preparation,
   706–707
Streptococci
   electrotransformation, 418
   group A
      disease associations, 171, 286, 405,
         410
      M protein, 171, 286–294, 411
         acid extraction, 287–288
         antibodies, 286–287
         antiphagocytic property, 286
         pepsin extraction, 288–292
         preparation, 287–288
         purification, 291–292, 294
         recombinant, 292–294
         structure, 286–287, 289
      M typing, 171–173
      opacity factor
         inhibition typing, 173
         reaction, 172–173
      pathogenic strategies, 406
      serogrouping, on acid extracts, 172
      serotyping, 171–173
      T protein, 171
      T typing, 171–172
      virulence analysis, transposon muta-
         genesis methods, 411–412
   group B
      disease associations, 173–174, 410

meningitis, 94, 96, 173
pathogenic strategies, 406
serotypes, 174
serotyping, 174
virulence analysis, transposon mutagenesis methods, 411
β-hemolytic, serologic identification,
   159, 170–174
oral
   preparation from small samples, 420–
      421
   transformation
      protocols, 416–419
      temperature-sensitive vectors for,
         417–418
   virulence analysis, transposon muta-
      genesis methods, 414–416
   virulence analysis, transposon mutagene-
      sis methods, 410–412
   virulence determinants, 411
Streptococcus agalactiae
   disease associations, 173
   polysaccharide-based typing, 161
   types, 161
Streptococcus dysgalactiae, hyaluronidase,
   assay, 613
Streptococcus faecium, peptidoglycans,
   low-molecular-weight oligomers and
   monomers, isolation, 269
Streptococcus lactis, electrotransformation, 382
Streptococcus milleri, hyaluronidase,
   assay, 614
Streptococcus mitis
   disease associations, 414
   IgA$_1$ protease, 543
Streptococcus mutans
   disease associations, 405, 414
   pathogenic strategies, 406
   virulence analysis, transposon mutagene-
      sis methods, 415–416
Streptococcus oralis, IgA$_1$ protease, 543
   properties, 553
Streptococcus pneumoniae
   capsular polysaccharide, 159, 169–170
   capsular serotyping, 170
   disease associations, 168
   IgA$_1$ protease, 543
   properties, 553

laboratory hazards and biosafety recommendations, 23
meningitis, 94, 168
  animal models, 95, 102
  lytic antibiotic therapy, 104
otitis, animal models, 59
polysaccharide-based typing, 161
types, 159, 161, 169
vaccine, 169
Streptococcus pyogenes
  genomic DNA, preparation, 175–177
  host–parasite interactions, animal chamber models, 121, 138–139
  hyaluronidase, assay, 615–616
  laboratory hazards and biosafety recommendations, 23
  peptidoglycans
    hot formamide extraction, 258
    preparations, teichoic acid contamination, 285
  serogroups, 170–171
  streptolysins
    cytopathogenic effects, 683
    mechanism of action, 680
  types, 161
  typing, biochemical basis, 161
Streptococcus sanguis
  disease associations, 414
  IgA$_1$ protease, 543
    properties, 553
  virulence analysis, transposon mutagenesis methods, 415–416
Streptococcus sobrinus, periodontitis, rat model, 117
Streptokinase, 411
Streptomyces fradiae, protease, assay, 560
Streptomyces griseus, protease, assay, 559–560
Streptomyces hyalurolyticus, hyaluronidase, assay, 615–616
Streptomycin resistance, carried by transposon, selection for, 394
Subtilisin, 566
  zymography in nondissociating gels with copolymerized substrates, 588–590
Suicide vectors
  for *Campylobacter*, 475–480

conjugative transfer of vectors from *Escherichia coli*, 477–478
transfer by electroporation, 478–479
conditionally counterselectable, for allelic exchange, 458–465
for *Pseudomonas aeruginosa*, 467–471
for Tn*phoA* delivery, 433

T

Temperature-sensitive mutants, 448–457
  characterization, 453–454
  generation, 449–451
  isolation, 450–452
  quantitative clearance studies with, 454–455
Tetracycline resistance, carried by transposon, selection for, 394
Thiry–Vella rabbit ileal loop model, secretory IgA response to *Shigella* in, 140–143, 155
Toxic shock syndrome, rabbit model, 132–133
Toxins, *see also specific toxins*
  ADP-ribosylating, 617–631
    active site residues, photoaffinity labeling, 631–639
      calculations, 639
      photolysis, 635–639
  cell shape changes caused by, 683–684
  cell surface changes caused by, 683–684
  cytopathogenic, 679–690
    classification, 679–680
    in cultured cells
      antiproliferative action, assay, 690
      assay, 685–690
      biochemical assay, 690
      cytopathogenicity after binding, assay, 690
      detection, 687–688
      titration, 688–690
    mechanism of action, 679–681
    targets, 679
  cytopathogenic effects
    definition, 679
    types seen by light microscopy, 681–685
  cytostatic, 679
  hemolytic, assays, 657–667

inhibition of host protein synthesis, 647–656

assay
  cell-free method, 651–656
  in fractionated cell lysates, 653–656
  in unfractionated cell lysates, 651–653
  whole cell method, 650–651
internalization by mammalian cells, 705–715
  monitoring, 706
intracellular localization, 714–716
intracellularly visible changes caused by, 684–685
intracellular processing, monitoring, 716
ion channel formation, and toxicity, 700–705
noncytotoxic, cytopathogenic effects, 679
pore formation, 658–659
  assay, 677–678
  in lipid bilayer membranes, 691–705
ribosome-inhibiting, 648
  mechanism of action, 649
thiol-activated, cell detachment, 682–683
toxicity, biochemical monitoring, 710–712
types, 679–680
uptake by mammalian cells, 705–715
Trachoma, cynomolgus monkey model, 70–78
Transducin, ADP-ribosylation, 628
Transferrin, 316
iron bound to, acquisition by *Neisseria gonorrhoeae*, 356–357
as iron-chelating bacteriostatic agent, 327–328
Transferrin receptors
eukaryotic, 362–363
microbial, 345
neisserial, 360–363
Transfer RNA, genes, as chromometers, 184
Transglycosylase
lytic, 270
  preparation, 271–272
  peptidoglycan fragments released by, 270–271
Transposase, 387–388

Transposons, *see also* Minitransposons
composite, 387
conjugative, 409
  filter-mating with, 409, 418–419
delivery into gram-positive bacteria, 408–410
in genetic analysis, advantages, 386–387
Tn5, properties, 387–388
Tn10, properties, 387–388
Tn551, in genetic manipulations of *Staphylococcus aureus*, 413
Tn916, 409, 411–415, 426
  transformation of oral streptococci, 416–419, 421
Tn917, 413–414, 426
  hot spots for, in streptococci, 414–415
Tn918, 412
Tn1545, 426
TnphoA
  chromosomal insertions, conversion to deletion mutations, 444–446
  delivery, 431–437
    broad-host-range plasmid vectors for, 431–433
  functions, 427–429
  gene fusion, conversion to chloramphenicol acetyltransferase operon fusion, 446–447
  as probe for genes linked to virulence factors, 447–448
  structure, 427–429
  virulence analysis with, 426–448
    applications, 437–448
    methodological considerations, 443–444
for virulence analysis in gram-positive bacteria, 405–426
Treponema denticola
periodontitis, 107
virulence factors, 106
Treponema pallidum
host–parasite interactions, animal chamber models, 121, 125
hyaluronidase, zymographic analysis, 616
laboratory hazards and biosafety recommendations, 23
urogenital infection, 83
Tripyridyl-*s*-triazine, in iron assays, 320

Trypanosoma cruzi, hemolysin, 667
Trypsin, 566
   inhibition by soybean trypsin inhibitor,
      demonstration by reverse zymo-
      graphy, 582–583
Tuberculosis, epidemiology, investigation
   by *Mycobacterium tuberculosis* DNA
   fingerprinting, 196, 205
Tumor necrosis factor, in rabbit bacterial
   meningitis model, 105

U

Ureaplasma urealyticum, IgA₁ protease,
   543
Urogenital system, infection
   animal models, 83–93
   pathogens, 83
Ustilago spaerogena, ferrichrome, isolation, 342–343

V

Vacuolation, cytosolic, toxin-induced, 684
Vectors, *see* Plasmids; Shuttle vectors;
   Suicide vectors
Vertebrates, studies with pathogenic bacteria, biosafety levels, 2, 11–16
Vibrio, outer membrane vesicles, 233
Vibrio anguillarum
   anguibactin, 352
   iron-regulated proteins, 352
Vibriobactin, 352
   purification, 340
Vibrio cholerae
   colonization factors, 518
      accessory, 519
      analysis by Tn*phoA* mutagenesis, 437–
         438
   disease caused by, *see* Cholera
   extracellular protease, production,
      tryptophan effects, 523–524
   growth, laboratory conditions for, 520–
      521, 524–525
   heat-labile enterotoxin, cytopathogenic
      effects, 684
   iron-regulated proteins, 352
   iron-regulated virulence factor, analysis
      by Tn*phoA* mutagenesis, 440–441

laboratory hazards and biosafety recommendations, 24
lipopolysaccharide-based typing, 161
pathogenesis, 518
   lipoproteins involved in, identification,
      446
pilin, 527
TcpJ, amino acid sequence, 528, 530
toxin, *see* Cholera toxin
toxin-coregulated pilus, 519
ToxR regulon, 519–520
types, 161
vibriobactin, 352
   isolation, 340
virulence factors, expression, regulators,
   519
Vibrio parahaemolyticus, laboratory hazards and biosafety recommendations,
   24
Vibrio vulnificus, neutral metalloproteinase, activities, 554
Virulence
   attenuation during inoculum preparation,
      41
   cell-surface determinants, analysis with
      transposon Tn*phoA*, 426–448
   and culture conditions, 41
   determinants, 502
      definition, 458
      identification, 458
      regulation, 458
   factors
      environmental regulation, 482
      identification, 482
   gram-positive bacteria, transposon muta-
      genesis studies, 406–426
   monitoring, Sereny assay, 46–47
V8 protease, 566

W

Wolinella recta, periodontitis, 107

X

Xanthomonas campestris, macromolecular
   transport in, proteins required for,
   amino acid sequences, 529

Y

Yersinia pestis, laboratory hazards and biosafety recommendations, 24

Z

Zymograms, 563
  development
    after nondissociating electrophoresis, 567, 571–574
    after SDS electrophoresis, 569–570, 576
    SDS removal before, requirement for, 578–580
    stock solutions, 567, 570
Zymography
  hyaluronidase assay, 615–616
  microbial proteases, 563–594
    advantages, 565
    applications, 564, 574–593
      with detergent-containing gels, 567–571
      enzymes, 565–566
      in nondissociating gels, 571–574, 587–593
        after cationic electrophoresis, 589–593
        with copolymerized substrates, 588–589
        with SDS-containing gels, 574–576
      copolymerized protein substrates for, 574–576
      limitations, 579–580
      with substrate diffusion into gels after electrophoresis, 583–584
    origins, 563–564
    principles, 563–564
    reverse, HMW protease inhibitor identification method, 581–583
    in SDS-containing polyacrylamide gels, 567, 569
    with substrate electrophoresis in gels containing acrylamide-conjugated substrates, 586
    with substrate electrophoresis in polyacrylamide gradient gels, 586
    zymogen activation studies, 586–587

ISBN 0-12-182136-6

90038